Theory and Design of Charged Particle Beams

WILEY SERIES IN BEAM PHYSICS AND ACCELERATOR TECHNOLOGY

Series Editor MEL MONTH

BROWN-	FUNDAMENTALS OF ION SOURCES*
CHAO-	PHYSICS OF COLLECTIVE BEAM INSTABILITIES IN HIGH ENERGY ACCELERATORS
EDWARDS AND SYPHERS-	AN INTRODUCTION TO THE PHYSICS OF HIGH ENERGY ACCELERATORS
MICHELOTTI-	INTERMEDIATE CLASSICAL DYNAMICS WITH APPLICATIONS TO BEAM PHYSICS*
REISER-	THEORY AND DESIGN OF CHARGED PARTICLE BEAMS

*Forthcoming titles

Theory and Design of Charged Particle Beams

MARTIN REISER

Electrical Engineering Department and Institute for Plasma Research
University of Maryland
College Park, Maryland

A Wiley-Interscience Publication
JOHN WILEY & SONS, INC.
New York / Chichester / Brisbane / Toronto / Singapore

Library of Congress Cataloging in Publication Data:

Reiser, M. (Martin), 1931–
    Theory and design of charged particle beams/Martin Reiser.
      p. cm.
    Includes bibliographical references and index.
    ISBN 0-471-30616-9
    1. Particle beams. I. Title.
QC793.3.B4R45   1994
539.7'3--dc20   93-33110

Printed in the United States of America

10 9 8 7 6 5 4 3 2 1

To the memory of Ilya M. Kapchinsky
(1919–1993) and L. Jackson Laslett (1913–1993).

Contents

Preface   **xiii**

Acknowledgments   **xvii**

1 Introduction   **1**

1.1 Exposition / 1
1.2 Historical Developments and Applications / 4
1.3 Sources of Charged Particles / 7
References / 14

2 Review of Charged Particle Dynamics   **15**

2.1 The Lorentz Force and the Equation of Motion / 15
2.2 The Energy Integral and Some General Formulas / 19
2.3 The Lagrangian and Hamiltonian Formalisms / 23
    2.3.1 Hamilton's Principle and Lagrange's Equations / 23
    2.3.2 Generalized Potential and Lagrangian for Charged Particle Motion in an Electromagnetic Field / 25
    2.3.3 Hamilton's Equations of Motion / 28
    2.3.4 The Hamiltonian for Charged Particles and Some Conservation Theorems / 30
2.4 The Euler Trajectory Equations / 37
    2.4.1 The Principle of Least Action and the Euler Equations / 37
    2.4.2 Relativistic Euler Equations in Axially Symmetric Fields / 41
2.5 Analytic Examples of Charged Particle Motion / 43
    2.5.1 Planar Diode without Space Charge / 43
    2.5.2 Planar Diode with Space Charge (Child–Langmuir Law) / 45
    2.5.3 Charged-Particle Motion in a Uniform Magnetic Field / 46

2.5.4   Charged Particle Motion in a Radial Electric Field / 47
2.5.5   The Harmonic Oscillator / 49
Reference / 52
Problems / 52

3 Beam Optics and Focusing Systems without Space Charge   56

3.1   Beam Emittance and Brightness / 56
3.2   Liouville's Theorem / 62
3.3   The Paraxial Ray Equation for Axially Symmetric
       Systems / 66
       3.3.1   Series Representation of Axisymmetric Electric and Magnetic
                Fields / 66
       3.3.2   Derivation of the Paraxial Ray Equation / 69
       3.3.3   General Properties of the Solutions of the Paraxial Ray
                Equations / 76
3.4   Axially Symmetric Fields as Lenses / 80
       3.4.1   General Parameters and Transfer Matrix of a Lens / 80
       3.4.2   Image Formation and Magnification / 83
       3.4.3   Electrostatic Lenses / 86
       3.4.4   Solenoidal Magnetic Lenses / 98
       3.4.5   Effects of a Lens on the Trace-Space Ellipse and Beam
                Envelope / 103
       3.4.6   Aberrations in Axially Symmetric Lenses / 106
3.5   Focusing by Quadrupole Lenses / 111
3.6   Constant-Gradient Focusing in Circular Systems / 116
       3.6.1   Betatron Oscillations / 116
       3.6.2   The Trace-Space Ellipse and Beam Envelope in a Betatron-Type
                Field / 121
       3.6.3   Focusing in Axisymmetric $\mathbf{E} \times \mathbf{B}$ Fields / 126
       3.6.4   Energy Spread, Momentum Compaction, and Effective
                Mass / 130
3.7   Sector Magnets and Edge Focusing / 135
3.8   Periodic Focusing / 139
       3.8.1   Periodic Focusing with Thin Lenses / 139
       3.8.2   General Theory of Courant and Snyder / 145
       3.8.3   The FODO Quadrupole Channel / 153
       3.8.4   Sector-Focusing Cyclotrons / 157
       3.8.5   Strong-Focusing Synchrotrons / 162
       3.8.6   Resonances in Circular Accelerators / 166
3.9   Adiabatic Damping of the Betatron Oscillation
       Amplitudes / 171
       References / 174
       Problems / 175

4   Linear Beam Optics with Space Charge   183

4.1   Theoretical Models of Beams with Space Charge / 183
4.2   Axisymmetric Beams in Drift Space / 191
    4.2.1   Laminar Beam with Uniform Density Profile / 191
    4.2.2   Beam Envelope with Self Fields and Finite Emittance / 202
    4.2.3   Limitations of the Uniform Beam Model and Limiting Currents / 203
    4.2.4   Self-Focusing of a Charge-Neutralized Beam (Bennett Pinch) / 208
4.3   Axisymmetric Beams with Applied and Self Fields / 210
    4.3.1   The Paraxial Ray Equation with Self Fields / 210
    4.3.2   Beam Transport in a Uniform Focusing Channel / 212
4.4   Periodic Focusing of Intense Beams (Smooth Approximation Theory) / 221
    4.4.1   Beam Transport in a Periodic Solenoid Channel / 221
    4.4.2   Beam Transport in a Quadrupole (FODO) Channel / 234
    4.4.3   Envelope Oscillations and Instabilities of Mismatched Beams / 240
    4.4.4   Coherent Beam Oscillations due to Injection Errors and Misalignments / 252
4.5   Space-Charge Tune Shift and Current Limits in Circular Accelerators / 260
    4.5.1   Betatron Tune Shift due to Self Fields / 260
    4.5.2   Current Limits in Weak- and Strong-Focusing Systems / 265
    4.5.3   Effects of Image Forces on Coherent and Incoherent Betatron Tune / 267
4.6   Charge Neutralization Effects / 273
    4.6.1   Ionization Cross Sections for Electrons and Proton Beams in Various Gases / 273
    4.6.2   Linear Beam Model with Charge Neutralization / 278
    4.6.3   Gas Focusing in Low-Energy Proton and H⁻ Beams / 281
    4.6.4   Charge-Neutralization Effects in Intense Relativistic Electron Beams / 285
    4.6.5   Charge-Neutralization Effects in High-Energy Synchrotrons and Storage Rings / 289
    4.6.6   Plasma Lenses / 294
References / 296
Problems / 298

5   Self-Consistent Theory of Beams   304

5.1   Introduction / 304
5.2   Laminar Beams in Uniform Magnetic Fields / 306
    5.2.1   A Cylindrical Beam in an Infinitely Strong Magnetic Field / 306
    5.2.2   Nonrelativistic Laminar Beam Equilibria / 311

5.2.3 Relativistic Laminar Beam Equilibria / 323

5.2.4 Paraxial Analysis of Mismatched Laminar Beams in Uniform Magnetic Fields / 331

5.3 The Vlasov Model of Beams with Momentum Spread / 335

5.3.1 The Vlasov Equation / 335

5.3.2 The Kapchinsky–Vladimirsky (K–V) Distribution / 341

5.3.3 Stationary Distributions in a Uniform Focusing Channel / 347

5.3.4 RMS Emittance and the Concept of Equivalent Beams / 358

5.4 The Maxwell–Boltzmann Distribution / 364

5.4.1 Coulomb Collisions between Particles and Debye Shielding / 364

5.4.2 The Fokker–Planck Equation / 368

5.4.3 The Maxwell–Boltzmann Distribution for a Relativistic Beam / 372

5.4.4 The Stationary Transverse Distribution in a Uniform or Smooth Focusing Channel / 379

5.4.5 Transverse Temperature and Beam-Size Variations in Nonuniform Focusing Channels / 390

5.4.6 The Longitudinal Distribution and Beam Cooling due to Acceleration / 393

5.4.7 Stationary Line-Charge Density Profiles in Bunched Beams / 402

5.4.8 Longitudinal Motion in rf Fields and the Parabolic Bunch Model / 414

5.4.9 Longitudinal Beam Dynamics in Circular Machines / 428

5.4.10 Effects of Momentum Spread on the Transverse Distribution / 437

5.4.11 Coupled Envelope Equations for a Bunched Beam / 447

5.4.12 Matching, Focusing, and Imaging / 453

References / 461

Problems / 462

6 Emittance Growth   **467**

6.1 Causes of Emittance Change / 467

6.2 Free Energy and Emittance Growth in Nonstationary Beams / 470

6.2.1 Analytical Theory / 470

6.2.2 Comparison of Theory, Simulation, and Experiment / 479

6.3 Instabilities / 491

6.3.1 Transverse Beam Modes and Instabilities in Periodic Focusing Channels / 491

6.3.2 Longitudinal Space-Charge Waves and the Resistive-Wall Instability / 498

6.3.3 Longitudinal Instability in Circular Machines and Landau Damping / 515

6.4 Collisions / 525
      6.4.1 The Boersch Effect / 525
      6.4.2 Intrabeam Scattering in Circular Machines / 530
      6.4.3 Multiple Scattering in a Background Gas / 537
6.5 Beam Cooling Methods in Storage Rings / 541
      6.5.1 The Need for Emittance Reduction / 541
      6.5.2 Electron Cooling / 542
      6.5.3 Stochastic Cooling / 544
      6.5.4 Radiation Cooling / 546
6.6 Concluding Remarks / 554
      References / 557
      Problems / 560

Appendix 1: Example of a Pierce-Type Electron Gun with Shielded Cathode   **564**

   References / 566

Appendix 2: Example of a Magnetron Injection Gun   **567**

   References / 569

Appendix 3: Four-Vectors and Covariant Lorentz Transformations   **570**

Appendix 4: Equipartitioning in High-Current rf Linacs   **573**

   References / 580

Appendix 5: Radial Defocusing and Emittance Growth in High-Gradient rf Structures (Example: The rf Photocathode Electron Gun)   **581**

   References / 588

List of Frequently Used Symbols   **590**

Bibliography (Selected List of Books)   **598**

Index   **601**

Preface

This book evolved over many years from the material I have taught in a graduate course on charged particle beams at the University of Maryland since the late 1960s. It is also influenced by undergraduate courses on principles of particle accelerators, physical electronics, and fundamentals of charged particle devices that I taught intermittently during this time.

Most important, though, this book reflects my research interests and experience in accelerator design and beam physics: cyclotron and ion source design during the 1960s, collective ion acceleration in the 1970s, and since then the physics of intense, high-brightness beams.

Although the connection with particle accelerators is emphasized in the book, I have tried to present a broad synoptic description of beams that applies to a wide range of other devices such as low-energy focusing and transport systems and high-power microwave sources. The material is developed from first principles, basic equations, and theorems in a systematic and largely self-sufficient way. Assumptions and approximations are clearly indicated; the underlying physics and the validity of theoretical relationships, design formulas, and scaling laws are discussed. The algebra is often more detailed than in other books. This is a feature that I retained from my class notes in order to make the derivations more transparent, which the students found especially valuable.

The "theory" in this book is an experimentalist's theory. It tries to get away with a minimum of mathematical complexity, avoids topics that are only of academic interest, and stresses the essential physical features and the relevance to laboratory beams. A central theme, which has only recently become the focus of research, is the behavior of space-charge dominated beams and the thermodynamic description of beams by a Maxwell–Boltzmann distribution. Due to longitudinal cooling by acceleration, beams are usually not in 3-D thermal equilibrium and are best described by a Maxwell–Boltzmann distribution with different transverse and longitudinal temperatures. The analysis of the "equilibrium" properties of such distributions, including the transverse and longitudinal density profiles and the modeling by equivalent beams with linear space-charge forces, takes a major part

xiii

of Chapter 5. This includes a significant amount of very recent work with my collaborators, such as image effects and the Boltzmann line charge density profiles in bunched beams, that has not yet been published. Nonlinear forces, instabilities, and collisions drive the beam toward thermal equilibrium and thermalize the free energy when the beam is not perfectly matched to the focusing system or deviates from the equilibrium density profile. These effects and the resulting emittance growth are discussed in Chapter 6, which includes a review of cooling methods in storage rings.

This book is not intended to give an extensive review of the entire field, with a comprehensive list of references and discussion of all important past and current developments, like Lawson's encyclopedic *Physics of Charged Particle Beams*. It is written as a textbook for the student, researcher, or newcomer who wants to have a thorough and systematic introduction to the theory and design of charged particle beams. However, it also addresses the needs of the more experienced physicists and engineers involved in the design or operation of particle accelerators, low-energy beam systems, microwave sources, free electron lasers, and other devices. To these professionals, the book offers a broad review of focusing systems, a detailed and critical evaluation of theoretical models, a comprehensive list of definitions of fundamental beam parameters and their relationships, and theoretical guidance for the design of the high-quality beams required in modern devices where space charge plays an important role.

Our analysis is limited to the basic physical properties of beams and their behavior in various focusing and accelerating systems. Thus the topic of instabilities is covered only in an introductory way that includes a few examples of fundamental interest.

The references at the end of each chapter are limited to historical papers and to more recent work that I considered important for the beam theory discussed in the book. There are undoubtedly omissions in each category. In particular, my selection in the latter category is admittedly subjective, and I apologize to the many researchers whose contributions are not mentioned. The selectivity in the references is balanced by an extensive bibliography at the end of the book to which I frequently refer for further references or elaboration of a topic.

With regard to use as a textbook, my experience in teaching and lecturing at the University of Maryland, in U.S. Accelerator Schools, and elsewhere indicated to me that there is a demand for a more introductory presentation, in addition to the advanced beam theory. Consequently, I have organized the material in such a way that the first four chapters can be used for a senior-year, undergraduate, special-topics course. Such courses are also often attended by beginning graduate students interested in the field or searching for a research topic. A broadly based course of this type on fundamentals of charged particle beams and focusing methods is more useful, in my opinion, than a course devoted entirely to more specialized topics such as beam optics or accelerators. Some material could be omitted or supplemented by special lecture notes reflecting the research interests of the faculty or student demand at a particular institution. Similarly, some topics in Chapters 5 and 6 could be included in such an undergraduate course.

If the book is used only for a one-semester graduate-level course, the material in the first four chapters has to be condensed and presented selectively to leave enough time for the more advanced topics in Chapters 5 and 6. For example, the review in Chapter 2, while useful for the electrical engineering student who has had no advanced course in classical mechanics, could be largely omitted for the physics student who has studied classical mechanics or is taking it concurrently. Similarly, some topics in Chapters 3 and 4 could be omitted depending, for instance, on whether the emphasis is to be on low-energy devices or on high-energy circular accelerators.

As is desirable for a textbook of this type, I have tried to present the material in a uniform notation throughout. Nevertheless, it was unavoidable to use some letters of the Latin and Greek alphabet for different purposes. However, the "List of Frequently Used Symbols" and the explanations in the text, including frequent repetitions of definitions, should help to clarify the intended meaning and to avoid the confusion that often arises with regard to this issue.

In this age of the computer, a comment on my philosophy with regard to numerical-versus-analytical treatment is called for. First, I am a firm believer in basic analytical theory. This analysis is required to guide experiment and simulation and to provide the indispensable parameter scaling necessary for physics understanding and design. Second, the discussion of computational tools, computer codes, and simulation techniques—even on an introductory level—is beyond the scope of this book. Third, the interplay between analytical theory, particle simulation, and experiment is an absolute necessity for achieving progress in the field of multiparticle beam dynamics. The example discussed in Section 6.2 is an illustration of this important point: neither theory nor simulation nor experiment alone would have been sufficient to obtain the final understanding and quantitative interpretation of the behavior of such a space-charge dominated beam. Indeed, the conclusion from my long experience is not to rest until "full agreement" is achieved between theory, simulation, and experiment.

Acknowledgments

I am deeply grateful for the support, advice, criticism, and collaboration that I received in developing the material for this book over so many years. First, I want to thank the University of Maryland, which provided me with a supportive environment for combining and pursuing my teaching and research interests. The collaboration and interaction with my colleagues in the charged particle beam group, notably Bill Destler, Victor Granatstein, Wes Lawson, Moon-Jhong Rhee, and Chuck Striffler, and others in the Institute for Plasma Research, has been highly stimulating and beneficial.

The material in the book evolved gradually over many years from ever-changing class notes. The comments and questions from many students attending the course have contributed to eliminating errors and clarifying the presentation. More important, however, I must give major credit to the research associates, graduate students, and outside collaborators who worked with me in unraveling the physics of space-charge dominated beams that is the major theme in Chapters 5 and 6. In particular, I am most grateful to John Lawson for the many years of our highly productive scientific interaction and for his personal friendship.

The University of Maryland electron beam transport facility, which formed the focal point for much of the research discussed in this book, and the experiments themselves were developed with the close collaboration of John Lawson and the help of many research associates, graduate students, and others. My special thanks go to Won Namkung, John McAdoo, Ekkehard Boggasch, Jian Guang Wang, Josephus Suter, Eric Chojnacki, Peter Loschialpo, Keng Low, Tom Shea, Yinbao Chen, Jianmei Li, David Kehne, Dunxiong Wang, Weiming Guo, and Hyyong Suk. The help of Bill Herrmannsfeldt in the design of the electron gun, the interaction with Richard True on the operation of gridded cathodes, and the technical support of Paul Haldemann and the workshops in the Physics Department and the Institute for Plasma Research are also highly appreciated. I am particularly indebted to David Kehne for the highly successful multiple-beam experiments and to Jian Guang Wang and Dunxiong Wang for the pioneering longitudinal beam physics experiments. Equally important has been the long-standing collaboration with Irving

Haber, who together with Helen Rudd, obtained the impressive simulation results that were so crucial for the interpretation of our experiments and the check of the theoretical concepts.

I am also indebted to Patrick O'Shea for his contribution to our work on intense electron beams and for valuable discussions on photocathodes and other topics. Similarly, I wish to express my appreciation to Eric Horowitz, Chu Rui Chang, Samar Guharay, Chris Allen, Chung-Hsin Chen, and Victor Yun for the important contributions they made in establishing our research program on high-brightness H^- beams.

Special thanks go to Chris Allen and Nathan Brown for the important theoretical work on the effects of image charges in bunched beams and on the longitudinal and transverse Boltzmann distribution (N. Brown), the results of which are presented in Section 5.4. I am most grateful to Jian Guang Wang, who helped me with the material on the longitudinal instability in Chapter 6 and with clarifying many points in other parts of the manuscript, and to Dunxiong Wang, who made valuable contributions, most notably with the Lorentz transformation for the Boltzmann distribution in Section 5.4.3. I also appreciate the help from Girish Saraph with the description of the generalized K-V distribution in Section 5.3.2.

My collaboration on particle simulation studies with Jürgen Klabunde and Jürgen Struckmeier at GSI during a sabbatical visit in 1982–1983, sponsored by a U.S. Senior Scientist Award from the Alexander-von-Humboldt Foundation of Germany, led to the important discovery and correct interpretation of emittance growth due to space-charge nonuniformity in high-intensity, low-emittance beams. This discovery motivated my research during the last decade and the important multiple-beam experiments discussed in Section 6.2. Jürgen Struckmeier has been particularly helpful with his contributions to envelope instabilities in Section 4.4.3 and the numerical simulation work reviewed in Section 6.3.1, including several figures. I have also benefited from many stimulating discussions with Ingo Hofmann and Rolf Müller at GSI and from the interaction with Bob Gluckstern and Alex Dragt at the University of Maryland.

My research on emittance growth in space-charge dominated beams would have been unthinkable without Tom Wangler at Los Alamos. The close collaboration with Tom, our mutual friendship, and the synchronism in our thinking have been crucial to the success that we made together in advancing the theoretical understanding of the role of space charge in beam physics. I want to thank Tom Wangler, Bob Jameson, and other colleagues at the Los Alamos Laboratory for the long and very productive interaction that I have enjoyed. In this context I also want to thank George Schmidt of the Stevens Institute, with whom Tom Wangler and I wrote the report on relativistic beam relationships that I refer to in the book. I hope that George will forgive me for not fully sharing his view of temperature as a Lorentz invariant and for my extensive use of "laboratory temperature" in the text.

The final manuscript was the product of many changes and additions that I made in response to the reviews by several great pioneers and leaders in the field of charged particle beam physics. I am especially indebted to John Lawson, who patiently read every page of the many versions of the manuscript and helped me immensely with his very extensive and detailed comments and criticism. His critical evaluation of Chapters 5 and 6 in the penultimate version resulted in major additions and thus important improvements in the final version.

The penultimate version of the manuscript was also carefully reviewed by Ilya Kapchinsky and Pierre Lapostolle. Both gave me detailed lists of comments, corrections, and suggestions that I incorporated into the final version and for which I am very thankful. I was deeply shocked and saddened by Ilya's death shortly after I had talked to him about the manuscript.

Other outside researchers gave me their views on specific topics. Bob Jameson encouraged me to write something on equipartitioning in rf linacs, as did Ilya Kapchinsky. Since this is a somewhat special topic and time was running out, I decided to do so in the Appendix. Wahab Ali helped me with the review of ionization cross sections in Section 4.6.1. Anton Piwinski, Allen Sørensen, and Mario Conti reviewed Section 6.4.2 on intrabeam scattering and gave me very valuable information that I used in the final version and for which I wish to thank them.

My research work and the development of the manuscript would not have been possible without the long-time support and encouragement of David Sutter of the Department of Energy (DOE) and more recently, Charles Roberson of the Office of Naval Research and Mark Wilson of DOE. I want to thank them and their institutions for the financial support and the personal interest that they have shown for my own work and our research in the charged particle beam group.

In the final analysis, and despite what I said in the above acknowledgments, this book would never have been finished without the invention of the word processor and Carol Bellamy. Carol was the last but most important person involved in the typing and editing of the manuscript that began in the 1970s with Eleanor Fisher and continued for a few years with Ann Rehwinkel. However, for the final six years of new versions, improvements, and corrections, it was Carol Bellamy who had the word processing expertise, strength, patience, and tenacity to see that the book was finally completed. I am deeply indebted to her for staying in this with me and not giving up. My thanks are also due to Mel Month and Gregory Franklin for gently but firmly steering me toward publication in the Wiley Series in Beam Physics and Accelerator Technology. I also want to thank Joan Hamilton of the Physics Department for her professional help in preparing most of the figures in the text and to Janice Schoonover for alleviating the paperwork and administrative chores during the final busy period of completing the manuscript.

Most important, I want to thank my wife, Inge, for bearing with me during the writing of this book.

Theory and Design of
Charged Particle Beams

CHAPTER 1

Introduction

1.1 EXPOSITION

Charged particle dynamics deals with the motion of charged particles in electric and magnetic fields. More specifically, it implies the behavior of *free* charged particles in applied electric and magnetic fields (single-particle dynamics) or in the collective fields generated by the particle distribution if the density is high enough that the mutual interaction becomes significant (self-field effects). Many aspects of gas discharges and plasmas (microscopic motion) are also included in charged particle dynamics. The interaction of free particles with the electron shell of atoms or molecules or with the periodic electric potential of crystals (electron diffraction) as well as the physics of *bound* particles (solid-state theory) are excluded. The particles' behavior in these cases is described by quantum mechanics, not classical mechanics.

The electric and magnetic fields may be static or time dependent and the kinetic energy of the particles may be relativistic. In general, the particles will be treated as classical point charges. Quantum-mechanical effects may be of importance in some applications, for example, in determining the resolution of the electron microscope, but they are ignored in this book. We shall also neglect electromagnetic radiation by accelerated charged particles except for a brief treatment in connection with radiation cooling: *Synchrotron radiation* limits the achievable kinetic energy in circular accelerators, especially for electrons and positrons, but it can also be utilized in damping rings to *cool* these *lepton* beams, as discussed at the end of Chapter 6. On the other hand, we consider collisional effects, such as intrabeam scattering, and collisions between beam particles and gas molecules. They play a major role in charge neutralization due to collisional ionization of the background gas, discussed in Chapter 4; in the formation of the thermal equilibrium distribution, treated in Chapter 5; and as a cause of emittance growth, covered in Chapter 6.

When the self fields are taken into account, a charged particle beam behaves like a *nonneutral* plasma, that is, a special class of plasma having a drift velocity much greater than the random thermal velocity and lacking in general the charge neutrality of a regular plasma composed of particles with opposite charge. A *beam* is a well-defined flow of a continuous stream or a bunch of particles that move along a straight or curved path, usually defined as the *longitudinal* direction, and that are constrained in the transverse direction by either applied focusing systems or by self-focusing due to the presence of particles with opposite charge. The transverse velocity components and the spread in longitudinal velocities are generally small compared to the mean longitudinal velocity of the beam. Examples are the *straight* beams in linear accelerators, cathode ray tubes, or electron microscopes and the *curved* beams in circular accelerators, such as betatrons, cyclotrons, and synchrotrons.

Most particle accelerators employ radio-frequency (rf) fields to accelerate the particles. The beam in these cases consists of short bunches with a pulse length that is usually small compared with the rf wavelength. To prevent the bunch from spreading due to its intrinsic velocity distribution or due to space-charge repulsion, external focusing forces must be provided in both transverse and longitudinal directions. In rf accelerators, the axial component of the electric field provides focusing in the longitudinal direction, while magnetic fields must be used for transverse focusing. Throughout most of this book we deal with continuous, or *long*, beams and *linear* transverse focusing systems in which the external force on a particle is proportional to the displacement from the axis, or central orbit, of the beam. A brief introduction to the acceleration and focusing of bunched beams is given in Chapter 5.

Nonlinear beam optics, or more generally, nonlinear beam dynamics, which deals with the effects of nonlinear forces due to aberrations in the applied focusing systems, is a highly specialized field that cannot be treated comprehensively within the scope of a book like this. We therefore limit this topic to brief discussions of aberrations in axisymmetric lenses (Section 3.4.6), resonances in circular accelerators (Section 3.8.6), and nonlinear longitudinal beam dynamics in rf accelerators (Section 5.4.8). We do, however, analyze in some detail the generally nonlinear nature of space-charge forces in the thermal distribution, which provides a realistic description of the behavior of laboratory beams (Sections 5.4.4 to 5.4.7 and 6.2). An example of the nonlinear interaction between the aberrations of a solenoid lens and the space charge of an electron beam is presented in Section 5.4.12.

Overall, the material presented in our book is developed in a systematic, largely self-contained manner. We start, in Chapter 2, with a review of the basic principles and formalisms of classical mechanics as applied to charged particle dynamics; our treatment is more comprehensive than the usually brief discussions presented in other books. We then proceed to a broad, general review of beam optics and focusing systems in Chapter 3. The topic of periodic focusing is treated in some detail because of its importance to beam transport and particle accelerators.

A central theme is the role of space charge and emittance in high-intensity, high-brightness beams. In Chapter 4 we use the model of a uniform-density beam with linear self fields. This model allows us to extend the linear beam optics of Chapter 3 to include space charge without having to cope with the mathematically more complicated nonlinear forces. Special emphasis is given to periodic beam transport with space charge (Section 4.4), space-charge effects in circular accelerators (Section 4.5), and charge-neutralization effects (Section 4.6).

The self-consistent theory is developed systematically in Chapter 5 from laminar beams (Section 5.2) to the Vlasov model for beams with momentum spread (Section 5.3), and then to the Maxwell–Boltzmann distribution, which is treated very extensively in Section 5.4. The latter section represents an attempt to develop a unifying thermodynamic description of a beam and contains a considerable amount of new material that is not found in other books on charged particle beams.

The thermodynamic description is continued in Chapter 6, which deals with the fundamental effects causing emittance growth. The concept of free energy, created when a beam is not in equilibrium, and its conversion into thermal energy and emittance growth is treated in Section 6.2, which includes a comparison between theory, simulation, and experiment. Transverse beam modes and instabilities are reviewed in Section 6.3.1. Longitudinal space-charge waves are discussed in Section 6.3.2 since they are fundamental to an understanding of the behavior of perturbations in a beam. Two historically important illustrations of the destructive interaction between the space-charge perturbations and the beam's environment are selected. One is the *resistive wall instability* (Section 6.3.2) in straight systems (microwave devices, linear accelerators); the other is the *longitudinal instability* in circular machines due to *negative-mass* behavior and interaction with the wall represented by a complex impedance (Section 6.3.3). These cases, which are treated for pedagogical reasons on a fundamental level, are intended merely as two examples of the many instabilities that may limit the beam intensity and cause emittance growth. A more extensive discussion of waves and instabilities in beams, including wakefield effects at relativistic energies, is beyond the scope of this book. An excellent introduction and survey of these topics with a comprehensive list of references to the scientific literatures is provided by Lawson [C.17, Chap. 6]. Collective instabilities in high-energy accelerators are treated comprehensively and on an advanced level in terms of the beams' wakefields and the wall impedances in the book by Chao [D.11].

Coulomb collisions as a source of emittance growth and energy spread are treated in Section 6.4. Our analysis of the *Boersch effect* (Section 6.4.1) shows that *intrabeam scattering* is relevant not only in high-energy storage rings (Section 6.4.2) but may also be significant in low-energy beam focusing, transport, and acceleration devices. Scattering in a background gas is discussed in Section 6.4.3. As a natural, complementary addition to our review of emittance growth, we present in Section 6.5 a brief survey of the methods to reduce emittance (*beam cooling*) in storage rings. Finally, in Section 6.6, we summarize the key topics that were discussed, comment on some questions that were left open, and mention a few issues that need further research.

The application of the theory to the design of charged particle beams is stressed throughout the book. Many formulas, scaling laws, graphs, and tables are presented in the text to aid the experimentalists and the designers of charged particle beam devices. Similarly, many of the problems at the end of the chapters were chosen to be of practical interest. The main emphasis of this book, though, is on the physics and design of *beams*. Only those features of a particular device that are relevant to an understanding of the physics and/or necessary for theoretical analysis and design are treated. Some supplemental material is presented in the appendixes.

Charged particle dynamics and the theory of charged particle beams combine aspects of classical mechanics, electromagnetic theory, geometrical optics, special relativity, statistical mechanics, and plasma physics. A few selected texts covering these fields are listed in the bibliography at the end of the book.

1.2 HISTORICAL DEVELOPMENTS AND APPLICATIONS

Historically, the first and most prominent area of charged particle dynamics is the field of electron optics, where most of the early work and theoretical development took place and which is well documented in many books listed in part C of the bibliography. The birth of electron optics may be traced to 1926, when H. Busch showed that the action of a short axially symmetric magnetic field on electron rays was similar to that of a glass lens on light rays. Then in 1931 and 1932, Davidson and Calbrick, Brüche, and Johannson recognized that this is also true for axially symmetric electric fields. The first use of magnetic lenses was by Knoll and Ruska (1931) and of electric lenses by Brüche and collaborators (1934).

Up to 1939, electron optics experienced a rapid development stimulated by strong industrial needs, especially electron microscopes, cathode ray tubes, and television. The classic book, which is an encyclopedia of electron optics in this important period and even today is very useful, is that of Zworykin et al., *Electron Optics and the Electron Microscope* [C.1].

During World War II, electron optics received new impulses from war requirements: cathode ray tubes for radar and image-converter tubes for infrared vision, but most important, the development of microwave devices (klystron, magnetron, etc.) for the generation of high-power electromagnetic waves in the range above 1000 MHz. The need for improvement of these latter tubes stimulated interest and progress in the study of space-charge effects in high-intensity beams. The classic reference here is Pierce's book [C.3].

Another important impetus that significantly expanded the field of electron optics, or charged particle dynamics in the broader sense, came from the development of high-energy particle accelerators. This development started around 1930 with the invention of the linear accelerator and the betatron in 1928, the cyclotron in 1931, and the electrostatic accelerator in 1931–1932. This was followed by the large high-energy accelerators existing today, such as the two-mile electron linac at Stanford and the proton synchrotron at Fermilab, near Chicago, now operating at an energy of about 1 TeV and called the *tevatron*. Beam dynamics in particle accelerators is now a major branch of charged particle dynamics. Electron and ion optics was extended to include the focusing of beams in circular accelerators. New types of focusing systems, such as quadrupole lenses, edge focusing in sector-shaped magnets, alternating-gradient focusing, and so on, were invented and contributed to the successful development of accelerators with steadily increasing energies and improving performance characteristics. New interest in particle dynamics came also from space science, industrial applications of electron–ion beam devices (welding, micromachining, ion implantation, charged particle beam lithography), and thermonuclear fusion.

In the decade from 1965 to 1975 two new types of accelerator were developed for the generation of electron beams with high peak power and short pulse length; these are the relativistic diode and the linear induction accelerator. The former produces intense relativistic electron beams (IREB), with peak currents ranging from kiloamperes to mega-amperes and energies from hundreds of keV to more than 10 MeV. Such high intensity electron beams are created when short high voltage pulses from so-called *Marx generators* or pulse transformers impinge on the diode. The associated high electric fields cause field emission from the cathode and plasma formation. The plasma expansion leads to gap closure, which, in turn, limits the beam pulse length to between 10 and 100 nanoseconds. These pulsed-power IREB generators have found applications as strong x-ray sources, for studies of the collective acceleration of positive ions by the electric fields associated with intense electron beams, and for the generation of high-power microwaves and *free electron lasers*. More recently, pulsed diodes have been developed that produce high-power ion beams for research on inertial fusion. Miller's book [C.18] presents a very useful introduction to the physics and technology of such pulsed-power, intense particle beams.

Like the betatron, the linear induction accelerator uses inductive electric fields produced by the time-varying flux in magnetic cores. These fields are applied in a sequence of gaps to accelerate pulsed beams of charged particles. The charged particles traverse the gaps only during the time interval in which the magnetic flux is changing, and hence a voltage drop appears across the gaps. In contrast to the radio-frequency resonance accelerators, induction linacs can accelerate very high peak currents, ranging typically from several hundred amperes to several kiloamperes. The largest accelerator in this class was the Advanced Test Accelerator (ATA) at the Lawrence Livermore National Laboratory. It accelerated a 10-kA 70-ns electron beam to an energy of 47 MeV. Originally developed for relatively short electron beams (10 to 100 ns), induction linacs are now also being used for longer pulses (microseconds) of both electron and ion beams. The best example in the latter category is the ion induction linac being developed at the Lawrence Berkeley Laboratory. It is designed for acceleration of high-current heavy-ion beams with the aim of using them as drivers—like laser beams—to ignite the fuel pellets of future inertial fusion reactors. Present experiments are at relatively low energies of a few MeV and a current of $\lesssim$ 1 A. A full-scale heavy-ion fusion driver system would require currents of heavy ions (mass number $\gtrsim$ 100) in the range 20 to 30 kA with an energy of 5 to 10 GeV and a pulse length of about 10 ns.

The more traditional radio-frequency (rf) linear accelerators are also being developed for high-power applications such as heavy-ion fusion, electron–positron linear colliders for high-energy physics, and other purposes. The invention of the low-energy radio-frequency-quadrupole (RFQ) accelerator by Kapchinsky and Teplyakov in 1970 has revolutionized the field of rf linacs for ion beams. Today, practically all rf linacs in major laboratories and industry throughout the world use the RFQ as an injector.

Other recent developments involve the use of intense electron beams as electromagnetic radiation sources. Of particular interest in this regard is the gyrotron, a new high-power microwave source in the centimeter and millimeter range, and the free electron laser (FEL), which covers a very wide spectrum from centimeter to optical wavelengths. All of these applications have triggered new research in the physics of intense high-brightness charged particle beams such as transport through periodic-focusing systems, beam stability in the presence of high space-charge forces, interaction with a plasma background, and nonlinear effects responsible for beam deterioration (emittance growth) or particle loss.

This book deals primarily with the theory and design of charged particle beams, not with the design principles of accelerators and other devices which are found in many of the books listed in the bibliography. Thus it will be appropriate to close this historical review by highlighting some of the major early milestones in the development of charged particle beam physics with regard to the theoretical understanding and modeling of the effects of space charge.

The recognition that there are fundamental current limits in charged particle beams plays an important role in beam theory and design. Historically, the fact that the magnetic self field of a relativistic, charge-neutralized beam stops the propagation of the beam when the current exceeds a critical value was discovered by Alfvén (in 1939) for electron propagation through space and later applied to laboratory beams by Lawson (in 1958). The critical current associated with this effect is known in the literature as the *Alfvén current* or *Alfvén–Lawson current*. Closely related to this effect is the work on self-focused relativistic electron beams by Bennett (1934) and Budker (1956).

The current limit due to space charge (in the absence of charge neutralization) in a diode is known as the *Child–Langmuir law* and dates to the early work of Child (1911) and Langmuir (1913). However, the related limit for a beam propagating through a drift tube was studied much later, and the formula for a relativistic electron beam derived by Bogdankevich and Rukhadze in 1971 is probably the one cited most frequently in the literature.

The foundation for the mathematical treatment of beams with space charge was laid by Vlasov in 1945. Vlasov integrated Liouville's theorem, Maxwell's equations, and the equations of motion into a self-consistent theoretical model that has become an indispensable tool for the theoretical analysis of beams. In 1959, Kapchinsky and Vladimirsky proposed a special solution to the Vlasov equation, known in the literature as the *K–V distribution*, which has the property that the transverse space-charge forces are linear functions of the particles' positions in the beam. This was a major milestone in beam physics whose practical importance for analysis and design cannot be overemphasized. The K–V distribution gained additional significance when Lapostolle and Sacherer in 1971 introduced the description of beams in terms of the root-mean-square (rms) properties (rms width, divergence, and emittance). They showed that beams having the same rms properties are equivalent. This equivalency principle is used extensively in Section 5.4 for correlating the nonanalytical Maxwell–Boltzmann distribution with the analytical K–V distribution in the transverse direction and with the parabolic line-charge distribution in the longitudinal direction, and in Section 6.2 for our theoretical treatment of emittance growth.

Another important milestone in the development of beam physics is the detailed pioneering work by Laslett in 1963 on the space-charge tune shift of the betatron oscillations in circular accelerators. This effect, often referred to as the *Laslett tune shift*, is of fundamental importance, as it limits the achievable intensity in these machines. With regard to understanding the physics of space-charge-dominated beams, the simulation work by Chasman in 1968 for linear accelerators, the analysis of collective oscillation modes in uniformly focused beams by Gluckstern, and the stability analysis by Davidson and Krall in 1970 constitute important achievements which influenced future work.

This list of historical milestones is obviously quite subjective and incomplete and could be extended into many directions, such as the rich field of beam instabilities, where the theoretical analysis of the *negative-mass instability* in 1959 comes to mind as a major event. But this book is not about instabilities. Furthermore, we wanted to limit the list to "historical" milestones, defined somewhat arbitrarily as events that occurred more than 20 years ago.

1.3 SOURCES OF CHARGED PARTICLES

Although the main topic of this book is beam dynamics, it will be beneficial to review briefly the basic principles and performance limitations of typical particle sources. This is particularly important for intense beams, where physical and technological constraints of the source pose fundamental limits for the beam current and the emittance or brightness that can be achieved.

The simplest conceptual model of a source is the planar diode. One of the two electrodes emits the charged particles; in the case of electrons it is called a *cathode*. A potential difference of the appropriate polarity accelerates the particles to the other electrode, called the *anode* in the electron case. In practice, the emitter has, of course, a finite size, and usually a circular shape with radius r_s. The *anode* contains a hole or a mesh to allow the beam to propagate into the vacuum tube downstream, where it is focused or accelerated depending on the particular application. Furthermore, the electrode in which the emitter is embedded as well as the *anode* may have a special nonplanar design to provide initial focusing for the beam. In a *Pierce-type* geometry, for example, the electrodes form an angle of less than 90° with respect to the beam axis to produce a transverse electrostatic force that exactly balances the repulsive Coulomb force due to the space charge of the beam (see [C.3, Chap. X]).

A schematic illustration of a typical diode-type electron gun with thermionic cathode, Pierce-type focusing electrode, and anode mesh is shown in Figure 1.1.

The electron beam radius in this example remains practically constant within the gun and then increases due to space-charge repulsion when the beam enters the field-free region outside the anode. To prevent divergence due to space-charge forces or transverse velocity spread, the beam has to be focused with appropriate magnetic or electrostatic lenses, as discussed in this book. Other types of electron sources employ field emission or photocathodes; the cathodes may have the shape of an annulus (to form a hollow beam) or a sharp tip. Additional intermediate electrodes (triode or tetrode configurations) may be used to control the beam parameters.

Conduction electrons in a metal have an energy distribution that obeys the Fermi–Dirac statistics. The electrons emitted from a thermionic cathode belong to the *Maxwellian tail* of the Fermi–Dirac distribution, and the current density J_{th} is given by the *Richardson–Dushman equation* [1]

$$J_{th} = AT^2 e^{-W/k_BT}. \tag{1.1}$$

Here T is the cathode temperature, W the work function of the cathode material (typically a few eV), and k_B is Boltzmann's constant (8.6175×10^{-5} eV/K). The theoretical value for the constant A is

$$A = \frac{4\pi emk^2}{h^3} = 1.2 \times 10^6 \text{ Am}^{-2}\text{K}^{-2}, \tag{1.2}$$

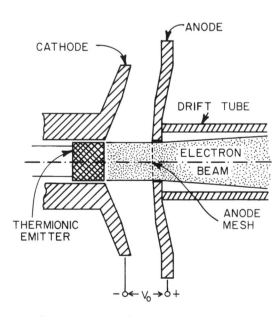

Figure 1.1. Schematic of an electron gun with thermionic cathode, Pierce-type-electrode geometry, and anode mesh. (See Appendix 1 for a discussion of such a gun without anode mesh.)

where $e = 1.6 \times 10^{-19}$ C is the electron charge, $m = 9.11 \times 10^{-31}$ kg the electron rest mass, and $h = 6.63 \times 10^{-34}$ Js is Planck's constant. Experimentally, one finds a value for A that is lower than (1.2) by a factor of about 2. Fabrication of thermionic cathodes is a highly specialized art where the choice and composition of materials is guided by requirements of low work function W, long lifetime (at high temperature), smoothness of emitting surface, and other factors. Pure tungsten has a work function of $W = 4.5$ eV, and tungsten cathodes operate at a temperature of 2500 K ($k_B T \sim 0.2$ eV), with a current density of about 0.5 A/cm^2. Considerably higher current densities of 10 to 20 A/cm^2 can be achieved with dispenser cathodes, which are used for high-power microwave generation. Dispenser cathodes use barium or strontium oxides impregnated in a matrix of porous tungsten (or similar metals). These cathodes operate at a typical temperature of 1400 K ($k_B T \sim 0.12$ eV) and have an effective work function of 1.6 eV.

A typical ion source with a diode configuration is shown schematically in Figure 1.2. The ions are extracted from the plasma of a gas discharge, and the accelerated beam passes through a hole in the *extraction electrode* into the vacuum drift tube. The emitting plasma surface area is not fixed as in the case of a cathode. Rather, it has a concave shape, called *meniscus*, which depends on the plasma density and the strength of the accelerating electric field at the plasma surface. The dashed lines in Figure 1.2 indicate the equipotential surfaces of the electric field distribution due to the applied voltage V_0 as well as the space charge of the beam. Note that there is a small potential drop between the plasma surface and the wall of the chamber that encloses the plasma. The concave shape of the meniscus and the aperture in the source electrode produce a transverse electric field component that results in a converging beam.

In general, ion sources are much more complex than electron guns. There are many different types of sources for the various particle species, such as light ions, heavy ions, or negative ions (e.g., H^-). Most of the sources employ magnetic fields to confine the plasma. Some have several electrodes at different potentials to better control the ion beam formation and acceleration process. A special problem with ion sources is the gas in which the plasma is formed and which leaks through the source aperture into the acceleration gap and the drift tube. Near the source the pressure is high enough that a plasma with density exceeding the beam density can be formed through ionizing collisions between the beam ions and gas molecules. This causes space-charge neutralization, which is advantageous for focusing but may also cause detrimental effects such as high-voltage breakdown and beam plasma instabilities. Another problem arises because ions with different charge state or mass are extracted from the plasma together with the desired species. In the case of negative ions such as H^-, for instance, electrons are also accelerated with the ion beam. Unless the number of contaminating particles is small, it is necessary in these cases to use deflecting magnetic fields to remove the undesired particle species from the beam.

For our purpose of illustrating the basic design concept of charged particle sources it suffices to consider the simple diode configurations of Figures 1.1 and 1.2. In such sources the space-charge electric field limits the amount of current that can be accelerated by a given voltage V_0. For a planar electrode geometry with a gap spacing d between the two plates, the limiting current density J (in the nonrelativistic limit and in MKS units) is given by the formula

$$J = 1.67 \times 10^{-3} \left(\frac{q}{mc^2} \right)^{1/2} \frac{V_0^{3/2}}{d^2} \quad [A/m^2], \qquad (1.3)$$

where q and m are the particle charge and mass, respectively, and c is the speed of light. The relation, first derived by Child and Langmuir [2], is known in the literature as *Child's law* or as the *Child–Langmuir law*. Applying this result to a uniform round beam emitted from a circular area with radius r_s yields for the beam current

$$I = 1.67\pi \times 10^{-3} \left(\frac{q}{mc^2} \right)^{1/2} V_0^{3/2} \left(\frac{r_s}{d} \right)^2 \quad [A]. \qquad (1.4)$$

However, in practical ion sources and electron guns with cylindrical geometry the beam current may be considerably lower than this limit, which is based on an ideal one-dimensional planar-diode geometry. The ratio $I/V_0^{3/2}$ is known as the *perveance* of the beam. A derivation of Child's law is given in Section 2.5.2.

An important figure of merit for a high-brightness beam is the *emittance*, which is basically defined by the product of the width and transverse velocity spread of

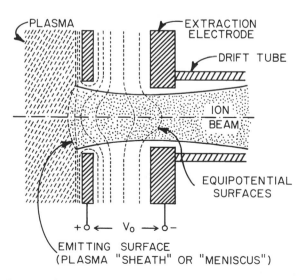

Figure 1.2. Schematic of a plasma ion source. The equipotential surfaces of the electric field distribution are indicated by dashed lines. The ions are emitted from the concave plasma sheath, which forms an equipotential surface.

the beam. The electrons in the tail of the Fermi–Dirac distribution inside a cathode and the ions in the plasma source have a Maxwellian velocity distribution given by

$$f(v_x, v_y, v_z) = f_0 \exp\left[-\frac{m\left(v_x^2 + v_y^2 + v_z^2\right)}{2k_B T} \right], \tag{1.5}$$

where T is the temperature of the cathode or the plasma. As a result, the particles emerge from the source with an intrinsic velocity spread. If x and y denote the two cartesian coordinates perpendicular to the direction of the beam, the rms values of the transverse velocity spread for the Maxwellian distribution are readily found to be

$$\tilde{v}_x = \tilde{v}_y = \left(\frac{k_B T}{m} \right)^{1/2}. \tag{1.6a}$$

If the emitting surface is a circle with radius r_s and with uniform current density, the rms width of the beam is

$$\tilde{x} = \tilde{y} = \frac{r_s}{2}. \tag{1.6b}$$

As explained in Section 3.2, an *effective normalized emittance* is defined nonrelativistically as

$$\epsilon_n = 4\tilde{x} \frac{\tilde{v}_x}{c} \tag{1.7a}$$

Substitution of (1.6a) and (1.6b) in (1.7a) yields

$$\epsilon_n = 2r_s \left(\frac{k_B T}{mc^2} \right)^{1/2} \quad \text{[m-rad]}. \tag{1.7b}$$

The normalized emittance measures the beam quality in two-dimensional *phase space*, which is defined by the space and momentum coordinates of the particle distribution (i.e., x, p_x, or x, v_x, nonrelativistically). From Liouville's theorem (discussed in Section 3.2) it may be shown that the normalized emittance remains constant if there are no nonlinear forces or coupling forces between different coordinate directions. Thus Equation (1.7b) constitutes a lower theoretical limit; in practice, nonlinear beam dynamics, instabilities, and other effects may cause emittance growth, so that the actual value is always larger than (1.7b).

For many high-power applications the output current of an electron gun is limited by the achievable current density J_c at the cathode rather than by the space-charge limit and by the high-voltage breakdown effect to be discussed below. In the widely used thermionic cathodes, current densities, in practice, are normally limited to 10 to 20 A/cm^2, and values as high as 100 A/cm^2 have been achieved in experimental studies, depending on the desired cathode lifetime, average beam power, and other factors. If the current density J_c is fixed, the desired beam current I determines the cathode radius r_s and hence also the emittance $\epsilon_n \pi$. Using $r_s = (I/J_c\pi)^{1/2}$, one can write Equation (1.7b) in the form

$$\epsilon_n = 2\left(\frac{I}{J_c\pi}\right)^{1/2}\left(\frac{k_B T}{mc^2}\right)^{1/2} \quad [\text{m-rad}], \tag{1.8}$$

which shows that the emittance increases with the square root of the product of beam current and cathode temperature and decreases with current density as $J_c^{-1/2}$. For $J_c = 10$ $A/cm^2 = 10^5$ A/m^2 and $k_B T = 0.1$ eV, one obtains $\epsilon_n = 1.6 \times 10^{-6} I^{1/2}$ m-rad.

In a new type of electron gun with photocathode that is being developed at various laboratories, a high-power laser beam is focused on the cathode surface and electron currents of several hundred A/cm^2 have been achieved. The photocathode is located inside the first cavity of an rf injector-linac structure, as shown in Figure 5.1 of Appendix 5. The strong axial electric field in this cavity (20–100 MV/m) rapidly accelerates the electrons to a high energy ($\gtrsim 1$ MeV). Timing and length of the laser pulse are chosen to produce a short electron bunch during a small phase interval within the accelerating part of the rf cycle. The high-brightness beams produced by the laser-driven *rf photocathode guns* are of particular interest for advanced particle accelerator applications such as high-energy e^+ e^- linear colliders and free electron lasers (FELs), which require beams with high intensity but very small emittance. The rf photocathode gun was first developed at Los Alamos, and the general concept is described in the early papers by Fraser et al. [3]. More recent reviews of the developments in this field can be found in References [4] and [5]. The problem of emittance growth in such electron guns due to rf defocusing and nonlinear space-charge forces is discussed briefly in Appendix 5.

The above scaling does not apply for high-intensity plasma-type ion sources with a simple diode geometry. In this case the achievable beam current is often limited by Child's law and by high-voltage breakdown. Several different empirical formulas for voltage breakdown have been developed over the years based on practical experience and theoretical models. According to these formulas the gap width d between the electrodes must not be smaller than a critical value that depends on the voltage V_0 between the electrodes as

$$d = CV_0^\alpha, \tag{1.9}$$

where C is a constant and the exponent α ranges between $1 < \alpha < 2$, depending on the model for breakdown. In one model, the electric field strength, V_0/d, is the parameter controlling breakdown, hence $\alpha = 1$. Another model assumes that the product of field strength and gap voltage (i.e., V_0^2/d) determines the breakdown condition, so that $\alpha = 2$. In a recent survey of experimental results with ion sources, Keller concluded that the relation

$$d_{[\text{mm}]} = 1.4 \times 10^{-2} V_{0[\text{kV}]}^{3/2} \tag{1.10}$$

(i.e., $\alpha = 1.5$) provided the best fit to the data [4]. This appears to be a reasonable compromise between the two extreme cases of $\alpha = 1$ and $\alpha = 2$. It should be pointed out, however, that such simple scaling laws have to be used with some caution. In practice, electrical breakdown is a very complicated phenomenon that depends on many details (other than gap spacing and voltage), such as gas flow from the source, geometry of the electrode structures, and surface cleanliness.

Another important constraint influencing the output characteristics (perveance and emittance) of high-current, low-emittance charged particle sources is imposed by considerations of beam optics. To minimize nonlinearities in the electrostatic field configuration, especially spherical aberrations, which would adversely affect the beam quality, the radius r_s of the beam at the emitter surface must not be larger than the gap width d. In most high perveance ion source designs, for instance, the ratio r_s/d is in the range

$$0.2 < \frac{r_s}{d} < 1.0. \tag{1.11}$$

It should be noted that this beam optics argument does not apply to intense relativistic electron beams and high-power ion diodes producing charge-neutralized beams with intensities far above the space-charge limit given in Equation (1.4).

The above set of equations and constraints defines the parameter space for high-perveance electron or ion sources. Thus, the intrinsic normalized emittance ϵ_n is determined by the beam radius r_s at the emitter surface and the source temperature $k_B T$ according to Equation (1.7). For electrons from thermionic cathodes, one typically has $k_B T_e \approx 0.1$ eV, while ion temperatures from plasma sources (e.g., protons or H^- ions) are usually an order of magnitude higher (i.e., $k_B T_i \approx 1$ to 5 eV). If ϵ_n, and thus r_s, are given (to meet the requirements of a particular application), the beam current and voltage are defined by Child's law (1.4) and the two constraints imposed by electrical breakdown (1.10) and beam optics (1.11).

For experiments in which a high-intensity beam is to be focused to a small spot size, the *unnormalized emittance* at the final beam energy, $\epsilon = \epsilon_n/\beta\gamma$, which represents the product of beam radius and divergence angle, is an important parameter. It is inversely proportional to the relativistic velocity and energy factors $\beta = v/c$ and $\gamma = (1 - \beta^2)^{-1/2}$, and hence decreases as the particles are accelerated to high energy. Emittance by itself is not sufficient to characterize the beam quality. A better figure of merit is the *brightness* B defined by the ratio of beam current I and the product of the two emittances, i.e., I/ϵ^2 for axisymmetric beams [See Equation (3.8)]. Since the emittance changes with energy, it is preferable to use the *normalized brightness* defined as $B_n = 2I/\pi^2 \epsilon_n^2$ [Equation (3.22)]. The normalized brightness, like the normalized emittance ϵ_n, is an invariant in an ideal system. Emittance growth due to nonlinear forces, instabilities, and other effects (discussed in Chapter 6) decreases the normalized brightness. By comparing the actual beam brightness with the ideal value one can assess the effectiveness of the design and performance characteristics of a particular device. As an example, let us consider a high-intensity electron beam from a dispenser-type cathode. Using Equation (1.8), one finds that the normalized brightness has an upper limit of

$$B_n = \frac{J_c}{2\pi} \frac{mc^2}{k_\mathrm{B} T}. \tag{1.12a}$$

This brightness limit depends on the ratio of the current density J_c at the cathode and the temperature T of the cathode, and it is independent of the current I. If one operates at a maximum current density of $J_c = 10$ A/cm² and at a cathode temperature of $k_\mathrm{B} T = 0.1$ eV, the brightness has an upper limit of

$$B_n = 8 \times 10^{10} \ \text{A/(m-rad)}^2. \tag{1.12b}$$

In practice, the brightness of the electron beam in the system downstream from the electron gun will always be less than this ideal value.

The preceding discussion was intended to provide an introductory overview of basic design principles of charged particle sources and of the fundamental performance limits of high-intensity beams due to constraints imposed by the physics and technology of source operation. Detailed descriptions can be found in the literature, such as the books C.15, C.16, and C.23 listed in the bibliography, and in the proceedings of accelerator conferences or topical meetings on low-energy beams and sources.

REFERENCES

1. O. W. Richardson, *Phil. Mag.* **28**(5), 633 (1914); S. Dushman, *Phys. Rev.* **21**(6), 623 (1923).

2. C. D. Child, *Phys. Rev. Ser. I* **32,** 492 (1911); I. Langmuir, *Phys. Rev. Ser. II*, **2,** 450 (1913).

3. J. S. Fraser, R. L. Sheffield, E. R. Gray, and G. W. Rodenz, *IEEE Trans. Nucl. Sci.* **NS-32,** 1791 (1985); J. S. Fraser and R. L. Sheffield, *IEEE J. Quantum Electron.* **23,** 1489 (1987).

4. P. O'Shea and M. Reiser, *AIP Conference Proc.* **279,** 579 (1993), ed. J. S. Wurtele.

5. I. Ben-Zvi, Conference Record of the 1993 IEEE Particle Accelerator Conference, 93CH3279-7, p. 2964.

6. R. Keller, in "High Current, High Brightness, and High Duty Factor Ion Injectors," *AIP Conf. Proc.* **139,** 1 (1986), ed. G. Gillespie, Y. Y. Kuo, D. Keefe, and T. P. Wangler.

CHAPTER 2

Review of Charged Particle Dynamics

2.1 THE LORENTZ FORCE AND THE EQUATION OF MOTION

In this chapter we present a brief review of the methods of relativistic classical dynamics for determining the motion of charged particles in electromagnetic fields. We begin with the force on a point charge q in an electromagnetic field, known as the *Lorentz force* and given by

$$\mathbf{F} = q(\mathbf{E} + \mathbf{v} \times \mathbf{B}). \tag{2.1}$$

Note that the *International System of Units* (SI), also referred to as the *mks system*, is used consistently throughout this book. Equation (2.1) is valid for static as well as time-dependent fields. The field vectors $\mathbf{E}$ and $\mathbf{B}$ obey Maxwell's equations, which in our case of charged particle motion in vacuum (where $\mathbf{D} = \epsilon_0\mathbf{E}, \mathbf{B} = \mu_0\mathbf{H}$) may be written in the form

$$\nabla \times \mathbf{E} = -\frac{\partial \mathbf{B}}{\partial t}, \qquad \nabla \cdot \mathbf{E} = \frac{\rho}{\epsilon_0}, \tag{2.2a}$$

$$\nabla \times \mathbf{B} = \mu_0\mathbf{J} + \frac{1}{c^2}\frac{\partial \mathbf{E}}{\partial t}, \qquad \nabla \cdot \mathbf{B} = 0. \tag{2.2b}$$

Here we used the relation $c^2 = 1/\epsilon_0\mu_0$ between the speed of light c, the permittivity ϵ_0, and the permeability μ_0 of free space. The current density $\mathbf{J}$ and the space-charge density ρ satisfy the continuity equation $\nabla \cdot \mathbf{J} + \partial\rho/\partial t = 0$. The motion of a particle due to the force of Equation (2.1) is determined by Newton's equation

$$\frac{d\mathbf{P}}{dt} = \mathbf{F} = q(\mathbf{E} + \mathbf{v} \times \mathbf{B}), \tag{2.3}$$

15

where **P** is the mechanical momentum. In nonrelativistic mechanics, **P** is simply the product of particle mass m and velocity **v** (i.e., $\mathbf{P} = m\mathbf{v}$). The above force equation is also correct relativistically. However, the relationship between **P** and the particle velocity is more complicated and, according to the theory of special relativity, given by

$$\mathbf{P} = \frac{m\mathbf{v}}{1 - v^2/c^2)^{1/2}},$$

or

$$\mathbf{P} = \gamma m\mathbf{v}, \tag{2.4}$$

where γ, also known as the *Lorentz factor,* is defined as

$$\gamma = \frac{1}{(1 - \beta^2)^{1/2}} \tag{2.5}$$

and $\beta = v/c$ is the ratio of the particle velocity v to speed of light in vacuum c. Substituting (2.4) into (2.3), one obtains

$$\gamma m \frac{d\mathbf{v}}{dt} + m\mathbf{v} \frac{d\gamma}{dt} = \mathbf{F} = q(\mathbf{E} + \mathbf{v} \times \mathbf{B}). \tag{2.6}$$

Solving for the acceleration $\mathbf{a} = d\mathbf{v}/dt$, one can write this equation in the form

$$\mathbf{a} = \frac{\mathbf{F} - (\mathbf{F} \cdot \boldsymbol{\beta})\boldsymbol{\beta}}{\gamma m}. \tag{2.7}$$

In the nonrelativistic limit where $\gamma = 1$ and $d\gamma/dt = 0$, the acceleration is parallel to the force and given by $\mathbf{a} = \mathbf{F}/m$. However, in the relativistic situation the acceleration and the force have, in general, different directions. As can be seen from Equation (2.7), only if the force is perpendicular or parallel to the velocity is **a** proportional to **F**. For $\mathbf{F} \perp \mathbf{v}$, one finds that

$$\mathbf{a}_\perp = \frac{\mathbf{F}_\perp}{\gamma m}, \quad \text{or} \quad \frac{d\mathbf{P}_\perp}{dt} = \gamma m \frac{d\mathbf{v}_\perp}{dt}, \tag{2.8a}$$

and for $\mathbf{F} \parallel \mathbf{v}$,

$$\mathbf{a}_\parallel = \frac{\mathbf{F}_\parallel}{\gamma^3 m}, \quad \text{or} \quad \frac{d\mathbf{P}_\parallel}{dt} = \gamma^3 m \frac{d\mathbf{v}_\parallel}{dt}. \tag{2.8b}$$

Thus in place of the mass m of nonrelativistic mechanics we have an *effective mass* that depends on the direction between the force and the velocity. The two effective masses of the two special cases (2.8a) and (2.8b) are known in the literature as the *transverse mass* m_t and the *longitudinal mass* m_l, respectively, and are defined by

$$m_t = \gamma m = \frac{m}{(1 - \beta^2)^{1/2}}, \qquad (2.9a)$$

$$m_l = \gamma^3 m = \frac{m}{(1 - \beta^2)^{3/2}}. \qquad (2.9b)$$

In addition, γm is also known as the *relativistic mass* and m as the *rest mass*, often written with a subscript as m_0. These various definitions of mass have led to considerable confusion, giving the impression that mass is a function of energy that also depends on the direction of the force. However, according to special relativity there is only one mass m that is independent of the frame of observation (i.e., invariant to a Lorentz transformation) [1].

The main task of charged particle dynamics is to determine the particle motion by solving Newton's equation for a given configuration of fields E and B. A special difficulty arises in high-intensity beams, where the fields depend also on the particles' electric and magnetic self fields, which in turn depend on the particles' motion. Known as the problem of self-consistency, this is addressed in Chapter 5.

Equation (2.3) is a vector equation that consists of a set of three second-order coupled differential equations. In cartesian coordinates we have

$$\frac{d}{dt}(\gamma m \dot{x}) = \dot{\gamma} m \dot{x} + \gamma m \ddot{x} = q(E_x + \dot{y} B_z - \dot{z} B_y), \qquad (2.10a)$$

$$\frac{d}{dt}(\gamma m \dot{y}) = \dot{\gamma} m \dot{y} + \gamma m \ddot{y} = q(E_y + \dot{z} B_x - \dot{x} B_z), \qquad (2.10b)$$

$$\frac{d}{dt}(\gamma m \dot{z}) = \dot{\gamma} m \dot{z} + \gamma m \ddot{z} = q(E_z + \dot{x} B_y - \dot{y} B_x), \qquad (2.10c)$$

Many of the cases treated in this book involve beams and field geometries with rotational symmetry which are best treated in cylindrical coordinates. By transformation from cartesian to cylindrical coordinates (r, θ, z), the velocity vector is given by $\mathbf{v} = \{\dot{r}, r\dot{\theta}, \dot{z}\}$, and the equations of motion take the form

$$\frac{d}{dt}(\gamma m \dot{r}) - \gamma m r \dot{\theta}^2 = q(E_r + r\dot{\theta} B_z - \dot{z} B_\theta), \qquad (2.11a)$$

$$\frac{1}{r}\frac{d}{dt}(\gamma m r^2 \dot{\theta}) = q(E_\theta + \dot{z} B_r - \dot{r} B_z), \qquad (2.11b)$$

$$\frac{d}{dt}(\gamma m \dot{z}) = q(E_z + \dot{r} B_\theta - r\dot{\theta} B_r). \qquad (2.11c)$$

It is immediately apparent that Equations (2.10) and (2.11) are rather complex second-order differential equations which permit rigorous analytical solutions in only a few simple cases. Furthermore, we see that the form in which our space variables enter into the equations depends on the coordinate system we choose. This is to say that we cannot write down a generalized form of scalar equation which applies to every component equation in any given coordinate system. This shortcoming of the Newtonian form of the equation of motion is avoided in the Lagrangian–Hamiltonian formalism, where generalized coordinates and generalized potentials are introduced. However, it should be recognized that the Newtonian equations of motion are a good starting point for many problems and that they are particularly useful in obtaining a simple physical picture of the forces and the resulting particle motion in complicated systems.

Of major interest in this book is the use of electric and magnetic fields as lenses to focus the beam along the desired path (in analogy to the focusing of the light rays in optics). In addition, electric and magnetic fields are also used to deflect the beams, as in cathode ray tubes or to bend them into circular orbits, as in cyclotrons and synchrotrons. For design purposes, it is interesting to compare the relative magnitude of electric and magnetic forces for the same amount of stored energy per unit volume and to understand the constraints imposed by technical limitations.

With $w_E = (\epsilon_0/2)E^2$ for the electrostatic energy and $w_M = (1/2\mu_0)B^2$ for the magnetostatic energy per unit volume, we find for $w_E = w_M$:

$$\frac{B}{E} = (\mu_0 \epsilon_0)^{1/2} = \frac{1}{c}$$

in free space. On the other hand, the ratio of magnetic and electric forces is given by

$$\frac{F_M}{F_E} = \frac{vB}{E},$$

and if we substitute the above relation for B/E, we get

$$\frac{F_M}{F_E} = \frac{v}{c}.$$

Since $v \lesssim c$, this implies that (except for extreme relativistic velocities) to achieve the same focusing or deflection force (e.g., in a cathode ray tube), more stored energy is needed if a magnetic field is used than with an electric field. However, in practice, one is severely limited by electrical breakdown problems to field strengths, which for static fields are below about 10 MV/m. Electromagnets with iron can produce fields of up to 2 tesla (T) limited by magnetic saturation of the iron. If we take a particle with velocity $v = 0.1c$ and compare the force in a magnetic field of $B = 2$ T with that in an electric field of $E = 10^7$ V/m, we find that the ratio of the forces $F_M/F_E = vB/E = 6$. The magnetic force is thus six times stronger than the electric force. On the other hand, for $v = 0.01c$, the force ratio is 0.6 and hence the electric field would be more effective at this lower velocity. For this reason, electric fields are limited to applications at low particle velocities. At relativistic energies, magnetic fields must be used for bending and focusing of particle beams.

In recent years, superconducting magnets producing magnetic fields of 4 T and higher have been developed for use in high-energy accelerators. A good example is the *tevatron* at Fermilab, where installation of 4-T superconducting bending magnets made it possible to double the proton energy to about 1 TeV.

2.2 THE ENERGY INTEGRAL AND SOME GENERAL FORMULAS

When **E** and **B** represent static fields that do not depend on time explicitly, the system is *conservative* and we can obtain a first integral of the equation of motion which may be identified with the total energy of the particles. To accomplish this, multiply each side of Equation (2.3) with **v**:

$$\frac{d}{dt}(\gamma m \mathbf{v}) \cdot \mathbf{v} = q\mathbf{E} \cdot \mathbf{v} + q(\mathbf{v} \times \mathbf{B}) \cdot \mathbf{v}.$$

Since $\mathbf{v} \perp (\mathbf{v} \times \mathbf{B})$, the last term on the right side is zero, and with $\mathbf{v} = d\mathbf{l}/dt$, where $d\mathbf{l}$ is the path element, we get

$$\frac{d}{dt}(\gamma m \mathbf{v}) \cdot \mathbf{v} = q\mathbf{E} \cdot \frac{d\mathbf{l}}{dt}. \tag{2.12}$$

The electric field **E** in a conservative system can be derived from a scalar potential ϕ:

$$\mathbf{E} = -\nabla\phi, \qquad \text{or} \qquad \phi = -\int \mathbf{E} \cdot d\mathbf{l}. \tag{2.13}$$

In the nonrelativistic case where $\gamma = 1$, integration of Equation (2.12) between two points along the particle's trajectory yields

$$\frac{m}{2}(v_2^2 - v_1^2) = q\int_1^2 \mathbf{E} \cdot d\mathbf{l} = -q(\phi_2 - \phi_1). \tag{2.14}$$

On the left-hand side, we have the change in kinetic energy of the particles, and we can interpret Equation (2.14) as follows:

1. The change in the particle's kinetic energy is given by the electrostatic potential difference between the two points considered.
2. The magnetic field does not affect the kinetic energy (i.e., it does not do any work even though it may change the direction of the particle's path).
3. If T denotes the kinetic energy and $U = q\phi$ the potential energy of the particles, we can state the physical contents of Equation (2.14) as $T + U = $ const (conservation of total energy).

To obtain the energy integral in the general relativistic case, we first differentiate the left side of Equation (2.12), which yields

$$\gamma m \frac{d\mathbf{v}}{dt} \cdot \mathbf{v} + m \frac{d\gamma}{dt} \mathbf{v} \cdot \mathbf{v} = q\mathbf{E} \cdot \frac{d\mathbf{l}}{dt}.$$

This may be written in the alternative form

$$\frac{\gamma m}{2} \frac{dv^2}{dt} + mv^2 \frac{d\gamma}{dt} = q\mathbf{E} \cdot \frac{d\mathbf{l}}{dt},$$

or

$$\frac{\gamma mc^2}{2} \frac{d\beta^2}{dt} + mc^2\beta^2 \frac{d\gamma}{dt} = q\mathbf{E} \cdot \frac{d\mathbf{l}}{dt}.$$

From $\beta^2 = 1 - 1/\gamma^2$ we have

$$\frac{d\beta^2}{dt} = \frac{2}{\gamma^3} \frac{d\gamma}{dt},$$

and since $1/\gamma^2 + \beta^2 = 1$, the preceding equation becomes

$$\frac{d}{dt}\left(\gamma mc^2\right) = q\mathbf{E} \cdot \frac{d\mathbf{l}}{dt} = -q\frac{d\phi}{dt}. \tag{2.15}$$

With $U = q\phi$, this result may be stated as

$$\frac{d}{dt}\left(\gamma mc^2 + U\right) = 0, \tag{2.16}$$

which is the law of conservation of energy in relativistic form. Binomial expansion in the velocity yields

$$\gamma mc^2 = \frac{mc^2}{(1 - \beta^2)^{1/2}} = mc^2\left(1 + \frac{1}{2}\beta^2 + \frac{3}{8}\beta^4 + \cdots\right). \tag{2.17}$$

For $v \ll c$, or $\beta \ll 1$, we obtain the nonrelativistic approximation

$$\gamma mc^2 \approx mc^2 + \frac{m}{2}v^2. \tag{2.18}$$

The first term on the right side of Equations (2.17) and (2.18) is the rest energy of the particles,

$$E_0 = mc^2, \tag{2.19}$$

which is the famous energy–mass equivalence principle of Einstein's special relativity theory. The remaining terms in Equation (2.17), which depend on the velocity v, can then be identified as the kinetic energy T. For the nonrelativistic approximation, we have

$$T = \frac{m}{2} v^2, \tag{2.20}$$

while in the relativistic case, we get

$$T = \gamma mc^2 - mc^2 = (\gamma - 1)mc^2 = E_T - E_0, \tag{2.21}$$

or

$$E_T = \gamma mc^2 = E_0 + T. \tag{2.22}$$

The total energy E_T of the particle is the sum of rest energy E_0 and kinetic energy T. The relationship between mechanical momentum $\mathbf{P} = \gamma m\mathbf{v}$ and energy E_T of a moving particle is, in view of $\beta^2 \gamma^2 = (\gamma^2 - 1)$, obtained as follows:

$$P^2 = m^2 c^2 \beta^2 \gamma^2 = m^2 c^2 (\gamma^2 - 1),$$

or

$$P^2 = \gamma^2 m^2 c^2 - m^2 c^2 = \frac{E_T^2 - E_0^2}{c^2}.$$

Thus,

$$P = \frac{(E_T^2 - E_0^2)^{1/2}}{c}, \quad \text{or} \quad \frac{P}{mc} = (\gamma^2 - 1)^{1/2}, \tag{2.23}$$

and

$$E_T^2 = c^2 P^2 + m^2 c^4 = \gamma^2 m^2 c^4. \tag{2.24}$$

Differentiating Equation (2.24) with respect to time t yields

$$2E_T \frac{dE_T}{dt} = 2c^2 \mathbf{P} \cdot \frac{d\mathbf{P}}{dt}, \quad \text{or} \quad \frac{dE_T}{dt} = \mathbf{v} \cdot \frac{d\mathbf{P}}{dt}.$$

In conclusion, we can write the equations of motion in the form

$$\frac{d\mathbf{P}}{dt} = \frac{d}{dt}(\gamma m\mathbf{v}) = q(\mathbf{E} + \mathbf{v} \times \mathbf{B}),$$

$$\frac{dE_T}{dt} = \frac{d\mathbf{P}}{dt} \cdot \mathbf{v} = q\mathbf{E} \cdot \mathbf{v}, \tag{2.25}$$

where $E_T = mc^2 + T = \gamma mc^2$. In the extreme relativistic limit ($E_T \gg E_0$), the relation in Equation (2.23) takes the approximate form $P = E_T/c$, with units of MeV/c often used in high-energy physics.

The two most important particles in this book are electrons and protons. Their charge, mass, and rest energy are given in Table 2.1. With $E_e = 0.511$ MeV being the rest energy of an electron, the rest energy of an ion can be calculated to good approximation as follows:

$$E_0 = AE_a - ZE_e = 931.481A - 0.511Z \quad [\text{MeV}]. \quad (2.26)$$

$E_a = 931.481$ MeV represents the atomic mass unit based on ^{12}C ($A = 12$ exactly). A is the atomic mass number of the element, and Z is the number of electrons removed from the atomic shell (i.e., the ionization state). This approximation is accurate to the extent that we can neglect the binding energy of the electrons that have been removed (i.e., for ions with a low charge state). Table 2.2 lists the values of A and E_0 for several light-ion species.

Table 2.1 Charge and mass of electron and proton

	Electron	Proton
Charge q	-1.602×10^{-19} C	1.602×10^{-19} C
Mass m	9.110×10^{-31} kg	1.673×10^{-27} kg
Rest energy E_0	0.511 MeV	938.259 MeV

Table 2.2 Rest energies of some isotopes and ions

Isotope	A (amu)	Rest Energy (MeV)	Ion	Rest Energy (MeV)
^{1}H	1.007825	938.770	^{1}H$^+$	938.259
			^{1}H$^-$	939.281
^{2}H	2.01410	1,876.096	^{2}H$^+$	1,875.585
^{3}He	3.01603	2,809.375	^{3}He$^+$	2,808.864
			^{3}He^{2+}	2,808.353
^{4}He	4.0026	3,728.346	^{4}He$^+$	3,727.835
			^{4}He^{2+}	3,727.324
^{6}Li	6.01512	5,602.970	^{6}Li$^+$	5,602.460
			^{6}Li^{3+}	5,601.437
^{12}C	12.000	11,177.772	^{12}C^{3+}	11,176.239
			^{12}C^{6+}	11,174.706
^{14}N	14.00307	13,043.594	^{14}N$^+$	13,043.083
			^{14}N^{7+}	13,040.017

2.3 THE LAGRANGIAN AND HAMILTONIAN FORMALISMS

2.3.1 Hamilton's Principle and Lagrange's Equations

The Newtonian equations of motion have the disadvantage that they differ in their form markedly when the coordinate system is changed. To circumvent this problem, one introduces in classical mechanics *generalized coordinates* q_i and the associated velocities of $\dot{q}_i$; one then defines a function $L(q_i, \dot{q}_i, t)$, the Lagrange function, from which the equations of motion can be generated in a form that is independent of the coordinate system.

We present a brief review of the main features of the Lagrangian formalism and refer to standard textbooks on classical mechanics, such as Goldstein [A.3], for a more detailed treatment. First, we have to recognize that a Lagrangian L can be defined only for systems with applied forces derivable from an ordinary or *generalized* potential. The simplest case is a *conservative* system where $\oint \mathbf{F} \cdot d\mathbf{l} = 0$ (work done around a closed path is zero). Moreover, if we take the *nonrelativistic* case of a conservative system and $\mathbf{B} = 0$, the Lagrange function is defined by the difference between kinetic and potential energy,

$$L = T - U, \tag{2.27}$$

where $\mathbf{F} = -\nabla U$.

We shall see below that we can define a Lagrange function also for the case that $\mathbf{B} \neq 0$ and the particle velocities are relativistic. For our application, we also note that the generalized coordinates q_i are normally cartesian, cylindrical, or spherical coordinates; however, in general, q_i can be any set of *coordinates* that uniquely defines the state of the system.

Suppose now that we are dealing with a system for which a Lagrangian can be defined. *Hamilton's variational principle* states that the motion of the system (in our case that of a charged particle in an electromagnetic field) from one fixed point at time t_1 to another point at time t_2 is such that the time integral of the Lagrangian, $\int L \, dt$, along the path taken is an extremum (actually, a minimum). Thus, if we compare different possible paths between the two points (i.e., consider small variations of the path taken), the actual path followed by the particle is defined by the condition that the variation of the time integral $\int L \, dt$ is zero, or

$$\delta \int_{t_1}^{t_2} L(q_i, \dot{q}_i, t) \, dt = \int_{t_1}^{t_2} \delta L(q_i, \dot{q}_i, t) \, dt = 0. \tag{2.28}$$

Since t_1 and t_2 remain fixed and we consider only virtual changes of the $q_i, \dot{q}_i$ such that $\delta q_i, \delta \dot{q}_i$ at the two endpoints of the path remain zero, we can take the δ under the integral. In classical mechanics, Hamilton's principle is often used as the starting point to derive Lagrange's equations of motion, and we shall now present this derivation.

The variation of L (i.e., the difference between L for the virtual coordinates $q_i + \delta q_i, \dot{q}_i + \delta \dot{q}_i$), and the unvaried original path $q_i, \dot{q}_i$ is (for a *conservative* system where $\partial L / \partial t = 0$)

$$\delta L = \sum_i \frac{\partial L}{\partial q_i} \delta q_i + \sum_i \frac{\partial L}{\partial \dot{q}_i} \delta \dot{q}_i, \tag{2.29}$$

where $\delta \dot{q}_i = d/dt(\delta q_i)$.

Now substitute Equation (2.29) into Equation (2.28) and perform partial integration of the second term involving $\dot{q}_i$:

$$\int_{t_1}^{t_2} \sum_i \frac{\partial L}{\partial \dot{q}_i} \frac{d}{dt} (\delta q_i) \, dt = \sum_i \frac{\partial L}{\partial \dot{q}_i} \delta q_i \Big|_{t_1}^{t_2} - \int_{t_1}^{t_2} \sum_i \frac{d}{dt} \frac{\partial L}{\partial \dot{q}_i} \delta q_i \, dt. \tag{2.30}$$

Since the variation at the endpoints is zero, the first term on the right-hand side is zero. Therefore, we may write Hamilton's principle in the form

$$\int_{t_1}^{t_2} \sum_i \left(\frac{\partial L}{\partial q_i} - \frac{d}{dt} \frac{\partial L}{\partial \dot{q}_i} \right) \delta q_i \, dt = 0. \tag{2.31}$$

In view of the fact that the δq_i are independent of each other, it follows that

$$\frac{d}{dt} \frac{\partial L}{\partial \dot{q}_i} - \frac{\partial L}{\partial q_i} = 0 \qquad (i = 1, 2, 3). \tag{2.32}$$

These are the Lagrange equations of motion. We will show that they are identical with the equations of motion in the Newtonian form [Equations (2.10) and (2.11)]. Consider a conservative system and assume that the motion is nonrelativistic and that $\mathbf{B} = 0$. In *cartesian coordinates*, the Lagrangian is defined as

$$L = T - U = \frac{m}{2}(\dot{x}^2 + \dot{y}^2 + \dot{z}^2) - q\phi(x, y, z),$$

and substitution in Equation (2.32) yields for the x-coordinate

$$\frac{d}{dt}(m\dot{x}) = -q\frac{\partial \phi}{\partial x} = qE_x.$$

This is identical with Equation (2.10a) when $\gamma = 1, \dot{\gamma} = 0$, and $\mathbf{B} = 0$. In *cylindrical coordinates*, we have

$$L = T - U = \frac{m}{2}\left(\dot{r}^2 + r^2\dot{\theta}^2 + \dot{z}^2\right) - q\phi(x, y, z),$$

and with

$$\frac{\partial L}{\partial \dot{r}} = m\dot{r}, \qquad \frac{\partial L}{\partial r} = mr\dot{\theta}^2 - q\frac{\partial \phi}{\partial r}$$

we obtain from Equation (2.32) for the radial motion

$$\frac{d}{dt}(m\dot{r}) - mr\dot{\theta}^2 = -q\frac{\partial \phi}{\partial r} = qE_r.$$

This equation is identical with Equation (2.11a) when $\gamma - 1, \dot{\gamma} - 0$, and $\mathbf{B} - 0$. In similar fashion one can derive the equations for the azimuthal and axial motion [Equations (2.11b) and (2.11c)].

(*Note:* In Hamilton's principle, the variation is taken between two unvaried points in space and time. The varied paths will *not* obey the equation of motion or the conservation laws; only the unvaried, actual path does.)

2.3.2 Generalized Potential and Lagrangian for Charged Particle Motion in an Electromagnetic Field

The Lagrange equations in the form of Equation (2.32) also apply for the more general case where forces can be derived from a *generalized potential*, or a velocity-dependent potential, U^*, Lagrange's function L is then defined (nonrelativistically) as $L = T - U^*$ and the generalized forces are obtained by the prescription

$$F_i = -\frac{\partial U^*}{\partial q_i} + \frac{d}{dt}\left(\frac{\partial U^*}{\partial \dot{q}_i}\right). \qquad (2.33)$$

The most important case of such a generalized potential is that of combined electric and magnetic forces on a moving charge. To derive U^* for this case, we go back to Maxwell's equations [Equations (2.2)] and introduce the vector potential $\mathbf{A}$:

$$\mathbf{B} = \nabla \times \mathbf{A}. \qquad (2.34)$$

The $\mathbf{E}$ vector can then be redefined for time-varying fields by

$$\mathbf{E} = -\nabla \phi - \frac{\partial \mathbf{A}}{\partial t}. \qquad (2.35)$$

With these two relations, the Lorentz force equation [Equation (2.1)] may be written in terms of the scalar potential ϕ and the vector potential $\mathbf{A}$ in the form

$$\mathbf{F} = q\left(-\nabla \phi - \frac{\partial \mathbf{A}}{\partial t} + \mathbf{v} \times \nabla \times \mathbf{A}\right). \qquad (2.36)$$

Note that ϕ and $\mathbf{A}$ are connected by the *Lorentz gauge condition*, which in free space is given by

$$\nabla \cdot \mathbf{A} + \frac{1}{c^2} \frac{\partial \phi}{\partial t} = 0. \tag{2.37}$$

The terms involving $\mathbf{A}$ on the right side of Equation (2.36) can be written in a more convenient form. Consider the x-component, for instance:

$$F_x = q\left[-\frac{\partial \phi}{\partial x} - \frac{\partial A_x}{\partial t} + v_y\left(\frac{\partial A_y}{\partial x} - \frac{\partial A_x}{\partial y} \right) - v_z\left(\frac{\partial A_x}{\partial z} - \frac{\partial A_z}{\partial x} \right) \right]. \tag{2.38}$$

The total time derivative of A_x is

$$\frac{dA_x}{dt} = \frac{\partial A_x}{\partial t} + v_x \frac{\partial A_x}{\partial x} + v_y \frac{\partial A_x}{\partial y} + v_z \frac{\partial A_x}{\partial z}, \tag{2.39}$$

and the x-component of $\mathbf{v} \times \nabla \times \mathbf{A}$ can therefore be written as

$$(\mathbf{v} \times \nabla \times \mathbf{A})_x = \frac{\partial}{\partial x}(\mathbf{v} \cdot \mathbf{A}) - \frac{dA_x}{dt} + \frac{\partial A_x}{\partial t}. \tag{2.40}$$

Furthermore, we recognize that

$$\frac{dA_x}{dt} = \frac{d}{dt}\left[\frac{\partial}{\partial v_x}(\mathbf{v} \cdot \mathbf{A}) \right]. \tag{2.41}$$

With these substitutions, we obtain for Equation (2.38)

$$F_x = q\left\{ -\frac{\partial}{\partial x}(\phi - \mathbf{v} \cdot \mathbf{A}) - \frac{d}{dt}\left[\frac{\partial}{\partial v_x}(\mathbf{A} \cdot \mathbf{v}) \right] \right\}. \tag{2.42}$$

Since the scalar potential ϕ is independent of velocity, this expression is equivalent to

$$F_x = -\frac{\partial U^*}{\partial x} + \frac{d}{dt}\left(\frac{\partial U^*}{\partial v_x} \right), \tag{2.43}$$

where

$$U^* = q\phi - q\mathbf{A} \cdot \mathbf{v} \tag{2.44}$$

is a generalized potential that has the form desired by the prescription of Equation (2.33). Consequently, the Lagrangian for a charged particle in an electromagnetic field can be written (nonrelativistically)

$$L = T - q\phi + q\mathbf{A} \cdot \mathbf{v}. \tag{2.45}$$

As an example, in cylindrical coordinates, we have

$$L = \frac{m}{2}\left(\dot{r}^2 + r^2\dot{\theta}^2 + \dot{z}^2\right) - q\phi(r,\theta,z) + q\left(\dot{r}A_r + r\dot{\theta}A_\theta + \dot{z}A_z\right). \tag{2.46}$$

If Equation (2.46) is substituted into the Lagrange equations [Equations (2.32)], one obtains the equations of motion in cylindrical coordinates and in the nonrelativistic approximation ($\gamma = 1$, $\dot{\gamma} = 0$), as is easily verified.

In order to obtain the equations of motion in the relativistically correct form, one has to modify the definition of the Lagrange function from $L = T - U^*$ to $L = T^* - U^*$, where

$$T^* = mc^2\left[1 - (1 - \beta^2)^{1/2}\right],$$

with $\beta = v/c$. Note that T^* is not the kinetic energy. Thus,

$$L = mc^2\left[1 - (1 - \beta^2)^{1/2}\right] - q\phi + q\mathbf{v} \cdot \mathbf{A} \tag{2.47}$$

is a suitable Lagrange function for a relativistic particle in an electromagnetic field. Let us check this for cartesian coordinates when $\mathbf{A} = 0$. Since $\beta^2 = (\dot{x}^2 + \dot{y}^2 + \dot{z}^2)/c^2$, we find that

$$\frac{\partial L}{\partial \dot{x}} = \frac{mc^2}{(1 - \beta^2)^{1/2}}\frac{\dot{x}}{c^2} = \gamma m\dot{x}, \quad\quad \frac{\partial L}{\partial x} = -q\frac{\partial \phi}{\partial x}.$$

Hence from Equation (2.32),

$$\frac{d}{dt}(\gamma m\dot{x}) = -q\frac{\partial \phi}{\partial x} = E_x,$$

in agreement with Equation (2.10a).

It should be noted that the definition of L in Equation (2.47) is not unique. We can add or subtract arbitrary constants. Thus, if we subtract mc^2, we get the form

$$L = -mc^2(1 - \beta^2)^{1/2} - q\phi + q\mathbf{v} \cdot \mathbf{A}, \tag{2.48}$$

which is more widely used (Goldstein [A.3], Panofsky and Phillips [A.1], Jackson [A.4], Septier [C.19], etc.).

2.3.3 Hamilton's Equations of Motion

The Lagrange equations are second-order differential equations, and hence, the motion of the particle is completely specified if the initial values for the generalized coordinates q_i and velocities $\dot{q}_i$ are given. In this sense, the q_i and $\dot{q}_i$ together form a complete set of independent variables necessary for describing the motion. For many applications, in particular for numerical techniques of calculating the particle motion, it is more convenient to replace the second-order differential equations by a set of twice the number of first-order differential equations. This is done in the Hamiltonian formulation of classical mechanics. However, rather than using the $(q_i, \dot{q}_i)$ pairs as independent variables, *generalized momenta, p_i,* also known as *canonical momenta* or *conjugate momenta,* are introduced in place of the $\dot{q}_i$. These momenta are defined as follows:

$$p_i = \frac{\partial L(q_i, \dot{q}_i, t)}{\partial \dot{q}_i}. \tag{2.49}$$

Thus, in cartesian coordinates we have from Equation (2.47)

$$p_x = \frac{\partial L}{\partial \dot{x}} = \gamma m\dot{x} + qA_x, \tag{2.50a}$$

$$p_y = \frac{\partial L}{\partial \dot{y}} = \gamma m\dot{y} + qA_y, \tag{2.50b}$$

$$p_z = \frac{\partial L}{\partial \dot{z}} = \gamma m\dot{z} + qA_z, \tag{2.50c}$$

or, in vector form,

$$\mathbf{p} = \gamma m\mathbf{v} + q\mathbf{A}. \tag{2.51}$$

The canonical momentum thus contains the added term of the vector potential $\mathbf{A}$ when a magnetic field is present. The mechanical momentum, which we denote with $\mathbf{P}$ (see Section 2.2), is then obtained from Equation (2.51) as

$$\mathbf{P} = \gamma m\mathbf{v} = \mathbf{p} - q\mathbf{A}. \tag{2.52}$$

In cylindrical coordinates, the Lagrangian of Equation (2.48) has the form

$$L = -mc^2 \left[1 - \frac{\dot{r}^2 + r^2\dot{\theta}^2 + \dot{z}^2}{c^2} \right]^{1/2} - q\phi + q\left(\dot{r}A_r + r\dot{\theta}A_\theta + \dot{z}A_z \right) \tag{2.53}$$

and the three canonical momentum components are

$$p_r = \frac{\partial L}{\partial \dot{r}} = \frac{mc^2}{(1 - \beta^2)^{1/2}} \frac{\dot{r}}{c^2} + qA_r = \gamma m\dot{r} + qA_r, \tag{2.54a}$$

$$p_\theta = \frac{\partial L}{\partial \dot{\theta}} = \frac{mc^2}{(1 - \beta^2)^{1/2}} \frac{r^2\dot{\theta}}{c^2} + qrA_\theta = \gamma mr^2\dot{\theta} + qrA_\theta, \tag{2.54b}$$

$$p_z = \frac{\partial L}{\partial \dot{z}} = \gamma m\dot{z} + qA_z. \tag{2.54c}$$

It is important to recognize that the *canonical angular momentum* p_θ does not have the same dimensions as p_r and p_z. The mechanical θ-momentum is therefore

$$P_\theta = \gamma mr\dot{\theta} = \frac{p_\theta - qrA_\theta}{r}. \tag{2.55}$$

The change in variables from the $(q_i, \dot{q}_i, t)$ set to the (q_i, p_i, t) set is accomplished by introducing the Hamiltonian $H(q_i, p_i, t)$ via the transformation

$$H(q_i, p_i, t) = \sum_i \dot{q}_i p_i - L(q_i, \dot{q}_i, t). \tag{2.56}$$

The total differential of H is

$$dH = \sum_i p_i\, d\dot{q}_i + \sum_i \dot{q}_i\, dp_i - \sum_i \frac{\partial L}{\partial q_i}\, dq_i - \sum_i \frac{\partial L}{\partial \dot{q}_i}\, d\dot{q}_i - \frac{\partial L}{\partial t}\, dt. \tag{2.57}$$

Since $p_i = \partial L/\partial \dot{q}_i$, the first and fourth terms on the right-hand side cancel, and we obtain

$$dH = \sum_i \dot{q}_i\, dp_i - \sum_i \frac{\partial L}{\partial q_i}\, dq_i - \frac{\partial L}{\partial t}\, dt = \sum_i \dot{q}_i\, dp_i - \sum_i \frac{dp_i}{dt}\, dq_i - \frac{\partial L}{\partial t}\, dt, \tag{2.58}$$

where we made the substitution [from Lagrange's equations of motion (2.32)]

$$\frac{\partial L}{\partial q_i} = \frac{d}{dt} \frac{\partial L}{\partial \dot{q}_i} = \frac{dp_i}{dt}.$$

On the other hand, $H = H(q_i, p_i, t)$; hence, the left-hand side may be written as

$$dH = \sum_i \frac{\partial H}{\partial q_i}\, dq_i + \sum_i \frac{\partial H}{\partial p_i}\, dp_i + \frac{\partial H}{\partial t}\, dt. \tag{2.59}$$

Since $\partial L/\partial q_i = dp_i/dt$, comparison of Equations (2.58) and (2.59) yields the equations

$$\frac{dq_i}{dt} = \frac{\partial H}{\partial p_i}, \qquad \frac{dp_i}{dt} = -\frac{\partial H}{\partial q_i}, \qquad -\frac{\partial L}{\partial t} = \frac{\partial H}{\partial t}. \tag{2.60}$$

These are Hamilton's canonical equations, which represent an alternative form of the equations of motion. They constitute a set of $2n$ first-order equations replacing the n Lagrange equations. In principle, the first step in solving a particular problem of charged particle motion in this canonical formulation is to set up the Lagrangian L as $L(q_i, \dot{q}_i, t)$. Using Equation (2.49), one then obtains the canonical momenta p_i, and with their aid, the Hamiltonian H is constructed according to the prescription of Equation (2.56). The equations of motion, which are now first order, then follow by substituting H in Equation (2.60).

2.3.4 The Hamiltonian for Charged Particles and Some Conservation Theorems

We shall now consider a *conservative* system in the *nonrelativistic* approximation and show that the Hamiltonian H in this case is the total energy of the particle. This will also explain the transformation [Equation (2.56)] which defined H. In a conservative system, the force is given by $\mathbf{F} = -\nabla U$, and the Lagrangian L is not an explicit function of time t. The total time derivative of L is then

$$\frac{dL}{dt} = \sum_i \frac{\partial L}{\partial q_i} \frac{dq_i}{dt} + \sum_i \frac{\partial L}{\partial \dot{q}_i} \frac{d\dot{q}_i}{dt}, \tag{2.61}$$

which in view of the Lagrange equations [Equation (2.32)] can be written as

$$\frac{dL}{dt} = \sum_i \frac{d}{dt} \frac{\partial L}{\partial \dot{q}_i} \dot{q}_i + \sum_i \frac{\partial L}{\partial \dot{q}_i} \frac{d\dot{q}_i}{dt} = \sum_i \frac{d}{dt} \left(\dot{q}_i \frac{\partial L}{\partial \dot{q}_i} \right). \tag{2.62}$$

It therefore follows that

$$\frac{d}{dt} \left(L - \sum_i \dot{q}_i \frac{\partial L}{\partial \dot{q}_i} \right) = \frac{d}{dt} \left(L - \sum_i \dot{q}_i p_i \right) = 0. \tag{2.63}$$

Considering the definition for the Hamiltonian H [Equation (2.56)], the last equation can be stated in the form

$$\frac{dH}{dt} = 0, \qquad \text{or} \qquad H = \text{const.} \tag{2.64}$$

Thus, for a conservative system, the Hamiltonian H is a constant of the motion.

Next, we prove that $H = T + U = $ total energy. For simplicity, let us consider cartesian coordinates, where the Lagrangian is given by

$$L = \frac{m}{2}\left(\dot{x}^2 + \dot{y}^2 + \dot{z}^2\right) - q\phi(x, y, z) + q\left(\dot{x}A_x + \dot{y}A_y + \dot{z}A_z\right). \qquad (2.65)$$

From the definition of H [Equation (2.56)] and using (2.50), we have

$$H = \dot{x}p_x + \dot{y}p_y + \dot{z}p_z - L;$$

hence,

$$H = \dot{x}(m\dot{x} + qA_x) + \dot{y}(m\dot{y} + qA_y) + \dot{z}(m\dot{z} + qA_z) \quad L. \qquad (2.66)$$

Substituting L in Equation (2.66) yields

$$H = \frac{m}{2}\left(\dot{x}^2 + \dot{y}^2 + \dot{z}^2\right) + q\phi = T + U. \qquad (2.67)$$

To obtain the Hamiltonian in the form $H(q_i, p_i, t)$ required for the equations of motion, we transform from the velocity components $(\dot{x}, \dot{y}, \dot{z})$ to the canonical momentum components $p_x = P_x + qA_x$, and so on. Then from Equation (2.67) with

$$\frac{m}{2}\dot{x}^2 = \frac{P_x^2}{2m} = \frac{(p_x - qA_x)^2}{2m},$$

and so on, we find that

$$H = \frac{1}{2m}\left[(p_x - qA_x)^2 + (p_y - qA_y)^2 + (p_z - qA_z)^2\right] + q\phi(x, y, z),$$

$$(2.68)$$

or

$$H = \frac{1}{2m}(\mathbf{p} - q\mathbf{A})^2 + q\phi(x, y, z)$$

in cartesian coordinates. In *cylindrical* coordinates, the nonrelativistic Hamiltonian is

$$H = \frac{1}{2m}\left[(p_r - qA_r)^2 + \left(\frac{p_\theta - qrA_\theta}{r}\right)^2 + (p_z - qA_z)^2\right] + q\phi(r, \theta, z).$$

$$(2.69)$$

By using the above Hamiltonian in Equation (2.60) one obtains Hamilton's equations of motion, which are first-order equivalents of the original force equations.

As an example, consider cartesian coordinates and no magnetic field ($\mathbf{A} = 0$); then, from Equations (2.68) and (2.60),

$$\frac{dp_x}{dt} = -\frac{\partial H}{\partial x} = -q \frac{\partial \phi}{\partial x}.$$

Now, in this case $p_x = P_x = m\dot{x}$; hence,

$$\frac{dp_x}{dt} = \frac{d}{dt}(m\dot{x}) = qE_x,$$

in agreement with Equation (2.10a).

The Hamiltonian in *relativistically* correct form is obtained by substituting the Lagrangian L of (2.47) in Equation (2.56) and changing to the (q_i, p_i, t) set of variables. Take the case $\mathbf{A} = 0$ in cartesian coordinates. Then with L from Equation (2.47):

$$H = \sum_i \dot{q}_i p_i - L = \sum_i \dot{q}_i \frac{\partial L}{\partial \dot{q}_i} - L,$$

$$H = \gamma m(\dot{x}^2 + \dot{y}^2 + \dot{z}^2) - mc^2\left[1 - (1 - \beta^2)^{1/2}\right] + q\phi,$$

$$H = mc^2\left[\frac{\beta^2}{(1 - \beta^2)^{1/2}} - 1 + (1 - \beta^2)^{1/2}\right] + q\phi,$$

$$H = mc^2\left[\frac{\beta^2 + 1 - \beta^2}{(1 - \beta^2)^{1/2}} - 1\right] + q\phi,$$

$$H = mc^2(\gamma - 1) + U = T + U,$$

as in the nonrelativistic case. Now, from Equation (2.24),

$$\gamma mc^2 = c(m^2 c^2 + P^2)^{1/2},$$

where $\mathbf{P} = \gamma m\mathbf{v} = \mathbf{p}$ since $\mathbf{A} = 0$; hence,

$$H = c(m^2 c^2 + P^2)^{1/2} + q\phi - mc^2.$$

To include the case where $\mathbf{A} \neq 0$, we simply replace $\mathbf{P}$ by $\mathbf{p} - q\mathbf{A}$ to get the relativistic Hamiltonian

$$H = c[m^2 c^2 + (\mathbf{p} - q\mathbf{A})^2]^{1/2} + q\phi - mc^2.$$

If we use the Lagrangian according to Equation (2.48), the last term, mc^2, on the right-hand side drops out, and the Hamiltonian has the simpler form $H = \gamma mc^2 + U$, or

$$H = c[m^2c^2 + (\mathbf{p} - q\mathbf{A})^2]^{1/2} + q\phi. \tag{2.70}$$

In cartesian coordinates the relativistic Hamiltonian is given by

$$H = c\left[m^2c^2 + (p_x - qA_x)^2 + (p_y - qA_y)^2 + (p_z - qA_z)^2\right]^{1/2} + q\phi. \tag{2.71}$$

For cylindrical coordinates one obtains

$$H = c\left[m^2c^2 + (p_r - qA_r)^2 + \left(\frac{p_\theta - qrA_\theta}{r}\right)^2 + (p_z - qA_z)^2\right]^{1/2} + q\phi. \tag{2.72}$$

We have shown that in a conservative system (where L and H do not depend explicitly on time), the Hamiltonian represents the total energy, which, in this case, is a constant of the motion. Another important conservation theorem is obtained for the case where the Hamiltonian does not depend on one of the space coordinates, for instance, q_j. The latter is then called a *cyclic variable* and from Hamilton's equations (2.60) follows

$$\frac{dp_j}{dt} = -\frac{\partial H}{\partial q_j} = 0, \tag{2.73}$$

or

$$p_j = \text{const} \tag{2.74}$$

(i.e., the canonical momentum variable p_j is a constant of the motion in this case).

The most important application of this theorem is for systems with cylindrical symmetry, where $\phi, \mathbf{A}$, and hence the Hamiltonian H are independent of the azimuth coordinate θ. Therefore,

$$\frac{dp_\theta}{dt} = -\frac{\partial H}{\partial \theta} = 0,$$

or

$$p_\theta - \gamma mr^2\dot\theta + qrA_\theta = rP_\theta + qrA_\theta = \text{const}. \tag{2.75}$$

Equation (2.75) represents the *conservation of canonical angular momentum,* which is very useful in the analysis of particle dynamics in axisymmetric fields. It is equivalent to *Busch's theorem,* which was originally derived from the equations of motion [Equation (2.11b)] and which states that

$$\gamma m r^2 \dot{\theta} + \frac{q}{2\pi} \psi = \text{const.} \tag{2.76}$$

Here $\psi = \int \mathbf{B} \, d\mathbf{S}$ is the magnetic flux enclosed by the particle trajectory (i.e., the flux inside a circle with radius r given by the radial distance r of the particle from the axis at a given position along the trajectory). The proof that Equation (2.76) is equivalent to Equation (2.75) simply follows from

$$\psi = \int \mathbf{B} \cdot d\mathbf{S} = \int (\nabla \times \mathbf{A}) \cdot d\mathbf{S} = \oint \mathbf{A} \cdot d\mathbf{l} = 2\pi r A_\theta , \tag{2.77}$$

hence, $r A_\theta = \psi/2\pi$.

As an example, consider a particle launched at point 1 in an axisymmetric magnetic field with $\dot{\theta}_1 = 0$. Then $\dot{\theta}$ at any other point (r, z) along the trajectory can be calculated from Equation (2.76):

$$\gamma m r^2 \dot{\theta} = -\frac{q}{2\pi} (\psi - \psi_1),$$

or

$$\dot{\theta} = -\frac{q}{2\pi \gamma m r^2} (\psi - \psi_1). \tag{2.78}$$

If the trajectory remains close to the axis so that to first-order approximation $B_z(r, z) \approx B_z(0, z) = B$, we find that

$$\dot{\theta} = -\frac{q}{2\pi \gamma m r^2} (B r^2 \pi - B_1 r_1^2 \pi) = -\frac{q}{2\gamma m} \left[B - B_1 \left(\frac{r_1}{r} \right)^2 \right]. \tag{2.79}$$

If, moreover, the magnetic field is uniform ($B = B_1$), we obtain

$$\dot{\theta} = -\frac{qB}{2\gamma m} \left[1 - \left(\frac{r_1}{r} \right)^2 \right] = \mp \frac{\omega_c}{2} \left[1 - \left(\frac{r_1}{r} \right)^2 \right], \tag{2.80}$$

where

$$\omega_c = \left| \frac{qB}{\gamma m} \right| \tag{2.81}$$

is the *cyclotron frequency.* The sign in (2.80) depends on the polarity of the charge and the direction of the magnetic field. It is negative when both q and B are either positive or negative, and it is positive when q and B have opposite signs.

Suppose now that the particle is launched in a region where $\mathbf{B} = 0$ with $\dot{\theta}_1 = 0$ (i.e., no initial velocity component in azimuthal direction). Then, since $\psi_1 = 0$, it follows from Equation (2.76) that

$$\gamma m r^2 \dot{\theta} = -\frac{q}{2\pi}\psi\,; \tag{2.82}$$

hence, we get for trajectories near the axis where $\psi \approx Br^2\pi$,

$$\dot{\theta} = -\frac{qB}{2\gamma m} = \mp\frac{\omega_c}{2} = \mp\omega_L\,, \tag{2.83}$$

where $\omega_L = \omega_c/2$ is known as the *Larmor frequency*.

Another important theorem that is used widely, especially in plasma physics, applies for *adiabatic* particle motion in magnetic fields. The motion of a particle is called adiabatic when the magnetic field varies so slowly along the particle's helical trajectory that the change of the field strength during one revolution or cyclotron period $\tau_c = 2\pi/\omega_c$ is negligibly small. In this case, the *theorem of adiabatic invariance* states that the magnetic flux encircled by the particle trajectory remains a constant of the motion.

The concept of adiabatic invariance is introduced by considering the *action integrals* of a system in terms of the generalized canonical coordinates q_i and momenta p_i. For each coordinate q_i, which is *periodic*, the action integral J_i is defined by

$$J_i = \oint p_i\,dq_i\,. \tag{2.84}$$

The integration is over a complete cycle of the periodic coordinate q_i. For a given system with specified initial conditions and with changes that are adiabatic, the action integral is invariant or a constant of the motion:

$$J_i = \text{const.} \tag{2.85}$$

As an example, let us consider an axially symmetric $\mathbf{B}$ field with particles moving adiabatically on spiraling trajectories that encircle the axis. The periodic coordinate is then θ (cylindrical coordinates) and with $p_\theta = \gamma m r^2 \dot{\theta} + qrA_\theta$, the action integral takes the form

$$J_\theta = \int_0^{2\pi} \gamma m r^2 \dot{\theta}\,d\theta + \int_0^{2\pi} qA_\theta r\,d\theta = \text{const.} \tag{2.86}$$

Now let

$$\dot{\theta} = -\frac{qB_z(r)}{\gamma m},$$

where $r = R$ is constant during one cyclotron period, and

$$\int_0^{2\pi} A_\theta r \, d\theta = 2\pi \int_0^R B_z(r) r \, dr = \psi.$$

Then we get for the first term in Equation (2.86),

$$-\int_0^{2\pi} \gamma m R^2 \left[\frac{qB_z(R)}{\gamma m} \right] d\theta = -2\pi R^2 q B_z(R),$$

and the action integral [Equation (2.86)] may be written as

$$2\pi B_z(R)R^2 - 2\pi \int_0^R B_z(r) r \, dr = \text{const},$$

or

$$2\pi B_z(R)R^2 - \psi = \text{const}. \tag{2.87}$$

If the particle orbit is confined to the region near the axis so that in first approximation $B_z(r) \approx B(0)$, we have $B_z(R) = B$ and

$$\psi \simeq BR^2 \pi. \tag{2.88}$$

Hence, $2\psi - \psi = \text{const}$, or

$$\psi = BR^2 \pi = \text{const}. \tag{2.89}$$

This is the adiabatic *theorem of magnetic flux conservation,* which can be generalized to arbitrary magnetic field configurations. Under adiabatic conditions, the particles spiral along magnetic flux lines such that the flux encircled remains constant (i.e., the radius R of the circle is proportional to $B^{-1/2}$). Since the mechanical momentum is defined by

$$P_\theta = \gamma m v_\theta = \gamma m R \dot{\theta} = -RqB, \tag{2.90}$$

we have the equivalent conservation law:

$$\frac{P_\theta^2}{B} = \text{const}, \qquad \text{or} \qquad P_\theta R = \text{const}. \tag{2.91}$$

The same law is also expressed as the *conservation of the magnetic moment*. Since the magnetic moment M is defined by the product of current I and enclosed area S (i.e., $M = I \cdot S = IR^2\pi$), we have for a circulating charge q,

$$I = \frac{q}{\tau} = \frac{q\omega_c}{2\pi} = \frac{q^2 B}{2\pi\gamma m},$$

and therefore, in view of Equation (2.89),

$$M - \frac{q\omega_c R^2}{2} = \frac{q^2 B R^2}{2\gamma m} = \text{const}. \tag{2.92}$$

We should note that this conservation law for the magnetic moment applies only when the particle's energy is either nonrelativistic ($\gamma \sim 1$) or does not change ($\gamma = \text{const}$), in contrast to the flux conservation law [Equations (2.87) and (2.89)]. As an example, for relativistic electron motion in a pulsed magnetic field, γ varies due to $\nabla \times \mathbf{E} = -(\partial \mathbf{B}/\partial t)$, and hence the magnetic moment may change although the flux conservation law, $\psi = \text{const}$, still holds

2.4 THE EULER TRAJECTORY EQUATIONS

2.4.1 The Principle of Least Action and the Euler Equations

In addition to Hamilton's principle, discussed in previous sections, the *principle of least action* plays an important role in classical mechanics. It holds for a conservative system where the Hamiltonian does not depend explicitly on time (i.e., where the total energy of the system is conserved). Applied to charged particle motion in a conservative electric and magnetic field, the principle of least action may be stated as follows: The line integral of the canonical momentum $\mathbf{p}$ along the path of motion between two given points in space is an extremum; that is,

$$\delta \int_1^2 \mathbf{p} \cdot d\mathbf{l} = 0. \tag{2.93}$$

Note that there is a distinct difference between the two variational principles. In *Hamilton's principle*, the variation of the path was taken between two unvaried points in *both space and time*; the varied paths do not obey the equations of motion or the conservation laws. In the *principle of least action*, on the other hand, the endpoints are fixed points in space but *not* in time; however, the varied paths do obey the law of conservation of energy. Thus, we are comparing all possible paths for a charged particle to go from one fixed point to another fixed point in space. The time of travel may differ, but along each path the sum of kinetic and potential energy remains a constant of the motion. The *actual* path followed by the particle in the real world is then uniquely determined by Equation (2.93).

The principle of least action is used to obtain equations for the trajectories directly rather than integrating the equations of motion with respect to time and then eliminating the time. With $\mathbf{p} = \gamma m \mathbf{v} + q\mathbf{A}$, we may write for Equation (2.93),

$$\delta \int_1^2 (\gamma m \mathbf{v} + q\mathbf{A}) \cdot d\mathbf{l} = \delta \int_1^2 (\gamma m v \, dl + q\mathbf{A} \cdot d\mathbf{l}) = 0. \qquad (2.94)$$

Since by definition $\mathbf{v}$ is in the direction of $d\mathbf{l}$, we have $\mathbf{v} \cdot d\mathbf{l} = v \, dl$. In cartesian coordinates $dl = [dx^2 + dy^2 + dz^2]^{1/2}$ and $\mathbf{A} \cdot d\mathbf{l} = A_x \, dx + A_y \, dy + A_z \, dz$.

Now, let us make one of the three coordinates, say x, the independent variable. Then, with $dy/dx = y'$ and $dz/dx = z'$ denoting the slopes of the trajectory, Equation (2.94) becomes

$$\delta \int_{x_1}^{x_2} \left[\gamma m v (1 + y'^2 + z'^2)^{1/2} + q(A_x + A_y y' + A_z z') \right] dx = 0, \qquad (2.95)$$

where the expression in brackets following the integral will be denoted by $F(x, y, z; y', z')$. [In general coordinates, the function F would be $F(q_i, q_i')$, where $q' = dq_i/dq_1$, for example.] In terms of the function F, Equation (2.95) can be stated in the form

$$\delta \int_{x_1}^{x_2} F(x, y, z; y', z') \, dx$$

$$= \int_{x_1}^{x_2} \left(\frac{\partial F}{\partial y} \delta y + \frac{\partial F}{\partial z} \delta z + \frac{\partial F}{\partial y'} \delta y' + \frac{\partial F}{\partial z'} \delta z' \right) dx = 0. \quad (2.96)$$

Since x is the independent variable and the variation is in the y and z directions, the term $(\partial F/\partial x)\delta x$ is zero and does not appear in the integral. Now

$$\delta y' = \frac{d(\delta y)}{dx}, \qquad \delta z' = \frac{d(\delta z)}{dx},$$

and partial integration of the terms involving y' and z' yields

$$\int_{x_1}^{x_2} \frac{\partial F}{\partial y'} \frac{d(\delta y)}{dx} \, dx = \left[\frac{\partial F}{\partial y'} \delta y \right]_{x_1}^{x_2} - \int_{x_1}^{x_2} \frac{d}{dx} \frac{\partial F}{\partial y'} \delta y \, dx.$$

The bracketed term equals zero, since $\delta y = 0$ at the endpoints. An analogous expression is obtained for the z' term. Using these results, we can write Equation (2.96) in the form

$$\delta \int_{x_1}^{x_2} F \, dx = \int_{x_1}^{x_2} \left[\left(\frac{\partial F}{\partial y} - \frac{d}{dx} \frac{\partial F}{\partial y'} \right) \delta y + \left(\frac{\partial F}{\partial z} - \frac{d}{dx} \frac{\partial F}{\partial z'} \right) \delta z \right] dx = 0,$$

(2.97)

which yields the two Euler equations:

$$\frac{\partial F}{\partial y} - \frac{d}{dx} \frac{\partial F}{\partial y'} = 0,$$

(2.98a)

$$\frac{\partial F}{\partial z} - \frac{d}{dx} \frac{\partial F}{\partial z'} = 0.$$

(2.98b)

In generalized coordinates, if q_1 is the independent variable, we can write

$$\frac{\partial F}{\partial q_2} - \frac{d}{dq_1} \frac{\partial F}{\partial q_2'} = 0,$$

(2.99a)

$$\frac{\partial F}{\partial q_3} - \frac{d}{dq_1} \frac{\partial F}{\partial q_3'} = 0.$$

(2.99b)

Returning now to Equation (2.95), we express v in terms of the kinetic energy and the rest energy. This is possible since in the principle of least action, the varied paths obey the law of conservation of energy. From Equation (2.23) we have

$$mc\gamma\beta = \frac{E_0}{c} (\gamma^2 - 1)^{1/2}.$$

Therefore, Equation (2.95) can be written as

$$\delta \int_{x_1}^{x_2} \left\{ [(\gamma^2 - 1)(1 + y'^2 + z'^2)]^{1/2} + \frac{qc}{E_0} (A_x + A_y y' + A_z z') \right\} dx = 0.$$

(2.100)

As a special example, let us consider the *nonrelativistic* case of a particle in an electrostatic field. In this situation, $\mathbf{A} = 0$ and

$$mv = [2qm\phi(x, y, z)]^{1/2},$$

where $\phi(x, y, z)$ is defined as the potential corresponding to the particle's kinetic energy at that point. Then Equation (2.95) becomes

$$\delta \int_{x_1}^{x_2} \left[\phi^{1/2}(1 + y'^2 + z'^2)^{1/2} \right] dx = 0. \tag{2.101}$$

Let us now further assume that the $x-y$ plane is a plane of symmetry. Then the force component in the z-direction is zero, and a particle with $z_0' = 0$ remains in this plane. For this case we have

$$F(x, y; y') = \phi^{1/2}(x, y)(1 + y'^2)^{1/2},$$

$$\frac{\partial F}{\partial y} = \frac{1}{2} \phi^{-1/2} \frac{\partial \phi}{\partial y}(1 + y'^2)^{1/2},$$

$$\frac{\partial F}{\partial y'} = \frac{1}{2} \phi^{1/2}(1 + y'^2)^{-1/2} 2y' = \phi^{1/2} \frac{y'}{(1 + y'^2)^{1/2}}.$$

If we differentiate the last expression with respect to x and substitute into the Euler equation (2.98a), we obtain, after some algebra, the following trajectory equation for the particle motion in the $x-y$ plane:

$$y'' = \frac{1 + y'^2}{2\phi} \left(\frac{\partial \phi}{\partial y} - y' \frac{\partial \phi}{\partial x} \right) \tag{2.102}$$

(see [C.1, p. 401]).

Equation (2.102) also holds, of course, for particle trajectories in the symmetry planes (meridional planes) of an axially symmetric system. In this case one uses cylindrical coordinates, and if z denotes the independent variable (i.e., $r' = dr/dz$, etc.), one gets

$$r'' = \frac{1 + r'^2}{2\phi} \left(\frac{\partial \phi}{\partial r} - r' \frac{\partial \phi}{\partial z} \right). \tag{2.103}$$

The equations for *skew* trajectories ($v_\theta \neq 0$) in axially symmetric systems that include both static electric and magnetic fields and in which the particle motion is relativistic are derived in the next section.

2.4.2 Relativistic Euler Equations in Axially Symmetric Fields

In axially symmetric fields, the vector potential $\mathbf{A}$ has only a θ-component, $A_\theta(r, z)$, and in place of Equation (2.100), we have (using polar coordinates)

$$\delta \int_{x_1}^{x_2} \left\{ [(\gamma^2 - 1)(r'^2 + r^2\theta'^2 + 1)]^{1/2} + \frac{qc}{E_0} A_\theta r\theta' \right\} dz = 0. \tag{2.104}$$

Here z is the independent variable and the path element dl is given by

$$dl = [(dr)^2 + (r\, d\theta)^2 + (dz)^2]^{1/2} = [r'^2 + r^2\theta'^2 + 1]^{1/2}\, dz.$$

The Euler equations (2.99a) and (2.99b) take the form

$$\frac{d}{dz} \frac{\partial F}{\partial r'} - \frac{\partial F}{\partial r} = 0, \tag{2.105a}$$

$$\frac{d}{dz} \frac{\partial F}{\partial \theta'} - \frac{\partial F}{\partial \theta} = 0. \tag{2.105b}$$

The electrostatic potential $\phi(r, z)$ is implicitly given in γ:

$$\gamma - \frac{E_T}{E_0} = \frac{T + E_0}{E_0} = \frac{q\phi(r, z) + F_0}{E_0}. \tag{2.106}$$

Note that ϕ is the *voltage equivalent* of the kinetic energy T, defined by the actual potential distribution set up by the electrodes plus the particles' initial *voltage* when entering the field.

We begin with Equation (2.105b). Since there is no θ dependence, we have $\partial F / \partial \theta = 0$. Thus

$$\frac{d}{dz} \frac{\partial F}{\partial \theta'} = \frac{d}{dz} \left[(\gamma^2 - 1)^{1/2}(r'^2 + r^2\theta'^2 + 1)^{-1/2} r^2\theta' + \frac{qc}{E_0} A_\theta r \right] = 0.$$

This equation can be integrated directly, leading to

$$\frac{r^2\theta'(\gamma^2 - 1)^{1/2}}{(r'^2 + r^2\theta'^2 + 1)^{1/2}} + \frac{qc}{E_0} A_\theta r = C. \tag{2.107}$$

It can be shown that Equation (2.107) is equivalent to the law of *conservation of canonical angular momentum*,

$$\gamma m r^2 \dot{\theta} + q A_\theta r = p_\theta = mcC. \tag{2.108}$$

It implies that a particle emitted from a source located in a magnetic field will attain mechanical angular momentum when leaving the field. A particle launched from a field-free region (i.e., $A_\theta = 0$) and passing through a magnetic field will experience a change in mechanical angular momentum, which, however, will be restored to its initial value when the particle leaves the field.

Equation (2.107) can be substituted into the first Euler equation for $r(z)$ in order to eliminate θ'. To simplify the algebra, we will introduce the following parameters [A.1, p. 432]:

$$\xi = \gamma^2 - 1, \qquad \eta = \frac{1}{(\xi)^{1/2}}\left(\frac{C}{r} - \frac{qc}{E_0}A_\theta\right), \qquad \lambda = \xi(1 - \eta^2). \qquad (2.109)$$

Dividing by $r(\gamma^2 - 1)^{1/2}$, we can write Equation (2.107) in the form

$$\frac{r\theta'}{(r'^2 + r^2\theta'^2 + 1)^{1/2}} = \frac{1}{(\xi)^{1/2}}\left(\frac{C}{r} - \frac{qc}{E_0}A_\theta\right) = \eta. \qquad (2.110)$$

From Equation (2.110) one finds that

$$r\theta' = \eta\frac{(1 + r'^2)^{1/2}}{(1 - \eta^2)^{1/2}} \qquad (2.111)$$

and

$$r'^2 + r^2\theta'^2 + 1 = \frac{1 + r'^2}{1 - \eta^2}. \qquad (2.112)$$

The function F is defined as

$$F = (\xi)^{1/2}(r'^2 + r^2\theta'^2 + 1)^{1/2} + \frac{qc}{E_0}\theta'rA_\theta. \qquad (2.113)$$

For the two derivative terms in the Euler equation (2.105a) one finds, after some manipulation, that

$$\frac{d}{dz}\frac{\partial F}{\partial r'} = \frac{\lambda^{1/2}}{(1 + r'^2)^{3/2}}r'' + \frac{1}{2\lambda^{1/2}}\frac{r'}{(1 + r'^2)^{1/2}}\left(\frac{\partial\lambda}{\partial z} + r'\frac{\partial\lambda}{\partial r}\right) \qquad (2.114)$$

and

$$\frac{\partial F}{\partial r} = \frac{(1 + r'^2)^{1/2}}{2\lambda^{1/2}}\frac{\partial\lambda}{\partial r}. \qquad (2.115)$$

Substitution of Equations (2.114) and (2.115) into Equation (2.105a) yields the differential equation for the radial motion $r(z)$ of a charged particle in axisymmetric, static electric and magnetic fields

$$r'' = \frac{1 + r'^2}{2\lambda}\left(\frac{\partial\lambda}{\partial r} - r'\frac{\partial\lambda}{\partial z}\right), \qquad (2.116)$$

where from (2.109)

$$\lambda = \gamma^2 - 1 - \left(\frac{C}{r} - \frac{qcA_\theta}{E_0}\right)^2.$$

This trajectory equation, in which the time t has been eliminated, is exact and in relativistic form; no simplifying assumptions have been made to this point. It agrees with Panofsky and Phillips [A.1, Equations (23), (24), p. 432] except for a factor $(1 + r'^2)$ associated with $\partial\lambda/\partial r$ in their result, which appears to be an error. For the *nonrelativistic* limit, Equation (2.116) is in agreement with Zworykin et al. [C.1, Equation (15.54), p. 504] if we make the substitutions $\xi \to \phi$, $D \to \eta$, and $\lambda \to \phi(1 - D^2)$. Specifically, if there is only an electric field (hence $A_\theta = 0$), and the particle moves within a fixed meridional plane (i.e., $C = 0$), we have $\xi = 2q\phi/E_0$, $\eta = 0$, and $\lambda = \xi = 2q\phi/E_0$. Then Equation (2.116) becomes identical with Equation (2.103).

It should be pointed out that the meridional plane in which the particle is located at any given time and in which $r(z)$ is measured is not the same throughout the motion, even if the particle crosses the axis. The meridional plane rotates about the z-axis, its azimuth being determined at any given point by integration of Equation (2.111):

$$\theta = \theta_0 + \int_{z_0}^{z} \frac{\eta}{r} \frac{(1 + r'^2)^{1/2}}{(1 - \eta^2)^{1/2}} dz. \tag{2.117}$$

2.5 ANALYTIC EXAMPLES OF CHARGED PARTICLE MOTION

2.5.1 Planar Diode without Space Charge

To illustrate the use of some of the concepts and of the equations of motion discussed earlier in this chapter, we now treat a few problems that yield relatively simple analytical solutions. As a first example, consider two infinite, parallel, conducting plates, one (the cathode) at $x = 0$ with potential $\phi = 0$, the other (the anode) at $x = d$ with potential $\phi = V_0$. Such configuration is known as a planar diode. Suppose that an electron (charge $q = -e$) is emitted from the cathode with a velocity $\mathbf{v}_0 = \{\dot{x}_0, \dot{y}_0, 0\}$, and determine its trajectory in the $x-y$ plane in the nonrelativistic limit. Ignore space-charge effects due to other electrons.

The static electric field between the two plates is given by $\mathbf{E} = -\nabla\phi$ and the potential $\phi(x)$ can be calculated from Laplace's equation (since $\rho = 0$):

$$\nabla^2\phi = \frac{d^2\phi}{dx^2} = 0, \tag{2.118}$$

with the solution

$$\phi(x) = \frac{V_0}{d} x \qquad (2.119)$$

and

$$E_x = E = -\frac{d\phi}{dx} = -\frac{V_0}{d} \qquad (2.120)$$

(i.e., the electric field is uniform).

The nonrelativistic equations of motion in Newton's form [Equations (2.10a) and (2.10b)] are readily integrated, yielding (with $\gamma = 1$)

$$m\ddot{x} = eE, \qquad (2.121)$$

$$\dot{x} = \frac{e}{m} Et + \dot{x}_0, \qquad (2.122)$$

$$x = \frac{eE}{m} \frac{t^2}{2} + \dot{x}_0 t, \qquad (2.123)$$

$$m\ddot{y} = 0, \qquad (2.124)$$

$$\dot{y} = \dot{y}_0, \qquad (2.125)$$

$$y = \dot{y}_0 t. \qquad (2.126)$$

Substituting $t = y/\dot{y}_0$ from (2.127) into Equation (2.123) gives the trajectory equation

$$x = \frac{1}{2} \frac{eE}{m} \frac{y^2}{\dot{y}_0^2} + \frac{\dot{x}_0}{\dot{y}_0} y, \qquad (2.127)$$

which is a parabola.

The kinetic energy gain is simply

$$\Delta T = T - T_0 = \frac{m}{2} (\dot{x}^2 - \dot{x}_0^2) = eV_0 \frac{x}{d}, \qquad (2.128)$$

and hence $\Delta T = eV_0$ when the electron arrives at the anode.

2.5.2 Planar Diode with Space Charge (Child–Langmuir Law)

Let us now include the effect of the space charge of the electron current in the diode on the potential distribution and electron motion. To simplify our analysis, we assume that all electrons are launched with initial velocity $v_0 = 0$ from the cathode (i.e., they are moving on straight lines in the x-direction). This is an example of *laminar flow* where electron trajectories do not cross and the current density is uniform. We try to find the steady-state solution ($\partial/\partial t = 0$) in a self-consistent form. The electrostatic potential is determined from the space-charge density ρ via Poisson's equation, with $\phi = 0$, at $x = 0$ and $\phi = V_0$ at $x = d$, as in the previous case. The relationship between ρ, the current density $\mathbf{J}$, and the electron velocity $\mathbf{v}$ follows from the continuity equation ($\nabla \cdot \mathbf{J} = 0$, or $\mathbf{J} = \rho\mathbf{v} = \text{const}$). The velocity in turn depends on the potential ϕ and is found by integrating the equation of motion. Thus we have the following three equations:

$$\nabla^2 \phi = \frac{d^2\phi}{dx^2} = -\frac{\rho}{\epsilon_0} \qquad \text{(Poisson's equation)}, \qquad (2.129)$$

$$J_x = \rho\dot{x} = \text{const} \qquad \text{(continuity equation)}, \qquad (2.130)$$

$$\frac{m}{2}\dot{x}^2 = e\phi(x) \qquad \text{(equation of motion)}. \qquad (2.131)$$

Substituting $\dot{x} = [2e\phi(x)/m]^{1/2}$ from (2.131) into (2.130) and $\rho = J_x/\dot{x}$ from (2.130) into (2.129) yields

$$\frac{d^2\phi}{dx^2} = \frac{J}{\epsilon_0(2e/m)^{1/2}} \frac{1}{(\phi)^{1/2}}, \qquad (2.132)$$

where the current density $J = -J_x$ is defined as a positive quantity. After multiplication of both sides of Equation (2.132) with $d\phi/dx$, we can integrate and obtain

$$\left(\frac{d\phi}{dx}\right)^2 = \frac{4J}{\epsilon_0(2e/m)^{1/2}} \phi^{1/2} + C. \qquad (2.133)$$

Now $\phi = 0$ at $x = 0$, and if we consider the special case where $d\phi/dx = 0$ at $x = 0$, we obtain $C = 0$. A second integration then yields (with $\phi = V_0$ at $x = d$)

$$\frac{4}{3}\phi^{3/4} = 2\left(\frac{J}{\epsilon_0}\right)^{1/2}\left(\frac{2e}{m}\right)^{-1/4} x,$$

or

$$\phi(x) = V_0\left(\frac{x}{d}\right)^{4/3}, \qquad (2.134)$$

with the relation

$$J = \frac{4}{9}\, \epsilon_0 \left(\frac{2e}{m} \right)^{1/2} \frac{V_0^{3/2}}{d^2}.$$ (2.135)

This equation is identical to Equation (1.3) and is known as *Child's law* or the *Child–Langmuir law*, referred to earlier (Section 1.3). For electrons, one gets

$$J = 2.33 \times 10^{-6} \frac{V_0^{3/2}}{d^2} \quad [\text{A/m}^2],$$ (2.136)

with V_0 in volts and d in meters.

By comparing the result (2.134) for $\phi(x)$ with the previous case, we see that the negative space charge of the electrons lowers the potential at any given point between the two electrodes of the planar diode. Equation (2.135) represents the space-charge limit (i.e., the maximum current density that can be achieved in the diode by increasing the electron supply from the cathode). The electric field at the cathode is zero in this case ($d\phi/dx = 0$ at $x = 0$). The current can be increased by either increasing the voltage or decreasing the cathode–anode spacing. The Child–Langmuir law is of fundamental importance for vacuum tubes, electron guns, and ion sources.

2.5.3 Charged Particle Motion in a Uniform Magnetic Field

We can solve this problem in either cartesian or cylindrical coordinates. If the magnetic field **B** is in the z-direction (i.e., $\mathbf{B} = \{0, 0, B\}$) and the velocity vector is in the $z = 0$ plane, the particle motion is a circle. If we choose a cylindrical coordinate system such that the center of the circle coincides with the origin, we find the solution

$$\dot{r} = 0, \qquad r_0 = \frac{\gamma m v}{qB},$$ (2.137)

$$\dot{\theta} = -\frac{qB}{\gamma m}, \qquad v = r_0 \dot{\theta},$$ (2.138)

as can be readily verified by substitution of these results in Equations (2.11a) and (2.11b).

The particle gyrates on a circle with constant radius r_0, known in the literature as the *cyclotron radius*, and with a constant angular velocity, the *cyclotron frequency* $\omega_c = |qB/\gamma m|$. If the center of the circle does not coincide with the origin, it is better to use cartesian coordinates (see Problem 2.2).

2.5.4 Charged Particle Motion in a Radial Electric Field

Let us consider two conducting, coaxial cylinders, the inner one with radius r_1 and at a potential $\phi(r_1) = V_1$, the outer one with radius r_2 and at a potential $\phi(r_2) = V_2$. By integrating Laplace's equation, we find for the electric field

$$E_r = -\frac{d\phi}{dr} = -\frac{V_2 - V_1}{\ln(r_2/r_1)}\frac{1}{r}. \tag{2.139}$$

The motion of a particle in this field depends on the charge q of the particle and the polarity of the electric field. Let us suppose that for a positive charge E_r is radially inward and for a negative charge radially outward, and furthermore, that the motion is nonrelativistic. Using cylindrical coordinates, we obtain the radial force equation

$$m\ddot{r} - mr\dot{\theta}^2 = qF_r, \tag{2.140}$$

and the azimuthal equation

$$\frac{1}{r}\frac{d}{dt}(mr^2\dot{\theta}) = 0, \quad \text{or} \quad mr^2\dot{\theta} = \text{const.} \tag{2.141}$$

It is readily verified that a special solution exists in which the particle moves on a circle of constant radius r_e, the *equilibrium radius*. In this case $\ddot{r} = 0, \dot{r} = 0, v = v_\theta = v_0 = r_e\dot{\theta}_0$, and the outward centrifugal force is exactly balanced by the inward electric force, hence

$$\frac{mv_0^2}{r_e} = qE_r(r_e) = qE_e, \tag{2.142}$$

or

$$r_e = \frac{mv_0^2}{qE_e}, \tag{2.143}$$

where $E_e = E_r(r_e)$ is the electric field at the equilibrium radius. The angular velocity is given by

$$\dot{\theta}_0^2 = \omega_0^2 = \frac{qE_e}{mr_e}. \tag{2.144}$$

What happens to a particle with the same velocity v that is displaced from the equilibrium radius r_e? To answer this question, we employ a technique widely used in similar problems which assumes that the displacement is small compared to the radius r_e so that the equations of motion can be *linearized.* Let the radial position of a particle be defined by

$$r(\theta) = r_e + x(\theta), \qquad \text{where } x \ll r_e. \tag{2.145}$$

Substitute this into the equations of motion, make a Taylor-series expansion about r_e, and keep only the linear terms. The azimuthal equation (2.141) then yields in the linear approximation

$$\dot{\theta} = \dot{\theta}_0 \frac{r_e^2}{(r_e + x)^2} \approx \dot{\theta}_0 \left(1 - 2 \frac{x}{r_e} \right). \tag{2.146}$$

The radial force equation (2.140) becomes

$$m\ddot{x} - mr_e \left(\frac{1 + x}{r_e} \right) \dot{\theta}_0^2 \left(\frac{1 - 4x}{r_e} \right) = q(E_e + E_r' x), \tag{2.147}$$

where

$$E_r' = \frac{dE_r}{dr} \bigg|_{r_e} = -\frac{E_e}{r_e}.$$

Using the force–equilibrium condition and keeping only the linear terms in x/r_e, we obtain the equation

$$\ddot{x} + \omega_e^2 x = 0, \tag{2.148}$$

where

$$\omega_e^2 = 2\dot{\theta}_0^2 = 2 \frac{qE_e}{mr_e}. \tag{2.149}$$

If $\omega_e^2 > 0$, the particles are performing harmonic oscillations about the equilibrium radius, which may be written in the form

$$x = x_m \sin\left[\sqrt{2}\, \dot{\theta}_0 (t - t_0) \right] = x_m \sin\left[\sqrt{2}\, (\theta - \theta_0) \right]. \tag{2.150}$$

The nodes of the oscillations are spaced at intervals of $\sqrt{2}\, \dot{\theta}_0 (t - t_0) = \sqrt{2}\, \Delta\theta = \pi$, or

$$\Delta\theta = \frac{\pi}{\sqrt{2}} = 127°17'. \tag{2.151}$$

An important application of this theory is the electrostatic analyzer. In this device, cylindrically shaped capacitor plates extending over a sector with an angle of $\pi/\sqrt{2}$ are used to deflect a beam and separate the particles with different velocity (velocity analyzer).

2.5.5 The Harmonic Oscillator

In the previous problems, the linearization of the equations of motion for small displacements from the equilibrium orbit led to Equation (2.148), which is the differential equation for an harmonic oscillator. Such a system is characterized by the fact that the forces acting on a particle are proportional to the displacement from the equilibrium position. We will now treat the harmonic-oscillator problem in the Hamiltonian framework. Consider the nonrelativistic motion of a particle with positive charge q in an electrostatic field defined by the potential

$$\phi(x) = V_0 \left(\frac{x}{a} \right)^2. \tag{2.152}$$

The canonical variables are $x, p = P = m\dot{x} = mv$, and the Hamiltonian H is given by

$$H(x, P) = T + U = \frac{P^2}{2m} + \frac{1}{2} kx^2 = H_0, \tag{2.153}$$

where $k = 2qV_0/a^2$. H_0 is the total energy and is constant for the conservative system being considered here. The two coordinates x and P define a two-dimensional space called *phase space*, and the particle motion in this two-dimensional space is an ellipse with semiaxes $(2mH_0)^{1/2}$ and $(2H_0/k)^{1/2}$, provided that k is a positive quantity.

Hamilton's equations of motion are

$$\frac{dP}{dt} = -\frac{\partial H}{\partial x} = -kx \tag{2.154}$$

and

$$\frac{dx}{dt} = \dot{x} = \frac{\partial H}{\partial P} = \frac{P}{m}. \tag{2.155}$$

Differentiating (2.155) and substituting in (2.154) yields the harmonic oscillator equation

$$\ddot{x} + \omega^2 x = 0, \tag{2.156}$$

with

$$\omega^2 = \frac{k}{m} = \frac{2qV_0}{ma^2}. \tag{2.157}$$

If $\dot{x} = v_0$, $x = 0$ at $t = 0$, the solution is

$$x = \frac{v_0}{\omega} \sin \omega t \tag{2.158}$$

and

$$\dot{x} = v_0 \cos \omega t, \qquad (2.159)$$

with

$$H_0 = \frac{m}{2} v_0^2. \qquad (2.160)$$

Using the radian frequency ω in place of the constant k, we can write the Hamiltonian in the alternative form

$$H(x, P) = \frac{P^2}{2m} + \frac{1}{2} m\omega^2 x^2. \qquad (2.161)$$

What happens if the potential is not constant but varies with time, that is, $V_0 = V_0(t)$, hence, $k = k(t)$ or $\omega = \omega(t)$? If the variation with time is slow enough that the potential change during one particle oscillation is negligibly small, we can make use of the adiabatic invariance of the action integral $J = \oint P \, dx$. In our case, using $\dot{x}$ in place of P and taking the integral over one-fourth of the oscillation cycle, from $x = 0$ to $x = x_m = v_0/\omega$, we can express this invariance as

$$J = \int_0^{x_m} \dot{x} \, dx = \text{const.} \qquad (2.162)$$

Now from (2.159) and (2.158) we have

$$\dot{x} = v_0(1 - \sin^2 \omega t)^{1/2} = (v_0^2 - \omega^2 x^2)^{1/2}, \qquad (2.163)$$

hence

$$J = \int_0^{x_m} (v_0^2 - \omega^2 x^2)^{1/2} \, dx = \frac{v_0^2}{\omega} \frac{\pi}{4} = \text{const.} \qquad (2.164)$$

In view of (2.160), we can express (2.164) in the equivalent form

$$\frac{H_0}{\omega} = \text{const.} \qquad (2.165)$$

The total energy H_0 is thus no longer constant but varies linearly with the frequency ω.

Recall that in two-dimensional phase space the particle trajectory is an ellipse with semiaxes $P_m = (2mH_0)^{1/2}$ in the P-direction and $x_m = (2H_0/m\omega^2)^{1/2}$ in the x-direction. The area of this ellipse is given by

$$A = P_m x_m \pi = 2\pi \frac{H_0}{\omega} = 4mJ. \qquad (2.166)$$

Consequently, the adiabatic invariance of the action integral implies that the area A of the ellipse traced by the particle trajectory in phase space remains constant.

Thus, for instance, if the potential increases adiabatically with time, the total energy H_0 increases proportional to the frequency ω. Furthermore, the amplitude x_m of the particle oscillation decreases as $\omega^{-1/2}$ while the momentum amplitude increases as $\omega^{1/2}$, so that $x_m P_m = $ const. It should be noted here that the phase-space area A remains constant even if the system is nonadiabatic. This more general conservation law follows from Liouville's theorem, which will be discussed in Section 3.2.

Another important relation for the harmonic oscillator system concerns the average values of kinetic energy and potential energy during one oscillation cycle. Since

$$\overline{\dot{x}^2} = \int_0^{2\pi} \dot{x}^2 \, d(\omega t) = v_0^2 \int_0^{2\pi} \cos^2 \omega t \, d(\omega t) = v_0^2 \pi \tag{2.167}$$

and

$$\overline{x^2} = \int_0^{2\pi} x^2 \, d(\omega t) = \frac{v_0^2}{\omega^2} \int_0^{2\pi} \sin^2 \omega t \, d(\omega t) - \frac{v_0^2}{\omega^2} \pi, \tag{2.168}$$

we find that

$$\overline{T} - \overline{U} = \frac{1}{2} H_0. \tag{2.169}$$

Thus, in the harmonic oscillator the average kinetic energy of the particle during one period is equal to the average potential energy or one-half of the total energy. This result also follows from the *virial theorem* of classical mechanics. For a system of pointlike particles with position vectors l_i and with applied forces $\mathbf{F}_i$ (including any constraints) acting in such a way that the coordinates and velocities of the particles remain finite, the virial theorem states that

$$\overline{T} = -\frac{1}{2} \overline{\sum_i \mathbf{F}_i \cdot \mathbf{l}_i}. \tag{2.170}$$

In the above harmonic-oscillator case, we have only one particle and the force can be derived from the potential energy U (i.e., $\mathbf{F} = -\nabla U$), hence

$$\overline{T} = \frac{1}{2} \overline{\frac{\partial U}{\partial x} x}. \tag{2.171}$$

Since $U = \frac{1}{2}kx^2$, one finds that $(\partial U/\partial x)x = kx^2 = 2U$, and therefore obtains the result $\overline{T} = \overline{U}$ of Equation (2.169).

REFERENCE

1. See Lev B. Okun, "The Concept of Mass," *Phys. Today* **42** (6), 31 (June 1989), for a discussion of this topic.

PROBLEMS

2.1 Derive the equations of motion in cylindrical coordinates [Equation (2.11)] from the cartesian form (2.10) by the appropriate transformations of the coordinates and the components of the velocity and field vectors.

2.2 Solve the equation of motion in cartesian coordinates for a charged particle moving in a uniform magnetic field $\mathbf{B} = B_0 \mathbf{a}_z$ and launched at $t = 0$ at the point $\{x_0, y_0, 0\}$ with velocity $\mathbf{v}_0 = \{\dot{x}_0, \dot{y}_0, 0\}$. Show that the trajectory is a circle described by

$$(x - x_c)^2 + (y - y_c)^2 = R^2$$

and determine the coordinates x_c and y_c of the center point and the radius R of the circle in terms of the initial conditions and the cyclotron frequency $\omega_c = |qB_0/\gamma m|$.

2.3 Solve the nonrelativistic equation of motion in cartesian coordinates for an electron $(q = -e)$ moving in a crossed electric and magnetic field given by $\mathbf{E} = \{0, -E, 0\}$ and $\mathbf{B} = \{0, 0, -B\}$. The initial conditions at $t = 0$ are $\mathbf{r} = \{x_0, y_0, 0\}$, $\mathbf{v} = \{\dot{x}_0, \dot{y}_0, 0\}$. The solution can be greatly simplified by transforming to a system $x' = x - (E/B)t$ moving in the x-direction with constant velocity E/B. The motion of the electron is cycloidal and can be traced by a wheel with radius a rolling in the x-direction to which a pencil is attached at a radial distance R from the center of the wheel. Show from the solution of the equation of motion that $a = E/B\omega$ and $R = 1/\omega[(\dot{x}_0 - E/B)^2 + \dot{y}_0^2]^{1/2}$. Sketch qualitatively the electron trajectory for the following four cases: (a) $R = a$ [i.e., $v_0 = 0$ (common cycloid)]; (b) $R > a$ (curtate cycloid or epicycloid); (c) $R < a$ (prolate cycloid or hypocycloid); and (d) $R = 0$, $y_0 = 0$.

2.4 Consider a charged particle moving in a uniform magnetic field. Let R denote the cyclotron radius, ω_c the cyclotron frequency, and assume that the center of gyration is displaced from the origin of a cylindrical coordinate system by a radial distance R_0. Show that the equation of motion for the radial position of the particle $r(t)$ is given by

$$\ddot{r} + r\frac{\omega_c^2}{4}\left[1 - \frac{(R_0^2 - R^2)^2}{r^4}\right] = 0.$$

2.5 Verify that the relativistic Hamiltonian $H = c[m^2c^2 + (\mathbf{p} - q\mathbf{A})^2]^{1/2} + q\phi$ yields the correct equations of motion in *cylindrical* coordinates [Equation (2.11)].

2.6 Find the Hamiltonian for charged particle motion in the region between the two coaxial conductors of a transmission line subject to the following conditions:

    (a) The inner conductor (radius r_1) is at an electrostatic potential $\phi = V_0$ with respect to the outer conductor (radius r_2).

    (b) A dc current I flows along the inner conductor in the positive z-direction, and a *return* current of the same magnitude flows in the outer conductor in the opposite direction.

    (c) The coaxial transmission line is located inside a solenoid which generates a static uniform magnetic field $\mathbf{B} = B_0\mathbf{a}_z$.

2.7 Solve the relativistic equation of motion for a positively charged particle moving in a uniform electric field $\mathbf{E} = (E,0,0)$ with initial conditions $t = 0: x = 0,\ y = 0,\ P_y = P_0$. Show that by elimination of time the trajectory is given as

$$x = \frac{U_0}{qE}\left(\cosh\frac{qEy}{P_0c} - 1\right), \qquad \text{where } U_0 = [(mc^2)^2 + c^2P_0^2]^{1/2}.$$

2.8 Two solenoids with current flow in opposite direction and an iron plate with infinite permeability in between form an ideal cusped magnetic field that can be approximated by the function

$$B_z(r,z) = \begin{cases} -B_1 & \text{for } z < 0 \\ +B_2 & \text{for } z > 0, \end{cases}$$

where B_1 and B_2 are independent of r, z and only linear functions of the currents I_1 and I_2 in the two solenoids. An electron with kinetic energy $T = eV_0$ is emitted from an electron gun at $z = -z_1$ with $r = r_0,\ \dot{r}_0 = 0$, and $\dot{\theta}_0 = 0$. It can pass into the region $z > 0$ (solenoid 2) through a suitably placed small hole in the iron plate. Determine the electron motion and plot qualitatively the trajectory for the following cases:

    (a) $B_1 = 0,\ B_2 = B_0$

    (b) $B_1 \neq B_2$

    (c) $B_1 = B_2 = B_0$

Show in case (c) that there exists a threshold field $B_0 = B_{\max}$ above which the electron cannot enter the region $z > 0$. Calculate $B_{\max}$ for an electron with kinetic energy of 2.5 MeV launched with $r_0 = 6$ cm.

2.9 A relativistic electron of kinetic energy $T = eV_1$ moves on a circle of radius $r = R_1$ in the midplane ($z = 0$) of an axisymmetric magnetic *mirror* field. The field at $r = R_1$ is $B_z(R_1, 0) = B_1$. At some instant, a switch is turned on and the magnetic field increases with time. The change in $\mathbf{B}(r, z, t)$ occurs slowly enough so that the electron motion is adiabatic and always stays in the midplane. Assume that $B_z(r, 0, t) = B(r)f(t)$.

    (a) How do the kinetic energy T, the momentum P, and the radius R change as the magnetic field rises? What requirements are imposed on the radial dependence $B(r)$ in order that $R^2 B(R) = $ const?

    (b) Find the condition for $B(r)$ where the radius of the electron orbit remains constant (i.e., $R = R_1 = $ const).

2.10 Consider the space-charge-limited planar diode (Child's law) treated in Section 2.5.2. Assume that the cathode and anode have a cross-sectional area A and that the current is I. Calculate the space-charge density $\rho(x)$, the total charge Q between anode and cathode, and the total surface charge Q_a on the anode. For comparison, calculate the surface charges on the anode, Q_a, and cathode, Q_c, when the space charge is negligible (planar diode without space charge).

2.11 Consider a static magnetic field $\mathbf{B} = \{0, 0, -B\}$ and a time-varying electric field $\mathbf{E} = \{0, -E, (1 - \alpha \cos \omega t), 0\}$ in cartesian coordinates. Solve the nonrelativistic equation of motion for an electron launched at time $t = t_0$ from the origin $\{0, 0, 0\}$ with zero initial velocity ($v_0 = 0$). Evaluate the integration constants and write the solutions $x(t)$ and $y(t)$ for the case $t_0 = 0$ in terms of the parameters ω, $\omega_c = eB/m$, $a = eE_1/m$, and α.

2.12 Provide the missing steps in the derivation of the relativistic Euler trajectory equation (2.116) for an axisymmetric system.

2.13 The betatron employs an axially symmetric magnetic field which is varied in time. Electrons are accelerated by the action of the induced azimuthal electric field associated with this time variation of $\mathbf{B}$ (Faraday's law!). The initial magnetic field is zero, and during the acceleration process the electrons are kept at a centered orbit of *constant* radius R. Consider motion in the midplane ($z = 0$) only and prove that the constant-radius condition implies that $B(R) = \frac{1}{2}\overline{B}$, where $\overline{B}$ is the average field inside the orbit.

2.14 In a magnetron, a radial electric field $E_r(r)$ is formed by two coaxial cylinders, the inner of radius r_1 (the cathode) and the outer of radius r_2 (the anode). The anode is at potential V_0 with respect to the cathode. A uniform magnetic field B_z exists in the axial direction. Consider an electron leaving the cathode with zero velocity. Calculate the relativistically correct radial dependence of the azimuthal velocity v_θ and the critical field value B_c versus V_0 for which the electron can just reach the anode (i.e., $B_z > B_c$:

electron misses anode, $B_z < B_c$: electron hits anode). Use conservation laws and either Lagrange's equation of motion or the Lorentz-force equations.

2.15 In a planar diode the voltage between the cathode (at $x = 0$) and the anode (at $x = d$) varies periodically with time as $V(t) = V_0 \cos \omega t$. Solve the non-relativistic equations of motion for an electron launched from the cathode at $t = t_0$ with initial conditions $y = 0$, $\mathbf{v}_0 = \{\dot{x}_0, \dot{y}_0, 0\}$. Find $x(t)$, $y(t)$, and the kinetic energy $T(t)$. Determine the transit time $\tau_d = t_d - t_0$ for an electron leaving the cathode with $v_0 = 0$ in the approximation $\omega(t_d - t_0) \ll 1$.

CHAPTER 3

Beam Optics and Focusing Systems without Space Charge

3.1 BEAM EMITTANCE AND BRIGHTNESS

The basic principles of producing charged particle beams in diode-type sources and performance limitations of such sources were discussed in Section 1.3. In the case of electron beams, the *source* is a piece of conducting material that forms the cathode; the electrons are accelerated across the potential difference in the diode and emerge through a hole in the anode. The cathode may be either heated (thermal emission) or cold (field emission) or the electrons may be produced by photoemission. Positive or negative ions, on the other hand, are usually formed in the plasma of a gas discharge; they are then extracted from this *ion source* by applying a potential difference (with appropriate polarity) between the source and an extractor electrode.

In view of the nature of the source, there is always a spread in kinetic energy and velocity in a particle beam. Each point on the surface of the source is emitting particles with different initial magnitude and direction of the velocity vector. This intrinsic *thermal velocity spread* remains present in the beam at any distance *downstream* from the source. In practice, the velocity spread of the beam from a given source may be considerably greater than the ideal thermal limit since many factors, such as temperature fluctuations in a plasma source, nonlinear forces (aberrations) due to the external or space-charge fields, and instabilities lead to a deterioration of the beam quality. The *emittance* provides a quantitative basis, or a *figure of merit*, for describing the quality of the beam. As we will see in the next section, it is closely related to two-dimensional projections of the volume occupied by the ensemble of particles in six-dimensional *phase space* as defined by the set of canonical coordinates q_i, p_i.

Most beams of practical interest have two planes of symmetry or are circularly symmetric. For the following discussion, assume that the beam propagates in the z-direction and has two planes of symmetry (x–z and y–z). The motion of each individual particle is defined by the three space coordinates (x, y, z) and the three mechanical momentum coordinates (P_x, P_y, P_z) at any given instant of time. An ensemble of particles forms a *beam* if their momentum component in the longitudinal direction is much larger than the momentum component in the transverse directions (i.e., in our case of cartesian coordinates if $P_x \ll P_z$, $P_y \ll P_z$). If the length of the beam is much greater than the diameter, we can treat the distribution as a *continuous beam*. On the other hand, if the length is comparable to the diameter, we are dealing with *bunched* beams.

Consider now a particle in the x–z plane with total momentum $P = (P_x^2 + P_z^2)^{1/2}$, where $P_x \le P_z \approx P$. The slope of the trajectory is by definition $x' = dx/dz = \dot{x}/\dot{z} \approx P_x/P$. At any given distance z along the direction of beam propagation, every particle represents a point in x–x' space, known as *trace space*. The area occupied by the points that represent all particles in the beam

$$A_x = \iint dx\, dx' \tag{3.1}$$

is related to (but, by our definition, not identical to) the *emittance* of the beam.

Unfortunately, there is no single definition of emittance that is consistently used in the literature, a fact that often causes confusion when results from different laboratories or publications are compared. Many authors, especially experimentalists, define the trace-space area A_x as the emittance. However, this definition does not distinguish between a well-behaved beam in a linear focusing system and a beam with the same trace-space area but a distorted shape due to nonlinear forces. To illustrate this point, let us consider Figure 6.1 in Chapter 6. This figure shows the progressive distortion of the trace-space boundary during the propagation of the beam through a periodic system of lenses with spherical aberrations. The area enclosed by this boundary remains constant in agreement with Liouville's theorem, which is discussed in Section 3.2. However, it is clear from this picture that the beam quality becomes progressively worse as the beam propagates through the focusing channel.

We prefer, therefore, a definition of emittance that measures the beam quality rather than the trace-space area. A measure of the beam quality is the product of the beam's width and divergence, where the divergence relates to the random (or thermal) velocity spread. To be mathematically more precise, we will use the moments of the particle distribution in x–x' trace space, $\overline{x^2}, \overline{x'^2}, \overline{xx'}^2$ to define an *rms emittance* $\tilde{\epsilon}_x$ (see Section 5.3.4) by

$$\tilde{\epsilon}_x = \left(\overline{x^2}\,\overline{x'^2} - \overline{xx'}^2 \right)^{1/2}, \tag{3.2a}$$

or, equivalently, by

$$\tilde{\epsilon}_x = \tilde{x}\tilde{x}'_{th} = \tilde{x}\,\frac{\tilde{v}_{x,th}}{v_0}\,. \tag{3.2b}$$

The term $\overline{xx'}^2$ in (3.2a) reflects a correlation between x and x' which occurs, for instance, when the beam is either converging (e.g., after passing through a lens) or diverging (e.g., after passing through a waist); it is zero at the waist of an ideal uniform beam. As discussed in Section 5.4.5, $\overline{xx'}^2$ represents an inward or outward *flow* term in the transverse kinetic energy. The difference between the total transverse kinetic energy and the flow energy is the *random,* or *thermal, kinetic energy* whose x-component is defined by $\tilde{v}_{x,th}$. Equation (3.2b) is therefore equivalent to (3.2a); $\tilde{x} = (\overline{x^2})^{1/2}$ is the rms width, $\tilde{x}'_{th} = (\overline{x'^2_{th}})^{1/2}$ the rms divergence, $\tilde{v}_{x,th} = (\overline{v^2_{x,th}})$ the rms velocity spread, and v_0 the mean axial velocity of the particle distribution. A rigorous derivation from the particle distribution function and detailed discussion of this relationship can be found in Sections 5.3.4 and 5.4.5 of Chapter 5.

The rms emittance provides the desired quantitative information on the quality of the beam. Thus in Figure 6.1, $\tilde{\epsilon}_x$ increases with distance through the focusing channel while the trace-space area A_x remains constant. The only problem with the rms emittance is that is gives more weight to the particles in the outer region of the trace-space area (e.g., the *halo* observed in some beams) as compared to those in the beam core. Removal of such particles can therefore significantly improve the rms emittance while the corresponding decrease of beam intensity may be comparatively very small.

In a system where all the forces acting on the particles are linear (i.e., proportional to the particle's displacement x from the beam axis), it is useful to assume an elliptical shape for the area occupied by the beam in x–x' trace space. If this ellipse has an upright position with major axes x_m and $(x')_m$, the trace-space area A_x, which is identical to the area of the ellipse, is given by

$$A_x = x_m(x')_m \pi \,. \tag{3.3}$$

In this special case we can define an emittance as the product of the width x_m and maximum divergence $(x')_m$ as

$$\epsilon_x = x_m(x')_m = \frac{A_x}{\pi}\,, \tag{3.4}$$

which is equal to the total trace-space area A_x divided by π. The definition $\epsilon_x = A_x/\pi$ also applies when the trace-space ellipse is tilted, and we will use it consistently in the chapters that deal with linear beam optics without space charge (Chapter 3) and with space charge (Chapter 4). Note that we use the brackets in $(x')_m$ to distinguish the maximum value of x' from the slope of the width $x'_m = dx_m/dz$ in a converging or diverging uniform beam.

As will be shown in Section 5.3.4, for a beam with uniform particle density where both space charge and external forces are linear, the relationships between $x_m, (x')_m, \epsilon_x$ and the corresponding rms quantities $\tilde{x}, \tilde{x}'_{th}, \tilde{\epsilon}_x$ are given by

$$x_m = 2\tilde{x} \tag{3.5a}$$

$$(x')_m = 2\tilde{x}'_{th} \tag{3.5b}$$

$$\epsilon_x = 4\tilde{\epsilon}_x . \tag{3.5c}$$

For the ideal uniform beams with linear focusing forces discussed in Chapters 3 and 4 we call x_m the *width*, $(x_m)'$ the *divergence*, and ϵ_x the *emittance* of the beam. For other beams having nonuniform particle distributions the total trace-space area comprising all particles is generally larger than $\epsilon_x \pi = 4\tilde{\epsilon}_x \pi$. However, in most cases of theoretical or practical interest one finds that the fraction of the beam outside of this area is relatively small. It is therefore meaningful to use ϵ_x, the *four-times rms emittance* as a measure for the overall beam quality, following a proposal by Lapostolle. [See Section 5.3.4 and the discussion at the end of Section 5.4.4 in connection with Equations (5.295a)–(5.298).] In the context of this more general application of the relations (3.5a)–(3.5c), we will call x_m the *effective width*, $(x')_m$ the *effective divergence*, and ϵ_x the *effective emittance*, of the beam.

As mentioned above, many authors identify the emittance with the trace-space area A_x of the beam by including the factor π implicitly or explicitly. In the first case, "emittance" is defined as $\epsilon_x = A_x$. In the second case, π is factored out and "emittance" is given as $\epsilon_x \pi = A_x$, where the factor π is often included with the units, e.g., $\epsilon_x = 20\pi$ mm-mrad. For the reasons stated, we prefer the definitions given in the above equations, and they are used consistently in this book.

For an axially symmetric beam, the above description of the beam properties in one trace-space plane $(x–x')$ is sufficient. However, in many cases one has two planes of symmetry, e.g., beams in quadrupole focusing channels. Hence, one also needs the effective width y_m, divergence $(y')_m$, and emittance ϵ_y for the $y–y'$ projection of the four-dimensional transverse trace-space distribution. In the case of bunched beams, the *longitudinal phase-space properties* have to be included to obtain a description of the overall beam quality in six-dimensional phase space. These properties and the associated definitions of longitudinal emittance, bunch width, divergence, and energy spread are presented in Sections 5.4.6–5.4.8 for linear accelerators and in 5.4.9 for circular machines.

The units of measurement for emittance are m-rad. However, since the typical widths and divergence angles of beams are in the range of mm (or cm) and milli-radians, respectively, it is customary to use units of mm-mrad or cm-mrad. Also, the normalized longitudinal emittance is often given in units of "electronvolt-seconds" (see Section 5.4.6).

The emittance, as defined here, is a somewhat incomplete description of the *quality* of the beam. For one thing, emittance depends on the kinetic energy of the

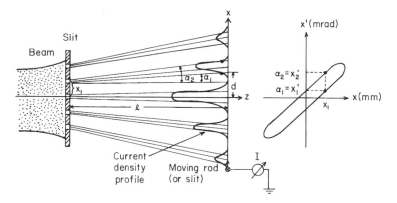

Figure 3.1. Method of measuring the trace-space distribution of a beam.

particles: according to Equation (3.2), the slope x' (and hence the area in $x-x'$ trace space) decreases as the longitudinal momentum P_z increases. One therefore has to *normalize* the emittance when one compares beams of different energy, as discussed in the next section. Another problem is due to the fact that the particle density across the beam as well as the density of the representative points in trace space is generally nonuniform, in practice, and decreases at the edges. Thus one has to specify what fraction of the beam particles lies within a given area. This is done by presenting the data in a contour map where different curves contain different fractions of the beam.

A method of measuring the trace-space distribution of laboratory beams is schematically illustrated in Figure 3.1. As shown in this figure, the beam is intercepted by a plate with a set of narrow slits (or, alternatively, with a single slit that can be moved across the beam). The particles passing through each slit form a narrow beamlet with a small divergence angle. At distance l downstream from the plate, the current density profile of each beamlet is scanned by a moving probe (thin tungsten rod, for instance) or by a second slit with a current collector behind it. Each slit position in the intercepting plate upstream defines an x-coordinate within the beam. The angular divergence of the beamlet passing through a slit is obtained from the width of the associated current density curve measured by the probe at distance l. For any given current density level, for example 10% of the highest peak, two points at distances d_1 and d_2 from the axis are defined, as indicated in Figure 3.1.The corresponding divergence angle α, or slope x', is given by the simple geometric relation $\tan \alpha_1 \approx \alpha_1 = (d_1 - x_1)/l = x_1'$, and likewise for α_2. If one plots the two angles (or slopes) for each slit position in an $x-x'$ coordinate system, one can construct a closed curve, as shown on the right side of the figure. The area enclosed by this curve is then the trace-space area A_x for the fraction of the beam defined by the given intensity level. Such emittance contours can be obtained for any fraction of the beam current distribution. Specifically, the contour corresponding to the zero current density points at the bottoms of each beamlet curve defines the total, or 100%, trace-space area of the beam. Similarly, one can construct trace-space contours for 95% or 90% of the beam current. This contour map then provides a good picture of the particle distribution in the $x-x'$ trace space and hence in the corresponding $x-p_x$ projection in phase space.

A useful description of the beam quality is obtained by measuring the trace-space area in both transverse directions. This is accomplished by either rotating the slits and probe of Figure 3.1 by 90° or by inserting a separate system for measuring the trace-space area A_y.

As stated earlier, emittance alone is not enough to define the quality of a beam. One can make the emittance as small as one desires for a particular application by use of collimating slits. What counts, however, is the number of particles, or the total beam current, with a given emittance. The figure of merit is therefore known as the *brightness* of the beam, commonly defined by [C.14, p. 160]

$$B = \frac{J}{d\Omega} = \frac{dI}{dS \, d\Omega},$$ (3.6)

which is the current density per unit solid angle. In this definition, brightness, like current density J, is a quantity that may vary across the beam. For many practical applications it is more meaningful to know the total beam current that can be confined within a given four-dimensional trace-space volume V_4 . The corresponding definition of average brightness is

$$B = \frac{I}{V_4},$$ (3.7)

where $V_4 = \int \int dS \, d\Omega$ represents the integral taken over the total trace-space volume. As pointed out previously in this section, the emittance is generally not directly proportional to the trace-space area. However, for any particle distribution whose boundary in four-dimensional trace space is defined by the hyperellipsoid

$$\frac{x^2}{a^2} + \left(\frac{ax'}{\epsilon_x}\right)^2 + \frac{y^2}{b^2} + \left(\frac{by'}{\epsilon_y}\right)^2 = 1,$$

one finds that $\int \int dS \, d\Omega = (\pi^2/2)\epsilon_x \epsilon_y$, and that the average brightness is accordingly given by

$$\overline{B} = \frac{2I}{\pi^2 \epsilon_x \epsilon_y} \qquad [\text{A/(m-rad)}^2].$$ (3.8)

The best known examples of this type are the *K–V distribution* and the *waterbag distribution*, which are treated in Chapter 5. In the waterbag case, the hyperellipsoidal volume is populated with uniform density. In the K–V beam, on the other hand, the particles occupy only the surface of the hyperellipsoid, and the volume in four-dimensional phase space is zero in a mathematical sense. However, the projections into any two-dimensional subspace $(x-x', y-y',$ etc.) are ellipses whose areas, $\epsilon_x \pi$ and $\epsilon_y \pi$, have uniform current densities; Equation (3.8) is therefore a valid definition of brightness for a K–V beam as well. Since Equation (3.8) is also consistent with our concept of emittance as a measure of beam quality, we will use this relation as the definition of brightness for any distribution provided that ϵ_x and ϵ_y denote effective ("four-times-rms") emittances. For the experimental determination of brightness it is necessary to specify the percentage of total beam current to which the effective emittance values being used in (3.8) apply. It should be noted that quite often in the literature the factor $2/\pi^2$ is left out, and brightness is simply defined as $I/\epsilon_x\epsilon_y$ or I/ϵ^2 if $\epsilon_x = \epsilon_y = \epsilon$. Sometimes the rms emittances are used in place of the effective emittances. Since $\epsilon_x^2 = 16\tilde{\epsilon}_x^2$, the brightness values associated with the rms emittances are 16 times higher than those calculated from Equation (3.8), and $I/\tilde{\epsilon}_x^2$ is almost 80 times greater than $2I/\pi^2\epsilon_x^2$. It is therefore important to clearly state which definition of brightness is used to avoid misunderstandings when brightness figures are reported in the literature or results from different experiments are compared.

In view of the energy dependence, one also has to normalize both emittance and brightness if one wants to compare the quality of different beams. These normalized quantities can be deduced from Liouville's theorem, which is discussed in the next section. According to Liouville's theorem, the *normalized emittance,* defined as $\epsilon_n = \beta\gamma\epsilon$, and the *normalized brightness,* $B_n = B/(\beta\gamma)^2$, are invariants under ideal conditions. A more thorough description of the concepts of beam emittance and brightness can be found in the article by C. Lejeune and J. Aubert in Supplement 13A, p. 159ff. of *Applied Charged Particle Optics* [C.19].

3.2 LIOUVILLE'S THEOREM

In Chapter 2 we introduced the canonical space and momentum coordinates q_i, p_i. If we construct a conceptual Euclidean space of six dimensions combining configuration space (q_i) and canonical momentum space (p_i), a particle is represented by a point, and all particles in a beam will occupy a volume in this six-dimensional hyperspace which is called *phase space*. We can define a particle density $n(x, y, z, p_x, p_y, p_z, t)$ in phase space, and the number dN of particles in a small volume element dV of phase space is then

$$dN = n\, dV = n\, dx\, dy\, dz\, dp_x\, dp_y\, dp_z. \tag{3.9}$$

Let us now consider a system of noninteracting particles. The motion of the ensemble of particles representing the beam in actual configuration space is associated with an equivalent motion of the representative points in phase space, and we can define a velocity vector $\mathbf{v} = \{\dot{q}_i, \dot{p}_i\}$ in phase space for each particle. As the ensemble moves, the volume it occupies in phase space also moves and changes its shape. Since the total number of particles and the associated representative points in phase space must remain constant, the motion in phase space must obey the continuity equation

$$\nabla \cdot (n\mathbf{v}) + \frac{\partial n}{\partial t} = 0, \tag{3.10}$$

or

$$n\nabla \cdot \mathbf{v} + \mathbf{v} \cdot \nabla n + \frac{\partial n}{\partial t} = 0. \tag{3.11}$$

Now, with $\mathbf{v} = \{\dot{q}_i, \dot{p}_i\}$, we have

$$\nabla \cdot \mathbf{v} = \sum_{i=1}^{3} \left(\frac{\partial \dot{q}_i}{\partial q_i} + \frac{\partial \dot{p}_i}{\partial p_i} \right). \tag{3.12}$$

If a Hamiltonian $H(q_i, p_i, t)$ can be defined for this system, Hamilton's equations hold; that is,

$$\dot{q}_i = \frac{\partial H}{\partial p_i}, \qquad \dot{p}_i = -\frac{\partial H}{\partial q_i},$$

$$\frac{\partial \dot{q}_i}{\partial q_i} = \frac{\partial^2 H}{\partial p_i \partial q_i}, \qquad \frac{\partial \dot{p}_i}{\partial p_i} = -\frac{\partial^2 H}{\partial q_i \partial p_i}. \tag{3.13}$$

As a result,

$$\nabla \cdot \mathbf{v} = \sum_{i=1}^{3} \left(\frac{\partial^2 H}{\partial p_i \partial q_i} - \frac{\partial^2 H}{\partial q_i \partial p_i} \right) = 0,$$

and (3.11) becomes

$$\frac{\partial n}{\partial t} + \mathbf{v} \cdot \nabla n = 0. \tag{3.14}$$

Since

$$\frac{dn}{dt} = \frac{\partial n}{\partial t} + \sum_i \frac{\partial n}{\partial q_i} \dot{q}_i + \sum_i \frac{\partial n}{\partial p_i} \dot{p}_i = \frac{\partial n}{\partial t} + \mathbf{v} \cdot (\nabla n), \tag{3.15}$$

we can write in place of (3.14)

$$\frac{dn}{dt} = 0, \quad \text{or} \quad n = n_0 = \text{const}; \tag{3.16}$$

that is, the density of points in phase space is a constant.

If n remains constant, the volume occupied by a group of particles in phase space also remains constant throughout the motion. If $\delta N = n\,\delta V$ is the number of particles in a small volume element δV, we have

$$\frac{d}{dt}\,(\delta N) = \frac{d}{dt}\,(n\,\delta V) = \frac{dn}{dt}\,\delta V + n\,\frac{d(\delta V)}{dt} = 0. \tag{3.17}$$

Thus, in view of (3.16),

$$\frac{d}{dt}\,(\delta V) = \frac{d}{dt}\int d^3 q_i\, d^3 p_i = 0. \tag{3.18}$$

Equations (3.16) and (3.18) are both two versions of *Liouville's theorem,* which states that the density of particles, n, or the volume occupied by a given number of particles in phase space remains invariant.

Liouville's theorem in the above form is, strictly speaking, valid only for noninteracting particles. However, it is still applicable in the presence of electric and magnetic self fields associated with the bulk space charge and current arising from the particles of the beam, as long as these fields can be represented by average scalar and vector potentials $\phi(x, y, z)$, $\mathbf{A}(x, y, z)$. This implies that a particle's interaction with its nearest neighbors can be neglected in comparison to the interaction with the average collective field produced by the other particles in the beam. Quantitatively, for this to be the case, the *Debye length* λ_D, discussed in Sections 4.1 and 5.4.1, must be large compared to the interparticle distance. If the fields of individual particles and particle–particle interactions become important, one must generalize the phase-space concept to a hyperspace of higher dimension. Thus, instead of 6, we have $6N$ independent space and canonical momentum coordinates in this case. Then a velocity vector $\mathbf{v} = \{\dot{q}_1, \dot{q}_2, \ldots, \dot{q}_{3N}; \dot{p}_1, \ldots, \dot{p}_{3N}\}$ may be defined, and Liouville's theorem applies to the particle distribution in $6N$-dimensional phase space but not in six-dimensional phase space. For a more detailed discussion, see Lawson's book [C.17, Sec. 4.2] and the references given there.

Returning now again to the six-dimensional phase space, we note the following:

1. Liouville's theorem also applies for the phase space defined by the *mechanical* momentum components P_i and spatial coordinates q_i; thus the conservation of the phase-space volume can be stated in the form (see Problem 5.7)

$$\iint d^3 q_i\, d^3 p_i = \iint d^3 q_i\, d^3 P_i = \text{const.} \tag{3.19}$$

2. While the *volume* in phase space remains constant, the *shape* generally does not. In fact, nonlinearities (aberrations) in the field configurations through which the particles move may cause considerable distortions (*filamentation*) in the shape of the phase-space volume; as a result, beam blowup and particle loss to nearby walls may occur.

3. The trace-space area A_x is related to the projection of the phase-space volume into the $x-P_x$ plane by

$$A_x = \frac{1}{P} \iint dx \, dP_x = \frac{1}{\gamma \beta mc} \iint dx \, dP_x. \tag{3.20}$$

If there is no coupling between the x-motion and the other directions (y and z), the area in x-P_x phase space defined by $\iint dx \, dP_x$ remains constant. Moreover, if there is no acceleration or deceleration ($\beta\gamma$ = const), the area A_x in x-x' trace space is also conserved. However, if there is an energy change (i.e., $\beta\gamma \neq$ const), A_x and, by implication, the emittance ϵ_x, do not remain constant, the change being inversely proportional to $\beta\gamma$ according to Liouville's theorem as stated in Equation (3.20). For this reason, one introduces the *normalized rms emittance*

$$\tilde{\epsilon}_n = \beta\gamma\tilde{\epsilon} \tag{3.21a}$$

or the equivalent *normalized (effective) emittance*

$$\epsilon_n = \beta\gamma\epsilon = 4\beta\gamma\tilde{\epsilon}. \tag{3.21b}$$

For beams in particle accelerators, the normalized emittance is a more useful quantity than the unnormalized emittance since in an ideal system (linear forces, no coupling) it remains constant. An increase of the normalized emittance is usually an indication that nonlinear effects causing a deterioration of beam quality are present in the system.

In similar fashion one can define a *normalized brightness*,

$$B_n = \frac{B}{\gamma^2\beta^2} = \frac{2I}{\pi^2\epsilon_n^2}, \tag{3.22}$$

which in an ideal system is also an invariant.

As an example of how the unnormalized emittance is changed by acceleration, consider a proton beam that is injected into a linear accelerator with a kinetic energy of 50 keV and an effective emittance of ϵ = 200 mm-mrad. It emerges from the accelerator with a kinetic energy of 80 MeV. What is the emittance of the accelerated proton beam if no particle loss and no distortions in the phase-space volume occur? We may treat the protons nonrelativistically since $T \ll E_0 = 938.25$ MeV. Now $v A_x =$ const, or $T^{1/2} A_x =$ const. Let ϵ_1 be the emittance at 50 keV, ϵ_2 at 80 MeV. Then $\epsilon_2 = \epsilon_1 (T_1/T_2)^{1/2} = \epsilon_1/40$, hence, $\epsilon_2 = 5$ mm-mrad. Thus, the emittance is reduced by a factor of 40 while the normalized emittance $\epsilon_n = \beta\gamma\epsilon$ remains constant.

We close this section with an example that illustrates how to calculate the rms emittance for a given theoretical distribution. Consider a round beam with uniform particle density $n(r) = n_0 =$ const in space and a thermal velocity spread defined by a Gaussian of the form $\exp[-m(v_x^2 + v_y^2)/2k_B T_\perp]$. This is a good approximation for a high-current electron beam with transverse temperature $T_\perp$ from a thermionic cathode. If $x_m = a$ is the beam radius, one finds for the rms width

$$\tilde{x} = \frac{\left(\overline{r^2}\right)^{1/2}}{\sqrt{2}} = \frac{1}{\sqrt{2}} \left[\frac{2\pi n_0 \int_0^a r^3 \, dr}{2\pi n_0 \int_0^a r \, dr} \right]^{1/2} = \frac{a}{2} ,$$

where $(\overline{r^2})^{1/2} = \tilde{r} = a/\sqrt{2}$ is the rms radius of the beam. The rms thermal velocity is defined by

$$\tilde{v}_{x,\text{th}} = \frac{\left(\overline{v_\perp^2}\right)^{1/2}}{\sqrt{2}} = \left(\frac{k_B T_\perp}{m} \right)^{1/2},$$

where $v_\perp^2 = v_x^2 + v_y^2$.

The effective normalized rms emittance is then given by

$$\epsilon_n = 4\tilde{\epsilon}_n = 2a \left(\frac{k_B T_\perp}{mc^2} \right)^{1/2},$$

which is identical to Equation (1.7b) if one replaces r_c with a. A method of determining the rms emittance from experimental data in the case of an axially symmetric beam can be found in the paper by Rhee and Schneider mentioned in Chapter 6 (Reference 14).

3.3 THE PARAXIAL RAY EQUATION FOR AXIALLY SYMMETRIC SYSTEMS

3.3.1 Series Representation of Axisymmetric Electric and Magnetic Fields

In the following, we consider particle motion in rotationally symmetric static fields $(\partial/\partial\theta = 0, \partial/\partial t = 0)$. We assume that the beam currents are low enough so that self fields generated by the particles may be neglected in comparison to the applied fields. These assumptions imply that

$$\nabla \times \mathbf{E} = 0, \qquad \nabla \cdot \mathbf{E} = 0, \qquad \mathbf{E} = -\nabla\phi,$$
$$\nabla \times \mathbf{B} = 0, \qquad \nabla \cdot \mathbf{B} = 0, \qquad \mathbf{B} = -\nabla\phi_m. \qquad (3.23)$$

Thus, both $\mathbf{E}$ and $\mathbf{B}$ may be derived from a scalar potential $f(r, z)$ which obeys a Laplace equation of the form

$$\nabla^2 f(r, z) = \frac{1}{r}\frac{\partial}{\partial r}\left(r\frac{\partial f}{\partial r}\right) + \frac{\partial^2 f}{\partial z^2} = 0. \qquad (3.24)$$

Since the potentials must be finite, continuous along the z-axis, and for symmetry reasons an even function of radius r , one can solve Equation (3.24) by means of a power series

$$f(r, z) = \sum_{\nu=0}^{\infty} f_{2\nu}(z) r^{2\nu} = f_0 + f_2 r^2 + f_4 r^4 + \cdots, \qquad (3.25)$$

where $f_0(z) = f(0, z)$ is the potential along the z-axis. Note that a linear term $f_1 r$ (and by implication any odd power of r) in the potential function $f(r, z)$ would lead to nonzero radial fields on the axis since $\partial f/\partial r = f_1 \neq 0$. This is inconsistent with the axial symmetry, which requires that $E_r = 0, B_r = 0$ at $z = 0$.

Differentiation of Equation (3.25) and substitution in (3.24) yields

$$\sum_{\nu=1}^{\infty}[2\nu + 2\nu(2\nu - 1)]f_{2\nu} r^{2\nu-2} + \sum_{\nu=0}^{\infty} f''_{2\nu} r^{2\nu} = 0, \qquad (3.26)$$

where $f''_{2\nu} = \partial^2 f_{2\nu}/\partial z^2$ and $[2\nu + 2\nu(2\nu - 1)] = (2\nu)^2$. From this equation, one obtains the recursion formula for the coefficients $f_{2\nu}$:

$$(2\nu + 2)^2 f_{2\nu+2} + f''_{2\nu} = 0, \qquad (3.27)$$

that is,

$$f_2 = -\frac{1}{4} f''_0 \qquad \text{for } \nu = 0,$$

$$f_4 = \frac{1}{64} f_0^{(4)} = \frac{1}{64}\frac{\partial^4 f_0}{\partial z^4} \qquad \text{for } \nu = 1, \text{ etc.}$$

The function of $f(r, z)$ can thus be written in the form

$$f(r,z) = f(0,z) - \frac{1}{4}\frac{\partial^2 f(0,z)}{\partial z^2}r^2 + \frac{1}{64}\frac{\partial^4 f(0,z)}{\partial z^4}r^4 - \cdots ,$$

or

$$f(r,z) = \sum_{\nu=0}^{\infty} \frac{(-1)^\nu}{(\nu!)^2} \frac{\partial^{2\nu} f(0,z)}{\partial z^{2\nu}} \left(\frac{r}{2}\right)^{2\nu}. \tag{3.28}$$

This shows that the potential distribution $f(r, z)$ in an axisymmetric system for any given point r, z off the axis may be determined from the potential distribution and its derivatives on the z-axis $(r = 0)$. In practice, it is relatively easy to obtain $f_0(z)$ with sufficient accuracy. However, small errors in the measurement (or numerical calculation) of $f_0(z)$ are strongly amplified in the derivatives so that the higher-order terms may become increasingly inaccurate. One therefore has to check carefully that the series representation can be applied within acceptable error bars.

The **E** and **B** fields are obtained by substituting ϕ or ϕ_m for $f(r, z)$ in Equation (3.28) and taking the gradients. Since the magnetic potential ϕ_m is not a measurable quantity like ϕ, one uses, in the case of magnetic fields, the axial field component $B_z(0, z) = B(z)$ on the axis rather than $\phi_m(0, z)$. The field components B_r and B_z off the axis are then obtained from $B(z)$ and its derivatives as shown below. First, one has

$$B_r = -\frac{\partial \phi_m}{\partial r}, \qquad B_z = -\frac{\partial \phi_m}{\partial z}. \tag{3.29}$$

From (3.28), with $f(r, z) = \phi_m$, one gets

$$B_z(r,z) = -\frac{\partial \phi_m(0,z)}{\partial z} + \frac{r^4}{4}\frac{\partial^3 \phi_m(0,z)}{\partial z^3} - \cdots , \tag{3.30}$$

Using $B_z(0, z) = B(z) = -\partial \phi_m(0, z)/\partial z$, this may be written in the form

$$B_z(r,z) = B - \frac{r^2}{4}\frac{\partial^2 B}{\partial z^2} + \frac{r^4}{64}\frac{\partial^4 B}{\partial z^4} - \cdots ,$$

or

$$B_z(r,z) = \sum_{\nu=0}^{\infty} \frac{(-1)^\nu}{(\nu!)^2} \frac{\partial^{2\nu} B}{\partial z^{2\nu}} \left(\frac{r}{2}\right)^{2\nu}. \tag{3.31}$$

Likewise,

$$B_r(r,z) = -\frac{r}{2}\frac{\partial B}{\partial z} + \frac{r^3}{16}\frac{\partial^3 B}{\partial z^3} - \cdots ,$$

or

$$B_r(r,z) = \sum_{\nu=1}^{\infty} \frac{(-1)^\nu}{\nu!(\nu-1)!} \frac{\partial^{2\nu-1} B}{\partial z^{2\nu-1}} \left(\frac{r}{2}\right)^{2\nu-1}. \tag{3.32}$$

3.3.2 Derivation of the Paraxial Ray Equation

With the series representation of the fields given in the preceding section, we are now able to calculate particle trajectories in axisymmetric systems to any degree of accuracy desired. The general approach is first to derive linear equations for the particle motion in which only terms up to first order in r and $r' = dr/dz$ are considered. One can then improve the accuracy by including higher-order terms in r and r'. These terms are necessary to study nonlinear effects, or aberrations, which are important in many applications, most notably the electron microscope. There are several approaches to tackling this problem. One is to start with the equations of motion in the Newtonian form or with Hamilton's equations. Another possibility is to use the Euler trajectory equations which were derived from the variational principle of least action.

The basic first-order optical equation which describes the motion of charged particles in an axisymmetric system is known as the *paraxial ray equation*. In the following we derive the paraxial ray equation from the equations of motion in the Newtonian form [Equations (2.11)].

The assumptions of paraxial motion are that the particle trajectories remain close to the axis; that is, r is very small compared to the radii of electrodes, coils, or iron pieces that produce the electric and magnetic fields. This also implies that the slopes of the particle trajectories remain small (i.e., $r' \ll 1$ or $\dot{r} \ll \dot{z}$). Furthermore, the azimuthal velocity v_θ must remain very small compared to the axial velocity (i.e., $r\dot{\theta} \ll \dot{z}$). Thus, in this linear approximation we have $\dot{z} = (v^2 - \dot{r}^2 - r^2\dot{\theta}^2)^{1/2} \approx v$. With these assumptions, only the first-order terms in the expansions of the fields need to be considered and the equations of motion can be linearized by expanding all quantities in terms of their values on the axis of the system and dropping all terms of order r^2, rr', r'^2, and higher.

The electric potential may be expressed in terms of the potential on the axis $(r = 0)$, which we denote with $V(z)$. From (3.28), we obtain with $f(r, z) = \phi(r, z), f(0, z) = \phi(0, z) - V(z)$:

$$\phi(r, z) = V - \frac{1}{4} V'' r^2 + \frac{1}{64} V^{(4)} r^4 - \cdots . \tag{3.33}$$

From this we obtain for the radial and axial electric field components the first-order relations

$$E_z = -V', \qquad E_r = \frac{1}{2} V'' r = -\frac{r}{2} \frac{\partial E_z}{\partial z} . \tag{3.34}$$

The first-order magnetic field terms are [from Equations (3.31) and (3.32)]

$$B_r = -\frac{1}{2} B' r, \qquad B_z = B . \tag{3.35}$$

Note that $E_\theta = 0$, $B_\theta = 0$, which follows from $\nabla \times \mathbf{E} = 0$ and $\nabla \times \mathbf{B} = 0$ with $\partial/\partial\theta = 0$. $V = V(z)$ and $B = B(z)$ are the electrostatic potential (corresponding to the kinetic energy of the particles) and magnetic field on the z-axis ($r = 0$), respectively.

If we substitute the above relations for the field components into Equations (2.11a) to (2.11c), using (2.76) in place of (2.11b), we obtain the following set of equations for the radial, azimuthal, and axial motion of the particles:

$$m \frac{d}{dt} (\gamma\dot{r}) - m\gamma r\dot{\theta}^2 = \frac{qr}{2} V'' + qr\dot{\theta}B, \qquad (3.36)$$

$$\gamma m r^2 \dot{\theta} = -\frac{q}{2\pi} \psi + p_\theta = -\frac{q}{2} Br^2 + p_\theta, \qquad (3.37)$$

$$m \frac{d}{dt} (\gamma\dot{z}) = -qV' + \frac{q}{2} r^2 \dot{\theta}B'. \qquad (3.38)$$

Since in the paraxial approximation $\dot{z} \approx v = \beta c$ (i.e., $v_\theta = r\dot{\theta} \ll v$), we can neglect the term $qr^2\dot{\theta}B'/2$ on the right-hand side of Equation (3.38). The differentiation with respect to time on the left-hand side of Equation (3.38) can be changed into differentiation with respect to the z-coordinate as follows:

$$\frac{d}{dt} (\gamma\dot{z}) = \frac{dz}{dt} \frac{d}{dz} (\gamma\dot{z}) = v \frac{d}{dz} (\gamma v) = \gamma' v^2 + \gamma v' v = c^2(\gamma'\beta^2 + \gamma\beta'\beta),$$

$$\qquad (3.39)$$

or, with $\beta'\beta = \gamma'/\gamma^3$,

$$\frac{d}{dt} (\gamma\dot{z}) = c^2(\beta^2 + \gamma^{-2})\gamma' = c^2\gamma'. \qquad (3.40)$$

Thus, (3.38) may be written as

$$mc^2\gamma' = -qV'. \qquad (3.41)$$

Integration of Equation (3.41) yields the energy conservation law $T + U = (\gamma - 1)mc^2 + qV = \text{const}$. If we define the potential such that $V = 0$ when $T = 0$, or $\gamma = 1$, the constant is zero and we get

$$\gamma(z) = 1 - \frac{qV(z)}{mc^2} = 1 + \frac{|qV(z)|}{mc^2}. \qquad (3.42)$$

Note that for positive q, the potential V is negative and vice versa; hence, $-qV$ is always positive, and $|V(z)|$ is the *voltage equivalent* of the particle's kinetic energy.

From Equation (3.37) one obtains for the angular velocity of the particles

$$\dot{\theta} = -\frac{qB}{2\gamma m} + \frac{p_\theta}{\gamma mr^2},$$

or

$$\theta' = \frac{\dot{\theta}}{\beta c} = -\frac{qB}{2\gamma m\beta c} + \frac{p_\theta}{mc\beta\gamma r^2}. \tag{3.43}$$

Integration of (3.43) with the initial condition $\theta = \theta_0$ and $z = z_0$ yields

$$\theta - \theta_0 = \int_{z_0}^{z} \left(-\frac{qB}{2\gamma m\beta c} + \frac{p_\theta}{mc\beta\gamma r^2} \right) dz. \tag{3.44}$$

By substitution of (3.43) into (3.36) one obtains for the radial motion

$$\frac{d}{dt}(\gamma\dot{r}) = \frac{qrV''}{2m} + \frac{r\dot{\theta}}{m}\left(m\gamma\dot{\theta} + qB \right)$$

$$= \frac{qrV''}{2m} + \frac{r}{m}\left(-\frac{qB}{2\gamma m} + \frac{p_\theta}{\gamma mr^2} \right)\left(\frac{qB}{2} + \frac{p_\theta}{r^2} \right),$$

or

$$\frac{d}{dt}(\gamma\dot{r}) = \frac{qrV''}{2m} - \frac{r}{4}\left(\frac{qB}{m} \right)^2 \frac{1}{\gamma} + \frac{p_\theta^2}{\gamma m^2 r^3}. \tag{3.45}$$

Now we have

$$\dot{\gamma} = \gamma'\dot{z} = \gamma'\beta c, \tag{3.46a}$$

$$\dot{r} = r'\dot{z} = r'\beta c, \tag{3.46b}$$

$$\ddot{r} = \beta c \frac{d}{dz}(r'\beta c) = r''\beta^2 c^2 + r'\beta'\beta c^2, \tag{3.46c}$$

Using these relations and $\beta'\beta = \gamma'/\gamma^3$, the left-hand side of Equation (3.45) may be written as

$$\frac{d}{dt}(\gamma\dot{r}) = c^2\left(\gamma\beta^2 r'' + \gamma'r' \right). \tag{3.47}$$

From (3.42), we have

$$qV'' = -mc^2\gamma''. \tag{3.48}$$

Substitution of (3.47) and (3.48) into (3.45) yields

$$c^2(\gamma\beta^2 r'' + \gamma'r') = -\frac{1}{2}c^2\gamma''r - \left(\frac{qB}{2m}\right)^2\frac{r}{\gamma} + \frac{p_\theta^2}{\gamma m^2 r^3},$$

or, after dividing by $c^2\gamma\beta^2$,

$$r'' + \frac{\gamma'}{\gamma\beta^2}r' + \frac{\gamma''}{2\gamma\beta^2}r + \left(\frac{qB}{2mc\beta\gamma}\right)^2 r - \frac{p_\theta^2}{m^2c^2\gamma^2\beta^2 r^3} = 0. \qquad (3.49)$$

This is the relativistically correct paraxial ray equation that defines the radial motion of the particles near the z-axis where the nonlinear force terms can be neglected. As explained earlier, $p_\theta = \gamma mr^2\dot{\theta} + qA_\theta r$ is the canonical angular momentum of the particles as determined by the initial conditions. The azimuthal position of the particles as a function of distance z can be determined from Equation (3.44).

Let us now discuss the physical contents of Equation (3.49). The first term, r'', represents the change of slope of the particle trajectory. The second term contains the effect of the axial electric field (acceleration or deceleration), the third term that of the radial electric field (focusing, defocusing), and the fourth term represents the magnetic force. The last term adds a *centrifugal potential* or an effective *repulsive core* when the canonical angular momentum is different from zero. In this case, the particle never crosses the axis (i.e., $r \neq 0$).

In the *nonrelativistic* limit, we can make the substitutions

$$\gamma \approx 1, \qquad \beta^2 = \frac{v^2}{c^2} \approx -\frac{2qV}{mc^2}, \qquad (3.50a)$$

$$\gamma' = -\frac{qV'}{mc^2}, \qquad \frac{\gamma'}{\beta^2} \approx \frac{V'}{2V}, \qquad (3.50b)$$

$$\gamma'' \approx -\frac{qV''}{mc^2}, \qquad \frac{\gamma''}{\beta^2} \approx \frac{V''}{2V}. \qquad (3.50c)$$

With these approximations we obtain from (3.49) the *nonrelativistic paraxial ray equation*

$$r'' + \frac{V'r'}{2V} + \frac{V''r}{4V} + \frac{q^2B^2r}{8mqV} - \frac{p_\theta^2}{2mqV}\frac{1}{r^3} = 0. \qquad (3.51)$$

Note that V and qV in the denominators are positive quantities representing the *voltage equivalent* of the particles' kinetic energy. For the angle θ the nonrelativistic approximation is [from Equation (3.44)]

$$\theta = \theta_0 - \int_{z_0}^{z}\left[\left(\frac{q^2B^2}{8mqV}\right)^{1/2} - \frac{p_\theta}{(2mqV)^{1/2}r^2}\right]dz. \qquad (3.52)$$

With respect to canonical angular momentum, $p_\theta = \gamma m r^2 \dot\theta + q A_\theta r$, three cases are of interest:

1. The particle starts in a field-free region ($A_\theta = 0$). In this case p_θ depends on the initial radius r_0, initial azimuthal velocity $\dot\theta_0$, and initial kinetic energy γ_0 of the particle and is given by

$$p_\theta = \gamma_0 m r_0^2 \dot\theta_0. \tag{3.53}$$

2. If in addition to $A_\theta = 0$ the initial angular velocity is zero (i.e., $v_\theta = r_0 \dot\theta_0 = 0$), one has

$$p_\theta = 0, \tag{3.54}$$

and the last term in the paraxial ray equation vanishes.

3. If a particle starts in a region where $A_\theta \neq 0$ with $\dot\theta_0 = 0$, or if we choose as the reference point a position along the trajectory inside a magnetic field where $\dot\theta_0 = 0$, we can introduce the magnetic flux $\psi_0 = 2\pi \int_0^{r_0} Br\, dr$. From (2.74) and (2.75) one then has

$$p_\theta = \frac{q\psi_0}{2\pi}, \tag{3.55}$$

and hence one can express the last term in the paraxial ray equation in the form

$$\frac{p_\theta^2}{m^2 c^2 \beta^2 \gamma^2} \frac{1}{r^3} = \left(\frac{q\psi_0}{2\pi m c \beta \gamma}\right)^2 \frac{1}{r^3} \qquad \text{(relativistic case)} \tag{3.56}$$

and

$$\frac{p_\theta^2}{2mqV} \frac{1}{r^3} = \frac{(q\psi_0)^2}{8\pi^2 mqV} \frac{1}{r^3} \qquad \text{(nonrelativistic case).} \tag{3.57}$$

It is often convenient to study the particle trajectories in a frame that rotates at the Larmor frequency ω_L and is therefore known as the *Larmor frame*. The angle θ_r between this rotating frame and the stationary laboratory system is given by

$$\theta_r = -\int_{z_0}^{z} \frac{qB}{2\gamma m \beta c}\, dz = \mp \int \frac{\omega_L}{\beta c}\, dz. \tag{3.58}$$

The angle θ_L of the particle in the Larmor frame is given by the difference between the angle θ in the laboratory frame and θ_r:

$$\theta_L = \theta - \theta_r = \int_{z_0}^{z} \frac{p_\theta \, dz}{\beta \gamma m c r^2} + \theta_0 . \tag{3.59}$$

When $p_\theta = 0$, or $\psi_0 = 0$, particle motion in this frame is in a plane through the axis which is known as the *meridional plane*. In this case, the trajectory $r(z)$ in the meridional plane may be found from Equation (3.49) alone by setting $p_\theta = 0$, and the rotation of the meridional plane is found from Equation (3.58). For nonmeridional motion when $p_\theta \neq 0$, one must solve Equation (3.49) first and then use the result to solve (3.44).

Although cylindrical coordinates are the natural choice for systems with axial symmetry, it is sometimes convenient to use cartesian coordinates (x, y) and obtain the projections of the trajectory on the two perpendicular planes. In this case we start with Equations (2.10a) and (2.10b). The first-order terms for the field components are [from Equations (3.34), (3.35)]

$$E_x = E_r \frac{x}{r} = \frac{1}{2} V'' x, \qquad E_y = E_r \frac{y}{r} = \frac{1}{2} V'' y , \tag{3.60}$$

$$B_x = -\frac{1}{2} B' x, \qquad B_y = -\frac{1}{2} B' y . \tag{3.61}$$

Transforming from time t to axial distance z as the independent variable [as in (3.46)] and introducing the Larmor frequency relations

$$\omega_L = \frac{qB}{2\gamma m} , \qquad \omega_L' = \frac{q}{2\gamma m} B' - \frac{\gamma'}{\gamma} \omega_L , \tag{3.62}$$

one obtains the two equations

$$x'' + \frac{\gamma' x'}{\beta^2 \gamma} + \frac{\gamma'' x}{2\beta^2 \gamma} + \frac{2\omega_L y'}{\beta c} + \frac{\omega_L' y}{\beta c} + \frac{\gamma' \omega_L y}{\beta \gamma c} = 0, \tag{3.63}$$

$$y'' + \frac{\gamma' y'}{\beta^2 \gamma} + \frac{\gamma'' y}{2\beta^2 \gamma} - \frac{2\omega_L x'}{\beta c} - \frac{\omega_L' x}{\beta c} - \frac{\gamma' \omega_L x}{\beta \gamma c} = 0. \tag{3.64}$$

Unlike the paraxial ray equation (3.49) that contains the r^{-3} term when $p_\theta \neq 0$, these are linear equations in x and y. However, they are coupled and thus have to be solved simultaneously. By introducing the complex variable $\zeta = x + iy = re^{i\theta}$, the two equations can be combined into one which can be transformed to the rotating Larmor frame via the transformation

$$\zeta = \zeta_L e^{i\theta_r} , \tag{3.65}$$

where

$$\zeta_L = x_L + iy_L = r_L e^{i\theta_L} \tag{3.66}$$

and θ_r is given by Equation (3.58).

The resulting differential equation describing the particle motion in the Larmor frame has the form

$$\zeta_L'' + g_1(z)\zeta_L' + g_2(z)\zeta_L = 0, \tag{3.67}$$

with

$$g_1(z) = \frac{\gamma'}{\beta^2 \gamma}, \qquad g_2(z) - \frac{\gamma''}{2\beta^2 \gamma} + \frac{\omega_L^2}{\beta^2 c^2}. \tag{3.68}$$

Thus one obtains for the trajectory coordinates $x_L = \mathrm{Re}\ \zeta_L$ and $y_L = \mathrm{Im}\ \zeta_L$ two identical equations [since $g_1(z)$ and $g_2(z)$ are real functions]. These equations are decoupled and linear in x_L and y_L, which explains the advantage of working in the Larmor frame. Note that $g_1(z)$ and $g_2(z)$ are the same functions as in the corresponding linear terms of the radial equation (3.49).

The three sets of equations for r and θ [Equations (3.49) and (3.44)], x and y [Equations (3.63) and (3.64)], or x_L and y_L [from Equation (3.67)] are, of course, equivalent forms of the paraxial ray equation and have the same physical content. To solve them, one must specify the initial conditions for r, r', θ, θ' (or p_θ) in the first case, or the corresponding initial values for x, x', y, y' or x_L, x_L', y_L, y_L' in the latter two cases. Particularly simple is the situation where the particles are launched with $v_\theta - 0$ in a region where the magnetic field is zero ($\mathbf{B} = 0$). In this case, $p_\theta = 0$; that is, the nonlinear term in the radial trajectory equation (3.49) vanishes and a particle stays in the meridional plane that is defined by the initial values of r and θ and that rotates with the Larmor frequency ω_L. One can choose one of the transverse planes in the Larmor frame, say the x_L–z plane by setting $\theta_L = 0$ to coincide with the meridional plane and thus needs only one of the two equations defined by (3.67) to describe the particle motion. The radial coordinate r is then identical with x_L. Thus we can use the equation

$$r'' + g_1(z)r' + g_2(z)r = 0 \tag{3.69}$$

and treat r like the cartesian coordinate x_L (i.e., it can be positive or negative, changing sign when a particle crosses the z-axis in the meridional plane). Note that (3.69) follows directly from (3.49) for $p_\theta = 0$. The transformation to the Larmor frame is a very important simplification. It allows us to apply Equation (3.69) and its properties, which will be discussed in the following sections, to both axisymmetric electric and magnetic fields.

Through the remainder of this section and in Section 3.4 we will, for the most part, restrict ourselves to axisymmetric systems described by Equation (3.69) with the understanding that the variable r behaves like a cartesian coordinate whether magnetic fields are present or not. This implies that we will consider only particle motion in a meridional plane. The rotation of this plane in the presence of a magnetic field can be calculated from Equation (3.58). Application of the results to the more general case of nonmeridional motion requiring the inclusion of angular momentum or the use of two equations is straightforward in view of the preceding discussion.

3.3.3 General Properties of the Solutions of the Paraxial Ray Equations

Let us now review some general mathematical properties of Equation (3.69) describing the linear beam optics in an axisymmetric system. In the case of magnetic fields this description is of course done in the rotating Larmor frame, as discussed in the preceding section. First, we recognize that equations of the form (3.69) [and, likewise, (3.67) in the complex variable ζ] are second-order, linear, ordinary differential equations. These have two independent solutions, say $u(z)$ and $v(z)$, from which the general solution can be constructed by linear superposition, that is,

$$r(z) = Au(z) + Bv(z), \tag{3.70}$$

$$r'(z) = Au'(z) + Bv'(z). \tag{3.71}$$

The constants A and B, and thus the solution $r(z)$, are uniquely determined by the initial conditions; for instance, if $r = r_0$, $r' = r_0'$ at $z = 0$, one has

$$r_0 = Au(0) + Bv(0), \tag{3.72}$$

$$r_0' = Au'(0) + Bv'(0). \tag{3.73}$$

Solving for the constants A and B yields

$$A = \frac{r_0 v' - r_0' v}{uv' - u'v}, \tag{3.74}$$

$$B = \frac{r_0' u - r_0 u'}{uv' - u'v}. \tag{3.75}$$

The denominator in the solutions for A and B is known as the Wronskian determinant,

$$W = uv' - u'v. \tag{3.76}$$

W is nonzero by virtue of the fact that u, v are linearly independent solutions. Differentiation of (3.76) yields

$$W' = uv'' - u''v. \tag{3.77}$$

From (3.69) we have $u'' = -g_1 u' - g_2 u$, $v'' = -g_1 v' - g_2 v$. When this is substituted in (3.77), one finds that the terms involving g_2 cancel, and one obtains the result

$$W' = -g_1(uv' - u'v) = -g_1 W. \tag{3.78}$$

Using $g_1 = \gamma'/\beta^2\gamma = \gamma\gamma'/(\gamma^2 - 1)$ from (3.68), we can integrate Equation (3.78) and obtain

$$W = W_0 \exp\left[-\int g_1(z)\,dz\right] = W_0 \exp\left(-\int \frac{d\gamma}{\beta^2\gamma}\right) = \frac{W_0}{[\gamma^2 - 1]^{1/2}} = \frac{W_0}{\beta\gamma}, \tag{3.79}$$

where W_0 is the integration constant that depends on the initial conditions. Thus, when $g_1 = \gamma'/\beta^2\gamma = 0$, the Wronskian is a constant, otherwise, it changes as $(\beta\gamma)^{-1}$.

The term $g_1 r'$ in Equation (3.69) [or in Equation (3.67)] can always be eliminated by introducing the *reduced variable*

$$R = r\left(\frac{W_0}{W}\right)^{1/2} = r\left[\exp\left(\int g_1(z)\,dz\right)\right]^{1/2} = (\beta\gamma)^{1/2} r, \tag{3.80}$$

which results in the equation

$$R'' + G(z)R = 0, \tag{3.81}$$

where

$$G(z) = g_2 - \frac{1}{4}g_1^2 - \frac{1}{2}g_1',$$

or

$$G(z) = \frac{\gamma'^2(\gamma^2 + 2)}{4\beta^4\gamma^4} + \frac{\omega_L^2}{\beta^2 c^2}. \tag{3.82}$$

In the nonrelativistic case, we have $R = (V)^{1/4}r$,

$$g_1 = \frac{V'}{2V}, \qquad g_2 = \frac{V''}{4V} + \frac{qB^2}{8mV}, \qquad g_1' = \frac{V''2V - 2V'^2}{4V^2},$$

and the function $G(z)$ takes the form

$$G(z) = \frac{3}{16}\left(\frac{V'}{V}\right)^2 + \frac{qB^2}{8mV}. \tag{3.83}$$

Furthermore, since

$$\int g_1(z)\,dz = \frac{1}{2}\int \frac{dV}{V} = \frac{1}{2}\ln V,$$

one finds for the reduced variable the relation

$$R = r\,\exp\!\left(\frac{1}{4}\ln V\right) = rV^{1/4}. \tag{3.84}$$

The form (3.81) of the paraxial ray equation is of great importance insofar as it involves only V and its first derivative V' on the z-axis. Thus if V is measured or calculated with some error and, hence, the second derivative, V'', is not too accurate, one can obtain better results by using reduced variables and Equation (3.81). Also, the reduced variable R is much smoother than r, which varies more strongly when the energy changes. The calculation of the focal length of an electrostatic lens is therefore more accurate when R is used, as is done in Section 3.4.3.

Let us now return to the solution of the paraxial ray equation as presented in Equations (3.70) to (3.76). By substituting the results for the constants A and B into the first two equations, one obtains a linear relationship between $r(z), r'(z)$ and the initial conditions $r_0(z), r_0'(z)$, which can be written in matrix form as

$$\begin{pmatrix} r \\ r' \end{pmatrix} = \begin{pmatrix} a & b \\ c & d \end{pmatrix}\begin{pmatrix} r_0 \\ r_0' \end{pmatrix} = \widetilde{M}\begin{pmatrix} r_0 \\ r_0' \end{pmatrix}. \tag{3.85}$$

The matrix $\widetilde{M}$ is known as the *transfer matrix* and the matrix elements depend on $u(z)$, $v(z)$ and the derivatives $u'(z)$, $v'(z)$. It is often convenient to choose two independent solutions having initial conditions $u(0) = 1$, $u'(0) = 0$, $v(0) = 0$, $v'(0) = 1$ and known as the *principal solutions*. In this case, the constants A and B are simply $A = r_0, B = r_0'$ [from Equations (3.74), (3.75)], and the transfer matrix is given by

$$\widetilde{M} = \begin{pmatrix} u(z) & v(z) \\ u'(z) & v'(z) \end{pmatrix}. \tag{3.86}$$

Note that the determinant of this matrix, $|\widetilde{M}|$, is defined by the Wronskian $W = uv' - u'v$ and hence changes its value as $(\beta\gamma)^{-1}$ when $g_1(z) \neq 0$ (i.e., when the particles are accelerated or decelerated). On the other hand, when $g_1(z) = 0$, or by operating with reduced variables (R, R'), one has the advantage that the determinant of the transfer matrix is always unity ($|\widetilde{M}| = W = 1$).

We recognize that a 2×2 matrix relation like (3.85) exists for any two points in a system described by linear equations of the form (3.69). Thus the radial coordinate and slope of a trajectory at three different positions, z_1, z_2, and z_3, are linked by the relation

$$\begin{pmatrix} r_2 \\ r_2' \end{pmatrix} = \tilde{M}_{21} \begin{pmatrix} r_1 \\ r_1' \end{pmatrix}, \tag{3.87}$$

$$\begin{pmatrix} r_3 \\ r_3' \end{pmatrix} = \tilde{M}_{32} \begin{pmatrix} r_2 \\ r_2' \end{pmatrix} = \tilde{M}_{31} \begin{pmatrix} r_1 \\ r_1' \end{pmatrix}, \tag{3.88}$$

where

$$\tilde{M}_{31} = \tilde{M}_{32} \tilde{M}_{21}. \tag{3.89}$$

This property of the solutions for individual particle trajectories relates to the concept of the emittance discussed in Sections 3.1 and 3.2. Suppose that we have a distribution of particles at some initial position z_1 such that the trace-space area is defined by an ellipse of the general form

$$a_1 r_1^2 + 2b_1 r_1 r_1' + c_1 r_1'^2 = 1. \tag{3.90}$$

The area occupied by this distribution at some other point z_2 is then readily found by solving the transfer matrix relation (3.87) for (r_1, r_1') in terms of (r_2, r_2') and substituting in (3.90). In view of the linear relationship between the two sets of variables, one obtains an equation of the form

$$a_2 r_2^2 + 2b_2 r_2 r_2' + c_2 r_2'^2 = 1. \tag{3.91}$$

This is again the equation of an ellipse where the coefficients a_2, b_2, c_2 are uniquely determined by the transfer matrix elements and the initial coefficients. Consequently, the motion of particles in this linear system is such that the trace-space area of the distribution remains an ellipse. Of course, the orientation, size, and shape of the ellipse are changing continuously as the beam propagates along the z-axis.

In the linear systems considered here, Equation (3.4) applies, and the emittance is defined by the area of the ellipse, which is given by

$$A_1 = \epsilon_1 \pi = \frac{\pi}{(a_1 c_1 - b_1^2)^{1/2}}, \qquad A_2 = \epsilon_2 \pi = \frac{\pi}{(a_2 c_2 - b_2^2)^{1/2}}, \tag{3.92}$$

for the two positions. Now, according to Liouville's theorem, if $mc\beta_1\gamma_1$ and $mc\beta_2\gamma_2$ denote the momentum of the particles at the two positions, respectively, the emittances are related as

$$\frac{\epsilon_2}{\epsilon_1} = \frac{A_2}{A_1} = \frac{\beta_1\gamma_1}{\beta_2\gamma_2} = \frac{W_2}{W_1}. \tag{3.93}$$

Thus, the emittance changes in the same ratio as the Wronskian determinant of the transfer matrix. It should be pointed out in this context that the beam description in the reduced variables R, R' also obeys Liouville's theorem.

There are several other forms in which paraxial ray equations and transfer matrices for axially symmetric systems may be written; these alternate formulations are discussed in Lawson's book [C.17]. Paraxial equations can be derived also for systems without axial symmetry, such as rectangular geometries (*strip beams*), systems with two symmetry planes (quadrupole fields), and beams in circular accelerators. Some of these equations are discussed in subsequent sections.

3.4 AXIALLY SYMMETRIC FIELDS AS LENSES

3.4.1 General Parameters and Transfer Matrix of a Lens

Whereas in the preceding section we did not make any assumptions about the axial distribution of the fields, we now focus our attention on fields that are of limited axial extent. These fields, which are zero outside a small interval $z_1 < z < z_2$, are generally employed as *ion-optical* or *electron-optical* lenses in the same way that glass lenses are used to focus light beams. Like glass lenses in optics, charged particle lenses can be used to form images of an object (electron microscope), to transport a beam from one point to another, or to focus a beam onto a small target.

We first discuss the *optical* properties of a single lens for charged particle beams neglecting space-charge effects. Let the electric or magnetic fields of such a lens be confined to the region $z_1 < z < z_2$. Outside this region the particle trajectories will be straight lines. As will be shown later, all fields of limited axial extent have a focusing action on a traversing beam; that is, they form converging lenses provided that the trajectories do not cross the axis within the lens region. Thus, if we choose a ray $u(v)$ which, prior to entering the lens, forms a straight line parallel to the axis [i.e., $(u, u') = (1, 0)$ for $z < z_1$], it will emerge from the lens with an angle of inclination toward the axis. Likewise, there will be a ray $v(z)$ converging toward the axis when entering the lens which will leave the lens on a straight line parallel to the axis with $(v, v') = (1, 0)$ for $z > z_2$ (see Figure 3.2).

Both rays will cross the axis at some point on the respective side of the lens. As in the case of a glass lens in optics, these two points are called *focal points*. The actual paths of the particles inside the lens need to be known, of course, to determine the displacement and slope of the trajectory emerging from the other side of the lens. If we extend the two straight lines of the trajectory on either side, they will intersect (dashed lines in Figure 3.2). The points of intersection define the two *principal planes* I and II. The four lens parameters (d_1, d_2, f_1, f_2) are defined in Figure 3.2 and are taken as positive numbers if they are as indicated. Thus, $d_1 > 0$ when plane I is to the right of the center of the lens, as shown. For a defocusing lens, f_2 would be negative.

The two principal planes and their respective focal lengths f_1 and f_2 completely determine the optical properties of the two particular solutions we have chosen, and

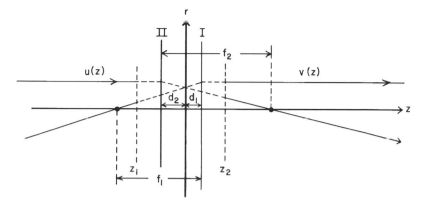

Figure 3.2. Principal planes and focal points of a lens.

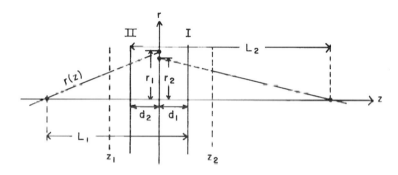

Figure 3.3. Trajectory crossing the axis at object and image plane.

since these are independent solutions, that of any other particle trajectory. Consider, for instance, the trajectory in Figure 3.3, which crosses the axis a distance L_1 from plane I to the left and at L_2 from plane II to the right. As we will see later, this ray defines the object and image planes of the system.

Any such ray can be described by linear superposition of two independent solutions like $u(z)$ and $v(z)$ defined above. If we project the slopes of the ray on each side through the lens, they intersect the midplane of the lens ($z = 0$) at distances r_1 and r_2, respectively, as shown in Figure 3.3. The action of a lens is thus seen to result in a change of the projected slope and displacement of the trajectory at the center, and it can be described by a lens transfer matrix relation

$$\begin{pmatrix} r_2 \\ r_2' \end{pmatrix} = \tilde{M} \begin{pmatrix} r_1 \\ r_1' \end{pmatrix} = \begin{pmatrix} m_{11} & m_{12} \\ m_{21} & m_{22} \end{pmatrix} \begin{pmatrix} r_1 \\ r_1' \end{pmatrix}. \tag{3.94}$$

This relation is valid for any trajectory, including the two principal solutions $u(z)$ and $v(z)$. Using $u(z)$ and $v(z)$ and the geometrical relations illustrated in Figure 3.2, we can find the transfer matrix elements in terms of the four lens parameters d_1, d_2, f_1, f_2. Starting with the parallel ray entering the lens from the left [i.e., $(u_1, u_1') = (1, 0)$], we get

$$\begin{pmatrix} u_2 \\ u_2' \end{pmatrix} = \begin{pmatrix} m_{11} & m_{12} \\ m_{21} & m_{22} \end{pmatrix}\begin{pmatrix} 1 \\ 0 \end{pmatrix} = \begin{pmatrix} m_{11} \\ m_{21} \end{pmatrix}. \tag{3.95}$$

From Figure 3.2 we have the relation

$$u_2' = -\frac{1}{f_2}, \qquad u_2 = 1 + u_2' d_2 = 1 - \frac{d_2}{f_2}.$$

Consequently,

$$m_{11} = 1 - \frac{d_2}{f_2}, \tag{3.96}$$

$$m_{21} = -\frac{1}{f_2}. \tag{3.97}$$

Using the other independent ray leaving the lens parallel to the axis [i.e., $(v_2, v_2') = (1, 0)$], we obtain the relation

$$\begin{pmatrix} v_2 \\ v_2' \end{pmatrix} = \begin{pmatrix} 1 \\ 0 \end{pmatrix} = \begin{pmatrix} m_{11} & m_{12} \\ m_{21} & m_{22} \end{pmatrix}\begin{pmatrix} v_1 \\ v_1' \end{pmatrix}. \tag{3.98}$$

From Figure 3.2 we have

$$v_1' = \frac{1}{f_1}, \qquad v_1 = 1 - \frac{d_1}{f_1},$$

hence

$$\begin{pmatrix} 1 \\ 0 \end{pmatrix} = \begin{pmatrix} m_{11} & m_{12} \\ m_{21} & m_{22} \end{pmatrix}\begin{pmatrix} 1 - \dfrac{d_1}{f_1} \\ \dfrac{1}{f_1} \end{pmatrix},$$

or

$$\begin{pmatrix} 1 - \dfrac{d_1}{f_1} \\ \dfrac{1}{f_1} \end{pmatrix} = (m_{11}m_{22} - m_{21}m_{12})^{-1}\begin{pmatrix} m_{22} \\ -m_{21} \end{pmatrix}. \tag{3.99}$$

Since m_{11} and m_{21} are known, we can solve for the other two matrix elements and obtain after some algebra the result

$$m_{22} = \frac{f_1}{f_2}\left(1 - \frac{d_1}{f_1}\right),$$ (3.100)

$$m_{12} = f_1\left(\frac{d_1}{f_1} + \frac{d_2}{f_2} - \frac{d_1 d_2}{f_1 f_2}\right).$$ (3.101)

In terms of the four lens parameters, the lens transfer matrix thus has the form

$$\tilde{M} - \begin{pmatrix} m_{11} & m_{12} \\ m_{21} & m_{22} \end{pmatrix} = \begin{pmatrix} 1 - \dfrac{d_2}{f_2} & f_1\left(\dfrac{d_1}{f_1} + \dfrac{d_2}{f_2} - \dfrac{d_1 d_2}{f_1 f_2}\right) \\ -\dfrac{1}{f_2} & \dfrac{f_1}{f_2}\left(1 - \dfrac{d_1}{f_1}\right) \end{pmatrix}.$$ (3.102)

When the two principal planes coincide, we may put $d_1 = d_2 = 0$, and the lens matrix becomes much simpler:

$$\tilde{M} = \begin{pmatrix} 1 & 0 \\ -\dfrac{1}{f_2} & \dfrac{f_1}{f_2} \end{pmatrix}.$$ (3.103)

The displacement of the trajectory, $\Delta r = r_2 - r_1$, is zero in this case and the lens changes only the slope of the trajectory. This is known as the *thin-lens approximation* or *weak-lens approximation*, which in many cases is sufficient to determine the focusing effects of a lens. Physically, a lens may be considered as *thin* or *weak* in the above sense when the width of the lens is short compared to the focal length. In this case the change of slope, $\Delta r' = r_2' - r_1'$, is small and the particle position $r(z)$ within the lens region may be regarded as constant to good approximation.

3.4.2 Image Formation and Magnification

As we know from light optics, one of the important features of a lens is the fact that it can form an image of an object. In our case of charged particle beam optics, the *object* can be an electron-emitting surface such as the cathode of an electron gun, a piece of material from which an electron beam is reflected into an electron microscope, or the plasma surface of an ion source. However, in a broader sense, any cross-sectional area of a particle beam can be an object. In this case, particles with different slope r' from a given point r within the beam are focused into a point at the image location. Indeed, from this point of view, there is no difference between an emitting surface, like a cathode, and a cross-sectional area of the beam. In either case, particles emerge from every point with different angles of their trajectories. An *image* is formed at the position where the trajectories emerging from a given object point are focused again into a point downstream from the lens.

To examine this image-forming property of a lens, let us return to Figure 3.3. The trajectory shown in this figure crosses the axis on the left side (*object side*) at distance L_1 from principal plane I and on the right (*image side*) at distance L_2 from the plane II. We know that this ray can be obtained by linear superposition of the two independent solutions u, v defined earlier, that is,

$$r(z) = Au(z) + Bv(z). \tag{3.104}$$

Now, on the object side of the lens ($z < 0$) we have $u_1 = 1$, $u_1' = 0$, $v_1 = 1/f_1$, $v_1 = v_1'[z + (f_1 - d_1)] = [z + (f_1 - d_1)]/f_1$, and hence

$$r(z) = A + B\frac{z + (f_1 - d_1)}{f_1}. \tag{3.105}$$

With the condition $r(z) = 0$ at distance L_1 from plane I [i.e., at $z = -(L_1 - d_1)$], one obtains from (3.105) the relation

$$\frac{f_1}{L_1} = \frac{B}{A + B}. \tag{3.106}$$

On the image side of the lens ($z > 0$), we have $v = v_2 = 1$, $v_2' = 0$, $u_2' = -1/f_2$, $u_2 = -[z - (f_2 - d_2)]/f_2$, and thus

$$r(z) = -A\frac{z - (f_2 - d_2)}{f_2} + B. \tag{3.107}$$

With $r(z) = 0$ at distance L_2 from plane II (i.e., at $z = L_2 - d_2$), one finds from (3.107) that

$$\frac{f_2}{L_2} = \frac{A}{A + B}. \tag{3.108}$$

Adding Equations (3.106) and (3.108) leads to the well-known lens equation relating object distance and image distance:

$$\frac{f_1}{L_1} + \frac{f_2}{L_2} = 1. \tag{3.109}$$

As we have made no restrictions concerning the slope r' of the ray at distance L_1 from plane I, our analysis implies that all rays emerging from the *object* point

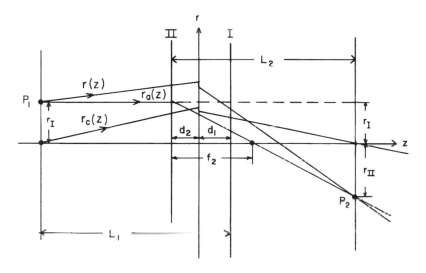

Figure 3.4. Relation between object point P_1 and image point P_2.

on the axis are *imaged* into a single point at distance L_2 from principal plane II on the axis.

Let us now consider the behavior of rays displaced from the axis at the object location L_1. As an example, consider the ray $r(z)$ emerging from point P_1 which is displaced by r_I in the object plane, $z = -(L_1 - d_1)$, as shown in Figure 3.4. We will show that all rays emerging from point P_1 are focused into point P_2 a distance r_{II} off the axis at the image plane, $z = L_2 - d_2$. The ray $r(z)$ can be represented by a linear superposition of the previous ray (starting from the axis), which we now denote with $r_c(z)$, and the parallel ray through point P_1, given by $r_a(z) = r_I u(z)$:

$$r(z) = A r_a(z) + B r_c(z). \tag{3.110}$$

At the object plane $r_c = 0$ and $r = r_I = r_a$, hence, $A = 1$, and thus

$$r(z) = r_a(z) + B r_c(z). \tag{3.111}$$

As $r_c(z) = 0$ at the image plane, all rays emerging from P_1 will have the same distance $r_{II} = r_a$ from the axis (i.e., they will be focused into point P_2). Q.E.D.

The *magnification* (i.e., the ratio r_{II}/r_I), can be found by obtaining r_a at $z = L_2 - d_2$. This ray crosses the axis at distance f_2 from plane II. From

Figure 3.4 we find that

$$\frac{r_I + r_{II}}{L_2} = \frac{r_I}{f_2},$$

that is,

$$\frac{r_{II}}{r_I} = \frac{L_2}{f_2} - 1. \tag{3.112}$$

Multiplying the right side with (3.109), which is unity, we obtain

$$\frac{r_{II}}{r_I} = \frac{L_2}{L_1}\frac{f_1}{f_2} - \frac{f_1}{L_1} + 1 - \frac{f_2}{L_2} = \frac{L_2}{L_1}\frac{f_1}{f_2} - \left(\frac{f_1}{L_1} + \frac{f_2}{L_2} - 1\right).$$

The last term on the right-hand side is zero and, hence,

$$\frac{r_{II}}{r_I} = \frac{f_1 L_2}{f_2 L_1}. \tag{3.113}$$

This is the magnification factor.

Equations (3.109) and (3.113) allow us to determine the image distance L_2 and the image magnification from the object distance L_1 and the two focal lengths f_1 and f_2. Thus, if the two focal points and the two principal planes of a lens are known, one can construct the image for an object at an arbitrary position along the axis, provided that both object and image lie in the field-free space outside the lens fields.

Since f_1/f_2 is a constant for any given lens, we see that $r_{II} : r_I = C(L_2 : L_1)$. If we consider f_1, f_2, L_1, L_2 as positive parameters and define the radial coordinate of an object point by r_o, that of an image point by r_i, then because of the image inversion, we have $r_o = r_I, r_i = -r_{II}$, and hence, we must write (3.113) in the form

$$r_i = -\frac{f_1 L_2}{f_2 L_1} r_o. \tag{3.114}$$

3.4.3 Electrostatic Lenses

We return now to the paraxial ray equation and consider first electrostatic lenses ($\mathbf{B} = 0$) for particles with zero azimuthal velocity ($p_\theta = 0$). In this case, the paraxial ray equation [Equation (3.49)] becomes

$$r'' + \frac{\gamma'}{\beta^2\gamma} r' + \frac{\gamma''}{2\beta^2\gamma} r = 0,$$

or

$$(\beta\gamma r')' + \frac{\gamma''}{2\beta} r = 0. \tag{3.115}$$

We shall now derive a simple relation between the two focal lengths f_1 and f_2. To do this, let us consider the Wronskian determinant of the two independent solutions, which from Equations (3.76) and (3.79) is given by

$$W = uv' - u'v = W_0(\gamma^2 - 1)^{-1/2} = \frac{W_0}{\gamma\beta}. \tag{3.116}$$

Now let us assume that to the left of the lens (in object space),

$$u = u_1 = 1, \qquad u_1' = 0, \qquad \gamma - \gamma_1 = \frac{qV_1}{E_0} + 1,$$

and to the right (in image space),

$$v - v_2 = 1, \qquad v_2' = 0, \qquad \gamma = \gamma_2 = \frac{qV_2}{E_0} + 1.$$

On the left side of the lens, we therefore get from (3.116)

$$v_1' = W_0(\gamma_1^2 - 1)^{-1/2},$$

whereas to the right,

$$u_2' = -W_0(\gamma_2^2 - 1)^{-1/2}.$$

From the preceding section, we know that $v_1' = 1/f_1$, $u_2' = -1/f_2$; hence, taking the ratio of the last two equations, we find that

$$\frac{f_1}{f_2} = \frac{(\gamma_1^2 - 1)^{1/2}}{(\gamma_2^2 - 1)^{1/2}} = \frac{P_1}{P_2} = \frac{\gamma_1\beta_1}{\gamma_2\beta_2}. \tag{3.117}$$

In the nonrelativistic limit, this relation takes the form

$$\frac{f_1}{f_2} = \left(\frac{V_1}{V_2}\right)^{1/2}. \tag{3.118}$$

Thus the ratio of the object-side and image-side focal lengths of an electrostatic lens is equal to the ratio of the mechanical momenta (or, in the nonrelativistic case, the velocities) on either side of the lens region. If the momenta in the field-free space on each side of the lens, or the *voltage equivalents*, are identical, we speak of a *unipotential lens*. In this case, the two focal lengths are the same (i.e., $f_1 = f_2$).

We will prove now that all axisymmetric electrostatic fields (with field-free regions on either side) form positive, or converging, lenses provided that the trajectories do not cross the axis within the lens region. First we see this qualitatively from Figure 3.5 for a *bipotential lens* (i.e., different potentials, or velocities, on each side) in which the electric field accelerates the particles. The radial force component on the left is focusing, that on the right is defocusing. The radius of the particle trajectory decreases and the velocity increases from left to right. Hence, converging action dominates over diverging, and we have a net focusing effect. A similar argument can be made if the field is decelerating. Quantitatively, this follows from Equation (3.81), which in the nonrelativistic case is

$$R'' + \frac{3}{16}\left(\frac{V'}{V}\right)^2 R = 0. \tag{3.119}$$

Integration yields

$$R_2' - R_1' = -\frac{3}{16}\int_{z_1}^{z_2}\left(\frac{V'}{V}\right)^2 R\,dz. \tag{3.120}$$

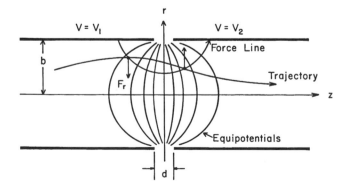

Figure 3.5. Trajectory through an electric lens (schematic).

As long as the particle does not cross the axis within the lens, $R > 0$, $(V'/V)^2 > 0$, and hence $R_2' - R_1' < 0$. From the definition $R = rV^{1/4}$, we have

$$R' = r'V^{1/4} + \frac{r}{4} V^{-3/4} V'.$$

As the limits of integration are assumed to be outside the lens field, V' is zero at these points, hence

$$r_2' V_2^{1/4} - r_1' V_1^{1/4} < 0.$$

In the special case where $r_1' = 0$, we find from this relation that $r_2' < 0$ (i.e., the parallel ray emerges from the lens with negative slope); hence, the lens is focusing. This argument is still valid if a magnetic field is considered (see next section) or added to the electric field. The reason is that in a solenoidal magnetic lens, the particle energy does not change; hence, $V_2 = V_1$, and therefore we have $r_2' - r_1' < 0$. A similar argument can be made for a two-dimensional system with planar electrodes separated by a gap. Thus we can generalize the conclusion and state that all axisymmetric or two-dimensional electrostatic and solenoidal magnetostatic lenses are focusing provided that the particle trajectory does not cross the axis within the lens field region.

We had seen in Section 3.4.1 that the action of a lens on the particle trajectory may be described by the transfer matrix M [Equation (3.102)]. When the thin-lens approximation can be applied, this matrix is greatly simplified and is given by Equation (3.103). In the case of electrostatic lenses, the accuracy of the thin-lens approximation can be improved by assuming that the reduced variable R, rather than r, is constant through the lens. One can show that R is uniformly concave toward the axis, irrespective of the character of the lens field; r, however, varies so that its value is larger in the converging parts of the lens than in the diverging part of it. Consequently, putting $r = $ const invariably leads to too low a value for $1/f$. Suppose that the path considered has a displacement r_1 and zero slope $(r_1' = 0)$ to the left of the field and a slope r_2' to the right of the field. The focal length f_2 is defined as

$$f_2 = -\frac{r_1}{r_2'}. \tag{3.121}$$

Now, in the nonrelativistic limit, one has, from Equation (3.84),

$$r_1 = \frac{R_1}{V_1^{1/4}}; \qquad r_2' = \frac{R_2'}{V_2^{1/4}}$$

since V_2' outside the field is zero. Then

$$\frac{1}{f_2} = -\frac{R_2'}{R_1} \left(\frac{V_1}{V_2} \right)^{1/4}. \tag{3.122}$$

Using (3.120) with $R_1' = 0$, this becomes

$$\frac{1}{f_2} = \frac{3}{16} \frac{1}{R_1} \left(\frac{V_1}{V_2} \right)^{1/4} \int_{z_1}^{z_2} \left(\frac{V'}{V} \right)^2 R \, dz .$$

Assuming that $R = R_1 = \text{const}$ leads to

$$\frac{1}{f_2} = \frac{3}{16} \left(\frac{V_1}{V_2} \right)^{1/4} \int_{z_1}^{z_2} \left(\frac{V'}{V} \right)^2 dz . \qquad (3.123)$$

Thus f_2 can be obtained by integration of $(V'/V)^2$ over the region of the lens, and f_1 is then determined by the relation

$$\frac{f_1}{f_2} = \left(\frac{V_1}{V_2} \right)^{1/2},$$

or

$$\frac{1}{f_1} = \frac{3}{16} \left(\frac{V_2}{V_1} \right)^{1/4} \int_{z_1}^{z_2} \left(\frac{V'}{V} \right)^2 dz . \qquad (3.124)$$

It should be noted that the actual value of R is always slightly smaller than the assumed constant value. Consequently, the focal length calculated by the above weak-lens formula is slightly shorter than the true focal lengths.

There are several types of electrostatic lenses, which may be classified as follows:

1. UNIPOTENTIAL OR EINZEL LENSES. These are characterized by equal constant potentials in object and image space. As mentioned before, the object-side and image-side focal lengths are then the same (i.e., $V_2 = V_1$; $f_2 = f_1$).

2. BIPOTENTIAL OR IMMERSION LENSES. The potentials in object and image space are different ($V_1 \neq V_2$). *Immersion lens* is derived from the analogy to the oil-immersion objectives of the light microscope, for which object and image are placed in media of different refraction index (i.e., oil and air, respectively). In this case, one has

$$\frac{f_2}{f_1} = \left(\frac{V_2}{V_1} \right)^{1/2} \qquad \text{(nonrelativistically)}.$$

3. SINGLE-APERTURE LENSES. These comprise the lens fields about an aperture in an electrode which separates two regions of different constant

electric field gradients (i.e., $V_1' \neq V_2'$). Since the electron paths in the object and image fields here are not straight lines but parabolas, the considerations leading to the above lens equations do not apply and, hence, these formulas cannot be used. One can show, however, that the image magnification and image distance can be found from the position of the focal points and principal planes if the points and planes are redefined in a suitable manner.

4. CATHODE LENSES. These are lenses which are terminated on one side by an emitting surface at zero potential, normal to the optical axis (e.g., the cathode of an electron gun). An example of this type of lens is a planar diode in which the anode has a hole through which the beam can pass. The electric field in the region of the anode hole has a defocusing radial component, and the system constitutes a lens with a diverging effect on the beam. The cathode lens can be considered as a single-aperture lens with the aperture in the anode and $V_2' = 0$.

Note that this classification is not unique. There is, for instance, an alternative classification that cuts across the above grouping and divides all electron lenses into long and short lenses.

An example of a bipotential lens is shown in Figure 3.6. Two coaxial cylinders with radii b_1, b_2, and separated a distance d, are at potentials V_1, V_2, respectively. The potential distribution for such a lens can be found by solving Laplace's equation and has the general form

$$\psi(r,z) = \frac{1}{2\pi} \int_{-\infty}^{\infty} a(k)J_0(ikr)e^{ikz}\, dk\,. \tag{3.125}$$

$J_0(ikr)$ is the normal Bessel function of the first kind of zero order and can be represented by the power series

$$J_0(ikr) = \sum_{n=0}^{\infty} \frac{1}{(n!)^2}\left(\frac{ikr}{2}\right)^{2n}\,. \tag{3.126}$$

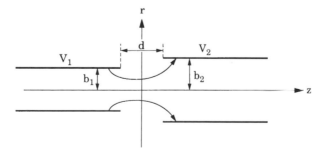

Figure 3.6. Bipotential lens formed by two coaxial conducting tubes at different potentials. (Typical lines of force are shown for $V_2 > V_1$.)

The coefficients $a_k(k)$ of the integral (3.125) must be determined from the boundary conditions for $\phi(r, z)$.

For the special case where the two cylinders have the same radius ($b_1 = b_2 = b$) and their separation is infinitesimally small ($d \rightarrow 0$), the potential functions may be written in the form

$$\phi(r, z) = \frac{V_1 + V_2}{2} + \frac{V_2 - V_1}{\pi} \int_0^\infty \frac{\sin\ kz}{k} \frac{J_0(ikr)}{J_0(ikb)} dk. \qquad (3.127)$$

On the axis ($r = 0$), this function becomes

$$\phi(0, z) = V(z) = \frac{V_2 + V_1}{2} + \frac{V_2 - V_1}{\pi} \int_0^\infty \frac{\sin\ kz}{k} \frac{dk}{J_0(ikb)}. \qquad (3.128)$$

Figure 3.7 shows $V(z)$ and the two first derivatives, $V'(z)$ and $V''(z)$. For convenience it was assumed that $V_1 = 0$, $V_2 = 1$, and that the distance z is given in units of the cylinder radius b.

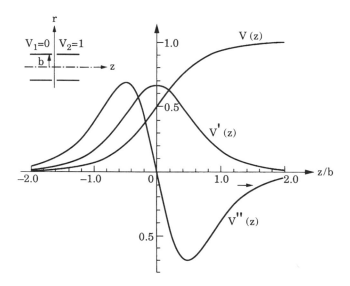

Figure 3.7. Potential distribution and derivatives on the axis of a two-cylinder lens with the same diameter and infinitesimally small separation d.

The potential distribution along the axis, $V(z)$, given in Equation (3.128) and plotted in Figure 3.7, can be approximated with a good degree of accuracy by the expression (see [C.11, Vol. I, p. 39])

$$V(z) = \frac{V_1 + V_2}{2} + \frac{V_2 - V_1}{2} \tanh\left(\frac{1.318}{b} z\right). \tag{3.129}$$

With this relatively simple analytical formula the integration of the paraxial ray equation and the determination of the focal properties of such a lens is obviously easier than using the Bessel function integrals. It is found that both f_2 and f_1 decrease (i.e., the refractive power of the lens increases) as the ratio V_2/V_1 increases. However, not much is to be gained if V_2/V_1 goes beyond about 10. Furthermore, both principal planes always lie on the low voltage side of this bipotential lens. The focal strength $1/f_2$ for a thin lens can be calculated analytically for the general relativistic case, and in the nonrelativistic approximation one finds with $\alpha = 1.318/b$ (see Problems 3.4 and 3.5)

$$\frac{1}{f_2} = \frac{3}{8} \alpha \left(\frac{V_1}{V_2}\right)^{1/4} \left(\frac{V_1 + V_2}{V_2 - V_1} \ln \frac{V_2}{V_1} - 2\right). \tag{3.130}$$

For the case where the two cylinders have the same radius $b_1 = b_2 = b$ but are separated by a distance d, the potential on the axis, $V(z)$, can be approximated by the analytical formula (see [C.11, Vol. I, p. 41])

$$V(z) = \frac{V_1 + V_2}{2} + \frac{V_2 - V_1}{2\alpha d} \ln \frac{\cosh \alpha z}{\cosh \alpha(z - d)}, \tag{3.131}$$

where $\alpha = 1.318/b$. In the limit $d \to 0$, one recovers Equation (3.129). If the diameters of the two cylinders differ, purely analytical methods cease to be effective and it is best to solve Laplace's equation, $\nabla^2 \phi = 0$, numerically (e.g., by the relaxation method) to obtain the potential distribution.

As a second example of electrostatic focusing we will now discuss the aperture lens illustrated in Figure 3.8. An electrode at potential V, located between two coplanar electrodes at potentials V_1 and V_2, has a small circular aperture of radius a through which the particle beam passes from one region to the other. In practice, the first electrode could be a cathode at $V_1 = 0$, the aperture plate could be an anode, and the third electrode could be absent. Alternatively, all three electrodes could have an aperture, as would be the case in an einzel lens. However, to analyze the effect of a single aperture, let us consider the geometry as shown in the figure.

First we note that the perturbation introduced by the aperture is confined to a region $z_1 < z < z_2$ whose width is comparable to the hole diameter, $2a$. Outside this small region the electric field on either side of the center electrode is practically uniform and the particle trajectory is either a straight line (for $\dot{r}_0 = 0$) or a parabola

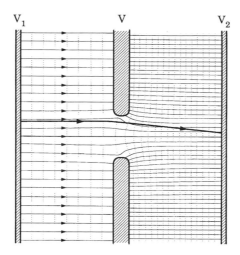

Figure 3.8. Field configuration and focusing action of a plane electrode with a circular aperture separating two regions of different field gradients.

(for $\dot{r}_0 \neq 0$). Thus we will assume that $E_{z1} = V_1' = $ const in the region $z \leq z_1$ and $E_{z2} = V_2' = $ const in the region $z \geq z_2$. For the configuration shown in the figure, we have $V_2 > V > V_1$ and $V_2' > V_1'$. Hence the radial force, qE_r, experienced by a particle passing through the aperture region is inward (i.e., the aperture acts like a focusing lens). Since electrostatic focusing is usually employed at low particle velocities, we can use the nonrelativistic force equation; hence, we obtain for the radial motion (with $\dot{\theta} = 0$),

$$m\ddot{r} = qE_r .$$

Integration through the aperture region yields

$$\dot{r}_2 - \dot{r}_1 = \frac{q}{m} \int_{t_1}^{t_2} E_r \, dt = \frac{q}{m} \int_{z_1}^{z_2} \frac{E_r}{v} \, dz . \qquad (3.132a)$$

Using the paraxial relation (3.34) between E_r and $\partial E_z / \partial z$, we can write

$$\dot{r}_2 - \dot{r}_1 = -\frac{q}{m} \int_{z_1}^{z_2} \frac{r}{2v} \frac{\partial E_z}{\partial z} \, dz . \qquad (3.132b)$$

Let us assume now that in the transition through the aperture region the radius r and velocity v of the particle remain approximately constant (thin-lens approximation).

The integral is then readily solved and we obtain for the change of the slope $r_2' - r_1' = (\dot{r}_2 - \dot{r}_1)/v$ the result

$$r_2' - r_1' = -\frac{qr}{2mv^2}(E_{z2} - E_{z1}) \tag{3.133a}$$

or alternatively, with $mv^2 = 2qV$, $E_z = V'$,

$$r_2' - r_1' = -\frac{r}{4}\frac{V_2' - V_1'}{V}. \tag{3.133b}$$

This result can also be obtained by applying Gauss's law for the electric flux to a cylinder of radius r and length $\Delta z = z_2 - z_1$. From $\int \mathbf{D} \cdot d\mathbf{S} = \int \epsilon_0 \mathbf{E} \cdot d\mathbf{S} = 0$ one obtains

$$E_{z1}r^2\pi + E_{z2}r^2\pi + 2\pi r \int_{z_1}^{z_2} E_r \, dz = 0,$$

and hence $\int_{z_1}^{z_2} E_r \, dz = -(r/2)(E_{z2} - E_{z1})$, which leads to (3.133a).

Setting $r_1' = 0$, one finds for the focal strength of the aperture lens

$$\frac{1}{f} = -\frac{r_2'}{r} = \frac{V_2' - V_1'}{4V}. \tag{3.134}$$

From these results we conclude that an aperture lens has a focusing effect if $V_2' > V_1'$, a defocusing effect if $V_2' < V_1'$, and no effect if $V_2' = V_1'$. Of special interest is the case where the beam emerges from a diode-type source, for instance an electron gun or an ion source, and propagates through an aperture into a field-free drift tube. In this configuration, the first electrode is a particle emitter (e.g., a thermionic cathode or a plasma surface), and the second electrode serves to extract and accelerate the beam. Since $V_2' = 0$, hence $f < 0$, the aperture in the *extractor* electrode has a defocusing effect. This single-aperture lens is then identical with the *cathode lens* of electron optics, as mentioned in the classification of electrostatic lenses given above.

In some applications the aperture may not be circular, but instead may have a rectangular shape having a width $\Delta x = 2a$ and a height $\Delta y = 2b$. For this case we can apply Gauss's law to a flux tube of length $\Delta z = z_2 - z_1$ and height y. Using the fact that

$$\int_0^a E_x \, dx = \overline{E}_x a = \int_0^b E_y \, dy = \overline{E}_y b,$$

where $\overline{E}_x$ and $\overline{E}_y$ are average field values, we find for the change of slopes $\Delta x'$, $\Delta y'$ in lieu of (3.133b) the relations

$$x_2' - x_1' = -\frac{x}{2(1 + a/b)}\frac{V_2' - V_1'}{V}, \tag{3.135a}$$

$$y_2' - y_1' = -\frac{y}{2(1 + b/a)}\frac{V_2' - V_1'}{V}. \tag{3.135b}$$

We must of course recognize that these linear relations are only approximately correct, as there are always aberrations (nonlinearities) associated with such rectangular apertures. From a practical point of view, two cases are of special interest. First, if $a = b$ (i.e., if the aperture has the shape of a square), the two equations are identical with (3.133b) if r is substituted by x or y. Second, if $b \rightarrow \infty$, the rectangular aperture becomes an infinitely long slit for a one-dimensional sheet beam. There is no force in the y-direction, and the change of slope for the x-motion is given by

$$x_2' - x_1' = -\frac{x}{2}\frac{V_2' - V_1'}{V}. \tag{3.136}$$

Using the above thin-lens approximations for the effect of an aperture and uniform fields for the regions between electrodes, one can calculate the focusing properties of more complicated electrode configurations such as cathode lenses or einzel lenses.

Finally, we note that the above theory also applies to cases where the electrode in question does not have a single aperture but consists of a configuration of parallel wires (*grids*) or a wire mesh. Such a configuration essentially subdivides the beam into many beamlets, each of which passes through a small aperture lens defined by the wire mesh. The formulas developed above must then be applied to each beamlet, and the net result is an effective increase of the emittance, as illustrated in Figure 3.9 (where a defocusing field geometry was assumed). As an example, consider an electron gun with a control grid at a potential V_g located a small distance from the thermionic cathode. Such a grid may be used to control the current and pulse width of the electron beam produced by the gun. Suppose that the grid is a mesh of thin wires crossing at right angles and forming square-shaped openings of width $\Delta x = \Delta y = 2a$. Each opening acts like an aperture lens; that is, the electrons passing through it experience a change of slope given by (3.135) with $a = b$ and with x and y measuring the distance from the center of the opening (see Figure 3.9). The maximum change of slope for each beamlet is then

$$\Delta x_{max}' = -\frac{a}{4}\frac{V_2' - V_1'}{V_g}, $$

where V_1' and V_2' are the potential gradients between grid and cathode, and grid and anode, respectively. Let R denote the total electron beam radius as defined by the

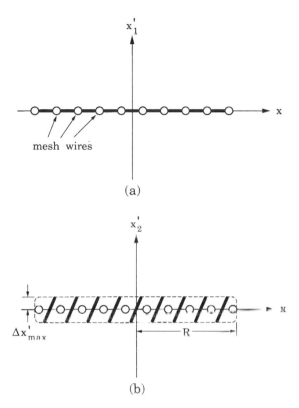

Figure 3.9. (a) Action of a defocusing grid on a laminar parallel beam in a trace-space diagram. Incoming particles (with $x_1' = 0$) occupy a straight line between mesh wires, and the emittance is zero. (b) Outgoing particles acquire slopes x_2' proportional to displacement from the center point of each wire mesh, straight lines representing particle distribution between mesh wires become tilted, and the effective emittance is $\epsilon_g \approx R \Delta x_{\max}'$.

cathode size. A parallel beam of electrons with zero initial emittance will acquire an effective emittance of $\epsilon_g \approx R|\Delta x_{\max}'|$ after passing through the wire mesh. The corresponding normalized emittance is

$$\epsilon_{n,g} = \frac{v}{c}\, \epsilon_g = \left(\frac{2eV_g}{mc^2}\right)^{1/2} \epsilon_g,$$

or

$$\epsilon_{n,g} = \frac{Ra}{4}\left(\frac{2eV_g}{mc^2}\right)^{1/2} \frac{|V_2' - V_1'|}{V_g}.$$

If the intrinsic thermal emittance, $\epsilon_{n,\text{th}}$, according to Equation (1.7), is included, one obtains for the total normalized emittance of an electron beam produced by a gun with cathode grid the result

$$\epsilon_n = \left(\epsilon_{n,\text{th}}^2 + \epsilon_{n,g}^2 \right)^{1/2}.$$

A numerical example illustrating the magnitude of this grid effect is given in Problem 3.6. Note that the emittance increase due to a cathode grid is proportional to the difference of the field gradients, $|V_2' - V_1'|$. An obvious conclusion, therefore, is to design and operate a gun such that this difference is as small as possible; ideally, the gradients on both sides should be the same, but this may not always be possible in practice.

3.4.4 Solenoidal Magnetic Lenses

In the case of a purely magnetic field, the paraxial ray equation (3.69) takes the form $r'' + g_2(z)r = 0$, or

$$r'' + k^2 r = 0, \tag{3.137}$$

where

$$k^2 = \left(\frac{qB}{2mc\gamma\beta} \right)^2 = \frac{\omega_L^2}{\beta^2 c^2}, \tag{3.138}$$

and $k^2 = q^2 B^2 / 8mqV$ in the nonrelativistic approximation. By integration, one gets for the change in the slope of the trajectory

$$r_2' - r_1' = - \int_{z_1}^{z_2} k^2 r \, dz. \tag{3.139}$$

Since k^2 is always positive, and if the particle does not cross the axis inside the lens field (i.e., $r > 0$), we see that $r_2' < r_1'$; hence, the lens is focusing, as stated earlier.

The major difference between solenoidal magnetic and electric lenses is that in the electric case the image is inverted, while in solenoidal magnetic lenses it is inverted and rotated by an angle θ_r given [from Equation (3.58)] by

$$\theta_r = - \int_{z_1}^{z_2} k \, dz. \tag{3.140}$$

Since, by Ampère's circuital law, the integral $\int_{-\infty}^{\infty} H \, dz = NI =$ number of ampere turns of the coil, we can also write

$$\theta_r = - \mu_0 \frac{q}{2mc\beta\gamma} NI, \tag{3.141}$$

or $\theta_r = -\mu_0 (q/8mV)^{1/2} NI$ nonrelativistically. This relation between the angle of rotation and the number of ampere turns of the coils exciting the solenoidal magnetic lens is accurate if we take the integral from a sufficiently field-free region on the left of the lens to the field-free space on the right of it. For electrons, one obtains the nonrelativistic relation

$$|\theta_r| = 0.1865 \times \frac{NI_{[A]}}{\left(V_{[V]}\right)^{1/2}} . \tag{3.142}$$

It the axial width of the magnetic lens is so short that the change in the radial coordinate r of a trajectory within the field region is negligibly small, the magnetic field in question constitutes a *thin* or *weak* magnetic lens, as defined earlier. In this case, r on the right side of (3.139) may be treated as a constant. Suppose that r_1' is zero to the left of the lens (parallel ray); then we obtain for the slope $r' = r_2'$ on the right side,

$$r' = -r \left(\frac{q}{2mc\beta\gamma} \right)^2 \int_{z_1}^{z_2} B^2 \, dz . \tag{3.143}$$

In a magnetic lens $f_2 = f_1$ since $V_2 - V_1$; that is, one deals with only one focal length, f, which in our case here is defined by

$$\frac{1}{f} = -\frac{r'}{r} = \left(\frac{q}{2mc\beta\gamma} \right)^2 \int_{z_1}^{z_2} B^2 \, dz . \tag{3.144}$$

Note that the focal length from this formula is shorter than the exact thick-lens result. The reason is that the trajectory radius r is not actually constant but decreases slightly through the lens due to the fact that $r'' = -(qB/2mc\beta\gamma)^2 r$ from Equation (3.137) is always negative.

As an example of a magnetic lens, let us consider the solenoid shown in Figure 3.10. The field produced by this arrangement may, in first approximation, be assumed to be uniform in the region $0 < z < l$ and zero outside this region if the diameter D of the aperture is small compared to the length l (i.e., $D/l \ll 1$).

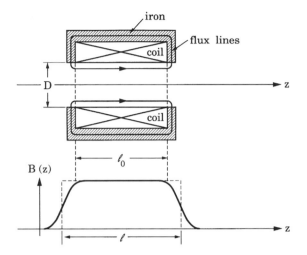

Figure 3.10. Solenoid lens with iron shield.

Mathematically, the *effective length* l of this equivalent uniform field (or *hard-edge*) approximation can be defined by

$$l = \frac{1}{B_0^2} \int_{-\infty}^{\infty} B^2(z)\, dz ,$$

where B_0 is the peak magnetic field (see Figure 3.10).

Since the paraxial ray equation involves only $B(z)$ on the axis and not the derivatives of $B(z)$, the treatment of solenoidal magnetic lenses is much simpler than that of electric lenses. In our particular case, using the hard-edge approximation we can integrate (3.137) with the assumption $k = \text{const}$ for $0 < z < l$ and $k = 0$ elsewhere. Thus, with initial conditions $r = r_0$, $r' = 0$ at $z = 0$, we obtain

$$r = r_0 \cos kz , \tag{3.145a}$$

$$r' = -r_0 k \sin kz . \tag{3.145b}$$

When the particle leaves the field, the radius and slope will be (with $z = l$ and $\phi = kl$)

$$r_l = r_0 \cos \phi , \tag{3.146a}$$

$$r_l' = \frac{r_0}{l} \phi \sin \phi , \tag{3.146b}$$

where

$$\phi = kl = \frac{qB_0l}{2mc\beta\gamma} = \frac{\mu_0 qNI}{2mc\beta\gamma},$$ (3.147a)

or

$$\phi = kl = \left(\frac{q}{8mV}\right)^{1/2} B_0l = \mu_0\left(\frac{q}{8mV}\right)^{1/2} NI \quad \text{(nonrelativistically)}.$$ (3.147b)

The image rotation is given by

$$\theta_r = -\phi\,;$$ (3.148)

that is, the parameter ϕ measures the amount of rotation of the meridional plane by the solenoidal magnetic field.

The focal length is obtained from (3.146) as

$$\frac{1}{f} = -\frac{r_i'}{r_0} = \frac{\phi \sin \phi}{l}.$$ (3.149)

The image-side principal plane is located at z_2, and from Figure 3.11 one has the relations

$$\frac{r_0 - r_i}{l - z_2} = -r_i', \quad \text{or} \quad z_2 = l + \frac{r_0 - r_i}{r_i'}\,;$$

hence,

$$z_2 = l\left(1 - \frac{1 - \cos \phi}{\phi \sin \phi}\right).$$ (3.150)

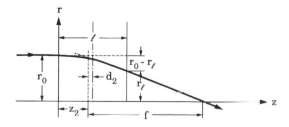

Figure 3.11. Trajectory entering solenoidal field with zero slope.

Expressing the location of the principal plane as a distance d_2 to the left of the center of the lens, one obtains

$$d_2 = \frac{l}{2} - z_2,$$

or

$$d_2 = l\left(\frac{1 - \cos \phi}{\phi \sin \phi} - \frac{1}{2}\right). \tag{3.151}$$

By comparison, the thin-lens approximation yields from Equation (3.144)

$$\frac{1}{f} = \left(\frac{q}{2mc\beta\gamma}\right)^2 B_0^2 l = \frac{\phi^2}{l} = k^2 l, \tag{3.152}$$

which follows also from the thick-lens formula (3.149) if one expands the $\sin \phi$ for $\phi \ll 1$. In the thin-lens approximation, the location of the principal plane, is of course, in the center of the lens; hence, $z_2 = l/2$ and $d_2 = 0$.

The comparison shows good agreement between thick-lens and thin-lens results for $\phi \leq 0.3$. The object-side principal plane is obtained by integrating (3.137) for a trajectory leaving the lens at $z = l$ with $r = r_l$ and $r' = r_l' = 0$. The solution in this case is

$$r = A \cos kz + B \sin kz, \tag{3.153a}$$

$$r' = -Ak \sin kz + Bk \cos kz. \tag{3.153b}$$

One finds that

$$A = r_l \cos kl, \qquad B = r_l \sin kl. \tag{3.154}$$

At $z = 0$, the particle's radial position and slope are

$$r_0 = r_l \cos kl, \tag{3.155a}$$

$$r_0' = r_l k \sin kl. \tag{3.155b}$$

The object-side focal length is

$$\frac{1}{f} = -\frac{r_0'}{r_l} = k \sin kl = \frac{\phi \sin \phi}{l},$$

which is the same as the image-side focal length, as expected for a magnetic lens. The location of the object-side principal plane is at $z_1 = (r_l - r_0)/r_0'$, or, with respect to the lens center, at $d_1 = z_1 - l/2$, that is,

$$d_1 = \left(\frac{1 - \cos \phi}{\phi \sin \phi} - \frac{1}{2} \right)l = d_2. \tag{3.156}$$

The two principal planes are thus located at an equal distance $(d_1 - d_2)$ upstream and downstream from the center of the lens.

We must now discuss the physical meaning of our results. At first glance, one would expect that a particle entering a uniform magnetic field on a straight path parallel to the field lines should not be deflected radially. This is certainly true if one disregards the transition from zero field to $B = $ const. However, our field $B_z = B_0$ for $0 < z < l$ and $B = 0$ outside this region has a B_r component in the fringe-field region, from $\nabla \cdot \mathbf{B} = 0$, and this condition has been utilized in the derivation of the paraxial ray equation. This means that in our equation (3.137), the linear force from the B_r component associated with the off-axis B_z component at the edges $(z = 0, z = l)$ is implicitly taken into account. In contrast, the simple, uniform-field equations in cartesian coordinates, $\ddot{x} = \omega_c \dot{y}$, $\ddot{y} = -\omega_c \dot{x}$, $\ddot{z} = 0$, where $\omega_c = qB_0/\gamma m$ is the cyclotron frequency, are valid only within the uniform field region $B_z = B_0$ and do not give a focusing action of this field unless the junction effect is considered separately.

The book by El-Kareh and El-Kareh [C.14] contains many examples of electric and magnetic lens design and tables of lens parameters for almost every type of lens used in practical applications. Also very useful are the books by Szilagyi [C.21] and Hawkes and Kaspar [C.22], where detailed treatments of charged particle beam optics, properties of lenses, and aberrations can be found, and the book by Wollnik [C.20], which contains special material on dipole magnets, electrostatic deflectors, quadrupole lenses, and the design of particle spectrometers.

3.4.5 Effects of a Lens on the Trace-Space Ellipse and Beam Envelope

In earlier sections we studied the motion of individual particles to determine the focal properties of a lens and its effects on the particle trajectories. If we want to know what happens to an entire distribution of particles comprising a beam it is convenient to use the trace-space ellipse discussed in Section 3.3.3. Suppose that at some initial position z_0 upstream from a lens the distribution of particles in $r - r'$ trace space fills an area bounded by an ellipse that is defined by an equation of the form (3.90). How does this ellipse change as the distribution of particles moves downstream through the lens?

In the field-free space on either side of the lens the particles' trajectories are straight lines determined by the slope r' and initial position r of each particle, and the motion can be described by the matrix

$$\widetilde{M} = \begin{pmatrix} 1 & z \\ 0 & 1 \end{pmatrix}. \tag{3.157}$$

Let the subscript 1 denote the particles' $r-r'$ coordinates at the entrance side of the lens center, 2 those at the exit side, and 3 at an arbitrary position downstream from the lens. Furthermore, let $\widetilde{M}_{01}$ be the free-space matrix between the initial position and the lens, $\widetilde{M}_{12}$ the lens matrix given by Equation (3.102), and $\widetilde{M}_{23}$ the free-space matrix between the lens and point 3 downstream. Then the relation between a particle's $r-r'$ coordinates at point 3 and the initial conditions (r_0, r_0') is given by

$$\begin{pmatrix} r_3 \\ r_3' \end{pmatrix} = \widetilde{M}_{23}\widetilde{M}_{12}\widetilde{M}_{01}\begin{pmatrix} r_0 \\ r_0' \end{pmatrix} = \begin{pmatrix} a_{11} & a_{12} \\ a_{21} & a_{22} \end{pmatrix}\begin{pmatrix} r_0 \\ r_0' \end{pmatrix}. \tag{3.158}$$

The equation of the trace-space ellipse at any of the four points is of the form

$$a_i r_i^2 + 2b_i r_i r_i' + c_i r_i'^2 = 1 \qquad (i = 0, 1, 2, 3). \tag{3.159}$$

Of particular interest is the case where the initial ellipse is upright (i.e., $b_0 = 0$). The motion of such an ellipse from object to image space is depicted schematically in Figure 3.12. In the free-space regions the slope of each particle trajectory remains constant and the beam is divergent. At the lens the shape and the radial position are changed and the beam converges until it reaches a waist. The location of this waist z_w is defined by the condition $b_3 = 0$ (upright ellipse). The image of the initial beam is located at z_i, a short distance past the waist position, as illustrated in the figure. It is determined by the condition that r_3 must be independent of r_0', hence $a_{12} = 0$. Note that the ellipse at the image position is tilted (not upright like the initial *object* ellipse).

The envelope of the beam, $R = r_{max}$, versus distance is also plotted schematically in Figure 3.12. For a tilted ellipse, the envelope radius is found by solving the ellipse equation for r'. Since this is a quadratic equation, there are generally two values of r' associated with each value of r, except for r_{max}, where r' is single-valued. From this condition one finds for the beam radius

$$r_{max} = R = \frac{A}{\pi}\sqrt{c} = \epsilon\sqrt{c}, \tag{3.160a}$$

while the slope of the envelope is defined by

$$R' = \frac{-\epsilon b}{\sqrt{c}}. \tag{3.160b}$$

Here $A = \epsilon\pi = \pi/(ac - b^2)^{1/2}$ is the area of the ellipse according to Equation (3.92) and a, b, c denote the three coefficients in the ellipse equation. For motion in free space, c depends on the initial conditions (R_0, R_0') and the distance z from the initial position. Differentiating (3.160) twice and using relations such as (3.92), one obtains the following differential equation for the beam envelope in free space:

$$R'' - \frac{\epsilon^2}{R^3} = 0. \tag{3.161}$$

This equation can be readily integrated, and one obtains for $R(z)$ the hyperbolic solution

$$R(z) = \left[R_0^2 + 2R_0 R_0' z + \left(\frac{\epsilon^2}{R_0^2} + R_0'^2 \right) z^2 \right]^{1/2}. \tag{3.162}$$

R_0 and R_0' denote the radius and slope of the envelope at the initial position, and ϵ the emittance.

Figure 3.12(b) depicts the change of the envelope as the beam propagates in the axial direction. The passage through the lens has a similar effect on the envelope as on an individual particle [i e , it changes the radius and slope (R, R')]. In general, this change must be calculated from (3.160a) and (3.160b) using the coefficients c_2, b_2 of the beam ellipse after the lens matrix has been applied. Only in the special

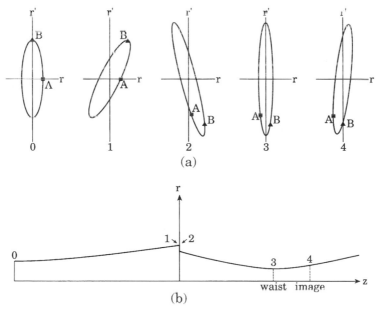

Figure 3.12. Qualitative change of the emittance ellipse in trace space and of the beam envelope as the beam travels from object space to image space of a lens.

case of a thin lens can the single-particle lens matrix $\tilde{M}_{12}$ also be used to obtain the change in the beam envelope directly via the relation

$$\begin{pmatrix} R_2 \\ R_2' \end{pmatrix} = \tilde{M}_{12} \begin{pmatrix} R_1 \\ R_1' \end{pmatrix}. \tag{3.163}$$

The reason for this is that a thin-lens transformation does not change the particle radius. Hence, a particle that is at the edge of the beam remains at the edge after the transformation has been applied. However, in a thick lens, a particle that coincides with the beam envelope on the upstream side of the lens is no longer at the beam edge after the lens transformation is applied, and vice versa. This can be seen in Figure 3.12 by following the motion of particle B from position 1 (entrance side of the lens) to position 2 (exit side of lens).

We must also keep in mind that the emittance does not necessarily remain constant when the lens transformation is applied. Thus, if the particle energy is changed, as is the case in a bipotential lens, the emittance also changes according to the relation

$$\frac{\epsilon_2}{\epsilon_1} = \frac{\beta_1 \gamma_1}{\beta_2 \gamma_2} = \frac{f_1}{f_2}. \tag{3.164}$$

In the free space downstream from the lens the beam envelope is again described by an equation of the form (3.162) with the new initial conditions R_2, R_2', and ϵ_2 and with z denoting the axial distance from the midplane of the lens.

3.4.6 Aberrations in Axially Symmetric Lenses

The paraxial ray equation was derived on the basis of idealizing assumptions (expanding the equations of motion and keeping only linear terms in r, r'). The lenses treated in this paraxial approximation are ideal in the sense that they produce sharp, faithful images of an object in a plane perpendicular to the beam axis. In practice, such perfect lenses do not exist as nonlinearities in the focusing fields, and other effects cause imperfections or aberrations. These aberrations can be classified according to the source by which they are caused, as follows: (1) *geometrical aberrations* (spherical aberration, coma, curvatures of field, astigmatism, and distortion of the *barrel, pin cushion,* or rotational type); (2) *chromatic aberrations* (due to energy spread in the beam); (3) *space-charge effects*; (4) *diffraction* (limits resolutions of electron microscopes); and (5) *imperfection* [such as mechanical misalignments, fluctuations (*ripple*) in the voltages and currents supplying the electric and magnetic lens elements, etc.].

In an ideal lens, all particles leaving a point r_o, θ_o in the object plane will arrive at the same point r_i, θ_i in the image plane. When aberrations are present, this is no longer the case, and particles emerging from an object point r_o, θ_o with different initial angles will arrive at different points $r_i + \Delta r_i, \theta_i + \Delta \theta_i$ in the image plane.

For a detailed discussion of the various types of aberrations, we must refer to the literature (e.g., the books on electron optics by Zworykin et al. [C.1] or Klemperer [C.2], or the more recent books by Wollnik [C.20], Szilagyi [21] or Hawkes and Kaspar [C.22]). We will, however, briefly discuss two types of aberrations that are of particular importance: the spherical aberration and the chromatic aberration.

The *spherical aberration* is a geometrical aberration that arises from third-order terms (r^3, $r^2 r'$, etc.) that are neglected in the paraxial ray equation. Note that r^2 terms are excluded by symmetry since the radial forces on a particle must change direction when the sign of r is changed. As an example, if one includes all terms up to third order in r and r' in the equations for solenoidal magnetic lenses, one obtains in place of the paraxial ray equation (3.137) the nonlinear equation

$$r'' + \kappa r + \kappa r'^2 r - \kappa \left(\frac{B'}{B} \right) r' r^2 + \left[\kappa^2 - \frac{1}{2} \kappa(B'') \right] r^3 = 0. \qquad (3.165)$$

Here it was assumed that $p_\theta = 0$ and κ is defined as $\kappa = k^2 = \omega_L^2 / \beta^2 c^2$; B' and B'' are the first and second derivatives of B_z on the axis with respect to z.

To illustrate the effect of spherical aberrations, consider the case of a thin lens shown in Figure 3.13. Two particle trajectories emerge from an *object* point on the axis with angles α_0 and $-\alpha_0$ and pass through the midplane of the lens at radial distance r_1 and $-r_1$. Due to the r_1^3 term, they will experience a stronger force than in the perfect lens and as a result, they will cross the axis at angles α_i and $-\alpha_i$ before reaching the image plane of the perfect lens. For small angles α_i, the displacement Δr_i at the image plane can be defined to good approximation be the relation

$$\Delta r_i = C_s \alpha_i^3, \qquad (3.166)$$

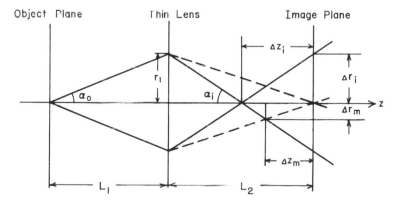

Figure 3.13. Effect of spherical aberration. The large-angle trajectories cross the axis at a distance Δz_i upstream from the ideal image plane.

where C_s is the spherical aberration coefficient, which depends on the initial conditions and the lens geometry. The crossing angle α_i depends on the initial angle α_0, or the object distance L_1. The crossover point is at a distance Δz_i upstream from the perfect image plane. If one considers the entire ensemble of trajectories within a beam, one finds that the minimum radius (waist of the beam envelope), which defines what is known in the literature as the *disk of least confusion*, is located at a distance of $\Delta z_m < \Delta z_i$ upstream from the perfect image plane, as indicated in Figure 3.13. If the object is at a large distance ($L_1 \gg L_2, L_2 \approx f_2$), the incident rays are practically parallel to the axis. In this case, defining the spherical aberration coefficient as $C_s(\infty)$, one can show that the radius Δr_m of least confusion and the associated distance Δz_m are given by

$$\Delta r_m \approx \frac{1}{4} C_s(\infty)\alpha_i^3, \qquad \Delta z_m \approx \frac{3}{4} \Delta z_i = \frac{3}{4} C_s(\infty)\alpha_i^2. \qquad (3.167)$$

For a unipotential or magnetic lens ($f_1 = f_2 = f$) , the relation between the spherical aberration coefficient for infinite and finite object distance is found to be

$$C_s(\infty) = \frac{f}{L_2} C_s(L_1), \qquad (3.168)$$

where L_2 defines the location of the ideal image plane, L_1 the object distance, and f the focal length of the lens. Spherical aberrations constitute a fundamental form of lens defects that, unlike the situation in light optics, cannot be eliminated completely. This is due to the constraints imposed on the field shapes by the conditions $\nabla \times \mathbf{B} = 0, \nabla \cdot \mathbf{E} = 0$ when space charge is neglected. (Unfortunately, space-charge effects tend to make things worse rather than better.) The ratio of the spherical aberration coefficient to the focal length, C_s/f, is used as a figure of merit defining the quality of a lens. Spherical aberration data for various types of lenses can be found in Septier [C.13, Vol. I], El-Kareh and El-Kareh [C.14], Szilagyi [C.21], and Hawkes and Kaspar [C.22].

Chromatic aberrations are due to the spread in kinetic energy that is inherent to some degree in any beam. They are different from geometrical aberrations in that they do not imply any nonlinear terms in the trajectory equations. Since the focal length f (or f_1 and f_2 for bipotential lenses) depends on the momentum, particles with different momentum or energy produce images at different distances from the lens. These images are perfect in the paraxial approximation, and the spread in the image locations, Δz_i, depends on the momentum spread ΔP in the beam. The variation of the focal length f with particle momentum responsible for this effect also produces a *circle of least confusion* of radius r_c. We can calculate this radius by considering a parallel beam consisting of trajectories that enter the lens with zero initial slope (i.e., $r_0' = 0$). Particles of momentum P will cross the axis downstream from the lens at the focal distance z_f. Those with a different momentum, say $P + \Delta P$, will be focused at a point $z_f + \Delta z_f$, where $\Delta z_f = (\partial f/\partial P) \Delta P$. If the angle of convergence for the particle with momentum P is α, then the radius of the circle of least confusion is (following Lawson, [C.14, p. 41])

$$r_c = \alpha \left(\frac{\partial f}{\partial P} \right) \Delta P = \alpha f \left(\frac{P}{f} \frac{\partial f}{\partial P} \right) \frac{\Delta P}{P} . \tag{3.169}$$

One now defines a chromatic aberration coefficient C_c for a lens by

$$\frac{C_c}{f} = \frac{1}{2} \left(\frac{P}{f} \frac{\partial f}{\partial P} \right), \tag{3.170}$$

and writes

$$r_c = 2\alpha C_c \frac{\Delta P}{P} = 2C_c \alpha \frac{\Delta \gamma}{\beta^2 \gamma} . \tag{3.171}$$

In the nonrelativistic limit one gets

$$r_c = C_c \alpha \frac{\Delta V}{V} , \tag{3.172}$$

where V is the voltage equivalent of the kinetic energy and ΔV represents half the total energy spread in the beam. For a thin, solenoidal magnetic lens we found that the focal length f is proportional to P^2 [Equation (3.144)], and hence we get for the chromatic aberration coefficient the value $C_c/f = 1$. In general, however, when one considers thick as well as bipotential lenses, the expressions for C_c can be rather complicated.

Although *space charge* is neglected in this chapter, we discuss briefly its *effect on spherical aberrations*. First, we note that the space charge associated with a beam acts like a defocusing lens. In an ideal beam with uniform charge density, the electric field, and hence the defocusing force, are proportional to the radius r and cause an increase in the focal length, which in turn changes the image location and magnification. Linear beam optics with space charge is treated in Chapter 4. In practice, the charge density is not uniform, and this nonuniformity causes spherical aberration, as we now show. Suppose that the charge density across the beam varies as

$$\rho(r) = \rho_0 \left(1 - \delta \frac{r^2}{a^2} \right) \qquad \text{for } r \leq a , \tag{3.173}$$

and $\rho(r) = 0$ for $r > a$, where $\delta = \Delta\rho/\rho_0$. From $\nabla \cdot \mathbf{E} = \rho/\epsilon_0$, one then gets for the radial electric field

$$E_r(r) = \frac{\rho_0}{\epsilon_0} \left(\frac{r}{2} - \frac{\delta r^3}{4a^2} \right) = \frac{\rho_0 r}{2\epsilon_0} \left(1 - \frac{\delta}{2} \frac{r^2}{a^2} \right). \tag{3.174}$$

As an illustration, consider a unipotential, thin lens with paraxial focal length f_0. Suppose that the linear part of the space-charge force increases the focal length to σf_0. A parallel ray entering the lens will therefore cross the axis at distance f_0 when no space charge is present and at distance σf_0 when the linear part of the space charge is taken into account (no aberration). The nonlinear (quadratic) term in the charge density reduces the defocusing force, and the trajectory will cross the axis at a distance $f_0 < z < \sigma f_0$, which depends on the incident radius of the particle. This is illustrated in Figure 3.14. For the outermost particle passing the lens at $r = a$, the angle of convergence is $\theta_0 = a/\sigma f_0$ when $\delta = 0$. However, for $\delta \neq 0$, the angle will be increased to

$$\theta = \frac{a}{\sigma f_0} \frac{1}{1 - \delta/2} \approx \theta_0 \left(1 + \frac{\delta}{2} \right). \tag{3.175}$$

At the focal plane ($z = \sigma f_0$), there will be a spot size of radius $\Delta r \approx a\delta/2$. Equating this with $C_s \theta^3$, where C_s is the spherical aberration coefficient of the combined lens and space-charge effect, we get

$$C_s \theta^3 = C_s \theta_0^3 \left(1 + \frac{3\delta}{2} \right) = \frac{a\delta}{2},$$

or

$$C_s = \frac{(\sigma f_0)^3}{a^2} \frac{\Delta \rho}{2\rho_0}. \tag{3.176}$$

From our previous discussion of spherical aberration, the circle of least confusion has a radius of

$$\Delta r_{\min} = \frac{1}{4} C_s \theta_0^3 = \frac{1}{8} a \frac{\Delta \rho}{\rho_0}. \tag{3.177}$$

Thus, as an example, if $a = 2$ cm, $\Delta\rho/\rho_0 = 0.2$, we get $\Delta r_{\min} = 0.5$ mm.

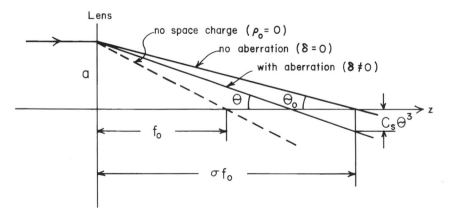

Figure 3.14. Spherical aberration due to space charge.

3.5 FOCUSING BY QUADRUPOLE LENSES

The focusing produced by axially symmetric systems is of second order since first-order fields of the form $E_r \propto r$ and $B_\theta \propto r$ are forbidden by the requirement of circular symmetry and the constraints that $\nabla \cdot \mathbf{E} = 0$, $\nabla \times \mathbf{B} = 0$. Thus, the focusing action of the axisymmetric lenses described in Section 3.4 is relatively weak. There are two alternative ways to provide stronger, first-order focusing: (1) utilizing charges and currents within the beam channel (i.e., $\nabla \cdot \mathbf{E} = \rho/\epsilon_0 \neq 0$, $\nabla \times \mathbf{B} = \mu_0 \mathbf{J} \neq 0$), as explained in Section 4.6; and (2) abandoning axial symmetry by introducing quadrupole fields or, as in circular accelerators, by using alternating-gradient ("strong") focusing.

In the present section we discuss quadrupole fields, which have two planes of symmetry. Following this, in the next section, we study the focusing of beams that propagate along a circular path in a magnetic guide field. Quadrupole fields are a special case of cylindrical multipole fields ("2n poles") which satisfy the condition $\nabla \cdot \mathbf{E} = 0$, $\nabla \times \mathbf{B} = 0$, and where the variation of the radial field component is proportional to $f(z)r^{n-1} \cos[2(n-1)\theta]$. In particular, a pure electric quadrupole field ($n = 2$) is given by

$$E_r = -\frac{E_0 r}{a} \cos 2\theta, \qquad E_\theta = \frac{E_0 r}{a} \sin 2\theta \qquad (3.178a)$$

in cylindrical coordinates, or

$$E_x = -E_0 \frac{x}{a}, \qquad E_y = E_0 \frac{y}{a} \qquad (3.178b)$$

in cartesian coordinates.

Such a two-dimensional field is produced by conducting boundaries shaped in hyperbolic form as shown in Figure 3.15 for the electric quadrupole. If potentials V_0 and $-V_0$ are applied, as shown in the figure, the potential distribution in the space between the electrodes is given by the expression

$$V(x, y) = \frac{V_0}{a^2} (x^2 - y^2), \qquad (3.179)$$

from which E_x and E_y given in (3.178b) are obtained, with E_0 defined as $E_0 = 2V_0/a$. In practice, the electric quadrupole potential distribution of Equation (3.179) can be approximated with good accuracy by using four cylindrical rods having a circular cross section of a radius a, rather than electrodes with hyperbolic shapes.

In similar fashion, a magnetic quadrupole field is described by

$$B_r = \frac{B_0 r}{a} \sin 2\theta, \qquad B_\theta = \frac{B_0 r}{a} \cos 2\theta, \qquad (3.180a)$$

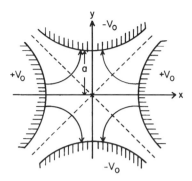

Figure 3.15. Electrodes and force lines in an electrostatic quadrupole.

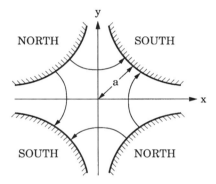

Figure 3.16. Field lines in a magnetic quadrupole. For a positively charged particle moving in the z-direction, the force components are focusing in the x-direction and defocusing in the y-direction.

or

$$B_y = B_0 \frac{x}{a}, \qquad B_x = B_0 \frac{y}{a}. \tag{3.180b}$$

Such a field is produced by a magnet configuration with hyperbolic pole shapes, as shown in Figure 3.16.

We will assume that in the electrostatic quadrupole the changes in kinetic energy remain negligibly small, so that for both magnetic quadrupoles and electric quadrupoles we have $\gamma = $ const or $\dot{\gamma} = 0$. The equations of motion for the electric quadrupole system are then

$$\gamma m \ddot{x} = q E_x = -\frac{q E_0}{a} x,$$

or

$$\ddot{x} + \frac{qE_0}{\gamma ma} x = 0,$$

and likewise

$$\ddot{y} - \frac{qE_0}{\gamma ma} y = 0.$$

In the case of a magnetic quadrupole one has

$$\gamma m\ddot{x} = -qv_z B_y = -qv_z \frac{B_0}{a} x,$$

or

$$\ddot{x} + \frac{qv_z B_0}{\gamma ma} x = 0,$$

$$\ddot{y} - \frac{qv_z B_0}{\gamma ma} y = 0.$$

These equations all have the same form, $\ddot{\xi} \pm \omega_0^2 \xi = 0$, with solutions $\xi = A\cos\omega_0 t + B\sin\omega_0 t$, or $\xi = A\cosh\omega_0 t + B\sinh\omega_0 t$, depending on the sign. Thus, we get focusing action in one plane of symmetry and defocusing in the other.

Let us now eliminate the time t and write the above equations as trajectory equations. With $z = v_z t$, $v_z \approx v \approx \mathrm{const}$, $d^2/dt^2 = v^2(d^2/dz^2)$, we obtain

$$x'' + \kappa x = 0, \tag{3.181a}$$

$$y'' - \kappa y = 0, \tag{3.181b}$$

where for magnetic quadrupoles

$$\kappa = \frac{qB_0}{\gamma mav}, \tag{3.182}$$

and for electric quadrupoles

$$\kappa = \frac{qE_0}{\gamma mav^2}. \tag{3.183a}$$

In the nonrelativistic case ($\gamma = 1$) the latter relation may be written in terms of the quadrupole voltage V_0 and beam voltage V_b as

$$\kappa = \frac{V_0}{V_b a^2} \tag{3.183b}$$

since $E_0 = 2V_0/a$ and $mv^2 = 2qV_b$. With initial conditions $x = x_0$, $x' = x_0'$, $y' = y_0'$ at $z = 0$, one obtains the solutions

$$x = x_0 \cos \sqrt{\kappa} \, z + \frac{x_0'}{\sqrt{\kappa}} \sin \sqrt{\kappa} \, z \,,$$

$$x' = -\sqrt{\kappa} \, x_0 \sin \sqrt{\kappa} \, z + x_0' \cos \sqrt{\kappa} \, z$$

or

$$\begin{pmatrix} x \\ x' \end{pmatrix} = \begin{pmatrix} \cos \sqrt{\kappa} z & \dfrac{1}{\sqrt{\kappa}} \sin \sqrt{\kappa} z \\ -\sqrt{\kappa} \sin \sqrt{\kappa} z & \cos \sqrt{\kappa} z \end{pmatrix} \begin{pmatrix} x_0 \\ x_0' \end{pmatrix} \tag{3.184a}$$

and

$$\begin{pmatrix} y \\ y' \end{pmatrix} = \begin{pmatrix} \cosh \sqrt{\kappa} z & \dfrac{1}{\sqrt{\kappa}} \sinh \sqrt{\kappa} z \\ \sqrt{\kappa} \sinh \sqrt{\kappa} z & \cosh \sqrt{\kappa} z \end{pmatrix} \begin{pmatrix} y_0 \\ y_0' \end{pmatrix} . \tag{3.184b}$$

A quadrupole field of a short axial width $\Delta z = l$, where l is normally greater than the semiaperture a, but less than $1\sqrt{\kappa}$, constitutes a quadrupole lens. In practice, the quadrupole field does not end abruptly. There is a fringe field which forms a transition from the ideal quadrupole field to the field-free region. However, as in the solenoid case of Section 3.4.4, we can replace the actual gradient profile, $\kappa(z)$, by an equivalent hard-edge approximation. If κ_0 denotes the peak gradient in the flat part of the profile, the effective length l of the equivalent hard-edge function is given by

$$l = \frac{1}{\kappa_0} \int_{z_1}^{z_2} \kappa(z) \, dz \,.$$

Hence, we have $\kappa = \kappa_0 = \text{const}$ for $0 \le z \le l$ and $\kappa = 0$ elsewhere. The hard-edge assumption is sufficient for a paraxial (first-order) analysis. Nonlinear effects due to fringe fields and nonhyperbolic boundaries are discussed by Hawkes [C.10] and Wollnik [C.20].

Two quadrupole lenses, arranged as a focusing–defocusing pair, have a net focusing effect that is much stronger than the focusing action of an axisymmetric lens of comparable size and field strength. This is why magnetic quadrupole *doublets* are widely used for focusing of high-energy particles in accelerators and beam-handling systems.

Electrostatic quadrupoles are limited in their application to focusing and transport of low-energy ion beams. This can be seen by comparing the electrostatic and magnetic gradient functions. From (3.183a) and (3.182) we find that

$$\frac{\kappa_E}{\kappa_M} = \frac{E_0}{v B_0} = \frac{E_0}{\beta c B_0} \,.$$

High-voltage breakdown limits E_0 to about 10^7 V/m and saturation of ferromagnetic materials limits B_0 to about 2 T, so that

$$\frac{\kappa_E}{\kappa_M} = \frac{1.67 \times 10^{-2}}{\beta}.$$

Thus below $\beta = 1.67 \times 10^{-2}$ (i.e., for protons with energies less than 130 keV or for electrons with energies less than 70 eV), electrostatic quadrupoles are more efficient than magnetic quadrupoles. At higher energies, however, the magnetic lenses are superior. Furthermore, by use of superconductors the focusing capability of magnetic quadrupoles can be increased substantially beyond the 2-T limit of room-temperature magnets.

The transfer matrix $\tilde{M}$ and the four lens parameters (or *cardinal points*, as they are often called) for each symmetry plane of a single quadrupole lens can be calculated in the same way as for the axially symmetric lenses of Section 3.4. In the thin-lens approximation, one finds that

$$\frac{1}{f} = \pm \kappa l. \tag{3.185}$$

The negative sign applies for the diverging case (y–z plane). The properties of a doublet consisting of a focusing and defocusing pair of thin lenses separated by a short drift space of length $\Delta z = s$ are obtained by multiplication of the appropriate transfer matrices. If f_1 denotes the focal length of the first lens, f_2 that of the second lens, one finds in the thin-lens approximation that the combined action of the two lenses is equivalent to that of a single lens of focal length F given by

$$\frac{1}{F} = \frac{1}{f_1} + \frac{1}{f_2} - \frac{s}{f_1 f_2}. \tag{3.186}$$

For a quadrupole doublet with equal strength (i.e., $f_1 = f$, $f_2 = -f$), one obtains

$$\frac{1}{F} = \frac{s}{f^2} = \kappa^2 l^2 s. \tag{3.187}$$

It is interesting to compare a quadrupole doublet with a solenoid having the same total length L and a magnetic field strength equal to the quadrupole field B_0 (at the pole tips). Taking the case $s = l$ when the two quadrupoles are adjacent to each other and hence $L = 2l$, one obtains

$$\frac{1/F_{\text{doublet}}}{1/f_{\text{solenoid}}} = \frac{4l^3}{a^2 L} = \frac{L^2}{2a^2}.$$

This relation shows that both lenses have equal strength when $L = \sqrt{2}a$. In practice, however, $L \gg a$, and hence the quadrupole doublet is stronger than a solenoid. As an example, when $L = 10a$, one finds that the doublet is 50 times stronger than the equivalent solenoid.

Other important quadrupole lens systems are the *triplet*, which consists of a lens of length l with a shorter lens of length $l/2$ on either side, and the periodic FODO channel discussed in Section 3.8.3.

Further details on quadrupole lenses can be found in Reference C.13 (article by Regenstreif, Vol. I) and in Livingood [D.1]. A thorough treatment of magnetic quadrupoles, including aberrations, is given by Hawkes [C.10].

3.6 CONSTANT-GRADIENT FOCUSING IN CIRCULAR SYSTEMS

3.6.1 Betatron Oscillations

So far, we have considered beams that move along a straight path. Let us now discuss the focusing of beams that move on circular orbits as is the case in high-energy accelerators with magnetic guide fields. In betatrons, classical cyclotrons, and synchrocyclotrons, the magnetic fields employed are axially symmetric and the orbits of the particles are circles. On the other hand, modern sector-focusing cyclotrons and high-energy synchrotrons have magnetic fields that vary azimuthally, and the orbit shape departs from a circle. In all cases, the fields are designed in such a way that the particles comprising the beam perform oscillations in the radial and axial directions about a *closed orbit* or *equilibrium orbit*. These oscillations are known as *betatron oscillations* since they were first investigated theoretically by Kerst and Serber [1] in connection with the betatron.

In Section 3.6 we restrict our analysis to an axially symmetric field. First, we define the equilibrium orbit of a particle of momentum $P = \gamma m v$ as the circle with radius R_0 centered on the axis. This orbit is in the *median plane* ($z = 0$), which is the plane where the radial component of **B** is zero (see Figure 3.17). The radius R_0 is found by equating the outward centrifugal force and the inward Lorentz force, which yields the well-known result

$$R_0 = \frac{\gamma m v}{q B_0} = \frac{\beta \gamma m c}{q B_0}, \qquad (3.188)$$

where $B_0 = B_z(R_0, 0)$.

A particle of the same momentum, which is displaced from this equilibrium orbit by a small amount $x = r - R_0$, experiences a radial force which will either drive it back toward R_0 (focusing) or farther away from R_0 (defocusing). Likewise, a particle that is displaced from the equilibrium orbit in the axial direction ($z \neq 0$) will experience a focusing or defocusing force. To determine whether focusing can be obtained simultaneously in both the radial and axial directions, we make a first-order analysis, as in the case of the paraxial ray equation (i.e., we assume that the displacements and slopes of the nonequilibrium trajectories are small).

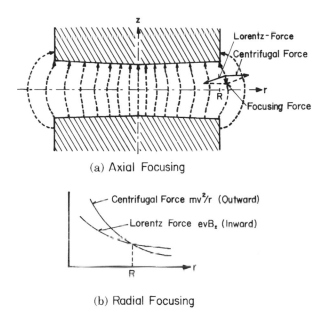

(a) Axial Focusing

(b) Radial Focusing

Figure 3.17. (a) Axisymmetric magnetic field configuration and focusing forces between pole shoes of a cyclotron-type magnet; (b) radial forces acting on particles neart equilibrium radius R.

Let $\mathbf{B} = \{B_r(r,z), 0, B_z(r,z)\}$, $\gamma m = $ const, and consider first the radial motion of a particle with velocity $\mathbf{v} = \{\dot{r}, r\dot{\theta}, 0\}$ moving in the median plane ($z = 0$). The radial force equation in this case is with $B_r(r,0) = 0, B_z = B_z(r,0)$:

$$\gamma m\ddot{r} - \gamma m r\dot{\theta}^2 = qr\dot{\theta}B_z. \tag{3.189}$$

If B_z is positive, a particle with positive charge will move in the negative θ-direction, and in the linear approximation we have $r\dot{\theta} = v_\theta \approx -v$. Thus we can write

$$\gamma m\ddot{r} - \frac{\gamma m v^2}{r} = -qvB_z. \tag{3.190}$$

Let

$$r = R_0\left(1 + \frac{x}{R_0}\right), \qquad x \ll R_0, \tag{3.191}$$

and

$$B_z(r) = B_z(R_0) + \frac{\partial B_z}{\partial r}x,$$

or

$$B_z = B_0\left(1 - n\,\frac{x}{R_0}\right),$$

(3.192)

where

$$n = -\frac{R_0}{B_0}\,\frac{\partial B_z}{\partial r}$$

(3.193)

is the *field index* and $\partial B_z/\partial r$ is evaluated at the equilibrium radius ($r = R_0$). We then obtain by substitution in (3.191) the first-order equation

$$\gamma m\ddot{x} - \frac{\gamma m v^2}{R_0}\left(1 - \frac{x}{R_0}\right) + qvB_0\left(1 - n\,\frac{x}{R_0}\right) = 0.$$

From the equilibrium-orbit condition ($x = 0$, $\ddot{x} = 0$), we have $\gamma m v^2/R_0 = qvB_0$, and the two corresponding terms in the last equation cancel. Thus one gets

$$\ddot{x} + \frac{v^2}{R_0^2}\,(1 - n)x = 0.$$

(3.194a)

But $v/R_0 = \omega_c$ is the cyclotron frequency at the equilibrium radius R_0; hence,

$$\ddot{x} + \omega_c^2(1 - n)x = 0,$$

(3.194b)

or

$$\ddot{x} + \omega_r^2 x = 0,$$

(3.194c)

where

$$\omega_r^2 = \omega_c^2(1 - n).$$

(3.195)

Let $s = R\theta = R\dot{\theta}t = vt$ denote the distance along the equilibrium orbit. Then, with $x'' = d^2x/ds^2 = (1/R_0^2\omega_c^2)\ddot{x}$, we may write Equation (3.194) in the alternative form

$$x'' + k_r^2 x = 0.$$

(3.196)

$\omega_r = \omega_c(1 - n)^{1/2}$ is the *radial betatron frequency*, $k_r = 2\pi/\lambda_r = \nu_r/R_0$ is the *betatron wave number*, λ_r the *betatron wavelength*, and

$$\nu_r = \frac{\omega_r}{\omega_c} = (1 - n)^{1/2}$$

(3.197)

is the number of radial betatron oscillation periods per revolution, also known as the *betatron tune*.

As we see from Equation (3.195), the orbits are unstable (exponential growth of x) when $\omega_r^2 < 0$ or $n > 1$, and they are stable (periodic solution for x) when $\omega_r^2 > 0$ or $n < 1$. In the latter case, we have (for $x = 0$ at $t = 0$)

$$x = x_m \sin \omega_r t = x_m \sin \left[(1 - n)^{1/2} \omega_c t \right],$$

or

$$x = x_m \sin k_r s. \tag{3.198}$$

Next, we examine the motion of a particle displaced from the equilibrium orbit in axial direction. The equation of motion is

$$\gamma m \ddot{z} = -q r \dot{\theta} B_r = -q v B_r. \tag{3.199}$$

In this case, $\mathbf{v} = \{0, r\dot{\theta}, \dot{z}\}$, $\dot{z} \ll |r\dot{\theta}|$, and again $r\dot{\theta} \approx -v$. Now $B_r = (\partial B_r / \partial z)_z +$ higher-order terms. But $\partial B_r / \partial z = \partial B_z / \partial r$ from $\nabla \times \mathbf{B} = 0$. Hence

$$\ddot{z} = \frac{qv}{\gamma m} \frac{\partial B_z}{\partial r} z = \frac{q R_0}{\gamma m} \frac{v}{R_0} \frac{R_0}{B_0} \frac{\partial B_z}{\partial r} z = -\omega_c^2 n z,$$

that is,

$$\ddot{z} + \omega_c^2 n z = 0, \tag{3.200a}$$

or

$$\ddot{z} + \omega_z^2 z = 0, \tag{3.200b}$$

where

$$\omega_z^2 = n \omega_c^2. \tag{3.201}$$

This may also be written in the form

$$z'' + k_z^2 z = 0, \tag{3.202}$$

with $k_z = 2\pi/\lambda_z = v_z/R_0$ and

$$v_z = n^{1/2}. \tag{3.203}$$

Thus, to get focusing in the axial direction (periodic solution in z), we must have $n > 0$. Orbit stability in both the radial and axial directions imposes the requirement

$$0 < n < 1. \tag{3.204}$$

In view of the definition (3.193) for the field index n, the stability condition implies that $B_z(r)$ must be a decreasing function of radius r [i.e., $(\partial B_z/\partial r) < 0$]; however, the gradient may not be greater than is allowed by the $n = 1$ limit or else there is no radial focusing. In the special case $n = 0.5$, the focusing strength is the same in both directions (i.e., $\nu_r = \nu_z$).

Figure 3.18 shows the ν_r and ν_z curves versus field index n for the respective ranges where the oscillation frequencies are real (focusing). Only in the region $0 < n < 1$, where the two curves overlap, does one get orbit stability simultaneously in both directions.

The differential equations (3.194) and (3.200) for the radial and axial motion of the particles about the equilibrium orbit are also known as the *Kerst–Serber equations*.

Note that the condition (3.204) also implies that $0 < \nu_r < 1$ and $0 < \nu_z < 1$. This means that it takes more than one revolution to complete a radial or axial betatron oscillation. The amplitude of a betatron oscillation is inversely proportional to the betatron tune (ν_r or ν_z, respectively). This follows from (3.198), which yields

$$x' = x_m k_r \cos k_r s \, ; \tag{3.205}$$

hence, with $x' = x_0'$ at $s = 0$, $k_r = \nu_r/R_0$:

$$x_m = \frac{x_0' R_0}{\nu_r} \, . \tag{3.206}$$

Since $x_0' = \dot{r}_0/v = \sin \alpha \approx \alpha$, we can also relate x_m to the orbit radius R_0 and the angle α between particle trajectory and equilibrium orbit at $x = 0$:

$$x_m = \frac{R_0 \alpha}{\nu_r} \, . \tag{3.207}$$

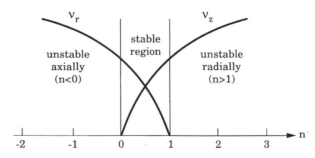

Figure 3.18. Betatron tunes versus field index n.

One betatron oscillation period is $\nu_r \, \Delta\theta = 2\pi$, and the corresponding change in azimuth angle is thus

$$\Delta\theta = \frac{2\pi}{\nu_r} . \tag{3.208}$$

The number of turns it takes for a particle to complete a betatron oscillation cycle [i.e., to return to its oscillation phase (or x, x' values) at a given azimuth angle] is then simply

$$N = \frac{\Delta\theta}{2\pi} = \frac{1}{\nu_r} . \tag{3.209}$$

As an example, suppose that $n = 0.36$. In this case, $\nu_r = (1 - n)^{1/2} = 0.8$, and it takes $N = 1/0.8 = 1.25$ revolutions to complete an oscillation period.

3.6.2 The Trace-Space Ellipse and Beam Envelope in a Betatron-Type Field

In $x - x'$ trace space, a particle moves on an ellipse which from (3.198) and (3.205) is given by

$$x^2 + \frac{x'^2}{k_r^2} = x_m^2 ,$$

or

$$\frac{x^2}{x_m^2} + \frac{x'^2}{k_r^2 x_m^2} = 1 . \tag{3.210}$$

This is an upright ellipse with major axes x_m and $k_r x_m$ in the x and x' directions, respectively, as shown in Figure 3.19. A particle starting on the equilibrium orbit $(x = 0)$ with $x' = x_m' = k_r x_m$ (point 1 in Figure 3.19) moves clockwise as s increases. If $\nu_r < 1$, it will be at point 2 after one revolution $(s = 2\pi)$, at point 3 after two revolutions, and so on.

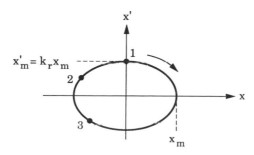

Figure 3.19. Ellipse representing particle motion in $x - x'$ space ($\nu_r < 1$).

If we consider a group of particles with different initial conditions x_0, x_0' at $s = 0$, they will each move on similar upright ellipses with constant ratio of the major axes, $x_m'/x_m = k_r$, as determined by the equations

$$x = x_0 \cos k_r s + \frac{x_0'}{k_r} \sin k_r s, \tag{3.211a}$$

$$x' = -x_0 k_r \sin k_r s + x_0' \cos k_r s, \tag{3.211b}$$

or, since $s = R_0\theta$, $k_r = \nu_r/R_0$,

$$x = x_m \cos \nu_r(\theta - \theta_m), \tag{3.212a}$$

$$x' = -k_r x_m \sin \nu_r(\theta - \theta_m). \tag{3.212b}$$

It follows that

$$x^2 + \frac{x'^2}{k_r^2} = x_0^2 + \frac{x_0'^2}{k_r^2} = x_m^2 \tag{3.213}$$

and

$$\tan \nu_r\theta_m = -\frac{R_0 x_0'}{\nu_r x_0}. \tag{3.214}$$

Suppose that all particles comprising a beam make a distribution of initial conditions x_0, x_0' which fills an elliptic area in $x-x'$ trace-space defined by the equation

$$a_1 x_1^2 + 2b_1 x_1 x_1' + c_1 x_1'^2 = 1 \tag{3.215}$$

and pictured in Figure 3.20. As each particle moves on an ellipse defined by its initial conditions according to Equation (3.213), the trace-space ellipse will rotate.

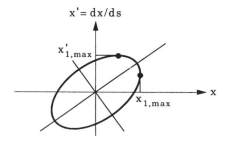

Figure 3.20. Ellipse representing beam trace-space area at $\theta = 0$.

Let us calculate the shape and orientation of this ellipse and the envelope point x_{max} as a function of azimuth angle θ. As we know from Section 3.3.3, all particles whose initial conditions correspond to points on the circumference of our *trace-space* ellipse (3.215) will remain on the circumference as the ellipse is changing. If x_1, x_1' denote the initial conditions of any such particle at $\theta = 0$, we obtain the coordinates x, x' at any other azimuth angle from the transfer matrix corresponding to Equation (3.211):

$$\begin{pmatrix} x \\ x' \end{pmatrix} = \begin{pmatrix} \cos \nu_r \theta & \dfrac{R_0}{\nu_r} \sin \nu_r \theta \\ -\dfrac{\nu_r}{R_0} \sin \nu_r \theta & \cos \nu_r \theta \end{pmatrix} \begin{pmatrix} x_1 \\ x_1' \end{pmatrix} = \begin{pmatrix} \alpha_{11} & \alpha_{12} \\ \alpha_{21} & \alpha_{22} \end{pmatrix} \begin{pmatrix} x_1 \\ x_1' \end{pmatrix}. \tag{3.216}$$

Since $\det \tilde{M} = 1$, we obtain

$$\begin{pmatrix} x_1 \\ x_1' \end{pmatrix} = \begin{pmatrix} \cos \nu_r \theta & -\dfrac{R_0}{\nu_r} \sin \nu_r \theta \\ \dfrac{\nu_r}{R_0} \sin \nu_r \theta & \cos \nu_r \theta \end{pmatrix} \begin{pmatrix} x \\ x' \end{pmatrix} = \begin{pmatrix} \alpha_{22} & -\alpha_{12} \\ -\alpha_{21} & \alpha_{11} \end{pmatrix} \begin{pmatrix} x \\ x' \end{pmatrix}. \tag{3.217}$$

Substituting this result in Equation (3.215) yields the equation of an ellipse of the form

$$ax^2 + 2bxx' + cx'^2 = 1 \tag{3.218}$$

The coefficients a, b, c depend on the initial coefficients (a_1, b_1, c_1), the betatron tune ν_r, and the azimuth angle θ. The calculation yields the result

$$a = a_1 \alpha_{22}^2 - 2b_1 \alpha_{22}\alpha_{21} + c_1 \alpha_{21}^2, \tag{3.219a}$$

$$b = -a_1 \alpha_{22}\alpha_{12} + b_1(\alpha_{12}\alpha_{22} + \alpha_{11}\alpha_{22}) - c_1 \alpha_{21}\alpha_{11}, \tag{3.219b}$$

$$c = a_1 \alpha_{12}^2 - 2b_1 \alpha_{12}\alpha_{11} + c_1 \alpha_{11}^2. \tag{3.219c}$$

We are particularly interested in the envelope of the beam which is defined by $x_m = \epsilon_x / \sqrt{c}$ [Equation (3.160a)], where ϵ_x is the emittance, which remains constant according to Liouville's theorem.

Using the relation (3.219c), one obtains for the envelope

$$x_m = \epsilon_x [a_1 \alpha_{12}^2 - 2b_1 \alpha_{12}\alpha_{11} + c_1 \alpha_{11}^2]^{1/2},$$

or

$$x_m = \epsilon_x \left[\frac{a_1 R_0^2}{\nu_r^2} \sin^2 \nu_r \theta - 2b_1 \frac{R_0}{\nu_r} \sin \nu_r \theta \cos \nu_r \theta + c_1 \cos^2 \nu_r \theta \right]^{1/2}. \tag{3.220}$$

Consider the special case where the initial ellipse is upright (i.e., $b_1 = 0$). Equation (3.220) then may be written in the form

$$x_m = x_{m1}\left[1 + h\sin^2 \nu_r\theta\right]^{1/2} = x_{m1}\left[1 + h\sin^2 k_r s\right]^{1/2}, \tag{3.221}$$

where x_{m1} is the beam envelope at $\theta = 0$ and

$$h = \frac{a_1 R_0^2}{c_1 \nu_r^2} - 1. \tag{3.222}$$

If $h \ll 1$, we can expand the expression on the right-hand side of Equation (3.221) and obtain the first-order relation

$$x_m = x_{m1}\left[1 + \frac{h}{4}(1 - \cos 2\nu_r\theta)\right]. \tag{3.223}$$

The beam envelope thus varies as a function of θ as shown qualitatively in Figure 3.21 and with a frequency that is twice the betatron frequency. The *ripple*, represented by the parameter h, is seen to depend on the ratio a_1/c_1 of the initial (upright) ellipse and on ν_r^2. Note that $1/\sqrt{a_1}$ and $1/\sqrt{c_1}$ are the major axes of the ellipse in the x and x' directions, respectively; that is, $1/\sqrt{a_1} = x_{m1}$ is the maximum displacement (envelope point), $1/\sqrt{c_1} = x'_{m1}$ the maximum slope in the particle distribution comprising the beam at the initial position ($\theta = 0$).

If the initial (upright) ellipse is chosen such that $h = 0$, hence

$$a_1 = \frac{c_1 \nu_r^2}{R_0^2},$$

or

$$x'_{m1} = \frac{\nu_r}{R_0} x_{m1}, \tag{3.224}$$

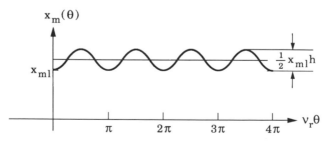

Figure 3.21. Beam envelope as function of θ if ripple is small.

then the beam envelope x_m remains a constant (no variation with angle θ), that is,

$$x_m = x_{m1} = \text{const.} \tag{3.225}$$

In this special case the beam is said to be *matched*. The beam ellipse then is identical with the ellipse on which the particle with the maximum amplitude moves in the $x-x'$ diagram (see Figure 3.19). Consequently, there is no rotation of the beam ellipse and x_m remains constant.

Let us now derive the differential equation that determines the beam envelope x_m as a function of angle θ or time t. By differentiation of Equation (3.221) with respect to s we get

$$\frac{dx_m}{ds} = x_{m1} h k_r \frac{\sin k_r s \cos k_r s}{[1 + h \sin^2 k_r s]^{1/2}} . \tag{3.226}$$

Differentiating again, one obtains from (3.226) after some algebra the envelope equation

$$\frac{d^2 x_m}{ds^2} - \frac{\epsilon_x^2}{x_m^3} + k_r^2 x_m = 0 , \tag{3.227}$$

or, in terms of time t, with $\theta = \omega_c t$, $\omega_r^2 = \omega_c^2 \nu_r^2$:

$$\frac{d^2 x_m}{dt^2} - \epsilon_x^2 \frac{\omega_o^2 R_0^2}{x_m^3} + \omega_r^2 x_m = 0 . \tag{3.228}$$

Using $X = x_m$, $X' = dX/ds$ for the envelope, one obtains from Equation (3.227) the alternative form

$$X'' + k_r^2 X - \frac{\epsilon_x^2}{X^3} = 0 . \tag{3.229}$$

An analogous equation may be obtained for the beam envelope in the z-direction. Thus, with $Z = z_m$, one gets

$$Z'' + k_z^2 Z - \frac{\epsilon_z^2}{Z^3} = 0 , \tag{3.230}$$

where $k_z = 2\pi/\lambda_z = \nu_z/R_0$, and ϵ_z represents the beam emittance in the z-direction. λ_z is the axial betatron wavelength.

The above differential equations for the beam envelope were derived from an initial upright ellipse, but they are also valid for a tilted initial ellipse. The only difference is that with a tilted ellipse, the initial slope (X'_1 or Z'_1) enters in the solutions of the envelope equations.

In the absence of a focusing force the second term in the envelope equation vanishes. Thus, for instance, when $k_z^2 = 0$ the axial envelope equation reduces to the form

$$Z'' - \frac{\epsilon_z^2}{Z^3} = 0, \tag{3.231}$$

which is the equation of the beam envelope in free space derived earlier for an axisymmetric beam [Equation (3.161)]. Free-space envelope equations like (3.231) apply to sections of a circular beam path with no focusing forces in one or both transverse directions. Examples are bending magnets with uniform field where $\nu_z^2 = 0$ and beam propagation along a straight path between magnets where both $\nu_r^2 = 0$ and $\nu_z^2 = 0$.

Note that the above envelope equations can be obtained from the Kerst–Serber equations simply by substituting X and Z for x and z and adding the emittance terms (negative sign!), which are proportional to $1/X^3$ or $1/Z^3$, respectively. While a single particle in its motion can have values $x = 0$, $z = 0$, the envelope of the beam can never approach the beam axis due to the repulsive emittance term.

3.6.3 Focusing in Axisymmetric E×B Fields

The linear theory of focusing of circular beams in axisymmetric magnetic fields can be generalized to include applied electric fields as well as the effects of the beam's magnetic and electric self field [2]. Let us assume a combination of electric and magnetic fields with axial symmetry and a *median plane* as defined by the field vectors $\mathbf{B} = \{B_r(r,z), 0, B_z(r,z)\}$, $\mathbf{E} = \{E_r(r,z), 0, E_z(r,z)\}$. The fields may be produced by charges and currents in conductors outside the beam as well as by the fields arising from the charges and the currents of the particles that constitute the beam. Although self-field effects will be treated in subsequent chapters, we will include them in this generalized theory for later reference. The only difference in the analysis as compared to the previous situation is that we can no longer use the condition $\nabla \cdot \mathbf{E} = 0$, $\nabla \times \mathbf{B} = 0$. Rather, we have to consider two contributions to the fields acting on an individual particle, namely, the applied fields $(\mathbf{E}_a, \mathbf{B}_a)$ and the self fields $(\mathbf{E}_s, \mathbf{B}_s)$; that is, we write

$$\mathbf{E} = \mathbf{E}_a + \mathbf{E}_s, \qquad \mathbf{B} = \mathbf{B}_a + \mathbf{B}_s.$$

For the steady state $(\partial/\partial t = 0)$ being considered here, we have from Maxwell's equation

$$\nabla \times \mathbf{E} = 0, \qquad \nabla \cdot \mathbf{E}_a = 0, \qquad \nabla \cdot \mathbf{E}_s = \frac{\rho}{\epsilon_0}, \tag{3.232}$$

$$\nabla \cdot \mathbf{B} = 0, \qquad \nabla \times \mathbf{B}_a = 0, \qquad \nabla \times \mathbf{B}_s = \mu_0 \mathbf{J} = \mu_0 \rho \mathbf{v}. \tag{3.233}$$

The radial force equation in this case takes the form

$$\frac{d}{dt}(\gamma m \dot{r}) - \gamma m r \dot{\theta}^2 = qE_r + qr\dot{\theta}B_z. \tag{3.234}$$

The equilibrium orbit is defined by the condition $d/dt = 0$, $r = R_0$, $\dot{\theta} = \dot{\theta}_0 = \omega_0 = -v_0/R_0$, $E_r = E_0$, $B_z = B_0$; that is,

$$\gamma m \frac{v_0^2}{R_0} = q(v_0 B_0 - E_0), \tag{3.235}$$

from which follows for the equilibrium radius

$$R_0 = \frac{\gamma m v_0}{q B_0} \frac{1}{1 - E_0/v_0 B_0}. \tag{3.236}$$

Introducing the cyclotron frequency $\omega_c = -qB_0/\gamma m$ and the frequency ω_e associated with the electric field and defined by

$$\omega_e^2 = \frac{qE_0}{\gamma m R_0}, \tag{3.237}$$

we can write Equation (3.235) in the form

$$\omega_0^2 - \omega_c \omega_0 + \omega_e^2 = 0. \tag{3.238}$$

Solving this equation for the angular frequency of rotation ω_0, one obtains

$$\omega_0 = \frac{1}{2}\left[\omega_c \pm (\omega_c^2 - 4\omega_e^2)^{1/2}\right]. \tag{3.239}$$

When the electric field is zero ($\omega_e^2 = 0$), we recover $\omega_0 = \omega_c =$ cyclotron frequency. In the case of zero magnetic field ($\omega_c = 0$), we need $\omega_e^2 < 0$ [i.e., $E_r < 0$ (inward radial electric field) for a rotating positively charged particle and get $\omega_0 = \omega_e$]. In the general case, we see from the above equations that the presence of an electric field changes both the equilibrium radius and the orbital frequency from the well-known cyclotron values R_c and ω_c.

The analysis for the motion of particles that are displaced from the equilibrium orbit in either the radial or axial direction follows the derivation given in Section 3.6.1. However, due to the presence of the electric field, the azimuthal velocity is no longer a constant but may change to first order.

In the relativistic case, the change in energy experienced by the particle, which is a radial distance $x = r - R_0$ off the equilibrium orbit, may be expressed in the form

$$\frac{d\gamma}{\gamma} = \frac{qE_0R_0}{\gamma mc^2} \frac{x}{R_0}. \tag{3.240}$$

The radian frequency is then found to be

$$\dot{\theta} = \omega_0 \left[1 - \left(1 - \frac{qE_0R_0}{mc^2\gamma^3\beta^2} \right) \frac{x}{R_0} \right], \tag{3.241}$$

where γ and β are the values at the equilibrium radius ($x = 0$).

Substituting (3.240) and (3.241) in the radial force equation, (3.234), and expanding all terms to first order leads to an equation of the form (3.196), where ν_r^2 can be expressed in terms of the fields, E_0, B_0, field gradients, $\partial B_z/\partial r, \partial E_r/\partial r$, and velocity, $v_0 = \beta c$, at the equilibrium radius R_0 as follows:

$$\nu_r^2 = 1 + \frac{E_0^2(1 - \beta^2)}{(\beta c B_0 - E_0)^2} + \frac{R_0\beta c(\partial B_z/\partial r) - [E_0 + R_0(\partial E_r/\partial r)]}{\beta c B_0 - E_0}. \tag{3.242}$$

If neither E_0 nor B_0 is zero, we can introduce the electric and magnetic field index

$$k_e = \frac{R_0}{E_0} \frac{\partial E_r}{\partial r}, \qquad k_m = \frac{R_0}{B_0} \frac{\partial B_z}{\partial r} = -n. \tag{3.243}$$

Furthermore, we define

$$\beta_0 = \frac{E_0}{cB_0}. \tag{3.244}$$

Then the expression for ν_r^2 may be written in the form

$$\nu_r^2 = 1 + \frac{\beta_0^2(1 - \beta^2)}{(\beta - \beta_0)^2} + \frac{\beta k_m - \beta_0(1 + k_e)}{\beta - \beta_0}. \tag{3.245}$$

The following limits are of interest:

1. MAGNETIC-FIELD CASE: $E_0 = 0$, $k_e = 0$

$$\nu_r^2 = 1 + k_m = 1 - n. \tag{3.246}$$

This is identical with solution (3.197) derived previously.

2. ELECTRIC-FIELD CASE: $B_0 = 0$, $k_m = 0$

$$\nu_r^2 = 3 - \beta^2 + k_e. \tag{3.247}$$

In the nonrelativistic limit, β^2 can be neglected, and if the electric field is produced between coaxial cylinders by an external voltage source, the field index $k_e = -1$. Hence $\nu_r^2 = 2$, in agreement with Equation (2.149).

In similar fashion, the axial motion of a particle can be analyzed. From the force equation in the z-direction,

$$\frac{d}{dt}(\gamma m \dot{z}) = qE_z - qr\dot{\theta}B_r,$$ (3.248)

one obtains by linearization an equation of the form (3.202), where v_z^2 is found to be

$$v_z^2 = \frac{R_0[\beta c(\partial B_r/\partial z) + \partial E_z/\partial z]}{\beta c B_0 - E_0}.$$ (3.249)

$\partial B_r/\partial z$ and $\partial E_z/\partial z$ are the axial gradients of the fields at the equilibrium radius.

For the special case where the self fields of the beam can be neglected (or for that part of v_z^2 that is due to the applied fields only), we can use $\nabla \cdot \mathbf{E} = 0$, $\nabla \times \mathbf{B} = 0$, from which follows

$$\frac{\partial E_z}{\partial z} = -\left(\frac{E_0}{R_0} + \frac{\partial E_r}{\partial r}\right).$$ (3.250)

Under these conditions, Equation (3.249) takes the form

$$v_z^2 = \frac{E_0 + R_0(\partial E_r/\partial r) - R_0\beta c(\partial B_z/\partial r)}{\beta c B_0 - E_0}.$$ (3.251)

With the definitions (3.243) and (3.244), this may also be written as

$$v_z^2 = \frac{(1 + k_e)\beta_0 - k_m\beta}{\beta - \beta_0}.$$ (3.252)

Note that for negatively charged particles ($q = -e$), we have to change the signs of all terms associated with the electric field and gradients. Thus, the equilibrium condition (3.235) is $\gamma m v_0^2/R_0 = e(v_0 B_0 + E_0)$. Furthermore, $\dot{\theta}_0 = v_0/R_0$, $\omega_c = eB_0/\gamma m$. Equation (3.249), for instance, has to be written as

$$v_z^2 = -\frac{R_0[\beta c(\partial B_r/\partial z) - \partial E_z/\partial z]}{\beta c B_0 + E_0}.$$ (3.253)

3.6.4 Energy Spread, Momentum Compaction, and Effective Mass

So far, we have considered only monoenergetic beams, where all particles have the same total momentum or kinetic energy. This even includes the $\mathbf{E} \times \mathbf{B}$ field case discussed in the preceding section, where particles do gain or lose energy when they depart from the equilibrium orbit. However, the assumption was that they all have the same kinetic energy at the equilibrium orbit $(r = R)$, and in that sense, we can still speak of a *monoenergetic* beam in the $\mathbf{E} \times \mathbf{B}$ case. What happens if there is a true momentum spread ΔP in the beam (i.e., if the particles at the radius R as well as at any other position have a difference in kinetic energy)? The most important effect is that particles with different momentum have different equilibrium radii about which they oscillate. Consider first the case of particles in a magnetic field (i.e., $\mathbf{E} = 0$). If R is the equilibrium radius of a particle with momentum P, defined by

$$R = \frac{P}{qB}, \tag{3.254}$$

a particle with momentum $P + dP$ will have a different equilibrium radius $R + dR$. For small fractional changes, one has to first order

$$\frac{dR}{R} = \frac{dP}{P} - \frac{dB}{B}. \tag{3.255}$$

The *momentum compaction factor* α, defined by

$$\alpha = \frac{dR/R}{dP/P}, \tag{3.256}$$

is a measure for the change in equilibrium radius due to a change in momentum. From (3.255) we obtain

$$\frac{dR}{R}\left(1 + \frac{R}{B}\frac{dB}{dR}\right) = \frac{dR}{R}(1 - n) = \frac{dP}{P}. \tag{3.257}$$

Thus,

$$\alpha = \frac{1}{1 - n} = \frac{1}{v_r^2}. \tag{3.258}$$

This expression holds for axisymmetric magnetic fields. When radial electric fields are present, B has to be replaced by the total guide field B_g, which from Equation (3.236) is defined as

$$B_g = B_z - \frac{E_r}{v_\theta} = B_0 - \frac{E_0}{v_0}. \tag{3.259}$$

In this case the momentum compaction factor α is given by

$$\alpha = \frac{1}{1 - n_g} . \tag{3.260}$$

where n_g is the effective gradient that includes the self fields.

Due to the change in radius R, there is also a change in the angular frequency ω and the revolution time $\tau = 2\pi R/\beta c$ of the particles. This relative change is readily found to be

$$\frac{d\tau}{\tau} = -\frac{d\omega}{\omega} = \left(\alpha - \frac{1}{\gamma^2}\right)\frac{dP}{P} = \eta\frac{dP}{P}, \tag{3.261a}$$

and it may be related to an equivalent velocity difference of

$$\frac{dv}{v} = -\frac{d\tau}{\tau} = -\eta\frac{dP}{P}, \tag{3.261b}$$

where the factor η is defined by

$$\eta - \alpha - \frac{1}{\gamma^2} = \frac{1}{v_r^2} - \frac{1}{\gamma^2}. \tag{3.262a}$$

In conventional weak focusing machines (e.g., betatron, cyclotron) the radial betatron tune is less than unity ($v_r < 1$); hence η is always positive ($\eta > 0$). However, in modern strong-focusing machines, the betatron tune is greater than unity ($v_r > 1$). Thus there will be a critical energy $\gamma_t mc^2$, known as the *transition energy*, where $\eta = 0$. Replacing v_r in Equation (3.262a) by γ_t, we can write the relation for η in the alternative form

$$\eta = \frac{1}{\gamma_t^2} - \frac{1}{\gamma^2}. \tag{3.262b}$$

For $\gamma > \gamma_t$, $\eta > 0$, as in the weak-focusing case, while for $\gamma < \gamma_t$, $\eta < 0$. The different operating regimes will be discussed further in connection with Equations (3.266) and (3.267).

Relation (3.261) gives the fractional change in the revolution time or frequency of a particle with momentum $P + dP$ as compared to a particle with momentum $P = \gamma m\beta c$, which is why η is also known as the *frequency slip factor*. If the path is straight rather than circular (i.e., if $\alpha = 0$ or $\eta = -1/\gamma^2$), then

$$\frac{d\tau}{\tau} = -\frac{1}{\gamma^2}\frac{dP}{P} \tag{3.263}$$

measures the difference in travel time of the two particles for a given distance. The minus sign indicates that, as expected, the travel time decreases when the momentum increases. By multiplying with the momentum $P = \gamma m v$, we can rewrite Equation (3.261b) in the form

$$dP = -\frac{\gamma m}{\eta} dv = m^* dv .$$

(3.264)

Here m^* is an *effective mass*, defined by

$$m^* = \frac{dP}{dv} = -\frac{\gamma m}{\eta} ,$$

(3.265)

which determines the relationship between the *momentum difference* and the *velocity difference* of the two neighboring particles in circular orbits.

In the case of a *straight* path ($\alpha = 0$), the effective mass is seen to be

$$m^* = m_l = \gamma^3 m ,$$

(3.266)

which is known as the *longitudinal mass*, m_l, and is a positive quantity. This relation is identical with (2.9b) that was obtained directly from the equation of motion. For the transverse motion the effective mass is $m_t = \gamma m$. If we apply the definition (3.265) to the relation (3.261) for particle motion in a *circular orbit* ($\alpha \neq 0$), we find that

$$m^* = -\frac{\gamma m}{\eta} = -\frac{\gamma^3 m}{\alpha \gamma^2 - 1} .$$

(3.267)

From this expression, we conclude the following: When $\eta < 0$, the effective mass is positive, as in the case of straight motion. However, when $\eta > 0$ (i.e., $\gamma > \gamma_t$ in strong-focusing machines), m^* is negative; and at the transition energy, where $\gamma = \gamma_t$ or $\eta = 0$, m^* goes to infinity. *Negative mass* means that the particle's revolution time increases when its momentum or kinetic energy is increased in contrast to the straight motion. At the transition point, the revolution time remains unaffected by a change in momentum. The sector-focusing cyclotron discussed in Section 3.8.4 operates in this way and is therefore also known as an *isochronous cyclotron*. The revolution time in this case is constant at all radii or energies (i.e., the particle's effective mass is infinite). In view of (3.258) and the condition $0 < n < 1$ for focusing in the axial and radial directions, we see that $\alpha > 1$, and hence $\alpha \gamma^2 > 1$ in this case. Thus, all devices with axisymmetric magnetic fields ($\nu_r < 1$) and all circular accelerators above the transition energy ($\gamma > \gamma_t$) are in the negative-mass region. This peculiarity of particle motion in a magnetic field is responsible for the so-called *negative-mass instability*, which poses a fundamental limit to the particle intensity in circular accelerators and is treated in Section 6.3.3.

Let us now examine how we can incorporate the changes due to energy spread into our first-order theory of betatron oscillations. We already pointed out that particles with different momentum oscillate about different equilibrium orbits. In general, we may suspect that the oscillation frequencies, ω_r and ω_z, are also functions of the momentum since the field index $n = -(R/B)(dB/dR)$ may vary with radius R. This is in fact the case, and one defines this effect by the *chromaticity parameters*

$$\xi_r = \frac{d\nu_r/\nu_r}{dP/P}, \qquad \xi_z = \frac{d\nu_z/\nu_z}{dP/P}. \tag{3.268}$$

Using (3.258), we may write these definitions in the alternative form

$$\xi_r = \frac{R\alpha}{2\nu_r^2} \frac{d(\nu_r^2)}{dR}, \qquad \xi_z = \frac{R\alpha}{2\nu_z^2} \frac{d(\nu_z^2)}{dR}. \tag{3.269}$$

It is easy to show that for a *scaling field* where $B/B_0 = (R/R_0)^{-n}$, the chromaticity parameters are zero. In a field where dB/dR is constant, n varies with radius and one obtains

$$\frac{dn}{dR} = -\frac{1}{B}\frac{dB}{dR} + \frac{R}{B^2}\left(\frac{dB}{dR}\right)^2 = \frac{n(1+n)}{R}. \tag{3.270}$$

In this particular case, with $\nu_r^2 = 1 - n$, $\nu_z^2 = n$, one finds that

$$\xi_r = -\frac{n(1+n)}{2(1-n)^2}, \qquad \xi_z = \frac{1+n}{2(1-n)}. \tag{3.271}$$

The variation of the equilibrium radius R with momentum P is a first-order effect, while the changes in ν_r^2 and ν_z^2 are of second order. We therefore neglect the latter in our first-order theory of betatron oscillations. The difference in momentum is incorporated in the theory in the following way: Let R_0 denote the equilibrium radius for particles with the average momentum P_0; that is, a particle with this momentum will perform betatron oscillations about R_0 in accordance with Equation (3.211) for the radial motion (and, likewise, for the axial motion). A particle with momentum $P_0 + \Delta P$ will perform similar oscillations about a displaced equilibrium radius $R_0 + \Delta R$. When the momentum spread is included, the equation of motion (3.190) must be modified by expanding the velocity as $v = v_0(1 + \Delta v/v_0) = v_0(1 + \Delta P/P_0)$. One then obtains, in lieu of (3.196),

$$x'' + k_r^2 x = \frac{1}{R_0}\frac{\Delta P}{P_0}. \tag{3.272}$$

The general solution consists of the linear superposition of the betatron oscillation amplitude x_b, which satisfies the homogeneous part of Equation (3.272), and the equilibrium orbit displacement x_e due to the momentum difference $\Delta P/P_0$, which is a special solution of the inhomogeneous equation; that is, we have

$$x(s) = x_b(s) + x_e(s).$$ (3.273)

The betatron oscillation obtained for $\Delta P = 0$ (homogeneous solution) may be written in the form

$$x_b(s) = x_b(0) \cos k_r s + \frac{x_b'(0)}{k_r} \sin k_r s,$$ (3.274a)

$$x_b'(s) = -x_b(0) k_r \sin k_r s + x_b'(0) \cos k_r s,$$ (3.274b)

where

$$k_r = \frac{2\pi}{\lambda_r} = \frac{\nu_r}{R_0} = \frac{\sqrt{1-n}}{R_0}.$$ (3.274c)

The special solution of the inhomogeneous equation ($\Delta P \neq 0$) is

$$x_e(s) = \frac{\Delta P}{P_0} \frac{1}{R_0 k_r^2} (1 - \cos k_r s),$$ (3.275a)

$$x_e'(s) = \frac{\Delta P}{P_0} \frac{1}{R_0 k_r} \sin k_r s.$$ (3.275b)

The general solution of Equation (3.272) for an off-momentum particle ($\Delta P \neq 0$) can be conveniently expressed with the aid of a 3×3 matrix as

$$\begin{pmatrix} x(s) \\ x'(s) \\ \frac{\Delta P}{P_0} \end{pmatrix} = \begin{pmatrix} \alpha_{11} & \alpha_{12} & \alpha_{13} \\ \alpha_{21} & \alpha_{22} & \alpha_{23} \\ 0 & 0 & 1 \end{pmatrix} \begin{pmatrix} x_0 \\ x_0' \\ \frac{\Delta P}{P_0} \end{pmatrix},$$ (3.276a)

where the matrix is given by

$$\tilde{M}_r = \begin{pmatrix} \cos k_r s & k_r^{-1} \sin k_r s & (R_0 k_r^2)^{-1}(1 - \cos k_r s) \\ -k_r \sin k_r s & \cos k_r s & (R_0 k_r)^{-1} \sin k_r s \\ 0 & 0 & 1 \end{pmatrix}.$$ (3.276b)

The last row in the matrix indicates that in the static magnetic field considered here, the momentum of the particles does not change. Note that the above expression is valid if $x \ll R_0$ and $\Delta P \ll P_0$. The fact that the radial motion of a particle depends on $\Delta P/P_0$ is known as *dispersion*. By contrast, the axial motion is nondispersive to first order. With the definition

$$k_z = \frac{2\pi}{\lambda_z} = \frac{v_z}{R_0} = \frac{\sqrt{n}}{R_0}, \qquad (3.277a)$$

we can write the axial matrix either as a 3×3 of the form

$$\tilde{M}_z = \begin{pmatrix} \alpha_{11} & \alpha_{12} & 0 \\ \alpha_{21} & \alpha_{22} & 0 \\ 0 & 0 & 1 \end{pmatrix}, \qquad (3.277b)$$

or simply as a 2×2 matrix

$$\begin{pmatrix} z \\ z' \end{pmatrix} = \begin{pmatrix} \cos k_z s & k_z^{-1}\sin k_z s \\ -k_z\sin k_z s & \cos k_z s \end{pmatrix}\begin{pmatrix} z_0 \\ z_0' \end{pmatrix}. \qquad (3.277c)$$

Momentum spread increases the effective radial width of a beam in a circular system. In practice, beams without space charge are found to have a Gaussian distribution in their betatron and momentum-spread amplitudes. If $\bar{x}_b = (x_b^2)^{1/2}$ denotes the rms value of the distribution in the betatron amplitudes and $\tilde{x}_e = (x_e^2)^{1/2}$ the rms value of the amplitude variation due to the momentum spread, the total rms half-width of the beam is given by

$$\bar{x} = (x_b^2 + x_e^2)^{1/2}.$$

We discuss dispersion further in Section 5.4.10.

3.7 SECTOR MAGNETS AND EDGE FOCUSING

Many magnets used in practice for deflecting a beam (*bending magnets*), as *momentum analyzers* (to separate ions or electrons of different momenta), or *mass spectrometers* (to separate charged particles of different mass) are sector shaped, as illustrated in Figure 3.22. Outside the magnetic field the particles move on a straight trajectory, inside the field on a circular path. In first approximation, one can neglect the fringe-field region and employ the formalism developed in Chapter 2 for particle motion in axisymmetric fields. Thus, one defines a central ray which inside the magnetic field is a circle of radius R_0. Particle motion with regard to this central ray is then described by a drift-space matrix in free space:

$$\begin{pmatrix} x \\ x' \\ \dfrac{\Delta P}{P} \end{pmatrix} = \begin{pmatrix} 1 & s & 0 \\ 0 & 1 & 0 \\ 0 & 0 & 1 \end{pmatrix}\begin{pmatrix} x_0 \\ x_0' \\ \dfrac{\Delta P}{P} \end{pmatrix}. \qquad (3.278)$$

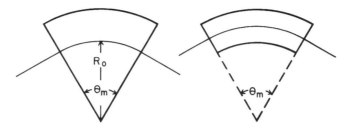

Figure 3.22. Two examples of sector magnets.

Inside of the magnet, the matrix (3.275) can be applied for the radial (or horizontal motion. For the vertical motion, a 2 × 2 drift-space matrix and the matrix (3.277) suffice. The trace-space coordinates of a particle at any point along the central trajectory are then obtained by matrix multiplication.

The above procedure is applicable as long as the particles enter the edge of the magnetic sector field at right angles. If the angle differs from 90°, the particle will experience a Lorentz force which will either be focusing or defocusing. This effect is known as *edge focusing*. To explain it, consider Figure 3.23, which shows a magnet with slanted edge so that a particle enters and leaves the field region at an angle α with respect to the normal. The magnetic field in the fringe region then has a component B_h normal to the edge for any point at a distance z outside the median plane. In the median plane, B_h will be zero by reason of symmetry. Now B_h can be decomposed into a component $B_{\parallel} = B_h \cos \alpha$ in the direction of the particle trajectory and a component $B_{\perp} = B_h \sin \alpha$ perpendicular to it. Only the latter exerts a force on the particle. Consider first the axial motion perpendicular to the plane of the figure when the particle passes through the fringe region as it leaves the sector magnet. The axial component of the Lorentz force is $q v B_{\perp}$, and one obtains the equation

$$\gamma m \ddot{z} + q v B_h \sin \alpha = 0. \tag{3.279}$$

Introducing the path length s along the trajectory, we get

Figure 3.23. Magnet with slanted edge and components of magnetic fringe field B_h. A particle with positive charge enters from the right side and leaves on the left side. The direction of the axial magnetic field lines is into the plane of the figure.

$$v = \frac{ds}{dt}, \qquad \frac{d^2z}{dt^2} \approx v\frac{dv_z}{ds}, \tag{3.280}$$

and thus,

$$\frac{dv_z}{ds} = -\frac{q}{\gamma m}B_h\sin\alpha. \tag{3.281}$$

Now integrate between point s_1, which is sufficiently far from the edge so that the magnetic field is zero, and point s_2, which is inside the magnet where there is only an axial component B_0 of the magnetic field:

$$\int_{s_1}^{s_2} dv_z = -\frac{q}{\gamma m}\sin\alpha\int_{s_1}^{s_2} B_h\, ds = -\frac{q}{\gamma m}\tan\alpha\int_{s_1}^{s_2} B_h\cos\alpha\, ds. \tag{3.282}$$

Assume that in the transition through the fringe region the axial displacement z of the particle from the median plane remains approximately constant (thin-lens approximation). Apply Stokes's theorem,

$$\int \nabla\times\mathbf{B}\cdot d\mathbf{S} = \oint \mathbf{B}\cdot d\mathbf{l} = 0,$$

for a rectangular closed path from point P_1 $(s_1,0)$ to P_2 $(s_2,0)$ in the median plane, up to point P_3 (s_2,z), then parallel to the median plane to point P_4 (s_1,z) and down to P_1 $(s_1,0)$. The nonzero contributions to the line integral are

$$\int_{P_4}^{P_3} B_h\cos\alpha\, ds \qquad \text{and} \qquad \int_{P_3}^{P_2} B_z\, dz = -B_0 z.$$

Consequently,

$$\int_{P_4}^{P_3} B_h\cos\alpha\, ds = -B_0 z, \tag{3.283}$$

and Equation (3.282) becomes

$$v_{z2} - v_{z1} = -\frac{q}{\gamma m}B_0 z\tan\alpha. \tag{3.284}$$

This may be written, with $v_z = dz/dt = v(dz/ds) = vz'$, as

$$z_2' - z_1' = -\frac{qB_0}{\gamma m v}z\tan\alpha. \tag{3.285}$$

Introducing the orbit radius $R_0 = \gamma m v/qB_0$ of the particles inside the magnetic field, we can write

$$z_2' - z_1' = -\frac{z}{R_0} \tan \alpha .$$ (3.286)

By definition, the focal length f_z is obtained (with $z_1' = 0$) from $1/f_z = -z_2'/z$, which yields

$$f_z = \frac{R_0}{\tan \alpha} .$$ (3.287)

This shows that there is focusing in the axial motion of the particle when $\alpha > 0$ and defocusing when $\alpha < 0$. The magnet edge can thus be considered as a thin magnetic lens that may be described by the transfer matrix

$$\begin{pmatrix} z_2 \\ z_2' \end{pmatrix} = \begin{pmatrix} 1 & 0 \\ -\dfrac{\tan \alpha}{R_0} & 1 \end{pmatrix} \begin{pmatrix} z_1 \\ z_1' \end{pmatrix} .$$ (3.288)

A similar analysis can be made for the particle motion in the median (radial) plane. One finds in this case for the focal length in the radial direction (see [D.1, Sec. 4.3])

$$f_r = -\frac{R_0}{\tan \alpha} .$$ (3.289)

Hence for the radial motion the edge is defocusing for $\alpha > 0$ and focusing for $\alpha < 0$, and the radial thin-lens matrix in for the edge region may be written in 3×3 form as

$$\begin{pmatrix} x_2 \\ x_2' \\ \dfrac{\Delta P}{P_0} \end{pmatrix} = \begin{pmatrix} 1 & 0 & 0 \\ \dfrac{\tan \alpha}{R_0} & 1 & 0 \\ 0 & 0 & 0 \end{pmatrix} \begin{pmatrix} x_1 \\ x_1' \\ \dfrac{\Delta P}{P_0} \end{pmatrix} .$$ (3.290)

The same results apply for the other side where the beam passes through the edge region on entering the magnet. In general, the edge angles at the entrance and exit side may differ (i.e., $\alpha_2 \neq \alpha_1$). The first-order properties of a sector magnet with such *thin-lens* (or *hard*) edges and two different angles α_1 and α_2 can thus be obtained by multiplication of three matrices, $\tilde{M}_2 \tilde{M}_s \tilde{M}_1$, where $\tilde{M}_1$ and $\tilde{M}_2$ represent the two edge-focusing lenses and $\tilde{M}_s$ the magnet sector.

3.8 PERIODIC FOCUSING

3.8.1 Periodic Focusing with Thin Lenses

A periodic-focusing system for charged particle beams consists of an array, or *lattice*, of periodically spaced lenses and other beam manipulation devices. Important applications of periodic focusing are microwave devices such as traveling-wave tubes, high-current beam transport over large distances, linear accelerators, sector-focusing cyclotrons, synchrotrons and storage rings, racetrack microtrons, and other devices for recirculating electron beams. One of the simplest cases of periodic focusing is a beam transport system with a periodic configuration of identical short solenoids. Circular accelerators constitute more complicated periodic systems in which the particles are bent around and traverse the same lattice of focusing lenses and deflecting magnets many times. In such systems one has practically two fundamental periods. One is the length S of a unit cell, and the other is the circumference C of the ring. If the ring lattice contains an integral number of N unit cells, then $C - NS$. In a perfect system, the forces acting on the particles have a repetition period of length S. However, if there are errors and misalignments in the system of lenses and bending magnets, the particles experience the resulting perturbation forces once in every revolution (i.e., with a repetition period of length C). For this reason, a circular focusing lattice is sometimes called a *doubly periodic system*.

As an introduction to the theory of periodic focusing, let us now consider the simplest case of a periodic system, which is a straight array of thin lenses (e.g., short solenoids), depicted in Figure 3.24. The basic building block, or unit cell, of such a periodic array consists of a lens and a drift space of length S. All lenses have the same focal length f. The particle trajectories between lenses are

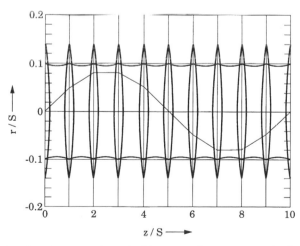

Figure 3.24. Particle trajectory and matched-beam envelope in a periodic thin-lens array with focal length $f = 2.618S$, where S is the cell length. The phase advance per cell is $\sigma = 36°$ (i.e., the particle performs one oscillation in 10 lens periods).

straight lines. Each lens changes the slope of the trajectory according to the relation $\Delta r' = -r/f$. The relation between the trajectory parameters at the input (n) and output $(n + 1)$ of each cell can be written in matrix form as the product of the lens and the drift-space matrices, that is,

$$\begin{pmatrix} r_{n+1} \\ r'_{n+1} \end{pmatrix} = \begin{pmatrix} 1 & 0 \\ -\dfrac{1}{f} & 1 \end{pmatrix} \begin{pmatrix} 1 & S \\ 0 & 1 \end{pmatrix} \begin{pmatrix} r_n \\ r'_n \end{pmatrix} = \begin{pmatrix} a & b \\ c & d \end{pmatrix} \begin{pmatrix} r_n \\ r'_n \end{pmatrix}. \tag{3.291}$$

In equation form this becomes

$$r_{n+1} = a r_n + b r'_n, \tag{3.292a}$$

$$r'_{n+1} = c r_n + d r'_n, \tag{3.292b}$$

where

$$a = 1, \tag{3.293a}$$

$$b = S, \tag{3.293b}$$

$$c = -\frac{1}{f}, \tag{3.293c}$$

$$d = 1 - \frac{S}{f}. \tag{3.293d}$$

From (3.292a) we get

$$r'_n = \frac{1}{b}(r_{n+1} - a r_n) \tag{3.294}$$

and thus

$$r'_{n+1} = \frac{1}{b}(r_{n+2} - a r_{n+1}). \tag{3.295}$$

Using relation (3.292b) for r'_{n+1} and substituting for r'_n from (3.294), we obtain the difference equation

$$r_{n+2} - (a + d)r_{n+1} + (ad - bc)r_n = 0, \tag{3.296}$$

which determines the change of the particles' radial position through the lens array. From the relations given in Equation (3.293) we can show that $ad - bc = 1$, hence we can rewrite (3.296) as

$$r_{n+2} - 2A r_{n+1} + r_n = 0, \tag{3.297}$$

where

$$A = \frac{1}{2}(a + d) = 1 - \frac{1}{2}\frac{S}{f} \qquad (3.298)$$

represents the trace of the matrix in (3.291).

In Figure 3.24 a particle trajectory was traced through a periodic system of thin lenses using a computer program. The ratio of the focal length f of the lenses to the cell length S was chosen to be $f/S = 2.618$. In this case the particle makes a full oscillation in approximately 10 lens periods. The result shown in this figure suggests that we can approximate the particle trajectory through the lens array by a sinusoidal oscillation. Indeed, this conclusion follows also from the difference equation (3.297), which is the equivalent of the harmonic oscillator equation $r'' + k^2 r = 0$. Thus we are led to try a solution of the form

$$r_n = r_0 e^{in\sigma}, \qquad (3.299)$$

which, when substituted in (3.297), leads to

$$e^{2i\sigma} - 2Ae^{i\sigma} + 1 = 0. \qquad (3.300)$$

From this equation we obtain

$$e^{\pm i\sigma} = \cos \sigma \pm i \sin \sigma = A \pm i\sqrt{1 - A^2}, \qquad (3.301)$$

where, in view of (3.298),

$$\cos \sigma = A = \frac{1}{2}(a + d) = 1 - \frac{1}{2}\frac{S}{f}. \qquad (3.302)$$

The general solution of (3.297) can be expressed as a linear superposition of $\exp(in\sigma)$ and $\exp(-in\sigma)$, or by the equivalent sinusoidal form

$$r_n = r_{max} \sin(n\sigma + \theta), \qquad (3.303)$$

where r_{max} and θ are determined by the initial conditions, r_0 and r'_0, at the entrance of the focusing system.

For the particle trajectory to be stable, the parameter σ must be a real number so that $|\cos \sigma| \leq 1$, or $|A| \leq 1$. From (3.302), the stability condition implies that

$$0 \leq S \leq 4f, \qquad \text{or} \qquad \frac{f}{S} \geq \frac{1}{4}. \qquad (3.304)$$

In accelerator theory the parameter σ is known as the *phase advance*, or *phase shift*, of the particle oscillation in one cell length of the periodic lattice, and it is usually given in degrees. If λ is the wavelength of the oscillation, we can write

$$\sigma = 360° \frac{S}{\lambda}. \tag{3.305}$$

In the example of Figure 3.24 we have a wavelength of $\lambda = 10S$, and the advance per cell is therefore $\sigma = 36°$. If the stability criterion (3.304) is not satisfied (i.e., if $f \leq S/4$), we obtain for (3.297) solutions of the form

$$r_n = C_1 e^{\delta n} + C_2 e^{-\delta n}, \tag{3.306}$$

where

$$e^{\pm \delta} = A \pm \sqrt{A^2 - 1}, \tag{3.307}$$

with the constants C_1 and C_2 being determined by the initial conditions. Since the magnitude of both $e^{+\delta}$ and $e^{-\delta}$ exceeds unity, the trajectory radius will increase exponentially. Such a case is illustrated in Figure 3.25, where we chose a value of $f/S = 0.246$ which is slightly below the stability threshold of $f/S = 0.25$. As can be seen, the maximum trajectory radius increases very rapidly. In an actual experiment, the beam simply blows up within a few lens periods when such "overfocusing" occurs.

Examination of the stability requirement (3.302) shows that the phase advance σ has the stability range $0 < \sigma < 180°$. For $\sigma = 0$ ($f \to \infty$), there is no focusing.

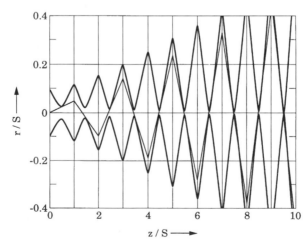

Figure 3.25. Particle trajectory and beam envelope in a periodic thin-lens array with focal length $f = 0.246S$, slightly below the stability threshold ($f = 0.25S$). The particle motion is unstable in this case.

On the other hand, when $\sigma > 180°$, particles cross the axis within a lens period and the trajectory becomes unstable, as illustrated in Figure 3.25.

So far we have treated the motion of a single particle through the periodic lens system. If we now consider a beam with emittance ϵ, we know from Section 3.4.5 that the envelope radius $R(z)$ in the drift space between lenses has the hyperbolic shape (3.162)

$$R(z) = \left[R_0^2 + 2R_0 R_0' z + \left(\frac{\epsilon^2}{R_0^2} + R_0'^2 \right) z^2 \right]^{1/2},$$

where R_0, R_0' are the initial radius and slope at the beginning of the drift section. The thin lenses merely change the slope of the envelope, as in the case of single trajectories.

For a *matched beam*, the envelope must be periodic with period length S, as can be seen in Figure 3.24. The waist R_w must occur at the center ($S/2$) of each cell. Furthermore, the slope at the entrance and exit side of each lens must have the same magnitude, $|R_0'|$, so that the slope R_0' at the lens exit is given by

$$2R_0' = -\frac{R_0}{f} . \tag{3.308}$$

Using these symmetry properties of the beam envelope one finds for the maximum radius

$$R_m = \sqrt{\epsilon S} \left[\frac{4(f/S)^2}{4(f/S) - 1} \right]^{1/4} \tag{3.309}$$

and for the waist radius

$$R_w = R_m \left(1 - \frac{S}{4f} \right)^{1/2} . \tag{3.310}$$

Figure 3.24 shows the matched-beam envelope for the case $f/S = 2.618$ or $\sigma_0 = 36°$; the emittance ϵ was chosen to yield $R_m = 0.1S$. As can be seen, the envelope ripple $\Delta R = R_m - R_w$ is relatively small in this example. On the other hand, in the example of Figure 3.25, the instability of single-particle motion also results in a rapid blowup of the beam envelope. It is not possible to obtain a matched, periodic solution for $R(z)$ in this instability case.

Equation (3.309) shows the scaling relationship between the maximum beam radius R_m, emittance ϵ, cell length S, and focal length f of the periodic thin-lens array. We can use this relation to define an important quantity known as the *acceptance* of the focusing system. Suppose that the aperture radius, $r_{max} = a$, of the lens channel is fixed. What is the maximum emittance, ϵ_{max}, that a matched beam can have for a given value of f/S to fit into the available aperture? This maximum emittance is identical with the acceptance; that is, we can make the identification $\epsilon_{max} = \alpha$ where α denotes the acceptance. A mathematical relation for the acceptance of a thin-lens array can be obtained by solving (3.309) for ϵ, substituting α for ϵ, and setting $R_m = a$. The result is

$$\alpha = \frac{a^2}{S}\left[\frac{4(f/S) - 1}{4(f/S)^2}\right]^{1/2}. \tag{3.311}$$

As can be seen from this relation, the acceptance of the focusing channel scales linearly with the product of aperture radius a and the ratio of the aperture to the cell length, a/S. In addition, it is a function of the focal length to the cell length f/S. The ratio a/S must be significantly less than unity (i.e., $a/S \ll 1$), to avoid nonlinear forces (lens aberrations) that would adversely affect the beam quality. A prudent choice might be $a/S = 0.2$, for instance. The focusing capability of the system is represented by the acceptance function α given in (3.311). This function is zero when $f = \infty$ (no focusing) and $f = S/4$ (stability limit); it has a maximum at $f = 0.5S$, corresponding to a phase advance of $\sigma = 90°$, in which case it is $\alpha = a^2/S$. The maximum acceptance of a periodic channel of thin lenses (with identical focusing length f) is therefore $\alpha_{max} = a^2/S$. If a/S is fixed for the reasons mentioned, the only way to match a beam of a given emittance into such a channel is to make the aperture radius $r_{max} = a$, and hence also S, sufficiently large. The alternative is to replace the array of *weak-focusing* lenses discussed here by *strong-focusing* lens configurations such as quadrupoles, as discussed in Section 3.5. Magnetic quadrupole lenses, for instance, are used in high-energy accelerators since they provide stronger focusing than solenoids at high particle kinetic energies. The derivation of the various relations for a channel consisting of thin quadrupole lenses is left as a problem (3.16) at the end of this chapter.

So far we have not discussed how to match the beam into a given focusing channel. The first requirement, of course, is that the emittance fit into the acceptance (i.e., $\epsilon \leq \alpha$). Assuming that this condition is satisfied, we must inject the beam into the first lens with initial condition given by $R_0 = R_m$ and $R_0' = -R_m/2f$. These two conditions can be met by using two matching lenses placed before the entrance into the focusing channel. If the beam is not properly matched, it will perform envelope oscillations that are generally not desirable since they may lead to deterioration of beam quality and particle losses. Envelope oscillations, including the effects of space charge, are discussed in Section 4.4.3.

With regard to the optimum choice of parameters, several considerations indicate that one should not operate in the region where the phase advance σ is greater than 90° even though the stability requirement permits higher values. First, as discussed above, the channel acceptance is a maximum, or, conversely, the beam radius R_m a minimum, when $\sigma = 90°$ in the thin-lens system being considered. Therefore, nothing is gained by increasing the focusing strength beyond this point. Second, the effects of random lens misalignments increase as $\sin(\sigma/2)$, as shown in Section 4.4.4; hence it is desirable to have as small a value of σ as possible. Finally, space charge perturbations cause envelope instabilities when $\sigma > 90°$, as we will see in Section 4.4.3. It should be noted that the $\sigma_0 < 90°$ rule is not restricted to the thin-lens array treated here; it applies also to other periodic focusing systems.

In the next section we present the general mathematical theory of periodic focusing. Following this we discuss three examples: a quadrupole focusing channel, the sector-focusing cyclotron, and the strong-focusing synchrotron. In addition, we briefly review the topic of resonances in circular accelerators.

3.8.2 General Theory of Courant and Snyder

Let us now consider the general case of a periodic-focusing system with two planes of symmetry and without the thin-lens restriction made in the preceding section. Mathematically, the linear, or paraxial, motion of charged particles in periodic systems is described by two differential equations of the form

$$x''(s) + \kappa_x(s)x = 0, \tag{3.312a}$$

$$y''(s) + \kappa_y(s)y = 0, \tag{3.312b}$$

Here x, y are the displacements from the beam axis and s is the independent variable measuring the distance along the beam axis (or equilibrium orbit in a circular accelerator). $\kappa_x(s)$, $\kappa_y(s)$ are the periodic-focusing functions, which satisfy the periodicity relation

$$\kappa(s + S) = \kappa(s), \tag{3.313}$$

where S is the length of one period. In a circular accelerator with circumference C and N focusing periods, or *unit cells*, we have the additional periodicity relation

$$\kappa(s + C) = \kappa(s), \tag{3.314}$$

where $C = NS$.

Many systems, such as quadrupole channels, have two planes of symmetry where the forces $\kappa_x(s)$ and $\kappa_y(s)$ may differ in phase or in both phase and amplitude. However, as long as there is no coupling between these two forces, the theory is the same, and in the following, we consider only motion in the x-direction.

Linear second-order differential equations with periodic coefficients of the form (3.312) are known as Mathieu–Hill equations. The properties of these equations and their solutions have been treated extensively in the literature [3]. The standard reference for periodic focusing of charged particles in the accelerator field is the theory by Courant and Snyder [4]. Although this theory deals with circular accelerators, the method and results apply equally to linear accelerators or beam transport systems. In this general treatment of periodic focusing we follow closely the Courant–Snyder theory.

First we recall from the paraxial theory that the solution of any linear second-order differential equation of the form (3.312), whether or not $\kappa(s)$ is periodic, is determined uniquely by the initial values (x_0, x_0') or (y_0, y_0'). Thus, for (x, x') at a distance s, one gets

$$x(s) = ax_0 + bx_0', \tag{3.315a}$$

$$x'(s) = cx_0 + dx_0', \tag{3.315b}$$

or, in matrix notation,

$$X(s) = \begin{pmatrix} x(s) \\ x'(s) \end{pmatrix} = \tilde{M}(s|s_0)X(s_0) = \begin{pmatrix} a & b \\ c & d \end{pmatrix}\begin{pmatrix} x(s_0) \\ x'(s_0) \end{pmatrix}. \tag{3.316}$$

Let us now examine the motion in a periodic system. In this case the matrix $\tilde{M}$ has the property

$$\tilde{M}(s + S|s) = \tilde{M}(s), \tag{3.317}$$

which is to say that the matrices describing particle motion through any one period of length S are identical. The matrix for passage through N periods is then obtained by multiplication of the matrices for a *unit cell*, that is,

$$\tilde{M}(s + NS|s) = [\tilde{M}(s)]^N. \tag{3.318}$$

The motion of the particle can be stable or unstable depending on whether $x(s)$ remains finite or increases indefinitely with distance x. For the motion to be stable it is necessary and sufficient that the elements of the matrix $\tilde{M}^N$ remain bounded for any number of periods N. To find this condition for stable motion, let us consider the eigenvalues for the characteristic matrix equation

$$\tilde{M}X = \lambda X, \tag{3.319}$$

which changes only the length, but not the direction of the vector $X = (x_0, x_0')$. Writing it out, we have

$$ax_0 + bx_0' = \lambda x_0, \tag{3.320a}$$

$$cx_0 + dx_0' = \lambda x_0', \tag{3.320b}$$

This system of linear equations have nonvanishing solutions only when

$$\begin{vmatrix} a - \lambda & b \\ c & d - \lambda \end{vmatrix} = 0, \tag{3.321}$$

or

$$\lambda^2 - \lambda(a + d) + (ad - cb) = 0. \tag{3.322}$$

The last term in Equation (3.322) represents the Wronskian of the matrix $\tilde{M}$, which is unity since we do not consider changes in the particle's kinetic energy, that is,

$$ad - bc = 1, \tag{3.323}$$

and hence Equation (3.322) becomes

$$\lambda^2 - \lambda(a + d) + 1 = 0. \tag{3.324}$$

Let us now introduce the parameter σ, already defined in the thin-lens case (3.302), by

$$\cos \sigma = \frac{1}{2}(a + d) - \frac{1}{2} \text{Tr}\, \tilde{M}. \tag{3.325}$$

The two solutions of the quadratic equation (3.324) are then

$$\lambda_{1,2} = \cos \sigma \pm i \sin \sigma. \tag{3.326}$$

The parameter σ will be real if $|a + d| \leq 2$ and imaginary or complex if $|a + d| > 2$. It will be advantageous to write the matrix $\tilde{M}$ in a form that contains $\cos \sigma$ and $\sin \sigma$. To do this we introduce the parameters $\hat{\alpha}, \hat{\beta}, \hat{\gamma}$ defined by the relations

$$a - d = 2\hat{\alpha} \sin \sigma, \tag{3.327a}$$

$$b = \hat{\beta} \sin \sigma, \tag{3.327b}$$

$$c = -\hat{\gamma} \sin \sigma. \tag{3.327c}$$

The matrix $\tilde{M}$ may now be written as

$$\tilde{M} = \begin{pmatrix} \cos \sigma + \hat{\alpha} \sin \sigma & \hat{\beta} \sin \sigma \\ -\hat{\gamma} \sin \sigma & \cos \sigma - \hat{\alpha} \sin \sigma \end{pmatrix}, \tag{3.328}$$

or

$$\tilde{M} = \tilde{I} \cos \sigma + \tilde{J} \sin \sigma, \tag{3.329}$$

where

$$\tilde{I} = \begin{bmatrix} 1 & 0 \\ 0 & 1 \end{bmatrix}, \quad \tilde{J} = \begin{bmatrix} \hat{\alpha} & \hat{\beta} \\ -\hat{\gamma} & -\hat{\alpha} \end{bmatrix}. \tag{3.330}$$

The condition det $\tilde{M} = 1$ implies that

$$\hat{\beta}\hat{\gamma} - \hat{\alpha}^2 = 1. \tag{3.331}$$

Note that det $\tilde{J} = 1$, Tr $\tilde{J} = 0$.

The representation (3.329) of $\tilde{M}$ has properties similar to the complex exponential $e^{i\sigma} = \cos \sigma + i \sin \sigma$. Thus one can show that

$$\tilde{M}^N = (\tilde{I} \cos \sigma + \tilde{J} \sin \sigma)^N = \tilde{I} \cos N\sigma + \tilde{J} \sin N\sigma. \tag{3.332}$$

The particle motion through a periodic array of N lenses (where N can be arbitrarily large) is stable when the parameter σ is real, or, in view of (3.325), when

$$|\mathrm{Tr}\tilde{M}| = |a + d| < 2, \tag{3.333}$$

and it is unstable when $|a + d| > 2$. The parameters $\hat{\alpha}, \hat{\beta}, \hat{\gamma}$ are also known in the literature as the *Courant–Snyder* or *Twiss parameters*, and $\hat{\beta}$ as the *amplitude function* or *betatron function*. We depart from the traditional notation by adding a hatch (^) to avoid confusion with the relativistic velocity and energy factors, β, γ. Furthermore, we use σ in place of μ to denote the phase advance per cell. (Courant and Snyder used the symbol μ. However, in more recent work on periodic focusing it has become customary to use σ instead, and we therefore adopt this notation [5].)

Returning now to Equation (3.316), we note that the two eigenvalues of the characteristic matrix equation, $\lambda_1 = \cos \sigma + i \sin \sigma$ and $\lambda_2 = \cos \sigma - i \sin \sigma$, obey the relation

$$\lambda_1 \lambda_2 = 1, \quad \text{or} \quad \lambda_2 = \frac{1}{\lambda_1}. \tag{3.334}$$

Consequently, there will be a unique pair (u, v) of independent solutions of the Mathieu–Hill equation fulfilling the condition

$$u(s + S) = \lambda_1 u(s), \quad v(s + S) = \frac{1}{\lambda_1} v(s). \tag{3.335}$$

This statement is called *Floquet's theorem*, and a pair of solution (u, v) satisfying (3.335) are called *Floquet functions*. The Floquet functions allow one to express any other solution of the Mathieu–Hill equation in the simplest way. In the special case where the force parameter κ is constant, Equation (3.312) reduces to the harmonic-oscillator equation and the Floquet functions are given by $u(s) = \exp[i\sqrt{\kappa}s]$, $v = \exp[-i\sqrt{\kappa}s]$. In the case of a periodic system, the Floquet function can be written in the phase-amplitude form

$$u = w(s)e^{i\psi(s)}, \tag{3.336a}$$

$$v = w(s)e^{-i\psi(s)}, \tag{3.336b}$$

which reduces to the harmonic-oscillator solution when $\kappa(s) = $ const. Actually, (3.336a) and (3.336b) represent two fundamental solutions even if $\kappa(s)$ is not periodic. In the following we will initially make no particular assumptions concerning $\kappa(s)$. The constraints imposed on $w(s)$ and $\psi(s)$ if $\kappa(s)$ is periodic will be discussed subsequently. First, we note that the use of (3.336) makes it possible to express any solution $x(s)$ as a linear combination of u and v in the form

$$x(s) = Aw(s)\cos[\psi(s) + \phi], \tag{3.337}$$

where the amplitude A and the phase ϕ are determined by the initial conditions. The Wronskian $W = uv' - u'v$, which is a constant, is given by

$$W = -2iw^2\psi' = W_1. \tag{3.338}$$

Choosing for the constant the value $W_1 = -2i$, one obtains the relation

$$\frac{d\psi}{ds} = \psi' = \frac{1}{w^2}, \tag{3.339}$$

which the two functions $w(s)$ and $\psi(s)$ have to satisfy if u, v are to be fundamental solutions.

Differentiation of either u or v and substitution into Equation (3.312) yields the differential equation

$$w'' + \kappa w - \frac{1}{w^3} = 0 \tag{3.340}$$

for the amplitude function $w(s)$. This equation has the form of the envelope equation for a beam with an emittance of elliptic shape as discussed in Section 3.4.5. Let us explore this analogy by deriving the equation of the trace-space ellipse for a particle whose trajectory is given by (3.337). Differentiation of (3.337) yields

$$x' = A[w'\cos(\psi + \phi) - w\psi'\sin(\psi + \phi)]. \tag{3.341}$$

By eliminating $\psi + \phi$, using $\cos^2(\psi + \phi) + \sin^2(\psi + \phi) = 1$, one obtains from (3.337) and (3.341) the equation

$$\frac{x^2}{w^2} + (wx' - w'x)^2 = A^2. \tag{3.342a}$$

We can write this in the form

$$\hat{\gamma}x^2 + 2\hat{\alpha}xx' + \hat{\beta}x'^2 = A^2, \tag{3.342b}$$

by using the definitions

$$\hat{\beta} = w^2, \tag{3.343a}$$

$$\hat{\alpha} = -ww', \tag{3.343b}$$

$$\hat{\gamma} = \frac{1}{w^2} + w'^2 = \frac{1 + \hat{\alpha}^2}{\hat{\beta}}. \tag{3.343c}$$

Equation (3.342) is the equation of an ellipse whose shape and orientation at any given s are determined by the amplitude factor A and the coefficients $\hat{\alpha}$ and $\hat{\beta}$, which in turn are defined by $w(s)$ and $w'(s)$. We conclude that all particles with the same initial value of A but different ϕ lie on the ellipse described by (3.342). To follow the motion of an ensemble of particles with the same A, we need to know how w varies with distance s, that is, we need to solve Equation (3.340) subject to the appropriate initial values $w(0), w'(0)$ at $s = 0$. As we pointed out above, the functions $w(s)$ and $\psi(s)$ are the same for all particles in the beam. Thus, particles with different A lie on different ellipses scaled in size but otherwise similar. The largest ellipse is defined by the maximum value of A, which we denote by A_0.

The area of each ellipse is $\pi A^2(\hat{\beta}\hat{\gamma} - \hat{\alpha}^2)^{-1/2} = \pi A^2$ since $\hat{\beta}\hat{\gamma} - \hat{\alpha}^2 = 1$ in view of (3.343c) (i.e., it is a constant through the motion of the beam). Specifically, the area of the largest ellipse is given by $A_x = \epsilon_x \pi$, where ϵ_x is the emittance of the beam, and we can write

$$A_x = A_0^2 \pi, \quad \text{or} \quad A_0 = \sqrt{\frac{A_x}{\pi}} = \sqrt{\epsilon_x}. \tag{3.344}$$

Thus, with $A^2 = A_0^2 = \epsilon_x$, we obtain from (3.342b) the equation of the beam ellipse in $x-x'$ trace space, that is,

$$\hat{\gamma}x^2 + 2\hat{\alpha}xx' + \hat{\beta}x'^2 = \epsilon_x. \tag{3.345}$$

This ellipse is illustrated in Figure 3.26, which also shows the relations for the intercepts with the two axes and for the maximum values of x and x'. The

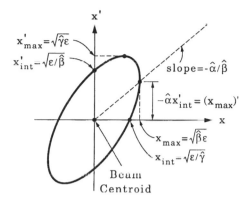

Figure 3.26. Trace-space ellipse described by equation $\hat{\gamma}x^2 + 2\hat{\alpha}xx' + \hat{\beta}x'^2 = \epsilon$ and relations for several important points on the circumference of the ellipse.

derivation of these relationships is left as a problem (3.17). The beam envelope x_m is characterized by the maximum value of x which occurs when $\psi + \phi = 0$ or $\cos(\psi + \phi) = 1$ and $A = A_0$. Thus, we have the relation

$$x_m(s) = A_0 w(s) = \sqrt{\epsilon_x}\, w(s) = \sqrt{\epsilon_x \beta(s)}. \tag{3.346}$$

Substitution of (3.346) into (3.340) yields the envelope equation

$$x_m'' + \kappa x_m - \frac{\epsilon_x^2}{x_m^3} = 0. \tag{3.347}$$

We note that the Courant–Snyder form (3.345) for the trace-space ellipse, which is widely used in the accelerator literature, differs from our previous notation in Sections 3.4.5 and 3.6.2. The relations between the coefficients $\hat{\alpha}, \hat{\beta}, \hat{\gamma}$ in Equation (3.345) and a, b, c in Equations (3.159) or (3.218)ff. are given by $a = \hat{\gamma}/\epsilon_x$, $b = \hat{\alpha}/\epsilon_x$, $c = \hat{\beta}/\epsilon_x$. Using these relationships, it is readily verified that the previous equations for the beam envelope, such as (3.160a) and (3.160b), are identical with the Courant–Snyder expressions in (3.346) and in Figure 3.26.

Comparison of Equations (3.340), (3.346), and (3.347) shows that the amplitude $w(s)$ of the fundamental solutions represents the characteristic envelope for a particular system defined by a given force function $\kappa(s)$. We can obtain $w(s)$ by integration of (3.340) subject to initial conditions $w(0), w'(0)$. The actual beam envelope then depends on the emittance ϵ_x and is found by relation (3.346). Furthermore, once we know $w(s)$, we can obtain the phase function $\psi(s)$ by integrating Equation (3.339).

Since u, v are linearly independent solutions, we can write the matrix $\widetilde{M}$ that determines the change in a particle's position and slope between two points s_2 and

s_1 in terms of u, v and hence in terms of the functions $w(s), \psi(s)$. The calculations for the matrix $\tilde{M}(s_2|s_1)$ lead to the following results:

$$
\left[
\begin{array}{cc}
\dfrac{w_2}{w_1}\cos\psi - w_2 w_1' \sin\psi & w_1 w_2 \sin\psi \\[2ex]
-\dfrac{1 + w_1 w_2' w_2 w_2'}{w_1 w_2}\sin\psi - \left(\dfrac{w_1'}{w_2} - \dfrac{w_2'}{w_1}\right)\cos\psi & \dfrac{w_1}{w_2}\cos\psi + w_1 w_2' \sin\psi
\end{array}
\right],
$$

$$(3.348)$$

where $\psi = \psi(s_2) - \psi(s_1), w_1 = w(s_1), w_2 = w(s_2)$.

The phase-amplitude solution (3.336), and hence the matrix (3.348), is valid whether or not $\kappa(s)$ periodic, as pointed out earlier. Let us now examine the constraints imposed by the condition that $w(s)$ be a periodic function of s with period S. In this case, if $s_2 - s_1 = S$, we have $w_2 = w_1 = w, w_2' = w_1' = w'$, and the matrix (3.348) takes the simpler form

$$
\tilde{M} = \left(
\begin{array}{cc}
\cos\psi - ww'\sin\psi & w^2\sin\psi \\[2ex]
-\dfrac{1 + w^2 w'^2}{w^2}\sin\psi & \cos\psi + ww'\sin\psi
\end{array}
\right).
$$

$$(3.349)$$

Furthermore, we recognize that the matrix $\tilde{M}$ is now identical with the matrix (3.328) provided that we use the relations (3.343) and make the additional identification

$$\psi = \psi(s_2) - \psi(s_1) = \sigma.$$

$$(3.350)$$

We note that the relations between $\hat{\alpha}, \hat{\beta}, \hat{\gamma}$, and w, w' given in Equation (3.343) are valid even if $\kappa(s)$ is not a periodic function. The periodicity condition is contained in (3.350), and in view of (3.339), we have for σ the relation

$$\sigma = \int_s^{s+S} \frac{ds}{w^2} = \int_s^{s+S} \frac{ds}{\hat{\beta}}.$$

$$(3.351)$$

As already mentioned in Section 3.8.1, σ is the *phase advance* or *phase shift* per period or cell. The particle motion in a periodic focusing channel is basically a pseudoharmonic oscillation with frequency $\omega = (\bar{\kappa})^{1/2}v$, or wavelength $\lambda = 2\pi v/\omega = 2\pi/(\bar{\kappa})^{1/2}$, and an amplitude modulation or *ripple* that varies with the period S of the force function. Here $\bar{\kappa}$ is the mean value of κ in one period and v is the velocity of the particles. The phase advance σ measures the fraction of the wavelength in one oscillation period [see Equation (3.305)]; thus, $\sigma = 90°$ implies that the particle completes one oscillation in four periods of the focusing channel. Note that the condition for stable motion is $|\cos\sigma| < 1$ (i.e., $\sigma < 180°$), in agreement with our result for the thin-lens channel treated in the preceding section.

In the case of circular accelerators with circumference $C = NS$, one uses in place of σ the parameter

$$\nu = \frac{N\sigma}{2\pi} = \frac{1}{2\pi} \int_s^{s+C} \frac{ds}{\hat{\beta}}, \qquad (3.352)$$

which is the number of betatron oscillations in one revolution, also known as the betatron *tune*.

From the form (3.337) of the solution of the equation of motion, we see that the largest displacement x_{max} is obtained where $w(s)$, and hence the betatron function $\hat{\beta}(s)$, attains its maximum value. In a given accelerator or focusing channel, the particle motion is usually restricted by the vacuum chamber or other structures. If x_{max} denotes the maximum excursion of the particle trajectory permitted by these aperture constraints, the *acceptance* or *admittance* of the system in the x-direction is defined by the quantity

$$\alpha_x = \frac{a^2}{w_{max}^2} = \frac{a^2}{\hat{\beta}_{max}}. \qquad (3.353)$$

A beam is considered *matched* when the emittance is equal to the acceptance (i.e., $\epsilon_x = \alpha_x$) and when the maximum displacement of the beam envelope $x_{max} = a$ occurs at the points (usually the center of the focusing lenses) where the betatron function $\hat{\beta}(s)$ has its maximum value $\hat{\beta}_{max}$. In this case no particles are lost to the walls. Note that the function $\hat{\beta}(s)$ has the same period S as the focusing function $\kappa(s)$, and in the matched case the beam envelope $x_m(s)$ therefore also varies with period S according to the relation (3.346). As an example, comparing Equation (3.353) with (3.311) of the preceding section, we see that $\hat{\beta}_{max}$ for an array of axisymmetric thin lenses is given by $\hat{\beta}_{max} = S[4(f/S)^2]^{1/2}/[4(f/S) - 1]^{1/2}$. Relations for a periodic axisymmetric channel consisting of thick lenses (e.g., solenoids) can be found in Section 4.4.1, Equations (4.163) to (4.170) and Figures 4.4 to 4.6.

3.8.3 The FODO Quadrupole Channel

As a first example of a general periodic-focusing system and to illustrate the use of the Courant–Snyder theory, let us consider a quadrupole channel in the FODO configuration. One period of such a channel is defined by a quadrupole of length l that is focusing in x and defocusing in y, a quadrupole that is defocusing in x and focusing in y, and two drift sections of length L each, as illustrated in Figure 3.27. The force function $\kappa(s)$, which we will assume to be piecewise constant, and the qualitative variation of the amplitude function $w(s) = \sqrt{\hat{\beta}(s)}$, and thus

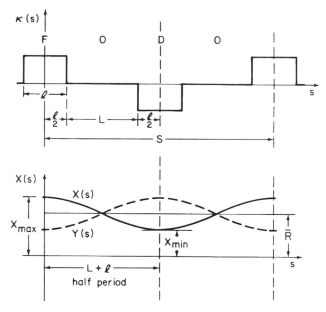

Figure 3.27. Gradient and envelope functions in a periodic quadrupole channel of the FODO type. (From Reference 2.)

of the matched-beam envelopes $x_m(s) = X(s), y_m(s) = Y(s)$, are indicated in the figure.

The transfer matrix $\tilde{M}$ for one period of such a FODO channel can be calculated by multiplication of the appropriate matrices for the four sections of one channel period using (3.184a), (3.184b), and the matrices for the two drift spaces. Comparing this with the matrix $\tilde{M}$ in the form (3.328), one finds for a FODO section the following results:

$$\cos \sigma = \cos \theta \cosh \theta + \frac{L}{l}\theta(\cos \theta \sinh \theta - \sin \theta \cosh \theta)$$
$$- \frac{1}{2}\left(\frac{L}{l}\right)^2 \theta^2 \sin \theta \sinh \theta, \tag{3.354}$$

$$\hat{\alpha} \sin \sigma = -\sin \theta \sinh \theta - \frac{L}{l}\theta \sin \theta \cosh \theta$$
$$- \frac{1}{2}\left(\frac{L}{l}\right)^2 \theta^2 \sin \theta \sinh \theta, \tag{3.355}$$

$$\hat{\beta} \sin \sigma = \frac{l}{\theta}\left[(\sin \theta \cosh \theta + \cos \theta \sinh \theta) + \frac{L}{l}(2\cos \theta \cosh \theta + \sin \theta \sinh \theta)\right]$$
$$+ \left(\frac{L}{l}\right)^2 \theta^2 \cos \theta \sinh \theta. \tag{3.356}$$

The parameter θ in these equations represents the focusing strength of the lenses and is defined by

$$\theta = \kappa^{1/2} l. \tag{3.357}$$

Note that κ is given by the relation (3.182) for magnetic quadrupoles and (3.183) for electric quadrupoles. The relations (3.354) to (3.356) are transcendental equations that must be solved by computer or presented in graphical form for practical use. A plot of the relationship between σ and θ for different values of L/l of a FODO channel, for instance, is shown in Section 4.4 (Figure 4.7), where periodic focusing in both FODO and solenoid channels, including space-charge effects, will be treated. Fortunately, there are several rather simple approximate formulas that are very useful for design and scaling purposes and therefore worthy of being presented here. First, one finds for $\theta \ll \pi/2$ from (3.354) for $\cos \sigma$ the approximate result

$$\cos \sigma = 1 - \frac{\theta^4}{6}\left[1 + 4\frac{L}{l} + 3\left(\frac{L}{l}\right)^2\right]. \tag{3.358}$$

If in addition to $\theta \ll \pi/2$ the drift length L is much larger than the lens width l (i.e., $L/l \gg 1$), we can neglect $1 + 4L/l$ compared to $3(L/l)^2$ and obtain

$$\cos \sigma = 1 - \frac{1}{2}\theta^4\left(\frac{L}{l}\right)^2 = 1 - \frac{\kappa^2 l^2 L^2}{2},$$

or

$$\cos \sigma - 1 - \frac{\eta^2}{2}, \tag{3.359}$$

where $\eta - \theta^2(L/l) - \kappa lL$. This is known as the *thin-lens approximation*, and for $\hat{\beta}_{\max}$ and acceptance α_x it yields the results

$$\hat{\beta}_{\max} = \frac{2L}{\eta}\left(\frac{2 + \eta}{2 - \eta}\right)^{1/2}, \tag{3.360}$$

$$\alpha_x = \frac{a^2}{\hat{\beta}_{\max}} = \frac{a^2\eta}{2L}\left(\frac{2 - \eta}{2 + \eta}\right)^{1/2}. \tag{3.361}$$

Finally, in the *smooth approximation*, where the phase advance per period is small ($\sigma \ll \pi/2$), one gets from (3.358) and (3.359) the simple relations

$$\sigma = \frac{\theta^2}{\sqrt{3}}\left[1 + 4\frac{L}{l} + 3\left(\frac{L}{l}\right)^2\right]^{1/2} \quad \text{for} \quad \theta \ll \frac{\pi}{2}, \tag{3.362a}$$

and

$$\sigma = \eta = \theta^2\left(\frac{L}{l}\right) = \kappa lL \qquad \text{for} \qquad \theta \ll \frac{\pi}{2}, L \gg l. \tag{3.362b}$$

The maximum and minimum values of the amplitude function $w(s)$ in the FODO channel can be obtained from the transfer matrix for half a period. By comparing this matrix with the form (3.348) and using the fact that $w_1 = w_{max}$, $w_2 = w_{min}$, $w_1' = w_2' = 0$, one finds that

$$\frac{w_{max}^2}{w_{min}^2} = \frac{\hat{\beta}_{max}}{\hat{\beta}_{min}} = \frac{1 + \tanh(\theta/2)[\tan(\theta/2) + (L/l)\theta]}{1 - \tan(\theta/2)[\tanh(\theta/2) + (L/l)\theta]} = n_1^2, \tag{3.363}$$

$$w_{max}^2 w_{min}^2 = \frac{l^2}{\theta^2} \frac{1 + \coth(\theta/2)[\tan(\theta/2) + (L/l)\theta]}{\tan(\theta/2)[\coth(\theta/2) + (L/l)\theta] - 1} = \frac{l^2}{\theta^2} n_2^2. \tag{3.364}$$

Solving for $\hat{\beta}_{max}$, $\hat{\beta}_{min}$ yields

$$\hat{\beta}_{max} = \frac{l}{\theta} n_1 n_2 = \frac{1}{\sqrt{\kappa}} n_1 n_2, \tag{3.365}$$

$$\hat{\beta}_{min} = \frac{l}{\theta} \frac{n_2}{n_1} = \frac{1}{\sqrt{\kappa}} \frac{n_2}{n_1}, \tag{3.366}$$

The above relations define the properties of a FODO channel such as the phase advance σ and the maximum of the amplitude function. If the maximum aperture available for the beam (e.g., the diameter of the beam pipe), is $2a$, the *acceptance* of the FODO channel is given by

$$\alpha_x = \frac{a^2}{\hat{\beta}_{max}} = \frac{a^2\sqrt{\kappa}}{n_1 n_2}. \tag{3.367}$$

As mentioned before, the acceptance of the channel is identical with the maximum emittance that a perfectly matched beam could have without particle loss to the wall of the beam pipe (provided that space-charge effects are negligible). Note that in the more general case of a FODO channel, the two quadrupoles and the two drift regions could have different lengths l_1, l_2 and L_1, L_2, and/or different strength $|\kappa_1|, |\kappa_2|$. The example presented here is a symmetric quadrupole system with $|\kappa_1| = |\kappa_2| = \kappa, l_1 = l_2 = l, L_1 = L_2 = L$. Further discussions of periodic beam transport systems—solenoids as well as quadrupoles—that include space-charge effects and several useful graphs can be found in Section 4.4.

3.8.4 Sector-Focusing Cyclotrons

In this section we discuss the sector-focusing cyclotron as a first example of a doubly periodic-focusing system in a circular accelerator. To appreciate the advantage of sector focusing in this case, we must first understand the basic operating principles and limitations of the classical cyclotron with axisymmetric magnetic field.

The cyclotron concept was invented by Lawrence in 1929; the first model was constructed a year later by Lawrence and Edlefsen, and the proof of principle was established by Lawrence and Livingston in 1931 [6]. The concept is based on the fact that a magnetic field B forces charged particles into circular orbits with angular frequency $\omega_c = qB/\gamma m$ and orbit radius $R = v/\omega$. During each revolution the particles pass through an even number of acceleration gaps across which an rf voltage $V = V_m \cos \omega_e t$ is maintained. When the radio frequency is in *resonance* with the circulating ions (i.e., when $\omega_e = \omega_c$), continuous acceleration occurs, and the ions travel on an expanding spiraling orbit from the center of the magnetic field to a maximum energy and radius determined by the size of the pole shoes of the magnet.

The magnetic field in the conventional cyclotron must decrease slightly with radius to produce the required force component toward the median plane, which serves to focus the beam during the many revolutions from center to maximum radius. The theory of gradient, or *betatron*, focusing was discussed in Section 3.6.1.

The focusing requirement of $dB/dr < 0$ implies that the orbital frequency $\omega_c = qB/\gamma m$ of the ions is not a constant, but decreases with radius. As a result, the resonance condition $\omega_e = \omega_c$ is violated, particles get out of step with the rf voltage, and after a certain number of turns the phase slip is large enough so that deceleration occurs. This dilemma is enhanced still further by the increase in the *relativistic mass*, γm, which also decreases the orbital frequency.

The maximum energy attainable in this type of cyclotron depends on the phase slip between the particles and the rf voltage; it is greater if the voltage is higher. The largest conventional machine was the 86-inch cyclotron at Oak Ridge National Laboratory, which accelerated protons to 24 MeV (with a peak accelerating voltage of $V_m = 500$ kV).

In 1945, McMillan and Veksler independently proposed the *synchrotron principle* [7] which made it possible to go beyond the energy limits of the conventional cyclotron and which led to the development of synchrocyclotrons and synchrotrons. The two basic ingredients in this new accelerator concept are (1) the modulation of the electric frequency (and in the synchrotron also the magnetic field) with time to maintain synchronism between radio frequency and circulating particle during accelerration; and (2) the existence of phase stability, which assures the continuous acceleration of nonsynchronous particles within certain limits.

The synchrocyclotron employs a cyclotron-type rf system with frequency ω_e modulated by the use of a rotating capacitor, tuning fork, or other means, such that ω_e is a function of time, decreasing in synchronism with the orbital frequency of the ions. After a group of ions is accelerated to full energy, the radio frequency returns to its starting value and begins another cycle of acceleration. The major drawback of this scheme is that beam intensities are down by a factor of 10^2 to 10^4 compared to those of the fixed-frequency cyclotrons. Many synchrocyclotrons were built throughout the world, the largest machined producing protons of about 1 GeV.

In 1938, L. Thomas had shown in a theoretical study that it should be possible to build an *isochronous cyclotron* with constant ion frequency ω_c by employing a magnetic field that varies sinusoidally with azimuth angle [8]. The average magnetic field increases with radius to compensate for the increase in the *relativistic mass*, γm, thus keeping $\omega_c = qB/\gamma m$ a constant while at the same time vertical focusing is provided by the azimuthal field variation (called *flutter*). Because of World War II and the invention of the synchrotron, this idea was not acted upon until 1950, when a group at the Lawrence Radiation Laboratory began a study and built an electron model that proved the feasibility of the new cyclotron concept [9]. Similar studies were soon started at other places in the United States and Europe, and since then a large number of sector-focusing cyclotrons have been built.

All sector-focusing cyclotrons employ magnets with wedge-shaped pole shoes producing a square-wave rather than a sinusoidal variation in azimuth. Recall that edge focusing in single-sector magnets was discussed in Section 3.7. A simple three-sector magnet configuration with straight radial wedges or *hills* of 60° in azimuthal width is illustrated in Figure 3.28. The equilibrium orbit deviates from a circle having a small radius of curvature in the *hills* and a large radius in the *valleys*. As indicated in the figure, a positive ion moving in a clockwise direction will have a radial velocity component pointing outward when the ion enters the hill sector and inward when it leaves it. A vertical cut through the magnet system along the equilibrium radius illustrates the magnetic field lines and the azimuthal field component B_θ in the edge regions of the pole-shoe sectors. For an ion displaced vertically from the midplane, there will be a force component $F_z = qv_rB_\theta$ that focuses the particle toward the midplane when it passes through the edge regions. This is, of course, the same edge-focusing effect that we discussed in Section 3.7. A low energies, this focusing force is sufficient to overcome axial defocusing due to the radially increasing average magnetic field. In most high-energy isochronous cyclotrons, however, the pole-shoe sectors are spiral-shaped rather than straight, as illustrated schematically in Figure 3.29.

The spiral configuration introduces an alternating-gradient focusing effect, marked in the figure by arrows that show the direction of the local magnetic field index, denoted here by $k = (r/B_z)(\partial B_z/\partial r)$. As we know from the discussion of quadrupole focusing in Section 3.5, a combination of focusing and defocusing lenses provides a net focusing effect [see Equations (3.186) and (3.187)]. Thus the alternating-gradient configuration of the spiral-sector system produces an additional net focusing force. This effect supplements edge focusing, thereby providing adequate axial stability for proton energies of several hundred MeV. The median-plane magnetic field of a configuration consisting of N sectors can be written in the form

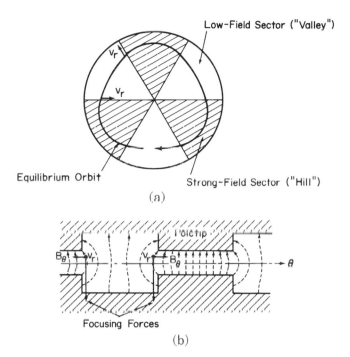

Figure 3.28. Three-sector magnet configuration with straight radial wedges, or *hills*, for an isochronous cyclotron. Equilibrium orbit (a), magnetic field lines, and axial focusing (b) are illustrated schematically.

$$B(r, \theta) = B(r)\left[1 + \sum_n f_n(r) \cos n(\theta \quad \phi_n(r)) \right] \quad (n - N, 2N, 3N, \text{ etc.}).$$

$$(3.368)$$

The phase angle ϕ_n, where the azimuthal variation of the nth field harmonic reaches its maximum value, varies with radius in accordance with the spiral shape of the pole sectors. Radial stability requires that the number of sectors be at least three or larger (i.e., $N \geq 3$). The average magnetic field $\overline{B}(r)$ increases with radius according to the relativistic energy change $\gamma = (1 - \beta^2)^{-1/2}$, that is,

$$\overline{B}(r) = \gamma B_0 = B_0(1 - \beta^2)^{-1/2} = B_0\left[1 - \left(\frac{r\omega_0}{c} \right)^2 \right]^{-1/2}, \quad (3.369)$$

where B_0 is the magnetic field at the center ($r - 0$). The orbital frequency is then $\omega_c = \omega_0 = qB_0/m$ and is thus constant. Calculation of the radial and axial betatron frequencies for such a sector field leads to rather complicated analytical

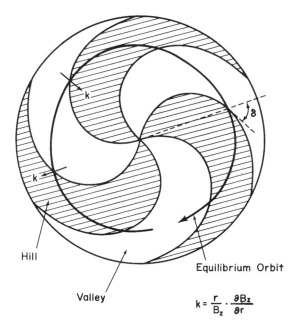

Figure 3.29. Sector-focusing cyclotron with a spiral-ridge magnetic field. Additional axial focusing is provided by the alternating-gradient forces in such a field configuration so that higher energies can be achieved than in a straight radial-sector machine.

expressions. For high accuracy, numerical orbit integration by computer is required. Neglecting a number of less important terms, first-order theory gives the following approximate results:

$$\nu_r^2 = 1 + \overline{k}, \tag{3.370}$$

$$\nu_z^2 = -\overline{k} + \frac{N^2}{N^2 - 1} F(1 + 2\tan^2 \delta), \tag{3.371}$$

where $\overline{k} = (r/\overline{B})(d\overline{B}/dr)$ is the index of the average magnetic field. The parameter F, given by

$$F \approx \frac{1}{2} \sum_n f_n^2, \tag{3.372}$$

represents the *flutter* of the magnetic field variation; δ is the (effective) spiral angle defined by

$$\tan \delta = \frac{r \, d\phi}{dr}, \tag{3.373}$$

where $\phi = \phi(r)$ is the azimuth angle of the peak field in the sectors.

Equation (3.370) for the radial frequency is identical with Equation (3.197) except that in this case $\nu_r > 1$, as $\bar{k}$ is positive. With respect to the vertical frequency [Equation (3.371)], the spiral angle δ and the flutter amplitude F must be large enough to compensate for the defocusing average field and, in addition, provide a net focusing effect such that $\nu_z > 0$. At small radii, sector focusing ceases to be effective since the azimuthal field amplitude, measured by $F(r)$, goes to zero as $(r/g)^N$, where g is the magnet gap width and N the number of sectors. To achieve good focusing at small radii, the number of sectors should be small (i.e., $N = 3$ or $N - 4$). As mentioned, fields with fewer than three sectors are unstable for the radial motion. The problem is alleviated in large cyclotrons which employ separated sectors in a ring-type configuration with beam injection from a small machine.

Sector-focusing cyclotrons are limited in energy by resonances in the radial motion which arise whenever the betatron frequency ν_r passes through certain critical values. Under the condition of isochronism, one finds from Equation (3.369) that

$$\bar{k} = \gamma^2 - 1 = \left(\frac{1 + T}{E_0}\right)^2 - 1 \qquad (3.374)$$

and hence,

$$\nu_r = \gamma = 1 + \frac{T}{E_0} = 1 + \frac{T}{mc^2} . \qquad (3.375)$$

Thus ν_r starts at unity and increases linearly with kinetic energy. According to the theory of resonances in circular accelerators discussed in Section 3.8.5, an *instability stop band* occurs in the radial motion whenever $\nu_r = N/2$, where N is the number of sectors [see Equation (3.404)]. A two-sector field is therefore intrinsically unstable. Using Equation (3.375), one finds that in a three-sector cyclotron $(N - 3)$, the stop band $\nu_r = 3/2$ occurs at a proton energy of 469 MeV, while $N = 4(\nu_r = 2)$ leads to a limit of 938 MeV. If terms neglected in Equations (3.370) and (3.371) are taken into account, the stop-band energy limits are found to be considerably lower than these values.

The largest sector-focusing cyclotron with a spiral-ridge magnetic field configuration of the type illustrated in Figure 3.29 is the TRIUMF machine at the University of British Columbia [10]. It has six ridges ($N = 6$) rather than the three shown in the figure, and accelerates H^- ions to a maximum energy of 500 MeV. Extraction from the cyclotron is achieved by passing the H^- beam through a stripper foil, thereby converting the H^- ions to protons ($H^- \rightarrow H^+ + 2e^-$) which escape from the sector field on trajectories with curvature in the outward direction. In a new project that has been proposed recently, the TRIUMF machine is to be used as an injector for a kaon factory [11].

Even higher proton energies, namely 590 MeV, have been achieved in the S.I.N. ring cyclotron in Switzerland [12]. It consists of a configuration of separated magnet sectors and uses a low-energy sector-focusing cyclotron as an injector for the ring machine. A separated sector magnet design has also been employed in the Indiana University Cyclotron Facility. This is a variable-energy, multiparticle machine that accelerates protons up to about 215 MeV and heavy ions of charge state Z and mass number A to energies up to 220 Z^2/A MeV. An interesting new development at Indiana University is the addition of a storage ring. In this ring, the beams injected from the cyclotron are further accelerated and cooled with a superimposed electron beam to very high phase-space densities that are particularly useful for some nuclear physics experiments [13].

Yet another development in the field of sector-focusing cyclotrons is the use of superconducting coils. The much higher magnetic fields that can be achieved make the *superconducting cyclotron* especially attractive for the cost-effective acceleration of heavy ions. The highest energies in a machine of this type, namely 1400 Z^2/A MeV, can be achieved with the facility at the National Superconducting Cyclotron Laboratory at Michigan State University [14].

3.8.5 Strong-Focusing Synchrotrons

For acceleration of protons to energies above 1 GeV, both linear accelerators and synchrocyclotrons are impractical, as the size of such machines would become prohibitively large. The only type of accelerator that has been capable so far of generating protons in the gigavolt (10^9 V) energy range is the synchrotron, which is based on the principle of phase-stable synchronous acceleration proposed by Veksler and McMillan. Fundamentally, the synchrotron is related to the synchrocyclotron, the main difference being that the orbit radius is kept constant, and the guiding magnetic field is provided by a number of individual dipole magnets placed along the orbit. The particles are first preaccelerated in a linear accelerator and then injected into the synchrotron ring. To keep the orbit radius constant in the synchrotron, the magnets are pulsed such that $B = B(t)$ increases from a minimum value at injection to the maximum given by the final energy of the particles.

In the early synchrotrons orbit stability was provided by constant-gradient focusing as in conventional cyclotrons. The focusing forces in constant-gradient synchrotrons are inherently weak, and consequently, the amplitudes of the betatron oscillations are relatively large. This necessitates the use of magnets with large gap dimensions to contain the beam and makes an accelerator of this kind prohibitively expensive if the energy exceeds more than a few GeV. The invention of the *alternating-gradient* or *strong-focusing principle* was, therefore, a major breakthrough in high-energy accelerator design. Alternating-gradient, or strong-focusing synchrotrons can be built with smaller magnets and have better beam quality and higher beam intensities than those of constant-gradient machines.

The principle of strong focusing was independently proposed first in 1950 by Christofilos [15], who did not publish his idea but applied for a U.S. patent in that year, and in 1952 by Courant, Livingston, and Snyder [16]. This new concept is most easily understood in terms of its well-known optical analog, the combination of focusing and defocusing lenses, that was discussed in Section 3.5. If two lenses of focal lengths f_1 and f_2 are combined, with a separation s between them, the focal length F of this system is, according to Equation (3.186), given by

$$\frac{1}{F} = \frac{1}{f_1} + \frac{1}{f_2} - \frac{s}{f_1 f_2}.$$

In the special case of a converging and diverging lens of equal, but opposite, strength, one has $f_2 = -f_1$, and hence [Equation (3.187)]

$$F = \frac{f_1^2}{s}, \qquad \text{or} \qquad \frac{1}{F} = \frac{s}{f_1^2}.$$

The focal length of such a two lens system is thus always positive (focusing). The application of this idea to synchrotrons implies the combination of strongly focusing and defocusing magnets. According to the theory of betatron oscillations, Equations (3.197) to (3.203), a magnet with negative gradient $dB/dr < 0$ is focusing vertically while defocusing radially if the field index $n > 1$. A radially increasing field ($n < 0$), on the other hand, focuses the particles only in the radial direction and is defocusing with respect to the vertical motion. The first alternating-gradient synchrotron ring consisted of a succession of magnets arranged in such a way that a magnet with large positive gradient is followed by one with a negative gradient of equal strength. The absolute values of n are typically in the range 10 to 100, compared to 0.5 in the conventional weak-focusing machines. Consequently, the frequencies of the corresponding radial and vertical oscillations are an order of magnitude larger than in constant-gradient accelerators.

The strong-focusing forces reduce the beam width and hence the size of the magnets and the vacuum tube, which results in substantial reduction of costs. All modern synchrotrons use arrays of quadrupole magnets for strong focusing, usually in a FODO configuration, and separate dipole magnets for bending of the particle orbits. This separate-function system provides better control, and is less expensive, than the use of alternating-gradient magnets, in which focusing and bending was combined. In addition, sextupole magnets placed at appropriate intervals are used to minimize the effects of chromaticity (i.e., the energy dependence of the betatron oscillations) (see Sections 3.6.4 and 5.4.10). The array of dipole, quadrupole, and sextupole magnets in a synchrotron is called a *lattice*. Such a lattice consists of a periodic configuration of N cells each of which contains identical sets of bending and focusing magnets [17].

Let us denote the length of one cell by S and the phase advance per cell by σ, as defined previously [Equation (3.351)]. In a synchrotron lattice with N cells, the circumference, or length of the closed equilibrium orbit, will be $C = NS$. The phase change of the betatron oscillations per revolution is then simply $N\sigma$, and the number of betatron wavelengths in one revolution, also known as the betatron frequency or *tune*, is given by Equation (3.352), (i.e., $\nu = N\sigma/2\pi$). We note that in the European literature the betatron tune is denoted by Q. The focusing functions

$\kappa(s)$ in Equation (3.312) for such a synchrotron lattice are thus doubly periodic functions with small period S and large period C.

According to the Courant–Snyder theory, the two fundamental solutions of Equation (3.312) may be written in the form

$$u = \hat{\beta}^{1/2}(s)e^{i\nu\psi(s)}, \qquad v = \hat{\beta}^{1/2}(s)e^{-i\nu\psi(s)}, \tag{3.376}$$

where

$$\psi(s) = \int \frac{ds}{\nu\hat{\beta}} \tag{3.377}$$

is a function that increases by 2π every revolution and whose derivative, $\psi' = d\psi/ds$, is periodic. The general solution of the differential equations (3.312) for such a doubly periodic lattice is, analogous to (3.337), of the form

$$x(s) = A\hat{\beta}^{1/2}(s)\cos[\nu\psi(s) + \phi], \tag{3.378}$$

where the constants A, ϕ are determined by the initial conditions. This is a pseudo-harmonic oscillation with varying amplitude $\hat{\beta}^{1/2}(s)$ and varying instantaneous wavelength

$$\lambda = 2\pi\hat{\beta}(s). \tag{3.379}$$

The orbital frequency of the particles in the synchrotron is determined by the ratio of the particle velocity v and the circumference C of the orbit (i.e., $\omega = 2\pi v/C = 2\pi\beta c/C$). In terms of the relativistic energy factor $\gamma = (1 - \beta^2)^{-1/2}$, we can write

$$\omega = \frac{2\pi c}{C}\left(\frac{\gamma^2 - 1}{\gamma^2}\right)^{1/2}. \tag{3.380}$$

At extreme-relativistic energies ($\gamma \gg 1$), the orbital frequency approaches the constant value $\omega = 2\pi c/C$.

The particle momentum P can be related to the average radius $\overline{R} = C/2\pi$ and the average bending magnetic field $\overline{B}$ by

$$P = \gamma\beta mc = (\gamma^2 - 1)^{1/2}mc = q\overline{B}\,\overline{R}. \tag{3.381}$$

The particles are accelerated by rf resonators located in the straight sections between magnets. From Equation (3.381) the rate of energy increase, $dE/dt = mc^2\,d\gamma/dt$, is determined by the rate of change of the average magnetic field, that is,

$$\frac{d\gamma}{dt} = \left(\frac{\gamma^2 - 1}{\gamma^2}\right)^{1/2}q\overline{R}c\,\frac{d\overline{B}}{dt}. \tag{3.382}$$

The corresponding energy gain per turn, $\Delta E = (2\pi/\omega)\, dE/dt$, is then obtained from Equations (3.380) and (3.381) and is given by

$$\Delta E = 2\pi q \overline{R}^2 \frac{d\overline{B}}{dt}. \tag{3.383}$$

To assure that the energy gain $qV_m \cos \omega_e t$ of the particles in the rf cavities during one revolution remains in step with the rising magnetic field, several conditions have to be met: (1) the electric frequency ω_e must be a multiple of the orbital frequency so that the particles remain in phase with the rf voltage; (2) the amplitude V_m of the accelerating voltage and the phase $\phi_s = \omega_e t_s$ of the *synchronous particle* during rf gap crossing must be correlated to satisfy the energy increase $\Delta E = qV_m \cos \phi_s$ as required in Equation (3.383); and (3) the acceleration process must be stable against phase oscillations (phase stability). The second condition can be satisfied for only one phase of gap crossing called the synchronous phase ϕ_s. Particles in the bunch whose phase differs from ϕ_s will gain less or more energy than the synchronous particle. However, the principle of phase stability mentioned earlier assures that the particles whose phase and energy differ slightly from the ideal values perform stable oscillations about the synchronous phase ϕ_s. To provide phase stability, the synchronous phase ϕ_s must be chosen to be within the proper quarter cycle of the rf voltage as defined by the relation (3.261) between revolution time and momentum, which depends on the *frequency slip factor* η. In strong-focusing synchrotrons the radial betatron tune, denoted by ν_r, is always greater than 1. Hence, as discussed in connection with Equation (3.262), at low energies, where $\gamma \ll \gamma_t$, an increase in momentum causes a decrease in revolution time. This implies that stability exists if the synchronous particle crosses the accelerating gaps when the voltage is rising. As γ increases, a critical *transition energy* occurs where $\gamma = \gamma_t$. Above that energy ($\gamma > \gamma_t$), particles behave as in the synchrocyclotron and constant-gradient synchrotron (i.e., the synchronous phase must be in a region of falling voltage). This means that in strong-focusing synchrotrons provisions must be made to shift the phase of the accelerating voltage at the point where the particles pass through the transition energy. If the injection energy is, however, higher than the transition energy, this phase shift can be avoided.

A major problem in the design of strong-focusing synchrotrons is the existence of unstable resonances in the betatron oscillations due to nonlinearities and imperfections in the magnet lattice. The operating point in $\nu_x - \nu_y$ parameter space must be carefully chosen, to be safely away from the nearest resonance; in typical machine designs one aims for a separation of $\Delta \nu \approx 0.25$. An introductory treatment of resonances will be given in the next section (3.8.6). The electric and magnetic self fields of the circulating beam produce a net defocusing force that is equivalent to an effective decrease $\Delta \nu$ of the betatron tune. The condition that this tune shift due to self fields cannot exceed the value $\Delta \nu = 0.25$ to avoid dangerous resonances imposes a fundamental limit to the current, or number of particles, in the circulating beam. This tune depression due to space-charge forces is treated in Section 4.5.

3.8.6 Resonances in Circular Accelerators

As was pointed out in the preceding sections on periodic focusing, circular accelerators are very sensitive to field errors or misalignments since the particles traverse the focusing lattice many times. Resonant-type instabilities occur when the errors or misalignments are encountered at the same phase of the betatron oscillations during each revolution (i.e., whenever there is an integral relationship between betatron frequency and orbital frequency). In this section we present a brief review of this important topic.

To illustrate a resonant-type instability in the particle motion, let us consider as a first example the effect of an error ΔB in one of the dipole magnets that guide the particles around the circular path. In the ideal system (i.e., when $\Delta B = 0$), the particles perform radial betatron oscillations about the equilibrium orbit. The number of these oscillations per turn is defined by the tune $\nu = N\sigma/2\pi$, where N is the number of unit cells in the circumference, $C = 2\pi R$, of the circular lattice, and R is the mean orbit radius. For the following analysis we will use the *smooth approximation* which ignores the ripple in the oscillation amplitude and assumes that the particle oscillation is sinusoidal; that is, it can be described by a differential equation of the form $x''(s) + k^2 x = 0$, where $k^2 = \nu^2/R^2$. (A formal treatment of the smooth approximation, including the effects of space charge, is presented in Section 4.4.) It will be convenient to introduce the azimuth angle $\theta = s/R$, in lieu of the distance, as the independent variable.

A field error ΔB in a short interval $\Delta\theta$ along the circumference can be analyzed as a function of θ in terms of a Fourier series, that is,

$$\Delta B(\theta) = \sum_n \Delta B_n \cos(n\theta + \theta_n), \tag{3.384}$$

where n is an integer and θ_n is the phase angle for the nth harmonic. The equation of motion for the perturbed trajectory in the radial direction takes the form

$$\frac{d^2 x}{d\theta^2} + \nu^2 x = R\frac{\Delta B}{B}, \tag{3.385}$$

where B is the average unperturbed magnetic field.

Let us now consider the Fourier component with the largest amplitude, denoted by the integer n. Using $\delta = \delta_n = R\,\Delta B_n/B$ and assuming that $\theta_n = 0$, we can write

$$\frac{d^2 x}{d\theta^2} + \nu^2 x = \delta \cos n\theta. \tag{3.386}$$

The solution of this differential equation is the sum of the solution x_h for the homogeneous equation and the particular solution x_p of the inhomogeneous equation, that is,

$$x = x_h + x_p. \tag{3.387}$$

The homogeneous solution is of the harmonic-oscillator form

$$x_h = c_1 \cos \nu\theta + c_2 \sin \nu\theta , \tag{3.388}$$

and for the particular solution one has

$$x_p = \frac{\delta}{\nu^2 - n^2} \cos n\theta + c_3 \cos \nu\theta + c_4 \sin \nu\theta . \tag{3.389}$$

A resonance with unlimited amplitude growth occurs for $\nu = n$, where the first term on the right-hand side of Equation (3.389) goes to infinity. Before further analyzing this resonance condition, however, let us first consider the situation where $\nu \neq n$. In this case one obtains a closed orbit solution, that is,

$$x_p(2\pi) = x_p(0), \qquad \frac{dx_p}{d\theta}(2\pi) = \frac{dx_p}{d\theta}(0), \tag{3.390}$$

by setting the two constants c_3 and c_4 to zero so that

$$x_p = \frac{\delta}{\nu^2 - n^2} \cos n\theta \qquad (\text{for } \nu \neq n). \tag{3.391}$$

The general solution (3.387) for this case can therefore be interpreted as a regular betatron oscillation with unperturbed tune ν about a new closed equilibrium orbit that takes into account the field error ΔB.

Returning now to the resonance where

$$\nu = n = \text{integer}, \tag{3.392}$$

we infer from (3.389) that there is no longer a stable, closed-orbit solution. Physically, what happens is that the "frequency" n of the driving force due to the field error is in resonance with the "frequency" ν of the radial betatron oscillations. This causes unlimited amplitude growth and thus instability of the radial motion. The rate of increase of the radial amplitude near the resonance can be calculated by considering a particle with initial conditions

$$x_p(0) = 0, \qquad \frac{dx_p}{d\theta}(0) = 0, \tag{3.393}$$

which results in the particular solution

$$x_p = \frac{\delta}{\nu^2 - n^2} (\cos n\theta - \cos \nu\theta), \tag{3.394}$$

This result is obviously different from the closed-orbit case (3.391), and using trigonometric relations it can be written in the alternative form

$$
x_p = \frac{2\delta}{\nu^2 - n^2} \sin\left(\frac{\nu + n}{2}\theta\right) \sin\left(\frac{\nu - n}{2}\theta\right),
\tag{3.395a}
$$

or

$$
x_p = \frac{\delta\theta}{\nu + n} \sin\left(\frac{\nu + n}{2}\theta\right) \sin\left(\frac{\nu - n}{2}\theta\right) \Big/ \left(\frac{\nu - n}{2}\theta\right).
\tag{3.395b}
$$

In the limit $\nu \to n$, where $\sin\alpha/\alpha \to 1$, one obtains

$$
x_p = \frac{\delta\theta}{2\nu} \sin\nu\theta \qquad \text{(for } \nu = n\text{)}.
\tag{3.396}
$$

This relation shows that at the resonance ($\nu = n$) the amplitude of the radial oscillation grows linearly with azimuth angle θ. Furthermore, the $\sin\nu\theta$ factor indicates that the maximum radial displacement occurs at a phase angle of 90° with respect to the field perturbation. In practice, of course, the radial amplitude of the particle motion does not grow indefinitely since the particles may get out of the resonance or hit the wall of the vacuum chamber after a number of revolutions.

As a second example of a resonant-type instability, let us consider the effect of a gradient error $\Delta\kappa$ in one of the quadrupole magnets of the circular focusing lattice. In the equation of motion such an error appears as a linear driving force. Again using the azimuth angle θ as the independent variable, we can write this equation in the smooth-approximation form

$$
\frac{d^2x}{d\theta^2} + \nu^2 x = R^2 \Delta\kappa x,
\tag{3.397}
$$

where ν is the betatron tune in the absence of the error. As in the previous example, the error $\Delta\kappa$ can be expressed in terms of a Fourier series of the angle θ. Considering again the Fourier component with the largest amplitude and defined by the integer n, we can rewrite (3.397) as

$$
\frac{d^2x}{d\theta^2} + (\nu^2 - \alpha\cos n\theta)x = 0.
\tag{3.398}
$$

This is a Mathieu-type equation that can be transformed to the standard Mathieu form

$$
\frac{d^2x}{d\phi^2} + (a - 2q\cos 2\phi)x = 0
\tag{3.399}
$$

by defining

$$n\theta = 2\phi; \qquad \left(\frac{2\nu}{n}\right)^2 = a; \qquad \frac{4\alpha}{n^2} = 2q. \qquad (3.400)$$

The analysis of Mathieu's equation shows the existence of stable and unstable solutions depending on the values of the parameters a and q. A Mathieu stability diagram is shown in Figure 3.30. The region in q versus a parameter space where the solutions $x(\phi)$ are periodic, and hence stable, are shaded in the figure. Outside these stability regions the solutions $x(\phi)$ are quasiperiodic function of ϕ with increasing amplitude (i.e., they are unstable).

As noted in Section 3.8.2, particle motion in a periodic-focusing channel is governed by a Mathieu equation. The stability condition $-1 < \cos\sigma < 1$ is consistent with the Mathieu diagram. From Figure 3.30 and Equation (3.399) we see that Mathieu's equation degenerates into the equation of a harmonic oscillator when $q = 0$, $a = m^2$ ($m = 1, 2, 3, \ldots$). In this case the solution is stable and of the form $x = \cos m\phi$. The period of the oscillation is therefore defined by

$$\Delta\phi = \frac{2\pi}{m} \qquad (m = 1, 2, 3, \ldots). \qquad (3.401)$$

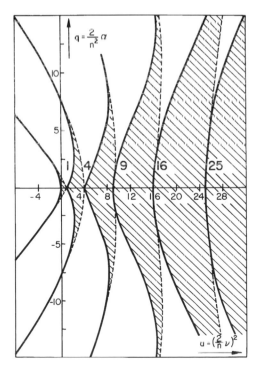

Figure 3.30. Mathieu stability diagram; particle motion is stable in shaded regions, unstable outside.

The analysis shows that the solutions in the unstable regions between the stability lines passing through the points $q = 0$, $a = m^2$ maintain the periodicity $\Delta\phi = 2\pi/m$, but the amplitudes increase without bounds. From (3.400) the azimuthal period $\Delta\phi$ and the corresponding betatron tune ν are then also preserved. Thus we can draw the important conclusion that a solution of Equation (3.398) is unstable if the oscillation period obeys the relation

$$\Delta\theta = \frac{2\Delta\phi}{n} = \frac{4\pi}{mn}. \tag{3.402}$$

The equivalent statement for the betatron tune ν is

$$\nu = \frac{2\pi}{\Delta\theta} = \frac{mn}{2}, \tag{3.403}$$

where m and n are both integers, with n denoting the harmonic of the gradient error responsible for the instability and m the periodicity. Since both m and n are integers it follows from (3.403) that the forbidden values of the betatron tune ν are either integers or half-integers, depending on the values of m and n. If one chooses a given value of the amplitude parameter q and increases the frequency parameter a, one passes through bands of instability or *stop bands* between stable regions. Each stop band is defined by the corresponding value of m and the associated half-integral or integral resonance according to Equation (3.403).

The above analysis of the effects of a gradient error also has important applications for the ideal lattice of a circular machine. The N unit cells of such a periodic structure constitute a strong variation of the gradient function with harmonic $n = N$. In view of (3.403) we conclude therefore that

$$\nu = p\frac{N}{2} \quad (p = 1, 2, 3, \dots) \tag{3.404}$$

should be forbidden values for the betatron tune, as we discussed in Section 3.8.4 for the case of sector-focusing cyclotrons.

In addition to the two examples presented above, there are many other effects leading to resonant-type instability of the betatron oscillations, such as perturbations resulting in nonlinear forces (e.g., $a_2 x^2$) or coupling between the two transverse directions (e.g., $b_1 xy$, etc.). All of these effects appear as driving terms on the right-hand side of the equation of motion, which can be put into the form

$$\frac{d^2 x}{d\theta^2} + \nu_x^2 x = a_0 + a_1 x + a_2 x^2 + b_1 xy + \cdots . \tag{3.405}$$

Strictly speaking, the parameters in this equation are not constants but vary with the angle θ; thus $\nu_x^2 = \kappa_0(\theta)$, $a_0 = a_0(\theta)$, and so on. The first term on the right-hand side, a_0, represents a dipole field error, the second term, $a_1 x$, a quadrupole field error, both of which were treated above. The next term, a_2x^2, represents a sextupole force. The general theoretical analysis of Equation (3.405) shows that the forbidden resonances in the horizontal and vertical betatron tunes, denoted by ν_x and ν_y, respectively, can be expressed in the general form

$$m\nu_x + n\nu_y = p. \tag{3.406}$$

Here m, n, and p are integers (i.e., $m, n, p = 0, \pm1, \pm2, \pm3$, etc.), that should all have the same sign (either $+$ or $-$). $|m| + |n| = l$ defines the *order of the resonance*. For $l > 4$, both theory and experiments show that the resonances are harmful only if the amplitudes of the associated errors are very large. Note that (3.406) contains all the resonances, including the dipole and quadrupole errors treated above. In a diagram plotting ν_y versus ν_x, the resonances appear as forbidden lines. Such diagrams are calculated for every circular accelerator, and the operating point (ν_x, ν_y) is chosen to be at the center of a stable region bounded by the nearest resonance lines that are considered dangerous, say $l \leq 4$.

The theory of resonances is of fundamental importance for the design of circular accelerators such as sector-focusing cyclotrons, synchrotrons, and storage rings. In practice, analytical theory must be complemented by detailed numerical computations to obtain information on the nonlinear properties and stability limits of a focusing lattice. This is particularly important for storage rings where all effects that may limit the lifetime of the beam must be understood. Fast particle tracking codes using *Lie operators* [18] and differential algebra techniques [19] have been developed to investigate the long-term behavior of the beam in such circular machines. This is a highly specialized and active field for which we have to refer to the appropriate literature such as the review articles in the *AIP Conference Proceedings* **249** (1992) (see D.8) by Symon (p. 278), Yan (p. 378) and Berz (p. 456) and references therein. A very exciting new development in this field is the experimental investigation of nonlinear beam-dynamics effects at Fermilab [20] and Indiana University [21].

3.9 ADIABATIC DAMPING OF THE BETATRON OSCILLATION AMPLITUDES

In the preceding sections on transverse focusing in accelerators such as betatrons (3.6.1), cyclotrons (3.8.4), and synchrotrons (3.8.5), we tacitly assumed that the particle energy, magnetic field, and focusing strength of the lenses remain constant. This assumption is justified if the changes in the pertinent parameters occur on a time scale that is very long compared to a betatron oscillation period. If, however, we are interested in determining how the oscillation amplitudes of individual particles, or the transverse dimensions of the beam, vary during the entire acceleration process, we must take that time variation of energy and focusing conditions into account. In betatrons and synchrotrons, for instance, the orbit radius remains constant during the acceleration cycle. However, the particle energy and the magnetic field increase with time and, as a result, the betatron oscillation frequency may also change.

As an example, let us consider the focusing in a betatron as discussed in Section 3.6.1. The axial motion of the particles was described by the Kerst–Serber equation (3.200), where $\omega_z^2 = \omega_c^2 n$ was considered to be constant. For evaluating the long-time behavior we must modify this equation to include the change of energy, $\gamma m c^2$, and of the axial oscillation frequency, ω_z. The appropriate equation of motion in place of (3.200) is then

$$\frac{d}{dt}(\gamma m \dot{z}) = \gamma m \ddot{z} + \dot{\gamma} m \dot{z} = F_z,$$

or, with $F_z/\gamma m = \omega_z^2 z$,

$$\ddot{z} + \frac{\dot{\gamma}}{\gamma}\dot{z} + \omega_z^2 z = 0, \tag{3.407}$$

where both $\dot{\gamma}/\gamma$ and ω_z^2 vary with time. The coefficient $\dot{\gamma}/\gamma$ of the $\dot{z}$ term is positive, indicating that the motion is damped, the decrease in amplitude being dependent on the rate of energy change. If the changes in γ and ω_z are adiabatic (i.e., if they occur slowly with respect to a betatron oscillation period), one can solve (3.407) by the approximate relation

$$z(t) = z_0 f(t) \exp i \int \omega_z(t)\, dt. \tag{3.408}$$

Differentiating (3.408) and substituting in (3.407) yields the following differential equation for the amplitude function $f(t)$:

$$\ddot{f} + 2\dot{f} i \omega_z + f i \dot{\omega}_z + \frac{\dot{\gamma}}{\gamma}\dot{f} + \frac{\dot{\gamma}}{\gamma} f i \omega_z = 0. \tag{3.409}$$

Now we assumed that the time variations of $f(t)$ and $\omega_z(t)$ are adiabatic with respect to the betatron period, $2\pi/\omega_z$. Therefore, we can neglect the first and fourth terms ($\ddot{f}$ and $\dot{\gamma}\dot{f}/\gamma$) and write Equation (3.409) in the form

$$\frac{\dot{f}}{f} = -\frac{\dot{\omega}_z}{2\omega_z} - \frac{\dot{\gamma}}{2\gamma}. \tag{3.410}$$

This can be readily integrated to yield

$$f = \frac{c_1}{(\omega_z \gamma)^{1/2}}, \tag{3.411}$$

where c_1 is an integration constant.

Substituting (3.411) in (3.408), we obtain

$$z = \frac{z_0 c_1}{(\omega_z \gamma)^{1/2}} \exp i \int \omega_z \, dt \, . \tag{3.412}$$

A similar result is obtained for the radial motion, where one has

$$x = \frac{x_0 c_1}{(\omega_r \gamma)^{1/2}} \exp i \int \omega_r \, dt \, . \tag{3.413}$$

The amplitudes of the axial and radial betatron oscillations are thus seen to damp with increasing energy as

$$z \sim (\omega_z \gamma)^{-1/2} \quad \text{and} \quad x \sim (\omega_r \gamma)^{-1/2}, \tag{3.414}$$

respectively. In this general form, the above result is applicable not only to betatrons, but to cyclotrons, synchrotrons, and to linear accelerators as well. One merely needs to change the notation appropriately. In linear accelerators and synchrotrons the two transverse directions are usually taken as x and y; hence one has $x \sim (\omega_r \gamma)^{1/2}$ and $y \sim (\omega_y \gamma)^{1/2}$ for the horizontal and vertical motion, respectively, in these cases.

For betatrons and synchrotrons with time-varying magnetic fields, one can use the relation $\omega = \nu \omega_c = \nu q B / \gamma m$ for the oscillation frequencies and write (3.414) in terms of the tune ν and magnetic field strength B as

$$z \sim (\nu_z B)^{-1/2}, \qquad x \sim (\nu_r B)^{-1/2}. \tag{3.415}$$

In the conventional betatron, the tunes are simple functions of the field index n, namely $\nu_z = (n)^{1/2}$ and $\nu_r = (1 - n)^{1/2}$ [see Equations (3.203) and (3.197)], and we can present our results in yet another form as

$$z \sim n^{-1/4} B^{-1/2}, \qquad x \sim (1 - n)^{-1/4} B^{-1/2}. \tag{3.416}$$

Thus if the change of the magnetic field B and the gradient $\partial B / \partial r$ at the equilibrium radius are known as functions of time during the acceleration cycle, one can calculate the change in the betatron amplitudes for (3.416).

Note that the damping laws for the betatron oscillations apply also to the beam width in the two transverse directions. In fact, we can derive these results in a much more elegant fashion from the conservation of the normalized beam emittance, $\epsilon_n = \beta \gamma \epsilon$, which follows from Liouville's theorem for linear focusing systems.

Let x_m denote the half-width of the beam, x_m' the maximum slope. For a matched beam, $\epsilon_x = x_m x_m' = x_m^2 k_x$, using $x_m' = k_x x_m$. Thus

$$\epsilon_{nx} = \beta\gamma\epsilon_x = \beta\gamma k_x x_m^2 = \text{const.} \tag{3.417}$$

But $k_x = \omega_x/v = \omega_x/\beta c$, hence

$$\gamma\omega_x x_m^2 = c\epsilon_{nx} = \text{const,} \tag{3.418}$$

or

$$x_m = \frac{(\epsilon_{nx} c)^{1/2}}{(\omega_x \gamma)^{1/2}} = \frac{\text{const}}{(\omega_x \gamma)^{1/2}}, \tag{3.419}$$

in agreement with (3.414).

The damping of the betatron oscillations and the beam cross section is a very important general effect in particle accelerators. It applies also when the space-charge forces reduce the net focusing forces. One merely has to use the space-charge depressed frequency, or tune, in this case. However, one must bear in mind that the above scaling relations apply, strictly speaking, only to a system in which all forces are linear in the transverse space coordinates and change adiabatically with time. Such systems also preserve the normalized emittance. If nonlinear effects increase the normalized emittance, we must use the more general scaling laws

$$x_m \sim \left(\frac{\epsilon_{nx}}{\omega_x \gamma}\right)^{1/2}, \qquad y_m \sim \left(\frac{\epsilon_{ny}}{\omega_y \gamma}\right)^{1/2} \tag{3.420}$$

for the beam size in the two transverse directions. Thus, if one wants to compare the matched-beam size at two different times or locations in the acceleration process, one must know not only the two energies and betatron frequencies, but also the change in the normalized emittance. Conversely, by measuring the matched-beam width at two locations, one can infer from (3.420) the emittance change that may have occurred.

REFERENCES

1. D. W. Kerst and R. Serber, *Phys. Rev.* **60**, 53 (1941).
2. M. Reiser, *Part. Accel.* **4**, 239 (1973).
3. See, for instance, N. W. McLachlan, *Theory and Applications of Mathieu Functions,* Clarendon Press, Oxford, 1951.
4. E. D. Courant and H. D. Snyder, *Ann. Phys.* **3**, 1 (1958).
5. J. Struckmeier, J. Klabunde, and M. Reiser, *Part. Accel.* **15**, 47 (1984).

6. E. O. Lawrence and N. F. Edlefsen, *Science* **72**, 376 (1930); E. O. Lawrence and M. S. Livingston, *Phys. Rev.* **38**, 834 (1931).

7. E. M. McMillan, *Phys. Rev.* **68**, 143 (1945); V. Veksler, *J. Phys. (Sov.)* **9**, 153 (1945).

8. L. H. Thomas, *Phys. Rev.* **54**, 580 (1938).

9. F. L. Kelly, P. V. Pyle, R. L. Thornton, J. R. Richardson, and B. T. Wright, *Rev. Sci. Instrum.* **27**, 493 (1956).

10. J. R. Richardson, E. W. Blackmore, G. Dutto, C. J. Kost, G. H. Mackenzie, and M. K. Craddock, *IEEE Trans. Nucl. Sci.* **NS-22**, 1402 (1975).

11. M. K. Craddock, *Part. Accel.* **31**, 183 (1990).

12. W. Joho, *IEEE Trans. Nucl. Sci.* **NS-22**, 1397 (1975).

13. R. E. Pollock, *Annu. Rev. Nucl. Part. Sci.* **41**, 357 (1991).

14. H. Blosser and F. Resmini, *IEEE Trans. Nucl. Sci.* **NS-26**, 3653 (1979).

15. N. Christophilos, "Focusing System for Ions and Electrons," U.S. Patent 2,736,799 (filed March 10, 1950, issued February 28, 1956).

16. E. Courant, M. S. Livingston, and H. Snyder, *Phys. Rev.* **88**, 1190 (1952).

17. For a review of lattice design in strong-focusing synchrotrons, see K. L. Brown and R. V. Servranckx, in *Physics of High Energy Accelerators* (ed. M. Month, P. F. Dahl, and M. Dienes), *AIP Conf. Proc.* **127**, AIP, New York, 1985, p. 62.

18. A. Dragt, *Rev. Nucl. Part. Sci.* **38**, 455 (1988).

19. M. Berz, *Part. Accel.* **24**, 109 (1989).

20. A. Chao et al., *Phys. Rev. Lett.* **61**, 2752 (1988).

21. M. Syphers et al., *Phys. Rev. Lett.* **71**, 719 (1993); S. Y. Lee, 1993 *IEEE Particle Accelerator Conference Record* 93CH3279-7, pp. 6–10.

PROBLEMS

3.1 Prove that the scalar potential of an axisymmetric field can be represented in the form (known as *Laplace's formula*)

$$\phi(r,z) = \frac{1}{\pi} \int_0^\pi \phi(z + ir\cos\theta)\,d\theta \,.$$

3.2 (a) Derive the nonrelativistic paraxial equations (3.51) and (3.52) for $B = 0$ and $p_\theta = 0$ (electrostatic field) from Euler's equations.
(b) Derive the relativistic paraxial equations (3.44) and (3.49) from the Euler equations (2.115) and (2.116).

3.3 Find the magnetic field $B(z)$ on the axis of a solenoid of length l, radius a, total number of turns N, and current I. Determine $B_r(r,z)$ and $B_z(r,z)$ from $B(z)$ by the power-series expansion up to third order in r .

3.4 Show that for nonrelativistic particles and in the thin-lens approximation the image-side focal strength of the bipotential lens of Figure 3.7 is given by

$$\frac{1}{f_2} = \frac{3}{8} \alpha \left(\frac{V_1}{V_2} \right)^{1/4} \left(\frac{V_1 + V_2}{V_2 - V_1} \ln \frac{V_2}{V_1} - 2 \right),$$

where $\alpha = 1.318/b$.

3.5 Repeat Problem 3.4 for relativistic particles, and show that in this case the image-side focal length is given by

$$\frac{1}{f_2} = \frac{\alpha}{4} \left(\frac{\gamma_1^2 - 1}{\gamma_2^2 - 1} \right)^{1/4}$$

$$\cdot \left[\frac{\gamma_1 + \gamma_2 + \frac{1}{2}(\gamma_1 \gamma_2 - 5)}{\gamma_2 - \gamma_1} \ln \frac{\gamma_2^2 - 1}{\gamma_1^2 - 1} - \frac{\gamma_1 \gamma_2 - 5}{\gamma_2 - \gamma_1} \ln \frac{\gamma_2 + 1}{\gamma_1 + 1} - 5 \right],$$

where $\alpha = 1.318/b$ [see Y. Chen and M. Reiser, *J. Appl. Phys.* **65**, 3324 (1989)].

3.6 An electron gun with a 1-cm-diameter thermionic cathode uses a rectangular wire mesh as a control grid. The mesh is located at a distance of $d_g = 0.11$ mm from the cathode. The wires have a diameter of 0.025 mm and are spaced at identical intervals of 0.182 mm in both the x and y directions. The anode is at a distance of $d_a = 15.4$ mm from the grid. Suppose that the gun operates with a cathode temperature of $k_B T = 0.08$ eV, a grid voltage of $V_g = 40$ V, and an anode voltage of $V_a = 5$ kV. Determine the total normalized emittance of the electron beam due to both the cathode temperature and the grid effect.

3.7 A symmetrical electrostatic *einzel lens* consists of three electrodes each of which has a circular aperture of diameter $2a$. The center electrode is at a potential V_2; the two outer ones have the same potential V_1 and separation $\Delta z = l$ from the center plate.

(a) Neglecting the electrode thickness, show that the focal length of this three-aperture lens is given by

$$\frac{l}{f} = \frac{3}{8} \left(\frac{V_2}{V_1} - 1 \right) \left[4 - \left(\frac{V_2}{V_1} \right)^{1/2} - 3 \left(\frac{V_1}{V_2} \right)^{1/2} \right],$$

and that the principal plane is located at a distance

$$\frac{d}{l} = \frac{4}{[3 - (V_2/V_1)^{1/2}][(V_2/V_1)^{1/2} + 1]} - 1,$$

from the center plate.

(b) Using the reduced variable $R = rV^{1/4}$, treat the three-aperture system as a thin lens and show that in this thin-lens approximation the focal length is given by

$$\frac{l}{f} = \frac{3}{8} \frac{[(V_2/V_1) - 1]^2}{V_2/V_1}.$$

(c) Plot l/f and d/l from (a) and l/f from (b) versus $V_2/V - 1$ for the range $0.5 \le V_2/V_1 \le 2$. How good is the accuracy of the thin-lens approximation in this range?

3.8 The field on the axis of a magnetic lens is given by

$$B_z(z) = B_0[1 + (z/a)^2]^{-1}.$$

(a) Determine the focal length, f, and the Larmor rotation, θ_r, in the *thin-lens* approximation.

(b) A 40-keV electron beam is emitted from an object at a distance of 20 cm from the lens; one wants to have a focal length of $f = 5a$, and an image with magnification 3:1 is to be formed downstream of the lens. Find a, B_0, f, image distance l_2, and the Larmor rotation angle of the image with respect to the object.

3.9 A beam of charged particles is focused by a thin unipotential lens. At a distance S_1 upstream of the lens, the particles in the beam occupy an upright ellipse in r–r' trace space with semiaxes $R_{max} = a_1$ and $r'_{max} = b_1$.

(a) Determine the equations for the beam ellipse, the maximum slope r'_{max}, and the envelope radius r_{max} at an arbitrary point downstream from the lens in terms of given parameters (S_1, f_1, a_1, b_1, and z). Specify the ellipse parameters at the image locations z_i and at the position z_w where the beam waist occurs.

(b) As an example, suppose that $a_1 = 0.5$ cm, $S_1 = 9$ cm, $b_1 = 2.19 \times 10^{-2}$ rad, and $f = 5.0$ cm. Find z_i, z_w, and the ellipse coefficients at z_i and z_w.

3.10 Two thin unipotential lenses with focal strength f_1 and f_2 are separated by a distance l. Obtain the transfer matrix relating r_2, r'_2 at the exit of lens 2 to r_1, r'_1 at the entrance of lens 1. What is the focal length F of the combined lens system? What is F if both lenses form a *quadrupole doublet* (one focusing, the other defocusing, of the same strength)?

3.11 Consider an electric quadrupole field that is focusing in the x-direction and defocusing in the y-direction (see the figure below). Assume that the field extends a distance l (from $z = -l/2$ to $z = +l/2$) in the z-direction and that it is strictly two-dimensional [i.e., $V = V(x, y)$ for $-l/2 \le z \le l/2$

and $V = 0$ outside]. Calculate the lens parameters (f_1, f_2, d_1, d_2) of this field for a particle of kinetic energy qV_1 traveling in the z-direction close to the axis both for motion in the $x-z$ plane [case (a)] as well as in the $y-z$ plane [case(b)]. Determine the length l for which the *convergence* $1/f_2$ is a maximum in the $x-z$ plane.

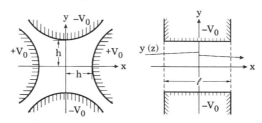

3.12 Derive the relationships between object and image of a lens [i.e., Equations (3.109) and (3.114)], from the transfer matrix $\tilde{M}_{30} = \tilde{M}_{32}\tilde{M}_{21}\tilde{M}_{10}$ between an *object* point (r_0, r_0') at a distance z_1 upstream and a point (r_3, r_3') at a distance z_2 downstream from the lens center. Here $\tilde{M}_{10}$ and $\tilde{M}_{32}$ are the drift-space matrices on either side of the lens and $\tilde{M}_{21}$ is the lens matrix as given in (3.102). (*Hint:* Make use of the fact that the image point is independent of the initial slope of the particle trajectory.)

3.13 Derive the relationship between (R_2, R_2') and (R_1, R_1') of the beam envelope in a thick-lens transformation. Show that (3.163) is not valid except for the thin-lens approximation.

3.14 Consider a source of ions that emits a beam of total radial (or horizontal) width $w_0 = 2x_0$. Let the starting condition of an ion that leaves the source at an arbitrary point be x_1, x_1' and $dP/P_0, P_0$ being the momentum of the *central-ray particle*. After traveling a distance S_0, the beam passes through a sector magnet with focusing edges (angles θ_1 at entrance, θ_2 at exit), as shown in the figure below. The magnet extends over an angle θ. After leaving the magnetic field, the beam travels again on a straight path with distance from the magnet edge given by S_1.

(a) Prove that the image formed of the source is at distance S_i downstream from the magnet given by

$$S_i = S_1(\text{image})$$

$$= \frac{S_0[\cos \phi + (t_1 \sin \phi)/\nu_r] + (R_0 \sin \phi)/\nu_r}{(S_0/R_0)[\nu_r \sin \phi - (t_1 + t_2)\cos \phi - (t_1 t_2 \sin \phi)/\nu_r] - \cos \phi - (t_2 \sin \phi)/\nu_r},$$

where

$$t_{1,2} = \tan \theta_{1,2},$$

$$\phi = \nu_r \theta,$$

$$R_0 = \frac{P_0}{qB_0}.$$

(b) Determine the image magnification $M_x - w_i/w_0 = w_i/2x_0$ and prove that M_x equals $-D^{-1}$, where D is the denominator of the expression for S_i.

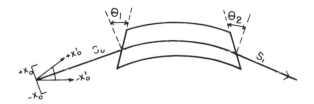

3.15 Consider a sector magnet with uniform magnetic field and normal entry and exit (edge angles $\theta_{1,2} = 0$). The source of a beam is at object distance S_0 from the edge of the magnet, the image at distance S_i downstream (see the figure below). Prove that the object point O, the center of curvature C of the bending radius R_0, and the image point I lie on a straight line, that is, that

$$\theta_0 + \theta + \theta_i = \pi.$$

This relation is known as *Barber's rule*. (*Hint:* You may make use of the expression for the image distance S_i given in Problem 3.14.)

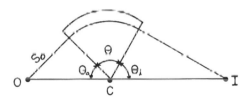

3.16 Derive the relations (3.309) to (3.311) for a periodic thin-lens array.

3.17 Following the procedure of Section 3.8.1, determine the unit-cell matrix, stability criterion, phase advance σ, matched-beam envelope parameters, and acceptance α_x for a periodic channel consisting of thin quadrupole lenses in a FODO array. Compare the results with the thin-lens approximations of Section 3.8.3.

3.18 Show that the Courant–Snyder amplitude function $\hat{\beta}(s)$ satisfies the differential equations

$$2\hat{\beta}\hat{\beta}'' - \hat{\beta}'^2 + 4\kappa\hat{\beta}^2 = 4 \tag{1}$$

and

$$\hat{\beta}''' + 4\kappa\hat{\beta}' + 2\kappa'\hat{\beta} = 0, \tag{2}$$

where $\kappa = \kappa(s)$ represents the focusing force function in the equation of motion. Show furthermore that in regions where $\kappa = $ const, the solution of (2) must be one of the three forms

$$\hat{\beta}(s) = A + Bs + Cs^2, \tag{3a}$$

$$\hat{\beta}(s) = A\cos 2\sqrt{\kappa}s + B\sin 2\sqrt{\kappa}s + C, \tag{3b}$$

$$\hat{\beta}(s) = A\cosh 2\sqrt{|\kappa|}\,s + B\sinh 2\sqrt{|\kappa|}\,s + C. \tag{3c}$$

Evaluate the constants A, B, C for each of these three cases in terms of the initial Courant–Snyder parameters $\hat{\alpha}_0$ and $\hat{\beta}_0$ at $s = 0$.

3.19 According to Equation (3.345), the equation of the beam ellipse in x–x' trace space may be written in the form

$$\hat{\gamma}x^2 + 2\hat{\alpha}xx' + \hat{\beta}x'^2 = \epsilon,$$

where, from (3.343c), the Courant–Snyder parameters $\hat{\alpha}, \hat{\beta}, \hat{\gamma}$ are related by $\hat{\beta}\hat{\gamma} - \hat{\alpha}^2 = 1$. Using these two equations, prove the validity of the relations for the intercepts $(x_{\text{int}}, x'_{\text{int}})$ and maximum values $(x_{\text{max}}, x'_{\text{max}})$ given in Figure 3.26.

3.20 Derive the relations (3.354) to (3.356) for a FODO channel.

3.21 Find the phase advance σ and the maximum of the amplitude function, $\hat{\beta}_{\text{max}}$, for a periodic-focusing channel consisting of solenoid lenses of length l and spacing L (between lenses) in terms of the focusing parameter $\theta = \sqrt{\kappa}\,l$ and the ratio L/l.

3.22 Determine the acceptance area A_x (i.e., the area $\int dx\,dx'$ in trace space) of a cylindrical pipe with radius a and length $L \gg a$ for a beam of charged particles for the following cases:

  (a) No focusing field exists.
  (b) A uniform magnetic field is applied along the pipe (in this case, take x, x' to be the radial coordinates in the Larmor frame).
  (c) A thin magnetic lens with focal length f is placed at the entrance of the pipe, and it produces a waist in the beam envelope at the center of the pipe.

Neglect self-field effects and sketch the acceptance area in a phase-space diagram for each case. What is the ratio of acceptance (b) to acceptance (a), and how does it vary with particle energy?

3.23 A sector-focusing cyclotron has an axial magnetic field $B_z = B$ in the median plane ($z = 0$) of the form

$$B(r, \theta) = \overline{B}(r)[1 + f_3(r)\cos 3\theta]$$

in cylindrical coordinates.

(a) Determine the radial variation of the average field $\overline{B}(r)$ and the field index $\overline{k}(r)$ that is required to keep the average cyclotron frequency ω_c constant. Show that the radial betatron tune ν_r increases linearly with the relativistic energy factor γ (i.e., $\nu_r = \gamma$).

(b) Suppose that we want to accelerate protons to a final energy of $T = 100$ MeV using an rf system with frequency $f = 20$ MHz. Determine the central magnetic field B_0, maximum average field $\overline{B}_{max}$, maximum orbit radius R_{max}, and the value of the azimuthal field variation factor f_3 necessary to achieve a vertical betatron tune of $\nu_z = 0.1$ at R_{max}.

3.24 Consider a simple periodic-focusing lattice of a synchrotron that can be treated like a quadrupole array in a FD (or DF) configuration without drift space. Assume that the two lenses comprising a unit cell have different lengths l_1 and l_2 and focusing strengths κ_1 and κ_2. Using the focusing parameters $\theta_1 = \sqrt{\kappa_1}\, l_1$ and $\theta_2 = \sqrt{\kappa_2}\, l_2$ determine $\cos \sigma_{FD}$ and $\cos \sigma_{DF}$ for the two orthogonal directions. Plot the stability boundaries for each direction in a θ_1^2 versus θ_2^2 diagram and indicate the region in which the motion is stable for both directions. This is known as a *necktie diagram* since the stable region resembles the shape of a necktie.

3.25 Let

$$\tilde{M}_0 = \begin{pmatrix} a_0 & b_0 \\ c_0 & d_0 \end{pmatrix}$$

define the matrix for one revolution and ν_0 the tune in an ideal circular machine. A gradient error $\Delta\kappa$ of a quadrupole magnet would be expected to change the tune from ν_0 to ν. Suppose that the single gradient error is equivalent to a thin-lens quadrupole with focal length f. The perturbed matrix for a single turn, $\tilde{M}$, may then be represented as the product of $\tilde{M}_0$ and the thin-lens matrix for the quadrupole with gradient error.

(a) Prove that

$$\cos 2\pi\nu = \cos 2\pi\nu_0 - \frac{1}{2}\frac{\hat{\beta}}{f}\sin 2\pi\nu_0,$$

where $\hat{\beta}$ is the value of the Courant–Snyder amplitude function at the perturbation.

(b) Assume that the tune change $\Delta\nu = \nu - \nu_0$ is small compared to ν_0 and show that

$$\Delta\nu = \frac{1}{4\pi}\frac{\hat{\beta}}{f}.$$

(c) If there is a distribution of gradient errors $\Delta\kappa(s)$ around the ring circumference, show that the tune change may be expressed by

$$\Delta\nu = \frac{1}{4\pi}\int_0^C \hat{\beta}(s)\,\Delta\kappa(s)\,ds.$$

3.26 Let $\tilde{M}$ represent the transport matrix between two points s_1 and s_2 in a periodic-focusing lattice. Prove that the phase advance from s_1 to s_2 is given by

$$\Delta\psi = \tan^{-1}\left(\frac{c}{\hat{\beta}_1 a - \hat{\alpha}_1 b} \right),$$

where a, b, c, d are the matrix elements of $\tilde{M}$ and $\hat{\alpha}_1, \hat{\beta}_1$ denote the values of the Courant–Snyder parameters at s_1.

CHAPTER 4

Linear Beam Optics with Space Charge

4.1 THEORETICAL MODELS OF BEAMS WITH SPACE CHARGE

In this chapter we include the effects of space charge in the transverse beam optics by using a uniform particle distribution in which both the charge and current density, ρ and $J_z = J$, are independent of the transverse coordinates. This uniform beam model assumes that the beam is continuous in the direction of propagation and has a sharp boundary with $\rho = $ const, $J = $ const inside and $\rho = 0$, $J = 0$ outside the boundary. The uniformity of charge and current density assures that the transverse electric and magnetic self fields and the associated forces are linear functions of the transverse coordinates. Thus, the uniform beam model allows us to extend the linear beam optics of Chapter 3 to include the space-charge effects or, more generally, the forces due to the *self fields*, in a straightforward manner. For nonneutral beams, the terms *space-charge fields* and *self fields* are interchangeable since the moving space charge of the particle distribution is the source for both the electric and the magnetic self fields. However, if the space charge of a beam is completely neutralized by a background of particles with opposite charge polarity, the electric self field is zero. Yet the beam current and hence the associated magnetic self field are still the same as without the charge-neutralization effect. For this reason it would be preferable to speak of *self-field* effects rather than of just *space-charge* effects. But we will use the two terms interchangeably.

Before proceeding with linear beam optics we first present in this section a general discussion of the theoretical beam models and the problems of including the space-charge forces in the beam dynamics. This discussion will also provide some clarification with regard to the limitations and usefulness of the uniform beam model. Specifically, it will be pointed out that real beams, in general, have nonuniform density profiles and that uniformity is approached only as the beam *temperature* (random transverse velocity spread) goes to zero. Nevertheless, the linear model yields valuable information on the *average* beam behavior (e.g., the rms width or divergence) and has become an indispensable tool in the design and operation of accelerators and other devices. A justification for the use of the uniform model is presented in Chapter 5, which deals with self-consistent beam theory. In particular, the concept of *equivalent beams*, discussed in Section 5.3.4, will establish a correlation between rms quantities of nonuniform beams and the equivalent uniform beam.

When the beam currents are high enough that self fields can no longer be neglected in comparison to the applied fields, the mathematical analysis becomes substantially more difficult and complex than the single-particle dynamics discussed in previous chapters. The self fields are functions of the charge and current distribution of the particle beam. At the same time, this distribution is affected by the total external and internal forces acting on the particles. Thus one has a closed loop in which the particle distribution changes the forces and the forces change the particle distribution.

It is straightforward to calculate with any desired degree of accuracy the motion of a single charged particle in an applied field by solving the Lorentz force equation. However, it is practically impossible to find exact solutions of the equations of motion for the enormous number of interacting particles in an intense beam. Even the most advanced computer codes can use only a relatively small number of *macroparticles* to represent the actual particle distribution in a beam and to thereby "simulate" the effects of the mutual interaction between the particles. Such codes, tracing thousands of macroparticles, have become indispensable tools for the study of beam physics and for the design of charged-particle beam devices in which self-field effects are important.

The mutual interaction of the charged particles in a beam can be represented by the sum of a "collisional" force and a "smooth" force. The collisional part of the total interaction force arises when a particle "sees" its immediate neighbors and is therefore affected by their individual positions. This force will cause small random displacements of the particle's trajectory and statistical fluctuations in the particle distribution as a whole. In most practical beams, however, this is a relatively small effect, and the mutual interaction between particles can be described largely by a smoothed force in which the "graininess" of the distribution of discrete particles is washed out. The space-charge potential function in this case obeys Poisson's equation, and the resulting force can be treated in the same way as the applied focusing or acceleration forces acting on the beam.

A measure for the relative importance of collisional versus smoothed interaction, of single-particle versus collective effects, is the *Debye length,* λ_D, a fundamental parameter in plasma physics that can also be applied to charged particle beams. If a test charge is placed into a neutral plasma having a temperature T and equal positive ion and electron densities n, the excess electric potential set up by this charge is effectively screened off in a distance λ_D by charge redistribution in the plasma. This effect is known as *Debye shielding.*

The Debye length λ_D in a nonrelativistic plasma is defined by the ratio of the rms random velocity $\tilde{v}_x = (\overline{v_x^2})^{1/2}$ and the plasma frequency ω_p,

$$\lambda_D = \frac{\tilde{v}_x}{\omega_p}, \tag{4.1a}$$

where $\omega_p = (q^2 n/\epsilon_0 m)^{1/2}$ and where the plasma ions and electrons are assumed to have the same charge q with opposite polarity. For a nonrelativistic isotropic plasma that is in thermal equilibrium at temperature T (*thermal distribution*), the rms thermal velocity is $\tilde{v}_x = (k_B T/m)^{1/2}$ and hence

$$\lambda_D = \left(\frac{\epsilon_0 k_B T}{q^2 n} \right)^{1/2}. \tag{4.1b}$$

A charged particle beam can be viewed as a *nonneutral plasma* that exhibits collective behavior (e.g., instabilities and electromagnetic wave propagation), similar to a neutral plasma (see [B.3]). Thus a local perturbation in the equilibrium charge distribution of a beam with transverse temperature T and density n, confined by external focusing fields, will be screened off in a distance corresponding to the Debye length λ_D.

However, in a charged particle beam moving at relativistic velocity, the nonrelativistic definitions of plasma frequency and Debye length implied in Equation (4.1) must be modified. As shown in Section 4.2, the force on a particle due to the self fields of the beam is proportional to ω_p^2. If transverse motion is considered, we have $\ddot{x} = \omega_p^2 x = F_s/\gamma m$, and since the electric Coulomb repulsion is reduced by magnetic attraction [i.e., $F_s - qE_s(1 - \beta^2) = qF_s/\gamma^2$], we obtain, with $E_s \sim qnx$ (assuming uniform density),

$$\omega_p^2 = \frac{q^2 n}{\epsilon_0 \gamma^3 m},$$

or

$$\omega_p = \left(\frac{q^2 n}{\epsilon_0 \gamma^3 m} \right)^{1/2}. \tag{4.2}$$

The same results also follows from a Lorentz transformation of the nonrelativistic plasma frequency in the beam frame to the laboratory frame.

It should be noted that most authors in the beam literature use either the nonrelativistic definition of ω_p, leaving out the factor γ^3, or (like this author in his past work) change only m to γm and leave out γ^2. However, it is argued here that (4.2) is the more logical relativistic definition with which the equations involving the plasma frequency become simpler and less confusing, as will be discussed in the appropriate context. The relation (4.2) is also correct for longitudinal motion in bunched beams, where there is no relativistic magnetic reduction of the electric space-charge force but where the longitudinal mass $\gamma^3 m$ takes the place of nonrelativistic mass.

Using (4.2) for the plasma frequency and assuming that the random transverse motion in the beam is nonrelativistic (i.e., $\tilde{v}_x \ll c$), we obtain for the Debye length in a relativistic beam the definition

$$\lambda_D = \frac{\tilde{v}_x}{\omega_p} = \left(\frac{\epsilon_0 \gamma^3 \tilde{v}_x^2}{q^2 n} \right)^{1/2}. \tag{4.3a}$$

For a thermal distribution in this case we can use the relation $\gamma m \tilde{v}_x^2 = k_B T$ to define the transverse temperature in the laboratory frame. With the substitution $\tilde{v}_x = (k_B T / \gamma m)^{1/2}$ we can then write the Debye length in the form

$$\lambda_D = \left(\frac{\epsilon_0 \gamma^2 k_B T}{q^2 n} \right)^{1/2}. \tag{4.3b}$$

It should be noted in this context that the relativistic definition of temperature is somewhat controversial. Many authors in the recent literature on relativistic gases and plasmas use only the temperature in the rest frame and treat it—like mass m—as a relativistic invariant. However, as will be shown in Section 5.4.3, one can also justify the use of a laboratory temperature. The relationship between temperature T_b in the beam (rest) frame and temperature T in the laboratory frame is $T = T_b / \gamma$. Hence, the above expression for the Debye length can be written in terms of the beam-frame temperature T_b as

$$\lambda_D = \left(\frac{\epsilon_0 \gamma k_B T_b}{q^2 n} \right)^{1/2}, \tag{4.3c}$$

where all quantities except T_b are defined in the laboratory frame. We shall use both definitions and indicate by the appropriate subscript which temperature is implied wherever necessary to avoid confusion.

If the Debye length is large compared with the beam radius ($\lambda_D \gg a$), the screening will be ineffective and single-particle behavior will dominate. On the other hand, if the Debye length is small compared to the beam radius ($\lambda_D \ll a$), collective effects due to the self fields of the beam will play an important role. It follows from (4.2) and (4.3) that the plasma frequency decreases with particle energy and that the Debye length increases so that at sufficiently high energy the space-charge forces become insignificant in comparison to the external forces acting on a beam.

Smooth functions for the charge and field distributions can be used as long as the Debye length remains large compared to the interparticle distance l_p, that is, as long as the number N_D of particles within a Debye sphere of radius λ_D remains very large ($N_D \gg 1$). On the other end of the spectrum, when λ_D becomes comparable to l_p, a particle will be affected by its nearest neighbors more than by the collective field of the beam distribution as a whole. In this limit, which occurs at extremely low temperature or very large density, the mutual interaction of single particles leads to configurations in the particle distribution in which crystal-like, "grainy" structures may develop. Such structures have been observed in particle simulation studies and are an important new topic of current beam research [1]. As mentioned, for most beams of practical interest the collisional forces are very small compared to the smooth forces and can be neglected. Notable exceptions are the *Boersch effect* at low energies and *intrabeam scattering* in high-energy storage rings, which are treated in Section 6.4. Collisions also play a fundamental role in driving a beam toward a Maxwell–Boltzmann distribution, as discussed below and in Section 5.4. We should also mention in this context that collisions between macroparticles used in computer simulation may cause artificial "numerical" emittance growth. Such effects can be avoided by judicious choice of mesh size and number of particles, by charge-averaging in a mesh, and by better modeling of the system being investigated.

If collisions can be neglected (i.e., if $\lambda_D \gg l_p$), the single-particle Hamiltonian, the particle distribution, and Liouville's theorem can be defined in six-dimensional phase space $(\mathbf{r}, \mathbf{P})$. This is possible because the smoothed space-charge forces acting on a particle can be treated like the applied forces. Thus, the six-dimensional phase-space volume occupied by a charged-particle distribution remains constant during propagation or acceleration. If, furthermore, all forces are linear functions of the particle displacement from the beam center, the normalized emittance associated with each direction remains a constant of the motion. For a matched beam in a linear focusing channel we can express this conservation law in terms of the rms beam width $\tilde{x}$, rms tranverse momentum $\tilde{P}_x$ and rms velocity $\tilde{v}_x$ as

$$\tilde{\epsilon}_n = \frac{\tilde{x}\tilde{P}_x}{mc} = \frac{\tilde{x}\gamma\tilde{v}_x}{c} = \text{const.} \tag{4.4a}$$

In the case of a thermal beam this relation can be written in terms of the laboratory or beam-frame temperatures as

$$\tilde{x}^2\gamma k_B T = \text{const,} \tag{4.4b}$$

or

$$\tilde{x}^2 k_B T_b = \text{const,} \tag{4.4c}$$

respectively.

Thus, as in a gas, if the beam in such an ideal system is compressed adiabatically by the applied focusing forces, its transverse temperature increases. Likewise, in an expanding beam the temperature decreases (i.e., the beam cools). On the other hand, nonlinear external or space-charge forces existing in the real world tend to increase the normalized emittance and the temperature of the beam and may also produce a temperature variation across the beam. The actual behavior is then more complicated than implied by relation (4.4).

As in thermodynamics and plasma physics, one of the fundamental questions of particle beam theory concerns the existence of equilibrium states in which the particle distribution remains stationary (i.e., it does not change with distance along the focusing channel). When collisions are negligible, the possible equilibria can be found with the help of the *Vlasov theory*. As shown in Section 5.3, many particle distribution functions can be constructed mathematically that are stationary solutions of the *Vlasov equation*, which combines the equations of motion for the particles and Maxwell's equations for the fields. Such distribution functions are useful tools for mathematical analysis or computer simulation, but the correlation with actual beams may often not be readily apparent. The notable exception is the *Maxwell–Boltzmann distribution*, also known as the *thermal distribution*, defined by $f(\mathbf{x}, \mathbf{P}) = f_0 \exp(-H/k_B T)$, where H is the single-particle Hamiltonian. It not only satisfies the Vlasov equation, but also represents the natural thermodynamic equilibrium state when collisions are included, as discussed in Section 5.4. Laboratory beams are usually not in thermal equilibrium. They have different *transverse* and *longitudinal temperatures, $T_\perp$ and $T_\parallel$,* with $T_\parallel \ll T_\perp$ due to longitudinal cooling (see Section 5.4.3) and other effects. If temperature relaxation (equipartitioning) due to collisions and nonlinear forces is slow compared to the lifetime of the beam, we can have a quasistationary state, or *metaequilibrium* (as defined in B.1, section 1.1 for a plasma). Thus, a perfectly matched, continuous beam in an axisymmetric uniform or smooth focusing channel or acceleration system can be treated as a transverse Maxwell–Boltzmann distribution for which the particle density profile obeys the Boltzmann relation (see Section 5.4.4).

$$n(r) = n_0 \exp\left[-\frac{q\phi(r)}{k_B T_\perp} \right]. \tag{4.5}$$

Here, n_0 is the density at $r = 0$, $T_\perp$ represents the transverse laboratory temperature of the beam, k_B is the Boltzmann constant, and $\phi(r)$ is the sum of the effective external potential, $\phi_e(r)$, and the effective potential due to the self fields, $\phi_s(r)(1 - \beta^2)$; that is,

$$\phi(r) = \phi_e(r) + \phi_s(r)(1 - \beta^2). \tag{4.6}$$

The space-charge potential $\phi_s(r)$ must obey Poisson's equation. The external potential in the linear focusing channel considered here is given by $q\phi_e(r) = \gamma m \omega_0^2 r^2/2$, where ω_0 is the particle oscillation frequency (when self fields are neglected). It follows from (4.5) that as $k_B T_\perp \rightarrow 0$ or, alternatively, as the repulsive self force becomes equal in magnitude to the external focusing force [i.e., $q\phi_s(1 - \beta^2) \rightarrow -q\phi_e$], the beam density profile becomes uniform with a sharp radius, a, hence

$$n(r) = n_0 = \text{const} \quad \text{for } r \leq a,$$

$$n(r) = 0 \quad \text{for } r > a. \tag{4.7}$$

On the other hand, at high temperature or, alternatively, when the self force becomes negligible compared to the external focusing force, we obtain a Gaussian profile,

$$n(r) = n_0 \exp\left[-\frac{\gamma m \omega_0^2 r^2}{2 k_B T_\perp} \right]. \tag{4.8}$$

In the first case [Equation (4.7)], the Debye length approaches zero (i.e., $\lambda_D \rightarrow 0$), while in the second case [Equation (4.8)], $\lambda_D \rightarrow \infty$.

Strictly speaking, these considerations are valid only for a beam in a long, uniform focusing channel. However, as shown in Section 4.4, a periodic-focusing system can often be described in terms of an equivalent uniform channel by using the *smooth approximation theory*. Hence, the argument that beams tend to approach a Boltzmann distribution if collisional and other effects have time to thermalize the distribution also applies in an approximate sense to periodic-focusing systems that are more commonly used in practice.

Since the potential function due to the self fields decreases with increasing energy [i.e., $\phi_s(1 - \beta^2) = \phi_s/\gamma^2 \rightarrow 0$ for $\gamma \gg 1$], we conclude that at sufficiently high energy, charged particle beams tend to have the Gaussian density distribution of Equation (4.8). By contrast, at low energy when the self force is comparable in strength to the external force, beams in smooth, linear focusing channels tend to have uniform density profiles.

A stationary distribution represents a state of minimum total energy. As shown in Chapter 6, deviations from this stationary state, such as beam mismatch, off-centering, and nonstationary particle density profiles, are associated with higher total energy. The difference in energy represents *free energy* that can be converted into random, or *thermal*, particle energy, thereby increasing the temperature and emittance of the beam. The mechanisms converting the free energy into emittance growth are collisional processes, instabilities, nonlinear space-charge forces, and any forces of a stochastic (random) nature acting on the particle distribution.

For the theoretical modeling of beams we can distinguish three regimes that can be characterized by the ratio of the Debye length to the effective beam radius, λ_D/a. When self-field effects dominate the beam physics (i.e., when $\lambda_D \ll a$), it is convenient for the mathematical analysis to neglect the thermal velocity spread altogether and use a laminar-flow model for the beam ($T_\perp = T_\| = 0$). In laminar flow, all particles at a given point are assumed to have the same velocity, so that particle trajectories do not cross. As we know from Equation (4.7), the particle density for a stationary laminar beam in a linear focusing channel is uniform. Like the external focusing force, the space-charge force is therefore a linear function of position within the beam. As a result, the linear beam optics techniques of Chapter 3 can be extended in a straightforward manner to include the self-field effects.

When the transverse thermal velocity spread becomes comparable to self field effects, so that $\lambda_D \sim a$, the density profile of a stationary beam becomes nonuniform, according to the Boltzmann relation (4.5). The forces due to the self fields of the beam are therefore nonlinear, a nonlaminar treatment of the beam is required, and the analysis becomes more complicated. A nonlaminar beam can be represented by the distribution of particles in phase space, $f(\mathbf{x}, \mathbf{P})$. The stationary state and the evolution of nonstationary distributions can be analyzed with the aid of the Vlasov equation, as mentioned above. Analytical techniques are rather limited in usefulness—except for the $K-V$ *model* discussed below—and must be complemented or replaced by particle simulation.

The third regime is characterized by $\lambda_D \gg a$, which implies that the self fields of the beam can be ignored. According to the Boltzmann relation (4.6), the steady-state density profile is Gaussian. However, the particle motion is entirely governed by the external fields; that is, the beam optics techniques and results of Chapter 3 are valid in this regime.

From a mathematical as well as a practical point of view, the most useful theoretical model satisfying the Vlasov equation is the distribution of Kapchinsky and Vladimirsky, known in the literature as the $K-V$ *distribution*. For the spatially uniform focusing channel discussed above, the K–V distribution is defined as a delta function of the transverse Hamiltonian [i.e., $f(x, y, P_x, P_y) = f_0 \delta(H_\perp - H_0)$]. Alternatively, the K–V distribution can be defined as a delta function of the transverse emittances [i.e., $f(x, x', y, y') = f_0 \delta(\epsilon_x, \epsilon_y)$]. In the latter case, it is also applicable to spatially varying focusing systems, consisting of discrete lenses, acceleration gaps, and so on, where the transverse Hamiltonian is not a constant of the motion. The K–V beam has the property that the density profile is uniform with sharp boundaries and the external forces are linear. Since the self fields of a uniform-density beam are linear functions of position, the density remains uniform and sharply bounded as the beam propagates through the focusing or accelerating system. Thus it is possible to extend the linear beam optics to include the space-charge forces in a straightforward way, and this will be done in subsequent sections of this chapter. It should be noted that the K–V model covers the entire range from space-charge-dominated laminar beams ($\lambda_D \ll a$) to the emittance-dominated beams ($\lambda_D \gg a$), where self-field forces are negligible. The self-consistent theory of beams, including the K–V distribution, is treated in Chapter 5. We show there that correlations between average beam parameters (rms width, divergence, emittance, etc.) of the K–V model, and other, more realistic distributions can be established. These correlations justify retroactively the extensive use of the uniform beam profile for the analysis of beam optics with self field presented in Chapter 4.

The major shortcoming of a linear model like the K–V beam is the fact that it does not provide any information on emittance growth due to nonlinear external or self forces. The determination of emittance growth requires additional tools such as nonlinear theory, particle simulation, and experiment, as discussed in Chapter 6.

In many practical devices, a background gas or plasma may affect the beam behavior in a substantial way. Depending on the gas density, ionizing collisions between beam particles and gas molecules may lead to partial or full charge neutralization of the beam. Secondary particles created in these collisions having the same charge polarity as the beam particles are expelled, while those with opposite charge polarity remain trapped in the potential well of the beam. The resulting charge neutralization, called *gas focusing*, is of great practical importance.

It occurs naturally in regions where the vacuum pressure is not low enough (e.g., in the low-energy beam transport lines near ion sources); or it may be deliberately utilized to transport high-current beams that could not be handled in conventional focusing systems. On the other hand, such charge-neutralization effects may lead to instabilities resulting in emittance growth and beam loss. Charge neutralization will be represented in our uniform-beam model by a partial neutralization factor f_e, and some special neutralization effects are discussed in Sections 4.2.4 (Bennett pinch) and 4.6 (neutralization in a background gas).

4.2 AXISYMMETRIC BEAMS IN DRIFT SPACE

4.2.1 Laminar Beam with Uniform Density Profile

We start our study of self-field effects in beams with a simplified model of a cylindrical beam propagating in a drift tube (i.e., with no applied fields present). Assume that a *laminar*, parallel beam of particles with uniform density is injected into a conducting drift tube. For vanishingly small currents, all particles would continue on trajectories parallel to the beam axis and the diameter of the beam would remain constant. As the current is increased, however, the space charge will produce a defocusing outward electric force and the beam will spread radially. At high velocities, the beam current produces a magnetic self field which exerts an attractive force that reduces the net defocusing effect. If the beam propagates through a region with a low-density background gas (rather than ideal vacuum), collisional ionization effects may result in partial neutralization of the beam space charge. Thus, in the case of an electron beam, the secondary, low-energy electrons created by the collisions are ejected and the positive ions remain inside the beam. Due to their heavy mass, these ions remain almost stationary compared to the fast beam electrons. If f_e is the ratio of positive ion charge to electron charge per unit volume, the electric field due to the space charge will be reduced by a factor $(1 - f_e)$. The magnetic field, however, remains unaffected as the stationary ions do not contribute to current flow. The ions, of course, do oscillate radially in the potential well of the electron beam, but the oscillation periods are long compared to the electron oscillation periods. On the other hand, if we are dealing with a positive ion beam, the ions from the collisions are ejected and the secondary electrons remain in the beam. These electrons are very mobile and oscillate rapidly across the beam in the transverse direction. The net effect is a partial charge neutralization of the ion beam which, as in the case of an electron beam, does not affect the beam current and the associated self magnetic field. As we will see, the combined effect of self magnetic field and partial charge neutralization may not only balance the repulsive electric force but may result in a net focusing or pinching of the beam.

Let us now list the assumptions that we will make in our simple uniform beam model:

1. The beam has a circular cross section with radius a and propagates within a concentric drift tube of radius b, and the variation of beam radius with axial

distance z is slow enough that axial electric field components E_z and radial magnetic field components B_r can be neglected.

2. The potential difference $\Delta\phi$ between beam axis and the drift-tube wall due to the space charge of the beam is small compared to the voltage equivalent of the particles' kinetic energy.

3. The beam particle density, as well as the density of charge-neutralizing particles of opposite polarity, is uniform inside the beam and zero outside. In view of assumption 2, the axial velocity of all beam particles is approximately the same, and we can therefore assume that the current density is uniform.

4. The flow is laminar (i.e., all beam particles move on trajectories that do not cross).

5. We consider a steady-state situation; that is, $\partial/\partial t = 0$ and the beam cross section at any given position along the direction of travel does not change with time.

6. The particle trajectories obey the paraxial assumption that the angle with the axis (slope) is small. This follows implicitly from assumption 1.

For the mathematical treatment that follows, we write the equations of motion for a positive charge q, as in previous chapters. The factor f_e represents a stationary charge distribution of opposite sign which results in a partial neutralization of the space charge of the primary particles. The results can be applied to electrons by setting $q = -e$. The current remains unaffected by the stationary particles. It should be pointed out that, strictly speaking, the assumption of uniform charge density is valid only when there are no charge-neutralizing particles present in the beam ($f_e = 0$). For a beam with *gas focusing* by secondary particles of opposite charge polarity, the density profiles of both species tend to become nonuniform. The uniform-density model is still useful, though, in describing the *average* behavior of the beam in this case.

First we note that in the steady state considered here the volume charge density ρ and the current density $\mathbf{J}$ at any point within the beam, or alternatively, the line charge density ρ_L and beam current I, are related by the continuity equation, that is,

$$\mathbf{J} = \rho\mathbf{v}, \tag{4.9a}$$

$$I = \rho_L v, \tag{4.9b}$$

where $\mathbf{v}$ is the velocity of a charge element at that point, and $v_z \approx v$ has been assumed in (4.9b). Due to the space charge of the beam, there exists a potential difference between the beam axis ($r = 0$) and the beam surface ($r = a$) and (for $b > a$) between the beam edge and the wall of the drift tube. If the total energy of the particles is a constant, the kinetic energy of a particle on the axis will be less than that of a particle on the beam edge. Thus, in general, we have a velocity distribution $v(r)$, and if $\rho = \text{const}$, $\mathbf{J}$ must be a function of radius, or vice versa. In principle, we could specify any one of the three functions, and the other two are then determined self-consistently by Equation (4.9), Maxwell's equations, and the equations of motion. However, in our uniform beam model, we abandon self-consistency to avoid mathematical complexity. As long as the paraxial assumption holds (i.e., $v_r \ll v$, $v_\theta \ll v$, $v_z \approx v$) and the difference in potential energy across the beam is small compared to the kinetic energy of the particles, the error will be small. Our major objective at this point is to gain physical insight into the behavior of the beam with a minimum of mathematical effort. Thus, we will assume that J, ρ, and $v_z \approx v$ are all constant across the beam (i.e., independent of radius r). Hence, with $\rho_0 = I/a^2\pi v$ denoting the charge density of the primary beam particles, we obtain

$$J_z = J = \frac{I}{a^2\pi}, \tag{4.10a}$$

$$\mu = \rho_0(1 - f_e) = \frac{I(1 - f_e)}{a^2\pi v} \qquad \text{for } 0 \le r \le a, \tag{4.10b}$$

and $J = 0$, $\rho = 0$ for $r > a$. In view of assumption 1, the electric field has only a radial component, which is readily found by application of Gauss's law, $\int \epsilon_0 \mathbf{E} \cdot d\mathbf{S} = \int \rho \, dV$, to a cylinder of radius r and unit length in the z-direction:

$$E_r = \frac{\rho_0(1 - f_e)r}{2\epsilon_0} - \frac{I(1 - f_e)r}{2\pi\epsilon_0 a^2 v} \qquad \text{for } r \le a, \tag{4.11a}$$

$$E_r = \frac{I(1 - f_e)}{2\pi\epsilon_0 v r} \qquad \text{for } r > a. \tag{4.11b}$$

When charge neutralization is absent ($f_e = 0$), we obtain

$$E_r = \frac{\rho_0 r}{2\epsilon_0} = \frac{Ir}{2\pi\epsilon_0 a^2 v} \qquad \text{for } r \le a \tag{4.11c}$$

and

$$E_r = \frac{I}{2\pi\epsilon_0 v r} \qquad \text{for } r > a. \tag{4.11d}$$

The magnetic field, which has only an azimuthal component, is obtained by applying Ampère's circuital law, $\int \mathbf{B} \cdot d\mathbf{l} = \mu_0 \int \mathbf{J} \cdot d\mathbf{S}$, which yields

$$B_\theta = \mu_0 \frac{Ir}{2\pi a^2} \qquad \text{for } r < a, \tag{4.12a}$$

$$B_\theta = \mu_0 \frac{I}{2\pi r} \qquad \text{for } r > a. \tag{4.12b}$$

By integrating Equations (4.11) we obtain for the electrostatic potential distribution (with $\phi = 0$ at $r = b$)

$$\phi(r) = V_s\left(1 + 2\ln\frac{b}{a} - \frac{r^2}{a^2}\right) \qquad \text{for } r \le a, \qquad (4.13a)$$

$$\phi(r) = 2V_s\ln\frac{b}{r} \qquad \text{for } a \le r \le b, \qquad (4.13b)$$

where

$$V_s = \frac{\rho_0(1 - f_e)a^2}{4\epsilon_0} = \frac{I(1 - f_e)}{4\pi\epsilon_0\beta c} \approx \frac{30I}{\beta}(1 - f_e) \qquad (4.14a)$$

and

$$V_s = \frac{\rho_0 a^2}{4\epsilon_0} = \frac{I}{4\pi\epsilon_0\beta c} \approx \frac{30I}{\beta} \qquad (4.14b)$$

when charge neutralization is absent ($f_e = 0$).

The peak potential on the beam axis ($r = 0$) is thus [from (4.13a)] $\phi(0) = V_0 = V_s[1 + 2\ln(b/a)]$, and the maximum electric field at the beam edge is $E_a = 2V_s/a \approx 60I(1 - f_e)/(\beta a)$, or $E_a \approx 60I/(\beta a)$ when $f_e = 0$.

We now examine the motion of a beam particle in this field using only the radial force equation

$$\frac{d}{dt}(\gamma m\dot{r}) = \gamma m\ddot{r} = qE_r - q\dot{z}B_\theta,$$

where we dropped the force term $qr\dot{\theta}B_z$ on the grounds that $r\dot{\theta}$ is negligibly small and $\gamma = $ const since there is no external acceleration. Substituting for E_r from (4.11a) and for B_θ from (4.12a), we get with $\epsilon_0\mu_0 = c^{-2}$, $\dot{z} = v = \beta c$,

$$\gamma m\ddot{r} = \frac{qIr}{2\pi\epsilon_0 a^2\beta c}(1 - f_e - \beta^2), \qquad (4.15a)$$

or

$$\gamma m\ddot{r} = \frac{qIr}{2\pi\epsilon_0 a^2\beta c}(1 - \beta^2) \qquad \text{for } f_e = 0. \qquad (4.15b)$$

With

$$\ddot{r} = v_z^2\frac{d^2r}{dz^2} = \beta^2 c^2 r'',$$

Equation (4.15) becomes

$$r'' = \frac{qIr(1 - f_e - \beta^2)}{2\pi\epsilon_0 a^2 mc^3\beta^3\gamma}. \qquad (4.16a)$$

or, with $f_e = 0$, $1 - \beta^2 = \gamma^{-2}$,

$$r'' = \frac{qIr}{2\pi\epsilon_0 a^2 mc^3 \beta^3 \gamma^3} . \tag{4.16b}$$

We will now introduce several parameters used in the literature on beams with space charge. First, we define a *characteristic current* I_0 by

$$I_0 = \frac{4\pi\epsilon_0 mc^3}{q} \approx \frac{1}{30} \frac{mc^2}{q} , \tag{4.17}$$

which is approximately 17 kA for electrons and $31(A/Z)$ MA for ions of mass number A and charge number Z. Next we introduce the *Budker parameter* [2] ν_B defined as the product of the number of primary beam particles per unit length, $N_L = \rho_L/q$, and the *classical particle radius* r_c. The latter is obtained by equating the rest energy mc^2 and potential energy $q^2/4\pi\epsilon_0 r_c$ of a point charge with mass m and charge q; hence, $r_c = q^2/4\pi\epsilon_0 mc^2$, and we find

$$\nu_B = N_L r_c = \frac{I}{I_0 \beta} . \tag{4.18}$$

Thus, for ultrarelativistic particles ($\beta \sim 1$), the Budker parameter is simply given by the ratio of the beam current I to the characteristic current I_0.

A third important beam physics parameter, the *plasma frequency*, was already introduced in Equation (4.2) for a relativistic, unneutralized beam ($f_e = 0$). For the more general case where charge neutralization is not zero ($f_e \neq 0$), our definition $\omega_p^2 = F_s/\gamma m$, using $1 - f_e = \beta^2 - \gamma^{-2}(1 - \gamma^2 f_e)$, yields

$$\omega_p^2 = \frac{q^2 n}{\epsilon_0 \gamma^3 m} (1 - \gamma^2 f_e), \tag{4.19}$$

or, in terms of the beam current I,

$$\omega_p^2 = \frac{qI}{\pi\epsilon_0 mc\beta\gamma^3 a^2} (1 - \gamma^2 f_e), \tag{4.20a}$$

and

$$\omega_p^2 = \frac{qI}{\pi\epsilon_0 mc\beta\gamma^3 a^2} \tag{4.20b}$$

when $f_e = 0$. Equation (4.15) for the radial motion of a particle may then be written in the form

$$\ddot{r} = \frac{\omega_p^2}{2} r . \tag{4.21}$$

The advantage of our generalized definition for the plasma frequency is now apparent: Equation (4.21) has the same mathematical form whether the motion is nonrelativistic or relativistic, and whether charge neutralization is present or not. From (4.19) we see that $\omega_p^2 > 0$; that is, the net space charge is defocusing when $\gamma^2 f_e < 1$. On the other hand, $\omega_p^2 < 0$ (i.e., the self fields produce a net focusing force) when

$$f_e > \frac{1}{\gamma^2}. \tag{4.22}$$

This relation is known as the *Budker condition of self focusing* [2]. It is of particular importance for intense relativistic electron beams where a small fraction of stationary positive ions is sufficient to focus the beam.

From Equations (4.19) and (4.20) we note that the plasma frequency is inversely proportional to the beam radius, a. As the radius a changes in a diverging or converging beam, ω_p will also change. Thus, we cannot integrate Equation (4.21) unless independent information is provided on the variation of ω_p with time or distance. Furthermore, it is desirable to eliminate time and introduce the distance along the direction of beam propagation as the independent variable, as was done in Equation (4.16). Before we proceed with solving this trajectory equation, we introduce another important parameter, the *generalized perveance K*, a dimensionless quantity, defined by Lawson [3] as

$$K = \frac{I}{I_0} \frac{2}{\beta^3 \gamma^3} (1 - \gamma^2 f_e). \tag{4.23}$$

As can be seen, the generalized perveance—unlike the plasma frequency—does not depend on the beam radius. It is solely defined by the beam current and particle energy and, where applicable, by the charge-neutralization factor f_e. When charge neutralization is absent (i.e., $f_e = 0$), the relationship among generalized perveance, Budker parameter, and the plasma frequency is given by

$$K = \frac{I}{I_0} \frac{2}{\beta^3 \gamma^3} = \frac{2\nu_B}{\beta^2 \gamma^3} = \frac{\omega_p^2 a^2}{2\beta^2 c^2}. \tag{4.24}$$

In terms of the generalized perveance, as defined in Equation (4.23), the equation (4.16) for the particle trajectories can be expressed as

$$r'' = \frac{K}{a^2} r. \tag{4.25}$$

Note that unlike Equation (4.21), this equation shows the explicit dependence on the beam radius. It applies to the trajectory of any particle within the beam $(r \leq a)$ and can be solved if the radius a is known as a function of distance. Now, under the conditions of laminar flow, the trajectories of all particles are similar and scale with the factor r/a. Specifically, the particle at $r = a$ will always remain at the boundary of the beam. Thus, by setting $r = a = r_m$ in Equation (4.25), we obtain the equation for the beam radius $r_m(z)$ in drift space, which may be written in the form

$$r_m r_m'' = K .$$ (4.26)

We should point out here that our paraxial beam model is valid only for $|K| \ll 1$, as discussed in Section 4.2.3. With regard to the solution of Equation (4.26), several special cases are of interest:

1. $f_e = 0$ (no stationary, neutralizing particles)

$$K = \frac{2\nu_B}{\beta^2 \gamma^3} = \frac{\omega_p^2 r_m^2}{2c^2 \beta^2} = \frac{I}{I_0} \frac{2}{\beta^3 \gamma^3} ,$$ (4.27a)

2. $f_e = 1$ (full charge neutralization by the stationary particles)

$$K = -\frac{2\nu_D}{\gamma} = -\frac{I}{I_0} \frac{2}{\beta\gamma} .$$ (4.27b)

3. $f_e = 0, \gamma \approx 1$ (nonrelativistic approximation)

$$K = -\frac{2\nu_D}{\beta^2} = -\frac{I}{I_0} \frac{2}{\beta^3} = \frac{qI}{2\pi\epsilon_0 m v^3} .$$ (4.27c)

Substituting $v = (2qV/m)^{1/2}$, where V denotes the beam voltage, Equation (4.27c) may be written as

$$K = \frac{I}{V^{3/2}} \left[\frac{1}{4\pi\epsilon_0 (2q/m)^{1/2}} \right].$$ (4.28)

For nonrelativistic, unneutralized ($f_e = 0$) beams, the ratio $I/V^{3/2}$ is known as the *perveance*. The *generalized perveance* K thus differs from the *perveance* $I/V^{3/2}$ by the factor in brackets in Equation (4.28). In the case of *electron beams* (and stationary positive ions), the above formulas for the generalized perveance K become numerically

$$K = 1.174 \times 10^{-4} \frac{I}{(\gamma^2 - 1)^{3/2}} \quad \text{for } f_e = 0,$$ (4.29a)

$$K = -1.174 \times 10^{-4} \frac{I}{(\gamma^2 - 1)^{1/2}} \quad \text{for } f_e = 1,$$ (4.29b)

and for a nonrelativistic electron beam with $f_e = 0$,

$$K = 1.515 \times 10^4 \, \frac{I}{V^{3/2}}. \tag{4.29c}$$

Finally, we note that the case $f_e = 1/\gamma^2$ represents a force-neutral beam where $K = 0$ and the particles move on straight lines.

Let us now integrate Equation (4.26) for the radius of the beam in the general case where $K \neq 0$ assuming that the beam has an initial radius $r_m = r_0$ and slope $r'_m = r'_0$ (at $z = 0$). First, we introduce dimensionless variables

$$R = \frac{r_m}{r_0}, \qquad R' = \frac{r'_m}{r_0}, \qquad Z = (\pm 2K)^{1/2} \frac{z}{r_0} = (2|K|)^{1/2} \frac{z}{r_0}. \tag{4.30}$$

Here and in the equations that follow, the plus sign applies when $K > 0$ (defocusing) and the minus sign when $K < 0$ (focusing). In terms of the new variables, Equation (4.26) can be written in the form

$$\frac{d^2R}{dZ^2} = R'' = \pm \frac{1}{2R}, \tag{4.31}$$

or, alternatively,

$$2R' \, dR' = \pm \frac{dR}{R}. \tag{4.32}$$

Integration of Equation (4.32) with $R = R_0 = 1$ and $R' = R'_0$ at $z = 0$ yields

$$R'^2 - R_0'^2 = \pm \ln R, \tag{4.33}$$

or

$$R = e^{\pm(R'^2 - R_0'^2)}. \tag{4.34}$$

By integration of Equation (4.33) one obtains

$$Z = \int_1^R \frac{dR}{R'} = \int_1^R \frac{dR}{\left(R_0'^2 \pm \ln R\right)^{1/2}}. \tag{4.35}$$

If we use R' as the independent variable, we get from Equations (4.32) and (4.34)

$$dR = 2R' e^{\pm(R'^2 - R_0'^2)} \, dR'. \tag{4.36}$$

Integration of Equation (4.36) then yields the alternative expression

$$z = 2e^{\mp R_0'^2} \int_{R_0'}^{(\pm \ln R + R_0'^2)^{1/2}} e^{\pm R'^2} dR',$$ (4.37)

or

$$\frac{z}{r_0} = \left(\left|\frac{2}{K}\right|\right)^{1/2} e^{\mp R_0'^2} \int_{R_0'}^{(\pm \ln R + R_0'^2)^{1/2}} e^{\pm R'^2} dR'.$$ (4.38)

The factor in front of the integral is always positive. As for the integral itself, the plus sign applies when $R' > 0$ (diverging beam) and the minus sign applies when $R' < 0$ (converging beam). If the initial slope is zero ($R_0' = 0$), for instance, the beam will diverge when $K > 0$, and we have then

$$\frac{z}{r_0} = \left(\frac{2}{K}\right)^{1/2} \int_0^{[\ln(r_m/r_0)]^{1/2}} e^{R'^2} dR'.$$ (4.39)

On the other hand, a converging beam results when $K < 0$, and we can write (for $R_0' = 0$)

$$\frac{z}{r_0} = \left(\left|\frac{2}{K}\right|\right)^{1/2} \int_0^{[\ln(r_0/r_m)]^{1/2}} e^{-R'^2} dR'.$$ (4.40)

Even if $K > 0$, we can still get a converging profile by passing the beam through a focusing lens. In this case, the initial slope of the beam profile will be negative ($R_0' < 0$). The beam radius r_m will decrease until a minimum is reached where $R' = 0$. Beyond that, the radius will increase again (diverging beam profile) as a result of the defocusing self-field forces. The minus sign in the upper limit of the integral in Equation (4.38) applies for the region from $z = 0$, $R = 1$ to the point z_m, where R is a minimum and $\ln R + R_0'^2 = 0$. For $z > z_m$, the plus sign in the integral applies. The integral in Equation (4.38) is of the type

$$\int_0^{\pm x} e^{\pm y^2} dy,$$

which is tabulated in handbooks of mathematical functions. For different values of the "reduced" initial slope R_0', one obtains for the case $K > 0(f_e = 0)$ the curves shown in Figure 4.1, where R is plotted versus the "reduced distance" $Z = (2K)^{1/2}(z/r_0)$. These curves might, for instance, represent the behavior of a charged particle beam after passage through a focusing lens which changes the slope R_0' as the beam enters the drift region following the lens. As we see, the beam diameter goes through a minimum (waist) which varies with the initial slope R_0'.

In some high-current applications one wants to pass as much current as possible through a tube of diameter D and length L with the help of a focusing lens at the tube entrance. In this case one has to focus the beam such that the waist is at the center of the tube, the beam having equal diameters at the entrance and exit of the tube ($R = R_0 = 1$). As we can see from Figure 4.1, there is a maximum value of Z, and hence for given values of beam voltage V, tube length L, and radius $r_0 = D/2$, a maximum current that one can get through the tube. This maximum Z value is 2.16, and the corresponding slope is about $R_0' = -0.92$ [i.e., $Z = (2K)^{1/2}(L/r_0) = 2.16$].

Letting $r_0 = D/2$, $z = L$, and $Z = 2.16$, we obtain from Equations (4.30) and (4.27a) for the maximum current:

$$I_m = I_0 \frac{\beta^3 \gamma^3}{4} (2.16)^2 \left(\frac{r_0}{L}\right)^2 = 1.166 I_0 \beta^3 \gamma^3 \left(\frac{r_0}{L}\right)^2. \tag{4.41}$$

For electrons the maximum current (in amperes) is

$$I_m = 0.496 \times 10^4 (\gamma^2 - 1)^{3/2} \left(\frac{D}{L}\right)^2 \tag{4.42}$$

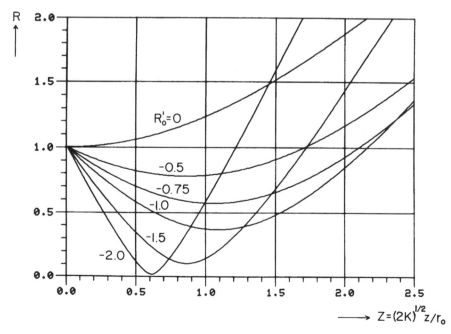

Figure 4.1. Beam radius R versus "reduced" distance Z for different initial slopes $R_0' = dR/dZ$ (laminar flow).

in the relativistic case, and

$$I_m = 38.5 \times 10^{-6} V^{3/2} \left(\frac{D}{L} \right)^2 \tag{4.43}$$

in the nonrelativistic approximation.

Returning again to Figure 4.1, let us consider now the upper curve, where the initial slope is zero ($R_0' = 0$). For this particular case, the beam diameter doubles in a reduced distance of about $Z = 2.12$, and one can show that the curve Z versus R can be approximated by the relation

$$Z = 2 \int_0^{[\ln R]^{1/2}} e^{R'^2} \, dR' \approx 2(R - 1)^{1/2}, \tag{4.44}$$

which is accurate to better than 3% for $1 \le R \le 2$. In this approximation, the beam radius is a quadratic function of distance given by

$$R = \frac{r_m}{r_0} = 1 + 0.25Z^2 = 1 + 0.5K \left(\frac{z}{r_0} \right)^2. \tag{4.45}$$

For an *electron* beam with no ions ($f_e = 0$), we obtain

$$R = \frac{r_m}{r_0} = 1 + 5.87 \times 10^{-5} \frac{I}{(\gamma^2 - 1)^{3/2}} \left(\frac{z}{r_0} \right)^2 \tag{4.46}$$

in the relativistic case, and

$$R = \frac{r_m}{r_0} = 1 + 7.58 \times 10^3 \frac{I}{V^{3/2}} \left(\frac{z}{r_0} \right)^2 \tag{4.47}$$

in the nonrelativistic approximation.

From these relations one can calculate the distance in which the beam radius doubles due to the space-charge repulsion. Setting $r_m = 2r_0$, one obtains for an electron beam from (4.46) for the doubling distance the relativistic relation

$$z = z_d = 1.31 \times 10^2 (\gamma^2 - 1)^{3/4} I^{-1/2} r_0$$

and from (4.47) the nonrelativistic formula

$$z = z_d = 1.15 \times 10^{-2} V^{3/4} I^{-1/2} r_0.$$

As an example, for an electron beam with a kinetic energy equal to the rest energy of 511 keV and a current $I = 200$ A, one finds $z_d = 21.0 r_0$ or $z_d = 52.6$ cm for $r_0 = 2.5$ cm. In this case, the beam expansion is slow enough ($z_d/r_0 \gg 1$) that the assumptions of our uniform beam model are well satisfied.

As was pointed out earlier, the defocusing self-field forces in a relativistic electron beam can be compensated by a background of stationary positive ions. Thus, when $f_e = 1/\gamma^2$ (i.e., $K = 0$), we get from Equation (4.26) $r_m'' = 0$, or $r_m = r_0'z + r_0$; and if the initial slope is zero ($r_0' = 0$), the beam diameter remains constant. The positive ions thus provide uniform focusing, like a long solenoidal magnetic field.

When $f_e > 1/\gamma^2$, K is negative and the beam pinches. The beam diameter decreases until it approaches zero at

$$(0.5|K|)^{1/2}\left(\frac{z}{r_0}\right) \approx 0.8 \, ,$$

as discussed by Lawson [C.17]. As $r_m \rightarrow 0$, our beam model breaks down. The motion is no longer paraxial as the slopes of the trajectories become very large; at the same time, the flow becomes nonlaminar. In practice, of course, all beams have transverse temperature, or nonzero emittance, which prevents such a collapse of the beam radius. The effects of finite emittance on the beam envelope are discussed in the next section, and then we explore the limits of our model and the concept of *limiting currents*.

4.2.2 Beam Envelope with Self Fields and Finite Emittance

The derivation leading to the trajectory equation (4.16) or (4.25) is valid for a uniform density profile whether the beam is laminar or nonlaminar. It was only when we derived the equation for the beam radius [Equation (4.26)] that we introduced the assumption of laminar flow. Let us now assume that the particles have a distribution in r, r' trace space that corresponds to an area of elliptical shape, as discussed in Chapter 3. If the radius r_m represents the envelope of the beam (i.e., if we set $a = r_m$), Equation (4.16) or (4.25) describes the motion of any particle at radius r within the beam. Specifically, when the current is negligibly small ($I \rightarrow 0$), we get $r'' = 0$, or $r = r_0 + r_0'z$, which is the motion of a particle in free space with zero self fields. For this case we found that the beam envelope obeys a differential equation of the form of Equation (3.161), which we rewrite with $R = r_m$ as

$$r_m'' = \frac{\epsilon^2}{r_m^3} \, , \tag{4.48}$$

where ϵ denotes the emittance of the beam.

Since the trajectory equation (4.16) is linear in r, we can obtain the general envelope equation by linear superposition of the two special solutions: Equation (4.26) for zero emittance and Equation (4.48) for zero space charge; that is, we can write

$$r_m'' = \frac{\epsilon^2}{r_m^3} + \frac{K}{r_m}. \tag{4.49}$$

A mathematically more rigorous derivation of this equation is presented in Section 5.3.2.

Integration of Equation (4.49) with the initial condition $r_m = r_0$ and $r_m' = r_0'$ at $z = 0$ and assuming that $K > 0$ yields the result

$$r_m' = \left[r_0'^2 + \epsilon^2 \left(\frac{1}{r_0^2} - \frac{1}{r_m^2} \right) + 2K \ln \frac{r_m}{r_0} \right]^{1/2} \tag{4.50}$$

and

$$z = \int_{r_0}^{r_m} \left[r_0'^2 + \epsilon^2 \left(\frac{1}{r_0^2} - \frac{1}{r_m^2} \right) + 2K \ln \frac{r_m}{r_0} \right]^{-1/2} dr_m. \tag{4.51}$$

When $K = 0$, the integral can be evaluated and we obtain the result (3.162) for the beam envelope without space charge. Likewise, for $\epsilon = 0$, we recover the result (4.35) for our uniform, laminar beam model. In the general case of a beam with space charge and finite emittance, the integral in (4.51) has to be evaluated numerically and one obtains envelope curves that are qualitatively similar to those shown in Figure 4.1.

4.2.3 Limitations of the Uniform Beam Model and Limiting Currents

It was noted earlier that our paraxial beam model becomes invalid when the particle trajectories either strongly converge or strongly diverge (i.e., when the assumptions $|K| \ll 1$, $\dot{r} \ll v$ are no longer satisfied). To explore this limitation of our model, let us assume that we have a fully charge-neutralized relativistic electron beam ($f_e = 1$), where $K = -(I/I_0)(2/\beta\gamma) = -2\nu_B/\gamma$ and the particles are affected only by the magnetic self force of the beam. From (4.25), one obtains for a particle within the beam the trajectory equation

$$r'' + \left(\frac{|K|}{a^2} \right) r = 0. \tag{4.52}$$

Now let us treat the beam radius as a constant. With $r = r_0$ and $r' = r_0' = 0$, the solution of Equation (4.52) at $z = 0$ is then

$$r = r_0 \cos \left(\sqrt{|K|} \, \frac{z}{a} \right) \tag{4.53}$$

and

$$r' = -\frac{r_0}{a} \sqrt{|K|} \sin\left(\sqrt{|K|}\,\frac{z}{a}\right). \tag{4.54}$$

The radial velocity is $v_r = \dot{r} = r'v$, or

$$v_r = -\frac{r_0}{a} v \sqrt{|K|} \sin\left(\sqrt{|K|}\,\frac{z}{a}\right). \tag{4.55}$$

It has a maximum value (for $r_0 = a$) of $v|K|^{1/2} = v(2\nu_B/\gamma)^{1/2}$. We had assumed that $v_r \ll v$, and therefore we require that $\nu_B/\gamma \ll 1$ or $|K| \ll 1$ in order for our model to be valid. This implies that the electron current must be substantially less than a critical current I_A, which follows from Equation (4.18) by setting $\nu_B/\gamma = 1$:

$$I_A = I_0 \beta\gamma \tag{4.56}$$

(i.e., $I_A \approx 1.7 \times 10^4 \beta\gamma$ amperes for electrons). This fundamental current limit was first derived in 1939 by H. Alfvén [4], who studied the propagation of electrons through a plasma in space. If one does not assume that $|v_r| \ll v$, it is still possible to solve the equation of motion since γ is a constant in the charge-neutral beam, where only the magnetic self field is present. The solution was obtained by Alfvén in terms of elliptical integrals, and it indicates that beam propagation essentially stops when the limit (4.56) is reached since most of the electrons are reflected back in the strong magnetic self field.

The factor ν_B/γ is a measure of the effects of the self fields on the beam dynamics. In terms of the beam current, we can write

$$\frac{I}{I_A} \approx \frac{\nu_B}{\gamma}. \tag{4.57}$$

Note that, in general, β and γ are functions of radius and that the beam current relates to the mean velocity $\overline{\beta}c$. In our simple model, β is uniform across the beam, and the velocity of individual particles is identical with the mean velocity. However, when $\beta = \beta(r)$, we have to use $\overline{\beta}$ in the definition [Equation (4.18)] of the Budker parameter ν_B.

We conclude from this analysis of a fully charge-neutralized beam that the uniform beam model in paraxial approximation is good only as long as $\nu_B/\gamma \ll 1$. When $\nu_B/\gamma \to 1$, the assumptions of uniform current density across the beam and $v_r \ll v$ are violated. The particles acquire increasingly larger values of radial velocity; as the more accurate trajectory calculations indicate, at $\nu_B/\gamma \approx 1$ the particles pass through the axis with no remaining axial velocity and are reflected backward. The beam therefore ceases to propagate in the forward direction, which explains why we speak of a *limiting* current, I_A. The stopping of a beam by its own magnetic self field was also studied by Lawson [3]. The relation (4.56) is often referred to as the *Alfvén current,* or as the *Alfvén–Lawson current,* or simply as the *magnetic current limit.*

A similar fundamental current limit also exists for a beam that is not charge-neutralized. In this case, without external focusing, the beam would expand radially due to the repulsive space-charge forces. For the following derivation, let us assume that an infinitely strong applied magnetic field prevents such expansion and keeps the beam radius constant. Due to the space-charge field, part of the kinetic energy of a particle inside the beam is converted into electrostatic potential energy. The potential difference between center $(r = 0)$ and the wall $(r = b)$ of a uniform-density, cylindrical beam of radius a inside a conducting tube of radius $b \geq a$ is obtained from Equations (4.13a) and (4.14b):

$$\phi(0) = V_0 = V_s\left(1 + 2\ln\frac{b}{a}\right) = \frac{I}{4\pi\epsilon_0\beta c}\left(1 + 2\ln\frac{b}{a}\right), \qquad (4.58)$$

where I is the beam current and βc the average (axial) velocity in the beam. Suppose that all particles are injected into the conducting drift tube with the same kinetic energy $(\gamma - 1)mc^2 = qV$. For a particle on the axis $(r = 0)$ inside the tube, the kinetic energy is then reduced by qV_0, so that [with $\gamma(0) = \gamma_0$]

$$\gamma_0 mc^2 = \gamma mc^2 - qV_0. \qquad (4.59)$$

One sees that a particle is stopped when all its kinetic energy is converted into potential energy [i.e., when $(\gamma - 1)mc^2 = qV_0$]. From (4.58), this happens when the current reaches the limit [5]

$$I = I_0\frac{\beta(\gamma - 1)}{1 + 2\ln(b/a)}. \qquad (4.60)$$

This value is actually a little too high since the current maximum is reached before the potential energy on the axis equals the kinetic energy. Solving (4.58) for the current I and expressing V_0 and β in terms of γ and γ_0, one finds from the condition $\partial I/\partial\gamma_0 = 0$, the more accurate limit (with $f_e = 0$)

$$I_L = I_0\frac{(\gamma^{2/3} - 1)^{3/2}}{1 + 2\ln(b/a)}, \qquad (4.61)$$

which was first derived by Bogdankevich and Rukhadze and independently by Nation and Read [6]. Note that this space-charge limiting current I_L is lower than the Alfvén–Lawson current I_A.

So far, we have considered the two extreme cases where the beam was either fully charge neutralized $(f_e = 1)$ or had no charge-neutralizing particles $(f_e = 0)$. In the first situation, the beam pinches and, as a result, stops propagating when

$\nu_B/\gamma \approx 1$ or $I = I_A$. In the latter case, the beam would blow up radially unless it is confined by a strong external magnetic field, and propagation stops when the potential energy of a particle in the beam becomes comparable to the kinetic energy at injection (i.e., when $I = I_L$).

Let us now examine what happens when the beam is partially neutralized and no external magnetic field is present. As we discussed in connection with Equation (4.26), the repulsive electric force exceeds the magnetic attraction when $f_e < 1/\gamma^2$ and the beam spreads radially ($r_m'' > 0$). On the other hand, for $f_e > 1/\gamma^2$, the beam pinches due to a net inward focusing force ($r_m'' < 0$). A special "force-free" state exists when $f_e = 1/\gamma^2$. When $f_e \neq 0$, the analysis leading to the space-charge limiting current (4.61) can be extended simply by including the factor $(1 - f_e)$ in the denominator; that is, one may write

$$I_L = I_0 \frac{(\gamma^{2/3} - 1)^{3/2}}{[1 + 2\ln(b/a)](1 - f_e)}. \tag{4.62}$$

This formula suggests that an arbitrarily large current can be achieved by using a large amount of charge neutralization; in fact, $I_L \to \infty$ when $f_e \to 1$. However, we have seen that for a charge-neutralized beam ($f_e = 1$), magnetic pinching leads to the Alfvén–Lawson limit I_A. Consequently, as $f_e \to 1$, Equation (4.62) must be modified by the constraint

$$I_L \leq I_A = I_0 \beta \gamma \qquad (\text{for } f_e \to 1). \tag{4.63}$$

From this discussion it appears that I_A constitutes a fundamental upper limit for particle beams. But this conclusion is not correct since I_A applies only to a beam that is fully charge neutralized by *stationary* particles of opposite charge. If the self-magnetic field of the primary beam becomes neutralized by an opposite current of *moving* secondary particles, pinching no longer occurs, and in principle, the beam current can become arbitrarily high. Such a *current neutralization* can be achieved by injecting co-moving particles of opposite charge into the primary beam, as, for example, in ion propulsion where electrons are used to neutralize the positive ion beam. Another mechanism is the generation of a return current due to the inductive fields associated with short beams propagating through a gas or plasma. The intense relativistic electron beams produced by high-power pulse generators have relatively short time durations with typical pulse widths of 10 to 100 ns. The rise time of these beams produces a time-varying magnetic field $\partial B_\theta/\partial t$ and hence an electric field E_z (from $\nabla \times \mathbf{E} = -\partial \mathbf{B}/\partial t$) as the beam front enters the gas region. As soon as the ionizing collisions of the electrons with the gas molecules produce a plasma, this $\mathbf{E}$ field generates a current in the opposite direction to the incoming electron beam. The magnetic field B_θ associated with this return current is opposite to the B_θ of the primary beam. Consequently, one gets a partial *magnetic neutralization* for which we introduce the factor $1 - f_m$. Our original equation for the magnetic field (4.12a) must then be written in the form

$$B_\theta = \frac{\mu_0 I(1 - f_m)r}{2\pi a^2} \qquad \text{for } r \le a. \tag{4.64}$$

Assuming that the beam is completely charge-neutralized ($f_e = 1$), the equation of motion (4.15) has to be modified and will now be

$$\gamma m \ddot{r} = -\frac{qIr\beta^2(1 - f_m)}{2\pi\epsilon_0 a^2 \beta c}. \tag{4.65}$$

The force is still inward, but it is weakened by the factor $1 - f_m$. The magnetic current limit in this case will then be increased by a factor $(1 - f_m)^{-1}$, that is,

$$I_A^* = \frac{17,000\beta\gamma}{1 - f_m}. \tag{4.66}$$

Since, by assumption, the beam is electrically neutral ($f_e - 1$), the consideration that kinetic energy is transformed into potential energy does not apply. Therefore, I_A^* can be considerably greater than the Alfvén current I_A if there is a large degree of magnetic neutralization [see Miller, C 18, Sections 4.3.3 and 5.5.] Thus, we can substitute I_A^* for I_A whenever a current in the direction opposite the primary beam current can be produced that results in partial magnetic neutralization.

The existence of upper limits for the beam current can also be understood from an energy-conservation or "power-balance" argument. As a charged particle beam propagates, kinetic energy has to be spent to build up the electric and magnetic self-field energy along the path of the beam. Suppose that a beam of length L and constant radius $r_m = a$ propagates inside a conducting drift tube of radius b. Let the pulse duration $\tau = L/v$ be short enough that the magnetic self field does not penetrate through the conducting wall. The total field energy associated with the beam is then given by

$$W = 2\pi L \left[\frac{\epsilon_0}{2} \int_0^b E_r^2 r \, dr + \frac{1}{2\mu_0} \int_0^b B_\theta^2 r \, dr \right]. \tag{4.67}$$

Substitution for E_r and B_θ from Equations (4.11) and (4.12) then yields the expression

$$W = \frac{I^2 L}{4\pi\epsilon_0 c^2} \left(\frac{1}{4} + \ln \frac{b}{a} \right) \left[\frac{(1 - f_e)^2}{\beta^2} + (1 - f_m)^2 \right], \tag{4.68}$$

where we have added the parameters f_e and f_m to include partial charge or current neutralization. This field energy must be supplied from the kinetic energy of the particles at the beam front, $(\gamma_f - 1)mc^2$. If $(\gamma_i - 1)mc^2$ represents the kinetic energy of the beam front particles at injection into the drift tube, one obtains the following energy conservation law in the form of a power-balance equation (assuming that the current I remains unchanged):

$$I(\gamma_f - 1)\frac{mc^2}{e} = I(\gamma_i - 1)\frac{mc^2}{e} - \frac{W}{L}\beta_f c, \tag{4.69}$$

where $\beta_f c$ is the final beam-front velocity. It is obvious from the last two equations that there is an upper limit for the beam current (unless $f_e = f_m = 1$) where the field energy is comparable to the kinetic energy of the particles and hence the beam can no longer propagate.

4.2.4 Self-Focusing of a Charge-Neutralized Beam (Bennett Pinch)

In our discussion of the laminar beam model (Section 4.2.1), we found that charge neutralization leads to self-focusing, or pinching, of the beam when $f_e > 1/\gamma^2$ and hence $K < 0$. We concluded, however, that the paraxial assumptions of our model are no longer valid when the beam radius approaches zero. A real beam is not perfectly laminar but always has a finite transverse temperature, or emittance, that prevents the collapse to zero radius. Let us now examine the pinch effect and the role of finite temperature more closely by returning to the beam envelope equation (4.49), which includes both the space charge and the emittance.

In the following, let us assume that we are dealing with a relativistic electron beam with stationary positive ions. If K is negative due to the fact that the inward magnetic force exceeds the repulsive space-charge force, there will be an equilibrium beam radius $r_m = a$, where this net inward force is just balanced by the outward "pressure" due to the thermal velocity spread, or emittance, of the beam. Setting $r_m'' = 0$ in (4.49), we find that

$$\epsilon^2 = -Ka^2 = |K|a^2. \tag{4.70}$$

Solving for the beam radius a yields

$$a = \frac{\epsilon}{\sqrt{|K|}}. \tag{4.71}$$

An alternative form of this equilibrium relation can be obtained by considering a thermal distribution and introducing the transverse electron beam laboratory temperature T_e in place of the emittance. With $\tilde{v}_x = (k_B T_e/\gamma m)^{1/2}, a = 2\tilde{x}$, we obtain from (4.4a) an effective emittance ($\epsilon = 4\tilde{\epsilon}$) of

$$\epsilon = \frac{\epsilon_n}{\beta\gamma} = 2a\left(\frac{k_B T_e}{\gamma m v^2}\right)^{1/2}. \tag{4.72}$$

Substitution of (4.72) in (4.70) yields

$$k_B T_e = \frac{1}{4}|K|\gamma m v^2 = \frac{1}{4}|K|\gamma \beta^2 mc^2. \tag{4.73}$$

For a fully charge-neutralized beam ($f_e = 1$, $f_m = 0$), K is given by Equation (4.27b); hence,

$$k_B T_e = \frac{1}{4}\frac{I}{I_0}\frac{2}{\beta\gamma}\gamma\beta^2 mc^2$$

or, with $I_0 = 4\pi\epsilon_0 mc^3/q$,

$$k_B T_e = \frac{1}{2}\frac{Iq\beta c}{4\pi\epsilon_0 c^2}. \tag{4.74}$$

If both sides of (4.74) are multiplied by the number of electrons per unit length, N_L, one obtains, with $N_L q\beta c = I$ and $1/\epsilon_0 c^2 = \mu_0$ on the right-hand side, the relation

$$2N_L k_B T_e = \frac{\mu_0}{4\pi}I^2. \tag{4.75}$$

In a real situation, the positive ions are not stationary and the ion distribution also has a transverse temperature, T_i, which we can assume to be nonrelativistic. If this ion temperature is included, one obtains the more general expression

$$2N_L k_B(T_e + T_i) = \frac{\mu_0}{4\pi}I^2, \tag{4.76}$$

which is known as the *Bennett pinch* relation [7].

The physical meaning of the pinch relation becomes clear when we recognize that $N_L k_B(T_e + T_i)$ represents the mean transverse kinetic energy per unit length of the beam, while the right-hand side relates to the field energy. From (4.68), we find that the magnetic field energy stored inside the electron beam (with $f_e = 1$, $f_m = 0$, $b = a$) per unit length is given by $W_m/L = (1/4)(\mu_0/4\pi)I^2$. Consequently, the Bennett pinch condition may be stated in the form

$$N_L k_B(T_e + T_i) = 2\frac{W_m}{L}; \tag{4.77}$$

that is, the mean transverse kinetic energy per unit length is equal to two times the magnetic field energy stored per unit length inside the beam region ($r \leq a$). In practice, it is difficult to produce such an ideal equilibrium state for a significant length of time.

4.3 AXISYMMETRIC BEAMS WITH APPLIED AND SELF FIELDS

4.3.1 The Paraxial Ray Equation with Self Fields

The results obtained for the self-field effects on a beam propagating inside a drift tube can be applied to the paraxial ray equation. To the extent that our uniform beam model is valid ($\nu/\gamma \ll 1$, etc.), Equation (4.25) is linear in radius r, and since the paraxial ray equation is linear, we can simply add the two force terms representing the applied fields and the self fields. Thus, we can amend Equation (3.49) by (4.25) and obtain the modified paraxial ray equation

$$
r'' + \frac{\gamma' r'}{\beta^2 \gamma} + \frac{\gamma''}{2\beta^2 \gamma} r + \left(\frac{qB}{2mc\beta\gamma} \right)^2 r - \left(\frac{p_\theta}{mc\beta\gamma} \right)^2 \frac{1}{r^3} - K \frac{r}{r_m^2} = 0,
$$

(4.78)

where K is the generalized perveance and r_m is the beam radius.

It should be pointed out in this context that in axisymmetric beams propagating through coaxial boundaries there are no electric or magnetic image fields. From Gauss's and Ampère's laws the self fields are entirely determined by the charge and current inside the radius r. This is no longer true, however, when the beam is displaced from the axis, as discussed in Section 4.4.4. Note also that we assumed the mean azimuthal beam rotation in the **B** field to be small enough that the axial self-magnetic field is negligible; otherwise, K would have to include a corresponding term. The above paraxial approximation thus implies that $\beta_r \ll \beta_z$ and $\beta_\theta \ll \beta_z$. We will see in Chapter 5 how the beam can be treated self-consistently when these paraxial restrictions are relaxed.

To solve the modified paraxial ray equation, we need to know the beam envelope r_m in the space-charge term as a function of axial distance z. From our previous studies we know that the envelope equation can be obtained from the trajectory equation by making the substitution $r = r_m$ and adding the emittance term. When electrostatic lenses are present in which a change of particle energy occurs (i.e., $\gamma' \neq 0$), the normalized emittance $\epsilon_n = \beta\gamma\epsilon$ must be used since ϵ is no longer constant. The envelope equation then takes the form

$$
r_m'' + \frac{\gamma' r_m'}{\beta^2 \gamma} + \frac{\gamma'' r_m}{2\beta^2 \gamma} + \left(\frac{qB}{2mc\beta\gamma} \right)^2 r_m -
$$

$$
\left(\frac{p_\theta}{mc\beta\gamma} \right)^2 \frac{1}{r_m^3} - \frac{\epsilon_n^2}{\beta^2\gamma^2 r_m^3} - \frac{K}{r_m} = 0. \quad (4.79)
$$

For laminar flow, the normalized emittance ϵ_n would be zero in our linear-beam model.

The paraxial approximation demands that the generalized perveance is substantially less than unity (i.e., $|K| \ll 1$) It is interesting to note that the emittance term in (4.79) has the same r_m^{-3} dependence as the angular momentum term. Both represent repulsive forces tending to diverge the beam. In fact, if we equate the two terms, we see that

$$\frac{p_\theta}{mc} = \epsilon_n = \beta\gamma\epsilon. \tag{4.80}$$

Thus, a nonzero canonical angular momentum, which gives rise to a rotation of the particle trajectories and hence a centrifugal force, has the same effect as the normalized emittance, $\beta\gamma\epsilon$. As an example, the magnetic field produced by the heating current for a thermionic cathode or the earth magnetic field may generate canonical angular momentum that in effect increases the normalized emittance of the electron beam.

The solution of the envelope equation is relatively simple when the applied fields acting on the beam can be represented by the thin-lens approximation. In this case, one can use the thin-lens matrix neglecting self-field forces when the beam passes through a lens, and Equation (4.51) for the beam envelope in the free space between lenses. If the self fields are defocusing ($K > 0$), for instance, the position of the image plane (upright ellipse) downstream from the lens is shifted farther away. One can balance this effect by increasing the focusing strength of the lens, thereby assuring that the image occurs in the same plane as in the absence of self-field forces. When the self fields are negligible, the envelope of the beam was found to have a hyperbolic shape as given by Equation (3.162), and the individual trajectories of the particles are straight lines. For repulsive self forces ($K > 0$), the beam envelope is obtained from Equation (4.39), which near the waist can be approximated by the quadratic function (4.45). The individual trajectories can be calculated from Equation (4.25) putting $a = r_m$, $f_e = 0$, that is,

$$r'' - K\frac{r}{r_m^2} = 0. \tag{4.81}$$

In general, $r_m = r_m(z)$ and the solution of (4.81) is complicated. However, when r_m varies rather slowly so that we can assume it to be piecewise constant, the solution of (4.81) is of the form

$$r = A\cosh\frac{\sqrt{K}z}{r_m} + B\sinh\frac{\sqrt{K}z}{r_m} \quad \text{for } K > 0. \tag{4.82}$$

Similarly, one finds that in the case of attractive self fields ($K < 0$, $f_e > 1/\gamma^2$) the solutions for the individual trajectories are oscillatory (assuming again that r_m may be considered as piecewise constant), that is,

$$r = A\cos\frac{\sqrt{|K|}z}{r_m} + B\sin\frac{\sqrt{|K|}z}{r_m} \quad \text{for } K < 0. \tag{4.83}$$

The constants A and B are determined by the initial position and slope of the particle. Obviously, if K is very small, the self-field effect represents only a small correction to the straight-line trajectory solution ($r'' = 0$, $r = r_0 + r_0'z$).

4.3.2 Beam Transport in a Uniform Focusing Channel

Let us now consider the case of a beam propagating through a long, uniform focusing channel. We will assume that there is no applied accelerating electric field ($\dot{\gamma} = 0$) and that the canonical angular momentum p_θ is zero. The restriction $p_\theta = 0$ implies that the particles are launched from a magnetically shielded source (i.e., $B = 0$ at source) with $\dot{\theta}_0 = 0$. For our further analysis we define the beam envelope by R. Setting $r_m = R$, the paraxial ray equation (4.78) may then be written as

$$r'' + k_0^2 r - \frac{K}{R^2} r = 0 \tag{4.84a}$$

In the literature on microwave sources, time t (rather than distance z) is preferred as the independent variable. The ray equation then takes the alternative form

$$\ddot{r} + \omega_0^2 r - \frac{\omega_p^2}{2} r = 0, \tag{4.84b}$$

as can readily be verified.

The terms $k_0^2 r$ and $\omega_0^2 r$ represent the linear external focusing force. The parameters $k_0 = 2\pi/\lambda_0$ and $\omega_0 = k_0 v$ ($v =$ beam velocity) define the wavelength λ_0 and oscillation frequency of the transverse particle oscillations due to the applied focusing force alone (i.e., when $K = 0$, or $\omega_p = 0$). The plasma frequency ω_p is defined by

$$\omega_p^2 = \frac{q^2 n}{\epsilon_0 \gamma^3 m} = \frac{2K v^2}{R^2}$$

according to Equations (4.2) and (4.24).

For the corresponding beam envelope equation one obtains from (4.79)

$$R'' + k_0^2 R - \frac{K}{R} - \frac{\epsilon^2}{R^3} = 0, \tag{4.85a}$$

which in the time domain becomes

$$\ddot{R} + \omega_0^2 R - \frac{\omega_p^2}{2} R - \frac{\epsilon^2 v^2}{R^3} = 0. \tag{4.85b}$$

The best known example of a uniform focusing channel is a long solenoid, for which case the oscillation frequency ω_0 is identical with the Larmor frequency ω_L, that is,

$$\omega_0 = \omega_L = \frac{|qB|}{2\gamma m}, \tag{4.86a}$$

and the wave number is given by

$$k_0 = \frac{\omega_L}{v} = \frac{|qB|}{2mc\beta\gamma}. \tag{4.86b}$$

Both ω_0 and k_0 are constants since the magnetic field B is uniform (i.e., independent of distance along the channel). Furthermore, we recall that the above equations describe the particle and envelope motion in the rotating Larmor frame.

Another example of a uniform transport channel that is often used for mathematical convenience is the case where the focusing force is provided by a transparent stationary cylinder of opposite charge with uniform density ρ_e. This is basically identical to the charge-neutralization effects discussed previously, for instance in connection with the *Bennett pinch* (Section 4.2.4). However, in our present context we treat the effect of the cylindrical channel of opposite charge like an external focusing force. We will see in Section 4.4 that in the *smooth approximation*, where only the average forces are considered, a periodic-focusing channel behaves like a cylinder of opposite charge. This equivalence is particularly apparent in the case of a periodic electrostatic quadrupole channel, where the transverse focusing forces are electrical in nature. However, the analogy also applies to magnetic quadrupole channels, or axisymmetric channels consisting of periodic arrays of short solenoids or electrostatic einzel lenses. Mathematically, the treatment of the *average* behavior of the particle motion or beam envelope in such periodic systems is identical with that in a uniform cylinder of opposite charge. The radial electric field due to a uniform charge distribution of density ρ_e is [from Equation (4.11c)] $E_r = \rho_e r/2\epsilon_0$, and one can readily show that the corresponding frequency ω_0 and wave number k_0 for the particle motion in such a field are given by

$$\omega_0 = \left[\frac{q\rho_e}{2\epsilon_0\gamma m}\right]^{1/2} \tag{4.87a}$$

and

$$k_0 = \frac{\omega_0}{v} = \left[\frac{q\rho_e}{2\epsilon_0\gamma m}\right]^{1/2}\frac{1}{v}. \tag{4.87b}$$

Let us now return to the envelope equation in the form (4.85a). The solution for the beam envelope will depend on the initial conditions, that is, on the beam radius $R(0)$ and the slope $R'(0)$ at the entrance ($z = 0$) of the uniform channel. In view of the fact that the force is constant (i.e., k_0 is independent of distance z), there will be a special solution where $R(z) = a = $ const, $R'(z) = 0$, and $R''(z) = 0$, and hence the beam envelope is a straight line. This special case is known as the *matched beam*, and from (4.85a) it is defined by the algebraic equation

$$k_0^2 a - \frac{K}{a} - \frac{\epsilon^2}{a^3} = 0. \tag{4.88a}$$

This can be written in the alternative forms

$$k^2 a - \frac{\epsilon^2}{a^3} = 0,$$ (4.88b)

or

$$ka^2 = \epsilon$$ (4.88c)

by introducing the wave number k defined as

$$k^2 = \left(\frac{2\pi}{\lambda}\right)^2 = k_0^2 - \frac{K}{a^2}.$$ (4.89a)

The last relation can be expressed in terms of the plasma wave constant $k_p = \omega_p/v$ as

$$k^2 = k_0^2 - \frac{1}{2}k_p^2,$$ (4.89b)

or in terms of the frequencies as

$$\omega^2 = \omega_0^2 - \frac{\omega_p^2}{2}.$$ (4.89c)

The parameters k and ω define the wavelength $\lambda = 2\pi/k$ and oscillation frequency of the particle oscillation due to the action of both the applied focusing force and the space-charge force. Since the space-charge force of the beam is defocusing, we have $k < k_0$, $\lambda > \lambda_0$, $\omega < \omega_0$, as can be seen from Equations (4.89). The ratio k/k_0, or alternatively, ω/ω_0, is known as the *tune depression* due to space charge.

The algebraic matched-beam envelope equation (4.88a) relates the four quantities a, k_0, K, and ϵ and can be solved for any quantity if the other three are given. We will first solve it for the beam radius a by assuming that k_0, K, and ϵ are known. First, we will consider the two extreme cases where either the emittance ϵ or the space-charge force (represented by the generalized perveance K) is zero. In the limit of zero emittance ($\epsilon = 0$), the flow is laminar and the beam radius is given by

$$a_B = \frac{K^{1/2}}{k_0}.$$ (4.90)

This type of flow, first identified by Brillouin [8] in 1945, is known as *Brillouin flow*.

When the space charge is negligible $(K = 0)$, on the other hand, the beam radius is

$$a_0 = \left(\frac{\epsilon}{k_0} \right)^{1/2}. \tag{4.91}$$

By comparison with (3.346) we see that the amplitude function w for the uniform transport channel in the case of zero space charge is $w_0 = k_0^{-1/2}$ and is independent of z. If we introduce the dimensionless parameter

$$u = \frac{K}{2k_0\epsilon}, \tag{4.92}$$

we can write the general solution of Equation (4.88a) for the beam radius in the form

$$a = a_B \left(\frac{1}{2} + \frac{1}{2} \sqrt{1 + u^{-2}} \right)^{1/2}, \tag{4.93a}$$

or

$$a = a_0 \left(u + \sqrt{1 + u^2} \right)^{1/2}. \tag{4.93b}$$

[See also Equation (5.293), which represents a useful approximation for practical design and scaling.]

From Equation (4.91) we see that without space charge, a beam with emittance ϵ has a radius a_0. As the current, and hence the space-charge parameter u increases, the beam radius expands to the value given in (4.93b), and the diameter of the beam pipe has to be increased accordingly. Conversely, we can say that a pipe with radius $a > a_0$ could accommodate a beam with zero space charge but larger emittance. From Equation (4.91), setting $\epsilon = \alpha$, $a_0 = a$, we can define a trace-space *acceptance* α of the pipe for zero space charge given by

$$\alpha = a^2 k_0 = a^2 \frac{\omega_0}{v}. \tag{4.94}$$

In many cases the beam radius a, or acceptance α, is given. For instance, the diameter of the vacuum pipe may be fixed, or the beam size may not exceed the *linear aperture* of the focusing system to avoid nonlinear effects. One can then calculate the maximum perveance or beam current from (4.88a), using (4.94), as

$$K = k_0^2 a^2 - \frac{\epsilon^2}{a^2} = k_0 \alpha \left[1 - \left(\frac{\epsilon}{\alpha} \right)^2 \right] \tag{4.95}$$

or

$$I = \frac{I_0}{2} \beta^3 \gamma^3 k_0 \alpha \left[1 - \left(\frac{\epsilon}{\alpha} \right)^2 \right], \tag{4.96}$$

where I_0 is the characteristic current defined in (4.17). We see that the current that can be transported through the focusing channel increases rapidly with the particle energy; furthermore, the acceptance α has to be larger than the emittance ϵ of the beam as indicated by the factor $1 - (\epsilon/\alpha)^2$. The transportable current reaches a maximum when the emittance ϵ becomes negligibly small compared to the acceptance α [i.e., when $\epsilon/\alpha \to 0$ (laminar beam limit)]. For such a laminar beam ($\epsilon = 0$), one gets the condition

$$K = k_0^2 a^2, \tag{4.97a}$$

or, in terms of frequencies,

$$\omega_0^2 = \frac{\omega_p^2}{2}. \tag{4.97b}$$

The second expression agrees with the well-known nonrelativistic relation $\omega_L = \omega_p/\sqrt{2}$ for ideal *Brillouin flow* in a long solenoid.

Let us now consider the motion of individual particles within the matched beam. We can write the particle trajectory equations in the alternative forms (space and time domains)

$$r'' + k^2 r = 0, \tag{4.98a}$$

or

$$\ddot{r} + \omega^2 r = 0, \tag{4.98b}$$

where the parameters k and ω are defined in Equations (4.89a) and (4.89c), respectively. Note that $k_0 = 2\pi/\lambda_0 = \omega_0/v$ and that λ and λ_0 are the wavelengths of the particle oscillations with and without self fields. Focusing requires that $\omega_0^2 \geq \omega_p^2/2$, so that ω is real. For ideal Brillouin flow, we have $\omega = 0$ in the Larmor frame and $|\dot{\theta}| = \omega_0 = \omega_L$ in the laboratory frame. The *tune depression* k/k_0 or ω/ω_0 can be related to the emittance ϵ, the acceptance α, and the parameter u by

$$\frac{k}{k_0} = \frac{\omega}{\omega_0} = \frac{\epsilon}{\alpha} = \sqrt{1 + u^2} - u. \tag{4.99}$$

For negligible space charge (i.e., $u = 0$), the particle oscillation frequency is equal to ω_0, and for the long solenoid $\omega_0 = \omega_L$, which is in accordance with the paraxial theory in Section 3.4.4. In the laminar-beam limit ($\epsilon = 0$), on the other hand, we have $\omega = \omega_p/\sqrt{2}$, as stated earlier.

Let us return now to the matched-beam envelope equation (4.88a) and compare the second and third terms representing the space charge and emittance. Clearly, when $Ka^2 > \epsilon^2$, we can say that the beam is *space-charge dominated*, while $Ka^2 < \epsilon^2$ implies an *emittance-dominated* regime. The transition between the two regimes occurs when

$$Ka^2 = \epsilon^2. \tag{4.100}$$

Using the three relations in (4.88) and the definition (4.89a), we can express Equation (4.100) in terms of the tune depression as

$$\frac{k}{k_0} = \frac{\omega}{\omega_0} = \sqrt{0.5} \approx 0.71. \tag{4.101}$$

Thus, when $k/k_0 < \sqrt{0.5}$, the beam is dominated by space charge, and when $k/k_0 > \sqrt{0.5}$, emittance dominates. For now, this distinction between the two regimes merely indicates which of the two terms in the envelope equation is more significant in determining the beam radius. We will see in the following discussion of a mismatched beam that there is also a difference in the internal dynamics of the particle motion.

To obtain the matched beam solution ($R = a = $ const) treated above, the beam must be properly matched into the focusing channel. In the solenoid case, where the source is in a region of zero magnetic field, the starting conditions must be chosen such that $R = a$ and $R' = 0$ when the beam reaches the uniform-focusing plateau inside the channel after passing through the fringe-field region. There are several possibilities for satisfying these matching requirements, each of which involves at least two parameters to control both the radius and slope of the beam envelope. In practice, the beam emerges from the source with an initial radius and slope that depend on the source design and operating conditions. By judicious choice of the source location with respect to the channel entrance, one can achieve the desired matched beam inside the channel provided that the generalized perveance K is fixed. A better solution is to place a small matching lens (e.g., a short solenoid) between the source and the focusing channel. By varying the focal length of this lens and its location between the source and the channel, one can achieve the desired matching even if K is not constant. A third possibility is to use two lenses at fixed positions and to vary the focusing strength of the two lenses to control the beam radius and slope for proper matching.

When the beam is not matched, the envelope radius becomes a periodically varying function of distance z. There are basically three major possibilities of beam mismatch:

1. The initial envelope radius and slope are not matched [i.e., $R(0) \neq a$, $R'(0) \neq 0$].

2. The beam is not axisymmetric (e.g., it has an elliptic cross section).
3. The density is not uniform.

The second case requires the use of two transverse coordinates $X(z)$ and $Y(z)$ for the envelope; this is treated in Section 4.4.3 and leads to two fundamental eigenmodes of the envelope oscillations. The third case lies outside the framework of the uniform-beam model and will be treated in Section 6.2.

In the first case, which we will now analyze, the beam remains axisymmetric. For small-amplitude oscillations, we can linearize Equation (4.85a) and find an approximate solution. Let

$$R = a + x, \tag{4.102}$$

where $|x| \ll a$ and a is the matched-beam radius. Then one obtains from (4.85a), using (4.88a) to cancel zero-order terms, the equation

$$x'' + \left(k_0^2 + \frac{K}{a^2} + 3\frac{\epsilon^2}{a^4} \right)x = 0. \tag{4.103}$$

This can be expressed in the equivalent forms

$$x'' + k_e^2 x = 0, \tag{4.104a}$$

$$\ddot{x} + \omega_e^2 x = 0. \tag{4.104b}$$

The parameter $k_e = 2\pi/\lambda_e$ is the wave number, and ω_e is the radian frequency of the envelope oscillations. By elimination of ϵ in Equation (4.103) with the aid of (4.88a) and by introducing the wave number k and frequency ω of the single-particle oscillations with space charge, as defined in (4.89a) and (4.89c), one obtains

$$k_e = [2k_0^2 + 2k^2]^{1/2} = \sqrt{2}\, k_0 \left[1 + \left(\frac{k}{k_0} \right)^2 \right]^{1/2}, \tag{4.105a}$$

$$\omega_e = [2\omega_0^2 + 2\omega^2]^{1/2} = \sqrt{2}\, \omega_0 \left[1 + \left(\frac{\omega}{\omega_0} \right)^2 \right]^{1/2}. \tag{4.105b}$$

The relationship between the wavelength λ_e associated with the beam envelope oscillation and the oscillation frequency ω_e is given by

$$\lambda_e = \frac{2\pi}{k_e} = \frac{2\pi v}{\omega_e}. \tag{4.106}$$

We note that Equations (4.105a) and (4.105b) represent the solutions of the envelope oscillations for the axisymmetric case, which is called the *in-phase mode*.

In Section 4.4.3 we also obtain the solutions for the quadrupole (ellipsoidal) case known as the *out-of-phase mode*. The relations (4.105a) and (4.105b) can be expressed in terms of the plasma wave number k_p and plasma frequency ω_p [using (4.89)] as

$$k_e = \left[4k_0^2 + k_p^2 \right]^{1/2} = 2k_0 \left[1 + \frac{1}{4} \left(\frac{k_p}{k_0} \right)^2 \right]^{1/2}, \tag{4.107a}$$

$$\omega_e = \left[4\omega_0^2 + \omega_p^2 \right]^{1/2} = 2\omega_0 \left[1 + \frac{1}{4} \left(\frac{\omega_p}{\omega_0} \right)^2 \right]^{1/2}, \tag{4.107b}$$

where

$$k_p = \frac{2\pi}{\lambda_p} = \frac{\omega_p}{v} = \left[2k_0^2 - 2k^2 \right]^{1/2}. \tag{4.108}$$

As we see from this analysis, the frequency ω_e associated with the ripple of the beam envelope differs from the frequency ω of the particle oscillations within the beam. In the limit of zero intensity ($\omega_p = 0$ or $K = 0$), we have $\omega = \omega_0$ and $\omega_e = 2\omega_0$. If the channel is a long solenoid, this implies that individual particles oscillate with the Larmor frequency, $\omega_0 = \omega_L$, while the envelope of the mismatched beam oscillates with the cyclotron frequency, $\omega_e = 2\omega_L = \omega_c$. For ideal *Brillouin flow* ($\epsilon = 0$), on the other hand, we have $\omega = 0$ and $\omega_e = \sqrt{2}\,\omega_0 = \omega_p$ (i.e., the envelope oscillates with a frequency given by the plasma frequency).

When the above linear approximation ($|x| \ll a$) is valid (i.e., for a small mismatch), the envelope oscillations are sinusoidal. For a large mismatch when $|x|$ is no longer small compared to the matched-beam radius a, numerical solution of the envelope equation (4.85a) is required. Such a solution [9] is shown in Figure 4.2(a,b), where the beam was injected into the long channel with an initial mismatch radius of $R_0 = 0.5a$ and a slope of $R_0' = 0$. The beam parameters in this case were chosen such that the single-particle tune depression was $k/k_0 = 0.8$. It is interesting to note that the wavelength of the envelope oscillation predicted by linear theory [$\lambda_e = 0.55\lambda_0$ from Equation (4.105a)] is in relatively good agreement with the numerical result [$\lambda_e \simeq 0.53\lambda_0$ from the plot in Figure 4.2(a,b)]. On Figure 4.2(b) we plotted the trace-space ellipse at 24 positions during one envelope oscillation. The dashed ellipse represents the matched beam. To obtain the equations for the ellipse, one must choose values for the perveance K, emittance ϵ, and wave constant k_0 that are consistent with the tune depression of $k/k_0 = 0.8$ using Equations (4.88) and (4.89a). The Courant–Snyder parameters $\hat{\alpha}, \hat{\beta}, \hat{\gamma}$, as defined in Section 3.8.2, are readily found from the envelope radius R and slope R' using Figure 3.26. Thus

$$\hat{\beta} = \frac{R^2}{\epsilon} \tag{4.109}$$

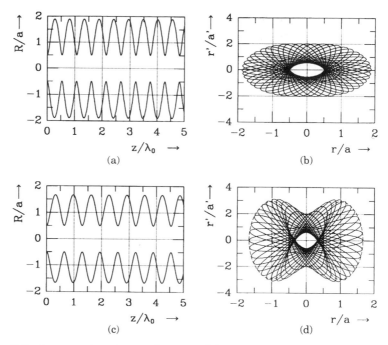

Figure 4.2. Beam envelope (a,c) and motion of the trace-space ellipse during one envelope oscillation period (b,d) for a mismatched beam with $a_0/a = 0.5$ and tune depression of $k/k_0 = 0.8$ (a,b) and $k/k_0 = 0.3$ (c,d). The matched-beam ellipses are indicated by a dashed line in each case. (From Reference 9; © 1991 IEEE.)

and since $R' = -\hat{\alpha}x_{\text{int}} = -\hat{\alpha}\sqrt{\epsilon/\hat{\beta}}$,

$$\hat{\alpha} = -R'\sqrt{\frac{\hat{\beta}}{\epsilon}} = -\frac{R'R}{\epsilon}. \qquad (4.110)$$

As we can see from the plot of Figure 4.2(b), the tips of the trace-space ellipse describe an ellipse that is concentric with the matched-beam ellipse, as expected from single-particle theory without space charge. This simple picture changes, however, when one studies the case of a beam with large tune depression, as we will see next.

In Figure 4.2(c, d) we have shown the envelope and trace-space ellipse for a tune depression of $k/k_0 = 0.3$, keeping the beam mismatch the same (i.e., $a_0/a = 0.5$). The value for the envelope oscillation wavelength predicted by linear theory ($\lambda_e \simeq 0.68\lambda$) in this case is also relatively close to the numerical result [$\lambda_e \simeq 0.65\lambda$ from Figure 4.2(c, d)]. However, the trace-space ellipse reveals an oscillatory pattern markedly different from that of Figure 4.2(b). We attribute this difference to the fact that the first case ($k/k_0 = 0.8$) is in the emittance-dominated regime while the second case ($k/k_0 = 0.3$) is in the space-charge-dominated regime. In the first case, the single-particle trajectories in the beam are simple betatron oscillations crossing the axis every half betatron period. The second case shows a beam in which the single-particle motion is dominated by the plasma oscillation. We can understand this behavior by looking at the extreme case of a laminar beam where emittance is zero ($\epsilon = 0$) and where single-particle trajectories do not cross the axis of the system. The trajectories in such a laminar beam are self similar and show the same behavior as the beam envelope [i.e., they oscillate between a maximum and minimum (at the waist) without crossing the axis]. As mentioned earlier, the transition between these two regimes occurs at $k/k_0 = \sqrt{0.5}$ [from Equation (4.101)].

It is interesting to note that a typical particle trajectory in the mismatched beam is no longer sinusoidal. This is due to the fact that the beam envelope R and hence the net force seen by the particle vary periodically with period λ_e. The trajectory is quasi-periodic with a wavelength $\lambda = k/2\pi$ that is approximately the same as in the matched-beam case in our example. Thus the mismatched beam represents a special case of periodic focusing in which the external force is uniform and the space-charge force varies periodically with distance. The more general case where both applied forces as well as the self forces are periodic is treated in the next section. Finally, we want to point out that if we apply the preceding analysis of beam transport to a long solenoid, we must keep in mind that the radial oscillations of the particles are in the Larmor frame (i.e., in the meridional plane which rotates with the Larmor frequency ω_L). The actual three-dimensional trajectories of the particles in the solenoid system have a helical shape which is obtained by a superposition of the radial oscillations, the Larmor rotations, and the axial velocity When the space charge is zero, for instance, the projections of the trajectories in the $x-y$ ($r-\theta$) plane are off centered circles. In this case, a particle that was launched with $v_r = 0$, $v_\theta = 0$ will describe a helix that will touch the z-axis without crossing it, as discussed in connection with Busch's theorem (Section 2.3.4). On the other hand, when the flow is laminar [i.e., entirely dominated by the space-charge fields (ideal Brillouin case)], the trajectory projections in the $x-y$ plane are centered circles. The particles rotate around the axis with the Larmor frequency in this case, whereas the trajectories in the Larmor frame are straight lines (since $\omega = 0$). In between these two extremes, the trajectory pattern is more complicated and depends on the ratio of the plasma frequency to the Larmor frequency, or, conversely, on ω/ω_L.

4.4 PERIODIC FOCUSING OF INTENSE BEAMS (SMOOTH-APPROXIMATION THEORY)

4.4.1 Beam Transport in a Periodic Solenoid Channel

In many practical applications the beams are focused by a periodic array, or *lattice*, of lenses rather than a uniform field. If the space-charge forces are linear, as assumed here, the theory of periodic focusing discussed in Section 3.8 can be amended to include the self fields in a straightforward way [10]. We begin our analysis with an axisymmetric channel consisting of periodically spaced short solenoid lenses. (A similar analysis can be applied to periodic arrays of electrostatic einzel lenses.) If $R(z)$ denotes the envelope of the beam and $\kappa_0(z) = [qB(z)/2mc\beta\gamma]^2$ denotes the periodic-focusing function of the lens system, the paraxial trajectory equation (4.84) can be written in the form

$$r'' + \kappa_0(z)r - \frac{K}{R^2(z)} r = 0. \tag{4.111}$$

To solve this equation, one must first find the beam radius $R(z)$ from the envelope equation

$$R'' + \kappa_0(z)R - \frac{K}{R} - \frac{\epsilon^2}{R^3} = 0. \tag{4.112}$$

When $R(z)$ is known, one can write (4.111) in the alternative form

$$r'' + \kappa(z)r = 0, \tag{4.113}$$

where

$$\kappa(z) = \kappa_0(z) - \frac{K}{R^2(z)}. \tag{4.114}$$

If S denotes the length of one focusing period, we have the periodicity condition

$$\kappa_0(z + S) = \kappa_0(z). \tag{4.115}$$

For the case where the beam is matched, $R(z)$ and $\kappa(z)$ are also periodic with period S, that is,

$$R(z + S) = R(z), \tag{4.116}$$

$$\kappa(z + S) = \kappa(z). \tag{4.117}$$

According to the theory discussed in Section 3.8.2, the solutions of Equation (4.113) can be written in the phase-amplitude form

$$r(z) = Aw(z)\cos[\psi(z) + \phi], \tag{4.118}$$

where A and ϕ are determined by the initial conditions and $w(z)$, $\psi(z)$ obey the relation

$$\frac{d\psi}{dz} = \psi' = \frac{1}{w^2} = \frac{1}{\hat{\beta}}. \tag{4.119}$$

Equations (4.118) and (4.119) are valid whether or not $\kappa(z)$ is periodic.

For a matched beam in a periodic channel, when both $R(z)$ and $\kappa(z)$ are periodic functions with period S, the particle trajectories are pseudoharmonic oscillations with a period or wavelength

$$\lambda = \frac{2\pi S}{\sigma}. \tag{4.120}$$

The parameter σ represents the *phase advance* of the particle oscillation per period *with space charge* and is given by the change of the phase function ψ in one channel period; that is, according to Equations (3.350), (3.351), and (4.119),

$$\sigma = \psi(z + S) - \psi(z) = \int_z^{z+S} \frac{dz}{w^2(z)} = \int_z^{z+S} \frac{dz}{\hat{\beta}(z)}. \tag{4.121}$$

When the space-charge term is absent ($K = 0$) [i.e., $\kappa(z) = \kappa_0(z)$], we will denote the phase and amplitude functions by $\psi_0(z)$ and $w_0(z)$, respectively. The *phase advance* of the particle oscillations *without space charge* is then defined as

$$\sigma_0 = \psi_0(z + s) - \psi_0(z) = \int_z^{z+S} \frac{dz}{w_0^2(z)} = \int_z^{z+S} \frac{dz}{\hat{\beta}_0(z)}, \tag{4.122}$$

and the wavelength of the particle oscillations is

$$\lambda_0 = \frac{2\pi S}{\sigma_0}. \tag{4.123}$$

As we will see below, the phase advances with and without space charge, σ and σ_0, are the key parameters that determine the beam physics. They take the place of the frequencies ω and $\omega_0 = \omega_L$ of the uniform solenoidal focusing system studied in Section 4.3.2.

In accordance with Equation (3.346), the beam envelope $R(z)$ is defined by the product of the amplitude function $w(z)$ and the square root of the emittance, that is,

$$R(z) = \sqrt{\epsilon}\, w(z) = \sqrt{\epsilon \hat{\beta}(z)}, \tag{4.124}$$

and likewise for zero space charge ($K = 0$)

$$R_0(z) = \sqrt{\epsilon}\, w_0(z) = \sqrt{\epsilon \hat{\beta}_0(z)}, \tag{4.125}$$

where $\hat{\beta} = w^2$, $\hat{\beta}_0 = w_0^2$, as defined in (3.343a). Note that $w_0(z)$ depends only on the focusing function $\kappa_0(z)$ and is found by solving the envelope equation (3.340), that is,

$$w_0'' + \kappa_0 w_0 - \frac{1}{w_0^3} = 0. \tag{4.126}$$

Thus, $w_0(z)$, or alternatively, $\hat{\beta}_0(z)$, describes the properties of the periodic-focusing lattice and is independent of the beam emittance ϵ and the generalized perveance K. [Note that this is not true for the amplitude functions $w(z)$, $\hat{\beta}(z)$ with space charge.]

In general, the solutions for the beam envelope and particle trajectories in a periodic-focusing channel with space charge must be obtained by numerical integration of the equations (4.111) and (4.112). Figure 4.3 shows such a numerical solution for a matched beam in the periodic solenoidal channel used in the University of Maryland electron-beam transport experiment. Each solenoid produces a field of the form

$$B(z) = B_0 \frac{\exp(-z^2/d^2)}{1 + z^2/b^2}, \qquad (4.127)$$

where $d = 3.24$ cm and $b = 4.40$ cm. The length of one period is $S = 13.6$ cm, and the actual field used in the computation is obtained by superposition of the lens fields. Note that $\kappa_0(z) \sim B^2(z)$ according to Equation (4.86b). For purposes of illustrating the nature of the periodic envelope and trajectory solutions, the peak value B_0 [hence the maximum of $\kappa_0(z)$], the beam emittance ϵ and the generalized perveance K were chosen to yield $\sigma = 36°$ and $\sigma_0 = 72°$. As Figure 4.3 indicates, the matched envelope of the beam is a periodic function $R(z)$ that has the same periodicity S as the focusing system. The particle trajectory shown in the figure is a pseudoharmonic function in which the period is determined by σ, or σ_0, and where the ripple in the amplitude reflects the periodicity S of the focusing system. When the space-charge term is set to zero ($K = 0$, $\sigma = \sigma_0$), the particle trajectory performs one oscillation in a distance $\lambda_0 = 2\pi S/\sigma_0$ that corresponds to five lens periods ($2\pi/\sigma_0 = 360°/72° = 5$). On the other hand, when space charge is included ($\sigma = 36°$), the particle oscillation wavelength increases to 10 periods since $2\pi/\sigma = 360°/36° = 10$.

When the variation of the beam radius during one focusing period is small compared to the mean radius in the period, one can use the *smooth-approximation theory* to solve the envelope and trajectory equations for the average values of the quantities involved. In effect, this implies replacing the periodic force $\kappa_0(z)$ by the constant average force $\overline{\kappa}_0$ and hence reducing the problem to the uniform focusing system treated in the preceding section. As we will see below, the smooth-approximation results (with suitable corrections to account for the envelope modulation) are fairly accurate for beams of practical interest.

The general derivation of the smooth-approximation theory for intense beams in periodic-focusing channels can be found in Reference 10. In the following analysis we consider a special case of a matched beam in an axisymmetric channel. (Quadrupole focusing is treated in Section 4.4.2 and envelope oscillations of mismatched beams in Section 4.4.3.)

The envelope $R(z)$ of the matched beam in a periodic channel can be written in terms of the mean radius $\overline{R}$, which is constant, and a modulation function $\delta(z)$ as

$$R(z) = \overline{R}[1 + \delta(z)]. \qquad (4.128)$$

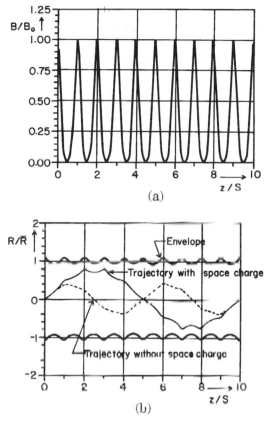

Figure 4.3. (a) Axial magnetic field variation $B(z)$ in a periodic focusing channel with solenoid lenses; (b) matched beam envelope and single particle trajectories without space charge ($\sigma_0 = 72°$) and with space charge ($\sigma = 36°$) in this channel.

Correspondingly, one can represent the amplitude function $w(z)$ by

$$w(z) = \overline{w}[1 + \delta(z)]. \tag{4.129}$$

The ripple function $\delta(z)$ has the period S; thus

$$\delta(z + S) = \delta(z), \tag{4.130}$$

and, by definition, the average of $\delta(z)$ over one period is zero, that is,

$$\int_z^{z+S} \delta(z)\, dz = 0. \tag{4.131}$$

By substituting (4.128) into the envelope equation (4.112), Taylor expanding, and keeping only the linear terms in δ, we obtain

$$\overline{R}\delta'' + \kappa_0(1 + \delta)\overline{R} - \frac{K(1 - \delta)}{\overline{R}} - \frac{\epsilon^2}{\overline{R}^3}(1 - 3\delta) = 0. \tag{4.132}$$

If we average over one period, using the fact that, from (4.131), $\overline{\delta} = 0$ and therefore also $\overline{\delta''} = 0$, we find that

$$\overline{\kappa_0}\overline{R} + \frac{\overline{R}}{S}\int_z^{z+S} \kappa_0(z)\delta(z)\,dz - \frac{K}{\overline{R}} - \frac{\epsilon^2}{\overline{R}^3} = 0. \tag{4.133}$$

The equivalent equation for the case where space charge is negligible (i.e., $K = 0$) is

$$\overline{\kappa_0}\overline{R}_0 + \frac{\overline{R}_0}{S}\int_z^{z+S} \kappa_0(z)\delta_0(z)\,dz - \frac{\epsilon^2}{\overline{R}_0^3} = 0. \tag{4.134}$$

$\overline{R}_0$ and $\delta_0(z)$ denote the average beam radius and ripple function for zero space charge, as defined by the relation

$$R_0(z) = \overline{R}_0[1 + \delta_0(z)], \tag{4.135}$$

analogous to (4.128).

Numerical studies indicate that the modulation function $\delta(z)$, defined in (4.128), has only a very weak dependence on the perveance K, as long as $|\delta(z)| \ll 1$, which is the case for $\sigma_0 \lesssim 90°$. Thus we have to good approximation

$$\delta(z) = \delta_0(z); \tag{4.136}$$

that is, we can replace $\delta(z)$ in the integral term of Equation (4.133) by $\delta_0(z)$.

From (4.124) and (4.125), the emittance can be related to the average amplitude functions with and without space charge as

$$\epsilon = \frac{\overline{R}^2}{\overline{\beta}} = \frac{\overline{R}^2}{\overline{w}^2} \tag{4.137a}$$

and

$$\epsilon = \frac{\overline{R}_0^2}{\hat{\beta}_0} = \frac{\overline{R}_0^2}{\overline{w}_0^2}. \tag{4.137b}$$

On the other hand, we have, from (4.121),

$$\sigma = \int_z^{z+S} \frac{dz}{\hat{\beta}(z)} = \frac{S}{\hat{\beta}} = \frac{S}{\overline{w}^2} \tag{4.138a}$$

and

$$\sigma_0 = \int_z^{z+S} \frac{dz}{\hat{\beta}_0(z)} = \frac{S}{\hat{\beta}_0} = \frac{S}{\overline{w}_0^2}. \tag{4.138b}$$

By substituting (4.137b) in (4.134) and using (4.138b), we obtain for the focusing force averaged along the envelope radius over one period of the solenoid array the result

$$\kappa_0 + \frac{1}{S} \int_z^{z+S} \kappa_0(z)\delta_0(z) \, dz = \frac{w_0^2}{S^2}. \tag{4.139}$$

Using (4.136), (4.137a), and (4.138a) in (4.133), we find for the net average force with space charge

$$\overline{\kappa}_0 + \frac{1}{S} \int_z^{z+S} \kappa_0(z)\delta_0(z) \, dz - \frac{K}{R^2} = \frac{\sigma^2}{S^2},$$

or, in view of (4.139),

$$\frac{\sigma^2}{S^2} = \frac{\sigma_0^2}{S^2} - \frac{K}{R^2}. \tag{4.140}$$

By substituting this result in (4.133) we obtain for the average beam radius $\overline{R}$ in our solenoidal channel the algebraic equation

$$\frac{\sigma_0^2}{S^2}\overline{R} - \frac{K}{R} - \frac{\epsilon^2}{R^3} = 0, \tag{4.141a}$$

which, in view of (4.140), may be written in the alternative form

$$\frac{\sigma^2}{S^2}\overline{R} - \frac{\epsilon^2}{R^3} = 0. \tag{4.141b}$$

This first major result of our smooth-approximation theory is equivalent to Equation (4.88) for the uniform focusing channel. Indeed, by comparing the two equations and considering relation (4.138b), we can make the important identification

$$k_0 = \frac{2\pi}{\lambda_0} = \frac{\sigma_0}{S} = \frac{1}{\hat{\beta}_0} \tag{4.142a}$$

for the beam without space charge and, likewise,

$$k = \frac{2\pi}{\lambda} = \frac{\sigma}{S} = \frac{1}{\hat{\beta}} \qquad (4.142b)$$

when space charge is included. Thus, we see that the average values of the amplitude functions $\hat{\beta}$ and $\hat{\beta}_0$ define the wavelengths λ and λ_0 of the particle oscillations with and without space charge.

The algebraic equation (4.141) can readily be solved in the same way as (4.88a). First, we obtain for the average radius $\overline{R}_0$ without space charge $(K = 0)$

$$\frac{\sigma_0^2}{S^2} \overline{R}_0 - \frac{\epsilon^2}{\overline{R}_0^3} = 0, \qquad (4.143)$$

which yields

$$\overline{R}_0 = \sqrt{\frac{\epsilon S}{\sigma_0}}. \qquad (4.144)$$

The same result can be obtained from Equation (4.125) by using $\overline{\hat{\beta}}_0 = S/\sigma_0$ from Equation (4.138b).

In analogy to Equation (4.92) of the uniform focusing case we define the dimensionless parameter

$$u = \frac{KS}{2\sigma_0\epsilon}, \qquad (4.145)$$

and obtain from Equation (4.141) for the average beam radius the result

$$\overline{R} = \overline{R}_0\left(u + \sqrt{1 + u^2}\right)^{1/2}. \qquad (4.146)$$

$\overline{R}_0$ is the average radius for zero space charge $(u = 0)$, as defined in Equation (4.144).

Likewise, we find for the phase advance with space charge

$$\sigma = \sigma_0\left(\sqrt{1 + u^2} - u\right). \qquad (4.147)$$

When space-charge effects are negligible, $u = 0$ and $\overline{R} = \overline{R}_0$, $\sigma = \sigma_0$. As space charge increases $(u > 0)$, the beam radius $\overline{R}$ becomes larger while the phase advance σ decreases $(\sigma \to 0$ as $u \to \infty)$.

Alternatively, we can solve for the generalized perveance K if the average radius and other parameters are given. Thus, we get from Equation (4.140)

$$K = \frac{\overline{R}^2}{S^2}(\sigma_0^2 - \sigma^2).$$ (4.148)

We can introduce the acceptance α defined as the maximum beam emittance ϵ_{max} for given radius $\overline{R}$ when space charge is zero ($\sigma = \sigma_0$). Equation (4.141) may then be written in the form

$$\frac{\sigma_0^2}{S^2}\overline{R} - \frac{\alpha^2}{\overline{R}^3} = 0,$$ (4.149)

which yields

$$\alpha = \sigma_0 \frac{\overline{R}^2}{S}.$$ (4.150)

Also, by comparison of the two expressions (4.141b) and (4.149) we find that

$$\frac{\epsilon}{\alpha} = \frac{\sigma}{\sigma_0}.$$ (4.151)

Using the last two relations we obtain for the generalized perveance (4.148) the alternative form

$$K = \frac{\sigma_0 \alpha}{S}\left[1 - \left(\frac{\epsilon}{\alpha}\right)^2\right].$$ (4.152)

The beam current that can be transported through a channel with acceptance α is then [10]

$$I = \frac{I_0}{2}\beta^3\gamma^3\frac{\sigma_0\alpha}{S}\left[1 - \left(\frac{\epsilon}{\alpha}\right)^2\right].$$ (4.153)

For transport of large currents, the emittance must be significantly less than the acceptance. The maximum current is obtained when $\epsilon/\alpha \to 0$ (laminar-flow limit), in which case one gets

$$I_{max} = \frac{I_0}{2}\beta^3\gamma^3\frac{\sigma_0\alpha}{S} = \frac{I_0}{2}\beta^3\gamma^3\sigma_0^2\left(\frac{\overline{R}}{S}\right)^2.$$ (4.154)

As we will see in Section 4.4.3, envelope instabilities limit the phase advance to $\sigma_0 \leq 90°$. In addition, the aspect ratio of beam radius to lens period must not be too large, say $\overline{R}/S < 0.2$, to avoid nonlinear effects, especially spherical aberrations, in the lenses.

The above set of equations for the average beam radius $\overline{R}$, for the phase advance σ, and for the generalized perveance K, or the beam current I, represent the essential results of the smooth approximation. The accuracy of these results depends on the geometrical configuration of the periodic lattice and on the phase advance σ_0. The latter, also known as the *zero-current phase advance*, can be calculated for a given periodic focusing function $\kappa_0(z)$ by the method described in Section 3.8 and below. A general criterion for the validity of the smooth approximation is that the variation of the beam envelope in one focusing period must be small compared to the average radius $\overline{R}$. This is usually satisfied when σ_0 is not too large.

From a practical point of view, the maximum beam radius, R_{max}—rather than the average radius, $\overline{R}$—is the quantity of interest since it relates directly to the channel aperture available to the beam. If we define $R_{max} = a$ and consider the relations (4.128), (4.129), (4.136), (4.124), and (4.125), we can write

$$\frac{a}{\overline{R}} = \frac{a_0}{R_0} = \frac{w_{0,max}}{w_0} = 1 + \delta_{0,max};\qquad(4.155)$$

hence, in view of (4.138b),

$$R_{max} = a = \overline{R}\left(\frac{\sigma_0}{S}\right)^{1/2} w_{0,max}.\qquad(4.156)$$

From these equations we can define a *ripple factor* G by

$$G = \left(\frac{\overline{R}}{a}\right)^2 = \frac{1}{\left(1 + \delta_{0,max}\right)^2} = \frac{S}{\sigma_0 w_{0,max}^2}.\qquad(4.157)$$

Let us now introduce for the acceptance α in lieu of (4.150) the exact definition (3.353) in terms of the maximum beam radius $R_{max} = a$, that is,

$$\alpha = \frac{a^2}{w_{0,max}^2} = \frac{a^2}{\hat{\beta}_{0,max}}.\qquad(4.158)$$

By comparison with (4.157) we then find that

$$\frac{\sigma_0 \alpha}{S} = \frac{\sigma_0^2 a^2}{S^2}G.\qquad(4.159)$$

Thus the equations (4.152) and (4.153) may be written in the alternative forms

$$K = \sigma_0^2 \frac{a^2}{S^2}G\left[1 - \left(\frac{\epsilon}{\alpha}\right)^2\right],\qquad(4.160)$$

and

$$I = \frac{I_0}{2} \beta^3 \gamma^3 \sigma_0^2 \frac{a^2}{S^2} G \left[1 - \left(\frac{\epsilon}{\alpha} \right)^2 \right].$$ (4.161)

This relation shows the explicit dependence of the transportable current on the semiaperture of the channel (or the allowed maximum beam radius) $R_{max} = a$, which is more useful than the average radius $\overline{R}$. The ripple factor G depends on σ_0 and on the shape of the focusing function $\kappa_0(z)$. Note that all quantities on the right-hand side of the last two equations are independent of space-charge forces and represent directly "measurable" parameters of the beam and the focusing channel.

In general, for periodic focusing functions $\kappa_0(z)$ of arbitrary shape, the quantities $w_{0,max}$, σ_0, G, and so on, must be calculated by numerical integration of Equations (4.126) and (4.122). However, in most cases of practical interest one can use for $\kappa_0(z)$ a *hard-edge approximation* that yields quite accurate analytical results and hence scaling relations not readily obtained from numerical studies. If $\kappa_{0,max}$ denotes the maximum of the focusing force, we can define the equivalent hard-edge function for each lattice period S of the channel by

$$\kappa_0 = \begin{cases} \kappa_{0,max} & \text{for } 0 \le z \le l, \\ 0 & \text{for } l \le z \le S, \end{cases}$$ (4.162)

where the effective length l of the lens is given by

$$l = \frac{1}{\kappa_{0,max}} \int_0^S \kappa_0(z)\, dz.$$ (4.163)

If L denotes the field-free region between the lenses, we have the relation

$$l + L = S.$$ (4.164)

To find the phase advance without space charge, σ_0, and other quantities for such a hard-edge periodic channel we define the focusing-strength parameter

$$\theta = \sqrt{\kappa_0}\, l$$ (4.165)

and follow the procedure discussed in 3.8.3 (see also Problem 3.21). From the transfer matrix $\tilde{M}$ for one channel period one finds that

$$\cos \sigma_0 = \cos \theta - \frac{1}{2} \frac{L}{l} \theta \sin \theta\, z.$$ (4.166)

The maximum value of the amplitude function is obtained from the transfer matrix for a half period (from $z = l/2$ to $z = L/2$):

$$\hat{\beta}_{0,\max} = w_{0,\max}^2 = \frac{l}{\theta}\left[\frac{1 + (L/l)(\theta/2)\cot(\theta/2)}{1 - (L/l)(\theta/2)\tan(\theta/2)}\right]^{1/2}. \tag{4.167}$$

From these two equations one can calculate the ripple function $G = G(\theta, L/l)$. Figure 4.4 shows the phase advance σ_0 versus θ for a periodic solenoid channel for different values of L/l. In Figures 4.5 and 4.6 we plotted $w_{0,\max}/\sqrt{l}$ and G versus σ_0 (rather than θ), with L/l as a parameter [11]. For thin lenses, where $\theta \ll \pi/2$, we have the approximation

$$\cos\sigma_0 = 1 - \frac{\theta^2}{2}\left(1 + \frac{L}{l}\right). \tag{4.168}$$

If, in addition, $\sigma_0 \ll \pi/2$, we obtain

$$\sigma_0 \approx \theta\left(1 + \frac{L}{l}\right)^{1/2}. \tag{4.169}$$

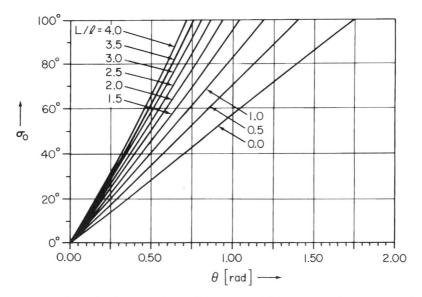

Figure 4.4. Relationship between phase advance σ_0 and focusing strength parameter θ in an axisymmetric periodic-focusing channel. (From Reference 11.)

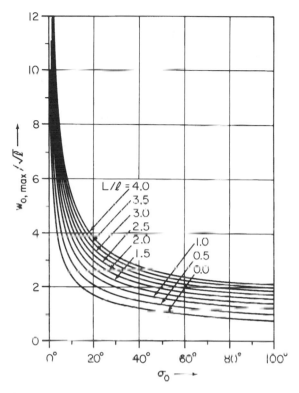

Figure 4.5. Amplitude function $w_{0,\,max}/l^{1/2}$ versus σ_0 in an axisymmetric periodic focusing channel for different values of L/l. (From Reference 11.)

In the latter case one finds for the ripple factor the approximation

$$G\left(\sigma_0, \frac{L}{l}\right) \approx \left(1 - \frac{\upsilon_0^2}{4}\frac{L/l}{1 + L/l}\right)^{1/2}\left(1 - \frac{\sigma_0^2}{12}\frac{L/l}{(1 + L/l)^2}\right)^{-1/2}. \qquad (4.170)$$

The above theory and the various relations for a periodic channel are discussed in more detail in References 10 and 11.

Let us now discuss two examples of periodic transport to illustrate the application of the theory and the accuracy of the approximation involved. First, we consider the case shown in Figure 4.3. Since $\kappa_0(z) \propto B^2(z)$, the hard-edge approximation (4.162) yields an effective length of

$$l = \frac{1}{B_0^2}\int_0^S B^2(z)\,dz\,,$$

which is found to have a value of $l = 3.34$ cm for the solenoidal field (4.127) of each lens. The period length is $S = 13.6$ cm, hence $L = 10.26$ cm and $L/l = 3.08$. To simplify the calculation, let us take $L/l \approx 3$ and use the plot for $G(\sigma_0, L/l)$ in

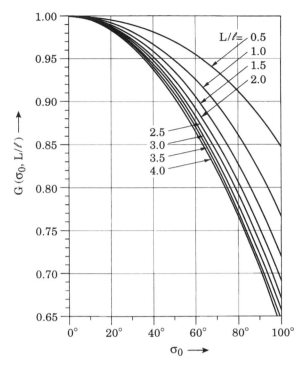

Figure 4.6. Ripple factor $G = (\overline{R}/a)^2$ versus σ_0 in an axisymmetric periodic focusing channel for different values of L/l. (From Reference 11.)

Figure 4.6, from which we find (for $\sigma_0 = 72°$) that $\theta \simeq 0.6$, $G = (\overline{R}/a)^2 \approx 0.82$, hence $a = R_{max} \simeq 1.1\overline{R}$. This is in good agreement with the numerical result of Figure 4.3, from which one infers an envelope modulation of slightly less than 10%.

As a second example, let us calculate the maximum electron-beam current that can be transported in the periodic solenoid channel of Figure 4.3 if the electron energy is 5 keV and the aperture radius is $a = 1$ cm. Assuming that $\epsilon \ll a$ and using the values $\sigma_0 = 72° = 0.4\pi$, $G = 0.82$, $a/S = 1/13.6$, $\beta\gamma = 0.14$, and $I_0 = 1.7 \times 10^4$ A, we find from Equation (4.161) a beam current of $I = 0.164$ A. Suppose now that the emittance of this beam is 8×10^{-5} m-rad; what is the phase advance with space charge σ? First, we find that the generalized perveance is $K = (I/I_0)(2/\beta^3\gamma^3) = 7.03 \times 10^{-3}$. Then we obtain for the parameter u the result $u = KS/(2\sigma_0\epsilon) = 4.76$. This yields from Equation (4.147) $\sigma \approx 0.104\sigma_0$, or $\sigma = 7.5°$. The phase advance due to the depression of the external focusing force by the space-charge repulsion is thus almost a factor 10 smaller in this case than the zero-current value of $\sigma_0 = 72°$.

4.4.2 Beam Transport in a Quadrupole (FODO) Channel

The foregoing theory of beam transport in a periodic system with axisymmetric lenses can be applied to a quadrupole channel in a straightforward way [10,11].

The major difference is that the focusing system, and hence the beam, has two planes of symmetry. Consequently, we need a set of two equations to describe the beam envelopes and the particle trajectories in the two planes. As we will see below, these two equations are coupled through the self-field terms. The case where self fields are negligible was treated in Section 3.8.3, where we considered a periodic system of hard-edge quadrupole lenses arranged in a FODO sequence. Such a system with the beam envelopes in the x and y directions is depicted in Figure 3.27. When the quadrupole lens is focusing in x and defocusing in y, the envelope function for a matched beam has a maximum in the x direction and a minimum in the y-direction. The beam cross section is then an ellipse with major axis in the x-direction and minor axis in the y-direction. Half a period later this ellipse has rotated by 90°. This picture also applies when linear self fields of a beam with uniform density are included.

If $X(z)$ denotes the x-envelope, $Y(z)$ the y-envelope, the ellipse describing the boundary of the beam obeys the equation

$$\frac{x^2}{X^2} + \frac{y^2}{Y^2} = 1, \tag{4.171}$$

and the charge density is defined by

$$\rho(z) = \begin{cases} \rho_0 & \text{for } \dfrac{x^2}{X^2} + \dfrac{y^2}{Y^2} \leq 1, \\ 0 & \text{for } \dfrac{x^2}{X^2} + \dfrac{y^2}{Y^2} > 1, \end{cases} \tag{4.172}$$

where

$$\rho_0 = \rho_0(z) = \frac{I}{\pi v X(z) Y(z)} \tag{4.173}$$

is constant inside the beam at any given position but varies with distance z. The electric field for such a charge distribution can be calculated from Poisson's equation, and one obtains

$$E_x = \frac{I}{\pi \epsilon_0 v} \frac{x}{X(X + Y)}, \tag{4.174}$$

$$E_y = \frac{I}{\pi \epsilon_0 v} \frac{y}{Y(X + Y)}. \tag{4.175}$$

In the case of a round beam, with $X = Y = a$, these results agree with Equation (4.11c) for the radial electric field E_r.

Similar expressions are obtained for the magnetic self-field component B_x, B_y. If $\kappa_{x0}(z)$ and $\kappa_{y0}(z)$ represent the external focusing functions in the two planes of symmetry, one obtains with the above self fields the following trajectory equations:

$$x'' + \kappa_{x0}x - \frac{2K}{X(X + Y)}x = 0, \tag{4.176}$$

$$y'' + \kappa_{x0}y - \frac{2K}{Y(X + Y)}y = 0. \tag{4.177}$$

These two equations are linear in x and y, but coupled through the self-field terms which can be determined from the two corresponding equations for the beam envelopes $X(z)$, $Y(z)$, that is,

$$X'' + \kappa_{x0}X - \frac{2K}{X + Y} - \frac{\epsilon_x^2}{X^3} = 0, \tag{4.178}$$

$$Y'' + \kappa_{y0}X - \frac{2K}{X + Y} - \frac{\epsilon_y^2}{Y^3} = 0. \tag{4.179}$$

The focusing functions κ_{x0} and κ_{y0} are periodic with period S. For the hard-edge approximation of a FODO system one has

$$S = 2(l + L), \tag{4.180}$$

where l is the length of a quadrupole lens and L the length of the drift space between lenses. We will show in Section 5.3.2 that the above equations—like all other linear beam-optics equations in Chapter 4—follow naturally from the self-consistent K–V beam model mentioned in Section 4.1. For our analysis we will assume that the two focusing functions have the same amplitudes (i.e., $|\kappa_{x0}| = |\kappa_{y0}| = \kappa_0$) and that the emittance is the same in both directions, hence $\epsilon_x = \epsilon_y = \epsilon$. The envelopes for a matched beam can then be written in terms of the mean radius $\overline{R}$, which is constant, and a modulation function $\delta(z)$ as

$$X(z) = \overline{R}[1 + \delta_x(z)] = \overline{R}[1 + \delta(z)], \tag{4.181}$$

$$Y(z) = \overline{R}[1 + \delta_y(z)] = \overline{R}[1 - \delta(z)], \tag{4.182}$$

where we used the fact that in a quadrupole channel $\delta_y(z) = -\delta_x(z)$. These relations are analogous to Equation (4.128) for the axisymmetric case and we can apply all equations of the preceding section to our quadrupole channel. The major difference is that we have two lenses of opposite polarity in each channel period. If these two lenses are identical in length and focusing strength, as is the case for the ideal symmetrical FODO channel that we consider below, the average value of $\kappa_0(z)$ will be zero; that is,

$$\overline{\kappa_0}(z) = \frac{1}{S} \int_z^{z+S} \kappa_0(z)\, dz = 0.$$

However, the second term in Equation (4.139) is not zero since $\delta(z)$ is positive $(X > \overline{R})$ when $\kappa_0(z)$ is positive (focusing lens) and negative $(X < \overline{R})$ when $\kappa_0(z)$ is negative (defocusing lens).

The calculations of σ_0 and $w_{0,\,\mathrm{max}}$ versus the quadrupole focusing parameter $\theta = \sqrt{\kappa_0}\, l$ for a FODO channel was carried out in Section 3.8.3. The related plots for different ratios of l/L are shown in Figures 4.7 and 4.8, and the ripple factor is plotted in Figure 4.9. These plots are from Reference 11, where more detailed information is given. Here we note only that the ripple factor is almost independent of the ratio of the quadrupole length to the drift space, l/L. In fact, one finds that for the region $\sigma_0 < 90°$, G can be approximated with reasonable accuracy by the relation [11]

$$G\left(\sigma_0, \frac{L}{l}\right) \approx 1 - \frac{1.2}{\pi}\sigma_0 \tag{4.183}$$

From Equation (4.161) of the smooth approximation theory, one then obtains for the maximum transportable beam current in a FODO channel with aperture $X_{\mathrm{max}} = a$ and period S the result

$$I_{\mathrm{max}} = \frac{I_0}{2}\,\beta^3\gamma^3 v_0^2\left(1 - \frac{1.2}{\pi}\sigma_0\right)\left(\frac{a}{S}\right)^2 \tag{4.184}$$

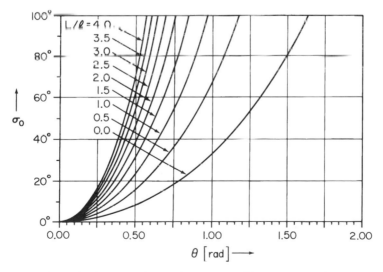

Figure 4.7. Relationship between phase advance σ_0 and focusing strength parameter θ in a periodic quadrupole (FODO) channel. (From Reference 11.)

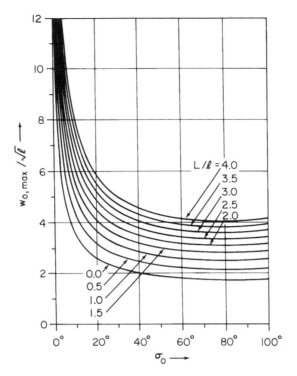

Figure 4.8. Amplitude function $w_{0,\max}/l^{1/2}$ versus σ_0 in a FODO channel for different values of L/l. (From Reference 11.)

The corresponding generalized perveance is

$$K_{\max} = \sigma_0^2\left(1 - \frac{1.2}{\pi}\sigma_0\right)\left(\frac{a}{S}\right)^2. \tag{4.185}$$

The ratio a/S should not be too large to avoid nonlinear forces in the fringe fields of the lenses. If we assume that $a/S = 0.1$ and a maximum phase advance of $\sigma_0 = 90°$ to avoid envelope instabilities, we find that $K_{\max} \approx 10^{-2}$. The phase advance σ_0 depends on the focusing parameter θ, which for magnetic quadrupoles is defined by

$$\theta = \left(\frac{qB_0}{mc\beta\gamma a_q}\right)^{1/2} l. \tag{4.186}$$

B_0 is the field strength at the pole shoe surface, a_q the quadrupole "radius" (i.e., the distance between the tip of the pole shoe and the axis), and l the effective

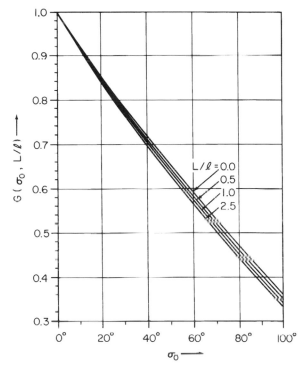

Figure 4.9. Ripple factor $G = (\bar{R}/a)^2$ in a FODO channel for different values of L/l. (From Reference 11.)

width of a quadrupole. In the case of electrostatic quadrupole lenses, the focusing parameter is given by

$$\theta = \left(\frac{2qV_0}{\gamma m v^2 a_q^2} \right)^{1/2} l, \tag{4.187}$$

where V_0 is the electrode potential and a_q the electrode "radius" as in the magnetic case. At nonrelativistic energies ($\gamma = 1$) where electrostatic quadrupoles are mostly used, one can introduce the beam voltage V_b from the kinetic energy relation

$$\frac{m v^2}{2} = q V_b \tag{4.188}$$

and obtain the simple formula

$$\theta = \left(\frac{V_0}{V_b} \right)^{1/2} \frac{l}{a_q}. \tag{4.189}$$

As mentioned in Section 3.5, magnetic quadrupoles provide stronger focusing than solenoid lenses for a given magnetic field strength, and they are used in all modern high-energy accelerators. Room-temperature electromagnets with iron pole shoes are limited to a field strength of 1 to 2 T due to iron saturation. To overcome this limitation, superconducting magnets producing fields in the range of 3 to 7 T have been developed. At low energies, designers for beam transport systems can choose between magnetic quadrupoles, solenoids, or axisymmetric electrostatic lenses. The choice depends on the application, particle species, kinetic energy, beam current, emittance, and on past experience at a particular laboratory. The stringent brightness and intensity requirements of such advanced accelerator applications as free electron lasers, heavy-ion inertial fusion, and high-current light-ion beams (p, H^-, etc.) pose great challenges for beam transport design. In many cases, such as low-energy transport of p or H^- beams from the source to the linear accelerator, charge neutralization via beam particle collisions in the background gas (known as *gas focusing*) is utilized to confine the beam. This topic is discussed in Section 4.6.

4.4.3 Envelope Oscillations and Instabilities of Mismatched Beams

The amount of beam current that can be transported through a periodic-focusing channel with a given aperture is a maximum when the beam is perfectly matched (i.e., when the mean beam radius is constant and the envelope is a periodic function with the same period as the lens system). Also, for space-charge-dominated beams, it is important that the particle density profile be as uniform as possible. In practice, perfect matching is often difficult to achieve. For instance, the beam current or emittance may differ from the design value. In pulsed beams the current may vary between front and tail. Matching lenses may not have the correct focusing strength or may not be in the right position. Indeed, beam matching between various components of an accelerator/transport system is one of the most important problems for the design and operation of any facility. Conversely, one must have an understanding of beam behavior when matching conditions are not perfect. As we know from our analysis of mismatch in a continuous (uniform) focusing channel, the beam envelope performs oscillation about the equilibrium (or matched beam) radius. We expect similar behavior for a periodic-focusing channel. However, due to the periodic nature of the focusing force acting on the beam, we have the possibility of parametrically excited instabilities that do not occur in uniform channels. As we will see, such instabilities do occur when $\sigma_0 > 90°$ and the beam intensity is sufficiently high.

Following the analysis by Struckmeier and Reiser [12], we will first calculate the envelope oscillation frequencies for small deviations from the matched-beam conditions in the smooth approximation which replaces the periodic channel by the equivalent uniform focusing channel. Next, we present a more rigorous analysis that takes into account the periodic variation of the focusing force and that leads to predictions of instabilities. We will carry out this study for the more general problem of a quadrupole channel which includes the axisymmetric system as a special case.

Let us start with the two envelope equations (4.178) and (4.179) and assume again that $\epsilon_x = \epsilon_y = \epsilon$. When the beam is not perfectly matched, the mean values of the envelope functions $\overline{X}(z)$ and $\overline{Y}(z)$ will differ from the matched radius $\overline{R}$ and will be functions of z. In the smooth approximation, we can replace the periodic-focusing functions $\kappa_{x0}(z)$ and $\kappa_{y0}(z)$ by σ_0^2/S^2, as shown in Section 4.4.1. Introducing the wave number k_0 defined as

$$k_0^2 = \frac{\sigma_0^2}{S^2},\tag{4.190}$$

we then obtain for $\overline{X}(z)$ and $\overline{Y}(z)$ the equations

$$\overline{X}'' + k_0^2\overline{X} - \frac{2K}{\overline{X} + \overline{Y}} - \frac{\epsilon^2}{\overline{X}^3} = 0,\tag{4.191}$$

$$\overline{Y}'' + k_0^2\overline{Y} - \frac{2K}{\overline{X} + \overline{Y}} - \frac{\epsilon^2}{\overline{Y}^3} = 0.\tag{4.192}$$

This is a set of coupled differential equations which, in contrast to (4.178), (4.179), has constant coefficients and can therefore be solved analytically. First, we note that in the matched-beam case we have $\overline{X} = \overline{Y} = \overline{R} = $ const, where the mean radius obeys the algebraic equation

$$k_0^2\overline{R} - \frac{K}{\overline{R}} - \frac{\epsilon^2}{\overline{R}^3} = 0,\tag{4.193}$$

which is identical with Equation (4.141a) for the axisymmetric channel. Introducing the space-charge depressed wave number, k or phase advance σ, defined by the relation

$$k^2 = \frac{\sigma^2}{S^2} = k_0^2 - \frac{K}{\overline{R}^2} = \frac{\sigma_0^2}{S^2} - \frac{K}{\overline{R}^2}\tag{4.194}$$

and substituting into Equation (4.193) one obtains for the average radius of the matched beam in the presence of space charge the result

$$R = \sqrt{\frac{\epsilon}{k}} = \sqrt{\frac{\epsilon S}{\sigma}}.\tag{4.195}$$

When the beam mismatch is small, the envelopes $\overline{X}(z)$ and $\overline{Y}(z)$ will not deviate very much from the mean radius $\overline{R}$. Defining the deviations by $\xi(z)$ and $\eta(z)$, we can write

$$\overline{X}(z) = \overline{R} + \xi(z),\tag{4.196}$$

$$\overline{Y}(z) = \overline{R} + \eta(z),\tag{4.197}$$

where $\xi, \eta \ll \overline{R}$.

By substituting (4.196), (4.197) into (4.191), (4.192), Taylor expanding, keeping only linear terms, and using the matched-beam relations (4.193), (4.194) to eliminate K and $\overline{R}$, we obtain

$$\xi'' + A_1 \xi + A_2 \eta = 0, \tag{4.198}$$

$$\eta'' + A_1 \eta + A_2 \xi = 0, \tag{4.199}$$

where

$$A_1 = \frac{3\sigma_0^2 + 5\sigma^2}{2S^2}, \qquad A_2 = \frac{\sigma_0^2 - \sigma^2}{2S^2}. \tag{4.200}$$

These coupled equations are identical in form with the set of second-order linear differential equations describing the behavior of two coupled harmonic oscillators. There are two fundamental modes of oscillation, which we define by

$$\zeta_1(z) = \xi(z) - \eta(z), \qquad \zeta_2(z) = \xi(z) + \eta(z). \tag{4.201}$$

The first mode, defined by ζ_1, corresponds to the case where the two oscillations in the x and y directions are 180° out of phase (antiparallel). By subtracting Equations (4.198) and (4.199), one obtains

$$\zeta_1'' + k_1^2 \zeta_1 = 0, \tag{4.202}$$

with

$$k_1 = (k_0^2 + 3k^2)^{1/2}, \qquad \text{or} \qquad \phi_1 = k_1 S = (\sigma_0^2 + 3\sigma^2)^{1/2}. \tag{4.203}$$

The second fundamental mode, defined by ζ_2, corresponds to the case where both oscillations are in phase (parallel) and is given by

$$\zeta_2'' + k_2^2 \zeta_2 = 0, $$

with

$$k_2 = (2k_0^2 + 2k^2)^{1/2}, \qquad \text{or} \qquad \phi_2 = k_2 S = (2\sigma_0^2 + 2\sigma^2)^{1/2}. \tag{4.204}$$

Any other case can be expressed as a superposition of these two fundamental modes. Suppose, for instance, that the initial conditions are $\xi_0 \neq 0$, $\xi_0' = 0$, $\eta_0 = 0$, $\eta_0' = 0$; then the envelope oscillations in the x and y directions are given by

$$\xi(z) = \xi_0 \cos\left[\frac{1}{2}(k_1 - k_2)z\right] \cos\left[\frac{1}{2}(k_1 + k_2)z\right], \quad (4.205)$$

$$\eta(z) = \xi_0 \sin\left[\frac{1}{2}(k_1 - k_2)z\right] \sin\left[\frac{1}{2}(k_1 + k_2)z\right]. \quad (4.206)$$

The envelope oscillations in this special case are characterized by a fast frequency variation, $\frac{1}{2}(k_1 + k_2)$, and a slow variation given by $\frac{1}{2}(k_1 - k_2)$.

When the space charge, or current, is negligibly small ($K \to 0$), we have from (4.194) $k = k_0$, or $\sigma = \sigma_0$, and hence the two fundamental modes converge, that is,

$$k_1 = k_2 = 2k_0, \quad \text{or} \quad \phi_1 = \phi_2 = 2\sigma_0. \quad (4.207)$$

Thus, in this limit, the envelopes of a mismatched beam oscillate with a frequency that is twice as fast as the single-particle oscillation frequency defined by the phase advance per period, σ_0.

On the other hand, when the space charge is very high and $k \to 0$, or $\sigma \to 0$, the envelope oscillation frequencies approach the lower limits of

$$k_1 = k_0, \quad \text{or} \quad \phi_1 = \sigma_0, \quad (4.208)$$

for the antiparallel mode, and

$$k_2 = \sqrt{2}\,k_0, \quad \text{or} \quad \phi_2 = \sqrt{2}\,\sigma_0 \quad (4.209)$$

for the parallel (in-phase) mode.

Note that the result (4.204) for the in-phase mode is identical with Equation (4.105a) for the axisymmetric envelope oscillation of a mismatched beam in a uniform focusing channel. This is not surprising since the smooth-approximation theory replaces the periodic-focusing force by the smoothed average force. Apart from this agreement, however, the above analysis for the periodic channel is more general than our previous calculation in that it includes two transverse degrees of freedom yielding two fundamental oscillation modes. This is of particular interest for quadrupole channels where small mismatch errors are more likely to produce the out-of-phase mode or a mixed mode. By analyzing the envelope perturbations, we arrive at a system of coupled linear differential equations with periodic rather than constant coefficients, which must be solved numerically. The starting point is again the nonlinear (coupled) system of the envelope equations (4.178), (4.179), in which we substitute the perturbed envelope functions directly:

$$X(z) = X_0(z) + \xi(z), \quad Y(z) = Y_0(z) + \eta(z). \quad (4.210)$$

Here, X_0 and Y_0 denote that matched envelope functions [i.e., periodic solutions of (4.178), (4.179)] and ξ, η denote the small perturbations:

$$\xi(z) \ll X_0(z), \qquad \eta(z) \ll Y_0(z).$$

Due to these conditions, we can linearize the differential equations for the perturbation functions $\xi(z)$ and $\eta(z)$ and obtain

$$\xi''(z) + a_1(z)\xi(z) + a_0(z)\eta(z) = 0, \tag{4.211}$$

$$\eta''(z) + a_2(z)\eta(z) + a_0(z)\xi(z) = 0, \tag{4.212}$$

with three S-periodic coefficients:

$$a_0(z) = \frac{2K}{[X_0(z) + Y_0(z)]^2}, \tag{4.213a}$$

$$a_1(z) = \kappa_{x0}(z) + \frac{3\epsilon^2}{X_0^4(z)} + a_0(z), \tag{4.213b}$$

$$a_2(z) = -\kappa_{y0}(z) + \frac{3\epsilon^2}{Y_0^4(z)} + a_0(z). \tag{4.213c}$$

To solve this system, we need the matched envelope functions $X_0(z)$ and $Y_0(z)$. The two second-order equations (4.211), (4.212) are equivalent to a system of four first-order differential equations. With $\zeta = (\xi, \xi', \eta, \eta')$, we may write in matrix notation

$$\zeta'(z) = \tilde{A}(z) \cdot \zeta(z), \tag{4.214}$$

with the S-periodic matrix

$$\tilde{A}(z) = \begin{pmatrix} 0 & 1 & 0 & 0 \\ -a_1(z) & 0 & -a_0(z) & 0 \\ 0 & 0 & 0 & 1 \\ -a_0(z) & 0 & -a_2(z) & 0 \end{pmatrix}. \tag{4.215}$$

If $\tilde{Z}(z)$ denotes the 4×4 solution matrix of (4.214) with $\tilde{Z}(0) = \tilde{E}$ ($\tilde{E}$ = unit matrix), we may write Floquet's theorem as follows:

$$\tilde{Z}(z + nS) = \tilde{Z}(z) \cdot \tilde{Z}(S)^n, \tag{4.216}$$

where n is an arbitrary integer number. The solution of (4.214) at any value $z + nS$ can be expressed as a product of the solution matrix $\tilde{Z}(z)$, $0 \le z \le S$, and the matrix $\tilde{Z}(S)$ at the end of the first focusing period. If we evaluate the eigenvalues and eigenvectors of $\tilde{Z}(S)$, we obtain a 4×4 matrix of eigenvectors C and a diagonal matrix of eigenvalues (denoted by Λ):

$$\tilde{Z}(S) \cdot C = \Lambda \cdot C.$$

If we define the matrix $\tilde{Y}(z)$ by

$$\tilde{Y}(z) = \tilde{Z}(z) \cdot C,$$

it follows from (4.216) that

$$\tilde{Y}(z + nS) = \Lambda^n \cdot \tilde{Y}(z). \tag{4.217}$$

Since $\tilde{Y}(z)$ is a solution matrix of (4.214), every special solution $\zeta(z)$ of (4.214) can be expressed as a linear combination of the column vectors of the matrix $\tilde{Y}(z)$. It is now obvious that a solution of (4.214) can be stable only if Λ^n remains finite for $n \to \infty$. One can readily prove that $Z(S)$ is symplectic and real, so the four eigenvalues occur both as reciprocal and as complex-conjugate pairs. Therefore, Λ^n can remain bounded only if all eigenvalues lie on the unit circle in the complex plane. Mathematically, this problem is identical to the two-dimensional linear oscillator without space charge treated by Courant and Snyder (Chapt. 3, [4]). Thus, if we express the eigenvalues in polar coordinates, that is,

$$\lambda = |\lambda| \cdot e^{i\phi}, \tag{4.218}$$

we arrive at only four possibilities for the four eigenvalues [12], assuming them to be distinct, as shown in Figure 4.10:

(a) All four eigenvalues lie on the unit circle, forming two complex conjugate and reciprocal pairs (no instability).
(b) One reciprocal pair is complex with $|\lambda| = 1$ (stable); the other pair is real with $|\lambda| \neq 1$ (unstable).
(c) Both reciprocal pairs are real with $|\lambda| \neq 1$ (unstable).
(d) Both reciprocal pairs are complex and are not on the unit circle, so that $\lambda_2 = 1/\lambda_1$, $\lambda_3 = \lambda_1^*$, $\lambda_4 = 1/\lambda_1^*$ ("confluent-resonance" instability).

By using relation (4.218) we can identify the growth rate (damping rate) $|\lambda|$ of the appropriate eigenvector passing through one focusing period and the phase shift ϕ of the corresponding envelope oscillation. A growth rate that is not equal to unity is an indication of instability. As we can see from Figure 4.10, this instability can occur only if $|\lambda| \neq 1$, and if:

1. One or both eigenvalue pairs lie on the real axis, that is, $\phi_{1,2} = 180°$ [Figure 4.10(b) or 4.10(c)].
2. The phase shift angles obey the relation $(\phi_1 + \phi_2) = 360°$, or are equal $(\phi_1 = \phi_2)$ [Figure 4.10(d)].

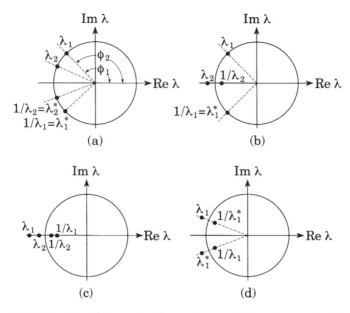

Figure 4.10. Location of eigenvalues for envelope oscillations (see text for discussion).

Case 1 can be seen as a half-integer resonance between the focusing structure and the envelope oscillation mode [i.e., half an oscillation occurs per period (*parametric resonance*)]. Case 2 is a resonance between both envelope oscillation frequencies, since they are equal (*confluent resonance*).

To illustrate the effects of envelope oscillations and instabilities computations were performed [12] for both a solenoid and a quadrupole (FODO) channel with hard-edge focusing functions as shown in Figure 4.11. The results of numerical integrations of Equation (4.214) are plotted in Figure 4.12 for the solenoid channel. The left side of each figure shows the ϕ_1, ϕ_2 values versus σ for several values of σ_0; the right side shows the growth rates $|\lambda|$ versus σ. Instability is indicated by $|\lambda|$-values differing from unity. The solid ϕ-lines are the perturbation phase shifts obtained by numerical integration and eigenvalue analysis; the dashed lines show the results obtained from Equations (4.203), (4.204) for a uniform or smooth channel. As can be seen from the figures, for $\sigma_0 = 90°$ these results are nearly identical. Above $\sigma_0 = 90°$, instability occurs in some specific regions. Note that σ is plotted on the abscissa as a decreasing function so that beam intensity increases from left to right. The value of σ at the origin corresponds to σ_0 (zero intensity), while $\sigma \to 0$ represents the laminar beam limit ($\epsilon = 0$).

For the solenoid channel, only parametric resonances occur, namely when a ϕ-curve reaches the 180°-line. In that case, the smooth approximation results differ from the exact periodic ones at and near the regions of instability, as one would expect.

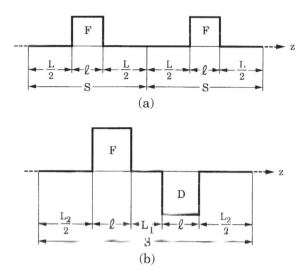

Figure 4.11. Hard-edge focusing functions $\kappa_0(z)$ used in the computations: (a) solenoid channel with $L/l = 3.0$, period length $S = 0.136$ m; (b) asymmetric quadrupole (FODO) channel with $L_1/l = 0.821$, $L_2/l = 2.858$, and period length $S = 1.238$ m. (From Reference 12.)

In the case of the quadrupole channel, shown in Figure 4.13, we are dealing with confluent resonances where phase locking occurs between the two modes so that $\phi_1 = \phi_2$. The instability occupies a certain range of σ values. As σ_0 increases, this patch gets wider and wider, extending over the entire region below $\sigma = 90°$ when σ_0 exceeds 120°.

For both types of beam transport channels, the instability growth rate increases with increasing σ_0, and at sufficiently high values of σ_0 there is an intensity threshold beyond which the beam is unstable for all values of $\sigma \to 0$. The results obtained here from the perturbation theory of the K–V envelope equations are equivalent with those obtained from the Vlasov equation perturbation analysis [13] for the special case of the *second-order even* mode.

As a check of the above linearized envelope perturbation theory and to further illustrate the beam behavior in the case of mismatch conditions, the envelope equations (4.112) and (4.178), (4.179) for the solenoid and quadrupole channels of Figure 4.11 were integrated numerically. By choosing the appropriate initial conditions, one can excite either one of the two fundamental modes or a mixed mode. Figure 4.14 shows a pure *in-phase* mode for the solenoid case. The phase advance without and with space charge is $\sigma_0 = 60°$, $\sigma = 21.2°$, resulting in a theoretical phase shift of

$$\phi_2 = (2\sigma_0^2 + 2\sigma^2)^{1/2} = 90°.$$

As can be seen in the figure, the envelope oscillation exhibits a pattern with a wavelength of four periods (i.e., $\phi_2 = 90°$), in excellent agreement with the

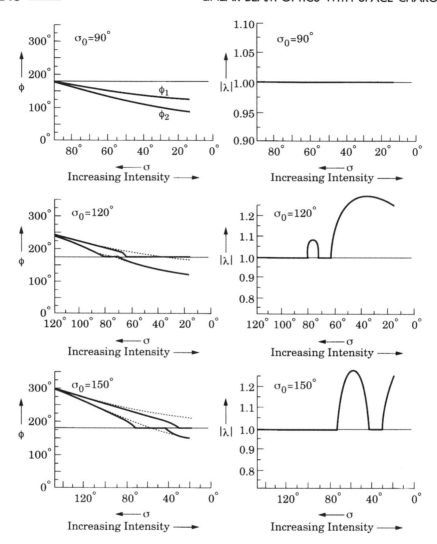

Figure 4.12. Phase shifts and growth rates of envelope perturbations versus decreasing σ (increasing beam intensity) for $\sigma_0 = 90°$, $120°$, and $150°$ for the solenoid channel. Dashed curves represent the smooth-approximation results. (From Reference 12.)

theory. A particle trajectory showing the oscillation period of about 17 lenses (in agreement with $\sigma = 21.2°$) is also plotted for comparison in the figure. An example of unstable behavior is illustrated in Figure 4.15 for the solenoid channel. The parameters in this case are $\sigma_0 = 120°$ and $\sigma = 34.6°$. As can be seen from Figure 4.12, the in-phase oscillation mode is unstable due to a parametric resonance ($\phi_1 = 180°$) with a growth rate of $|\lambda| = 1.283$, whereas the 180° *out-of-phase* mode is stable ($\phi_2 = 134°$, $|\lambda| = 1$). Figure 4.15 shows the increasing oscillation

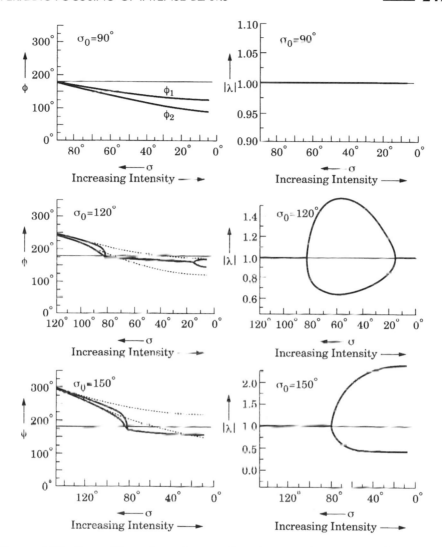

Figure 4.13. Phase shifts and growth rates of envelope perturbations versus decreasing σ (increasing beam intensity) for $\sigma_0 = 90°$, $120°$, and $150°$ for the quadrupole channel. Dashed curves represent the smooth-approximation results. (From Reference 12.)

amplitude for the unstable in-phase mode. A typical particle trajectory in the beam which starts out with pseudoharmonic motion is seen to lose its periodicity quickly as the envelope becomes unstable.

Similar results are obtained for mismatched beams in the quadrupole channel. Figure 4.16 shows the envelope oscillation in the case $\sigma_0 = 60°$, $\sigma = 21.2°$, for the in-phase mode whose wavelength extends over four cells in agreement with the linear theory. The particle trajectory also behaves as expected. A case of unstable

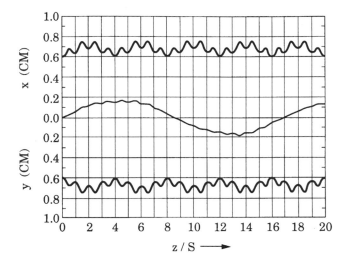

Figure 4.14. Solenoid channel, stable in-phase mode ($\sigma_0 = 60°$, $\sigma = 21°$). (From Reference 12.)

behavior (i.e., exponential growth of the beam radius) is demonstrated in Figure 4.17. The chosen parameter values of $\sigma_0 = 120°$, $\sigma = 35°$ are in the region of a confluent resonance where, according to Figure 4.13, one has

$$\phi_1 = \phi_2 = 162°, \qquad |\lambda| = 1.395.$$

Thus in this case both modes are unstable.

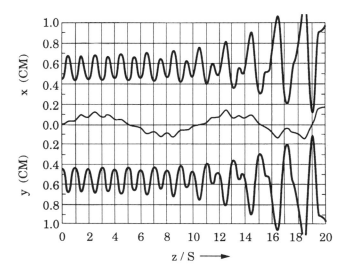

Figure 4.15. Solenoid channel, unstable in-phase mode ($\sigma_0 = 120°$, $\sigma = 34.6°$). (From Reference 12.)

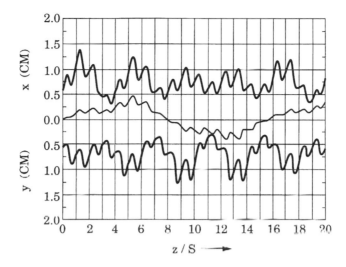

Figure 4.16. Quadrupole channel, in-phase mode ($\sigma_0 = 60°$, $\sigma = 21°$). (From Reference 12.)

The detrimental effects of envelope instabilities in the focusing region above $\sigma_0 - 90°$ is also observed in computer simulation studies as well as in experiments. As a consequence, periodic transport channels for high beam currents must be designed to operate at values of σ_0 below $90°$. In this region, the smooth approximation theory can be applied to design the focusing systems, as was pointed out previously.

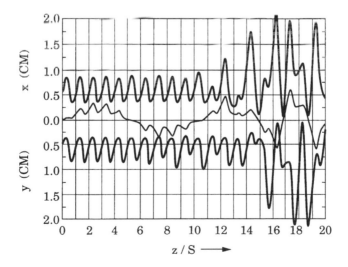

Figure 4.17. Quadrupole channel, slightly mismatched beam ($\sigma_0 = 120°$, $\sigma = 35°$). (From Reference 12.)

4.4.4 Coherent Beam Oscillations due to Injection Errors and Misalignments

In our analysis of periodic focusing so far we have assumed that the lenses are perfectly aligned and that the center of the beam coincides with the optical axis of the focusing channel. Since such an ideal system cannot be realized in practice, it is very important to analyze and understand the effects of injection errors and misalignments on the beam. One type of injection error that results in beam mismatch has already been discussed in the preceding section. There, the beam remains centered, but the envelope performs oscillations that may lead to beam loss when the beam strikes the drift-tube wall or becomes unstable. In the present section we will be concerned with errors leading to displacements of the beam center from the ideal optical axis of the channel. Such displacements are caused by injection errors and misalignments of lenses or other hardware components, and they lead to coherent oscillations of the beam centroid about the optical axis. These oscillations are called *coherent* since they are performed by the beam as a whole (i.e., the beam behaves very much like an oscillating rigid body). By contrast, the single-particle oscillations about the beam axis are called *incoherent* since they are not in phase (i.e., at any given position different particles in the distribution have different phase angles).

Let us first consider the case where the beam is injected into an ideal, perfectly aligned focusing channel with a small aiming error; that is, the beam centroid is displaced from the optical axis or makes a small angle with the axis at the channel entrance. As a result of this injection error, we expect that the beam will perform a coherent oscillation about the channel axis. To determine the frequency or wavelength of this oscillation, we must bear in mind that the *centroid* is defined as the center of mass of the particle distribution. Thus the self fields are zero at the centroid position (at least to the extent that the effects of conducting boundaries can be ignored), and the motion of the centroid is therefore governed by the external focusing force alone. This implies that the trajectory of the centroid is identical with that of a single particle in the absence of space charge. Consequently, we expect that the wavelength of the coherent beam oscillation is given by $\lambda_0 = 2\pi/k_0 = 2\pi S/\sigma_0$, where σ_0 is the phase advance without space charge. However, this description is correct only as long as the space-charge forces are small. When the self fields of the beam are not negligible in comparison with the external focusing fields, the effect of the image charges induced in the conducting drift-tube wall by the off-centered beam must be taken into account. As we will see later in this section, this image effect will increase the oscillation wavelength by an amount that depends on the beam current (or generalized perveance) and the drift-tube radius. We will proceed with our analysis by first neglecting the image force and then adding it later as a correction.

Returning now to our discussion of injection errors, let us suppose that the beam centroid at the channel entrance ($z = 0$) has a displacement x_0 and a slope x_0' with respect to the optical axis of the ideal channel. With these initial conditions and ignoring the image force, as stated, the coherent oscillation of the beam in the focusing channel will be given by the "single-particle" equation

$$x(z) = x_0 \cos k_0 z + \frac{x_0'}{k_0} \sin k_0 z .$$ (4.219)

The amplitude of this oscillation (i.e., the maximum displacement from the axis) is defined by

$$x_m = \left[x_0^2 + \left(\frac{x_0'}{k_0} \right)^2 \right]^{1/2} .$$ (4.220)

As an example, a beam injected with an error of $x_0 = 1$ mm, $x_0' = 20$ mrad into a periodic channel with $S = 15$ cm and $\sigma_0 = 60°$ will perform a coherent oscillation with a wavelength $\lambda_0 = 90$ cm and an amplitude of $x_m = 3.0$ mm. By increasing the phase advance to $\sigma_0 = 90°$, one would decrease the oscillation wavelength to $\lambda_0 = 60$ cm and the amplitude to $x_m = 2.2$ mm.

Next, let us consider the effects of lens misalignments. Suppose first that only one lens, with period S and located at $z_i \leq z \leq z_f$ in an otherwise perfect channel, is translationally offset a distance Δ from the channel axis. If the beam centroid within this lens has a displacement $x(z)$ from the channel axis, its transverse position with regard to the center of the misaligned lens is $x(z) - \Delta$. Since the force experienced by the centroid is proportional to its distance from the lens axis, the equation of motion for the centroid trajectory is given by

$$x'' - \kappa_0(z)(x - \Delta),$$

where $\kappa_0(z)$ represents the focusing force of the lens. Again using the smooth approximation theory, we can replace $\kappa_0(z)$ by the constant average focusing force for the lens period [i.e., $\kappa_0(z) \rightarrow k_0^2 = \sigma_0^2/S^2$], and write the equation of motion in the form

$$x'' + k_0^2 x = k_0^2 \Delta \qquad \text{for } z_i \leq z \leq z_f .$$ (4.221)

Note that $z_j = (z_i + z_f)/2$ defines the center of the misaligned lens period and $z_f - z_i = S$ the length of the period. If we assume that the beam is perfectly centered when it enters the misaligned lens (i.e., $x_i = 0$ and $x_i' = 0$ at $z = z_i$), then the solution of (4.221) is readily obtained as

$$x(z) = \Delta[1 - \cos k_0(z - z_i)]$$ (4.222a)

and

$$x'(z) = k_0 \Delta \sin k_0(z - z_i)$$ (4.222b)

for $z_i \leq z \leq z_f$.

The displacement and slope of the centroid trajectory at the end of the misaligned lens period are then

$$x_f = \Delta[1 - \cos k_0(z_f - z_i)] = \Delta[1 - \cos k_0 S], \qquad (4.223a)$$

$$x_f' = k_0 \Delta \sin k_0(z_f - z_i) = k_0 \Delta \sin k_0 S. \qquad (4.223b)$$

The misaligned lens thus produces a beam offset that is equivalent to an injection error with regard to the motion through the remaining part of the focusing channel. Thus we can use Equation (4.219), with (4.223) as initial conditions, to describe the centroid trajectory in the perfectly aligned channel section beyond the displaced lens. The resulting equation is then given by

$$x(z) = \Delta[1 - \cos k_0(z_f - z_i)] \cos k_0(z - z_f)$$
$$+ \Delta \sin k_0(z_f - z_i) \sin k_0(z - z_f) \qquad \text{for } z > z_f.$$

This may be written in the simpler forms

$$x(z) = \Delta[\cos k_0(z - z_f) - \cos k_0(z - z_i)], \qquad (4.224)$$

or

$$x(z) = 2\Delta \sin \frac{\sigma_0}{2} \sin \frac{\sigma_0}{S}(z - z_j) \qquad \text{for } z > z_f, \qquad (4.225)$$

where we used $k_0 = \sigma_0/S$ and $z_j = (z_i + z_f)/2$.

The above analysis can readily be extended to more than one misaligned lens. Thus if two neighboring lenses are misaligned, one with offset Δ_1 at $z_j = z_1$, the other with offset Δ_2 at $z_j = z_2 = z_1 + S$, one finds for the centroid trajectory in the channel downstream of the two lenses

$$x(z) = 2\Delta_1 \sin \frac{\sigma_0}{2} \sin \frac{\sigma_0}{S}(z - z_1) + 2\Delta_2 \sin \frac{\sigma_0}{2} \sin \frac{\sigma_0}{S}(z - z_2)$$
$$\text{for } z > z_2 + \frac{S}{2}. \qquad (4.226)$$

Generalizing this linear superposition to N successive misaligned lenses, one gets

$$x(z) = \sum_{j=1}^{N} 2\Delta_j \sin \frac{\sigma_0}{2} \sin \frac{\sigma_0}{S}(z - z_j) \qquad \text{for } z > z_N + \frac{S}{2}. \qquad (4.227)$$

Thus if the misalignment offsets Δ_j of the N lenses are known, one can calculate both the displacement and slope of the beam centroid at the end of the Nth lens period ($z = z_N + S/2$) or at any position z in an ideal channel section following the N misaligned lenses.

In practice, the alignment state of a focusing channel is known only within a certain accuracy limit. The remaining alignment errors below this accuracy limit are usually statistical (i.e., random) in nature. The deflections experienced by the beam in such a system of lenses with random misalignment are analogous to the problem of *random walk*. This problem is exemplified, for instance, by the scattering of a particle passing through a gas and suffering deflections from its path in collisions with the randomly distributed gas molecules. We can apply the statistical analysis of random walk to estimate an expectation value for the deflection amplitude of the centroid after the beam passed through a system of N lenses with random alignment errors. To carry this out, we will rewrite Equation (4.227) using the trigonometric identity $\sin(\alpha - \beta) = \sin \alpha \cos \beta - \cos \alpha \sin \beta$, which yields for the displacement and slope of the centroid trajectory at position $z = z_N + S/2$

$$
x(z) = \sin \frac{\sigma_0}{S} z \left[\sum_{j=1}^{N} 2\Delta_j \sin \frac{\sigma_0}{2} \cos \frac{\sigma_0}{S} z_j \right]
$$

$$
- \cos \frac{\sigma_0}{S} z \left[\sum_{j=1}^{N} 2\Delta_j \sin \frac{\sigma_0}{2} \sin \frac{\sigma_0}{S} z_j \right] \tag{4.228a}
$$

and

$$
x'(z) = \frac{\sigma_0}{S} \cos \frac{\sigma_0}{S} z \left[\sum_{j=1}^{N} 2\Delta_j \sin \frac{\sigma_0}{2} \cos \frac{\sigma_0}{S} z_j \right]
$$

$$
+ \frac{\sigma_0}{S} \sin \frac{\sigma_0}{S} z \left[\sum_{j=1}^{N} 2\Delta_j \sin \frac{\sigma_0}{2} \sin \frac{\sigma_0}{S} z_j \right]. \tag{4.228b}
$$

The square of the amplitude, $A^2 = x_m^2$, of the coherent beam oscillation after passage through the N lenses is then given by

$$
A^2 = x^2 + \left(\frac{x'}{k_0} \right)^2 = \left[\sum_{j=1}^{N} 2\Delta_j \sin \frac{\sigma_0}{2} \cos \frac{\sigma_0}{S} z_j \right]^2
$$

$$
+ \left[\sum_{j=1}^{N} 2\Delta_j \sin \frac{\sigma_0}{2} \sin \frac{\sigma_0}{S} z_j \right]^2. \tag{4.229}
$$

This may be written as

$$
A^2 = 4 \sin^2 \frac{\sigma_0}{2} \left[\sum_{j=1}^{N} \Delta_j^2 + \sum_{k=1}^{N} \sum_{j \neq k} \Delta_j \Delta_k \cos \frac{\sigma_0}{S} (z_j - z_k) \right]. \tag{4.230}
$$

If the alignment errors Δ_j for the N lenses were known, Equation (4.230), which was derived from (4.227), would allow us to calculate the exact value of the amplitude A. However, if the errors are not known and are statistically random in nature, we can only calculate an expectation value for A. To accomplish this, let us assume that $(\Delta_1, \Delta_2, \ldots, \Delta_N)$ represents a set of N independent, identically distributed random variables. This implies that we consider an infinite number of possible alignment states in which the displacement of each lens can assume any random value within a continuum of values over a given range. We will assume that the averages (first and second moments) of the distribution for each lens are identical and given by $\overline{\Delta_1} = \overline{\Delta_2} = \overline{\Delta_j} = 0$ and $\overline{\Delta_1^2} = \overline{\Delta_2^2} = \overline{\Delta_j^2} = \overline{\Delta^2}$ (for $j = 1, 2, \ldots, N$). Each possible set of the N variables will yield a different value for A^2 in Equation (4.230). If we take the average over all sets, we find that

$$\overline{\Delta_j \Delta_{j'}} = \begin{cases} 0 & \text{for } j \neq j', \\ \overline{\Delta^2} \neq 0 & \text{for } j = j'. \end{cases} \tag{4.231}$$

Hence the average (expectation value) of A^2 is

$$\langle A^2 \rangle = 4 \sin^2 \frac{\sigma_0}{2} N \langle \Delta^2 \rangle, \tag{4.232}$$

or, taking the square root,

$$\tilde{A} = 2\tilde{\Delta} \sin \frac{\sigma_0}{2} \sqrt{N}, \tag{4.233}$$

where $\tilde{A} = A_{\text{rms}} = (\overline{A^2})^{1/2}$, $\tilde{\Delta} = \Delta_{\text{rms}} = (\overline{\Delta^2})^{1/2}$.

Thus we obtain the very important result that the maximum rms displacement, $\tilde{A}$, of the beam centroid from the axis of a focusing channel with N randomly misaligned lenses is proportional to the rms value of the misalignments, $\tilde{\Delta}$, and increases with the square root of the number of lenses. In addition, it also increases with the zero-current phase advance σ_0 of the channel as $\sin(\sigma_0/2)$. As an example, if $\sigma_0 = 90°$, $N = 50$, and $\tilde{\Delta} = 0.2$ mm, we find that $\tilde{A} = 2.0$ mm. It should be noted that the above analysis of the effects of lens displacements in one transverse coordinate, x, can readily be extended to include misalignments in the other directions, y and z, or tilt angles. Such generalization still leads to a relation of the form (4.233). However, $\tilde{A}$ then represents the total transverse rms amplitude $\tilde{A}_r = (\tilde{A}_x^2 + \tilde{A}_y^2)^{1/2}$ and $\tilde{\Delta}$ the rms sum of all random misalignment errors.

Let us now discuss the effect of image forces on the coherent motion of a beam that is displaced from the axis. Figure 4.18 shows a beam of radius a, horizontally offset by an amount ξ, in a conducting pipe of radius b. When the beam is centered (i.e., $\xi = 0$), the image charges induced on the inner surface of the wall are distributed uniformly in azimuth, and there will be no net electric field at the center. However, when the beam is offset, as shown in the figure,

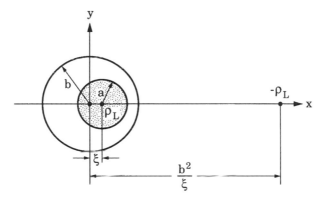

Figure 4.18. Off-centered beam with radius a and displacement ξ in conducting drift tube with radius b can be treated as a line charge, ρ_L, if $a \ll b$. Image $-\rho_L$, is located at b^2/ξ.

the image charge varies with azimuth along the pipe surface. From the geometry of Figure 4.18, we infer that the image charge has a maximum at $x = b, y = 0$ and a minimum at $x = -b, y = 0$, and that there should be a defocusing force in positive x-direction on the centroid of the beam. The potential distribution and electric field produced by this image charge can be calculated in a straightforward way by adding the free-space potential $\phi_f(x,y)$ of the beam (in the absence of the tube wall) and the potential $\phi_i(x,y)$ due to the image charge. The free-space potential can be found from Poisson's equation or Gauss's law. It varies as $\phi_f(R_1) = \phi_a - (\rho_0 a^2/4\epsilon_0)R_1^2/a^2$ with distance R_1 from the beam center inside the beam ($R_1 \leq a$) and as $\phi_f(R_1) = -(\rho_L/2\pi\epsilon_0) \ln(R_1/R_{10})$ outside the beam ($R_1 > a$). ϕ_a and R_{10} are constants determined by the boundary conditions, ρ_0 is the uniform charge density, and $\rho_L = \rho_0 a^2 \pi$ the line charge density of the beam. Note that outside the beam the charge distribution can be replaced by a line charge ρ_L. The image potential can be found by placing a line charge of opposite polarity, $-\rho_L$, at a distance x_i from the center of the tube. It varies with distance R_2 from the image location as $\phi_i(R_2) = (\rho_L/2\pi\epsilon_0) \ln(R_2/R_{20})$, where R_{20} is a constant. At the wall of the conducting tube the total potential must be zero (i.e., $\phi = \phi_f + \phi_i = 0$ at $x^2 + y^2 = b^2$). From this calculation one finds that the image location is given by $x_i = b^2/\xi$, as indicated in Figure 4.18 (see Problem 4.19).

The electric field produced by the image charge at the center of the beam is then found to good approximation (for $\xi \ll b$) as

$$E_x = \frac{\rho_L}{2\pi\epsilon_0} \frac{1}{(b^2/\xi) - \xi} \approx \frac{\rho_L}{2\pi\epsilon_0} \frac{\xi}{b^2}. \tag{4.234}$$

The corresponding force on the particle of charge q, $F_x = qE_x$, is directed away from the axis; hence it is defocusing (i.e., it reduces the net restoring force on a centroid particle).

In addition to the electric image force there is also a magnetic image force. The main difference here is that we must distinguish between the ac case and the dc case. When the beam consists of a pulse, or a sequence of pulses, whose time duration is short compared to the magnetic diffusion time, the situation is similar to the electric image case. The ac currents induced in the conducting wall surrounding the beam produce a magnetic image field that must be tangential to the boundary surface. The associated force reduces the electrostatic image force by the factor $1 - \beta^2 = \gamma^{-2}$, and we obtain for the net image force in this ac case the result

$$F_i = \frac{q\rho_L(1 - \beta^2)}{2\pi\epsilon_0} \frac{\xi}{b^2} = \frac{qI}{2\pi\epsilon_0 c \beta\gamma^2} \frac{\xi}{b^2}. \tag{4.235}$$

If we add this force to the external focusing force and use x in place of ξ, we obtain the following equation of motion for a particle representing the centroid of the beam:

$$x'' + (k_0^2 - k_i^2)x = 0, \tag{4.236}$$

where

$$k_i^2 = \frac{K}{b^2}. \tag{4.237}$$

$K = (I/I_0)(2/\beta^3\gamma^3)$ denotes the generalized perveance of the beam, as before. The image effect thus reduces the focusing force, and the corresponding effective phase advance is given by

$$\sigma_{\text{eff}} = \left(\sigma_0^2 - \frac{K}{b^2} S^2\right)^{1/2}. \tag{4.238a}$$

When the emittance term is negligible so that $K = k_0^2 a^2 = (\sigma_0^2/S^2)a^2$ this relation can be written in the form

$$\sigma_{\text{eff}} = \sigma_0\left(1 - \frac{a^2}{b^2}\right)^{1/2}, \tag{4.238b}$$

which shows a simple dependence on the ratio of the beam radius a to tube radius b.

The above analysis is valid only for beams with short pulse lengths, or for the early part of a long pulse (i.e., for times that are short compared to the magnetic diffusion time, τ_m).

In the dc case (i.e., for continuous beams or, more generally, for beams whose pulse duration is large compared to τ_m), the conducting wall is no longer a boundary for the magnetic field produced by the beam. The currents induced in the wall decay exponentially with a characteristic time constant τ_m, and at time $t \gg \tau_m$ the magnetic field of the beam has completely penetrated, or diffused, through the tube walls surrounding the beam. Thus there is no magnetic image effect due to the conducting walls in this case. However, there still can be another magnetic image force if there is magnetic material outside the beam tube, such as the ferromagnetic poles of dipole or quadrupole magnets. The dc magnetic field of the beam will then be modified to satisfy the boundary condition requiring that the field lines are perpendicular to the pole surface. For simplicity, let us assume that this second image effect can be neglected. Then the dc case is equivalent to free space as far as the magnetic self field is concerned. The factor $1 - \beta^2 = \gamma^{-2}$ introduced for the ac case must then be taken out again, and Equation (4.238) becomes

$$\sigma_{\text{eff}} = \left(\sigma_0^2 - \frac{K\gamma^2}{b^2} S^2 \right)^{1/2} \qquad \text{when } t \gg \tau_m. \qquad (4.239)$$

From standard electromagnetic theory one finds for the magnetic diffusion time the relation

$$\tau_m = \frac{4d^2 \sigma \mu}{\pi^2}, \qquad (4.240)$$

where d is the width of the conducting drift-tube wall, σ the conductivity (not to be confused with the phase advance σ), and μ the magnetic permeability of the wall material. As an example, for a copper wall with $d = 2.5 \times 10^{-3}$ m thickness, using $\sigma \approx 6 \times 10^7 \ (\Omega \cdot \text{m})^{-1}$ and $\mu = \mu_0 = 4\pi \times 10^{-7}$ H/m, one finds that $\tau_m = 1.9 \times 10^{-4}$ s. Thus, in this case one would use Equation (4.238) for beams with pulse length $\tau_p \ll 200$ μs and (4.239) when $\tau_p \gg 200$ μs. In between these two limits one must take into account the penetration of the magnetic field into the wall as a function of time.

To evaluate the significance of the image effect let us calculate σ_{eff} for two examples. First consider a 100-mA 100-kV proton beam in a focusing channel with $\sigma_0 = 60°$, lens period $S = 0.2$ m, and a drift-tube radius of $b = 0.02$ m. For these parameters one finds that $K = 2 \times 10^{-3}$. Since the beam is nonrelativistic ($\gamma \approx 1$), the magnetic image is negligible (i.e., the pulse length is unimportant), and one obtains for the effective phase advance the result $\sigma_{\text{eff}} \approx 54°$. This corresponds to a decrease of 10% from $\sigma_0 = 60°$.

As a second example, consider an electron beam of 20 A and 100 keV with a pulse length of $\tau_p = 2$ μs propagating through a periodic solenoid channel with $\sigma_0 = 80°$, $S = 0.2$ m, $b = 0.02$ m. Since $\tau_p \ll \tau_m$, Equation (4.238) is valid, the generalized perveance is $K = 8.4 \times 10^{-3}$, and one obtains $\sigma_{\text{eff}} = 60°$ (i.e., a decrease of the phase advance by 25%).

These two examples illustrate that the image force can be significant for beams with high perveance. The effect can readily be incorporated into the theory of coherent beam oscillations by using σ_{eff} in place of σ_0 in Equations (4.219) through (4.233) of this section. Thus Equation (4.233) may be written as $\tilde{A} = 2\tilde{\Delta} \sin(\sigma_{\text{eff}}/2)\sqrt{N}$. Since $\sigma_{\text{eff}} < \sigma_0$, we see that the image effect in the line-charge approximation appears actually to be benign, as it reduces the amplitudes of the coherent beam oscillations. However, one must bear in mind that in practice the image effect may give rise to nonlinear forces and hence emittance increase. This occurs when either the beam size is not significantly smaller than the drift-tube diameter and the particle distribution is not exactly uniform or when the conducting boundaries are nonaxisymmetric, as in the case of electrostatic quadrupole lenses. Such conditions would warrant further analysis and numerical simulation studies that are beyond the scope of this book.

Coherent beam oscillations are a particular problem in linear accelerators. In a frame moving with the particles, the lattice of acceleration gaps and focusing elements is seen as a periodic array of lenses with regard to the transverse motion. If the phase advance σ_0 is constant in the accelerator, our analysis can be applied to this problem. Otherwise, the theory can be modified appropriately. Except near injection, the image effects can be neglected since $K \sim I/\beta^3\gamma^3$ rapidly decreases with increasing energy. In place of the image force, however, a much more serious effect arises that is especially worrisome in high-current electron linacs. An electron beam displaced from the axis excites electromagnetic waves with transverse electric field components in the accelerator waveguide or drift-tube structures. The transverse electric fields of these waves then interact with the particles arriving later in the beam pulse, thereby increasing the centroid displacement from the axis. Due to the increasing oscillation amplitude, even more energy is fed into these unwanted electromagnetic modes, leading to further beam off-centering. This process, which is intrinsically unstable, is known as the *beam breakup instability*. It occurs in both electron induction linacs with relatively long beam pulses and in rf linacs with short bunches. In the latter case, the instability is also known as the *transverse wakefield effect*. This is because the effect can be described in terms of the *wakefield* generated by a short relativistic electron bunch passing through an aperture in a disk-loaded waveguide structure, or through any other discontinuity. The transverse component of the wakefield produced by the bunch head can displace the tail of the same bunch or affect other bunches trailing behind. The effect of coherent beam oscillations and associated instability can be minimized by careful design, such as precision alignment, use of dipole magnets for beam steering at periodic intervals, programming of rf phase history in an rf linac, and other measures. The variation in rf phase introduces an energy spread in the bunch. This, in turn, produces a spread in the transverse oscillations (σ_0) which destroys the coherence in the interaction with the transverse electromagnetic field components, thereby damping the instability.

4.5 SPACE-CHARGE TUNE SHIFT AND CURRENT LIMITS IN CIRCULAR ACCELERATORS

4.5.1 Betatron Tune Shift due to Self Fields

So far in this chapter on linear beam optics with self fields we have restricted our analysis to straight beams such as beam transport through periodic-focusing channels. However, the uniform beam model with linear forces can also be applied to circular accelerators. The main difference is that in circular systems the particles pass through the same focusing lattice repeatedly in many revolutions. Therefore, any field errors, misalignments, and nonlinearities have a much more serious effect than in straight transport lines. In particular, the change in the betatron oscillation frequency due to space charge that can be tolerated is considerably smaller in circular machines than in linear accelerators or transport lines. As discussed in Section 3.8.6, imperfections and nonlinear forces cause resonance-like amplitude growth for the transverse betatron oscillations that must be avoided. These forbidden resonances are summarized in Equation (3.406) by the relation $m\nu_x + n\nu_y = p$, where ν_x, ν_y are the two betatron tunes and m, n, p are integers. In the design of synchrotrons and storage rings, one therefore chooses an operating point of ν_x and ν_y that is not near any dangerous resonances. However, the resonance lines in the ν_x versus ν_y diagram are so closely spaced that relatively small changes of the betatron tune may drive the beam into instability. By far the most important effect in this regard is the tune shift caused by the defocusing self-field forces.

Another important difference between circular beams and straight beams is the effect of dispersion due to the momentum spread in the particle distribution [see Section 3.6.4, Equations (3.272) to (3.276)]. For the calculations of space-charge effects in circular accelerators presented in this section, we will, however, neglect the momentum spread. The inclusion of dispersion into the tune-shift formulas is given in Section 5.4.7. When dispersion is neglected, calculation of the betatron tune shift is a straightforward extension of the smooth-approximation theory for a FODO channel presented in Section 4.4.3. Let us assume a focusing lattice with $\nu_x = \nu_y = \nu_0$ without space charge and a beam with identical emittance $\epsilon_x = \epsilon_y = \epsilon$ in both directions. Furthermore, assume that the beam is matched, having a mean cross-sectional radius $\overline{X} = \overline{Y} = a$ and that it extends uniformly along the entire circumference C of the synchrotron. The relationship between beam radius a, average focusing strength k_0^2, generalized perveance K, and emittance ϵ is then determined by the envelope equation (4.193), which, if $\overline{R}$ is replaced by a, takes the form

$$k_0^2 a - \frac{K}{a} - \frac{\epsilon^2}{a^3} = 0. \tag{4.241}$$

The wave number k_0 is defined by the ratio of the phase advance per period without space charge, σ_0, and the length of one period, S, that is,

$$k_0 = \frac{2\pi}{\lambda_0} = \frac{\sigma_0}{S}.$$

Introducing the betatron tune $\nu_0 = N\sigma_0/2\pi$ [Equation (3.352)], where N is the number of focusing cells along the circumference, C, and defining the mean radius

R of the equilibrium orbit by $C = 2\pi R$, we get the alternative relation

$$k_0 = \frac{2\pi}{\lambda_0} = \frac{\nu_0}{R}.$$

(4.242)

Likewise, we can define the wave number k and depressed tune ν due to self fields as

$$k = \frac{2\pi}{\lambda} = \frac{\nu}{R}.$$

(4.243)

The envelope equations (4.241) may then be written in the alternative form

$$k^2 a - \frac{\epsilon^2}{a^3} = 0,$$

(4.244)

where

$$k^2 = k_0^2 - \frac{K}{a^2}.$$

(4.245)

Subtracting (4.244) from (4.241), we obtain

$$k_0^2 - k^2 = \frac{\nu_0^2 - \nu^2}{R^2} = \frac{K}{a^2}.$$

(4.246)

Since the allowed tune shift $\Delta\nu = \nu - \nu_0$ is very small compared to the tune ν_0 (i.e., $\Delta\nu \ll \nu_0$), we have

$$\nu_0^2 - \nu^2 = -(\nu_0 + \nu)\Delta\nu \approx -2\nu\Delta\nu,$$

hence

$$\Delta\nu = -\frac{KR^2}{2\nu a^2} = -\frac{IR^2}{I_0\beta^3\gamma^3\nu a^2}.$$

(4.247)

From (4.244) we obtain $k^2 = \nu^2/R^2 = \epsilon^2/a^4$, or

$$\epsilon = \frac{\nu}{R}a^2.$$

(4.248)

The corresponding normalized emittance is then

$$\epsilon_n = \beta\gamma\epsilon = \frac{\beta\gamma\nu a^2}{R}.$$

(4.249)

Substitution of (4.249) in (4.247) yields

$$\Delta \nu = -\frac{IR}{I_0 \epsilon_n \beta^2 \gamma^2}. \tag{4.250}$$

The above analysis for a continuous, or unbunched, beam can readily be extended to the case where the beam consists of discrete bunches. The current I in Equation (4.250) must then be replaced by the peak current in the bunch, $\hat{I}$. It is customary to use the bunching factor B_f, which is defined by the ratio of the average current, $\bar{I}$, to the peak current, $\hat{I}$, that is,

$$B_f = \frac{\bar{I}}{\hat{I}}, \quad \text{or} \quad \hat{I} = \frac{\bar{I}}{B_f}. \tag{4.251}$$

Note that B_f has the range $0 < B_f \leq 1$, and that $B_f = 1$ represents the unbunched beam treated above.

Introducing $\hat{I} = \bar{I}/B_f$ in place of the current I, we obtain from (4.250) the relation

$$\Delta \nu = -\frac{\bar{I}R}{I_0 \epsilon_n \beta^2 \gamma^2 B_f}. \tag{4.252}$$

An alternative form often found in the literature uses in place of the average current the total number of particles $N_t = 2\pi R I/q\beta c$ and in place of I_0 the classical particle radius r_c. This yields

$$\Delta \nu = -\frac{N_t r_c}{2\pi \epsilon_n \beta \gamma^2 B_f}, \tag{4.253}$$

where

$$r_c = \frac{q^2}{4\pi \epsilon_0 mc^2}. \tag{4.254}$$

For the proton, $r_c = 1.535 \times 10^{-18}$ m, and for the electron, $r_c = 2.818 \times 10^{-15}$ m. As mentioned earlier, the effect of dispersion on the space-charge tune shift will be discussed in Section 5.4.7. Solving (4.252) for the average current, one gets

$$\bar{I} = \frac{I_0 \epsilon_n \beta^2 \gamma^2 B_f \Delta \nu}{R}, \tag{4.255}$$

where $I_0 \approx 3.1 \times 10^7$ A for protons and 1.7×10^4 A for electrons.

As the above formulas indicate, the tune shift is proportional to the beam intensity (i.e., current I or particle number N_t) and is inversely proportional to the normalized emittance ϵ_n. Furthermore, it decreases with increasing energy and is therefore most pronounced at the injection point of a synchrotron. As a general conservative rule one tries to limit the tune shift at injection to a value of

$$|\Delta \nu| \leq 0.25. \tag{4.256}$$

As an example, consider the booster synchrotron for the Fermilab proton accelerator (see D.10, Table B.1 in Appendix A). It has a circumference of $C = 2\pi R \approx$ 470 m and an injection energy of 200 MeV ($\beta\gamma \approx 0.7$). The normalized emittance for the injected proton beam has a design value of $\epsilon_n = 8$ mm-mrad. Assuming a bunched beam with a bunching factor of $B_f = 0.25$, and taking $|\Delta\nu| = 0.25$, one finds for the allowed average current the value $\bar{I} \approx 98$ mA, which corresponds to a total number of particles of $N_t \approx 1.7 \times 10^{12}$.

In practice, tune shifts greater than $|\Delta\nu| = 0.25$ may be tolerated as long as the emittance dilution or particle losses that occur when a resonance is encountered remain within acceptable limits. The problem of resonance traversal of the beam due to the space-charge tune shift was first studied theoretically by M. Month and W. T. Weng, who concluded that $|\Delta\nu|$ can exceed the theoretical limit of 0.25 significantly. [See the review article by W. T. Weng in *AIP Conference Proceedings* **153**, pp. 349–389, (1987) listed in D.7, and references therein.] This finding is in agreement with experimental observations and has also been confirmed in numerical simulation studies by I. Hofmann [*Part. Accel.* **39**, 169 (1992)]. It should also be pointed out in this context that actual beams usually do not have the uniform density profile implied by the above theory. Take, for example, a Gaussian profile with standard deviation δ, that is,

$$n(r) = n_0 \exp\left(-\frac{r^2}{2\delta^2}\right). \tag{4.257}$$

The radial force due to the electric and magnetic self fields of such a Gaussian distribution is a nonlinear function of the radius r. Hence, in contrast to the uniform-beam case, the betatron oscillation frequency is a function of a particle's position in the beam [i.e., $\nu = \nu(r)$]. The betatron tune shift $\Delta\nu$ of particles near the center of the beam is larger than in the equivalent uniform beam, while particles in the Gaussian tail have a smaller tune shift. As a result, particles in the core of the distribution may encounter a resonance while the outer particles remain unaffected. This may lead to an increase of the effective emittance, thereby reducing the tune shift and moving the particle distribution away from the resonance.

From Equation (4.252) we see that the cure for emittance dilution and beam loss due to the betatron tune shift is higher injection energy and smaller ring size. The original booster synchrotron at Fermilab, for instance, has an injection energy of 200 MeV and a radius of 75 m. This was found to restrict the beam intensity severely, and an upgrade to 400 MeV has been undertaken.

4.5.2 Current Limits in Weak- and Strong-Focusing Systems

The theory of tune shift due to space charge presented in this section also applies to other circular accelerators, such as betatrons and cyclotrons, where the focusing is weaker than in the case of strong-focusing synchrotrons. As an example, let us consider a hypothetical betatron with a field index of $n = 0.5$, hence $\nu_r = \nu_z = \sqrt{0.5} \approx 0.7$. The main effect in this case is that the space charge decreases the already weak focusing force. This, in turn, increases the beam size and may lead to particle losses to the walls. An upper limit for the beam current exists where the depressed tune approaches zero (i.e., $\nu \to 0$), and hence the net focusing force is zero. We can calculate this space-charge limit for a betatron from Equation (4.246) and obtain

$$K_{max} = \nu_0^2 \frac{a^2}{R^2},$$

or

$$I_{max} = \frac{I_0 \nu_0^2 a^2 \beta^3 \gamma^3}{2R^2}. \tag{4.258}$$

As an example, suppose that the betatron has a radius of $R = 0.5$ m, a useful beam aperture of $a = 1$ cm, and operates at a tune of $\nu_0 = \sqrt{0.5}$. If the electron beam is injected at an energy of 100 keV ($\beta\gamma = 0.656$), one finds that the maximum current is 4.8 A.

In deriving the relation (4.258) we neglected the emittance of the beam. If the emittance is included, we obtain from (4.241) the more accurate results

$$K_{max} = \nu_0^2 \frac{a^2}{R^2} - \frac{\epsilon^2}{a^2},$$

or

$$I_{max} = \frac{I_0 \beta^3 \gamma^3}{2} \left(\frac{\nu_0^2 a^2}{R^2} - \frac{\epsilon_n^2}{\beta^2 \gamma^2 a^2} \right). \tag{4.259}$$

The maximum current that can be injected into the betatron is then reduced by an amount that depends on the emittance. On the other hand, the above relations show that the current can be increased substantially by raising the injection energy ($I \sim \beta^3 \gamma^3$). At an injection energy of 1 MeV ($\beta\gamma = 2.783$), for instance, the maximum current increases by a factor $(2.783/0.656)^3 = 76.5$: in the above example from 4.8 A to 367 A if the emittance term can be neglected.

We note in this context that various methods to improve the focusing, and hence to increase the space charge limit, in a betatron have been proposed or studied. The first idea, tried unsuccessfully in the 1950s, was to use charge neutralization (*plasma betatron*). More recent interests have focused on high-current acceleration schemes involving toroidal magnetic fields [14], *modified betatrons* with additional toroidal [15], or stellarator-type [16] fields, and other configurations. A good review of these various schemes, including a detailed list of relevant papers, can be found in Reference 16.

The net effect of all these schemes is to increase the effective betatron frequency significantly. In a sense, the betatron is converted into a strong-focusing device with $\nu_0 > 1$.

The difference between a weak-focusing circular accelerator, such as a conventional betatron or constant-gradient synchrotron, and a strong-focusing machine can be illustrated by comparing the beam current that can be handled by each device. Let $\nu_{0,w} = \sqrt{0.5}$ be the effective betatron tune in the weak-focusing machine, as in our previous example. The corresponding maximum perveance is then obtained from (4.246) by setting $\nu_{0,w}^2 = 0.5$ and $\nu^2 = 0$, that is,

$$K_w = 0.5 \frac{a^2}{R^2}. \tag{4.260}$$

For the strong-focusing device, we will assume that the allowed tune shift is $|\Delta \nu_s| = |\nu_s - \nu_{0,s}| = 0.25 \ll \nu_{0,s}$; hence, from (4.247),

$$K_s \approx 2\nu_{0,s} |\Delta \nu_s| \frac{a^2}{R^2} = 0.5 \nu_{0,s} \frac{a^2}{R^2}. \tag{4.261}$$

Assuming that both machines have the same energy, major radius R, and minor radius a, we find from the last two equations that

$$\frac{K_s}{K_w} = \frac{I_s}{I_w} = \nu_{0,s} \quad \text{(for } \nu_{0,s} \gg 1\text{)}. \tag{4.262}$$

Thus the current ratio scales linearly with the effective betatron tune ν_0 that can be achieved in the strong-focusing configuration.

As a further application of the theory of tune depression developed in this section, let us compare a circular, strong-focusing lattice with a linear FODO transport system having the same focusing strength per unit length, as defined by $k_0 = \sigma_0/S = \nu_0/R$. For the linear channel, there is no constraint on the tune depression (i.e., $\sigma \to 0$, or $k \to 0$). Hence, from Equation (4.246) we obtain for the maximum perveance

$$K_l = k_0^2 a^2. \tag{4.263}$$

In the circular system, the depressed tune is limited by an allowed shift of $|\Delta \nu| \ll \nu_0$, so that

$$K_c = \frac{2\nu_0 |\Delta \nu|}{R^2} a^2 = 2k_0^2 \frac{|\Delta \nu|}{\nu_0} a^2. \tag{4.264}$$

The ratio of the beam currents that can be handled by the two systems is then given by

$$\frac{I_l}{I_c} = \frac{1}{2} \frac{\nu_0}{|\Delta \nu|} = 2\nu_0 \qquad (4.265)$$

if we assume a tune shift of $|\Delta \nu| = 0.25$. Thus, as an example, for a tune ν_0 between 5 and 6, the linear channel can transport a 10- to 12-fold higher current than the equivalent circular machine.

4.5.3 Effects of Image Forces on Coherent and Incoherent Betatron Tune

In the foregoing analysis of self-field effects on the betatron oscillations and beam currents in circular machines we have not considered the image forces due to the boundaries surrounding the beam. For a vacuum tube with circular cross section, we could apply the results of Section 4.4.4 directly to our problem. However, in practice, many vacuum tubes in circular accelerators have a rectangular or elliptical shape, with the height, Δy, usually considerably smaller than the radial width, Δx. In general, the boundary problem can be very complicated, as both tube cross sections and wall materials may vary along the circumference of the accelerator. Also, as discussed in Section 4.4.4, the magnetic field case is much more complicated than the electric image problem and one must distinguish between short-time (ac) and long-time (dc) behavior.

The problem of electric and magnetic image forces in vacuum tubes with various geometries and boundary conditions was treated in detail by Laslett [17]. We limit our analysis to the simple model of a line charge ρ_L between two infinite conducting planes, illustrated in Figure 4.19. As shown in the figure, we assume that the beam is displaced in the vertical direction from the midplane by a distance η, which is considered to be small compared to the separation $2h$ of the two planes (i.e., $\eta \ll 2h$). This will allow us to evaluate the image effects for both the coherent oscillations of the beam as a whole and the incoherent betatron oscillations of the individual particles. Following Laslett [17], we will limit the analysis to the vertical motion (y-direction) where the image force is defocusing and the beam size is limited by the aperture $2h$ between the conducting planes.

The two-dimensional electrostatic field problem in Figure 4.19 can be treated either by conformal mapping or by the method of images. We choose the latter, which is well known from basic electrostatic theory. The two nearest images with respect to the upper and lower planes are shown in the figure. The first two have a negative line charge of $-\rho_L$ and are located at points $y_1 = 2h - \eta$ and $y_2 = -(2h + \eta)$. At a distance y from the origin in the $x = 0$ plane, these two image charges produce electric fields $E_{y1}^i = (\rho_L/2\pi\epsilon_0)(2h - \eta - y)^{-1}$ and $E_{y2}^i = -(\rho_L/2\pi\epsilon_0)(2h + \eta + y)^{-1}$. The two negative images generate positive images, which in turn produce negative images, and so on.

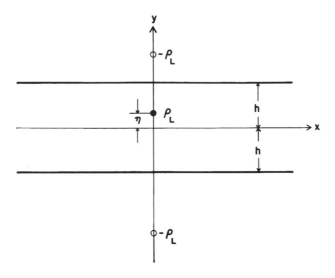

Figure 4.19. Beam represented by a line charge ρ_L between two conducting planes. The line charge is displaced from the midplane by a distance η. The first images above and below the conducting planes are indicated.

By summation of the contributions from this infinite series of images one obtains

$$E_y^i =$$

$$\frac{\rho_L}{2\pi\epsilon_0}\left[\frac{1}{2h - \eta - y} - \frac{1}{2h + \eta + y} + \frac{1}{4h - \eta + y} - \frac{1}{4h + \eta - y} + \cdots\right],$$

$$(4.266)$$

which can be written in the form

$$E_y^i =$$

$$\frac{2\rho_L}{2\pi\epsilon_0}\left[\frac{\eta + y}{4h^2 - (\eta + y)^2} + \frac{\eta - y}{16h^2 - (\eta - y)^2} + \frac{\eta + y}{36h^2 - (\eta + y)^2} + \cdots\right].$$

$$(4.267)$$

By assuming that $|\eta \pm y| \ll 2h$, one gets the first-order approximation

$$E_y^i \simeq \frac{\rho_L}{4\pi\epsilon_0 h^2}\left[\frac{\eta + y}{1} + \frac{\eta - y}{4} + \frac{\eta + y}{9} + \frac{\eta - y}{16} + \cdots\right],$$

or

$$E_y^i = \frac{\rho_L}{4\pi\epsilon_0 h^2}\left[\left(1 + \frac{1}{4} + \frac{1}{9} + \frac{1}{16} + \cdots\right)\eta + \left(1 - \frac{1}{4} + \frac{1}{9} - \frac{1}{16} + \cdots\right)y\right].$$
(4.268)

The numbers in brackets associated with η represent the series of expansion of $\pi^2/6$, and those associated with y represent $\pi^2/12$. Thus we obtain the result

$$E_y^i = \frac{\rho_L \pi}{48\epsilon_0 h^2}(y + 2\eta).$$
(4.269)

Note that in contrast with the beam in a cylindrical pipe of Section 4.4.4, there is an electrostatic image field in our case, even if the beam is centered in the midplane. For $\eta = 0$ we get, from (2.269),

$$E_y^i = \frac{\rho_L \pi}{48\epsilon_0 h^2} y \qquad (\text{for } \eta = 0).$$
(4.270)

On the other hand, the image field at the center of the displaced beam is found by setting $y = \eta$ in (2.269) and is given by

$$E_y^i = \frac{\rho_L \pi}{16\epsilon_0 h^2}\eta \qquad (\text{for } y = \eta).$$
(4.271)

As can be seen from the last three equations, the image fields produce defocusing forces in the y-direction. On the other hand, since $\nabla \cdot \mathbf{E}^i = 0$, the corresponding forces in the x-direction have opposite signs and are thus focusing.

Let us now evaluate the effect of the image forces on the betatron oscillations of the particles in the beam. If the net force in the y-direction is different from the net force in the x-direction, the matched beam will have an elliptical cross section even in the smooth approximation being used here. Let a and b denote the semiaxes in the radial (x) direction and axial (y) direction, respectively. The trajectory equation for particles in the $x = 0$ plane will then be of the smooth-approximation form

$$y'' + \left[k_{yo}^2 - \frac{2K}{b(a + b)} - k_{yi}^2\right]y = 0.$$
(4.272)

The first term in brackets represents the external focusing force, the second term the space-charge force without images, and the last term, $-k_{yi}^2$, the effect of the electrostatic image force. Using (4.270), one finds for the electrostatic image term

$$k_{yi}^2 = \frac{\pi^2}{24h^2}K\gamma^2,$$
(4.273)

where $K = (I/I_0)(2/\beta^3\gamma^3)$ is the generalized perveance.

The equation of motion (4.272) may be written in the alternative form

$$y'' + k_y^2 y = 0,$$
(4.274)

with

$$k_y^2 = k_{yo}^2 - \frac{2K}{b(a + b)} - k_{yi}^2.$$
(4.275)

The tune shift of the incoherent betatron oscillations is then, by analogy to Equations (4.246) and (4.247), given as

$$\nu_y^2 - \nu_{yo}^2 = -\frac{2KR^2}{b(a + b)}\left[1 + \epsilon_1 \gamma^2 \frac{b(a + b)}{h^2}\right],$$
(4.276a)

or

$$\Delta \nu_y \approx -\frac{KR^2}{\nu_y b(a + b)}\left[1 + \epsilon_1 \gamma^2 \frac{b(a + b)}{h^2}\right],$$
(4.276b)

where the electrostatic image coefficient ϵ_1 for our case has the value

$$\epsilon_1 = \frac{\pi^2}{48}.$$
(4.277)

Equation (4.276) includes in the perveance parameter K the focusing effect $(1 - \beta^2 = \gamma^{-2})$ due to the magnetic self field of the beam without magnetic image forces. It is thus equivalent to the dc case with no ferromagnetic boundaries. The bunched-beam result (ac case) is obtained by setting $\gamma^2 = 1$ in the bracketed term on the right-hand side of Equation (4.276). The correction factor due to image effects depends on the beam size and is most significant in the dc case (unbunched beam). Taking the example of the booster synchrotron for the Fermilab accelerator with an injection energy of 200 MeV ($\gamma = 1.21$) and assuming that $a \approx b = 0.1h$, one obtains

$$\epsilon_1 \gamma^2 \frac{b(a + b)}{h^2} = \frac{\pi^2}{48} \times 1.21^2 \times 2 \times 0.1^2 = 6 \times 10^{-3},$$

which is negligibly small. On the other hand, if the beam size increases (e.g., due to interaction with a resonance, mismatch, poor emittance, or other causes), these image effects may become significant. However, the above first-order results are valid only as long as $a \ll h, b \ll h$, and one must use nonlinear theory or particle simulation to compute the beam behavior in this situation.

To analyze the effect of image forces on the coherent motion of a displaced beam we set $y = \eta$ and $K = 0$ in Equation (4.272) and use relation (4.271) for calculating k_{yi}^2. One finds that

$$k_{yi}^2 = \frac{\pi^2}{8h^2} K\gamma^2,$$ (4.278)

and for the change in the coherent betatron tune

$$\nu_c^2 - \nu_{c0}^2 = -\frac{2KR^2}{h^2}\epsilon_{1c}\gamma^2,$$ (4.279a)

or

$$\Delta\nu_c \approx \frac{KR^2}{\nu_c h^2}\epsilon_{1c}\gamma^2,$$ (4.279b)

where

$$\epsilon_{1c} - \frac{\pi^2}{16}.$$ (4.280)

The result (4.279) corresponds to the dc case when ferromagnetic boundaries are neglected. For a bunched beam (ac limit) we obtain in lieu of (4.279)

$$\nu_c^2 - \nu_{c0}^2 = -\frac{2\overline{K}R^2}{B_f h^2}\epsilon_{1c} = -\frac{4\overline{I}R^2}{I_0 B_f h^2 \beta^3 \gamma^3}\epsilon_{1c},$$ (4.281a)

or

$$\Delta\nu_c \approx -\frac{\overline{K}R^2}{\nu_c B_f h^2}\epsilon_{1c} = -\frac{2\overline{I}R^2}{\nu_c I_0 B_f h^2 \beta^3 \gamma^3}\epsilon_{1c}.$$ (4.281b)

Taking the parameters for the Fermilab booster, namely $R = 75$ m, $B_f = 0.25$, $\nu_{c0} = 6.7$, $\beta\gamma = 0.7$, and using $\overline{I} = 100$ mA, $h = 5$ cm, we find from (4.281a) that

$$\Delta\nu_c^2 = \nu_c^2 - \nu_{c0}^2 = -\frac{4 \times 0.1 \times 75^2}{3.1 \times 10^7 \times 0.25 \times 0.05^2 \times 0.7^3} \times \frac{\pi^2}{16} = -0.2.$$

In this case the image forces are seen to have a noticeable effect on the coherent motion of a vertically displaced beam.

Laslett calculated the tune shift of the incoherent and coherent oscillations due to self-field effects for a variety of other boundary conditions such as rectangular and elliptical tube cross sections [17]. For the dc case, he introduced the ferromagnetic image force coefficient ϵ_2 in addition to the electrostatic image coefficient ϵ_1. When the beam passes through a dipole bending magnet, for instance, the ferromagnetic boundaries can be represented by two parallel planes separated by a distance $2g$. The dc magnetic image coefficient then has the same value as in the analogous electrostatic problem, namely $\epsilon_2 = \epsilon_1 = \pi^2/48$.

The tune-shift equations for the dc case must be modified when ferromagnetic images are present. Thus Equation (4.276) for the incoherent tune shift becomes

$$\nu_y^2 - \nu_{yo}^2 = -\frac{2KR^2}{b(a+b)}\left[1 + \epsilon_1\gamma^2\frac{b(a+b)}{h^2} + \epsilon_2\beta^2\gamma^2\frac{b(a+b)}{g^2}\right],$$

(4.282a)

or

$$\Delta\nu_y \approx -\frac{KR^2}{\nu_y b(a+b)}\left[1 + \epsilon_1\gamma^2\frac{b(a+b)}{h^2} + \epsilon_2\beta^2\gamma^2\frac{b(a+b)}{g^2}\right].$$

(4.282b)

Accordingly, for relativistic beams ($\beta \simeq 1, \gamma \gg 1$) and pole-shoe separation $2g$ comparable to the vacuum tube aperture $2h$, the ferromagnetic image term is seen to be of the same magnitude as the electrostatic image term.

The corresponding tune shift equation for a bunched beam (ac case), or for pulse durations short compared to the magnetic diffusion time τ_m, is

$$\nu_y^2 - \nu_{yo}^2 = -\frac{2KR^2}{b(a+b)}\left[1 + \epsilon_1\frac{b(a+b)}{h^2}\right],$$

(4.283a)

or

$$\Delta\nu_y \approx -\frac{KR^2}{\nu_y b(a+b)}\left[1 + \epsilon_1\frac{b(a+b)}{h^2}\right].$$

(4.283b)

At relativistic energies the difference between the dc and ac image terms can be very significant. As an example, consider the case of the high-current betatron with 1 MeV injection energy ($\gamma = 3, \beta\gamma = 2.78$) discussed in connection with Equation (4.259). Let $a = b, b/h \simeq 0.2$ and $b/g \simeq 0.18, \epsilon_1 = \epsilon_2 = \pi^2/48$. For a time $t < \tau_m$ after beam injection, the ac formula (4.283) applies and the image term is

$$\epsilon_1\frac{b(a+b)}{h^2} = \frac{\pi^2}{48}\,2 \times 0.2^2 = 1.64 \times 10^{-2}.$$

At later times when $t > \tau_m$, Equation (4.282) applies and the image term is

$$\epsilon_1\gamma^2\frac{b(a+b)}{h^2} + \epsilon_2\beta^2\gamma^2\frac{b(a+b)}{g^2} =$$

$$\frac{\pi^2}{48}(3^2 \times 2 \times 0.2^2 + 2.78^2 \times 2 \times 0.18^2) = 0.25.$$

This is a very significant change that increases the tune shift and reduces the total current that can be accelerated in the betatron. The frequency of the coherent motion of a displaced beam is also strongly affected by this time-varying behavior of the magnetic image forces.

The various formulas and examples presented in this section show that the correction factors for the space-charge tune shift due to the image forces depend on the geometry of the beam and vacuum chamber, the particle energy, and the time variation of the beam. Generally speaking, the larger the ratio of beam size to chamber height, $(a + b)/h$, and the larger the energy (i.e., γ), the greater is the image correction factor. It is interesting to note that this general trend is in just the opposite direction as the space-charge term without images, where one has the scaling $KR^2/[a(a + b)] \propto I[\beta^3\gamma^3 a(a + b)]^{-1}$. Finally, we note that the betatron tune shift due to self field effects, including images, is often referred to in the literature as the *Laslett tune shift*.

4.6 CHARGE NEUTRALIZATION EFFECTS

4.6.1 Ionization Cross Sections for Electron and Proton Beams in Various Gases

Since in practice it is not possible to obtain a perfect vacuum in the beam tubes of accelerators and other devices, there is always a finite probability that partial charge neutralization occurs due to ionizing collisions of the beam particles in the background gas. In fact, in some cases, such as the transport of low-energy H^+ or H^- beams, or of intense relativistic electron beams, the background gas is used deliberately to achieve better focusing or even self-focusing via charge neutralization, as will be discussed below. The degree of neutralization depends on the gas density n_g, the chemical composition of the gas, the ionization cross section σ_i for the production of electron–ion pairs, the velocity v of the beam particles, and the pulse length of the beam. The density increase with time, dn/dt, of the electrons or ions created in the collisions between the beam particles, and the number of gas molecules or atoms is given by

$$\frac{dn}{dt} = n_b n_g \sigma_i v, \qquad (4.284)$$

where n_b is the beam density.

The secondary particles created in the collisions that have the same charge polarity as the beam particles are expelled to the walls by the beam's space charge if no magnetic fields are present. (If there is a magnetic field, the situation becomes more complicated as the particles undergo $\mathbf{E} \times \mathbf{B}$ drift.) Those having opposite charge polarity are trapped and contribute to partial charge neutralization. In general, the transverse velocity and spatial distribution of the beam and the trapped particles are different, so that the charge-neutralization factor f_e is a function of radius r and distance z along the beam [i.e., $f_e = f_e(r, z)$]. However, in this section on charge-neutralization effects, we treat f_e as a constant since we are interested here in the gross features and linear beam theory rather than in a truly self-consistent description. An important parameter is the *charge-neutralization time*, τ_N, defined as the time it takes to obtain full charge neutralization of the beam ($f_e = 1$). Suppose that both beam and secondary particles are singly charged and that the gas and beam densities are constant. If we neglect recombination of electrons with ions, which is a valid assumption at low density and short times, then Equation (4.284) can be readily integrated, yielding $n(t) = n_b t / \tau_N$, with the relation

$$\tau_N = \frac{1}{n_g \sigma_i v} \tag{4.285}$$

for the charge-neutralization time (where $n = n_b$, or $f_e = 1$). The corresponding mean free path between ionizing collisions is $\bar{l}_i = \tau_N v = (n_g \sigma_i)^{-1}$.

For an ideal gas, the density at standard atmospheric pressure (760 torr) and standard temperature (0°C) is given by Loschmidt's number $n_L = 2.69 \times 10^{25}$ m^{-3}. Thus the relationship between n_g and the pressure p is

$$n_g[\text{m}^{-3}] = 3.54 \times 10^{22} \times p[\text{torr}] = 2.65 \times 10^{20} \times p[\text{Pa}], \tag{4.286}$$

where 1 pascal (Pa) = 7.5×10^{-3} torr and 10^5 Pa = 1 bar.

The ionization cross section depends on the velocity v of the beam particles and the atomic properties of the gas species. Experimental data and empirical scaling laws based on Bethe's theory [18] can be found in the literature. Good references for electron impact ionization cross sections in various gases are the papers by Kieffer and Dunn [19] for electron energies below 10 kV, and by Rieke and Prepejchal [20] for electron energies above 10 keV. Following Slinker, Taylor, and Ali [21], the general scaling law for electron impact ionization can be written in the form

$$\sigma_i = \frac{8a_0^2 \pi I_R A_1}{m_e c^2 \beta^2} f(\beta) \left(\ln \frac{2A_2 m_e c^2 \beta^2 \gamma^2}{I_R} - \beta^2 \right), \tag{4.287a}$$

or, numerically,

$$\sigma_{i[\text{m}^2]} = \frac{1.872 \times 10^{-24} A_1}{\beta^2} f(\beta)[\ln(7.515 \times 10^4 A_2 \beta^2 \gamma^2) - \beta^2], \tag{4.287b}$$

where $a_0 = 0.529 \times 10^{-10}$ m is the Bohr radius, $I_R = 13.6$ eV the ionization energy of atomic hydrogen (Rydberg energy), $\beta = v/c$, $\gamma = (1 - \beta^2)^{-1/2}$, and m_e is the electron mass. A_1 and A_2 are dimensionless empirical constants that depend on the gas species. $f(\beta)$ is a correction function for fitting the data at low energy near the threshold where the kinetic energy T of the electrons equals the ionization energy I_i. It is given by [21]

$$f(\beta) = \frac{I_i}{T}\left(\frac{T}{I_i} - 1\right) = \frac{2I_i}{m_e c^2 \beta^2}\left(\frac{m_e c^2 \beta^2}{2I_i} - 1\right). \tag{4.288}$$

Note that $f(\beta) = 0$ at $T = I_i$ where $\sigma_i = 0$, and $f(\beta) \rightarrow 1$ for $T \gg I_i$.

The constants A_1 and A_2 are related to the parameters M^2 and C of Rieke and Prepejchal [20] by

$$A_1 = M^2 \qquad \text{and} \qquad 7.515 \times 10^4 A_2 = e^{C/M^2}.$$

Specifically, for electron ionization of molecular hydrogen (H_2), where $I_i = 15.4$ eV, one finds (with $M^2 = 0.695$, $C = 8.115$ from Table IV of Reference 20)

$$A_1 = 0.695 \qquad \text{and} \qquad A_2 = 1.567,$$

By substitution of these values in Equation (4.287b), one obtains for the electron ionization cross section in H_2 the relation

$$\sigma_{i[m^2]} = \frac{1.301 \times 10^{-24}}{\beta^2} f(\beta)\left[\ln\left(1.177 \times 10^5 \beta^2 \gamma^2\right) - \beta^2\right], \tag{4.289a}$$

with

$$f(\beta) = \frac{6.027 \times 10^{-5}}{\beta^2}\left(1.659 \times 10^4 \beta^2 - 1\right). \tag{4.289b}$$

This cross section for electrons in H_2 is plotted in Figure 4.20 as a function of β (a) and of the electron kinetic energy T (b). Table 4.1 shows the values of the constants for H_2, He, Ne, and Ar from the data of Reference 20, where information on many other gas species can be found.

Impact ionization of proton beams at low energies between 0.3 keV and 5 MeV in various gases was studied by Rudd et al. [22]. The authors found that the experimental data could be fit with an empirical scaling law of the form

$$\sigma_i = \left(\frac{1}{\sigma_l} + \frac{1}{\sigma_h}\right)^{-1} = \frac{\sigma_l \sigma_h}{\sigma_l + \sigma_h}, \tag{4.290a}$$

where

$$\sigma_l = 4\pi a_0^2 C x^D, \tag{4.290b}$$

$$\sigma_h = \frac{4\pi a_0^2}{x}\left[A \ln(1 + x) - B\right], \tag{4.290c}$$

$$x = \frac{T_e}{I_R} = \frac{T_p}{1836 I_R} = \frac{m_e c^2 \beta^2}{2 I_R}, \tag{4.290d}$$

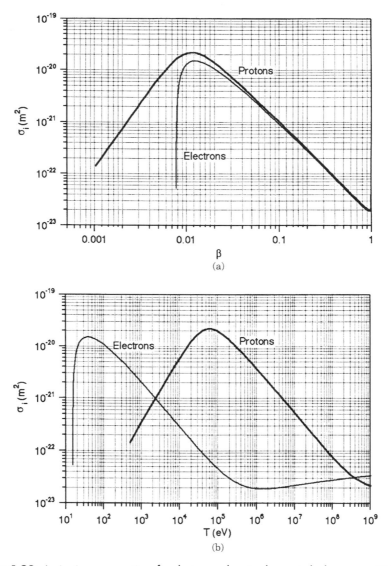

Figure 4.20. Ionization cross sections for electron and proton beams in hydrogen gas (H_2) as a function of $\beta = v/c$ (a) and kinetic energy T (b).

and A, B, C, and D are fitting constants that depend on the gas species. Note that $v = \beta c$ is the velocity of the protons, $T_p = m_p v^2/2$ is the (nonrelativistic) proton energy, and $T_e = m_e v^2/2$ is the kinetic energy of the electron with the same velocity as the proton ($m_p/m_e = 1836$). At the high-energy end of the range covered by this study, say $T_p > 1$ MeV, one has $x \gg 1$, $\ln(1 + x) \approx \ln x$,

Table 4.1 Values of fitting constants for ionization cross section of electrons in several gases

Gas Species	$M^2 = A_1$	C	$A_2 = 1.331 \times 10^{-5} e^{C/M^2}$
He	0.745	8.005	0.6174
Ne	2.02	18.17	0.1073
Ar	4.22	37.93	0.1066
H_2	0.695	8.115	1.5668

Source: Reference 20.

$\sigma_h \ll \sigma_l$, hence

$$\sigma_i \approx \sigma_h = \frac{1.872 \times 10^{-21}}{\beta^2} \left[A \ln(1.879 \times 10^4 \beta^2) + B \right]. \tag{4.290e}$$

On the other hand, for protons with kinetic energies $T_p > 1$ MeV the behavior of the ionization cross section is similar to that of electrons; that is, the relativistic formula (4.287) can be applied, setting $f(\beta) = 1$. Comparison of relations (4.290e) and (4.287b) in the nonrelativistic region near T_p of 1 to 5 MeV where they overlap and where $\beta^2 \ll 1$ shows that

$$A_1 = A \quad \text{and} \quad A_2 = 0.25 e^{B/A}.$$

As an example, let us take the ionization cross section for proton beams in molecular hydrogen (H_2). From Reference 22 one obtains for the constants in this case

$$A = 0.71, \quad B = 1.63, \quad C = 0.51, \quad D = 1.24;$$

$$A_1 = 0.71, \quad A_2 = 2.483.$$

The corresponding formulas for the cross section σ_i are then

$$\sigma_{l[m^2]} = 3.575 \times 10^{-15} \beta^{2.48}, \tag{4.291a}$$

$$\sigma_{h[m^2]} = \frac{1.872 \times 10^{-24}}{\beta^2} \left[0.71 \ln(1 + 1.879 \times 10^4 \beta^2) + 1.63 \right]$$
$$\text{for } T_p < 5\text{MeV}, \tag{4.291b}$$

$$\sigma_{i[m^2]} = \frac{1.329 \times 10^{-24}}{\beta^2} \left[\ln(1.866 \times 10^5 \beta^2 \gamma^2) - \beta^2 \right]$$
$$\text{for } T_p > 5\text{MeV}, \tag{4.291c}$$

Figure 4.20 shows the cross section for protons in hydrogen gas (H_2) as a function of kinetic energy based on Equations (4.291a) to (4.291b), for the

Table 4.2 Values of fitting constants for ionization cross section of protons in various gases

Gas Species	$A = A_1$	B	C	D	$A_2 = 0.25e^{B/A}$
H	0.28	1.15	0.44	0.907	15.193
He	0.49	0.62	0.13	1.52	0.886
Ne	1.63	0.73	0.31	1.14	0.391
Ar	3.85	1.98	1.89	0.89	0.418
Kr	5.67	5.50	2.42	0.65	0.659
Xe	7.33	11.10	4.12	0.41	1.136
H_2	0.71	1.63	0.51	1.24	2.483
N_2	3.82	2.78	1.80	0.70	0.518
O_2	4.77	0.00	1.76	0.93	0.250
CO	3.67	2.79	2.08	1.05	0.535
CO_2	6.55	0.00	3.74	1.16	0.250

Source: Reference 22.

energy range below 5 MeV and on Equation (4.291c) for energies above 5 MeV. Table 4.2 lists the values of the constants for H_2 and various other gases published in Reference 22, where additional information can be found. Equations (4.287) to (4.291) for the ionization cross section show that the dominant parameter is the particle velocity $v = \beta c$. As can be seen in the plot of σ_i versus the relative velocity β in Figure 4.20 (a) the two curves of σ_i for electrons and protons differ only at low velocities near the peak, $\sigma_{i,max}$, and are practically identical in the high-energy region. This "asymptotic" behavior of the cross section, independent of particle species, is an important feature of the Bethe theory [18]. It can be used to extrapolate experimental data to higher energies, as was done in Figure 4.20 for both electrons and protons or to estimate the cross sections for other species. Thus, the cross sections for positrons (e^-), on the one hand, and antiprotons ($\bar{p}$) and H^- ions, on the other hand, are basically identical with those of electrons and protons, respectively.

4.6.2 Linear Beam Model with Charge Neutralization

The charge-neutralization effects can be incorporated into our linear beam model by using the expression (4.23) for the generalized perveance, that is,

$$K = K_0[1 - \gamma^2 f_e(\tau)] \qquad (4.292)$$

where $K_0 = (I/I_0)(2/\beta^3\gamma^3)$ and $f_e(\tau)$ is the charge-neutralization factor. The parameter τ measures the "time into the beam pulse" at a given position in the beam tube. Thus, $\tau = 0$ defines the instant where the beam front passes through the location of interest. In most cases $f_e(\tau)$ can be approximated as a function of τ by (see [D.4, Fig. 10])

$$
f_e(\tau) =
\begin{cases}
\dfrac{\tau}{\tau_N} & \text{for } 0 \le \tau \le \tau_N \\
1 & \text{for } \tau > \tau_N.
\end{cases}
\tag{4.293}
$$

However, there are notable exceptions, such as beams of time duration $\tau_p < \tau_N$, where $f_{e,\max} = t/\tau_p < 1$, or overneutralized beams, where $f_{e,\max} > 1$ for $\tau > \tau_N$. Also, in practice, the $f_e(\tau)$ curve approaches the steady state $[f_e(\tau) = \text{const}, df_e/d\tau = 0]$ in exponential fashion [i.e., $f_e(\tau) \sim 1 - \exp(-\tau/\tau_N)$], rather than linearly with time. But these are minor details that can be easily accounted for if they become important. Charge neutralization increases the net focusing force acting on the particles and hence reduces the average beam radius. For simplicity, let us assume that the external focusing is provided by a symmetrical periodic-focusing lattice (i.e., $\overline{X} = \overline{Y} = a$) and that the round-beam smooth approximation can be used. The mean radius a then obeys the envelope equation

$$
a'' + k_0^2 a - \frac{K_0}{a}\left[1 - \gamma^2 f_e(\tau)\right] - \frac{\epsilon^2}{a^3} = 0.
\tag{4.294}
$$

The wave number k_0, which represents the external focusing strength, is defined by $k_0 = \sigma_0/S$ [Equation (4.190)]. In circular machines, it relates to the betatron tune without space charge, ν_0, and the mean orbit radius, R, by $k_0 = \nu_0/R$ [Equation (4.242)]. The second and third terms of Equation (4.294) can be combined into a single term, $k^2 a$, with

$$
k^2 = k_0^2 - \frac{K_0}{a^2}\left[1 - \gamma^2 f_e(\tau)\right],
\tag{4.295}
$$

where $k = \sigma/S = 2\pi/\lambda$ defines the effective particle oscillation wavelength, λ, or phase advance, σ, in the presence of self fields and charge neutralization.

Theoretically, a quasi-steady state (matched beam) exists where $a'' = 0$ and $a = \text{const}$ for any given value of the charge-neutralization factor f_e. If the beam is matched at a given time and $f_e(\tau)$ changes adiabatically, the radius will change adiabatically and the beam remains matched at all times (see Section 3.9). The increase of the charge-neutralization factor $f_e(\tau)$ is considered adiabatic if it occurs on a time scale that is long compared to a betatron oscillation period $T_b = \lambda/v$, so that $\Delta f_e/f_e \ll 1$ during the time interval $\Delta\tau = T_b$. If, on the other hand, the change of $f_e(\tau)$ occurs nonadiabatically (i.e., if the rise $df_e/d\tau = \Delta f_e/T_b$ during one betatron period is significant), the beam cannot be matched for the entire pulse duration. In the latter case, if the beam is matched for a particular value of $f_e(\tau)$ at a given time τ, it will be mismatched at other times during the pulse and hence perform envelope oscillations (see Section 4.4.3). Furthermore, as long as $f_e(\tau)$

varies with time, the frequency and amplitude of these oscillations will also change. An estimate of the degree of mismatch can be obtained by comparing the maximum and minimum values of the equilibrium radii for the range considered. As an example, suppose that a space-charge-dominated beam ($K_0 a^2 \gg \epsilon^2$) is matched at $\tau = \tau_1 = 0$ when it is unneutralized ($f_e = 0$). Its radius, $a = a_1$, is then obtained from (4.294) to good approximation as

$$a_1 = \left(\frac{K_0}{k_0^2} \right)^{1/2}. \tag{4.296}$$

At a later time, $\tau = \tau_2$, when $\gamma^2 f_e(\tau_2) = 1$, the space-charge term in the envelope equation is zero and one obtains a matched radius of

$$a_2 = \left(\frac{\epsilon}{k_0} \right)^{1/2}. \tag{4.297}$$

The ratio of these two radii is

$$\frac{a_2}{a_1} = \left(\frac{\epsilon k_0}{K_0} \right)^{1/2} = \left(\frac{I_0 \epsilon_n k_0}{2I} \right)^{1/2} \beta\gamma. \tag{4.298}$$

For space-charge-dominated beams, where $\epsilon k_0 \ll K_0$, the two radii can differ significantly (i.e., $a_2/a_1 \ll 1$). This is the case, for instance, in low-energy, high-brightness proton and H$^-$ beams ($\beta \ll 1$, $f_e \approx 1$) and in intense relativistic electron beams ($\beta \approx 1$, $f_e = 1 - \beta^2 \ll 1$). If such beams are matched into the focusing channel for $f_e = 0$ at the beginning of a pulse, there will be strong envelope oscillations in the later parts of the pulse, and vice versa.

If $f_e > \gamma^2$, the space-charge term in the envelope equation becomes positive (focusing). For sufficiently high partial neutralization, the positive space-charge term becomes equal in magnitude to the negative emittance term. In this case, the beam can be "self-focused" (i.e., external forces are not required to confine the beam). Setting $a'' = 0$ and $k_0 = 0$ in Equation (4.294), we find the equilibrium condition

$$\frac{K_0}{a} (\gamma^2 f_e - 1) = \frac{\epsilon^2}{a^3}, \tag{4.299}$$

from which follows that the neutralization factor f_e must satisfy the relation

$$\gamma^2 f_e - 1 = \frac{\epsilon^2}{K_0 a^2}. \tag{4.300}$$

Alternatively, for a given value of f_e we can solve for the equilibrium radius, which yields

$$a = \frac{\epsilon}{\sqrt{K}} = \epsilon [K_0 (\gamma^2 f_e - 1)]^{-1/2}. \tag{4.301}$$

When $f_e = 1$, this relation is identical with Equation (4.71) of Section 4.2.4 for a fully neutralized Bennett beam. The equilibrium state of a partially neutralized relativistic electron ring confined by an axial magnetic field was first discussed by Budker and formed the basis for the *electron ring accelerator* concept studied in the 1970s (see Schumacher [D.4 and references therein]).

It should be pointed out in this context that the above simplified model represents the effects of the secondary charge-neutralizing particles on the beam particles by a single parameter, $f_e(\tau)$, but it ignores the dynamical details of the mutual interaction between the two particle species. A more accurate theory would have to consider the distribution and motion of both particle species in a self-consistent manner. The main advantage of the linear beam model is its simplicity and the fact that it provides an adequate description of the average behavior of the beam (mean radius, net average charge density, etc.). Linear beam codes that include an empirically determined charge neutralization factor f_e are indispensable tools for the interpretation of experimental data and for the design of beam transport systems with a charge-neutralizing background gas.

In the following subsections we discuss the effects of charge neutralization for four cases of practical interest: low-energy proton and H^- beams, intense relativistic electron beams, high energy circular accelerators and storage rings, and plasma lenses.

4.6.3 Gas Focusing in Low-Energy Proton and H^- Beams

Space-charge effects are most severe at low energy. One of the most difficult problems, therefore, is to focus the high-brightness beam extracted from a proton or H^- source. Such beams must be transported over a distance of typically 30 to 100 cm and matched into the radio-frequency-quadrupole (RFQ) injector that constitutes the first stage of a typical RF linear accelerator facility. The conventional method is to use a combination of gas focusing and magnetic lenses (solenoids or quadrupoles) for this task [24] Magnetic lenses alone are not capable of handling the beam. Judicious use of charge neutralization in the background gas improves the focusing considerably. Even so, some loss of beam current and deterioration of beam quality appears to be unavoidable in such a system.

To understand the effects of charge neutralization in low-energy beam transport, let us first consider a proton beam. Suppose that the kinetic energy of the protons is 100 keV and that the beam passes through a vacuum tube with hydrogen gas (H_2) at a pressure of 10^{-5} torr (i.e., a gas density of $n_g = 3.5 \times 10^{17}\ m^3$). The collisions between the protons and the hydrogen molecules cause dissociation ($H_2 \rightarrow 2H$) and produce positive ions (H_2^+, H^+) which are expelled from the beam region and electrons which remain trapped and gradually neutralize the positive charge of the beam. From Figure 4.20 one finds a cross section of $\sigma_i \simeq 2 \times 10^{-20}\ m^2$ for ionizing collisions by 100-keV protons in H_2. With $v = 4.38 \times 10^6$ m/s for the proton velocity one obtains from Equation (4.285) a neutralization time of $\tau_N = 32.6\ \mu s$. Thus, if the proton beam has a pulse duration of $\tau_p > \tau_N$, the charge-neutralization factor f_e will rise linearly with time from zero at $\tau = 0$

to $f_e = 1$ at $\tau = 32.6$ μs and then remains constant. The proton beam radius will decrease with time from a maximum at $\tau = 0$ to a small value at the fully neutralized state ($\tau > \tau_N$). To get a rough idea of the magnitude of the effect, suppose that the proton current is $I = 200$ mA ($K_0 \approx 4 \times 10^{-3}$), that the normalized emittance $\epsilon_n = 6 \times 10^{-7}$ m-rad ($\epsilon = 4 \times 10^{-6}$ m-rad), and that the average focusing due to magnetic lenses is given by $k_0 = 2\pi/\lambda_0 = 2\pi$ m^{-1}. With these numbers one finds from Equations (4.296) and (4.297) that $a_1 \approx 10$ mm and $a_2 \approx 0.8$ mm, hence $a_2/a_1 = 0.08$. This is a very significant change in the beam radius, and one would expect that a large fraction of the beam current in the initial part of the beam pulse ($\tau < 32.6$ μs) could not be focused into the RFQ and would therefore get lost.

A similar situation exists for the transport of low-energy H$^-$ beams except that the physics is more complicated. Thus, in contrast to the proton case, the electrons produced in the ionizing collisions are repelled from the H$^-$ beams while the positive ions are trapped and provide the charge neutralization. In addition, one or both electrons can be stripped from the H$^-$ ion, converting it into either a neutral particle (H^0) or a proton (H$^+$). For a 100-keV H$^-$ beam passing through hydrogen gas (H$_2$), for instance, the ionization cross section is about the same as in the proton case ($\approx 2 \times 10^{-20}$ m^2). The cross section for stripping [23] one electron from the H$^-$ ion is 4×10^{-20} m^2. Particle losses due to electron stripping depend on the gas density and the length of the gas region and may become significant if the density is too high. A unique feature of low-energy H$^-$ beam transport through a background gas is the fact that a state of overneutralization ($f_e > 1$) can be achieved in which the potential in the beam region is positive with respect to the wall [25]. The positive ions then experience a repulsive force and escape from the beam region with a velocity v_i that depends on the potential at the radial position where they are created. Secondary electrons can also escape from the positive potential well since they are born with an energy distribution [26] that is practically Maxwellian with temperature $k_B T_e$. The temperature depends on the beam voltage, V_b, and the ionization potential of the background gas, V_i, by the approximate relation [26, 27]

$$\frac{k_B T_e}{e} \approx \frac{2}{3} \left(\frac{V_b V_i}{m_b/m_e} \right)^{1/2}, \qquad (4.302)$$

where m_b/m_e represents the mass ratio of beam ions and electrons with $m_b/m_e = 1836$ for protons and $m_b/m_e = 1838$ for H$^-$ ions. As an example, the ionization potential of molecular hydrogen (H$_2 \rightarrow$ H$_2^+ + e$) is $V_i = 15.4$ eV. Hence, 100-keV proton or H$^-$ beams produce electrons with a temperature of $k_B T_e \approx 19.3$ eV and a corresponding mean energy of $1.5 k_B T_e \approx 29.0$ eV.

According to the theory [27], a quasi-steady state exists in which the H$^-$ beam is overneutralized ($f_e > 1$) (i.e., self-focused) and in which the positive potential difference between the center and the edge of the beam is on the order of $\Delta\phi \approx k_B T_e/e$. In this state, the number of electrons and positive ions created by collisions of the H$^-$ beam in the background gas is exactly balanced by the number of electrons and ions escaping from the beam. This overneutralized state can occur only when the gas density exceeds a critical value, $n_{g,0}$, where the mean escape time of the positive ions, τ_i, is equal to the neutralization time, τ_N, and the beam is fully neutralized ($f_e = 1$). If one takes $\tau_i = \bar{x}/\bar{v}_i$, where $\bar{v}_i$ is the mean ion velocity and $\bar{x} = a/2$ is the mean escape distance equal to half the beam radius a, one obtains for the critical gas density the relation [25]

$$n_{g,0} = \frac{2}{a} \frac{\bar{v}_i}{v} \frac{1}{\sigma_i}. \qquad (4.303)$$

The mean escape velocity $\bar{v}_i$ depends on the temperature of the positive ions, which is assumed [25] to be on the order of $k_B T_i \sim 0.1$ eV. As the beam tends to become charge neutralized ($f_v \rightarrow 1$), the initially negative beam potential goes toward zero ($\Delta\phi \rightarrow 0$). The positive ions are then no longer trapped but can escape to the wall with a velocity $\bar{v}_i$. For a gas density $n_g < n_{g0}$, the ion escape time is faster than the neutralization time ($\tau_i < \tau_N$), hence the beam will not become fully neutralized. At the critical density, $n_g = n_{g0}$, the two effects (ion escape and creation) exactly balance each other; hence $f_e = 1$, $\Delta\phi = 0$. When the gas density exceeds the critical value, $n_g > n_{g0}$, the beam becomes overneutralized. The self-focused steady state described above is achieved when the net focusing force acting on the H$^-$ beam due to the positive space charge ($f_e > 1$) is sufficient to balance the beam divergence due to the emittance. The positive beam potential $\Delta\phi$ in this equilibrium state is determined by the electron temperature, which is considerably higher than the ion temperature ($\Delta\phi = k_B T_e \gg k_B T_i$). As a result, the ion escape velocity is greater than at the critical density, that is, we have to good approximation $\bar{v}_i = (2\Delta\phi/m_i)^{1/2}$ rather than $\bar{v}_i = (k_B T_i/m_i)^{1/2}$ at $n_g = n_{g0}$.

To minimize stripping losses and other problems in a low-energy H$^-$ beam transport system with gas focusing, one prefers to operate at as low a gas pressure as possible. Since, according to Equation (4.303), the critical gas density is proportional to the ion escape velocity ($n_{g0} \sim \bar{v}_i$) and $\bar{v}_i$ scales with the ion mass as $\bar{v}_i \sim m_i^{-1/2}$, it is advantageous to use a background gas with high atomic mass such as xenon (atomic mass number $A = 131.3$). The ionization cross section for a 100-keV H$^-$ beam in Xe gas is comparable to that for protons [22] (i.e., $\sigma_i \approx 11 \times 10^{-20}$ m^2). Compared to molecular hydrogen ($A \approx 2$), the critical density which scales as $n_{g0} \sim m_i^{-1/2}\sigma_i^{-1}$ is lower in xenon by a factor of $(131/2)^{1/2}(11/2) = 44.5$.

Let us now examine the steady state of a self-focused H$^-$ beam in the framework of the linear theory developed in Section 4.6.2. With $\gamma = 1$, we obtain from Equation (4.300) for the neutralization factor f_e that is required to focus a beam with emittance ϵ, perveance K_0, and radius a the result

$$f_e - 1 = \frac{\epsilon^2}{K_0 a^2} = \frac{I\epsilon_n\beta}{2I_0 a^2}, \qquad (4.304)$$

where $I_0 = 3.1 \times 10^7$ A is the characteristic current. Alternatively, one can solve for the beam radius and get

$$a = \frac{\epsilon}{[K_0(f_e - 1)]^{1/2}} = \epsilon_n \left[\frac{I_0 \beta}{2I(f_e - 1)} \right]^{1/2}. \tag{4.305}$$

At first sight, Equation (4.305) suggests that the beam radius decreases with increasing current and increasing charge overneutralization. However, the product $I(f_e - 1)$ is related to the positive potential difference in the overneutralized H^- beam by

$$\Delta\phi = V_s = \frac{I(f_e - 1)}{4\pi\epsilon_0 c\beta} = \frac{30I(f_e - 1)}{\beta}, \tag{4.306}$$

according to Equation (4.14). At the same time, $\Delta\phi$ is proportional to the electron temperature, and one has approximately

$$\Delta\phi \approx \frac{k_B T_e}{e} = \frac{2}{3} \left(\frac{V_b V_i}{m_b/m_e} \right)^{1/2}, \tag{4.307}$$

as was pointed out in the discussion following Equation (4.302). Substituting (4.306) and (4.307) into (4.305), one obtains for the beam radius

$$a = \epsilon_n \left(\frac{3}{4} \frac{m_b c^2}{e} \right)^{1/2} \left(\frac{m_b/m_e}{V_b V_i} \right)^{1/4} = 1.74 \times 10^5 \epsilon_n \frac{1}{(V_b V_i)^{1/4}}. \tag{4.308}$$

Thus, according to this rather crude model, the equilibrium radius of an over-neutralized, self-focused H^- beam scales as $a \sim \epsilon_n/V_b^{1/4}$; that is, it is linearly proportional to the normalized emittance ϵ_n, depends only weakly on the beam voltage and, most surprisingly, is independent of the beam current, I. For a 100-keV H^- beam propagating through xenon gas (ionization potential $V_i = 12.1$ V) and having a normalized emittance $\epsilon_n = 2 \times 10^{-7}$ m-rad, one obtains a radius of $a = 10^{-3}$ m = 1 mm. This constitutes a very strong focusing effect which cannot be achieved by external means (magnetic or electrostatic lenses) unless the beam current is very small, say $I \lesssim 50$ mA at 100 keV. This theoretical result explains the appeal of gas focusing. In practice, however, it is very difficult to achieve the ideal equilibrium state described here, and gas focusing in both H^- and proton beam is not well understood theoretically. Geometry effects, plasma-type instabilities, local pressure variations, and other factors make theoretical modeling extremely difficult. Charge neutralization alone is usually inadequate and must be supplemented by magnetic solenoid or quadrupole lenses to achieve better control, especially for matching the beam into the RFQ accelerator. Even so, substantial beam losses and emittance growth are typical for such systems.

Electrostatic quadrupole (ESQ) lenses, which have been used successfully for heavy ions [28], or combinations of ESQ and einzel lenses offer an attractive alternative for low-energy transport of high-brightness H^+ and H^- beams [27, 29, 30]. These provide strong focusing and at the same time prevent charge neutralization and plasma buildup since the ions and electrons created in collisions with the background gas are immediately accelerated to the ESQ electrodes with the appropriate voltage polarity. Ongoing research in this area will undoubtedly lead to improved designs for low-energy transport of both H^- and proton beams [24, 30].

4.6.4 Charge-Neutralization Effects in Intense Relativistic Electron Beams

For high-energy particle beams, the space-charge neutralization effects differ from the low-energy cases discussed in Section 4.6.3 in two ways. First, the ionization cross sections are considerably lower than near the peaks of the ionization curves (see Figure 4.20). Second, the magnetic self field of the beam reduces the space-charge defocusing force by the factor $1 - \beta^2 = \gamma^{-2}$ at relativistic velocities. This means that one can have self-focusing when the beam is only partially neutralized, as pointed out before.

Let us now examine the effects of charge neutralization in intense relativistic electron beams. Such beams are produced by applying high-voltage pulses in the range 0.1 to 10 MV with time durations of typically 10 to 100 ns across a diode. Electron currents ranging from 10^3 to 10^6 A are produced mostly by field emission; thermionic cathodes are also being used when high-brightness currents in the kiloampere range are desired. (For a general review of the physics of intense charged particle beams, see Miller [C 18].)

An important parameter in the theory and application of intense relativistic electron beams is the space-charge limiting current, I_L, discussed in Section 4.2.3. We first consider the case where the beam current is less than I_L. As a specific example, let us examine the self-focusing effect in a 1-MeV 5-kA electron beam. Suppose that this beam has an emittance of $\epsilon = 5 \times 10^{-4}$ m-rad, a pulse length of $\tau_p = 30$ ns, and that it is injected into a drift tube filled with hydrogen gas (H_2) at a pressure of $p = 50$ mtorr. Let the initial beam radius be $a = 1$ cm and the drift-tube radius be $b = 3$ cm. The geometry of this system is illustrated in Figure 4.21, which also shows the potential on the beam axis when the beam can propagate into the drift tube for $I < I_L$ (solid curve). For the case where $I > I_L$, which will be discussed later, the beam cannot propagate and forms a so-called *virtual cathode* near the injection plane; the potential then has the shape shown by the dashed curve.

From Figure 4.20 the ionization cross section for 1-MeV electrons in H_2 gas is 2×10^{-23} m^2, and the resulting neutralization time is $\tau_{N,e} = 100$ ns (i.e., considerably longer than the pulse length of the beam). However, the positive hydrogen ions produced in the collisions are accelerated in the potential well of

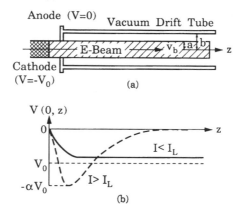

Figure 4.21. (a) Electron beam propagating through a vacuum drift tube and (b) potential on beam axis; the dashed curve indicates the potential when the beam current exceeds the space-charge limit $(I > I_L)$ and electrons are reflected back (virtual cathode formation).

the beam to energies in the range of 100 keV, where they are much better ionizers as the 1-MeV electrons. As mentioned earlier, the ionization cross section for 100-keV protons in H_2 gas is 2×10^{-20} m^2, and the resulting neutralization time at 50 mtorr is $\tau_{N,i} = 6.5$ ns. Olson, who studied this effect, estimated that the effective neutralization time due to the combined action of the relativistic electrons and the positive ions in H_2 gas can be approximated by the relation [D.4, p. 49]

$$\tau_{N[\text{ns}]}^{e,i} \approx \frac{1.0}{p[\text{torr}]}, \tag{4.309}$$

which for $p = 50$ mtorr yields $\tau_N^{e,i} = 20$ ns. The neutralization factor f_e thus increases linearly with time as $f_e(\tau) = \tau/\tau_N^{e,i}$ for $\tau \leq \tau_N^{e,i} = 20$ ns and then remains constant at $f_e = 1$ for the remaining 10 ns of the electron beam pulse. For the given values of the beam radius at injection $(a_0 = 1$ cm$)$ and drift-tube radius $(b = 3$ cm$)$, the space-charge limiting current from Equation (4.61) is $I_L \approx 6$ kA, which is above the beam current of 5 kA. The beam front will thus propagate into the drift tube, but the beam radius will blow up rapidly and hit the drift-tube wall due to lack of focusing. As the charge neutralization increases with time, the radial divergence will decrease until the equilibrium condition is reached at $a = a_0 = 1$ cm. Since $K_0 = 2.6 \times 10^{-2}$ and $\epsilon = 5 \times 10^{-4}$ m-rad, we obtain from (4.300) a neutralization factor of $f_e = 0.12$ that occurs at a time of 2.4 ns after injection of the beam front. If this value of the neutralization factor could be kept constant, the beam radius would remain matched, with $a = a_0 = 1$ cm, through the remainder of the pulse. In fact, however, f_e increases further, hence the beam will become overfocused and experience large envelope oscillations. The amplitude and wavelength of these oscillations for a given slice of beam within the pulse (defined by the time τ from the beam front) can be obtained by solving the envelope equation (4.294), with initial conditions $a_0 = 1$ cm, $a_0' = 0$ and using the value $f_e(\tau)$ for the neutralization parameter. The large envelope oscillations in this example of a relativistic electron beam are unavoidable since the change of $f_e(\tau)$ during the first 20 ns is nonadiabatic.

A special situation arises when the electron-beam current, I, exceeds the space-charge limiting value, I_L, given in Equations (4.61) and (4.62). In this case the beam front will not propagate into the drift tube and a virtual cathode forms until $f_e(\tau)$ becomes sufficiently large that I_L exceeds the beam current, so that $I < I_L$. The beam behavior depends very strongly on the pressure of the gas in the drift tube. An interesting feature is the fact that the positive ions formed in the collisions between electron beam and background gas experience *collective acceleration* as the electron beam propagates into the drift tube after it is sufficiently charge neutralized. The collective acceleration is attributed to the high electric field associated with the virtual cathode and its subsequent motion as the beam propagates. Figure 4.22 depicts the situation in the early stage where the beam enters into the drift region through the anode plane. A fraction of the beam corresponding to the limiting current, I_L, will propagate into the drift tube. The rest, corresponding to the difference $I - I_L$, will be reflected back into the anode and diode region. A virtual cathode forms at a short distance d_m beyond the anode plane. The associated potential variation along the z-axis is depicted at the bottom of

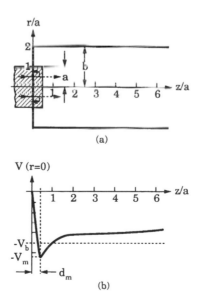

(a)

(b)

Figure 4.22. IREB injection into drift tube when current is above the space-charge limit ($I > I_L$). (a) Beam front stops a short distance d_m from the anode, electrons are reflected; (b) typical potential variation along the beam axis; the potential minimum V_m at virtual cathode exceeds the beam voltage V_b.

Figure 4.22. The potential drops almost linearly from zero at $z = 0$ to a minimum of $-V_m$ at $z = d_m$, whose magnitude can exceed the electron-beam voltage, V_b. Using a planar geometry the electric field at $z = 0$ can be calculated in terms of the injected current density J and the electron energy factor γ_b, and one obtains

$$E = \left(\frac{4mcJ}{e \epsilon_0} \right)^{1/2} (\gamma_b^2 - 1)^{1/4}. \tag{4.310}$$

With $J = I/a^2 \pi$ this can be expressed as

$$E_{[\text{MV/m}]} = \frac{2 \times 10^2}{a[\text{cm}]} \left(\frac{I}{I_0} \right)^{1/2} (\gamma_b^2 - 1)^{1/4}, \tag{4.311}$$

where a is the beam radius and $I_0 = 17$ kA the characteristic current. As an example, for $I = 2I_0 = 34$ kA and $a = 1$ cm one gets $E = 485$ MV/m. Theory and computer simulation show that the virtual cathode potential actually oscillates axially and in magnitude with roughly the plasma frequency, ω_p, about mean values of d_m and V_m, where the latter corresponds to the beam voltage, V_b. If a background gas is present and the beam becomes charge neutralized, the virtual cathode moves forward. Collective ion acceleration is observed in an intermediate gas pressure regime (typically, 50 to 100 mtorr). If the pressure is too low, the beam will not be neutralized during the pulse duration time (i.e., $\tau_N > \tau_p$); hence, it will not propagate and no collective ion acceleration is observed. If the pressure is too high, neutralization occurs so fast that there is no time to establish a virtual cathode, and hence no collective acceleration can occur. The positive ions that are accelerated in the intermediate pressure regime are found to have a broad energy spectrum, with a mean energy that is approximately equal to the electron kinetic energy for singly ionized particles (i.e., $\overline{T}_i \approx qV_b$) and a peak energy of $T_{i,\text{max}} \lesssim 1.5qV_b$. The peak energy can be considerably higher if the gas is confined to a small region near the anode and the drift-tube region downstream is vacuum. Simulation studies [31, 32] show that as a plasma is formed by collisional ionization in the localized gas region, the virtual cathode moves from the anode plane to the plasma surface on the vacuum side. A group of ions born near the anode gain an energy of qV_b as they fall down the potential well. When the well begins to move, they "surf" along and gain an additional energy of $\sim qE\,\Delta z$, so that

$$T_{i,\text{max}} = qV_b + qE\,\Delta z, \tag{4.312}$$

where Δz is the width of the localized gas/plasma region. In this configuration peak ion energies of three to eight times the electron energy have been observed.

Collective ion acceleration methods using the space-charge field of intense relativistic electron beams were studied extensively in the 1970s. General reviews of the work during this period can be found in the books by Olson and Schumacher

[D.4], Rostoker and Reiser [D.5], and Miller [C.18]. More recent studies have been concerned with the external control of the virtual cathode motion through localized gas channels to increase the ion energy. By injecting laser-produced H_2 gas clouds along the electron-beam path in the vacuum drift tube in a properly timed sequence, Destler et al. [33] were able to extend the surfing effect and obtained peak proton energies of $T_{i,\max} \approx 20qV_b$. However, so far this scheme has not yet been employed to build an inexpensive practical accelerator for isotope production or other applications.

Current interest in the collective accelerator field has shifted toward *laser beat-wave* and *wakefield acceleration* in dense plasmas, where, theoretically, electric fields in the range of $\gtrsim 1$ GeV/m are predicted, and other concepts. Most of these schemes are aimed at future linear c^+c^- colliders in the TeV range, which require very high gradients to be economic in cost. A major problem facing these novel schemes is the high luminosity requirement for such a linear collider. It is difficult to see how the high particle intensity and low emittance needed can be achieved. However, it is still too early to assess the ultimate feasibility of a collider based on one of these new methods. A recent review of the work in this new field can be found in [D.8].

4.6.5 Charge-Neutralization Effects in High-Energy Synchrotrons and Storage Rings

In high-energy synchrotrons and storage rings, partial charge neutralization in the residual gas background of the evacuated beam tubes may significantly alter the betatron tune. However, there is a major difference between continuous beams and bunched beams. The situation in the latter case is much more complicated, and in this section we consider only the continuous beam case. Since the residual vacuum pressure is usually much lower than in the cases considered in previous subsections, charge-neutralization effects in circular machines occur adiabatically (i.e., on a time scale that is large compared to a betatron oscillation). The beam is thus in a quasiequilibrium state at all times, except when a resonance instability is encountered. With $a'' = 0$ and $k_0 = \nu_0/R$, one obtains for the beam envelope from (4.294) the smooth-approximation equation

$$\frac{\nu_0^2}{R^2} a - \frac{K_0}{a}\left[1 - \gamma^2 f_e(\tau)\right] - \frac{\epsilon^2}{a^3} = 0. \tag{4.313}$$

The parameter τ measures the time after injection of the beam into the machine. If the emittance remains constant, the beam radius decreases with increasing charge neutralization. However, in circular machines the change in the betatron tune, $\Delta\nu$, is much more important than that of the beam size. To calculate this tune shift due to charge neutralization we can use the formulas derived in Section 4.5 by making the substitution $K = K_0[1 - \gamma^2 f_e(\tau)]$. In lieu of (4.247), (4.250), we then obtain, respectively,

$$\Delta\nu = -\frac{K_0 R^2[1 - \gamma^2 f_e(\tau)]}{2\nu a^2} \tag{4.314}$$

or

$$\Delta\nu = -\frac{IR[1 - \gamma^2 f_e(\tau)]}{I_0\epsilon_n\beta^2\gamma^2}.$$ (4.315)

Thus, if the machine operates in a nonaccelerating cycle where $\gamma = \text{const}$ and $I = \text{const}$, the tune depression $\Delta\nu$ will have a negative maximum at $\tau = 0$, $f_e(0) = 0$ and then $|\Delta\nu|$ will decrease linearly with time. If the cycle lasts long enough, $\Delta\nu$ will go through zero at $f_e(\tau) = \gamma^{-2}$ and then become positive at later times. Hence, the possibility exists that ν may be driven into a resonance above the single-particle design value ν_0 (rather than below when charge neutralization is negligible).

These effects are most significant when the beams are unbunched so that the neutralizing particles from the background gas can accumulate without interruption. A numerical example that illustrates the effect of charge neutralization on the betatron tune is given in Problem 4.14.

In storage rings where high-energy particles are trapped for many hours, special measures must be taken to prevent charge neutralization. Thus, an ultrahigh vacuum of $p \lesssim 10^{-10}$ torr is maintained in these machines. However, even at these low pressures the typical charge neutralization time is still less than a minute. Therefore, clearing electrodes have to be installed along the ring with sufficient electric field strength to extract the charge-neutralizing particles from the beam. Even so small uncleared pockets of ions (or electrons) remaining trapped in the electrostatic potential well of the beam may cause serious beam-quality deterioration and beam loss. These detrimental effects are caused by dipole or quadrupole-type instabilities which occur when the ion bounce frequency, ω_i, in the potential well of the beam is in resonance with a sideband frequency of the beam's betatron oscillations.

Dipole-type instabilities are excited when the center of mass of the beam particle distribution (beam centroid) and that of the trapped ion distribution are displaced from the equilibrium orbit and from each other in one or both transverse directions. In the following we present a brief linear analysis of this effect while referring to the original literature [34, 35] for further details. Let us consider a relativistic electron or antiproton beam with trapped positive ions in a synchrotron or storage ring. (The same analysis will, of course, also apply to a proton beam with trapped electrons.) To simplify the theory, we will use the smooth approximation and assume that the betatron tune and emittance is the same in both transverse directions. The beam then has circular cross section with effective radius $a = 2\tilde{x}$, where $\tilde{x} = \tilde{y}$ is the rms width in each transverse direction. The distribution of stationary positive ions trapped in the beam is also assumed to have a circular cross section with radius a. Suppose that the beam centroid as well as the center of the ion distribution are displaced in the x-direction from the equilibrium position by amounts $\bar{x}_b$ and $\bar{x}_i$, respectively, where $\bar{x}_b \ll a$ and $\bar{x}_i \ll a$. This displacement could be either in the horizontal or vertical direction. The coherent motion of the beam and ion distribution is then described by the coupled equations

$$\ddot{\bar{x}}_b + \nu_0^2\omega_0^2\bar{x}_b + \omega_{bi}^2(\bar{x}_b - \bar{x}_i) = 0,$$ (4.316a)

$$\ddot{\bar{x}}_i + \omega_i^2(\bar{x}_i - \bar{x}_b) = 0,$$ (4.316b)

where $\omega_0 = v/R$ is the angular revolution frequency, $\overline{R}$ the average orbit radius of the beam, and ν_0 the betatron tune without space charge. ω_i is the ion bounce frequency in the space-charge well of the beam and is given by

$$\omega_i = \left(\frac{q^2 n_b}{2\epsilon_0 m_i} \right)^{1/2}, \tag{4.317}$$

where q and m_i are the ion charge and mass, respectively, and n_b is the beam density. The frequency ω_{bi} represents the focusing effect of the positive ion distribution on the beam and is given by

$$\omega_{bi} = \left(\frac{q^2 n_i}{2\epsilon_0 \gamma m_b} \right)^{1/2} = \omega_i \left(\frac{m_i}{\gamma m_b} f_e \right)^{1/2}, \tag{4.318}$$

where $f_e = n_i/n_b$ is the charge-neutralization factor and m_b the mass of the beam particles.

Since the ions are stationary while the beam particles are moving with velocity v in the direction $s = \overline{R}\theta = (v/\omega_0)\theta$ along the circular orbit, the dot representing the total time derivative implies that

$$\dot{\overline{x}}_b = \frac{d\overline{x}_b}{dt} = \frac{\partial \overline{x}_b}{\partial t} + \frac{\partial \overline{x}_b}{\partial s} v = \frac{\partial \overline{x}_b}{\partial t} + \omega_0 \frac{\partial \overline{x}_b}{\partial \theta} \tag{4.319a}$$

for the beam, and

$$\dot{\overline{x}}_i = \frac{d\overline{x}_i}{dt} = \frac{\partial \overline{x}_i}{\partial t} \tag{4.319b}$$

for the ions.

Let us now try a solution of Equation (4.316) that is of the form $\exp[i(\omega t - k_s s)] = \exp[i(\omega t - l\theta)]$, where $k_s s = k_s(v/\omega_0)\theta = l\theta$ and l is an integer, that is,

$$\overline{x}_b = \overline{x}_{b0} \exp[i(\omega t - l\theta)], \tag{4.320a}$$

$$\overline{x}_i = \overline{x}_{i0} \exp[i(\omega t - l\theta)]. \tag{4.320b}$$

Differentiating and substituting in Equation (4.316) then yields

$$[-(\omega - l\omega_0)^2 + \omega_b^2]\overline{x}_{b0} = \omega_{bi}^2 \overline{x}_{i0}, \tag{4.321a}$$

$$(-\omega^2 + \omega_i^2)\overline{x}_{i0} = \omega_i^2 \overline{x}_{b0}, \tag{4.321b}$$

from which one obtains the dispersion relation

$$[(\omega - l\omega_0)^2 - \omega_b^2](\omega^2 - \omega_i^2) = \omega_{bi}^2\omega_i^2. \tag{4.322}$$

The frequency ω_b in the preceding two equations is defined as

$$\omega_b = (\nu_0^2\omega_0^2 + \omega_{bi}^2)^{1/2}. \tag{4.323}$$

Equation (4.322) is a fourth-order algebraic equation for the unknown frequency ω of the coherent oscillations of the beam and ion centroids. It has, in general, four roots that depend on the values of the frequencies $\omega_0, \omega_b, \omega_i, \omega_{bi}$, and the integer (space harmonic) l. The latter determines the number of spatial oscillation periods per revolution of the perturbation. The general analysis of the dispersion relation reveals large regions of resonant-type instability in parameter space. Graphical plots of ω_b versus ω_i for given values of the other parameters show that only a small region near the origin is stable while instability exists for all values of ω_b, ω_i outside this region [34]. When the right-hand side of Equation (4.322) is small (i.e., $\omega_{bi}\omega_i \ll \omega^2$) and, by implication, $\omega_{bi} \ll \nu_0\omega_0$, one can obtain in the neighborhood of the resonance the approximate solution [35] for low-frequency ω:

$$\omega = \omega_i + \delta \pm i\left(\frac{\omega_i^2\omega_{bi}^2}{4\omega_i\nu_0\omega_0} - \delta^2\right)^{1/2}, \tag{4.324}$$

with

$$\delta = \frac{1}{2}(l\omega_0 - \nu_0\omega_0 - \omega_i). \tag{4.325}$$

Resonance occurs when $\delta = 0$, that is,

$$\omega_i = (l - \nu_0)\omega_0 \tag{4.326}$$

and

$$\omega = \omega_i \pm i\left(\frac{\omega_i\omega_{bi}^2}{4\nu_0\omega_0}\right)^{1/2}. \tag{4.327}$$

The negative imaginary part of the frequency defines the growth rate τ of the instability, which is given by

$$\frac{1}{\tau} = \text{Im}(\omega_i),$$

or

$$\tau = 2\left(\frac{\nu_0\omega_0}{\omega_i\omega_{bi}^2}\right)^{1/2} = 2\frac{(\nu_0\omega_0)^{1/2}}{\omega_i^{3/2}}\left(\frac{\gamma m_b}{f_e m_i}\right)^{1/2}, \tag{4.328}$$

where we used relation (4.318) for ω_{bi}.

We conclude, therefore, that in the unstable parameter regime where the above approximation is valid (low-frequency ω, small partial charge neutralization, i.e., $f_e \ll 1$, etc.), the two beam centers oscillate coherently at the ion bounce frequency, ω_i. At the same time, the ion bounce frequency corresponds to a side band, $l - \nu_0$, of the betatron tune according to (4.326).

Quadrupole-type instabilities are caused by mutual excitation of oscillations in the shape, or envelope, of the beam and the shape of the ion distribution. They are qualitatively somewhat similar to the envelope oscillations discussed in Section 4.4.3 except that in the present case we have a resonant interaction between two particle species rather than between the beam and a periodic-focusing lattice. Linear analysis [34] shows that the quadrupole resonances occur when

$$\omega_i - \left(\frac{l}{2} - \nu\right)\omega_0, \tag{4.329}$$

where l is an integer, as before. In the ω_h versus ω_i diagrams, the resonant type instabilities of the quadrupole type have the form of narrow bands, with the first one occurring in the stable region of the dipole interaction. The most important aspect of both dipole and quadrupole instabilities is that they depend on the beam current, or total number of particles stored in the ring, and on the beam size, or emittance. Assuming a uniform-density round beam with radius a and a negligible degree of partial neutralization (say, $f_e \lesssim 0.01$), the ion bounce frequency in the potential well of the beam is readily calculated as

$$\omega_i = \left(\frac{2qV_s}{m_i a^2}\right)^{1/2}, \tag{4.330}$$

where q and m_i are the ion charge and mass, respectively, and V_s is the beam potential as defined in Equation (4.14) (see Problem 4.17). As the total number of beam particles stored in the ring is increased, the beam potential V_s rises and instability occurs when a dangerous resonance of the type predicted by (4.326) or (4.329) is encountered.

Dipole and quadrupole instabilities were found to severely limit the particle number in the antiproton ($\bar{p}$) accumulator rings at CERN and Fermilab [36]. At CERN, for instance, the betatron tune is $\nu_0 = 2.25$ in both the horizontal and vertical directions. Signals from pickup electrodes indicated when instability occurred. The main culprits are thought to be CO^+ ions, whose bounce frequency in the beam matches the resonant frequencies inferred from the measurement. By applying rf signals with appropriate frequencies and phases it was possible to damp the most prominent dipole instability with $l = 3$ at $\omega_i = (3 - \nu_0)\omega_0$. This technique, which introduces a spread in the particle oscillation frequencies, thereby detuning the resonance condition, is known as *Landau damping*. A second method that has been found effective in damping instabilities is called *ion shaking*. An rf field applied to the beam by electrodes induces coherent oscillations of the $\overline{p}$ beam about the equilibrium orbit with very small amplitudes of less than 0.01 mm. If the frequency of these kicker signals is chosen to be close to one of the dipole sideband resonances of $l \pm \nu_0$, the oscillations of the trapped ions are resonantly driven to large amplitudes so that they effectively escape from the beam's potential well. Using this technique it was possible to increase significantly the number of antiprotons stored in the accumulator ring [36].

4.6.6 Plasma Lenses

The charge neutralization effects discussed in the preceding sections occur "naturally" when a charged particle beam passes through a region of gas. If gas pressure and ionization cross sections are sufficiently high, the beam creates a plasma along its path. Plasma particles with the same charge polarity as the beam particles are expelled by the beam's space-charge force, and the remaining oppositely charged particles reduce the repulsive Coulomb force of the beam.

A major drawback of such "gas focusing" is the fact that the degree of charge neutralization varies along the beam according to Equation (4.293). Furthermore, for low-energy (nonrelativistic) ions even full charge neutralization is not sufficient to balance the outward pressure represented by the emittance term in the beam envelope equation. This has led to proposals and exploration of alternative methods such as forming the plasma by ionizing the gas prior to the beam arrival, using the magnetic force due to the discharge current in a z-pinch, or creating a nonneutral electron plasma for focusing of positive ions. Such *plasma lenses* are of particular interest for focusing intense beams to a small spot size. Notable examples are the focusing of the intersecting electron and positron beams in a linear e^+e^- collider, the final focusing of heavy ion beams for inertial fusion, and the matching of low energy proton, H^-, or other ion beams into the small aperture of an RFQ linac (discussed in Section 4.6.3). Theoretically, the focusing strength of a plasma lens can exceed the capability of conventional and even superconducting magnetic lenses by as much as several orders of magnitude depending on the particular application. Experimentally, many difficulties have been encountered in developing a practical device, and this is still a very active field of research. A detailed discussion is beyond the scope of this book; we will merely present a very brief review of the three methods mentioned above.

Historically, the first important event in this field was Gabor's proposal in 1947 to use a nonneutral electron plasma, confined in a magnetron-type trap, as an effective space-charge lens for focusing of positive ions beams [37]. This *Gabor lens,* as it became known in the literature, was investigated experimentally and theoretically in the former Soviet Union, at Livermore, Brookhaven, and more recently at Fermilab [38]. The Fermilab experiments were concerned with focusing proton and H$^-$ beams into an RFQ linac. None of this past work has led to a practical device. A theoretical comparison of the Gabor lens with conventional lenses and reference to important past studies can be found in [39].

In the z-pinch type of plasma lens, a high axial current is generated in the plasma. The Lorentz force, $F_r = qv_z B_\theta$, due to the azimuthal magnetic field produced by the current focuses the beam particles. Such a lens was used for the first time to focus the proton beam from the 184-inch cyclotron at Berkeley in 1950 [40] and for capturing 3-GeV muons and kaons at Brookhaven in 1964 [41]. More recently, a z-pinch was employed in a successful demonstration experiment at GSI Darmstadt, where a 460 MeV heavy-ion beam (Ar^{11+}) was focused to a small spot size of about 1 mm [42]. This experiment was motivated by the final focusing requirements for heavy ion inertial fusion.

The most active research work in recent years has been concerned with the use of plasma lenses for focusing the electron and positron beams to the extremely small spot size required at the interaction point of a linear e^+e^- collider [43]. Following the first proposal by Chen [44] in 1987 there have been a number of theoretical studies, e.g., by Rosenzweig and Chen [45], Whittum [46], Chen et al. [47], Katsoulcas and Lai [48]. There have also been proof-of-principle experiments [49,50] of a preliminary nature.

In the most recent theoretical studies it is proposed to use an adiabatic, tapered plasma channel with increasing density (focusing strength) to guide the beams into the interaction region of the linear collider [47,48]. In the case of the electron beam, for instance, the initial ion density n_i is considerably less than the beam density n_e ($f_e = n_i/n_e \lll 1$). It then increases adiabatically so that full charge neutralization ($f_e = 1$) is reached towards the end of the channel. In practice, the problem of forming such a plasma column with the desired properties needs to be solved. In the Japanese experiment an argon plasma was produced by a discharge and confined by the octupole field of permanent magnets [50]. Another possibility is to use a laser pulse for preionizing the gas, as was successfully demonstrated at Livermore [51] in ion-focusing experiments with an intense relativistic electron beam. (Laser gas preionization is also being considered for the final focusing of the heavy ion beams in inertial fusion.)

Let us now estimate the strength of ion focusing on a relativistic electron beam propagating through a plasma channel. We will assume a round beam and use the envelope equation (4.294), with $k_0 = 0$, to calculate the beam radius in the channel. Since for the highly relativistic energies of a linear collider $\gamma^2 f_e \gg 1$, we can write (4.294) in the form

$$a'' + \frac{K_0}{a}\gamma^2 f_e - \frac{\epsilon^2}{a^3} = 0. \qquad (4.331)$$

Using $I = en_e a^2 \pi c$, $I_0 = 4\pi\epsilon_0 mc^3/e$, $f_e = n_i/n_e$, we obtain the alternative expression

$$a'' + \frac{e^2 n_i(z)}{2\pi\epsilon_0 \gamma mc^2}a - \frac{\epsilon^2}{a^3} = 0. \qquad (4.332)$$

If we approximate the actual beam with an equivalent cylinder of uniform charge density, effective length l_b and total number of particle $N_e = n_e a^2 \pi l_b$, we can write the envelope equation in yet another form as

$$a'' + \frac{2r_c N_e f_e}{\gamma l_b a} - \frac{\epsilon^2}{a^3} = 0.$$ (4.333)

Here, r_c is the classical particle radius ($r_c = 2.82 \times 10^{-15}$ m for e^- and e^+). If $f_e(z) = n_i(z)/n_e(z)$ changes adiabatically with distance z and the beam is well matched we can set $a'' = 0$ and obtain from (4.333) for the radius

$$a(z) = \epsilon_n \left[\frac{l_b}{2r_c \gamma N_e f_e(z)} \right]^{1/2},$$ (4.334)

where $\epsilon_n = \gamma \epsilon$ is the normalized emittance of the relativistic beam ($\beta \approx 1$).

As an example, let $N_e = 10^{10}$, $\gamma = 10^6$ (500 GeV energy), $l_b = 4 \times 10^{-3}$ m, $\epsilon_n = 10^{-5}$ m-rad, and $f_e = 1$ (at the end of the column). With these numbers we find from (4.334) a final radius of $a = 84 \times 10^{-9}$ m $= 84$ nm. Such submicron beam radii are needed to meet the luminosity requirements of future TeV linear colliders. The focusing strength of plasma columns is orders of magnitude greater than that of superconducting magnetic quadrupoles. This explains the strong interest in plasma lenses and motivates the current research activity in this field. It will take a systematic research effort for several years to determine the practical feasibility and ultimate technological limitations of these focusing methods.

REFERENCES

1. See, for instance, R. W. Hasse and J. P. Schiffer, *Ann. Phys.* **203**, 419 (1990).
2. G. I. Budker, *CERN Symposium on High Energy Accelerators,* CERN, Geneva, 1956, Vol. I, p. 68.
3. J. D. Lawson, *J. Electron. Control* **5**, 146 (1958).
4. H. Alfvén, *Phys. Rev.* **55**, 425 (1939).
5. C. L. Olson and J. W. Poukey, *Phys. Rev.* **A9**, 2631 (1974).
6. L. S. Bogdankevich and A. A. Rukhadze, *Usp. Fiz. Nauk* **103**, 609 (1971) [*Sov. Phys.- Usp.* **14**, 163 (1971)]; J. A. Nation and M. Read, *Appl. Phys. Lett.* **23**, 426 (1973).
7. W. H. Bennett, *Phys. Rev.* **45**, 89 (1934).
8. L. Brillouin, *Phys. Rev.* **67**, 260 (1945).
9. M. Reiser, *1991 IEEE Particle Accelerator Conference Record 91CH3038-7,* pp. 2497–2499.
10. M. Reiser, *Part. Accel.* **8**, 167 (1978).
11. M. Reiser, *J. Appl. Phys.* **52**, 555 (1981).
12. J. Struckmeier and M. Reiser, *Part. Accel.* **14**, 227 (1984).

13. I. Hofmann, L. J. Laslett, L. Smith, and I. Haber, *Part. Accel.* **13**, 145 (1983).

14. N. Rostoker, *Part. Accel.* **5**, 93 (1973).

15. P. Sprangle and C. A. Kapetanakos, *J. Appl. Phys.* **49**, 1 (1978).

16. C. W. Roberson, A. Mondelli, and D. Chernin, *Part. Accel.* **17**, 79 (1985).

17. L. J. Laslett, *Proc. 1963 Summer Study on Storage Rings, Accelerators, and Experimentation at Super-High Energies* (ed. J. W. Bittner), BNL-7534, Brookhaven National Laboratory, Upton, NY, 1963, pp. 324–367; reprinted in *LBL Document PUB-6161*, 1987, Vol. III, pp. 4–30 ff.

18. H. Bethe, *Ann. Phys.* **5**, 325 (1930).

19. L. J. Kieffer and G. H. Dunn, *Rev. Mod. Phys.* **38**, 1 (1966).

20. F. F. Rieke and W. Prepejchal, *Phys. Rev. A* **6**, 1507 (1972).

21. S. P. Slinker, R. D. Taylor, and A. W. Ali, *J. Appl. Phys.* **63**, 1 (1988).

22. M. E. Rudd, Y.-K. Kim, D. H. Madison, and J. W. Gallagher, *Rev. Mod. Phys.* **57**, 965 (1985).

23. *Atomic Data for Controlled Fusion Research,* ORNL-5206, Vol. I, Oak Ridge National Laboratory, Oak Ridge, TN, February 1977.

24. M. Reiser, *Nucl. Inst. Methods Phys. Res. B* **56/57**, 1050 (1991).

25. M. D. Gabovich, I. S. Simonenko, and I. A. Soloshenko, *Zh. Tekh. Fiz.* **48**, 1389 (1978) [English translation *Sov. Phys. Tech. Phys.* **23**, 783 (1978)].

26. M. E. Rudd, *Phys. Rev. A* **20**, 787 (1979).

27. M. Reiser, C. R. Chang, D. Chernin, and E. Horowitz, "Microwave and Particle Beam Sources and Propagation," *SPIE* **873**, 172 ff. (1988).

28. M. G. Tiefenbach and D. Keefe, *IEEE Trans. Nucl. Sci.* **NS-32**, 2483 (1985).

29. O. A. Anderson et al., *Nucl. Instrum. Methods, B* **40/41**, 877 (1989).

30. S. Guharay, C. K. Allen, and M. Reiser, "High-Brightness Beams for Advanced Accelerator Applications," *AIP Proc.* **253**, 67–76 (1992), ed. W. W. Destler, and S. K. Guharay.

31. C. R. Chang and M. Reiser, *J. Appl. Phys.* **61**, 899 (1987).

32. R. L. Yao and C. D. Striffler, *J. Appl. Phys.* **67**, 1650 (1990).

33. W. W. Destler, J. Rodgers, and Z. Segalov, *J. Appl. Phys.* **66**, 2894 (1989).

34. D. G. Koshkarev and P. R. Zenkevich, *Part. Accel.* **3**, 1 (1972).

35. L. J. Laslett, A. M. Sessler, and D. Möhl, *Nucl. Instrum. Methods* **121**, 517 (1974).

36. J. Merriner, D. Möhl, Y. Orlov, A. Poncet, and S. van der Meer, *Part. Accel.* **30**, 13 (1990).

37. D. Gabor, *Nature* **160**, 89 (1947).

38. J. A. Palkovic, R. E. Mills, C. Schmidt, D. E. Young, *1989 IEEE Particle Accelerator Conference Record 89CH2669-0*, pp. 304–306; J. A. Palkovich, *1993 IEEE Particle Accelerator Conference Record 93CH3279-7*, pp. 21–25.

39. M. Reiser, *1989 IEEE Particle Accelerator Conference Record 89CH2669-0*, pp. 1744–1747.

40. W. K. H. Panofsky and W. R. Baker, *Rev. Sci. Instrum.* **21**, 445 (1950).

41. E. B. Forsyth, L. M. Lederman, and J. Sunderland, *IEEE Trans. Nucl. Sci.* **12**, 872 (1965).

42. E. Boggasch et al., *Phys. Rev. Lett.* **66,** 1705 (1991).

43. B. Richter, *IEEE Trans. Nucl. Sci.* **NS-32,** 3828 (1985).

44. P. Chen, *Part. Accel.* **20,** 171 (1987).

45. J. B. Rosenzweig and P. Chen, *Phys. Rev. D* **40,** 923 (1989).

46. D. Whittum, "Theory of the Ion-Channel Laser," University of California, Berkeley, Ph.D. Dissertation (1989).

47. P. Chen, K. Oide, A. M. Sessler, and S. S. Yu, *Phys. Rev. Lett.* **64,** 1231 (1990).

48. T. Katsouleas and C. H. Lai, *AIP Conf. Proc.* **279,** 551–564 (1993); ed. J. S. Wurtele.

49. J. B. Rosenzweig et al., Phys. *Fluids B* **2,** (6, part 2), 1376–1383 (1990).

50. H. Nakanaishi et al., *Phys. Rev. Lett.* **66,** 1870 (1991).

51. See R. J. Briggs, *Phys. Rev. Lett.* **54,** 2588 (1985) and references therein.

PROBLEMS

4.1 A uniform relativistic electron beam with constant radius $a = 1$ cm is propagating inside a conducting drift tube with radius $b = 2$ cm. The kinetic energy of the electrons is 1 MeV and the current is 2 kA.

    (a) Calculate the potential difference between the beam axis and the wall of the drift tube.

    (b) Determine the electric and magnetic field energy and the capacitance and inductance per unit length. Compare the capacitance with the case where the electron beam is replaced by a solid conductor having the same charge per unit length on its surface.

4.2 Derive the relation (4.61) for the space-charge limiting current. Calculate the electrostatic field energy per meter in the beam for this case when $b = a$ (beam fills drift tube) and compare it with the beam kinetic energy.

4.3 A toroidal relativistic electron beam is confined in an axisymmetric magnetic mirror field with the following assumptions: (1) the beam has a circular cross section of radius a; (2) the major radius R_0 of the beam is large enough that self fields resulting from the curvature of the beam can be neglected (i.e., the self fields can be calculated as if the beam were moving on a straight path); (3) both charge and current density within the beam may be considered uniform; and (4) a background of stationary positive ions of uniform density is present within the beam; the charge density of the ions is f_e times that of the electrons ($f_e < 1$).

    (a) Calculate the betatron tunes ν_r and ν_z.

    (b) State the condition for which the beam retains a circular cross section.

    (c) Derive a relation for f_e which assures that the beam remains focused in both radial and axial direction.

    (d) Determine the numerical values of ν_r and ν_z for the case where the beam parameters have the following values: electron energy, 2 MeV;

beam current, 2 kA; major radius R_0, 6 cm; minor radius a, 3 mm; fraction of ions, 5% of electrons.

4.4 A magnetic mirror for confinement of a toroidal electron beam with mean radius R is formed by a combination of a long solenoid and two coils, as shown in the figure below. The solenoidal field and the field due to the two mirror coils can be increased separately with time, and the combined field along the z-axis ($r = 0$) and close to the midplane ($z = 0$) may be described by the approximate formula

$$B_z(z, t) = B_0 F(t) + \left(B_1 + \frac{k^2}{2} z^2\right) f(t),$$

where the first term on the right side represents the long solenoid and the second term the two mirror coils. Assume that the field increase with time is adiabatic (i.e., the change of B_z is negligibly small during one period of the particle oscillations between the magnetic mirrors).

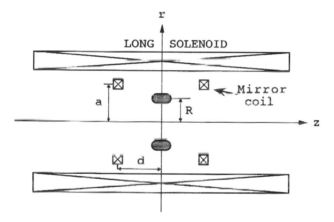

(a) List all constants of the motion that you can think of.
(b) Determine the constants B_1, k^2, and the field index n in terms of the coil radius a, coil separation $2d$, the peak field B_m that each mirror coil produces at its center ($t = 0$, $z = d$) and (in the case of n) the equilibrium orbit radius R. Assume that $F(t) = 1$, $f(t) = 1$ for this calculation.
(c) Consider a $T = 2$ MeV electron beam with major radius $R = 6$ cm and minor radii $x_m = 0.5$ cm, $z_m = 0.5$ cm in radial and axial directions at time $t = 0$. Let $a = d = 10$ cm. Suppose that $F(t) = f(t) = 1$ (i.e., no change with time), and that $B_m = 300$ G. Calculate B_0, B_1, n, v_r, and ω_c. Neglect the self fields of the beam.
(d) Suppose now that $F(t) = 1$ and $f(t) = 1 + 9(1 - e^{-t/\tau})$. Find n, v_r, v_z, x_m, and z_m when $t \to \infty$.

(e) Finally, let $F(t) = f(t) = 1 + 9(1 - e^{-t/\tau})$ with initial conditions at $t = 0$ as in (c). Calculate T, B_0, B, R, N, v_r, v_z, x_m, z_m when $t \to \infty$. Suppose that the electron-beam current at $t = 0$ is $I = 10^3$ A. What is the current when $t \to \infty$?

4.5 An axisymmetric beam of 100-keV electrons is injected into a drift tube of length $L = 100$ cm and diameter $D = 4$ cm. At the entrance of the drift tube, the beam is focused by a thin, solenoid lens with an effective width of $l = 4$ cm and magnetic field B on the axis. Calculate the following quantities:

(a) The maximum electron beam current that can be passed through the drift tube for laminar flow, the associated field B of the lens, the radius $a = r_{min}$ at the beam waist, and the value of the plasma frequency ω_p of the beam at the waist.

(b) The electrostatic potential difference between the beam axis and the drift-tube wall at the waist in case (a).

(c) The acceptance α of the drift tube (i.e., the maximum emittance that the beam could have) if the self fields are eliminated by charge and current neutralization.

4.6 The envelope equation (4.85a) for a beam in a uniform focusing channel ($k_0^2 = $ const) can be integrated once without any approximations. Obtain this first integral for r'_m if the initial conditions are $r_m = a$, $r'_m = r'_0$ at $z = 0$, where $r_m = a$ is the equilibrium radius of the matched beam. By setting $r_m = a + x$, assuming that $x \ll a$, and using Taylor expansion up to second-order terms in x/a, one can obtain a second integral. Find this integral for r_m, or x, as a function of r'_m and determine the maximum amplitude x_{max} of the envelope ripple as a function of r'_0 when the beam is mismatched.

4.7 Design a periodic hard-edge solenoidal focusing channel for a 10-keV, 0.6-A electron beam having a normalized emittance of $\epsilon_N = 9 \times 10^{-6}$ m-rad. The ratio L/l of the drift space to the width of a lens is to be 3, the desired phase advance without space charge is $\sigma_0 = 72°$, and the mean radius of the beam is $\overline{R} = 1$ cm. Using the smooth-approximation theory, determine the length of a focusing period S, the phase advance with space charge σ, and the maximum radius of the beam, R_{max}. Calculate the solenoidal magnetic field B_0, the mean plasma frequency $\overline{\omega}_p$, the associated plasma wavelength $\overline{\lambda}_p$, and the Debye length λ_D of this electron beam.

4.8 Consider an axisymmetric beam with current I, voltage V, and radius a propagating inside a drift tube with radius $b > a$. Suppose that the beam has a nonuniform density profile given by $n(r) = n_0[1 - (r/a)^2]$ for $r \ll a$ and $n(r) = 0$ for $a < r < b$. Calculate the following quantities:

(a) Self fields E_r, B_θ and the associated force F_r on a particle

(b) Potential distribution $V(r)$

    (c) Electrostatic energy per unit length
    (d) Rms beam radius $\tilde{a}$,
    (e) Radius r_m, where the force F_r is a maximum

4.9 Consider the thin beam (line charge ρ_L) displaced from the axis of a conducting drift tube by a distance $x_1 = \xi$, as shown in Figure 4.18. Prove that the electrostatic potential distribution within the drift tube satisfying the boundary condition can be obtained by adding the potential due to an image line charge $-\rho_L$ located at distance $x_2 = b^2/\xi$.

4.10 A round beam with particle energy γmc^2 and a Gaussian density profile $n(r) = n_1 \exp[-r^2/2\delta^2]$ propagates through a uniform, linear focusing channel defined by the single-particle oscillation frequency ω_0.

    (a) Determine the rms radius $\tilde{r}$ and rms width $\tilde{x}$ of this distribution.
    (b) Calculate the number of particles per unit length N_L and the beam current I in terms of n_1 and δ
    (c) Find the density n_0 and radius a_0 of the equivalent uniform density beam having the same rms radius $\tilde{r}$ and beam current I as the Gaussian beam.

4.11 Consider the Gaussian beam of Problem 4.10.

    (a) Calculate and plot schematically the radial force $F(r)$ due to the self fields and its derivative dF_r/dr versus radius r. Find the values of r/δ where F_r and dF_r/dr have a maximum.
    (b) Show that for small radii the force F_r is to first approximation a linear function of radius r. Determine the radius r/δ where the difference between the linear approximation and the actual force reaches 10%.
    (c) Using the result of (b), calculate the small-radius particle oscillation frequency ω that includes the self force as a function of ω_0, I, δ, and γ.

4.12 Derive the result (4.251) for the tune shift $\Delta\nu$ due to space-charge forces from the relation

$$\Delta\nu = \frac{1}{4\pi} \int_0^C \hat{\beta}(s)\,\Delta\kappa(s)\,ds$$

given in Problem 3.25(c). (Note that the space-charge force can be equated to a gradient error $\Delta\kappa$.)

4.13 Suppose that the radial force in a continuous-focusing channel has a nonlinear term due to spherical aberrations so that $F_r(r) = -\alpha_1 r - \alpha_3 r^3$ and $\alpha_3 r^3 = \alpha_1 r$ at radius r_1. Assuming a Boltzmann distribution of the form (4.5), find the density profile $n(r)$ versus radius in the laminar-flow limit $(k_B T_\perp \rightarrow 0)$ for a relativistic beam with radius $a = 0.5 r_1$.

4.14 Consider the 200 MeV low-energy ring with circumference $C = 470$ m of the Fermilab accelerator discussed in Section 4.5 following Equation (4.256). Assume that the beam is unbunched $(B_f = 1)$.

 (a) Determine the amount of fractional charge neutralization, f_e, due to ionizing collisions in the background gas that would exactly balance the tune shift $\Delta \nu$ due to the space-charge forces.

 (b) Suppose that the background gas is molecular hydrogen (H_2). Find the pressure for which the above value of f_e is reached in a time of 1 ms.

4.15 In an e^+e^- linear collider electron and positron bunches are accelerated to very high energy in two opposing accelerators. The bunches are then focused to a very small cross section and forced to pass through each other in a head-on collision at the so-called interaction point between the two linear accelerators. For the following calculations, consider an electron bunch (coming from the left side) passing through a positron bunch (coming from the right). Assume that each bunch can be represented by a cylinder of radius a and length $l \gg a$ having uniform charge density and the same total number of particles.

 (a) Calculate the radial force F_r on an electron (positron) at radius r in the midplane of the bunch *before* the collision.

 (b) Calculate the radial force F_r at the instant where the two bunches completely overlap each other *at* the intersection point. Compare the direction and magnitude of F_r at and before the intersection point.

 (c) The effect of one bunch on the other can be represented by an equivalent focusing lens. Determine the focal length f and the so-called disruption parameter $D = l/f$ at the intersection point as a function of N, γ, a using the thin-lens approximation.

 (d) Find the self-magnetic field $B_\theta(r)$ of the two bunches at the moment of overlap and the associated radius of curvature R for a particle traveling at the outermost radius $r = a$.

 (e) Calculate the disruption parameter D, the magnetic field $B_\theta(a)$, and the associated radius of curvature R at the intersection point for the following specific parameters: $a = 1$ μm, $l = 1$ mm, $N = 5 \times 10^{10}$, and a particle energy of 100 GeV.

4.16 One of the most important problems in accelerator and beam transport design is to match the beam from one focusing system into another. Consider a beam that is to be matched from a focusing channel with uniform (or smooth-approximation) wave number $k_{01} = 2\pi/\lambda_{01}$ *without* space charge into a channel characterized by $k_{02} = 2k_{01}$. Let R_1, R_2 and k_1, k_2 denote the matched beam radii and wave numbers with space charge in each channel, respectively. Note that the problem is similar to quarterwave matching between two transmission lines. Thus an appropriate uniform focusing element, characterized by constants k_0 without space charge and k with space charge, and length $\Delta s = d$, can be inserted between the two channels to achieve perfect matching.

(a) Neglecting space charge, determine k_0 and d in terms of given parameters.

(b) Repeat for a beam with space charge.

(c) Calculate k and d for a laminar beam ($\epsilon = 0$).

4.17 The antiproton ($\overline{p}$) accumulator ring at CERN operates at an ultrahigh vacuum of 0.75×10^{-10} torr and uses clearing electrodes to prevent any significant buildup of partial charge neutralization. The circumference of the ring is $2\pi\overline{R} = 150$ m, the $\overline{p}$ energy is 3 GeV, the total number of antiprotons is typically $N_b = 5 \times 10^{11}$, the betatron tune is $\nu_x = \nu_y = \nu_0 = 2.25$, and the $\overline{p}$ beam has an average circular cross section with radius $a = 3.3$ mm. Despite the clearing electrodes, dipole and quadrupole instabilities are encountered due to a small number of H^+ ions trapped in the $\overline{p}$ beam ($f_e = N_i/N_b \approx 0.01$).

(a) Calculate the cross section σ_i for ionization of H by a 3-GeV $\overline{p}$ beam.

(b) Calculate the charge neutralization time τ_N at a H pressure of 0.75×10^{-10} torr.

(c) Find the self potential V_s of the $\overline{p}$ beam assuming that f_e can be neglected.

(d) Determine the bounce frequency ω_i of the H^+ ions trapped in the p beam corresponding to the V_s value of (c).

(e) Suppose that the dipole instability is observed at the sideband of the betatron frequency where $l = 3$; determine the frequencies $f_i = \omega_i/2\pi$, f_{bi}, f_b, and the growth rate τ of the instability assuming that the neutralization factor is $f_e = 0.01$.

4.18 Prove that Equation (4.324) is an approximate solution of the dispersion relation (4.322).

CHAPTER 5

Self-Consistent Theory of Beams

5.1 INTRODUCTION

In the uniform-beam model of Chapter 4, we made the assumption that charge density ρ, particle velocity $\mathbf{v}$, and current density $\mathbf{J}$ are independent of the transverse coordinates (x, y) and that the external forces acting on the beam are linear. This allowed us to obtain the relatively simple paraxial trajectory equations, which are linear in x and y. However, the uniform-beam model does not in general satisfy Maxwell's equations and the equations of motion in a self-consistent manner when the paraxial approximations are violated. Furthermore, from a general theoretical point of view, the equilibrium state of a beam in a linear focusing channel tends to be more like a Boltzmann distribution which has a nonuniform density profile except in the zero-temperature limit, where the density is constant across the beam, as discussed in Section 4.1.

When the density is nonuniform, the forces associated with the self fields of the beam are nonlinear. In most laboratory systems nonuniformity of the density profile tends to be the rule rather than the exception; and, in addition, the applied focusing or accelerating forces also have an unavoidable minimum amount of associated nonlinearity. In other words, the general system that we are dealing with is intrinsically nonlinear, and we need to develop theoretical models (including particle simulation codes) that are self-consistent to a desired degree of accuracy. Such models are necessary to evaluate the nonlinear effects that are neglected in the uniform beam model and that cause emittance growth, halo formation, and particle loss.

To understand the self-consistency problem, we must recognize that on the one hand, the positions and velocity vectors of the particles in the beam determine the charge and current density, ρ and $\mathbf{J}$, at each point. On the other hand, ρ and $\mathbf{J}$ are the sources of the electric and magnetic self fields, which, together with the applied fields, determine the motion (i.e., the position and velocity of the particles). Thus one deals with a closed loop in which the motion of the distribution of particles changes the fields and the forces due to these fields change the particle distribution. A truly self-consistent theoretical model of the beam must close this self-interaction loop.

The mathematical difficulties involved in the self-consistency problem are quite formidable, and only relatively simple beam geometries can be solved by analytical techniques. With only one exception (the K–V distribution in Section 5.3.2), all of the self-consistent models presented in this chapter assume beams with cylindrical symmetry and applied focusing forces that are either uniform or "smoothed" over the lattice periods of a periodic focusing system. First, in Section 5.2, we discuss laminar beams in uniform magnetic fields; and we begin this analysis with a simple model of a cylindrical beam in an infinitely strong magnetic field to illustrate the self-consistency problem. Both nonrelativistic and relativistic descriptions of the stationary states (equilibria) of laminar beams will be treated. The nonrelativistic analysis closely follows the electron-beam theory developed in connection with microwave tube design (klystrons, traveling-wave tubes, etc.) during the period 1945–1970 after World War II. The relativistic theory, which is of interest for the intense relativistic electron beams, high-power microwave devices and electron beams developed more recently is mathematically more complex, and analytical results are available only for the simplest beam geometries.

In Section 5.3 we derive the self-consistent Vlasov equation, which allows us to treat nonlaminar beams (i.e., beams having an intrinsic velocity spread). We then discuss several important examples of stationary nonlaminar distributions in a linear focusing channel. These are distributions that satisfy the stationary (time-independent) Vlasov equation and hence represent matched beams. The best known examples are the K–V distribution and the Maxwell Boltzmann distribution, also known as the *thermal distribution*. Section 5.4 is devoted to a detailed analysis of the Maxwell–Boltzmann distribution, which will be shown to be the natural equilibrium state of a beam when the Coulomb collisions between the particles are taken into account. Beams with collisions are treated self-consistently by the Fokker–Planck equation, which reduces to the Vlasov equation when collisions are neglected. Although the charged particles are usually in thermal equilibrium inside the sources (thermionic cathode, plasma, etc.), acceleration results in a cooling of the longitudinal temperature. The typical laboratory beams are thus characterized by two different temperatures in the transverse and longitudinal directions. The relationships between temperature, emittance, and other beam parameters will be discussed for both the transverse and longitudinal Maxwell–Boltzmann distributions. The analysis will be restricted for the most part to external focusing fields with linear forces; the major exception is the longitudinal beam dynamics in rf fields (Section 5.4.8). However, the space-charge forces of a Maxwell–Boltzmann distribution in harmonic oscillator potentials are in general nonlinear. We will also investigate the effects of momentum spread on the transverse focusing and the dispersion that occurs in circular accelerators. The coupled envelope equations for bunched beams with space charge will be analyzed in Section 5.4.11, and in the final section (5.4.12) we discuss briefly the problems of matching, focusing, and imaging of beams.

5.2 LAMINAR BEAMS IN UNIFORM MAGNETIC FIELDS

5.2.1 A Cylindrical Beam in an Infinitely Strong Magnetic Field

We begin our analysis of a self-consistent laminar flow with a simple cylindrical beam model where we assume that both the source and the accelerated beam are immersed in an infinitely strong uniform magnetic field. The Lorentz force due to this applied field is infinitely stronger than the defocusing forces due to the electric and magnetic self fields. As a result, there is no transverse velocity component (i.e., $v_r = v_\theta = 0$) and the particle trajectories are straight lines with radius $r = \text{const.}$

Let us assume that the beam has a constant radius $r = a$ and that it propagates inside a conducting drift tube with radius $r = b$, as indicated in Figure 5.1. The drift tube is connected with the anode (or, in the case of ions, with the extraction electrode of the ion source). If the potential difference between the cathode (plasma surface) and the anode (extraction electrode) is ϕ_b, the particles will enter the drift tube with a kinetic energy of $T_b = q\phi_b$. Due to the beam space charge, however, there will be a potential distribution inside the drift tube which will reduce the kinetic energy of the particles in accordance with the energy conservation law. At a sufficient distance from the tube entrance, this potential (as well as all other

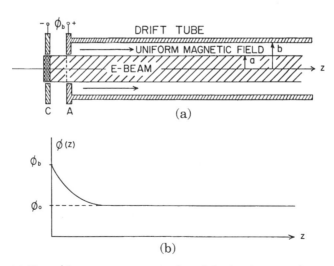

Figure 5.1. (a) Electron beam propagating inside a drift tube along a uniform magnetic field. The electrons are emitted from a thermionic cathode C and enter the drift tube through a mesh in the anode A. (b) Graph showing schematically the variation of the potential function along the axis.

parameters describing the beam) will be independent of the z-coordinate and vary only with radius r in view of the axial symmetry. If $\phi_e(r)$ denotes the electrostatic potential due to the space-charge field in this region of the tube, then from energy conservation the kinetic energy of the particles at a given radius r will be

$$T(r) = T_b - q\phi_e(r) = q\phi(r). \tag{5.1}$$

Here we introduced the function $\phi(r)$, which, as in previous contexts, represents the voltage equivalent of the kinetic energy. This definition implies that both the potential function $\phi(r)$ and the particle charge q are treated as positive quantities. In the following we use $\phi(r)$ rather than $\phi_e(r)$ as the potential function. The variation of the potential function ϕ (and hence of the kinetic energy of the particles) versus distance z along the axis of the beam is shown schematically in Figure 5.1. Note that $\phi = 0$ at the emitter surface of the source and $\phi = \phi_b$ at the anode mesh where the beam enters the drift tube. At a distance from the anode comparable to the drift tube diameter the potential ϕ (or kinetic energy) drops to a constant value ϕ_0. In this region, ϕ is only a function of radius r varying from $\phi = \phi_0$ on the axis to $\phi = \phi_a$ at the edge of the beam ($r = a$) and $\phi = \phi_b$ at the tube wall ($r = b$). It is important to recognize that the maximum kinetic energy of the particles at the beam edge, $q\phi_a$, is less than the injection energy, $q\phi_b$, when $b > a$, as discussed in Section 4.2.3. Since $T = (\gamma - 1)mc^2$, we can write Equation (5.1) in the alternative form

$$\gamma(r) - 1 + \frac{q\phi(r)}{mc^2}. \tag{5.2}$$

The variation of the potential function $\phi(r)$, and hence $\gamma(r)$, with radius is determined by Poisson's equation, $\nabla^2\phi_e = -\rho/\epsilon_0$, which relates the electrostatic potential ϕ_e to the charge density ρ in the beam. Replacing ϕ_e by $T_b - \phi$ from (5.1) and ρ by qn, where $n(r)$ is the particle density and where the charge q is treated as a positive quantity, and considering the fact that there is no azimuthal or axial variation, we can write Poisson's equation in the form

$$\nabla^2\phi = \frac{1}{r}\frac{d}{dr}\left(r\frac{d\phi}{dr}\right) = \frac{qn(r)}{\epsilon_0}. \tag{5.3}$$

Integration yields the radial electric field

$$E_r = -\frac{d\phi_e}{dr} = \frac{d\phi}{dr} = \frac{q}{\epsilon_0 r}\int_0^r n(r)r\,dr, \tag{5.4}$$

which can also be obtained directly from applying Gauss's law.

From the continuity equation one obtains for the current density $\mathbf{J} = J\mathbf{a}_z$ the relation

$$J(r) = \rho(r)v(r) = qc\,n(r)\,\beta(r),\tag{5.5}$$

with $\beta(r)$ related to $\phi(r)$ by the energy conservation law as

$$\beta(r) = \left[1 - \left(\frac{1}{\gamma(r)}\right)^2\right]^{1/2} = \frac{[2mc^2\,q\phi + (q\phi)^2]^{1/2}}{mc^2 + q\phi}.\tag{5.6}$$

The total beam current is given by

$$I = 2\pi \int_0^a J(r)r\,dr = 2\pi qc \int_0^a n(r)\beta(r)r\,dr.\tag{5.7}$$

Poisson's equation (5.3), the continuity equation (5.5), and energy conservation (5.6) represent three relationships between the four functions $\phi(r)$, $n(r)$, $J(r)$, and $\beta(r)$. Thus we can choose one of the four functions, and then the remaining three are self-consistently determined by these equations. As an example, let us choose the current density to be independent of radius r (i.e., $J = $ const). Then, from (5.7), $I = Ja^2\pi$, and from (5.5),

$$n(r) = \frac{I}{qc\,a^2\pi\beta(r)}.\tag{5.8}$$

Substituting (5.8) into (5.3) and using the relationship (5.6) for $\beta(r)$ yields the differential equation

$$\frac{1}{r}\frac{d}{dr}\left(r\frac{d\phi}{dr}\right) = \frac{I}{\epsilon_0 a^2\pi c}\frac{mc^2 + q\phi}{[2mc^2 q\phi + (q\phi)^2]^{1/2}},\tag{5.9}$$

which determines the potential function $\phi(r)$ in a self-consistent manner. If one uses γ in place of ϕ, one can write this equation in the alternative form

$$\frac{1}{r}\frac{d}{dr}\left(r\frac{d\gamma}{dr}\right) = \frac{qI}{\epsilon_0 a^2\pi mc^3}\left(1 - \frac{1}{\gamma^2}\right)^{-1/2}.\tag{5.10}$$

These equations have to be integrated numerically. In the nonrelativistic case, where $\beta c = (2q\phi/m)^{1/2}$, Equation (5.9) takes the form

$$\frac{1}{r}\frac{d}{dr}\left(r\frac{d\phi}{dr}\right) = \frac{I}{\epsilon_0 a^2\pi c(2q/mc^2)^{1/2}}\phi^{-1/2}.\tag{5.11}$$

For simplicity let us assume that the beam fills the drift tube so that $b = a$; that is, we can ignore the factor $1 + 2\ln(b/a)$, which arises when $b > a$, as we know from Section 4.2. This case is discussed by Pierce [C.3]. When $\phi = 0$ at $r = 0$ (i.e., when the kinetic energy of the particles on the axis is zero), this equation has the special solution

$$\phi(r) = \left(\frac{9'}{16\pi\epsilon_0 c\sqrt{2q/mc^2}} \right)^{2/3} \left(\frac{r}{a} \right)^{4/3}. \tag{5.12}$$

Putting $r = a$, $\phi(a) = \phi_a$, $\epsilon_0 c = \sqrt{\epsilon_0/\mu_0} = 1/120\pi$, one gets from (5.12) the *perveance* k of this beam

$$k = \frac{I}{\phi_u^{3/2}} = \frac{16\pi\sqrt{2}}{9 \times 120\pi\sqrt{mc^2/q}} - 20.95 \times 10^{-3} \left(\frac{mc^2}{q} \right)^{-1/2}. \tag{5.13}$$

For electrons, the perveance has the value

$$k = 29.34 \times 10^{-6} \text{A/V}^{3/2}. \tag{5.14}$$

On first thought, one would assume that the above current is the upper limit that can be obtained for a given potential ϕ_a; that is, as the beam current is gradually increased, the potential $\phi(0) = \phi_0$ at the center will gradually decrease until it becomes zero and the current reaches its maximum. This, however, is not the case. If one obtains the general numerical solution of Equation (5.11) for all possible values of ϕ_0 (namely, $0 \le \phi_0 < \phi_u$), one finds that the maximum current occurs at a value of $\phi_0 = 0.174\phi_a$. This is illustrated in Figure 5.2, where

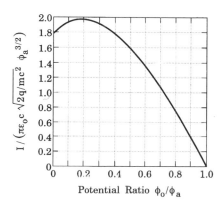

Figure 5.2. Graph showing dependence of current on potential ratio ϕ_0/ϕ_a.

$I/(\pi\epsilon_0 c\sqrt{2q/mc^2}\,\phi_a^{3/2}$, is plotted versus the potential ratio ϕ_0/ϕ_a. The maximum of the curve is 1.963, and the corresponding current is

$$I_m = 1.963\,\pi\epsilon_0 c\sqrt{\frac{2q}{mc^2}}\,\phi_a^{3/2}. \tag{5.15}$$

This yields a maximum perveance of

$$k_m = \frac{I_m}{\phi_a^{3/2}} = \frac{1.963\,\pi\sqrt{2}}{120\,\pi\sqrt{mc^2/q}} = 23.13 \times 10^{-3}\left(\frac{mc^2}{q}\right)^{-1/2}, \tag{5.16}$$

which for electrons has a value of

$$k_m = 32.4 \times 10^{-6}\,\text{A/V}^{3/2}. \tag{5.17}$$

This is 10% higher than the value for $\phi_0 = 0$ obtained in Equation (5.14). If one would try to inject currents higher than I_m, the potential at the center would drop to $\phi_0 = 0$, but particles would be reflected back toward the source from the *virtual cathode* that is formed in the beam and the net forward current would drop to the value of Equation (5.13). Actually, the situation is more complicated than that; one finds that in the region to the left of the current peak oscillations occur (i.e., one cannot achieve a stable steady-state operation).

The more general relativistic case, which requires a solution of Equation (5.10), was treated by Bogdankevich and Rukhadze in 1971. They obtained for the maximum current the approximate solution

$$I_m = I_L = I_0(\gamma_a^{2/3} - 1)^{3/2},$$

which was presented in Equation (4.61) (assuming that $b = a$). It is worth noting that in the extreme relativistic this space-charge current limit has the same value as the Alfvén–Lawson current (4.56) (i.e., $I_L = I_A = I_0\gamma_a$ for $\gamma_a \gg 1$, $\beta_a = 1$). This is true only when the beam fills the drift tube ($b = a$). If the drift-tube radius is greater than the beam radius, I_L is always lower than I_A by the factor $1 + 2\ln(b/a)$.

5.2.2 Nonrelativistic Laminar Beam Equilibria

In the preceding section we discussed the idealized model of a laminar beam in an infinitely strong magnetic field where all particles are forced to move on straight trajectories along the magnetic flux lines. Let us now consider the more realistic case where the beam is confined by a uniform axial magnetic field of finite strength. In this section we present the nonrelativistic theory where $\gamma = 1$, $\beta^2 \ll 1$, and hence the magnetic self fields can be neglected. The equilibrium state is characterized by exact force balance at every radial position within the beam. This implies that the particles must have an azimuthal velocity component, v_θ, so that the radially inward Lorentz force, $qv_\theta B_z$, can balance the outward electric force, qE_r, due to the space charge and the centrifugal force, mv^2/r, due to the rotation. The desired azimuthal motion is achieved when the magnetic field at the source is different from the field in the downstream equilibrium region so that the particles cross flux lines and experience a force $F_\theta - qv_z B_r$. In addition, the space-charge electric field, E, produces an azimuthal force, $E \times B$. The effects of the launching conditions on the beam equilibrium are discussed below. The particle trajectories in the equilibrium state are thus helices encircling the axis with constant radius r and with two components, v_z and v_θ, both of which may, in general, be functions of r. It follows that $v_r = \dot{r} = 0$ and $\ddot{r} = 0$ for all particles. The nonrelativistic radial force balance equation may then be written in the form

$$\frac{mv_\theta^2(r)}{r} + qE_r(r) + qv_\theta(r)B_0 = 0, \tag{5.18}$$

where B_0 is the applied axial magnetic field and E_r is the radial electric field due to the space charge of the beam. In some special cases there may also be an applied radial electric field, for instance when the beam is hollow and there is a coaxial inner conductor at an electrostatic potential with respect to the drift tube. In view of the cylindrical symmetry such an applied field is of the form C/r, where C is a constant. The inclusion of an applied electric field is straightforward. However, to simplify matters we will limit our analysis to the cases where no external electric field is present. These cases are also the more important ones from a practical point of view. It will be convenient to write the radial force-balance equation in terms of the angular frequency $\dot{\theta} = \omega$. By substituting $v_\theta - r\omega$ and introducing the cyclotron frequency $\omega_c = -qB_0/m$ in (5.18), one obtains

$$r\omega^2(r) + \frac{qE_r(r)}{m} - r\omega(r)\omega_c = 0. \tag{5.19}$$

The space-charge electric field, $E_r(r)$, is determined by the particle density, $n(r)$, via Poisson's equation or Gauss's law, as given in Equation (5.4). Using this relationship we can write (5.19) in the alternative form

$$\omega^2(r) + \frac{q^2}{\epsilon_0 mr^2} \int_0^r n(r)r\,dr - \omega(r)\omega_c = 0. \tag{5.20}$$

This equation can be used in two ways. If the particle density, $n(r)$, is known, one can solve for the angular frequency and obtains

$$\omega(r) = \omega_L \pm \left[\omega_L^2 - \frac{q^2}{\epsilon_0 m r^2} \int_0^r n(r) r \, dr \right]^{1/2}, \qquad (5.21)$$

where the Larmor frequency $\omega_L = \omega_c/2$ was introduced.

One the other hand, if $\omega(r)$ is known, one can solve Equation (5.20) for the particle density, which yields the equation

$$n(r) = -\frac{\epsilon_0 m}{q^2} \frac{1}{r} \frac{d}{dr} \{ r^2 [\omega^2(r) - \omega_c \omega(r)] \}. \qquad (5.22)$$

Note from Equation (5.21) that in order for ω to be real, the condition

$$\frac{q^2}{\epsilon_0 m r^2} \int_0^r n(r) r \, dr < \omega_L^2 \qquad (5.23)$$

must be satisfied. This implies that the magnetic restoring force exceeds the repulsive electrostatic force of the space charge.

The force-balance equation (5.20) and its alternative forms (5.21) and (5.22) contain only the two functions $n(r)$ and $\omega(r)$. It appears, therefore, that we can make an arbitrary choice of one of these two functions and then determine the other self-consistently. For example, if we assume that the density is uniform [i.e., $n(r) = n_0 = $ const], we find from (5.21) the solution

$$\omega = \omega_L \pm \left(\omega_L^2 - \frac{\omega_p^2}{2} \right)^{1/2}, \qquad (5.24)$$

where $\omega_p = (q^2 n_0 / \epsilon_0 m)^{1/2}$ is the plasma frequency. Thus, for a beam with uniform density n_0, where the plasma frequency ω_p is by definition constant (independent of radius r), all particles rotate around the axis with constant angular frequency ω. This state is known in the literature as a *rigid-rotor equilibrium* [B.3] or an *isorotational beam* [C.9]. Note that the frequency ω as defined here for the laboratory frame is not identical with the frequency ω used in Chapters 3 and 4 to describe particle motion in the Larmor frame ($\omega_{L.f.}$). The relationship between the two frequencies (with $\omega = \omega_{Lab}$) is $\omega_{L.f.} = \omega_{Lab} - \omega_L$.

Before proceeding with a more detailed discussion of this special result we must recognize that the force-balance equation alone does not completely describe the equilibrium state of the beam. In addition to force balance, the particles must also satisfy the two conservation laws for energy and canonical angular momentum.

The energy conservation law relates the total velocity of the particles, $v(r)$, to the potential function $\phi(r)$ according to the nonrelativistic formula

$$v(r) = \left[v_\theta^2(r) + v_z^2(r)\right]^{1/2} = \left[\frac{2q\phi(r)}{m}\right]^{1/2}. \tag{5.25}$$

The potential function $\phi(r)$, in turn, is determined from the particle density distribution $n(r)$ via Poisson's equation (5.3).

The conservation law for the canonical angular momentum p_θ, also known as *Busch's theorem*, given in Equation (2.76), implies that for each particle

$$mr^2\dot\theta + \frac{q}{2\pi}\psi = p_\theta = \text{const.}$$

Here ψ denotes the magnetic flux enclosed by the particle in the equilibrium region. The value of p_θ is determined by the magnetic field configuration at the emitter surface of the source. (We will assume that in all cases $\dot\theta = \omega = 0$ at the source.) Two such configurations are illustrated in Figure 5.3, where two solenoids, separated by an annular iron plate, are used to control the magnetic field at the source, B_s, and in the downstream region, B_0, independently. In the configuration shown on top of the figure, the source solenoid is turned off. All of the magnetic flux generated by the other solenoid passes radially outward through the iron plate, and the source is in the field-free region ($B_s = 0$). Thus, $p_\theta = 0$ for all particles. The other case illustrated in the figure shows a so-called cusp-field configuration in which both solenoids produce a magnetic field of the same strength but opposite polarity ($B_s = -B_0$).

The canonical angular momentum is determined by the magnetic flux ψ_s enclosed by the particle's initial radius r_s, that is,

$$p_\theta = \frac{q}{2\pi}\psi_s = \frac{q}{2}B_s r_s^2. \tag{5.26}$$

In each of the two field geometries the particles cross magnetic flux lines and rotate about the axis in the uniform field of the downstream equilibrium region, as illustrated in the figure. However, the flux change, and hence the rotation frequency ω, in the cusp case is twice as large as in the upper configuration of a magnetically shielded source. The variation of the axial magnetic field for the two cases is shown in Figure 5.3(c). By varying the current in the source solenoid one can achieve other configurations, such as the dashed curve in the graph where the field at the source has the same polarity but a lower value as in the downstream region.

From Busch's theorem one obtains for the angular frequency of the particles in the equilibrium region the relation

$$\dot\theta = \omega = \frac{1}{mr^2}\left(p_\theta - \frac{q}{2\pi}\psi\right). \tag{5.27}$$

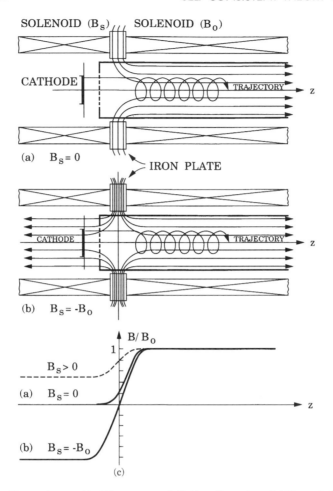

Figure 5.3. Electron beam with different magnetic field configurations. (a) Magnetic flux is zero at the cathode ($p_\theta = 0$); (b) cathode and downstream equilibrium region are immersed in a uniform magnetic flux with opposite polarity (cusp geometry); (c) magnetic field variation along the axis for cases (a) and (b) and for $0 < B_s < B_0$ (dashed curve). Case (a) corresponds to the Brillouin solid beam where the beam current I is a maximum. In case (b), on the other hand, there is no laminar-flow equilibrium, hence $I = 0$.

Using Equation (5.26) for p_θ and $\psi = B_0 r^2 \pi$ for the flux enclosed downstream, one can write this relation in the form

$$\omega(r) = \frac{qB_s r_s^2}{2mr^2} - \frac{qB_0}{2m} = -\frac{qB_0}{2m}\left(1 - \frac{B_s r_s^2}{B_0 r^2}\right),$$

or

$$\omega(r) = \omega_L\left(1 - \frac{\psi_s}{\psi}\right). \tag{5.28}$$

The force-balance requirement, Equation (5.20), energy conservation, Equation (5.25), Poisson's equation (5.3), and conservation of canonical angular momentum, Equation (5.28), form a complete self-consistent set of *equations of state* for the laminar-beam equilibrium. Note that these are four equations for the four functions $v_\theta(r) = r\omega(r)$, $v_z(r)$, $n(r)$, and $\phi(r)$. The only free parameter in this set is the canonical angular momentum, p_θ, or the magnetic flux ratio ψ_s/ψ. If $p_\theta(r)$ is given (i.e., if a particular field configuration for the source is chosen), the four functions are uniquely determined. Thus the free choice of either $n(r)$ or $\omega(r)$ implied by the force-balance equation does not exist when the conservation of energy and canonical angular momentum are included in the analysis. We will see below that the rigid-rotor beam ($n = $ const, $\omega = $ const) constitutes a special solution of the equations of state satisfying these conservation laws.

We can reduce the number of equations to two by eliminating $n(r)$ and $\phi(r)$. First, let us differentiate Equation (5.25) with respect to r using a prime (') to denote d/dr:

$$v_\theta(r)v_\theta'(r) + v_z(r)v_z'(r) = \frac{q\phi'(r)}{m} = \frac{qE_r(r)}{m}. \tag{5.29}$$

By substituting for E_r from Equation (5.19) and using $v_\theta = r\omega$, one obtains the differential equation

$$v_z(r)v_z'(r) + \omega(r)\{r^2[\omega(r) - \omega_L]\}' = 0. \tag{5.30}$$

Equations (5.30) and (5.28) uniquely determine the self consistent solutions for the two functions $v_z(r)$ and $\omega(r)$. The nature of these solutions, and hence the structure of the beam equilibrium, depends on p_θ (i.e., the magnetic field configuration and the launching conditions for the particles from the source).

As a first example, let us examine the rigid-rotor equilibrium obtained from the force-balance equation. Inspection shows that the solution $\omega = \omega_0 = $ const is compatible with Equation (5.30) and with Equation (5.28), provided that the flux ratio ψ_s/ψ is a constant for all particles. The latter condition is obviously satisfied for a magnetically shielded source where $p_\theta = 0$, hence $\psi_s/\psi = 0$. In this case, one has from (5.28) $\omega = \omega_L$, which is consistent with Equation (5.24) provided that $\omega_L^2 = \omega_p^2/2$. This is the condition of ideal Brillouin flow already encountered in Section 4.3.2 on beam transport in a uniform solenoidal magnetic field. We compare the results of the self-consistent theory with those from the linear (paraxial) theory of Section 4.3.2 in more detail below.

The flux ratio ψ_s/ψ is a constant when the source is in a uniform magnetic field, B_s, and the particles are launched from an emitter surface with a circular area to form a solid cylindrical beam. Under conditions of laminar flow, there is a correlation between the launching radius r_s of a particle and the radius r in the equilibrium state downstream. For each particle the ratio $r_s/r = \alpha$ is a constant, hence the flux ratio $\psi_s/\psi = B_s r_s^2/B_0 r^2$ is also constant, and the canonical angular momentum is a quadratic function of radius r:

$$p_\theta(r) = \frac{q}{2\pi} \psi_s = \frac{qB_s r_s^2}{2} = \frac{qB_0\alpha^2 r^2}{2}. \tag{5.31}$$

The constant rotation frequency ω of the rigid-rotor beam must satisfy the two equations (5.24) and (5.28). By comparison we find the following relation between the flux ratio ψ_s/ψ and the ratio ω_p/ω_L:

$$\frac{\psi_s}{\psi} = \mp\left(1 - \frac{\omega_p^2}{2\omega_L^2}\right)^{1/2}. \tag{5.32}$$

We see from this relation that ω_p/ω_L is a maximum, that is, the maximum amount of charge (or beam current) can be confined, when

$$\omega_p^2 = 2\omega_L^2, \tag{5.33}$$

which implies that $\psi_s = 0$ (shielded source) and $\omega = \omega_L$. On the other hand, one finds that $\omega_p/\omega_L = 0$ for either $\psi_s/\psi = 1$, $\omega = 0$, or $\psi_s/\psi = -1$, $\omega = 2\omega_L = \omega_c$. Thus, no rigid-rotor equilibrium exists for a finite space charge ($\omega_p > 0$) in the two cases where the source is immersed in the same uniform field as the downstream beam ($B_s = B_0$) or in the case of an ideal cusp ($B_s = -B_0$) illustrated in Figure 5.3(b). For flux ratios $0 < |\psi_s/\psi| < 1$, rigid-rotor confinement is possible, but the confined charge is always less than in the ideal Brillouin case ($\psi_s = 0$).

The variation of ω versus $\omega_p^2/2\omega_L^2$ according to Equation (5.24) is plotted in Figure 5.4. By substituting $\omega = $ const into Equation (5.30), one finds that the axial velocity v_z of the particles is a quadratic function of radius r except for $\omega = 0$ where $v_z = v = $ const and for $\omega = \omega_L$ where $v_z = v_0 = $ const. For the ideal cusp field one gets $v_z^2 = v^2 - \omega_c^2 r^2$, where the total velocity is the same for all particles since there is no space-charge potential. Note that in the Larmor frame used in Chapters 3 and 4, the frequency $\omega - \omega_{Lab}$ changes to $\omega_{L.f.} = \omega_{Lab} - \omega_L$. Thus when $\psi = 0$, $\omega_p^2 = 2\omega_L^2$, the frequency in the Larmor frame is zero ($\omega_{L.f.} = 0$) and the particle trajectories are straight lines.

We now proceed to study the so-called *isovelocity* solution $v_z = v_0$ of the equations of state in more detail. When v_z is a constant (independent of radius r), one obtains from Equation (5.30) for the angular frequency ω the solution

$$\omega(r) = \omega_L + \frac{C}{r^2}, \tag{5.34}$$

where C is an integration constant whose value depends on the canonical angular momentum at the source.

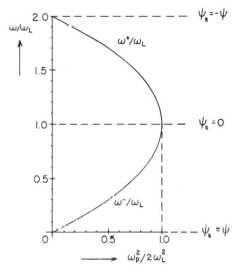

Figure 5.4. Relationship between angular frequency, ω, and plasma frequency squared, ω_p^2, in the rigid-rotor beam. The upper and lower branches of the curve (ω^+ and ω^-) represent the two solutions of Equation (5.24). The value of the magnetic flux at the source, ψ_s, is indicated on the right side for the three cases $\omega = 0$, $\omega = \omega_L$, and $\omega = 2\omega_L$.

The case $C = 0$ corresponds to the special rigid-rotor solution

$$\omega = \omega_L = \frac{\omega_p}{\sqrt{2}} \tag{5.35}$$

for a shielded source ($p_\theta = 0$). Thus, in this case, all particles within the equilibrium beam not only have a constant axial velocity but also a constant rotational frequency. This *isovelocity-isorotational* type of laminar flow is also known in the literature as a *Brillouin solid beam*.

The other solution ($C \neq 0$) of Equation (5.34) implies launching conditions such that the canonical angular momentum p_θ, or the flux ψ_s enclosed at the source, is the same for all particles within the beam. Such a configuration is illustrated in Figure 5.5, where a hollow beam is formed and both the source and the downstream equilibrium are immersed in the same uniform magnetic field ($B_s = B_0$). The radius, r_s, of the emitter surface is equal to the radius, r_0, of the inner edge of the downstream equilibrium beam. Therefore, $p_\theta = 0.5qB_0r_0^2$ and

$$\omega(r) = \omega_L\left(1 - \frac{r_0^2}{r^2}\right). \tag{5.36}$$

Thus the angular frequency is $\omega = 0$ at $r = r_0$ and then increases with the radius until it reaches the value $\omega = \omega_L(1 - r_0^2/a^2)$ at the outer edge of the beam. This

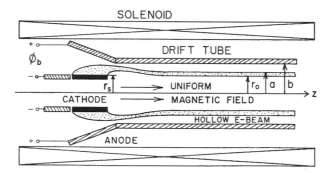

Figure 5.5. Schematic of electrode and magnetic field configuration to form a Brillouin hollow beam (used, for example, in magnetron injection guns). Both source and drift-tube region are immersed in the same uniform magnetic field. The magnetic flux enclosed at the source is the same for all particles ($p_\theta = 0.5qB_0r_s^2$ and $r_s = r_0$). A practical example of a magnetron injection gun is discussed in Appendix 2.

type of immersed flow is also known in the literature as a *Brillouin hollow beam,* and the electron source producing such a hollow beam is known as a *magnetron injection gun.* In practice, the cathode is located in the fringe region of the solenoid where the magnetic flux lines begin to diverge away from the axis. The cathode has a conical surface which coincides with the magnetic flux surface so that the total flux enclosed by the particles remains constant ($\psi = B_0r_0^2\pi$).

The particle density distribution for the various self-consistent beam configurations is obtained by substituting $\omega(r)$ from the canonical angular momentum equation (5.27) in (5.22), which yields

$$n(r) = \frac{\epsilon_0 m}{q^2}\left(\frac{2p_\theta^2}{mr^4} + \frac{\omega_c^2}{2}\right). \qquad (5.37)$$

In terms of the particles' plasma frequency $\omega_p = [q^2n/\epsilon_0 m]^{1/2}$, this relation may be written as

$$\omega_p^2 = \frac{\omega_c^2}{2} + \frac{2p_\theta^2}{m^2r^4}. \qquad (5.38)$$

For the Brillouin solid beam, where $p_\theta = 0$, $\omega = \omega_L$, we thus find the solution

$$n = n_0 = \frac{\epsilon_0 m\omega_c^2}{2q^2}, \qquad \text{or} \qquad \omega_p = \frac{\omega_c}{\sqrt{2}}. \qquad (5.39)$$

In the case of the Brillouin hollow beam, where $p_\theta = qB_0r_0^2/2$, $\omega = \omega_L(1 - r_0^2/r^2)$, on the other hand, we obtain

$$n = n_0\left[1 + \left(\frac{r_0}{r}\right)^4\right], \qquad \text{or} \qquad \omega_p = \frac{\omega_c}{\sqrt{2}}\left[1 + \left(\frac{r_0}{r}\right)^4\right]^{1/2}. \qquad (5.40)$$

Let us now examine the two types of beam separately.

(a) Brillouin Solid Beam   Figure 5.3(a) shows in schematic form the configuration for a Brillouin-type solid electron beam with uniform magnetic focusing. As discussed above, the magnetic field at the cathode is zero ($B_s = 0$) and the particles acquire an azimuthal velocity $v_\theta = r\omega_L$ when entering the uniform field region after crossing the flux lines.

The potential $\phi(r)$ in the beam is obtained from Equation (5.25) with $v_z = v_0$ and $v_\theta = r\omega = r\omega_c/2$; one gets

$$\phi(r) = \frac{mv_0^2}{2q} + \frac{\omega_c^2 m r^2}{8q} = \phi_0 + \frac{qB_0^2}{8m} r^2, \qquad (5.41)$$

where ϕ_0 is the potential on the axis ($r = 0$) related to the axial velocity by

$$v_z = v_0 - \sqrt{\frac{2q\phi_0}{m}}. \qquad (5.42)$$

The axial current density is given by

$$J_z = qn_0 v_z = \frac{\epsilon_0 m \omega_c^2}{2q} \sqrt{\frac{2q\phi_0}{m}}, \qquad (5.43)$$

and the total current carried by the beam in longitudinal direction is then simply

$$I = J_z a^2 \pi = \frac{a^2 \pi \epsilon_0 m \omega_c^2}{2q} \sqrt{\frac{2q\phi_0}{m}} - \frac{a^2 \pi \epsilon_0 q B_0^2}{2m} \sqrt{\frac{2q\phi_0}{m}}, \qquad (5.44)$$

where $r = a$ is the outer beam radius. If the potential ϕ_a at the beam edge is introduced, one has from (5.41) with $r = a$,

$$\phi_a = \phi_0 + \frac{\omega_c^2 m a^2}{8q} = \phi_0 + \frac{qB_0^2}{8m} a^2, \qquad (5.45a)$$

which can be solved for the applied magnetic field, yielding

$$B_0 = \frac{2}{a} \left(\frac{2m}{q} \right)^{1/2} (\phi_a - \phi_0)^{1/2}. \qquad (5.45b)$$

With the aid of these relations we can eliminate either B_0 or ϕ_0 in Equation (5.44) and obtain for the beam current the alternative expressions

$$I = 4\pi\epsilon_0 \sqrt{\frac{2q}{m}} (\phi_a - \phi_0)\phi_0^{1/2}, \qquad (5.46a)$$

or

$$I = \frac{\pi \epsilon_0 m a^2 \omega_c^2}{\sqrt{2}q} \left(\frac{q\phi_a}{m} - \frac{\omega_c^2 a^2}{8} \right)^{1/2}.$$ (5.46b)

The last equation has the advantage that it relates the beam current only to the experimental parameters ϕ_a, ω_c (or B_o), and a. Note that we can solve Equation (5.46b) for any one of the four quantities I, ϕ_a, B_o, a if the other three are given. For simplicity let us again assume that the beam fills the drift tube ($b = a$) so that $q\phi_a$ represents the initial kinetic energy of the particles. As we see from Equation (5.46), for a given potential ϕ_a, the current varies as the magnetic field, and hence ω_c, is changed. To find the maximum current, we write $\omega_c^2 a^2/8 = x$, $I = Cx(A - x)^{1/2}$. Differentiating and setting $dI/dx = 0$ yields

$$x = \frac{2}{3} A,$$

that is,

$$\frac{\omega_c^2 a^2}{8} = \frac{q^2 B_0^2 a^2}{8m^2} = \frac{2}{3} \frac{q\phi_a}{m},$$ (5.47a)

or

$$B_0 = \frac{4}{\sqrt{3}} \frac{1}{a} \left(\frac{m\phi_a}{q} \right)^{1/2}.$$ (5.47b)

Substituting (5.47a) in (5.45a) leads to the important relation

$$\phi_0 = \frac{1}{3} \phi_a.$$ (5.48)

Note that we get the same result by using Equation (5.46a), differentiating with respect to ϕ_0 and setting $\partial I/\partial \phi_0 = 0$. The maximum current I_m is thus obtained in the beam when the beam voltage on the axis is one-third of the voltage at the beam edge, which implies that the ratio of the azimuthal velocity v_θ on the outer radius to the constant axial velocity $v_z = v_0$ is $\sqrt{2}$. By substitution of (5.47) into (5.46), one finds that

$$I_m = \frac{16\pi\epsilon_0}{3\sqrt{6}} \sqrt{\frac{q}{m}} \phi_a^{3/2}.$$ (5.49)

This corresponds to a maximum perveance for an electron beam of

$$k_m = \frac{I_m}{\phi_a^{3/2}} = 25.4 \times 10^{-6} \text{A/V}^{3/2}.$$ (5.50)

In summary, this type of Brillouin flow represents the ideal case of a rigid-rotor beam in which all particles rotate about the axis at a constant angular frequency $\omega = \omega_L$ and have the same constant axial velocity $v_z = v_0 = (2q\phi_0/m)^{1/2}$. The particle density is uniform and the plasma frequency is 0.707 times the cyclotron frequency ($\omega_p = \omega_c/\sqrt{2}$). The current reaches a theoretical upper limit $I = I_m$ that is defined by Equation (5.49). The applied magnetic field required to focus I_m is given in Equation (5.47b) and is seen to be inversely proportional to the beam radius a.

In deriving the above relations for a Brillouin solid beam we assumed that the beam fills the drift tube ($a = b$) so that ϕ_a represents the beam voltage at injection. If the drift-tube radius is greater than the beam radius ($b > a$), Equations (5.41) through (5.46b) are still valid. However, the voltage ϕ_a is no longer constant, and one must use the relation

$$\phi_b - \phi_a = 2(\phi_a - \phi_0) \ln\frac{b}{a} , \tag{5.51a}$$

or

$$\phi_b - \phi_0 = (\phi_a - \phi_0)\left(1 + 2 \ln\frac{b}{a}\right) \tag{5.51b}$$

between the beam voltage ϕ_b and the values ϕ_a at the beam edge ($r = a$) and ϕ_0 on the axis ($r = 0$). By eliminating ϕ_a in Equation (5.46a) (i.e., by expressing I in terms of ϕ_b and ϕ_0, differentiating with respect to ϕ_0, and setting $\partial I/\partial \phi_0 = 0$), one finds that $\phi_0 = \phi_b/3$ and that $\phi_a^{3/2}$ in (5.49) must be replaced by $\phi_b^{3/2}/[1 + 2 \ln(b/a)]$. We note that the beam current I for given voltage, magnetic field, and beam radius in the general case ($b > a$) is always less than for the case where the beam fills the drift tube ($b = a$). The maximum current I_m is reduced by the geometry factor $1 + 2 \ln(b/a)$.

When the conditions for perfect Brillouin flow are not exactly satisfied, the force balance implied in Equation (5.18) is violated so that the particles no longer move with constant radius. As a result, both the particle trajectories as well as the beam radius vary periodically with distance z. We investigate this case of a *rippled* or *mismatched* beam with the aid of the paraxial ray equation in Section 5.2.4.

The Brillouin beam discussed here is of great importance from both a theoretical point of view and in regard to practical applications. Because of the relatively low magnetic field that is required, and since it is possible to achieve rather uniform current density and velocity profiles, this type of beam is used, for instance, in microwave tubes such as klystrons.

(b) Brillouin Hollow Beam   As was mentioned above, this type of beam corresponds to a source located inside the uniform magnetic field region with all particles starting from a surface that coincides with a magnetic flux tube and having zero initial angular velocity. The cathode can have a cylindrical shape, as illustrated in Figure 5.5 (if it is located in the uniform field region of the solenoid), or a conical shape (if it located in the fringe region), as is usually the case in practical designs, such as in magnetron injection guns. For simplicity, we take the cylindrical cathode shown in the figure and assume that the inner beam radius, r_0, downstream is equal to the cathode radius, r_s (i.e., $r_0 = r_s$).

According to Equation (5.36), the particles at the inner edge of the beam do not rotate ($\omega = 0$ for $r = r_s$) since they do not cross any flux lines, and from Equation (5.40) we see that the plasma frequency at this radius is equal to the cyclotron frequency, ω_c. As the radius increases, the angular frequency increases, while the plasma frequency decreases approaching the asymptotic value $\omega_p = \omega_c/\sqrt{2}$ at the outer beam edge when $a \gg r_0$.

Following the same procedure as in case (a), one finds for the total beam current

$$I = \frac{\pi \epsilon_0 \omega_c^2 a^2 m}{\sqrt{2}\,q} \left(1 - \frac{r_0^4}{a^4}\right) \left[\frac{q\phi_a}{m} - \frac{\omega_c^2 a^2}{8}\left(1 - \frac{r_0^2}{a^2}\right)^2\right]^{1/2}. \tag{5.52}$$

As before, ϕ_a denotes the potential on the outer beam edge and represents the *voltage equivalent* of the initial kinetic energy. For fixed values of ϕ_a, r_0, or a, the current reaches a maximum when

$$(\omega_c a)^2 \left(1 - \frac{r_0^2}{a^2}\right)^2 = \frac{16}{3} \frac{q\phi_a}{m}, \tag{5.53}$$

which leads to the expression for the maximum current

$$I_m = \frac{16\pi\epsilon_0}{3\sqrt{6}} \sqrt{\frac{q}{m}} \, \phi_a^{3/2} \frac{1 + r_0^2/a^2}{1 - r_0^2/a^2}. \tag{5.54}$$

In the case of electrons, the corresponding perveance is

$$k_m = 25.4 \times 10^{-6} \frac{1 + r_0^2/a^2}{1 - r_0^2/a^2} \quad [\text{A/V}^{3/2}]. \tag{5.55}$$

This is larger than the maximum perveance of the solid Brillouin beam by the geometry factor in Equation (5.55). If the ratio of inner to outer beam radius is 0.5, for instance, the factor is 1.67, and for $r_0/a = 0.75$, the perveance is increased by 3.57.

The potential distribution across the beam is given, in analogy to (5.41), by

$$\phi(r) = \phi_0 + \frac{r^2 \omega^2 m}{2q} = \phi_0 + \frac{q B_0^2 r^2}{8m}\left(1 - \frac{r_0^2}{r^2}\right)^2 \quad \text{for } r_0 \leq r \leq a. \tag{5.56}$$

Substituting (5.56) into (5.52) with $r = a$, $\phi(a) = \phi_a$, yields

$$I = \frac{\pi \epsilon_0 B_0^2 a^2}{\sqrt{2}} \left(\frac{q}{m} \right)^{3/2} \phi_0^{1/2} \left(1 - \frac{r_0^4}{a^4} \right). \tag{5.57}$$

This equation represents the relationship between beam current, I, magnetic field, B_0, inner and outer beam radii, r_0 and a, and the voltage ϕ_0 at $r = r_0$ that must be satisfied to obtain the self-consistent laminar equilibrium of the Brillouin hollow beam.

Similar relations between the parameters of the beam and the magnetic field can be obtained for the other types of laminar equilibria discussed in connection with the rigid-rotor solution. A comprehensive review of the various types of laminar beams can be found in the book by Kirstein, Kino, and Waters [C.9].

5.2.3 Relativistic Laminar Beam Equilibria

In an exact relativistic treatment of beam equilibria [1], the magnetic self field $B_{s\theta}$ and B_{sz} due to the axial and azimuthal current components must be included in the equations of state. As before, let us assume that the cylindrical relativistic beam with equilibrium radius a is injected from a diode into a conducting drift tube of circular cross section, with radius b. The tube is inside a solenoid which produces a uniform static magnetic field in the region downstream from the diode. For the general derivation, the emitting surface may be either disk-shaped (solid beam) or annular (hollow beam) and may be either in a magnetically shielded region or linked by magnetic flux lines. The injection conditions are such that the beam assumes a laminar-flow equilibrium state at a distance comparable to a few tube diameters downstream from the injection point.

The most relevant case is that of an intense relativistic electron beam (IREB). Normally, such a beam is a pulse of short time duration (10 to 100 ns); however, the length of the beam is usually considerably larger than the tube diameter so that the postulated equilibrium state can be reached after transient effects due to the beamfront have decayed. Considering such laboratory beam pulses, the equilibrium state can exist for only a short period of time during which the magnetic self fields of the beam do not penetrate through the walls of the conducting tube. However, by letting the magnetic boundary increase beyond the tube radius or to infinity, the solutions for any intermediate situation or for a long beam can readily be obtained from the equations. We assume that the conducting pipe is at anode potential ϕ_b and that all particles are injected with the same kinetic energy $(\gamma_b - 1)mc^2 = q\phi_b$. As before, the potential ϕ is defined as a positive quantity representing the *voltage equivalent* of the kinetic energy, and the particle charge is also taken as positive.

In the region of the equilibrium state, an electrostatic field is set up by the space charge such that $\phi = \phi_a$ at the surface of the beam and $\phi = \phi_0$ at the center ($r = 0$) or the inner edge of a hollow beam. The energy conservation law then implies that the kinetic energy of the particles is less than $q\phi_b$ and varies as a function of radius from a minimum of $(\gamma_0 - 1)mc^2 = q\phi_0$ at the center to the maximum of $(\gamma_a - 1)mc^2 = q\phi_a$ at the surface of the beam $(r = a)$. When the beam fills the entire pipe $(a = b)$, the kinetic energy of the outermost particles is equal to the injection energy (i.e., $q\phi_a = q\phi_b$). The relativistic radial force balance equation may be written in the form

$$\frac{\gamma(r)mv_\theta^2(r)}{r} + qE_r(r) + qv_\theta(r)B_z(r) - qv_z(r)B_\theta(r) = 0, \tag{5.58}$$

where E_r and B_θ are self fields, while $B_z = B_0 + B_{sz}$ includes both the uniform applied magnetic field, B_0, and the axial self field B_{sz}. Note that B_{sz} is due to the azimuthal current density, J_θ, and since it is in the opposite direction to the applied field, it is called *diamagnetic*. In addition to force balance, we have the conservation law for the canonical angular momentum for each particle, which in this relativistic case takes the form

$$\gamma(r)mrv_\theta(r) + qrA_\theta(r) = \gamma(r)mrv_\theta(r) + q\int_0^r B_z(r)r \, dr = p_\theta(r). \tag{5.59}$$

It implies that all particles at given radius r (which, under laminar-flow conditions, were emitted from the source at the same radius r_s) have the same canonical angular momentum p_θ. Particles at different radii in the equilibrium beam have different p_θ values if they are emitted from a source that is linked by magnetic flux lines such that the magnetic vector potential A_θ varies across the emitting surface [i.e., $A_\theta = A_\theta(r_s)$ at the source].

The relationship between azimuthal velocity v_θ, axial velocity v_z, energy factor γ, and the potential function ϕ in the equilibrium beam is defined by the relativistic energy conservation law:

$$\gamma(r) = [1 - \beta_\theta^2(r) - \beta_z^2(r)]^{-1/2} = 1 + \frac{q\phi(r)}{mc^2}. \tag{5.60}$$

The potential is determined by the particle density $n(r)$ via Poisson's equation, $\nabla^2\phi(r) = qn(r)/\epsilon_0$, which in this relativistic case may be written in the form

$$\frac{1}{r}\frac{d}{dr}\left(r\frac{d\phi}{dr}\right) = \frac{mc^2}{q}\frac{1}{r}\frac{d}{dr}\left(r\frac{d\gamma}{dr}\right) = \frac{qn}{\epsilon_0}. \tag{5.61}$$

The radial electric field is then given by

$$E_r = \frac{d\phi}{dr} = \phi' = \frac{mc^2}{q}\gamma'. \tag{5.62}$$

The magnetic self-field components are determined by the current density $\mathbf{J} = qn\mathbf{v}$
From Maxwell's equation, $\nabla \times \mathbf{B} = \mu_0\mathbf{J} = \mu_0qn\mathbf{v}$, we obtain

$$\frac{1}{r}(rB_\theta)' = \mu_0qnv_z \tag{5.63}$$

and

$$B'_z = -\mu_0qnv_\theta. \tag{5.64}$$

Equations (5.58) through (5.64) constitute a complete set of relations which allow
one to determine the field components and the beam properties in a self-consistent way.

In analogy to the nonrelativistic analysis, these relations can be reduced to two
equations of state which in addition to v_z, v_θ, and p_θ contain the relativistic
energy factor γ and which take the form

$$[r(\gamma v_z)']' - v_z(r\gamma')' + \frac{1}{m}\left(\frac{v_\theta p'_\theta}{v_z}\right)' = 0, \tag{5.65}$$

$$\frac{v_\theta}{r}\gamma' + (\gamma + \gamma'r)\left(\frac{v_\theta}{r}\right)' + (\gamma v'_\theta)' + \frac{1}{m}\left(\frac{p'_\theta}{r}\right)' = 0. \tag{5.66}$$

As in the nonrelativistic case, the possible solutions allowed by these two equations
depend on the launching conditions at the source, which are represented by the
canonical angular momentum p_θ. We will limit ourselves to a brief discussion of the
results for a shielded source where $p_\theta = 0$ and refer for the details of the analysis
to Reference 1. If $p_\theta = 0$, one obtains for a solid beam the solutions $v_z = v_0 =$
const, $\omega = \omega(r) \propto r_0/(r_0^2 + r^2)$, and $n = n(r) \propto (r_0^2 + r^2)/(r_0^2 - r^2)^3$, where
r_0 is an integration constant. This is in contrast to the rigid-rotor solution for the
nonrelativistic case where axial velocity v_z, angular frequency ω, and particle
density n are constant (i.e., independent of radius r).

The relativistic energy factor γ which determines the kinetic energy of the
particles in the equilibrium state is found to vary with radius r as

$$\gamma(r) = \gamma_0\frac{r_0^2 + r^2}{r_0^2 - r^2}, \tag{5.67}$$

where $\gamma_0 = (1 - v_0^2/c^2)^{-1/2}$ defines the kinetic energy on the beam axis ($r = 0$).
At the outer edge of the beam, defined by the radius $r = a$, we get from (5.67)
the equation

$$\frac{\gamma_a}{\gamma_0} = \frac{1 + a^2/r_0^2}{1 - a^2/r_0^2}. \tag{5.68}$$

This relates the integration constant r_0 to the beam radius a and the ratio γ_a/γ_0 and hence to the kinetic energies $q\phi_a = (\gamma_a - 1)mc^2$ and $q\phi_0 = (\gamma_0 - 1)mc^2$ defined by the potentials ϕ_a, ϕ_0 on the beam edge and on the axis, respectively.

The potential between the beam edge $(r = a)$ and the wall $(r = b)$ varies logarithmically with radius r, and by setting $r = b$ and $\gamma(b) = \gamma_b$ one finds that

$$\frac{\gamma_b}{\gamma_0} = \frac{\gamma_a}{\gamma_0} + \left(\frac{\gamma_a^2}{\gamma_0^2} - 1 \right) \ln \frac{b}{a}. \tag{5.69}$$

Note that $(\gamma_b - 1)mc^2 = q\phi_b$ is the kinetic energy of the beam particles at injection and corresponds to the diode voltage defined by ϕ_b.

For the experimentalist, the most important information is the relationship between total beam current, injection energy (or diode voltage), and applied magnetic field that has to be met in order to achieve laminar-flow equilibrium for a given beam and tube diameter. The axial beam current is readily obtained by integrating $2\pi qnv_0 r\, dr$ from $r = 0$ to $r = a$, which yields, in terms of γ_a/γ_0, the result

$$I = I_0 \frac{(\gamma_0^2 - 1)^{1/2}}{2} \left(\frac{\gamma_a^2}{\gamma_0^2} - 1 \right), \tag{5.70}$$

where $I_0 = 4\pi\epsilon_0 mc^3/q \approx 17{,}000$ A for electrons. The current thus depends on ϕ_a and the ratio of the potential on axis to the potential at the beam surface, ϕ_0/ϕ_a. In Figure 5.6 we plotted I/I_0 versus the potential ratio ϕ_0/ϕ_a for several values of the potential ϕ_a. Note that in the case $a = b$, ϕ_a represents the diode voltage. The current is seen to have a maximum at small values of the potential ratio and is zero at $\phi_0/\phi_a = 0$ and $\phi_0/\phi_a = 1$. The region to the left of the maximum, where the slope of the curve is positive $(\partial I/\partial\phi_0 > 0)$, is unstable, as is known from the nonrelativistic theory.

For the applied magnetic field, B_0, one obtains the expression

$$B_0 = \frac{4mc}{qr_0} \frac{(1 - a^4/r_0^2 b^2)}{(1 - a^2/r_0^2)^2}, \tag{5.71}$$

which, in view of (5.68), may also be written in the form

$$B_0 = \frac{mc}{qa} \left(\frac{\gamma_a^2}{\gamma_0^2} - 1 \right)^{1/2} \left[\left(\frac{\gamma_a}{\gamma_0} + 1 \right) - \left(\frac{\gamma_a}{\gamma_0} - 1 \right) \frac{a^2}{b^2} \right]. \tag{5.72}$$

Figure 5.7 shows how B_0 varies with the potential ratio ϕ_0/ϕ_a for the case $b = a$. The three curves correspond to the values of ϕ_a used in Figure 5.6 for the current. Given the beam current, potential ϕ_a, and beam radius a, one can thus determine

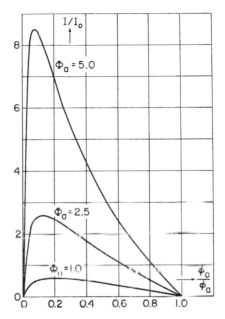

Figure 5.6. Beam current (in units of $I_0 = 4\pi\epsilon_0 mc^3/q$) versus potential ratio ϕ_0/ϕ_a for different values of $\Phi_a = q\phi_a/mc^2$. (From Reference 1.)

ϕ_0/ϕ_a and the required magnetic field B_0. Of the two values of ϕ_0/ϕ_a associated with a given current I, one must choose the larger one that corresponds to stable current flow.

If the beam does not fill the entire pipe ($b > a$), the procedure for determining the allowed combination of the parameters I, ϕ_b, a, b, B_0 is a little more complicated. One must then use Equation (5.69), which relates the potential on the beam edge, ϕ_a, to the diode voltage, ϕ_b, in combination with Equations (5.70)

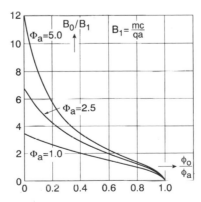

Figure 5.7. Applied magnetic field B_0 (in units of mc/qa) versus potential ratio ϕ_0/ϕ_a for different values of $\Phi_a = q\phi_a/mc^2$. (From Reference 1.)

and (5.72) for the current and the magnetic field, respectively. Note from these formulas that as b/a increases, both the current and the required applied magnetic field decrease (assuming a constant diode voltage).

The maximum value of the beam current, known in the literature as the *space-charge current limit*, plays an important role in many relativistic electron beam experiments and devices. We will therefore derive an analytical expression for this value from Equation (5.70). By differentiation with respect to γ_0, and setting $\partial I/\partial \gamma_0 = 0$, one obtains the equation

$$\gamma_0^4 + \gamma_a^2 \gamma_0^2 - 2\gamma_a^2 = 0, \tag{5.73}$$

which has the solution

$$\gamma_0^2 = \frac{\gamma_a^2}{2}\left[\left(1 + \frac{8}{\gamma_a^2}\right)^{1/2} - 1\right]. \tag{5.74}$$

Substitution of (5.74) into (5.70) then yields for the maximum current the expression

$$\frac{I_m}{I_0} = \frac{1}{2}\left\{\frac{\gamma_a^2}{2}\left[\left(1 + \frac{8}{\gamma_a^2}\right)^{1/2} - 1\right] - 1\right\}^{1/2}\left[\frac{2}{(1 + 8/\gamma_a^2)^{1/2} - 1} - 1\right]. \tag{5.75}$$

This is obviously a somewhat complicated functional form, especially if the case $b > a$ is considered, where Equation (5.69) has to be used to find the relationship between γ_a and the injection-energy parameter γ_b.

For the applied magnetic field required to focus the limiting current, one obtains in the case $b = a$, by substitution of (5.74) into (5.72), the result

$$B_0 = \frac{2mc}{qa}\left[\frac{2}{(1 + 8/\gamma_a^2)^{1/2} - 1} - 1\right]^{1/2}. \tag{5.76}$$

In the *ultrarelativistic* limit ($\gamma_a^2 \gg 1$), the potential on the beam axis approaches a maximum value which is independent of ϕ_a and given by

$$\gamma_0^2 = 2. \tag{5.77}$$

The corresponding relations for the limiting current, magnetic field, and energy factors (potentials) are

$$I_m = I_0 \frac{\gamma_a^2}{4}, \tag{5.78}$$

$$B_0 = \frac{2mc}{qa}\frac{\gamma_a}{\sqrt{2}}, \tag{5.79}$$

and

$$\gamma_b = \gamma_a + \frac{\gamma_a^2}{\sqrt{2}} \ln \frac{b}{a}. \tag{5.80}$$

Figure 5.8 shows the limiting current, I_m/I_0, versus the diode voltage $\Phi_b = q\phi_b/mc^2$ in the range $0 \le \Phi_b \le 7$ (0 to 3.5 MV for electrons) for several ratios of tube to beam radius. The corresponding curves for the applied magnetic field B_0 are presented in Figure 5.9. As an example, take an electron beam with $b = a$, a beam radius of $a = 1$ cm, and a diode voltage of 1 MV ($\Phi_b = 2$). With $I_0 \approx 17$ kA, one finds that $I_m = 30$ kA, $B_0 = 0.7$, and $T = 7$ kG. If the tube diameter is 25 percent larger than the beam diameter, these values drop to $I_m = 15$ kA and $B_0 = 6$ kG. In the latter case, the potential on the beam edge drops from 1 MV to 650 kV according to Equations (5.74) and (5.69).

In practice it is not possible to achieve the equilibrium state that is defined by the space-charge current limit of Equation (5.75), so that the current that can be propagated for a given beam voltage is usually lower than I_m. The design for an actual beam system then requires simultaneous solution of the three equations (5.69), (5.70), and (5.72) for the five experimental parameters I, γ_b, a, b, B_0 and the two theoretical parameters γ_0, γ_a. To find a solution, four of the seven parameters must be specified, and the other three are then calculated self-consistently

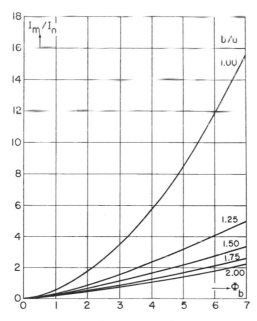

Figure 5.8. Limiting current I_m (in units of $I_0 = 4\pi\epsilon_0 mc^3/q$) versus potential ratio Φ_a for different values of b/a. (From Reference 1.)

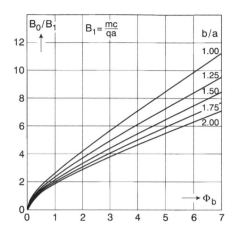

Figure 5.9. Applied magnetic field (in units of mc/qa) versus potential ratio Φ_a for different values of b/a in the limiting-current case $(I = I_m)$. (From Reference 1.)

from the equations for the equilibrium state. This procedure is best carried out by numerical solution.

The relationship for the nonrelativistic beam can be recovered from the above relativistic equations by assuming that $q\phi \ll mc^2$, and hence $\gamma^2 = (1 + q\phi/mc^2)^2 \approx 1 + 2q\phi/mc^2$. The proof is left as Problem 5.5.

When the source is immersed in the applied magnetic field and the canonical angular momentum p_θ varies with radius, the mathematical analysis becomes considerably more difficult. The axial velocity profile is no longer uniform, and in general, it is not possible to eliminate one of the two velocity components from Equations (5.65) and (5.66) to obtain a single equation that determines the equilibrium state. An inspection of the force balance equation (5.58) shows that as in the nonrelativistic case, the particles must acquire an angular velocity component, v_θ, in order to achieve a force equilibrium. As the particles leave the source (cathode), v_θ is initially zero and there is a net repulsive force $qE_r - qv_zB_\theta$ which increases the radius and results in a rotation of the trajectory due to the $\mathbf{E} \times \mathbf{B}$ effect of the E_r and B_z field components. At a short distance downstream from the source, an equilibrium state is reached in which the inward Lorentz force, $qv_\theta B_z$, balances the net defocusing action of the remaining force terms. In general, the azimuthal velocity component v_θ will be very small compared with v_z, and a relatively strong applied magnetic field B_0 is needed to confine the beam. Consequently (in contrast to the flow from a magnetically shielded source), the axial diamagnetic self field is very small and may be neglected. With the simplifying assumptions $B_0 \to \infty$, $v_\theta = 0$, and $J_z = qnv_z = $ const (uniform current density), Bogdankevich and Rukhadze obtained the solution for the limiting current given in Equation (4.61) and discussed in Section 5.2.1. It is interesting to compare this result, which applies to beams immersed in an infinitely strong magnetic field, with Equation (5.75) for the limiting current in the case of a magnetically shielded source. For convenience, consider the case $b = a$ and ultrarelativistic energies where $\gamma_a \gg 1$. A comparison then shows only the linear dependence $I_L = I_0\gamma_a$ of the Bogdankevich–Rukhadze current with voltage as compared with the quadratic dependence in Equation (5.78). Thus, one is led to conclude that injection from a shielded source should yield limiting currents, which, at relativistic energies, are substantially larger than the maximum currents achievable in a system where the source is located within the applied field. In the first case, the beam exhibits considerable rotational motion with the average v_θ comparable to v_z, while in the latter case, $v_\theta \approx 0$. Furthermore, due to the approximations made, the theory of immersed flow yields no practical relationship for the magnetic field strength required to focus the beam, other than the statement that B_0 should be very large. We do know, however, from the nonrelativistic theory of magnetically focused beams, that the magnetic field, required to focus a beam of a given current, voltage, and radius, is larger when the source is immersed in the field than in the case where the source is magnetically shielded.

5.2.4 Paraxial Analysis of Mismatched Laminar Beams in Uniform Magnetic Fields

In previous sections we considered self-consistent laminar beam equilibria where the beam radius a and the radius r of each particle trajectory were constant (i.e., independent of axial position z in the region downstream from the diode). If the beam is not launched with the correct initial conditions, the trajectories and the beam radius will perform oscillations about the respective equilibrium radii. Knowledge of the behavior of such a mismatched beam is very important for practical design and experiments. To study this problem without excessive mathematical difficulties, we will use the paraxial theory (i.e., we abandon the self-consistent approach). We will carry out the analysis for the general case of immersed flow ($p_\theta \neq 0$); the shielded flow is then obtained by setting $p_\theta = 0$ in the equations. As we shall see, the paraxial analysis leads to useful results, and the errors are small when the currents are not too high.

For laminar flow, the emittance is neglected ($\epsilon = 0$), and since there are no axial electric field components (i.e., $\gamma' = \gamma'' = 0$) the envelope equation (4.79) for immersed beams can be applied in the form

$$r_m'' + \left(\frac{qB_0}{2mc\beta\gamma}\right)^2 r_m - \left(\frac{p_\theta}{mc\beta\gamma}\right)^2 \frac{1}{r_m^3} - \frac{K}{r_m} = 0. \qquad (5.81)$$

Since p_θ is determined by the magnetic field B_s and the initial radius r_s at the source (i.e., $p_\theta = \frac{1}{2}qB_s r_s^2$), we may write

$$r_m'' + \kappa r_m - \kappa_s \frac{r_s^4}{r_m^3} - \frac{K}{r_m} = 0, \qquad (5.82)$$

where

$$\kappa = \left(\frac{qB_0}{2mc\beta\gamma} \right)^2, \qquad \kappa_s = \left(\frac{qB_s}{2mc\beta\gamma} \right)^2. \tag{5.83}$$

Note that Equation (5.82) is formally identical with (4.85a) except that we have the p_θ term in place of the emittance term. Indeed, as was already pointed out in Section 4.3.1, a nonzero canonical angular momentum p_θ has the same effect on the beam envelope as a finite emittance ϵ. Thus, the following mathematical analysis for laminar flow with $p_\theta \neq 0$ is directly applicable to a nonlaminar beam where $\epsilon \neq 0$. A special solution of Equation (5.82) is $r_m = a = $ const, which corresponds to a force balance condition where the outer radius of the beam is constant. For this case we have $r_m'' = 0$ and obtain the relation

$$a^4 - \frac{K}{\kappa} a^2 - \frac{\kappa_s}{\kappa} r_s^4 = 0. \tag{5.84}$$

Two special cases are of interest:

1. $K = 0$ (space-charge forces are negligible), in which case we obtain the solution

$$a = r_a = r_s \left(\frac{\kappa_s}{\kappa} \right)^{1/4} = r_s \left(\frac{B_s}{B_0} \right)^{1/2}. \tag{5.85}$$

   Note that the downstream equilibrium radius is greater or less than the cathode radius r_s if the magnetic flux density B_s at the source is greater or less than the uniform field B_0.

2. $B_s = 0$, that is, the source is shielded from magnetic flux ($p_\theta = 0$). In this case, the value of the equilibrium radius a is defined as r_b, where

$$a^2 = r_b^2 = \frac{K}{\kappa}. \tag{5.86}$$

In the nonrelativistic case, $K = (I/\phi_0^{3/2})/4\pi\epsilon_0(2q/m)^{1/2}$ and $\kappa = qB_0^2/8m\phi_0$, and we obtain from (5.86) the relation

$$I = \frac{\epsilon_0 \pi a^2 B_0^2 \phi_0^{1/2} (q/m)^{3/2}}{\sqrt{2}}, \tag{5.87}$$

which is identical with Equation (5.44). Thus, in the case of a solid non-relativistic Brillouin beam, the paraxial analysis yields the same results as the self-consistent theory provided that we use the potential on the axis, ϕ_0, in both cases.

With the two radii r_a and r_b, Equation (5.84) may be expressed in the form

$$a^4 - a^2 r_b^2 - r_a^4 = 0.$$ (5.88)

The general solution for the equilibrium beam radius a may then be written in the two equivalent forms

$$a = r_b \left[\frac{1 + (1 + 4r_a^4/r_b^4)^{1/2}}{2} \right]^{1/2},$$ (5.89)

$$a = r_a \left[\left(1 + \frac{r_b^4}{4r_a^4} \right)^{1/2} + \frac{r_b^2}{2r_a^2} \right]^{1/2}.$$ (5.90)

One can show from these relations that even if $I \neq 0$, the equilibrium beam radius r_e is larger or smaller than r_s when B_s is smaller or larger than B_0 (i.e., the beam is magnetically expanded when $B_s > B_0$ and magnetically compressed when $B_s < B_0$). Also, the equilibrium radius in the case where $B_s \neq 0$ will exceed the radius in the case $B_s = 0$ (solid Brillouin beam) for the same current I and voltage ϕ_0.

When the equilibrium conditions are not satisfied, the downstream beam radius is rippled [i.e., $r_m = r_m(z)$] and we must solve Equation (5.82) to obtain the solution for this case. If we define the beam radius r_m in terms of the unrippled equilibrium radius a by the substitution $R = r_m/a$ and the slope by $dR/dz = R'$, we obtain the following first integral of Equation (5.82) after multiplying by $2r_m'$:

$$R'^2 = R_0'^2 + \kappa \left(R_0^2 - R^2 + \frac{1}{R_0^2} - \frac{1}{R^2} \right) + \frac{K}{a^2} \left(\ln \frac{R^2}{R_0^2} + \frac{1}{R^2} - 1 \right),$$ (5.91)

where $R_0 = R(0)$, $R_0' = R'(0)$.

Putting $R' = 0$, one can find the maximum and minimum excursions of the beam radius ($R_{\max}$ and $R_{\min}$) by solving the transcendental relation for R as a function of R_0', κ, K, and a^2.

The second integral can be obtained from (5.91) in closed form only if the space-charge term K vanishes (zero current limit); otherwise, one has to resort to a numerical integration method. However, when the ripple amplitudes are relatively small compared with the equilibrium radius a, one can derive an approximate solution. By substituting $r_m = a(1 + x)$, where $|x| \ll 1$, into Equation (5.82), expanding, and keeping only first-order terms in x, one obtains the equation

$$x'' + 4\kappa \left(1 - \frac{r_b^2}{2a^2} \right) x = 0,$$ (5.92)

which is analogous to (4.103). This is the differential equation of the linear harmonic oscillator, which for $x = 0$, $x' = x'_0$ at $z = 0$ has a solution of the form

$$x = x'_0 \sin k_e z = x'_0 \sin \frac{2\pi}{\lambda} z, \tag{5.93}$$

where the wavelength λ_e of the beam envelope oscillation is given by

$$\lambda_e = \frac{2\pi}{k} = 2\pi \left[2\sqrt{\kappa} \left(1 - \frac{r_b^2}{2a^2} \right)^{1/2} \right]^{-1}. \tag{5.94}$$

In a solid Brillouin beam ($B_s = 0$), where $r_b = a$, the wavelength is

$$\lambda_e = \frac{2\pi}{\sqrt{2\kappa}}. \tag{5.95}$$

If we introduce the time $t = z/v_0$, we obtain the envelope oscillation frequency or the transverse resonant frequency, ω_e, given by $\omega_e = 2\pi v_0/\lambda_e$, or

$$\omega_e = \left(1 - \frac{r_b^2}{2a^2} \right)^{1/2} 2\sqrt{\kappa}\, v_0. \tag{5.96}$$

When $K = 0$ (no space charge), we find that

$$\lambda_e = \frac{2\pi}{2\sqrt{\kappa}} = \frac{2\pi v_0}{\omega_c}, \qquad \omega_e = \omega_c. \tag{5.97}$$

For the solid Brillouin beam ($B_s = 0$), the envelope oscillation wavelength and frequency are given by

$$\lambda_e = \frac{2\pi\sqrt{2}\, v_0}{\omega_c}, \qquad \omega_e = \frac{\omega_c}{\sqrt{2}} \quad \text{(in agreement with Section 4.3.3).} \tag{5.98}$$

Thus, in the absence of space charge, the frequency associated with the envelope oscillation of the beam is equal to the cyclotron frequency. In a solid Brillouin beam emitted from a magnetically shielded source, on the other hand, the envelope frequency is equal to the plasma frequency given by Equation (5.39) (i.e., $\omega_e = \omega_p = \omega_c/\sqrt{2}$). Interestingly, in the latter case, the paraxial analysis yields the same results as the self-consistent theory for a nonrelativistic beam.

5.3 THE VLASOV MODEL OF BEAMS WITH MOMENTUM SPREAD

5.3.1 The Vlasov Equation

When the effect of the velocity spread (temperature, emittance) of the beam is not negligible compared with the space-charge force, the flow is nonlaminar, and the theoretical model has to be modified. In the paraxial theory, the nonlaminar situation is represented by the emittance term in the envelope equation. As one might expect, a self-consistent theory of nonlaminar flow is not that simple; that is, one cannot merely add a temperature term to the self-consistent laminar equations discussed in preceding sections. The accepted method of describing self-consistent equilibria in this more general case is the Vlasov model [2]. It applies to all systems (nonneutral beams as well as neutral plasmas) for which Liouville's theorem is applicable and where collisions between particles can be neglected. A system of identical charged particles is defined by the distribution function $f(q_i, p_i, t)$ in six-dimensional phase space, where q_i and p_i represent the conjugate canonical space and momentum coordinates. Liouville's theorem states that

$$\frac{df}{dt} = \frac{\partial f}{\partial t} + \sum_{i=1}^{3} \left(\frac{\partial f}{\partial q_i} \dot{q}_i + \frac{\partial f}{\partial p_i} \dot{p}_i \right) = 0, \tag{5.99}$$

which is equivalent to the statement that the volume occupied by a given number of particles in phase space remains constant (see Section 3.2), that is,

$$\iint d^3q\, d^3p = \text{const.} \tag{5.100}$$

The phase-space coordinates q_i, p_i obey Hamilton's equations of motion (2.60), that is,

$$\dot{q}_i = \frac{\partial H}{\partial p_i}, \qquad \dot{p}_i = -\frac{\partial H}{\partial q_i}, \tag{5.101}$$

where $H(q_i, p_i, t) = c(m^2c^2 + (\mathbf{p} - q\mathbf{A})^2)^{1/2} + q\phi$ is the relativistic Hamiltonian. The scalar potential ϕ and the vector potential $\mathbf{A}$ represent the sum of the applied fields and the self fields generated by the particles. The self-field contributions are determined by the space charge ρ and current density $\mathbf{J}$, which are obtained by integrating the distribution function $f(q_i, p_i, t)$ in momentum space, namely,

$$\rho = q \int f(q_i, p_i, t)\, d^3p, \tag{5.102}$$

$$\mathbf{J} = q \int \mathbf{v} f(q_i, p_i, t)\, d^3p. \tag{5.103}$$

In the case of explicit time dependence, $\partial f / \partial t \neq 0$, one has to solve the wave equations for ϕ and $\mathbf{A}$ (subject to the boundary conditions):

$$\nabla^2 \phi - \mu_0 \epsilon_0 \frac{\partial^2 \phi}{\partial t^2} = -\frac{\rho}{\epsilon_0}, \tag{5.104}$$

$$\nabla^2 \mathbf{A} - \mu_0 \epsilon_0 \frac{\partial^2 \mathbf{A}}{\partial t^2} = -\mu_0 \mathbf{J}. \tag{5.105}$$

By substituting (5.101) into the Liouville equation (5.99), we obtain the relativistic *Vlasov equation* [2], also known as the *kinetic equation:*

$$\frac{\partial f}{\partial t} + \sum_{i=1}^{3} \left(\frac{\partial f}{\partial q_i} \frac{\partial H}{\partial p_i} - \frac{\partial f}{\partial p_i} \frac{\partial H}{\partial q_i} \right) = 0. \tag{5.106}$$

The set of equations (5.102) to (5.106) determine self-consistently the dynamics of an ensemble of charge particles that obey Liouville's theorem.

An alternative and often more convenient formalism specifies the distribution function in terms of the space coordinates q_i and the mechanical momentum components $P_i = p_i - qA_i$ [i.e., $f = f(q_i, P_i, t)$]. By transforming from (q_i, p_i) to other variables (Q_i, P_i), Equation (5.100) may be expressed as

$$\iint D \, d^3q \, d^3p = \iint d^3Q \, d^3P = \text{const}, \tag{5.107}$$

where

$$D = \frac{\partial(q_1, q_2, q_3, p_1, p_2, p_3)}{\partial(Q_1, Q_2, Q_3, P_1, P_2, P_3)} \tag{5.108}$$

is the *Jacobian* of the transformation. Obviously, when $D = 1$, Liouville's theorem also applies to the phase space defined by the other variables (Q_i, P_i). It is easy to show that the Jacobian determinant $D = 1$ for the transformation $(q_i, p_i) \rightarrow (q_i, P_i)$. As a result, Liouville's theorem may be stated in the alternative form

$$\iint d^3q \, d^3P = \text{const}, \tag{5.109}$$

or

$$\frac{\partial f}{\partial t} + \sum_{i=1}^{3} \left(\frac{\partial f}{\partial q_i} \dot{q}_i + \frac{\partial f}{\partial P_i} \dot{P}_i \right) = 0. \tag{5.110}$$

The $\dot{P}_i$ are determined by the electric field $\mathbf{E}$ and the magnetic field $\mathbf{B}$ via the Lorentz force equation

$$\frac{d\mathbf{P}}{dt} = q\mathbf{E} + q\mathbf{v} \times \mathbf{B}. \tag{5.111}$$

Substitution of (5.111) into (5.110) then yields the relativistic Vlasov equation in the alternative form

$$\frac{\partial f}{\partial t} + \sum_{i=1}^{3} \left[\frac{\partial f}{\partial q_i} \dot{q}_i + q(\mathbf{E} + \mathbf{v} \times \mathbf{B})_i \frac{\partial f}{\partial P_i} \right] = 0, \qquad (5.112)$$

where the velocity $\mathbf{v}$ has to be expressed in terms of the mechanical momentum, that is,

$$\mathbf{v} = \frac{\mathbf{P}}{m} \left(1 + \frac{P^2}{m^2 c^2} \right)^{-1/2}. \qquad (5.113)$$

The electric and magnetic field are determined self-consistently by Maxwell's equations.

$$\nabla \times \mathbf{E} = -\frac{\partial \mathbf{B}}{\partial t}, \qquad (5.114a)$$

$$\nabla \times \mathbf{B} = \mu_0 q \int \mathbf{v} f(q_i, P_i, t) \, d^3 P + \mu_0 \epsilon_0 \frac{\partial \mathbf{E}}{\partial t}, \qquad (5.114b)$$

$$\nabla \cdot \mathbf{E} = \frac{q}{\epsilon_0} \int f(q_i, P_i, t) \, d^3 P, \qquad (5.114c)$$

$$\nabla \cdot \mathbf{B} = 0. \qquad (5.114d)$$

In the nonrelativistic case (i.e., when $\gamma = 1$), one can express the distribution function in terms of the velocity $\mathbf{v}$ rather than the momentum $\mathbf{P}$. With $f(\mathbf{r}, \mathbf{v}, t)$, the Vlasov equation may then be written as

$$\frac{\partial f}{\partial t} + \sum_{i=1}^{3} \left[\frac{\partial f}{\partial q_i} \dot{q}_i + \frac{q}{m} (\mathbf{E} + \mathbf{v} \times \mathbf{B})_i \frac{\partial f}{\partial \dot{q}_i} \right] = 0, \qquad (5.115)$$

where $\mathbf{r} = \{q_i\} = \{x, y, z\}$ and $\mathbf{v} = \{\dot{q}_i\} = \{\dot{x}, \dot{y}, \dot{z}\}$ in the case of cartesian coordinates. If the beam is composed of different types of particles or ions, the distribution function for each species obeys a Vlasov equation of the form (5.112). The self fields are then obtained by summation of all contributions to total charge density ρ and current density $\mathbf{J}$ in Maxwell's equations (5.114).

The *equilibrium* states of a distribution of particles (i.e., in our case of a charged particle beam) are defined by time-independent solutions of the *Vlasov–Maxwell equations*. In this case, with $\partial/\partial t = 0$, Equations (5.112) and (5.114) take the form

$$\sum_{i=1}^{3} \left[\frac{\partial f}{\partial q_i} \dot{q}_i + q(\mathbf{E} + \mathbf{v} \times \mathbf{B})_i \frac{\partial f}{\partial P_i} \right] = 0, \qquad (5.116)$$

$$\nabla \times \mathbf{E} = 0, \tag{5.117a}$$

$$\nabla \times \mathbf{B} = \mu_0 q \int \mathbf{v} f(q_i, P_i, t) \, d^3P, \tag{5.117b}$$

$$\nabla \cdot \mathbf{E} = \frac{q}{\epsilon_0} \int f(q_i, P_i, t) \, d^3P, \tag{5.117c}$$

$$\nabla \cdot \mathbf{B} = 0. \tag{5.117d}$$

In principle, this set of Vlasov–Maxwell equations defining the stationary states of charged particle distributions obviously has many solutions which depend on the form of the distribution functions and the parameters characterizing the system. The main problem is to find a particular distribution function which permits a mathematical analysis without excessive difficulties and which, at the same time, represents a good model of a physically realizable system. The usual approach is to choose a distribution function which depends on the constants or integrals of the motion and which therefore, by definition, is a solution of the Vlasov equation. Suppose, for instance, that the constants or integrals of the motion (e.g., total energy, canonical angular momentum, etc.) of a system of particles are known and defined by I_1, I_2, and so on. Then, any distribution function which is an arbitrary function of these integrals, $f(I_1, I_2, \ldots)$, satisfies Liouville's theorem and therefore the Vlasov equation; that is (with $\partial/\partial t = 0$),

$$\frac{df}{dt} = \sum_j \frac{\partial f}{\partial I_j} \frac{dI_j}{dt} = 0, \tag{5.118}$$

since $dI_j/dt = 0$, for $j = 1, 2, \ldots$.

Unfortunately, it is not easy to find constants or integrals of the motion and appropriate distribution functions in the general cases of three-dimensional systems, especially when space-charge forces are involved. The most important class of problems that can be treated without excessive difficulties by the Vlasov method are those in which the Hamiltonian (i.e., the total energy of the particles) can be separated into a transverse and a longitudinal part. As an illustration, let us consider the Hamiltonian for particle motion in a nonrelativistic beam confined by an electrostatic potential, that is,

$$H = \frac{1}{2m} \left(P_x^2 + P_y^2 + P_z^2 \right) + q\phi(x, y, z), \tag{5.119}$$

where $\phi = \phi_a + \phi_s$ is the sum of the applied (focusing) potential ϕ_a and the space-charge (defocusing) potential ϕ_s. Now let us assume that we are dealing with a continuous beam and that the combined potential is of the form

$$\phi(x, y, z) = \phi(x, y)f(z), \tag{5.120}$$

where $f(z)$ is either constant (i.e., $\partial f/\partial z = 0$) or where its variation with distance z is so small that the effect of the axial electric field $E_z = -\partial \phi/\partial z$ on the longitudinal momentum is negligible. Under these conditions the axial velocity and momentum of the particles remain approximately constant, and the Hamiltonian can be separated into a transverse part, $H_\perp$, and a longitudinal part, $H_\parallel$, where

$$H_\perp = \frac{1}{2m}\left(P_x^2 + P_y^2\right) + q\phi(x,y)f(z), \qquad (5.121a)$$

$$H_\parallel = \frac{1}{2m}P_z^2. \qquad (5.121b)$$

We note that in this approximation $H_\parallel$ is a constant of the motion whereas $H_\perp$ is not because of the variation $f(z)$ in the potential function. Strictly speaking, since the total Hamiltonian H in the system described by Equation (5.119) is a constant of the motion, $H_\parallel$ would have to vary with distance z if $H_\perp = H_\perp(z)$. However, the approximation $H_\parallel \approx$ const is justified if the changes in the longitudinal momentum due to the axial force $F_z = qE_z$ are negligibly small (i.e., $\Delta P_z \ll P_z$). A good example for this case is an electrostatic quadrupole channel where the applied potentials (and hence also the beam's self potential) vary with distance z. In this case, the transverse Hamiltonian $H_\perp$ is not a constant, and hence, we cannot use it to construct a distribution function that would satisfy the stationary Vlasov equation. On the other hand, if we consider a uniform focusing channel, where the potential function does not vary with distance z (i.e., if $\partial f/\partial z = 0$), then both $H_\perp$ and $H_\parallel$ are also constants of the motion. In this case, any distribution function of $(H_\perp, H_\parallel)$ would satisfy the time-independent Vlasov equation and hence represent a stationary beam. A good example of such a system is the cylindrical beam in a long solenoid with uniform magnetic field. In the rotating Larmor frame, the applied radial Lorentz force $F_r = -qv_\theta B_z$ is equivalent to a focusing electrostatic force $qE_r = -q\partial\phi_a/\partial r$ which opposes the repulsive force due to the space-charge potential $\partial\phi_s/\partial r$. The steady state in this uniform magnetic field is characterized by a beam that has a constant radius and hence no z-variation of the self field (matched beam). A second, albeit somewhat more academic example is a beam that propagates through a "transparent" cylinder of a stationary opposite charge with uniform density in which collisions can be neglected and which acts like an applied focusing potential. The effects of charge neutralization discussed in Section 4.6 can be treated by such a model. Finally, the *smooth approximation* of beams in periodically varying focusing systems offers a third example that is of great importance from a practical point of view. The smoothed applied focusing force is independent of z and can be treated like a harmonic oscillator potential in the transverse direction, as discussed in Section 4.4. The corresponding average transverse Hamiltonian is then a constant of the motion and, mathematically, *average* stationary states can be constructed by distribution functions of the form $f(H_\perp, H_\parallel)$. In doing so one of course neglects the axial variation of both the applied potential ϕ_a and the space-charge potential ϕ_s (arising from the periodic ripple in the beam envelope). One of the most important examples of such a distribution function is

$$
f = f_0 \exp\left(-\frac{H_\perp}{k_B T_\perp}\right)\delta(P_z - P_0). \tag{5.122}
$$

The longitudinal part is a delta function, which means that all particles have the same axial momentum. The beam is therefore "cold" in the axial direction (i.e., it has zero longitudinal temperature). The transverse part corresponds to a two-dimensional *Maxwell–Boltzmann* distribution with constant transverse temperature $k_B T_\perp$, also known as a transverse *thermal distribution*. The particle density, which is only a function of the transverse coordinates, is obtained by integrating over the momentum components, yielding

$$
n(x,y) = n_0 \exp\left[-\frac{q\phi_a(x,y) + q\phi_s(x,y)}{k_B T_\perp}\right]. \tag{5.123}
$$

This equation relating the particle density to the potential and the temperature is known as the *Boltzmann relation*. When the applied potential function is given and the space-charge potential is negligible, the density variation is readily defined by Equation (5.123). However, when ϕ_s is not negligible the situation becomes mathematically much more complicated since ϕ_s and $n(x,y)$ are related by Poisson's equation, $\nabla^2\phi_s = -qn/\epsilon_0$. We discuss this complication further in Section 5.3.3, where several examples of stationary distributions are treated, and in Section 5.4.4, where the transverse Maxwell–Boltzmann distribution is analyzed in much more detail.

In the relativistic case, the Hamiltonian corresponding to Equation (5.119) has the form [see Equation (2.70)]

$$
H = c\left(m^2 c^2 + P_x^2 + P_y^2 + P_z^2\right)^{1/2} + q\phi(x,y,z), \tag{5.124}
$$

in which the transverse and longitudinal kinetic energy parts are not readily separated since they appear inside the square root. Fortunately, for most beams of practical interest, the transverse momentum components are small compared to the longitudinal momentum, so that (5.124) can be approximated by

$$
H = \gamma m c^2 + \frac{P_x^2 + P_y^2}{2\gamma m} + q\phi(x,y)f(z), \tag{5.125a}
$$

or, alternatively, in terms of the velocities,

$$
H = \gamma m c^2 + \frac{\gamma m}{2}\left(v_x^2 + v_y^2\right) + q\phi(x,y)f(z). \tag{5.125b}
$$

Thus, as in the nonrelativistic case, we can separate the transverse part of the Hamiltonian from the longitudinal term (represented by γmc^2). As we will see in the next sections, it will be convenient to redefine the transverse Hamiltonian in terms of the slopes x' and y' by dividing (5.125b) with γmv^2 and writing

$$H_\perp = \frac{1}{2}(x'^2 + y'^2) + \frac{q\phi(x,y)}{\gamma mv^2} f(z).$$ (5.126)

When $f(z)$ is constant, this relativistic Hamiltonian for the transverse motion is a constant, and any distribution function $f(H_\perp)$ will satisfy the stationary Vlasov equation. Before we discuss several functions of this type, we will first consider the more general case where $f(z)$ varies with distance z and where $H_\perp$ is not a proper constant of the motion. We recall from Chapters 3 and 4 that the emittances ϵ_x and ϵ_y (or the normalized emittances when acceleration is involved) are constant if all forces acting on the particles are linear in x and y. A distribution function $f(\epsilon_x, \epsilon_y)$ would therefore satisfy the time-independent Vlasov equation. As it happens there is only one self-consistent distribution where both the applied and space-charge forces are linear and where the emittances are preserved. This distribution is a delta function of the emittances. Known as the K–V distribution, it is treated in the next section.

5.3.2 The Kapchinsky–Vladimirsky (K–V) Distribution

In statistical mechanics, the distribution in which the forces are linear and the phase-space areas remain constant is known as the *microcanonical distribution*. Kapchinsky and Vladimirsky [3] used this distribution to study the effects of space charge on the transverse beam dynamics in a linear accelerator with magnetic quadrupole focusing elements. Their work, which was published in 1959, has been of major importance in accelerator theory and design, and their beam model is now generally referred to as the *K–V distribution*. For the forces to be linear in the transverse coordinates x and y, the conditions for paraxial motion must be satisfied (i.e., $v_x \ll v_z$, $v_y \ll v_z$, $v_z \approx v$). Furthermore, the changes in the beam size must occur slowly enough that longitudinal forces due to the beam's self-field components can be neglected. All particles then have the same axial velocity $v_z \approx v$; that is, the potential difference across the beam must be small compared to the kinetic energy.

Let us now trace the steps that lead to the K–V distribution function. The equations of motion for a continuous beam in which both the applied as well as the space-charge forces are linear functions of the transverse coordinates x and y have already been derived in Section 4.4.2 for a quadrupole focusing channel [Equations (4.176) and (4.177)]. We will write them in the form

$$x'' + \kappa_x(z)x = 0,$$ (5.127a)

$$y'' + \kappa_y(z)y = 0.$$ (5.127b)

The focusing functions $\kappa_x(z)$ and $\kappa_y(z)$ include the space-charge forces and are defined by

$$\kappa_x(z) = \kappa_{x0}(z) - \frac{2K}{X(X + Y)}, \tag{5.128a}$$

$$\kappa_y(z) = \kappa_{y0}(z) - \frac{2K}{Y(X + Y)}. \tag{5.128b}$$

The beam has in this general case an elliptical cross section; $X(z)$ and $Y(z)$ are the semiaxes of the ellipse, and they are found by solving the beam envelope equations (4.178) and (4.179). The two trajectory equations (5.127) have the same form as Equation (3.312), which we studied in Section 3.8.2. Thus we can represent the solutions in the phase-amplitude form

$$x(z) = A_x w_x(z) \cos\left[\psi_x(z) + \phi_x\right], \tag{5.129a}$$

$$y(z) = A_y w_y(z) \cos\left[\psi_y(z) + \phi_y\right]. \tag{5.129b}$$

The phase functions ψ_x and ψ_y satisfy the relations

$$\frac{d\psi_x}{dz} = \psi_x' = \frac{1}{w_x^2}, \qquad \psi_y' = \frac{1}{w_y^2}, \tag{5.130}$$

while w_x, w_y obey the equation

$$w_{x(y)}'' + \kappa_{x(y)} w_{x(y)} - \frac{1}{w_{x(y)}^3} = 0. \tag{5.131}$$

The parameters A_x, A_y, ϕ_x, ϕ_y depend on the initial conditions (x_0, x_0') and (y_0, y_0') and remain constant throughout the motion. Specifically, in analogy to Equation (3.342), we have the relations

$$A_x^2 = \frac{x^2}{w_x^2} + \left(w_x x' - w_x' x\right)^2, \tag{5.132a}$$

$$A_y^2 = \frac{y^2}{w_y^2} + \left(w_y x' - w_y' y\right)^2, \tag{5.132b}$$

which represent equations of ellipses. As discussed in Section 3.8.2, the maximum value of the amplitude parameter is related to the emittance. Thus, if ϵ_x denotes the emittance in $x-x'$ trace space, ϵ_y in $y-y'$ trace space, then

$$A_{x,\max}^2 = \epsilon_x, \qquad A_{y,\max}^2 = \epsilon_y. \tag{5.133}$$

This is in agreement with Liouville's theorem, which states that for a system of particles where the motion in each cartesian plane is decoupled from that in the other directions, the corresponding emittances remain constant during the motion.

Each value $A_x \le A_{x,\max}$, or $A_y \le A_{y,\max}$ defines an ellipse in $x-x'$ or $y-y'$ space whose area is conserved (i.e., A_x^2 and A_y^2 are integrals of the motion). Likewise, any linear combination, say $A_x^2 + CA_y^2$, is a conserved quantity. The constant C is given by the ratio of the emittances (i.e., $C = \epsilon_x/\epsilon_y$), and has the value $C = 1$ if the two emittances are the same. Thus we can define as a new integral of motion the quantity F given by

$$F = A_x^2 + \frac{\epsilon_x}{\epsilon_y} A_y^2, \qquad (5.134a)$$

or, alternatively, the dimensionless quantity G defined as

$$G = \frac{A_x^2}{\epsilon_x} + \frac{A_y^2}{\epsilon_y}. \qquad (5.134b)$$

Mathematically, any distribution function $f(F)$, or $f(G)$, would satisfy the time-independent Vlasov equation. However, only the special microcanonical distribution function proposed by Kapchinsky and Vladimirsky, known as the $K-V$ *distribution*, produces linear equations of motion with variables separated. It has the form

$$f = f_0^* \delta(F - F_0), \qquad (5.135a)$$

or

$$f = f_0 \delta(G - 1), \qquad (5.135b)$$

where $\delta(x)$ is the Dirac delta function with the property

$$\delta(x) = 0 \quad \text{for } x \ne 0, \qquad \int_{-\infty}^{\infty} \delta(x)\, dx = 1. \qquad (5.136)$$

Kapchinsky and Vladimirsky treated only the case $C = 1$ where $\epsilon_y = \epsilon_x = F_0$. However, we will not make this restriction and will treat the more general case of the K–V distribution, where $\epsilon_x \ne \epsilon_y$ (see the comment in Reference 3). For such a choice of the distribution function, the representation points of all particles in the beam lie on the surface of the hyperellipsoid

$$\frac{1}{\epsilon_x}\left[\frac{x^2}{w_x^2} + (w_x x' - w_x' x)^2 \right] + \frac{1}{\epsilon_y}\left[\frac{y^2}{w_y^2} + (w_y y' - w_y' y)^2 \right] = 1 \qquad (5.137)$$

in the four-dimensional phase space defined by the coordinates x, y, x', y'. The projection of this hyperellipsoid in the $x-x'$ plane gives the result

$$\frac{x^2}{w_x^2} + (w_x x' - w_x' x)^2 = \epsilon_x, \tag{5.138a}$$

which, from Equations (3.341) to (3.345), may be written in terms of the Courant–Snyder parameters $\hat{\alpha}$, $\hat{\beta}$, $\hat{\gamma}$ as

$$\hat{\gamma}_x x^2 + 2\hat{\alpha}_x xx' + \hat{\beta}_x x'^2 = \epsilon_x. \tag{5.138b}$$

These are the equations of the emittance ellipse in the $x-x'$ plane with the area $\epsilon_x \pi$. A similar relationship is obtained for the projection in the $y-y'$ plane, where on the right-hand side one obtains ϵ_y in lieu of ϵ_x.

The K–V distribution has the interesting property that all two-dimensional projections ($x-x'$, $x-y$, etc.) yield uniform particle densities in both the symmetric case $\epsilon_x = \epsilon_y$ and in the asymmetric case $\epsilon_x \neq \epsilon_y$ (see Reference 3 and Problem 5.8). Thus, the density in the ellipse (5.138) is uniform. Likewise, the density across the beam in the $x-y$ plane is uniform, as required to obtain linear space-charge forces. From Equation (5.137) we see that the coordinates of all particles obey the relation

$$\frac{x^2}{\epsilon_x w_x^2} + \frac{y^2}{\epsilon_y w_y^2} \le 1,$$

and, consequently, with $w^2 = \hat{\beta}$, the ellipse

$$\frac{x^2}{\hat{\beta}_x \epsilon_x} + \frac{y^2}{\hat{\beta}_y \epsilon_y} = 1 \tag{5.139}$$

represents the boundary of the beam outside of which there are no particles. The semiaxes of this ellipse, which represent the envelopes in the x and y directions, are given by

$$X(z) = \sqrt{\epsilon_x \hat{\beta}_x(z)}, \qquad Y(z) = \sqrt{\epsilon_y \hat{\beta}_y(z)}. \tag{5.140}$$

The distribution function $f(G)$ is a solution to the time-independent *Vlasov–Maxwell equations* and allows us to determine the charge density, current density, and associated fields in a self-consistent manner. First, the charge density can be calculated from the expression

$$\rho = qf_0 \int_{-\infty}^{\infty} \int_{-\infty}^{\infty} \delta(G - 1)\, dx'\, dy'. \tag{5.141}$$

The integration yields

$$\rho = q f_0 \pi \sqrt{\frac{\epsilon_x \epsilon_y}{\hat{\beta}_x \hat{\beta}_y}}, \tag{5.142}$$

which represents a uniform distribution in the beam cross section.

The total beam current I in the z-direction is given by

$$I = v_z \iint \rho(x, y, z) \, dx \, dy = v \rho X Y \pi \tag{5.143}$$

since ρ is independent of x and y and the beam cross section is an ellipse with area $XY\pi$. Thus, we can express ρ in terms of the current as

$$\rho(z) = \frac{I}{\pi v X(z) Y(z)}, \tag{5.144}$$

in agreement with Equation (4.173). With (5.140) and (5.144), the normalization constant f_0 in (5.135) and (5.142) is then found to be

$$f_0 = \frac{I}{\pi^2 q v \epsilon_x \epsilon_y}. \tag{5.145}$$

If the beam envelopes change along the path length z, the charge density ρ is a function of z. However, as was pointed out above, the changes occur along distances that are significantly greater than the beam width. The electrostatic potential ϕ may then be calculated from Poisson's equation for any given position z by approximating the beam as an infinite elliptical cylinder with semiaxes $X(z)$, $Y(z)$ and having uniform charge density; thus

$$\nabla^2 \phi = \frac{\partial^2 \phi}{\partial x^2} + \frac{\partial^2 \phi}{\partial y^2} = -\frac{\rho(z)}{\epsilon_0}, \tag{5.146}$$

where ρ is given in Equation (4.172). Note that we neglect any image charges from boundaries in this approximation.

The solution of Equation (5.146) for the potential distribution inside the beam is found to be

$$\phi(x, y, z) = -\frac{\rho(z)}{4\epsilon_0} \left[x^2 + y^2 - \frac{X(z) - Y(z)}{X(z) + Y(z)} (x^2 - y^2) \right] + \text{const.} \tag{5.147}$$

From this expression one obtains the electric field components inside the beam given in Equations (4.174) and (4.175). The magnetic self-field components are defined by $\mathbf{B} = (\mathbf{v} \times \mathbf{E})/c^2$, and the associated Lorentz force reduces the electric force by the factor $1 - \beta^2 = 1/\gamma^2$, as shown in Equations (4.15) and (4.16). Substitution of these force components into the equations of motion then leads to the trajectory equations (4.176) and (4.177), which can be written in the form (5.127). The associated envelope equations (4.178) and (4.179) can be obtained by substituting (5.140) into (5.131) using the definition (5.128) for the function $\kappa(z)$. The proof that the K–V distribution (5.135) yields the desired linear equations for beams with space charge is thus essentially completed.

To apply the K–V model to a specific beam channel, one must know the functions $\kappa_{x0}(z)$ and $\kappa_{y0}(z)$ representing the external focusing force, the generalized perveance K representing the space charge, and the emittances ϵ_x and ϵ_y. Then one must first find the envelopes $X(z)$, $Y(z)$. The envelope equations represent a system of two nonlinear, second-order coupled differential equations which, in general, must be solved numerically for given initial conditions. The results for $X(z)$ and $Y(z)$ can then be substituted into Equations (4.176), (4.177) to find the trajectories for individual particles in the beam.

If the channel consists of periodically spaced quadrupole lenses with period S such that $\kappa_{x0}(z + S) = \kappa_{x0}(z)$ and $\kappa_{y0}(z + S) = \kappa_{y0}(z)$, proper initial conditions will give periodic solutions for the envelopes with period S, as discussed in Section 4.4.2. In this case, the beam is said to be *matched* to the channel. For any other initial conditions in the periodic system one obtains the envelope oscillations of *unmatched* beams treated in Section 4.4.3. The K–V distribution $f = f_0 \delta(G - 1) = f(x, y, x', y')$ thus represents the phase-space function that generates the linear self-fields in coordinate space (x, y or r, θ) that were used in the paraxial theory of Chapter 4.

The above trajectory and envelope equations are applicable not only to straight focusing channels and linear accelerators, but also to beams in circular accelerators. In axisymmetric (*weak-focusing*) systems, for instance, the independent variable is the path length $s = \overline{R}\theta$, where $\overline{R}$ is the average radius of the equilibrium orbit; the functions κ_{x0} and κ_{y0} then represent the radial and axial betatron frequencies (see Section 3.6.1), $\kappa_{x0} = (1 - n)/\overline{R}^2$ and $\kappa_{y0} = n/\overline{R}^2$. Since κ_{x0} and κ_{y0} are independent of s in this case, a matched-beam solution exists where $X = $ const, $Y = $ const, which can be found by setting $X'' = 0$, $Y'' = 0$ in Equations (4.178), (4.179). In alternating-gradient (*strong-focusing*) synchrotrons, on the other hand, κ_{x0} and κ_{y0} are periodic functions of the path length s; the closed equilibrium orbit is not a circle in this case, and the periodicity is defined by the number of focusing lattice units along the circumference of the accelerator.

If we apply Equation (4.178) and (4.179) to a round beam ($X = Y = R$) and replace $\epsilon_x = \epsilon_y$ by ϵ and $\kappa_{x0} = \kappa_{y0}$ by $\kappa_0 = (qB_z/2mc\beta\gamma)^2$, we obtain the envelope equation (4.85a) for a paraxial beam in a solenoidal magnetic field, namely,

$$R'' + \kappa_0 R - \frac{K}{R} - \frac{\epsilon^2}{R^3} = 0.$$

At first glance, this analogy is somewhat surprising since the K–V equations were based on a self-consistent distribution function, whereas the paraxial theory did not use a self-consistent approach. However, as we did point out, the K–V distribution does in fact imply the paraxial assumption that the transverse velocity components v_x, v_y are small compared with the axial velocity. This assumption was necessary to linearize the equations of motion. The K–V theory may thus be called a *self-consistent paraxial* theory. It applies to beams with elliptic cross section in focusing fields with linear external forces that differ in the two perpendicular transverse directions as well as to round beams in axisymmetric focusing systems. A truly self-consistent nonparaxial theory comparable to that for relativistic laminar flow entails considerable mathematical difficulty and results in a rather complex set of nonlinear, complicated relations for the density profile and the associated self fields. An example of this type is the intense relativistic electron beam model of Hammer and Rostoker [4], in which the focusing force is produced by a background of stationary positive ions providing partial charge neutralization.

The K–V theory is a good and very useful approximation for beams where the current remains well below the space-charge limit. This is true for practically all accelerators and other devices. The most notable exception is the intense relativistic electron beam generator (IREB) which usually operates near the limiting current. It must be kept in mind, however, that the K–V model does not include nonlinear effects that increase the emittance. We deal with emittance growth effects in Chapter 6.

5.3.3 Stationary Distributions in a Uniform Focusing Channel

In the preceding section we have shown that the K–V distribution is a self consistent solution of the time-independent Vlasov equation when the external forces acting on the particles are linear functions of the transverse displacements x, y from the beam axis. In general, the amplitudes of these forces may be different in the two transverse directions and may vary with the distance along the path of the beam. The K–V distribution has the property that the electric and magnetic self forces due to space charge and beam current are also linear functions of x, y. This property is independent of the strength of the internal fields (and hence the generalized perveance K), and the particle density within the beam is always uniform.

In this section we consider the simplest case of a focusing system, namely, a uniform channel in which the external forces are linear, axisymmetric, and independent of the longitudinal distance z. A stationary particle distribution in such a channel has the property that the internal forces are also axisymmetric and independent of the distance z along the beam; however, with the exception of the K–V beam, the space-charge forces may be nonlinear functions of x, y. Mathematically speaking, a stationary distribution in a uniform channel is characterized by the fact that for all forces acting on the particles $\partial/\partial z = 0$ and $\partial/\partial t = 0$. We will assume that all particles have the same axial velocity $\dot{z}$ and that the transverse velocity components $\dot{x}, \dot{y}$ are very small compared to $\dot{z}$ (paraxial approximation). The relativistic energy factor γ can then be treated as a constant, with $\gamma = (1 - \beta^2)^{-1/2} \approx (1 - \dot{z}^2/c^2)^{-1/2}$. The total force acting on a particle in the transverse direction consists of the linear external focusing force; the outward force, qE_r, due to the space-charge electric field; and the inward force, $-q\dot{z}B_\theta = -qE_r\beta^2$, due to the magnetic self field of the beam. Note that there is also a force $q\dot{r}B_\theta$ in the z-direction due to the magnetic self field, but in our beam model the associated change in the axial velocity is so small that we can neglect it. It is easy to show that the equation of motion for a particle experiencing these forces can be derived from a Hamiltonian $H_\perp$, which is a function of r and $v_\perp = (\dot{x}^z + \dot{y}^z)^{1/2}$. If we use the trajectory slopes $x' = \dot{x}/v$, $y' = \dot{y}/v$ in place of $\dot{x}$, $\dot{y}$, the Hamiltonian can be defined in dimensionless form as

$$H_\perp(r, r') = \frac{1}{2} r_\perp'^2 + \frac{1}{2} k_0^2 r^2 + \frac{q\phi_s(r)}{\gamma m v^2} [1 - \beta^2], \qquad (5.148)$$

where $r^2 = x^2 + y^2$ and $r_\perp'^2 = x'^2 + y'^2$. Note that $r_\perp'^2$ stands for $v_\perp^2/v^2 = x'^2 + y'^2$ and is not to be confused with the square of the slope $r'^2 = (dr/dz)^2$ of $r(z)$, where r is the radial coordinate. The first term on the right-hand side represents the transverse kinetic energy, the second term is the potential energy due to the external focusing field, and $\phi_s(r)$ in the third term is the electrostatic potential due to the space charge of the beam. Note that (5.148) is identical with (5.126) except that the potential function is split into the applied part, ϕ_a, assumed to be of the harmonic oscillator form ($\propto r^2$) and the space-charge part, ϕ_s, in which the factor $1 - \beta^2$ represents the attractive magnetic self force arising from $\dot{z}B_\theta$.

Physically, such a uniform focusing system corresponds to particle motion through a long solenoid as seen in the Larmor frame or to a particle beam passing through a channel of stationary charges of opposite polarity and uniform density ρ_e, or to a smoothed periodic channel, as was pointed out earlier. In the first case the constant k_0^2 is given by $k_0^2 = \kappa = \omega_L^2/v^2$, where ω_L is the Larmor frequency, while in the second case it is proportional to the charge density ρ_e of the focusing background particle distribution, and in the third case it relates to the phase advance σ_0 and period length S by $k_0^2 = \sigma_0^2/S^2$. The Hamiltonian $H_\perp$ for particle motion in such a system is constant, and thus a stationary, self-consistent solution of the steady-state Vlasov equation can be represented by a properly chosen distribution function $f(H_\perp)$. Although the Hamiltonian depends only on r and $r_\perp'$, we should bear in mind that both $H_\perp$ and $f(H_\perp)$ are functions in four-dimensional phase space (x, y, x', y') or (r, θ, r', ψ), where θ and ψ denote the angles of the cylindrical coordinate system in x, y and x', y' space, respectively.

If a distribution function $f(H_\perp)$ is given, the self-consistent determination of the particle density $n(r)$ and the space-charge potential $\phi_s(r)$ follows the procedure outlined in Section 5.3.1. First, one obtains for the density

$$n(r) = \int_0^{a'(r)} \int_0^{2\pi} f(H_\perp(r, r_\perp')) \, d\psi \, r' \, dr' = \pi \int_0^{a'^2} f(H_\perp(r, r_\perp')) \, d\left(\frac{1}{2} r_\perp'^2\right),$$

$$(5.149)$$

where $a'(r)$ denotes the maximum value of $r'_\perp$ in the particle distribution at a given radius r.

The space-charge potential ϕ_s is then found by solving Poisson's equation,

$$\frac{1}{r} \frac{d}{dr}\left(r \frac{d\phi_s(r)}{dr}\right) = -\frac{\rho(r)}{\epsilon_0} = -\frac{qn(r)}{\epsilon_0}, \tag{5.150}$$

which, by substituting (5.149) for $n(r)$, takes the form

$$\frac{1}{r} \frac{d}{dr}\left(r \frac{d\phi_s(r)}{dr}\right) = -\frac{q\pi}{\epsilon_0} \int_0^{a'^2} f(H_\perp(r, r'_\perp)) d\left(\frac{1}{2} r'^2_\perp\right). \tag{5.151}$$

It will be convenient to introduce an effective potential $W(r)$ which represents the sum of the external focusing potential and the self-field potential, that is,

$$W(r) = \frac{1}{2} k_0^2 r^2 + \frac{q\phi_s(r)}{\beta^2 \gamma^3 mc^2}, \tag{5.152}$$

where $1 - \beta^2 = 1/\gamma^2$ has been used. The Hamiltonian may then be written as

$$H_\perp(r, r'_\perp) = \frac{1}{2} r'^2_\perp + W(r). \tag{5.153}$$

A particle reaching the outer edge of the beam, defined by $r_{max} = a$, will have zero slope at this point, and its Hamiltonian, or transverse total energy, will have the maximum value given by

$$H_0 = W(a), \qquad \text{with } r'_\perp(a) = 0. \tag{5.154}$$

At any radius $r < a$ inside the beam, this particle has the maximum value $a'(r)$ and its Hamiltonian is

$$H_0 = \frac{1}{2} a'^2(r) + W(r) = W(a). \tag{5.155}$$

It follows from these last two equations that

$$\frac{1}{2} a'^2(r) = W(a) - W(r). \tag{5.156}$$

We may thus write Equation (5.149) in the alternative form

$$n(r) = 2\pi \int_0^{W(a) - W(r)} f(H_\perp(r, r'_\perp)) d\left(\frac{1}{2} r'^2_\perp\right),$$

or

$$n(r) = 2\pi \int_{W(r)}^{W(a)} f(H_\perp)\, dH_\perp. \tag{5.157}$$

Likewise, the Poisson equation (5.151) may be written in the form

$$\frac{1}{r}\frac{d}{dr}\left(r\,\frac{d\phi_s(r)}{dr}\right) = -\frac{2\pi q}{\epsilon_0}\int_{W(r)}^{W(a)} f(H_\perp)\, dH_\perp. \tag{5.158}$$

The boundary of the particle distribution in four-dimensional space outside of which the density is zero is defined by the maximum value of the Hamiltonian [i.e., by setting $H_\perp(r, r'_\perp) = H_0 = W(a)$]. If $a' = a'(0)$ defines the maximum value of $a'(r)$ occurring at $r = 0$, we obtain from Equations (5.153) to (5.156) the equation for the boundary

$$\frac{r'^2_\perp}{a'^2} + \frac{W(r) - W(0)}{W(a) - W(0)} = 1. \tag{5.159}$$

In the following we discuss three distributions that have been treated in the literature and are known to be stationary self-consistent solutions of the steady-state Vlasov equation for a uniform focusing channel: the K–V distribution (introduced in the preceding section), the waterbag distribution, and the Gaussian distribution.

For a uniform focusing system, the K–V distribution can be represented as a delta function of the transverse Hamiltonian, that is,

$$f(H_\perp) = f_1 \delta(H_\perp - H_0), \tag{5.160}$$

where f_1 is a normalization constant. This distribution has the property that all particles in the beam have the same total transverse energy defined by H_0; hence, the particles populate the surface of a hypersphere in four-dimensional phase space uniformly. We note that for an asymmetric matched beam, where the focusing forces and/or the emittances differ, (5.160) is not valid and $F(H_\perp)$ has a rectangular shape. (See the report by Saraph and Reiser mentioned in Reference 3.)

The particle density is found from Equation (5.157) by substitution of (5.160), yielding

$$n(r) = 2\pi f_1 \int_{W(r)}^{W(a)} \delta(H_\perp - H_0)\, dH_\perp = 2\pi f_1 = n_0 \tag{5.161}$$

for $H_\perp \leq H_0 = W(a)$. The density is thus uniform, as expected for a K–V distribution, and the normalization constant is related to the density by

$$f_1 = \frac{n_0}{2\pi}. \tag{5.162}$$

With this result for the particle density, Poisson's equation becomes

$$\frac{1}{r}\frac{d}{dr}\left(r\frac{d\phi_s(r)}{dr}\right) = -\frac{qn_0}{\epsilon_0} = -\frac{\rho_0}{\epsilon_0},$$

which yields for the space-charge potential the solution

$$\phi_s(r) = \frac{\rho_0}{4\epsilon_0}(a^2 - r^2) - \frac{I}{4\pi\epsilon_0 v}\left(1 - \frac{r^2}{a^2}\right) \tag{5.163}$$

if $\phi_s = 0$ at $r = a$ is assumed and the relation (4.10b) between charge density ρ_0 and beam current I is used. The associated electric field is a linear function of radius r and is given by

$$E_r = -\frac{d\phi_s}{dr} = \frac{\rho_0}{2\epsilon_0}r = \frac{I}{2\pi\epsilon_0 a^2 v}r. \tag{5.164}$$

By substituting (5.163) for ϕ_s and introducing the generalized perveance K defined in (4.23), we obtain for the Hamiltonian (5.148) in the case of a K–V distribution the result

$$H_\perp(r,r_\perp') = \frac{1}{2}r_\perp'^2 + W(r) = \frac{1}{2}r_\perp'^2 + \frac{1}{2}k_0^2 r^2 + \frac{1}{2}K\left(1 - \frac{r^2}{a^2}\right). \tag{5.165}$$

We note that the distribution functions (5.135) and (5.160) are equivalent descriptions of a K–V beam. The latter was chosen for an axisymmetric beam in a uniform focusing system, while the former also includes the more general case of quadrupole fields that may vary with distance (although still linear in the two transverse directions). When the external forces vary with distance, a stationary distribution is characterized by the fact that the beam radius is no longer constant [i.e., $a = a(z)$]. Special cases of this type are the periodic-focusing systems discussed in Sections 4.4.1 to 4.4.3.

The phase-space boundary of the K–V distribution is obtained from Equation (5.159) and given by

$$\frac{r^2}{a^2} + \frac{r_\perp'^2}{a'^2} = 1, \tag{5.166}$$

where the relations $W(0) = \frac{1}{2}K$ and $W(a) = \frac{1}{2}k_0^2 a^2$ have been used.

Our second example of a stationary beam in a uniform focusing channel is the *waterbag distribution*, which is defined by the Heaviside step function

$$f(H_\perp) = f_2\theta(H_0 - H_\perp),\tag{5.167}$$

that is, $f(H_\perp) = \text{const}$ for $0 \le H_\perp \le H_0$ and $f(H_\perp) = 0$ for $H_\perp > H_0$. This distribution has the property that all transverse total energies between $H_\perp = 0$ and $H_\perp = H_0$ occur with equal probability and that the particles populate the interior of the hypersphere defined by H_0 uniformly.

Substitution (5.167) in (5.157) and integration yields for the particle density the result

$$n(r) = 2\pi f_2[W(a) - W(r)].\tag{5.168}$$

In view of the definition (5.152) for $W(r)$, the density thus depends linearly on the space-charge potential $\phi_s(r)$, which must be determined from Poisson's equation

$$\frac{1}{r}\frac{d}{dr}\left(r\frac{d\phi_s(r)}{dr}\right) = -\frac{2\pi q f_2}{\epsilon_0}\left[W(a) - \frac{1}{2}k_0 r^2 - \frac{q\phi_s(r)}{\beta^2\gamma^3 mc^2}\right].\tag{5.169}$$

This equation can be simplified by introducing the potential function

$$U(r) = W(r) - W(a) + \frac{2k_0^2}{k_1^2},\tag{5.170}$$

where the constant k_1^2 is defined by

$$k_1^2 = \frac{2\pi q f_2}{\epsilon_0}\frac{q}{\beta^2\gamma^3 mc^2}.\tag{5.171}$$

In place of Equation (5.169) one then obtains

$$\frac{1}{r}\frac{d}{dr}\left(r\frac{dU(r)}{dr}\right) - k_1^2 U(r) = 0.\tag{5.172}$$

This equation has the solution

$$U(r) = CI_0(k_1 r),\tag{5.173}$$

where C is an integration constant and $I_0(k_1 r)$ the modified Bessel function of order zero. If we let the space-charge potential take the value $\phi_s = 0$ at the edge of the beam ($r = a$), where $W(a) = \frac{1}{2}k_0^2 a^2$, the integration constant becomes

$$C = \frac{U(a)}{I_0(k_1 a)} = 2\frac{k_0^2}{k_1^2}\frac{1}{I_0(k_1 a)}.$$ (5.174)

The effective potential is then given by

$$W(r) = W(a) - 2\frac{k_0^2}{k_1^2}\left(1 - \frac{I_0(k_1 r)}{I_0(k_1 a)}\right).$$ (5.175)

For the particle density one finds that

$$n(r) = n_f\left(1 - \frac{I_0(k_1 r)}{I_0(k_1 a)}\right),$$ (5.176)

where

$$n_f = 4\pi f_2\frac{k_0^2}{k_1^2} = \frac{2k_0^2\epsilon_0\beta^2\gamma^3 mc^2}{q^2}$$ (5.177)

and where the density at $r = 0$ is given by

$$n(0) = n_0 = n_f\left(1 - \frac{1}{I_0(k_1 a)}\right).$$ (5.178)

In the extreme space-charge limit ($k_1 \to \infty$), the density is seen to be essentially uniform inside the beam and given by n_f. Thus n_f is the limiting density where the internal force due to the self fields is exactly equal and opposite to the external focusing force.

The beam current is defined by the integral

$$I = 2\pi\int_0^a vqn(r)r\,dr = 2\pi qvn_f\int_0^a\left(1 - \frac{I_0(k_1 r)}{I_0(k_1 a)}\right)r\,dr,$$

which yields

$$I = \pi qn_f va^2\left[1 - \frac{2}{k_1 a}\frac{I_1(k_1 a)}{I_0(k_1 a)}\right] = \pi qn_f va^2\frac{I_2(k_1 a)}{I_0(k_1 a)}.$$ (5.179)

By substituting relation (5.177) for n_f and introducing the characteristic current I_0, defined in Equation (4.13), we can write Equation (5.179) in the form

$$I = \frac{1}{2} I_0 \beta^3 \gamma^3 k_0^2 a^2 \frac{I_2(k_1 a)}{I_0(k_1 a)} . \tag{5.180}$$

From this we obtain the equivalent equation for the generalized perveance

$$K = k_0^2 a^2 \frac{I_2(k_1 a)}{I_0(k_1 a)} . \tag{5.181}$$

For given values of the generalized perveance K, the external focusing parameter k_0, and the beam radius a, the parameter k_1 can be calculated by solving Equation (5.181) numerically.

The space-charge potential $\phi_s(r)$ is found from (5.152) by substituting (5.175) for $W(r)$, and one obtains

$$\phi_s(r) = \frac{1}{2} I_0 \frac{\beta^2 \gamma^3}{4\pi\epsilon_0 c} k_0^2 a^2 \left[1 - \frac{r^2}{a^2} - \frac{4}{k_1^2 a^2} \left(1 - \frac{I_0(k_1 r)}{I_0(k_1 a)} \right) \right] . \tag{5.182}$$

By introducing the beam current I from (5.180), we can write this equation in the form

$$\phi_s(r) = \frac{I}{4\pi\epsilon_0 v} \frac{I_0(k_1 a)}{I_2(k_1 a)} \left[1 - \frac{r^2}{a^2} - \frac{4}{k_1^2 a^2} \left(1 - \frac{I_0(k_1 r)}{I_0(k_1 a)} \right) \right] , \tag{5.183}$$

which can be compared with the result (5.163) for the K–V distribution.

The Hamiltonian is given by

$$H_\perp(r, r'_\perp) = \frac{1}{2} r'^2_\perp + \frac{1}{2} k_0^2 a^2 \left[1 - \frac{4}{k_1^2 a^2} \left(1 - \frac{I_0(k_1 r)}{I_0(k_1 a)} \right) \right] . \tag{5.184}$$

For its maximum value $H_0 = W(a)$ one has $r'_\perp = a'$ at $r = 0$ and $r'_\perp = 0$ at $r = a$, which yields the relation

$$\frac{1}{2} a'^2 + \frac{1}{2} k_0^2 a^2 \left[1 - \frac{4}{k_1^2 a^2} \left(1 - \frac{I_0(0)}{I_0(k_1 a)} \right) \right] = \frac{1}{2} k_0^2 a^2 . \tag{5.185}$$

By substituting n_0/n_f from (5.178), we obtain the following relation for the parameter k_1:

$$k_1^2 = \frac{4 k_0^2 n_0}{a'^2 n_f} , \tag{5.186}$$

or, in view of (5.177),

$$k_1^2 = \frac{2q^2 n_0}{a'^2 \epsilon_0 \beta^2 \gamma^3 mc^2}. \tag{5.187}$$

Finally, by substituting (5.175) into (5.159), we obtain for the phase-space boundary of the waterbag distribution the equation

$$\frac{r_\perp'^2}{a'^2} + \frac{I_0(k_1 r) - 1}{I_0(k_1 a) - 1} = 1. \tag{5.188}$$

The shape of this boundary in the x–x' plane is shown in Figure 5.10 for several values of the parameter $k_1 a$, and the corresponding density profiles $n(r)/n_0$ are displayed in Figure 5.11. Of interest are two limiting cases. First, when the space-charge effects are negligible ($k_1 \rightarrow 0$), the phase-space area takes the well-known shape of an ellipse and is thus similar to the boundary of a K–V distribution. The density profile in this limit becomes

$$n(r) = \frac{2}{a^2 \pi} \left(1 - \frac{r^2}{a^2} \right). \tag{5.189}$$

Second, in the extreme space-charge limit ($k_1 \rightarrow \infty$), we see that the phase-space contour becomes rectangular and the density profile uniform. In fact, as discussed later, all nonuniform distributions in linear focusing channels become uniform in the laminar-flow limit, where the emittance is zero. The net potential in the interior of the beam is approaching the value zero in this limit due to the fact that the space-charge potential $\phi_s(r)$ is quadratic in r and exactly cancels the external focusing potential. The beam in this case resembles a rigid box in which the particles move

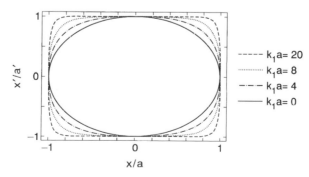

Figure 5.10. Phase-space boundary of stationary waterbag distributions for different values of the parameter $k_1 a$. (Courtesy of J. Struckmeier.)

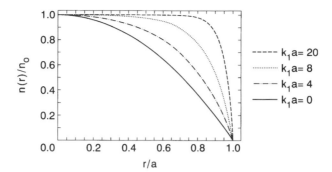

Figure 5.11. Density profile of stationary waterbag distributions for different values of the parameter $k_1 a$. (Courtesy of J. Struckmeier.)

freely on straight trajectories until they reach the "wall" at the edge ($r = a$), where they are reflected. Since the phase-space boundary becomes rectangular when $k_1 \rightarrow \infty$, we can represent the trace-space area by the approximate relation

$$\epsilon \pi = 4aa',$$ (5.190)

while the total beam current can be approximated by $I = q n_0 a^2 \pi v$. Using these relations, we obtain for $k_1^2 a^2$ the result

$$k_1^2 a^2 = \frac{32 q^2 n_0 a^4}{\epsilon^2 \pi^2 \epsilon_0 \beta^2 \gamma^3 m c^2} = \frac{64}{\pi^2} \frac{K a^2}{\epsilon^2},$$ (5.191)

where K is the generalized perveance of the beam. By comparing this result with the envelope equation (4.86), we see that $k_1^2 a^2$ is proportional to the ratio of the space-charge and emittance terms. If ω_0 denotes the particle oscillation frequency in the uniform focusing channel without space charge and ω the frequency with space charge, one can also show that $k_1^2 a^2 \propto \omega_0 / \omega$. This relation, as well as Equation (5.191), is of course valid only in the limit $k_1^2 a^2 \gg 0$ (i.e., $K a^2 \gg \epsilon^2$ or $\omega_0 \gg \omega$).

Finally, we note that the normalization constant f_2 for the waterbag distribution can be calculated from Equation (5.177) if k_1^2 is known. On the other hand, by substituting k_1^2 from (5.187) in (5.177), one obtains the simple relation

$$f_2 = \frac{2 q n_0}{a'^2 \pi},$$ (5.192)

indicating that f_2 is proportional to the density n_0 and inversely proportional to the square of the maximum slope a' at the center of the beam ($r = 0$).

The third example of a stationary beam in a uniform focusing channel is the *Gaussian distribution* defined by

$$f(H_\perp) = f_3 \exp\left(-\frac{H_\perp}{\alpha}\right), \qquad (5.193)$$

where f_3 and α represent normalization constants. Its form is identical to the Maxwell–Boltzmann distribution discussed in Section 5.3.1 [Equation (5.122)] except that in the present case the Hamiltonian is defined in terms of r' rather than $P_\perp$. Furthermore, the value of $H_\perp$ cannot be arbitrarily large since the particle beam has a finite radius ($r = a$). Thus the range of $H_\perp$ is given by $0 \le H_\perp \le H_0 = W(a)$, as in the case of the waterbag distribution; that is, the exponential function is truncated at some specified value $H_{\perp\,max} = H_0$ with $f(H_\perp) = 0$ for $H_\perp > H_0$. If H_0 is chosen to be in the "tail" of the distribution, one can take the upper limit of integrals involving the exponential function (5.193) at $H_\perp = \infty$ without significant loss of accuracy. The mean value of $H_\perp$, for instance, is given in this approximation by

$$\overline{H}_\perp = \frac{\int_0^{H_0} H_\perp \exp(-H_\perp/\alpha)\, dH_\perp}{\int_0^{H_0} \exp(-H_\perp/\alpha)\, dH_\perp} \approx \alpha \qquad (5.194)$$

and can be expressed in terms of the transverse beam temperature $T_\perp$ in the laboratory frame as

$$\overline{H}_\perp \approx \alpha = \frac{k_B T_\perp}{\gamma m v^2}. \qquad (5.195)$$

From Equations (5.193) and (5.157) we obtain for the particle density

$$n(r) = 2\pi f_3 \alpha \left[\exp\left(-\frac{W(r)}{\alpha}\right) - \exp\left(-\frac{W(a)}{\alpha}\right) \right]. \qquad (5.196)$$

Using $W(a) = H_0$, this result can be substituted into Poisson's equation, yielding

$$
\begin{aligned}
n(r) &= -\frac{\epsilon_0}{q}\frac{1}{r}\frac{d}{dr}\left(r\frac{d\phi_s(r)}{dr}\right) \\
&= 2\pi f_3 \alpha \left[\exp\left(-\frac{k_0^2 r^2}{2\alpha} - \frac{q\phi_s(r)}{\alpha\beta^2\gamma^3 mc^2}\right) - \exp\left(-\frac{H_0}{\alpha}\right) \right],
\end{aligned}
\qquad (5.197)
$$

which must be solved numerically to obtain either the space-charge potential ϕ_s or the particle density n as a function of radius. If the Gaussian function $f(H_\perp)$ is not truncated so that $\alpha = k_B T_\perp/\gamma m v^2$ and $\exp(-W(r)/\alpha) = \exp(-H_0/\alpha) \to 0$, it becomes identical to the transverse Maxwell–Boltzmann distribution. The density profile (5.197) then takes the form of the ideal Boltzmann relation $n(r) = n(0)\exp(-q\phi(r)/k_B T_\perp)$, where $\phi(r)$ includes both the applied focusing potential ϕ_a and the space-charge potential ϕ_s. We will study the Maxwell–Boltzmann distribution, including the behavior of the density profile with temperature $T_\perp$ in Section 5.4.4. Here we note only that at high temperature, when ϕ_s is negligible, the profile becomes truly Gaussian in radius, as can be seen from Equation (5.197). On the other hand, as $T_\perp \to 0$ and the space-charge potential ϕ_s balances the applied potential ϕ_a, the density profile becomes uniform, as was the case with the waterbag distribution (see Figure 5.11). This will be discussed in more detail in Section 5.4.4.

5.3.4 RMS Emittance and the Concept of Equivalent Beams

In the preceding section we discussed three examples of self-consistent, stationary particle distributions in a uniform focusing channel. Laboratory beams as well as distributions used in computer simulation studies may differ significantly from such stationary theoretical solutions of the Vlasov equation. Furthermore, most focusing systems consist of discrete lenses, and often these lenses are quadrupoles which do not exhibit the axial symmetry assumed in our theoretical models.

To analyze and compare the behavior of different stationary or nonstationary distributions, Lapostolle and Sacherer in 1971 introduced rms quantities (for beam radius, emittance, etc.) and the concept of equivalent beams [5, 6]. According to this concept, two beams composed of the same particle species and having the same current and kinetic energy are *equivalent* in an approximate sense if the second moments of the distribution are the same. This implies that the rms beam widths and rms emittances in the two orthogonal transverse directions are identical, assuming that the two beams are compared at identical positions in the same focusing systems.

Consider a normalized stationary or nonstationary distribution $f(x, y, x', y')$ in four-dimensional transverse trace space. The second moment in the particle coordinates x is defined by

$$\overline{x^2} = \iiiint x^2 f(x, y, x', y')\, dx\, dx'\, dy\, dy', \qquad (5.198)$$

and the *rms beam width* in the x-direction is then given by

$$x_{\text{rms}} = \tilde{x} = \left(\overline{x^2}\right)^{1/2}. \qquad (5.199)$$

In similar fashion the other second moments, such as $\overline{x'^2}$, $\overline{xx'}$, $\overline{y^2}$, and so on, and associated rms quantities ($\tilde{x}'$, $\tilde{y}'$, etc.) are defined. As an example, let us take a K–V distribution whose boundary in the x–x' plane is described by a tilted ellipse of the form (5.138b), that is,

$$\hat{\gamma}x^2 + 2\hat{\alpha}xx' + \hat{\beta}x'^2 = \epsilon_x.$$

Consider a position where the ellipse is upright (i.e., $\hat{\alpha} = 0$) and let $x_{max} = a$, $x'_{max} = a'$ denote the maximum x-position (radius or envelope) and maximum slope in the particle distribution. Then it is straightforward to show that

$$\overline{x^2} = \frac{\hat{\beta}\epsilon_x}{4} = \frac{a^2}{4}, \qquad \tilde{x} = \frac{a}{2} \tag{5.200}$$

and

$$\overline{x'^2} = \frac{\hat{\gamma}\epsilon_x}{4} = \frac{a'^2}{4}, \qquad \tilde{x}' = \frac{a'}{2}. \tag{5.201}$$

The *total* or *100% emittance* encompassing all particles in the K–V distribution is given by

$$\epsilon_x = \frac{aa'}{\sqrt{\hat{\beta}\hat{\gamma}}} = aa' \tag{5.202}$$

since $\sqrt{\hat{\beta}\hat{\gamma}} = 1$ in this case. The *rms emittance* can be defined from the relation $\overline{\epsilon_x^2} = \overline{x^2}\,\overline{x'^2}$ as

$$\left(\overline{\epsilon_x^2}\right)^{1/2} = \tilde{\epsilon}_x = \left[\overline{x^2}\,\overline{x'^2}\right]^{1/2} = \frac{aa'}{4} = \frac{\epsilon_x}{4}. \tag{5.203}$$

For the more general situation of a tilted ellipse ($\hat{\alpha} \neq 0$), the total emittance of a K–V beam in the x–x' plane is given by

$$\frac{\epsilon_x}{\left(\hat{\beta}\hat{\gamma} - \hat{\alpha}^2\right)^{1/2}} = \epsilon_x, \tag{5.204}$$

since $(\hat{\beta}\hat{\gamma} - \hat{\alpha}^2)^{1/2} = 1$ from (3.343c). The analogous definition of the rms emittance for this case is

$$\tilde{\epsilon}_x = \left[\overline{x^2}\,\overline{x'^2} - \overline{xx'}^2\right]^{1/2}. \tag{5.205}$$

Similar relations apply for the y-direction. Equation (5.205) represents the general definition of the rms emittance for an arbitrary distribution and is very useful in describing laboratory beams as well as theoretical or particle simulation results. The *normalized rms emittance* is obtained by multiplication with the factor $\beta\gamma$ (i.e., $\tilde{\epsilon}_{nx} = \beta\gamma\tilde{\epsilon}_x$) and it is in general not a constant. Nonlinear space-charge forces, instabilities, collisions, and other effects may lead to emittance growth. This topic is of great current interest for the generation and acceleration of high-intensity, high-brightness beams. Considerable progress has been made during the past years in obtaining a better understanding of the sources of emittance growth, discussed in Chapter 6.

Of particular interest is the behavior of distributions that are not of the stationary type treated in Section 5.3.3 and hence would be expected to change with distance along the linear uniform focusing channel. Examples of such cases are the nonstationary waterbag (WB), parabolic (PA), conical (CO), and nonstationary Gaussian (GA) distribution functions that are often used to represent the initial state of a beam in particle simulation studies, such as the investigation of beam transport in a magnetic quadrupole FODO system [7]. These distributions, which are listed in Table 5.1 (together with the stationary K–V beam), are defined as functions of the "radius" $r_4 = (x^2 + y^2 + x'^2 + y'^2)^{1/2}$ in four-dimensional trace space and not as functions of the Hamiltonian $H_\perp$. Consequently, they do not represent stationary solutions of the Vlasov equations, and as discussed in the next section, the simulation studies show that they do not retain their initial mathematical form. The normalization factors for each distribution in Table 5.1 have been chosen such that the integral of the distribution over the four-dimensional trace-space volume is unity. The ratio of total emittance to rms emittance, $\epsilon_t/\tilde{\epsilon}$, is a measure of the tail in the distribution (i.e., how far the particles are spread out in trace space compared to the rms area). For a K–V beam with its uniform density this ratio is 4, as discussed earlier. The ratio $\epsilon_t/\tilde{\epsilon}$ then increases in the order in which the distributions are listed, reaching a maximum value for a Gaussian distribution that has the form $\exp(-r_4^2/2\tilde{\delta})$. The constant $\tilde{\delta}$ represents the rms width of the Gaussian distribution (i.e., $\tilde{\delta} = \tilde{x} = \tilde{y}$ when $0 \leq x < \infty$ is assumed). For numerical simulation studies, however, the Gaussian tail is truncated at a finite radius $r_4 = n\tilde{\delta}$, where n is an integer. In the case $n = 4$ (i.e., $r_4 \leq 4\tilde{\delta}$), for instance, one finds that $\epsilon_t/\tilde{\epsilon} \approx 16$. However, the number of particles outside the K–V ellipse (i.e., outside the trace-space area defined by $4\tilde{\epsilon}$) represent only a small percentage of the total beam in the distributions of Table 5.1. For this reason, Lapostolle proposed to use an emittance defined by $4\tilde{\epsilon}$, rather than the rms emittance $\tilde{\epsilon}$, as a measure for the trace-space area of the beam (see our comment in Reference 5). We will simply call this quantity the *effective emittance* and use the symbols ϵ_x and ϵ_y. Thus, in agreement with the relations given in Section 3.1, we define

$$\epsilon_x = 4\tilde{\epsilon}_x = 4\left[\overline{x^2}\ \overline{x'^2} - \overline{xx'}^2\right]^{1/2}, \tag{5.206}$$

and likewise, $\epsilon_y = 4\tilde{\epsilon}_y$. In similar fashion we define an *effective beam radius*

$$X = 2\tilde{x} = 2\left(\overline{x^2}\right)^{1/2}. \tag{5.207}$$

For a K–V beam, these quantities are identical with the total emittance ϵ_t and the beam radius $r = a$, respectively.

Lapostolle [5] and Sacherer [6] have shown that K–V envelope equations can be derived for either the rms or effective beam radii of more general distributions such

Table 5.1 Definition and properties of K–V beam and four nonstationary distribution functions often used in computer simulation studies ($r_4^2 = x^2 + y^2 + x'^2 + y'^2$)

Distribution Function	Definition (Normalized), $f(r_4)$	Ratio of Total Emittance to rms Emittance, $\epsilon_t/\tilde\epsilon$	Particle Density in Real Space, $r^2 = x^2 + y^2$	Nonlinear Field Energy Factor U/w_0, $w_0 = \dfrac{I^2}{16\pi\epsilon_0 c^2\beta^2}$
Kapchinsky–Vladimirsky (K–V)	$\dfrac{1}{2\pi^2 a^3}\,\delta(r_4 - a)$	4	$\dfrac{1}{\pi a^2}$	0
Waterbag (WB)	$\dfrac{2}{\pi^2 a^4}$	6	$\dfrac{2}{\pi a^2}\left(1 - \dfrac{r^2}{a^2}\right)$	0.0224
Parabolic (PA)	$\dfrac{6}{\pi^2 a^4}\left(1 - \dfrac{r_4^2}{a^2}\right)$	8	$\dfrac{10}{3\pi a^2}\left(1 - 3\dfrac{r^2}{a^2} + 2\dfrac{r^3}{a^3}\right)$	0.0563
Gaussian (GA)	$\dfrac{1}{4\pi^2\delta^4}\exp\left(-\dfrac{r_4^2}{2\delta^2}\right)$ $\delta^2 = \overline{x^2}$	$\simeq n^2$ if truncated at $n\delta$, $n \geq 4$	$\dfrac{1}{2\pi\delta^2}\exp\left(-\dfrac{r^2}{2\delta^2}\right)$	0.1544

as the ones listed in Table 5.1. (Sacherer preferred to use rms quantities and did not adopt Lapostolle's notation.) This then led to the concept of *equivalent beams,* which says that two different phase-space distributions of a given particle species with the same kinetic energy and beam current are equivalent when they have the same first and second moments [i.e., when the rms (or effective) emittances and radii are identical]. Nonstationary distributions used in computer simulation work to study beam transport in a focusing channel do not retain their initial mathematical form, just like experimental nonstationary (e.g., mismatched) beams would not be expected to maintain their initial profile. The concept of equivalent beams allows one, however, to describe their average or rms behavior by solving the rms envelope equation or by studying the equivalent K–V beam. There is the implicit assumption in this concept that the rms emittance of two beams being compared remains the same or that the emittance change with time (or distance) is known a priori. This assumption is in general not correct, as we discuss in Section 6.2.

Let us now derive the rms envelope equations following the original work by Lapostolle and Sacherer. Consider a focusing channel with linear external forces and (generally nonlinear) space-charge forces acting on the particles. In cartesian coordinates the equations for the transverse motion of a particle in such a channel can be written in the form

$$x'' + k_{x0}^2 x - F_x = 0, \tag{5.208a}$$

$$y'' + k_{y0}^2 y - F_y = 0. \tag{5.208b}$$

The functions F_x and F_y represent the forces due to the electric and magnetic self fields of the beam and are defined by

$$F_{x(y)} = \frac{qE_{x(y)}(1 - \beta^2)}{\gamma mc^2\beta^2} = \frac{qE_{x(y)}}{mc^2\gamma^3\beta^2}, \tag{5.209}$$

where the factor $1 - \beta^2 = \gamma^{-2}$ takes into account the relativistic reduction of the electrostatic Coulomb repulsion by the force qv_zB_θ due to the self-magnetic field.

Note that the constant k_0^2, which represents the external focusing force, can have different magnitudes and signs in the two directions (i.e., the beam would have an elliptical shape in this general situation). For $k_{x0}^2 = k_{y0}^2 = k_0^2$, we recover the axisymmetric case (round beam) discussed so far.

If we multiply Equation (5.208a) by x and average over the distribution, we obtain

$$\overline{xx''} + k_{x0}^2\overline{x^2} - \overline{F_xx} = 0. \tag{5.210}$$

Now we have the following relations:

$$\tilde{x}^2 = \overline{x^2}, \qquad \tilde{x}'^2 = \overline{x'^2},$$

$$\left(\overline{x^2}\right)' = 2\overline{xx'} = (\tilde{x}^2)' = 2\tilde{x}\tilde{x}', \tag{5.211}$$

$$\left(\overline{x^2}\right)'' = (\tilde{x}^2)'' = 2(\tilde{x}\tilde{x}')' = 2(\tilde{x}\tilde{x}'' + \tilde{x}'^2). \tag{5.212}$$

Furthermore,

$$\left(\overline{xx'}\right)' = \overline{x'^2} + \overline{xx''} = \overline{x'^2} - k_{x0}^2\overline{x^2} + \overline{F_x x}, \tag{5.213}$$

where we substituted for $\overline{xx''}$ from Equation (5.210). The last equation may be written in the alternative form

$$\frac{1}{2}\left(\overline{x^2}\right)'' = \left(\overline{xx'}\right)' = \tilde{x}\tilde{x}'' + \tilde{x}'^2 = \overline{x'^2} - k_{x0}^2\overline{x^2} + \overline{F_x x} \tag{5.214}$$

or, with $\tilde{x}' = \overline{xx'}/\tilde{x}$ and $\tilde{x}^2 = \overline{x^2}$,

$$\tilde{x}\tilde{x}'' + \frac{\left(\overline{xx'}\right)^2}{\tilde{x}^2} - \frac{\overline{x'^2}\overline{x^2}}{\tilde{x}^2} + k_{x0}^2\tilde{x}^2 - \overline{xF_x} = 0.$$

By introducing the rms emittance, as defined in Equation (5.205) and dividing by $\tilde{x}$, we obtain the equation for the rms beam envelope:

$$\tilde{x}'' + k_{x0}^2\tilde{x} - \frac{\tilde{\epsilon}^2}{\tilde{x}^3} - \frac{\overline{xF_x}}{\tilde{x}} = 0. \tag{5.215}$$

It has been shown by Sacherer [6] that the term $\overline{xF_x}$ is independent of the form of the distribution and has the same value as for the equivalent K–V distribution, that is,

$$\overline{xF_x} = \frac{K}{2}\frac{\tilde{x}}{\tilde{x} + \tilde{y}}, \tag{5.216}$$

and furthermore,

$$\overline{xF_x} + \overline{yF_y} = \frac{K}{2}. \tag{5.217}$$

With (5.216) the rms envelope equation becomes

$$\tilde{x}'' + k_{x0}^2\tilde{x} - \frac{K}{2(\tilde{x} + \tilde{y})} - \frac{\tilde{\epsilon}_x^2}{\tilde{x}^3} = 0. \tag{5.218}$$

A similar equation can be derived for the y-envelope. By introducing the effective beam width $X = 2\tilde{x}$, $Y = 2\tilde{y}$, and the effective emittance $\epsilon_x = 4\tilde{\epsilon}_x$, $\epsilon_y = 4\tilde{\epsilon}_y$, we obtain the two equivalent equations for the effective beam envelopes:

$$X'' + k_{x0}^2X - \frac{2K}{X + Y} - \frac{\epsilon_x^2}{X^3} = 0, \tag{5.219a}$$

$$Y'' + k_{x0}^2Y - \frac{2K}{X + Y} - \frac{\epsilon_y^2}{Y^3} = 0. \tag{5.219b}$$

These equations are identical in form to the K–V envelope equations (4.178) and (4.179), but they apply to any other transverse phase-space distributions as well. If a given distribution is stationary, the effective emittance $\epsilon_{x(y)}$ does not change and the above equations can be solved for the effective envelopes (X, Y) of the beam. All stationary beams with the same perveance K and effective (or rms) emittance $\epsilon_{x(y)}$ have the same effective (or rms) radii. On the other hand, for nonstationary distributions one expects that the emittance will change, and one would have to know the evolution of this change to solve Equations (5.219a) and (5.219b) for X and Y as functions of distance along the focusing channel. As we will see in Section 6.2.1, one can in fact derive a differential equation relating the emittance growth to the rate of change of the free energy in a nonstationary beam. But its usefulness in determining the evolution of ϵ with distance is rather limited. However, by comparing the final equilibrium state with the initial nonstationary state, using the concept of equivalent beams, it is possible to obtain upper limits for the emittance growth. The analytical expressions for these limits also exhibit the scaling of emittance growth with the experimental parameters. On the other hand, to unravel the dynamical details of the emittance growth processes, one must rely on computer simulation and experiments, as we discuss in Section 6.2.2.

The Vlasov equation provides an extremely useful framework for the study of beam equilibria and for a stability analysis which shows whether a particular equilibrium is stable or unstable against various types of perturbations. With regard to the equilibrium state, we have seen that any distribution that depends only on the constants or integrals of the motion satisfies the time-independent Vlasov equation and hence represents an equilibrium beam. From a practical point of view, one would like to know which of the many possible theoretical distributions represents the best model for a real beam in the laboratory. Obviously, the Vlasov theory in itself does not give an answer to this question. However, as we mentioned in Section 4.1, based on thermodynamic arguments, the Maxwell–Boltzmann distribution provides the most physical description of a laboratory beam. To prove this assertion we need to go beyond the Vlasov equations and include the Coulomb collisions between the particles in our model. As we will see in the next section, Coulomb collisions play the key role in achieving the thermal equilibrium represented by the Maxwell–Boltzmann distribution.

5.4 THE MAXWELL–BOLTZMANN DISTRIBUTION

5.4.1 Coulomb Collisions between Particles and Debye Shielding

In our self-consistent theoretical models of both laminar and nonlaminar beams discussed so far we have made the assumption that the space-charge forces acting on the particles can be derived from smoothed potential functions. This means that a particle does not "see" its immediate neighbors but only the smooth collective field of the particle distribution as a whole. To the extent that this is correct, Liouville's theorem can be applied in six-dimensional phase space, which plays a central role in the theory and design of charged particle beams. In this section we examine the validity of this assumption. We do this in a coordinate system in which the centroid of the beam is at rest and we call this the *beam frame*. Furthermore, we assume that the particle motion in this beam frame is nonrelativistic, which is the case for most beams of practical interest, and that there is an effective three-dimensional applied potential $\phi_a(\mathbf{r}, t)$ which keeps the particles confined. In the laboratory frame this situation corresponds to a bunch of charged particles that are being acted upon by applied focusing forces in both the transverse and longitudinal directions. The continuous beams that we have treated so far would be obtained by letting the axial force go to zero and the bunch length go to infinity.

Let us now turn our attention to the Coulomb interactions between the particles in this bunch. Since the particle motion is nonrelativistic and the mean velocity of the distribution as a whole is zero in the beam frame, the magnetic force can be ignored completely. Suppose that we have N identical particles with charge q and mass m in the bunch whose location at a particular instant of time is given by the position vector $\mathbf{r}$. The force exerted by particle j on particle i is given by Coulomb's law as

$$\mathbf{F}_{ij} = \frac{q^2 \mathbf{r}_{ij}}{4\pi\epsilon_0 r_{ij}^3}, \qquad (5.220)$$

where $\mathbf{r}_{ij} = \mathbf{r}_i - \mathbf{r}_j$.

The total force on particle i is given by the sum of the forces exerted by all other particles, that is,

$$\mathbf{F}_i = \frac{q^2}{4\pi\epsilon_0} \sum_{j\neq i}^{N} \frac{\mathbf{r}_{ij}}{r_{ij}^3}. \qquad (5.221)$$

In addition, we have the applied external force $\mathbf{F}_a = -q\nabla\phi_a(\mathbf{r}, t)$, which, however, we omit from consideration temporarily since we are concerned with the space-charge interaction between the particles.

The long-range nature of the Coulomb forces implies that many particles will contribute to the total force $\mathbf{F}_i$ on our test particle. The many small contributions of the "distant" particles will add up to a smooth function whose effect on the particle trajectory can be described in terms of a space-charge potential $\phi_s(\mathbf{r}, t)$ that acts in a continuous fashion just like the external potential. On the other hand, the few particles in the immediate neighborhood of our test particle are seen as discrete point charges which will effectively change the curvature of the test particle's trajectories in very short distances. The encounters with these neighbors can be described as "collisions" that cause rapid fluctuations in the particle's motion. Thus, we can divide the total Coulomb interaction force on the test particle into two components. One represents the gradient of the smooth space-charge potential

$\phi_s(\mathbf{r}, t)$ of the large number of "distant" particles, the other the collisional force due to fewer neighbors:

$$\mathbf{F}_i = -\frac{\partial}{\partial \mathbf{r}} \phi_s(\mathbf{r}, t) + \frac{q^2}{4\pi\epsilon_0} \sum_j \frac{\mathbf{r}_{ij}}{r_{ij}^3}. \tag{5.222}$$

The smooth space-charge potential can be represented by the volume integral over the charge density function ρ by

$$\phi_s(\mathbf{r}, t) = \int \frac{\rho \, dV}{4\pi\epsilon_0 |\mathbf{r} - \mathbf{r}'|} = \int \frac{\rho \, d\mathbf{r}}{4\pi\epsilon_0 |\mathbf{r} - \mathbf{r}'|}. \tag{5.223}$$

In turn this can be related to the distribution function $f(\mathbf{r}, \mathbf{v}, t)$ as

$$\phi_s(\mathbf{r}, t) = \frac{q}{4\pi\epsilon_0} \iint \frac{f(\mathbf{r}', \mathbf{v}', t)}{|\mathbf{r} - \mathbf{r}'|} \, d\mathbf{v}' \, d\mathbf{r}', \tag{5.224}$$

where $\mathbf{r}', \mathbf{v}'$ denote the position and velocity vectors of the "field" particles over which the integration is taken. As a result of the collisional or fluctuating part of the interaction between the particles, the time variation of the distribution function will now include a collisional term $[\partial f / \partial t]_c$. Thus, in place of the nonrelativistic Vlasov equation (5.115), with $\mathbf{B} = 0$, we have

$$\frac{df}{dt} = \frac{\partial f}{\partial t} + \mathbf{v} \cdot \frac{\partial f}{\partial \mathbf{r}} + \frac{q}{m}(\mathbf{E}_a + \mathbf{E}_s) \cdot \frac{\partial f}{\partial \mathbf{v}} = \left[\frac{\partial f}{\partial t}\right]_c. \tag{5.225}$$

This is known as the *Boltzmann equation*. The two electric field vectors $\mathbf{E}_a$ and $\mathbf{E}_s$ represent the applied field and the smoothed part of the space charge field, respectively, while $[\partial f / \partial t]_c$ on the right-hand side stands for the effects of *Coulomb collisions* or *Coulomb scattering*, as it is often called. Clearly, when $[\partial f / \partial t]_c \neq 0$, the total derivative of the distribution function, df/dt, is not zero; hence Liouville's theorem in $(\mathbf{r}, \mathbf{v})$ phase space does not hold. In this case, the rate of change of the distribution function along a given trajectory in phase space is due entirely to the Coulomb collisions represented by $[\partial f / \partial t]_c$. On the other hand, if $[\partial f / \partial t]_c = 0$, we recover the Vlasov equation, as expected.

Before proceeding further with our analysis of the Boltzmann equation, we need to define a characteristic distance from a given point charge which separates the region near the charge where the collisional forces dominate from the "distant" region where the particles produce a smooth space-charge force on the point charge considered. This distance is known as the *Debye length*, λ_D, which we introduced in Section 4.1 and used in Section 5.3.3. Historically, the problem was first investigated by Debye and Hückel [8], who showed that the electric field of an ion in an electrolyte was effectively screened by the cloud of particles with opposite charge surrounding it. This concept applies to both a neutral plasma and to a nonneutral charged particle distribution.

Consider a plasma consisting of singly charged positive ions and electrons in thermal equilibrium at a temperature T. If we place a test charge q into this plasma (say at $r = 0$ of a spherical coordinate system), it will disturb charge neutrality and produce an electrostatic potential $\phi(r)$. Both electrons and ions have a Maxwellian velocity distribution, and their density functions will obey the Boltzmann relation

$$n(r) = n_0 \exp\left[-\frac{q\phi(r)}{k_B T} \right], \tag{5.226}$$

which one obtains from the Vlasov equation analogous to the derivation of (5.123). The potential function $\phi(r)$ must relate to the difference in charge density between electrons and ions (produced by the presence of the test charge) via Poisson's equation:

$$\nabla^2 \phi(r) = -\frac{q}{\epsilon_0} [n_i(r) - n_e(r)]$$

$$= -\frac{q}{\epsilon_0} n_0 \left\{ \exp\left[-\frac{q\phi(r)}{k_B T} \right] - \exp\left[\frac{q\phi(r)}{k_B T} \right] \right\}. \tag{5.227}$$

The positive sign in the argument of the second exponential function on the right-hand side results from the fact that the electrons have a negative charge. Assuming that $q\phi(r) \ll k_B T$, we can expand the exponential functions and obtain to first order

$$\nabla^2 \phi(r) = -\frac{q}{\epsilon_0} n_0 \left[\frac{-2q\phi(r)}{k_B T} \right] = \frac{2\phi(r)}{\lambda_D^2}, \tag{5.228}$$

where

$$\lambda_D = \left(\frac{\epsilon_0 k_B T}{q^2 n_0} \right)^{1/2} \tag{5.229}$$

is the Debye length. The latter can also be expressed in terms of the rms thermal velocity $\tilde{v}_x = (k_B T/m)^{1/2}$ and the plasma frequency $\omega_p = (q^2 n/\epsilon_0 m)^{1/2}$ as

$$\lambda_D = \frac{\tilde{v}_x}{\omega_p}. \tag{5.230}$$

The solution of Equation (5.228) is

$$\phi(r) = \frac{q}{4\pi\epsilon_0 r} \exp\left(-\frac{\sqrt{2} r}{\lambda_D} \right), \tag{5.231}$$

as is readily verified by differentiating and substituting into the Poisson equation.

Now one can see that for $r \ll \lambda_D$ the exponential term is close to unity and the potential $\phi(r)$ is essentially that of the unscreened test charge, that is,

$$\phi(r) = \frac{q}{4\pi\epsilon_0 r} \qquad \text{for } r \ll \lambda_D. \tag{5.232}$$

On the other hand, when $r \gg \lambda_D$, the exponential term dominates and the potential $\phi(r)$ goes toward zero much faster than without the shielding effect.

The condition $q\phi/k_B T \ll 1$ that we used in deriving the Poisson equation (5.228) implies that the average potential energy per particle should be small compared to the average kinetic energy per particle. If we substitute $r = \lambda_D$ in the point-charge potential (5.232), divide by $k_B T$, and use Equation (5.229), we obtain

$$\frac{q\phi(\lambda_D)}{k_B T} = \frac{q^2}{4\pi\epsilon_0 \lambda_D k_B T} = \frac{1}{4\pi\lambda_D^3 n_0} = \frac{1}{3N_D}, \tag{5.233}$$

where N_D is the number of particles inside the Debye sphere of volume $(4\pi/3)\lambda_D^3$. The condition

$$\frac{q\phi(\lambda_D)}{k_B T} \ll 1 \quad \text{implies that} \quad N_D \gg 1, \tag{5.234}$$

which means that the number of particles inside the Debye sphere is very large. When this is the case, the smooth part of the Coulomb interaction force in Equation (5.222) exceeds that of the collisional part. Thus, collisional effects under these conditions, which apply to most particle beams, are small, and our assumption that Liouville's theorem holds has now been validated. There are, however, exceptions, such as the Boersch effect and intrabeam scattering in high-energy synchrotrons and storage rings, that we discuss in Section 6.4.1 and 6.4.2. Furthermore, scattering in a background gas, time-varying nonlinear space-charge or applied forces of a stochastic nature, and instabilities have the same effect as Coulomb collisions, violating Liouville's theorem and causing emittance growth (see Section 6.2). It is therefore important that we pursue the thermodynamic treatment of collisions in further detail.

5.4.2 The Fokker–Planck Equation

Let us now return to the Boltzmann equation (5.225). To continue the analysis we need to evaluate the collisional term $[\partial f/\partial t]_c$ on the right-hand side of the equation. Modeling the physics of a particular problem, finding mathematical expressions for the collision term, and solving the Boltzmann equation is rather complicated even in relatively simple cases. The application to Coulomb interactions in charged particle beams was discussed by Jansen, whose recent book presents a very detailed review of this topic, with a comprehensive list of references [9].

Briefly, the effects of Coulomb interactions on the particle distribution in a beam or a plasma can be described as a diffusion process that is opposed by a dynamical friction ("drag") force. The thermal outward flow of particles from a given region due to diffusion is slowed down by collisions that reduce the forward momentum components. The rate of change of the distribution function at any given point $[\partial f/\partial t]_c$ due to these processes can be modeled as [Reference 9, Equation (4.3.10)]

$$\left[\frac{\partial f}{\partial t} \right]_c = \beta_f \frac{\partial(\mathbf{v}f)}{\partial \mathbf{v}} + D \frac{\partial^2 f}{\partial v^2}, \tag{5.235}$$

where β_f is the coefficient of dynamical friction and D the coefficient of diffusion (which for simplicity is assumed to be isotropic in this case). Substitution of (5.235) into Equation (5.225) yields

$$\frac{df}{dt} = \frac{\partial f}{\partial t} + \mathbf{v} \cdot \frac{\partial f}{\partial \mathbf{r}} - \frac{q}{m} \frac{\partial \phi}{\partial \mathbf{r}} \cdot \frac{\partial f}{\partial \mathbf{v}} = \beta_f \frac{\partial(\mathbf{v}f)}{\partial \mathbf{v}} + D \frac{\partial^2 f}{\partial v^2}. \tag{5.236}$$

This form of the Boltzmann equation is known as the *Fokker–Planck* equation. If the two coefficients β_f, D and the applied potential ϕ_a are given, Equation (5.236), together with Equation (5.224) for the self potential ϕ_s, represents a self-consistent description of the evolution of the particle distribution function $f(\mathbf{r}, \mathbf{v}, t)$ in space and time.

Without a focusing force and without friction (i.e., when $\phi_a = 0$ and $\beta_f = 0$), no stationary solution exists. The particle distribution then simply expands in space and time, and the density $n(r, t)$ of the core decreases monotonically, as can be shown mathematically. When a confining potential is present, an equilibrium exists even if $\beta_f = 0$. However, particles in the high-energy tail may leak out of the system if their kinetic energy exceeds the potential energy $q\phi_a$ due to the applied force. The distribution then becomes a truncated Maxwellian. A good example is the atmosphere in the earth's gravitational field. Due to collisions there is a slow leakage near the top of the atmosphere of particles whose velocity v exceeds the escape velocity $v_e (v > v_e)$. Another example are the beams in storage rings, where there is a continuous loss of particles whose energy exceeds the potential energy of the confining fields. In the case of the atmosphere, the escape of particles into space does not matter since there are always enough new particles entering from the surface of the earth to balance these losses. However, in the storage rings, the diffusion of particles is an important factor that contributes to the finite lifetime of the beam.

When there is no confining potential but friction exists (i.e., when $\phi_a = 0$ and $\beta_f \neq 0$), it is easy to see that a stationary solution of the Fokker–Planck equation exists. This solution is found by setting the time derivatives equal to zero, that is,

$$\frac{df}{dt} = 0; \qquad \frac{\partial f}{\partial t} = 0, \qquad \left[\frac{\partial f}{\partial t} \right]_c = 0, \tag{5.237}$$

and hence

$$\beta_f \frac{\partial(\mathbf{v}f)}{\partial \mathbf{v}} + D \frac{\partial^2 f}{\partial v^2} = 0. \qquad (5.238)$$

By integrating the corresponding equation for each velocity component, one finds that

$$f(\mathbf{v}) = C \exp\left[-\frac{\beta_f(v_x^2 + v_y^2 + v_z^2)}{2D} \right] = C \exp\left(\frac{-\beta_f v^2}{2D} \right), \qquad (5.239)$$

where the constant C is found from the normalization $\iiint f(\mathbf{v}) \, dv_x \, dv_y \, dv_z = 1$, which yields $C = (\beta_f/2\pi D)^{3/2}$.

As we see, the equilibrium distribution satisfying the Fokker–Planck equation is a Gaussian in the three velocity components, and it can be shown that it is identical to a *Maxwellian distribution* of statistical mechanics:

$$f(\mathbf{v}) = \left(\frac{m}{2\pi k_B T} \right)^{3/2} \exp\left[-\frac{m(v_x^2 + v_y^2 + v_z^2)}{2k_B T} \right]. \qquad (5.240)$$

By comparing the last two equations one finds that the ratio of the diffusion and friction coefficients relate to the temperature and particle mass as

$$\frac{D}{\beta_f} = \frac{k_B T}{m}. \qquad (5.241)$$

Assuming a Maxwellian distribution one can calculate the two coefficients by averaging over the statistical fluctuations of the particle velocities due to the Coulomb collisions. The results of the rather lengthy calculations are

$$\beta_f = \frac{16\sqrt{\pi}}{3} \frac{n r_c^2 Z^4 c \ln \Lambda}{(2k_B T/mc^2)^{3/2}}, \qquad (5.242)$$

$$D = \frac{8\sqrt{\pi}}{3} \frac{n r_c^2 Z^4 c^3 \ln \Lambda}{(2k_B T/mc^2)^{1/2}}, \qquad (5.243)$$

where n is the particle density, Z the charge state of the particles (in the case of multiply charged ions), and r_c the *classical particle radius* defined as

$$r_c = \frac{q^2}{4\pi\epsilon_0 mc^2} = \begin{cases} 2.8180 \times 10^{-15} \text{ m} & \text{for electrons,} \\ 1.5347 \times 10^{-18} \text{ m} & \text{for protrons.} \end{cases} \qquad (5.244)$$

The parameter $\ln \Lambda$ is known as the *Coulomb parameter*, and it is usually defined in terms of the Debye length λ_D, and the *impact parameter b* corresponding to a

90° deflection of a test particle's trajectory due to the Coulomb interaction with a single field particle:

$$\ln \Lambda = \ln \frac{\lambda_D}{b} . \tag{5.245}$$

The impact parameter b represents the distance between the test particle and the field particle at the point of closest approach ($r = b$), where the potential energy is equal to the initial kinetic energy. If v_r represents the relative velocity between two interacting particles and v_{th} the thermal velocity, one can show that $\overline{v_r^2} = 2v_{th}^2$. The impact parameter can then be defined as

$$b = \frac{q^2}{4\pi\epsilon_0 m v_{th}^2} , \tag{5.246a}$$

where

$$v_{th}^2 = 3\frac{k_B T}{m} \tag{5.246b}$$

and

$$T = \frac{1}{3} (T_x + T_y + T_z) = \frac{2}{3} T_\perp + \frac{1}{3} T_\| \tag{5.246c}$$

is the average temperature of the beam. If the beam is in thermal equilibrium, $T = T_x = T_y = T_z = T_{eq}$. If the beam is not in three-dimensional equilibrium (i.e., initially $T_x \neq T_y \neq T_z$), two possibilities exist: (1) the conditions are such that given enough time, equilibrium can be achieved, in which case $T = T_{eq}$ remains constant; or (2) equilibrium cannot be achieved in principle, as is the case in many storage rings (see Section 6.4.2), in which case T will be increasing with time.

Using λ_D from Equation (5.229), and Equations (5.246) for b and v_{th}, we obtain for the Coulomb logarithm

$$\ln \Lambda = \ln \frac{(\epsilon_0 k_B T)^{3/2} 12\pi}{q^3 n^{1/2}} = \ln \frac{3}{2\sqrt{\pi}} \frac{(k_B T/mc^2)^{3/2}}{r_c^{3/2} n^{1/2}} . \tag{5.247}$$

This expression is valid as long as the Debye length λ_D is less than the average beam radius a. If $\lambda_d > a$, one uses the radius a in place of λ_D in the Coulomb logarithm:

$$\ln \Lambda = \ln \frac{4\pi\epsilon_0 a m v_{th}^2}{q^2} = \ln \frac{12\pi\epsilon_0 a k_B T}{q^2} = \ln \left(\frac{3a}{r_c} \frac{k_B T}{mc^2} \right). \tag{5.248}$$

Some authors use the effective radius $a = \sqrt{2}\,\tilde{r}$, some use the rms radius $\tilde{r}$, and others the rms width $\delta_x = \tilde{r}/\sqrt{2}$ for a. However, because of the logarithmic dependence, the Coulomb parameter varies only slowly over a wide range of the parameters involved. Thus, for electrons, one finds that $6 < \ln \Lambda < 30$ for densities between 10^3 and 10^{24} m^{-3} and temperatures between 10^2 and 10^8 K. Finally, we note that Equations (5.242) and (5.243) do indeed satisfy the relation (5.241) that we obtained by comparing the Gaussian solution of the stationary Fokker–Planck equation with the Maxwellian distribution.

In our analysis so far of the steady-state solution (5.238) of the Fokker–Planck equation we have focused entirely on the effects of random Coulomb collisions. However, it should be pointed out that the Gaussian distribution is of much more general importance. Thus, according to the *central limit theorem* of statistical mechanics, any processes of a random, statistically independent nature acting on a particle distribution in a harmonic oscillator potential will lead to displacements in the particles' positions that obey a Gaussian distribution. Examples of this kind are the random misalignments treated in Section 4.4.4 (related to the problem of "random walk"); random fluctuations of the fields or field gradients in the focusing magnets or rf cavities of an accelerator or storage ring due to vibrations; rf noise from the acceleration cavities, or other sources of noise; collisions between beam particles and the molecules of a background gas; and nonlinear forces of a stochastical nature. Another, very important example that will be discussed in Section 6.2 are nonlinear space-charge forces. Since the particle distribution of a beam is confined by focusing potentials and the individual particles are performing oscillations, there is a continuous exchange between position (potential energy) and velocity (kinetic energy) so that displacements in position due to random processes translate into velocity changes, and vice versa. With a harmonic oscillator potential, given by $\phi(x, y, z) = \text{const}(x^2 + y^2 + x^2)$, the Gaussian distribution of the central limit theorem is of the form

$$f(\mathbf{r}, \mathbf{v}) = C_0 \exp\left[-C_1(x^2 + y^2 + x^2) - C_2(v_x^2 + v_y^2 + v_x^2) \right], \qquad (5.249)$$

where C_0, C_1, C_2 are constants.

This observation concerning the central limit theorem is very important for our following discussion of the Maxwell–Boltzmann distribution as the "natural" equilibrium state of a charged particle beam, which, as we will see, is identical to Equation (5.249). We elaborate more on this topic in connection with our discussions of the causes of emittance growth in Section 6.2.

5.4.3 The Maxwell–Boltzmann Distribution for a Relativistic Beam

We are now ready to integrate the results of the steady-state Vlasov equation (Section 5.3.3) with those of the Fokker–Planck equation (Section 5.4.2) to obtain a modified model for the "natural" thermodynamic equilibrium state of a charged particle beam. First, we note that the left-hand side of the Fokker–Planck equation is identical to the Vlasov equation and contains the potential function ϕ due to both the applied and self fields. The right-hand side, which contains the effects of Coulomb scattering, yielded a Maxwellian velocity distribution [Equation (5.240)] as the only steady-state solution. While the steady-state Vlasov equation is satisfied by any distribution that is an arbitrary function of the invariants of the motion, only one solution satisfies both sides of the time-independent Fokker–Planck equation: the *Maxwell–Boltzmann distribution*, also known as the *thermal distribution*, which in the beam frame considered here has the form

$$f(\mathbf{r}, \mathbf{v}) = f_0 \exp\left[-\frac{m(v_x^2 + v_y^2 + v_x^2)}{2k_B T} - \frac{q\phi(x, y, z)}{k_B T} \right]. \tag{5.250}$$

It can be written in terms of the nonrelativistic single-particle Hamiltonian

$$H = \frac{m}{2} (v_x^2 + v_y^2 + v_z^2) + q\phi(x, y, z) \tag{5.251}$$

as

$$f(H) = f_0 \exp\left(-\frac{H}{k_B T} \right), \tag{5.252}$$

where the constant $C = (\beta_f/2\pi D)^{3/2} = (m/2\pi k_B T)^{3/2}$ of Equation (5.240) has been absorbed in the normalization factor, f_0. In the case of a harmonic-oscillator potential, Equation (5.250) becomes identical in form to Equation (5.249).

By integrating the Maxwell–Boltzmann distribution over the spatial coordinates (x, y, z) we recover the Maxwellian velocity distribution (5.240). On the other hand, by integrating over the velocities (v_x, v_y, v_z) we recover the Boltzmann relation for the particle density (5.226).

The only remaining step is to transform the Maxwell–Boltzmann distribution from the beam frame to the laboratory frame in which the beam physics is usually described. Let us assume that the laboratory beam propagates in the s-direction, that the centroid position of the distribution is $s_0(t)$, and that a particle's position $s(t)$ is described relative to s_0 by $z_l = s - s_0$. If the beam considered is nonrelativistic in the laboratory frame, the transformation is an easy task. Using the subscript b for quantities measured in the beam frame and l for the counterparts in the laboratory, all we need to do is substitute in Equation (5.250)

$$z = z_b = z_l = s - s_0 \tag{5.253a}$$

for the longitudinal coordinate and

$$v_z = v_{bz} = v_{lz} - v_0 = \Delta v_{lz} \tag{5.253b}$$

for the longitudinal velocity, where v_0 is the centroid velocity of the distribution in the laboratory and $s_0 = v_0 t$. We then obtain for the distribution in the laboratory frame

$$f_l = f_0 \exp\left[-\frac{m\left[v_x^2 + v_y^2 + (\Delta v_{lz}^2) \right]}{2k_B T} - \frac{q\phi(x, y, z_l)}{k_B T} \right] \tag{5.254}$$

since the transverse velocity components (v_x, v_y), the transverse coordinates (x, y), and the temperature are unaffected by the nonrelativistic (Galilean) transformation.

Unfortunately, the Lorentz transformation for a relativistic Maxwell–Boltzmann distribution is not so straightforward; in fact, it is rather tricky, as we will see. We already noted in Section 5.3.1 that the transverse and longitudinal particle motion cannot be separated if the beam is relativistic in both the beam frame and the laboratory frame. Fortunately, for most beams of practical interest the motion in the beam frame is nonrelativistic. Thus, we will limit our discussion to relativistic laboratory beams that have a nonrelativistic Maxwell–Boltzmann distribution in the beam frame.

First, we note that the distribution function (5.250) is not in a covariant form suitable for the Lorentz transformation. In special relativity, space and time, momentum and energy, electric scalar potential and magnetic vector potential, are intricately linked and must be represented by appropriate four-vectors. Thus

$$P^i = \left(P_x, P_y, P_z, \frac{E}{c} \right) \tag{5.255}$$

is the momentum-energy four-vector, with the energy defined as $E = \gamma mc^2$. The four-vector potential is defined by

$$A^i = \left(A_x, A_y, A_z, \frac{\phi}{c} \right), \tag{5.256}$$

where ϕ represents the electric scalar potential.

The transformation of such four-vectors from one frame to another is presented for convenient reference in Appendix 3. If P_{lz} is the z-component of a particle's momentum in the laboratory frame, and E_l the energy, then the z-component of the momentum in the beam frame is given by

$$P_{bz} = \gamma_0 \left(P_{lz} - \frac{v_0}{c^2} E_l \right), \tag{5.257}$$

where v_0 is the beam velocity in the laboratory frame. Considering now the entire particle distribution, we note that by definition the average momentum in the beam frame is zero (i.e., $\overline{\mathbf{P}_b} = 0$, hence specifically $\overline{P_{bz}} = 0$). Using this result and taking the average of both sides of Equation (5.257), we obtain

$$v_0 = \frac{c^2 \overline{P_{lz}}}{\overline{E_l}} \tag{5.258}$$

and for the corresponding energy factor

$$\gamma_0 = \left(1 - \frac{v_0^2}{c^2} \right)^{-1/2} = (1 - \beta_0^2)^{-1/2} = \frac{\overline{E_l}}{(\overline{E_l}^2 - c^2 \overline{P_{lz}}^2)^{1/2}}. \tag{5.259}$$

These two equations uniquely define the *center-of-momentum frame*, that is, the velocity v_0 and energy $\gamma_0 mc^2$ of the "beam centroid" particle (in the laboratory frame), whose velocity and energy in the beam frame are $v_b = 0$ and $E_b = mc^2$, respectively. Here mc^2 represents, of course, the particle's rest energy.

A key question with regard to our Maxwell–Boltzmann distribution is, how does the temperature transform from one Lorentz frame to another? The literature contains conflicting answers to this question [10]. On the one hand, one might think that temperature, representing the average thermal energy of the particles, could be treated like energy, which would suggest a transformation of the form $T_l = \gamma_0 T_b$. Upon further thought one of course realizes that temperature is a measure of the random motion of the particles contained in the velocity or momentum distribution. So an appropriate four-momentum vector for temperature would appear to be the answer. However, the accepted convention is to treat temperature as a scalar. Much of the literature in the early part of the century [11] upon the transformation $T = T_0/\gamma_0$, which satisfics all relevant thermodynamic relations for a Lorentz transformation from the rest-frame temperature T_0 to a moving-frame temperature T). Applied to our problem, the corresponding transformation from the beam frame ($T_0 = T_b$) to the laboratory frame ($T = T_l$) is thus

$$T_l = \frac{T_b}{\gamma_0}. \qquad (5.260)$$

More recently, some authors [12] have preferred to use a temperature that is Lorentz invariant; that is, there is only one temperature, the temperature as measured by an observer in the rest frame of the system.

We take the position that both viewpoints are correct and that the temperature definition one should use depends on the situation being considered. Some problems, such as Coulomb scattering, are best described in the beam (rest) frame of the system and in terms of the beam-frame temperature T_b, as we did above in our discussions of this topic. On the other hand, we will see below that there is also justification for using a laboratory temperature as defined in Equation (5.260). In this regard, our position on temperature differs from that on mass, where we prefer to use the definition of mass as a Lorentz-invariant scalar to avoid the problems of different transverse and longitudinal masses (see Section 2.1). Fortunately, unlike mass, the temperature does not exhibit asymmetry with regard to longitudinal and transverse motion of the particles. The transformation (5.260) holds for both directions (i.e., for $T_\parallel$ as well as $T_\perp$), and the confusion regarding the various definitions of mass does not exist here.

Returning now to our problem of transforming from the beam frame to the laboratory frame, we will use the covariant form of the Maxwell–Boltzmann distribution given in Reference 12 (p. 46):

$$f(x^i, P^i) = A \, \exp\left[-\frac{(P^i + qA^i)U_i}{k_B T_b} \right]. \qquad (5.261)$$

The four-vectors P^i and A^i are defined in (5.255), (5.256), respectively. U_i is the covariant partner of the center-of-momentum four-velocity vector (see Appendix 3 for details), which is in the beam frame

$$U_{bi} = (0, 0, 0, c), \qquad (5.262)$$

and in the laboratory frame

$$U_{li} = (0, 0, -\beta_0 \gamma_0 c, \gamma_0 c). \qquad (5.263)$$

The temperature T_b is treated here as a Lorentz-invariant scalar. The transformation of the momentum and potential four-vectors P^i and A^i from the beam frame to the laboratory frame is given by

$$P_l^i = (Q_j^i)^{-1} P_b^j, \qquad (5.264a)$$

$$A_l^i = (Q_j^i)^{-1} A_b^j, \qquad (5.264b)$$

where

$$\left(Q_j^i \right)^{-1} = \begin{bmatrix} 1 & 0 & 0 & 0 \\ 0 & 1 & 0 & 0 \\ 0 & 0 & \gamma_0 & \beta_0 \gamma_0 \\ 0 & 0 & \beta_0 \gamma_0 & \gamma_0 \end{bmatrix}. \qquad (5.265)$$

Let us assume now that in the beam frame the particles have nonrelativistic velocities. The distribution (5.261) has the form

$$f_b = A_{b0} \exp\left(-\frac{E_b + q\phi_b}{k_B T_b} \right), \qquad (5.266a)$$

or since $E = \gamma m c^2 = m c^2 + (\tfrac{1}{2}) m v^2$,

$$f_b = A_{b1} \exp\left[-\frac{m(v_{bx}^2 + v_{by}^2 + v_{bz}^2)}{2 k_B T_b} - \frac{q\phi_b}{k_B T_b} \right], \qquad (5.266b)$$

where the factor $\exp(-mc^2/k_B T_b)$ has been included in the normalization constant A_{b1}. Clearly, Equation (5.266b) is identical to the Maxwell–Boltzmann distribution of Equation (5.250), as expected.

The transformation from the beam frame to the laboratory frame is somewhat lengthy and is left to be carried out in Problem 5.11(a). With $z_l = s - s_0$, $\Delta v_{lz} = v_l - v_0$, as in (5.253), the final result can be written in the form

$$f_l = A_l \exp\left[-\frac{\gamma_0 m(v_{lx}^2 + v_{ly}^2) + \gamma_0^3 m(\Delta v_{lz}^2)^2}{2 k_B T_b / \gamma_0} - \frac{q\phi_l(x, y, z_l)}{k_B T_b / \gamma_0} \right], \qquad (5.267)$$

where again the factor $\exp[-mc^2/k_B T_b]$ was absorbed in the constant A_l and where $k_B T_b$ denotes the beam-frame temperature. Also, the potential function ϕ_l includes the magnetic vector potential A_{lz} that is generated by the transformation. It represents the sum of the effective potential due to the applied focusing forces, ϕ_{la}, and the potential due to the self fields of the beam, ϕ_{ls}; that is, $\phi_l = \phi_{la} + \phi_{ls}/\gamma_0^2$, where the factor γ_0^2 in the second term represents the focusing effect of the beam's magnetic self field $(1 - \beta_0^2 = 1/\gamma_0^2)$.

Equation (5.267) is the desired Maxwell–Boltzmann distribution in the laboratory frame for a relativistic beam with nonrelativistic transverse and longitudinal velocities in the beam frame. We have written the equation in a form suggesting that we introduce a laboratory temperature $T_l = T_b/\gamma_0$. Otherwise, the factor γ_0 combined with $k_B T_b$ in the denominator would appear in the kinetic and potential energy terms in the numerator, causing considerable confusion. Thus, the Maxwell–Boltzmann equation (5.267) can be used to justify the transformation (5.260) for the temperature, rather than keeping the beam temperature as a Lorentz invariant. In terms of the laboratory temperature T_l, we can write Equation (5.267) as

$$f_l = A_l \, \exp\left[-\frac{\gamma_0 m(v_{lx}^2 + v_{ly}^2) + \gamma_0^3 m(\Delta v_{lz})^2}{2k_B T_l} - \frac{q\phi_l}{k_B T_l}\right], \qquad (5.268a)$$

which, if we introduce the *transverse mass* $m_t = \gamma_0 m$ and the *longitudinal mass* $m_l = \gamma_0^3 m$, can be put into the suggestive form

$$f_l = A_l \, \exp\left[-\frac{m_t(v_{lx}^2 + v_{ly}^2) + m_l(\Delta v_{lz})^2}{2k_B T_l} - \frac{q\phi_l}{k_B T_l}\right], \qquad (5.268b)$$

which resembles the nonrelativistic distribution (5.266b). We note that one can obtain the relativistic Maxwell–Boltzmann distribution from (5.266b) more directly by applying the Lorentz transformations for the velocities and the scalar potential [see Problem 5.11(b)].

As we will see, beams are usually not in the three-dimensional thermal equilibrium implied by the single-temperature expression given above. Acceleration tends to cool the beam longitudinally while keeping the transverse temperature unaffected. Thus it is useful to write the Maxwell–Boltzmann distribution in terms of a transverse temperature $T_\perp$ and a longitudinal temperature, $T_\parallel$, as

$$f_l = A_l \, \exp\left[-\frac{\gamma_0 m(v_{lx}^2 + v_{ly}^2)}{2k_B T_{l\perp}} - \frac{q\phi_{l\perp}}{k_B T_{l\perp}}\right]\exp\left[-\frac{\gamma_0^3 m(\Delta v_{lz})^2}{k_B T_{l\parallel}} - \frac{q\phi_{l\parallel}}{k_B T_{l\parallel}}\right], \qquad (5.269)$$

where we assumed that the potential function can be split into transverse and longitudinal parts.

The two temperatures for each frame are then defined by the second velocity moments of the distribution as

$$k_B T_{b\perp} = k_B T_{bx} = m\overline{v_{bx}^2} = k_B T_{by} = m\overline{v_{by}^2}, \tag{5.270a}$$

$$k_B T_{l\perp} = k_B T_{lx} = \gamma_0 m\overline{v_{lx}^2} = k_B T_{ly} = \gamma_0 m\overline{v_{ly}^2}, \tag{5.270b}$$

$$k_B T_{b\parallel} = k_B T_{bz} = m\overline{\Delta v_{bz}^2}, \tag{5.270c}$$

$$k_B T_{l\parallel} = k_B T_{lz} = \gamma_0^3 m\overline{\Delta v_{lz}^2}. \tag{5.270d}$$

We recognize that the distribution function (5.269) can be expressed in terms of the transverse and longitudinal Hamiltonians and temperatures in the laboratory frame as

$$f(H_\perp, H_\parallel) = A \, \exp\left(-\frac{H_\perp}{k_B T_\perp}\right)\exp\left(-\frac{H_\parallel}{k_B T_\parallel}\right), \tag{5.271}$$

where we dropped the subscript l. This two-temperature Maxwell–Boltzmann distribution provides the most realistic theoretical description for laboratory beams. It should be pointed out, however, that in a strict mathematical sense the separation of the Hamiltonian into a transverse and longitudinal part is possible only when the coupling due to the space-charge forces is negligible. This is the case for long or continuous beams and for bunched beams in either the high-temperature limit where emittance dominates or in the low-temperature limit $(T_\perp \to 0, T_\parallel \to 0)$ where the space-charge forces tend to be linear. Otherwise, there is coupling between the transverse and longitudinal motion via the space-charge potential, and the two-temperature distribution is only a crude approximation that does not satisfy the stationary Vlasov equation. The coupling may lead to a relatively rapid change of the distribution towards thermal equilibrium $(T_\perp = T_\parallel)$, with relaxation time depending on the strength of the space-charge coupling forces and the difference in the two temperatures. This *equipartitioning* process is particularly strong in high-current rf linacs and will be discussed in Appendix 4.

If we are dealing with a continuous beam rather than a bunch, the longitudinal potential term is zero $(\phi_\parallel \to 0)$ since there is no applied longitudinal focusing force. In this case, Equation (5.269) represents a Maxwell–Boltzmann distribution for a continuous beam with longitudinal temperature, that is,

$$f = A \, \exp\left(-\frac{H_\perp}{k_B T_\perp}\right)\exp\left[-\frac{\gamma_0^3 m(\Delta v_z)^2}{k_B T_\parallel}\right]. \tag{5.272}$$

Finally, if we let the longitudinal temperature go to zero (i.e., $\overline{\Delta v_z^2} = 0$) and assume a nonrelativistic energy, we recover the transverse two-dimensional Maxwell–Boltzmann distribution of Equation (5.122).

In practice, beam focusing and acceleration systems are designed to be "linear" in the applied forces as much as possible to avoid emittance growth from non-linearities. The corresponding applied potential functions are then quadratic in the displacements of the particles from the beam centroid.

5.4.4 The Stationary Transverse Distribution in a Uniform or Smooth Focusing Channel

The stationary Vlasov distributions discussed in Section 5.3.3 represent examples of continuous beams in which the transverse Hamiltonian $H_\perp$ is a constant of the motion and the longitudinal temperature is zero. In view of what we know now, the transverse Maxwell–Boltzmann or thermal distribution is the one that best describes the equilibrium state of a real beam in transverse phase space. Let us from now on again define all quantities in the laboratory frame unless stated otherwise, and drop the subscript l. As in Section 5.3.3, we assume a uniform focusing channel in which the applied potential function has the form

$$\phi_a(r) = \frac{1}{2}\,\gamma_0 m v_0^2 k_0^2 r^2 \tag{5.273}$$

so that the focusing force acting on the particles is linear and independent of the axial coordinate z. This description also applies, of course, to the average behavior of the matched beam in a periodic channel in the smooth approximation. Using $v_x = v_0 x'$, $v_y = v_0 y'$, $r^2 = (x^2 + y^2)$, $r'^2 = (x'^2 + y'^2)$, and Equation (5.273), one can express the transverse Maxwell–Boltzmann distribution for a matched beam in such a system as

$$f_\perp = f_0\,\exp\!\left(-\frac{H_\perp}{k_{\mathrm B}T_\perp}\right), \tag{5.274a}$$

or

$$f_\perp = f_0\,\exp\!\left[-\frac{\gamma_0 m v_0^2(r'^2 + k_0^2 r^2)}{2k_{\mathrm B}T_\perp} - \frac{q\phi_s(r)}{k_{\mathrm B}T_\perp \gamma_0^2}\right], \tag{5.274b}$$

which is identical to Equation (5.193) if one sets $\alpha = k_{\mathrm B}T_\perp/\gamma_0 m v_0^2$. By integrating Equation (5.274b) with respect to the transverse velocity, or $r'_\perp = v_\perp/v_0$, one obtains the well-known Boltzmann relation for the particle density profile, which in our case has the form

$$n(r) = n(0)\exp\!\left[-\frac{\gamma_0 m v_0^2 k_0^2 r^2}{2k_{\mathrm B}T_\perp} - \frac{q\phi_s(r)}{k_{\mathrm B}T_\perp \gamma_0^2}\right]. \tag{5.275}$$

The space-charge potential $\phi_s(r)$ is related to the density $n(r)$ via Poisson's equation. Thus, the Boltzmann density profile $n(r)$ is in general nonanalytic and must be determined by a numerical method. Only in the low- and high-temperature limits does one get analytic solutions for $n(r)$.

In the first case $(T_\perp \to 0)$, the beam is laminar and the space-charge potential ϕ_s exactly balances the applied focusing potential, so that

$$\frac{\gamma_0 m v_0^2 k_0^2 r^2}{2} = -\frac{q\phi_s(r)}{\gamma_0^2} = \frac{q^2 n_0 r^2}{4\epsilon_0 \gamma_0^2}, \tag{5.276}$$

where we used Poisson's equation. The density profile in this case is thus uniform inside the beam, that is,

$$n(r) = \begin{cases} n_0 & \text{for } 0 \le r \le a_0, \\ 0 & \text{for } r > a_0, \end{cases} \tag{5.277}$$

where a_0 is the beam radius at zero temperature. The relation (5.276) shows that we can replace the external focusing force by a background of stationary ions or electrons of opposite charge to that of the beam particles.

In the second analytic case, at high temperature $(T_\perp \to \infty)$ the space charge is negligible, and by setting $\phi_s(r) = 0$, we obtain from (5.275) the Gaussian density profile:

$$n(r) = n(0) \exp\left(-\frac{\gamma_0 m v_0^2 k_0^2}{2k_B T_\perp} r^2\right). \tag{5.278}$$

These results for the low- and high-temperature cases are in agreement with experimental observations: space-charge-dominated beams in high-intensity, low-energy devices tend to have a uniform density profile, whereas emittance-dominated beams in high-energy synchrotrons, for instance, tend to have a Gaussian shape.

To obtain the solutions for the general case where both space charge and emittance are important, we must integrate Equation (5.275) numerically. We will adopt the procedure used by Lawson [C.17, p. 203], who considered a nonrelativistic beam. The relativistic case can be treated in the same way except that we include the additional factor γ_0^2 in our equations.

Following Lawson, it will be convenient to use the self-electric field, $E_s(r) = -d\phi_s(r)/dr$, which from Poisson's equation or Gauss's law is related to the charge distribution by

$$E_s(r) = \frac{q}{\epsilon_0 r} \int_0^r rn(r)\,dr. \tag{5.279}$$

Substitution for $n(r)$ from (5.275) using $\phi_s(r) = -\int_0^r E_s(r)\,dr$ and relation (5.276) for the external force term yields

$$E_s(r) = \frac{qn(0)}{\epsilon_0 r} \int_0^r r \exp\left[\frac{-q^2 n_0 r^2}{4\epsilon_0 k_B T_\perp \gamma_0^2} + \frac{q}{k_B T_\perp \gamma_0^2} \int_0^r \right] dr. \tag{5.280}$$

Let us now introduce the Debye length λ_D, corresponding to the density on the axis, which by our definition (4.3) is

$$\lambda_D = \lambda_D(0) = \left[\frac{\epsilon_0 k_B T_\perp \gamma_0^2}{q^2 n(0)} \right]^{1/2}. \tag{5.281}$$

Then, by normalizing the radius as $x = r/\lambda_D$ and the electric field as $F = \epsilon_0 E_s / n(0) q \lambda_D$, we can write (5.280) as

$$F(x) = \frac{1}{x} \int_0^x x \, \exp\left[\int_0^x F(x) \, dx - \frac{n_0 x^2}{4n(0)} \right] dx. \tag{5.282}$$

The Boltzmann relation (5.275) for the particle density may then be written in the form

$$n(x) = n(0) \, \exp\left[\int_0^x F \, dx - \frac{n_0 x^2}{4n(0)} \right]. \tag{5.283}$$

We note that the two equations (5.282) and (5.283) that determine our relativistic Boltzmann distribution are identical with Lawson's nonrelativistic relations. The distributions for different temperatures are normalized by requiring that the number of particles per unit length N_L be the same, that is,

$$N_L = 2\pi \int_0^\infty rn(r) \, dr = \pi a_0^2 n_0. \tag{5.284}$$

By solving Equations (5.282) and (5.283) numerically, using the relations (5.281) and (5.284), we obtain [13] the density profiles shown in Figure 5.12. The eight curves labeled 1 to 8 in (a) are identical to Lawson's results [C.17, p. 203]. Curve 7a represents an additional profile close to the zero-temperature limit. In Figure 5.12(b) we normalized these curves so that the rms radius is the same. The normalization (5.284) defines the ratio $n(0)/n_0$ for any given value of $\lambda_D(0)/a_0$. Note that the radius r in the figure is given in units of the zero-temperature radius a_0 [i.e., the variable x is taken in the form $x = (r/a_0)(a_0/\lambda_D)$]. The density curves plotted in the figure for different values of the parameter λ_D/a_0 show the general behavior discussed above and in Section 4.1. The shape varies from the uniform distribution (5.277) in the laminar limit, where temperature and emittance are zero, to the Gaussian profile (5.278) as λ_D/a_0 increases toward infinity. Table 5.2 shows a list of relevant parameters for each curve in Figure 5.12. The first column gives the density ratios $n(0)/n_0$, which, except for case 7a, were chosen to agree with Lawson's eight profiles. The second column shows the ratio of the Debye length on the axis $\lambda_D(0)$ to the zero-temperature radius a_0; these values differ from Lawson's results, which appear to be incorrect, probably because of an error in the normalization procedure. Our values for $\lambda_D(0)/a_0$ in Table 5.2 converge toward the analytical result $\lambda_D(0)/a_0 = n_0/2n(0)$ for $T_\perp \rightarrow \infty$, whereas Lawson's numbers do not converge toward this limiting value. The third column lists the rms radius $\tilde{r}$ in units of the zero-temperature radius $\tilde{r}_0 = a_0/\sqrt{2}$, as calculated for each curve. Note that $\tilde{r}/\tilde{r}_0 = a/a_0$, where $a = \sqrt{2}\,\tilde{r}$ is the effective radius of the equivalent uniform beam. The next three columns show the values for the relevant parameters of the equivalent uniform beam calculated from Equations (5.286) and (5.287) below. The average Debye length $\overline{\lambda_D}$ was calculated using the relation $\overline{\lambda_D}/a = (\lambda_D(0)/a_0)(n(0)/n_0)^{1/2}$. The last column shows the ratio of the density on axis to the zero-temperature density, $n(0)/n_0$, for each of the rms normalized curves in (b). For $T_\perp \rightarrow \infty$, one gets $n(0)/n_0 = 2$.

Near the limit of laminar flow, the Maxwell–Boltzmann distribution, like the waterbag distribution treated in Section 5.3.3, has a uniform density profile, with a transition from $n(r) = n_0$ in the interior to $n(r) = 0$ for $r > a$ that depends on the ratio $\overline{\lambda_D}/a$. This transition at high space-charge density, where $\overline{\lambda_D} \ll a$, was studied by Hofmann and Struckmeier [14], who found that both distributions can be approximated in this case by the function

$$n(r) = n_0\left\{1 - \frac{\exp[(r - a)/\overline{\lambda_D}]}{\sqrt{r/a}}\right\}. \qquad (5.285)$$

The ratio of the average Debye length to the beam radius, $\overline{\lambda_D}/a$, can be related to ϵ^2/Ka^2 or to the ratio of the particle oscillation frequencies with and without self

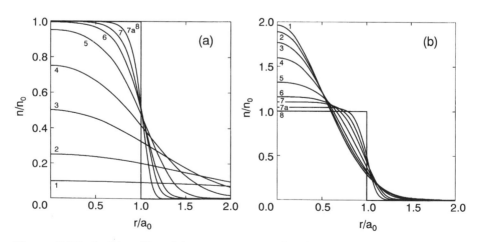

Figure 5.12. Radial profiles of the transverse Maxwell–Boltzmann distribution in a uniform focusing system for different temperatures. All beams have the same number of particles per unit length. (a) The focusing force is kept constant so that the beam width increases with increasing temperature. (b) The focusing force is increased to keep the rms radius constant. Table 5.2 lists the relevant parameter values for each curve. (From Reference 13.)

Table 5.2 Relevant parameters for the radial Boltzmann density profiles of Figure 5.12

Curve	$n(0)/n_0$	$\lambda_D(0)/a_0$	$\bar{r}/\bar{r}_0$	$\overline{\lambda_D}/a$	Ka^2/ϵ^2	k/k_0	$n(0)/n_0$ for $\bar{r} = r_0$
1	0.1	4.82	4.43	1.52	0.054	0.974	1.96
2	0.25	1.81	2.75	0.905	0.153	0.931	1.89
3	0.5	0.795	1.88	0.562	0.396	0.846	1.77
4	0.75	0.432	1.46	0.374	0.893	0.727	1.60
5	0.95	0.229	1.18	0.223	2.51	0.534	1.32
6	0.995	0.145	1.08	0.144	6.00	0.378	1.16
7	0.9995	0.107	1.04	0.107	10.9	0.290	1.08
7a	0.999995	0.0710	1.02	0.0710	24.8	0.197	1.04
8	1	0	1	0	∞	0	1

Source: Reference 13

fields, ω/ω_0. From Equations (4.88) and (4.89) one obtains, with $\omega_0/\omega = k_0/k$,

$$\frac{Ka^2}{\epsilon^2} = \frac{\omega_0^2}{\omega^2} - 1 = \frac{k_0^2}{k^2} - 1. \tag{5.286}$$

Using the relationship between normalized emittance $\epsilon_n = \beta\gamma\epsilon$ and temperature $T_\perp$ and Equation (5.286), one can show that

$$\frac{\overline{\lambda_D}}{a} = \left(\frac{\epsilon^2}{Ka^2}\right)^{1/2}\left(\frac{1}{8}\right)^{1/2} = \left(\frac{k^2}{k_0^2 - k^2}\right)^{1/2}\left(\frac{1}{8}\right)^{1/2}. \tag{5.287}$$

Thus, when $\overline{\lambda_D} \ll a$ or $k \ll k_0$, $\overline{\lambda_D}/a$ is directly proportional to the *tune depression* $k/k_0 = \omega/\omega_0$ of the particle oscillations of the beam.

The preceding two equations are very important from a practical point of view since they relate the experimental parameters K, a, and ϵ to the beam physics parameters λ_D and k/k_0. To obtain this correlation and to describe the behavior of the Maxwell–Boltzmann distribution in more detail, we will use the concept of equivalent beams introduced in Section 5.3.4. According to this concept, any distribution can be modeled to good approximation by an equivalent analytical beam having the same rms radius, rms transverse velocity, and rms emittance. For an axisymmetric stationary (i.e., matched) beam in a uniform focusing channel characterized by the focusing constant k_0, we obtain from (5.218) with $\bar{x}'' = \bar{y}'' = 0$ the rms envelope equation

$$k_0^2\bar{x} - \frac{K}{4\bar{x}} - \frac{\bar{\epsilon}^2}{\bar{x}^3} = 0, \tag{5.288a}$$

where $\tilde{x} = \tilde{y} = \tilde{r}/\sqrt{2}$ is the rms width in each of the two orthogonal directions and $\tilde{r}$ is the rms radius of the beam. This equation may be expressed in terms of the effective radius $a = \sqrt{2}\tilde{r} = 2\tilde{x}$ and the effective emittance $\epsilon = 4\tilde{\epsilon}$ of the equivalent uniform (K–V) beam as

$$k_0^2 a - \frac{K}{a} - \frac{\epsilon^2}{a^3} = 0, \tag{5.288b}$$

which is identical to Equation (4.88a). The unnormalized emittance in this case corresponds to an upright ellipse in x-x' trace space of area $\tilde{\epsilon}\pi$ and is defined by

$$\tilde{\epsilon} = \tilde{x}\frac{\tilde{v}_x}{v_0}, \qquad \text{or} \qquad \epsilon = 4\tilde{\epsilon} = a(x')_{\text{max}}. \tag{5.289a}$$

Introducing the laboratory temperature from (5.270) [i.e., using $\tilde{v}_x = (\overline{v_x^2})^{1/2} = (k_B T_\perp/\gamma_0 m)^{1/2}$], we can express the emittance by

$$\tilde{\epsilon} = \tilde{x}\left(\frac{k_B T_\perp}{\gamma_0 m v_0^2}\right)^{1/2}, \qquad \text{or} \qquad \epsilon = 2a\left(\frac{k_B T_\perp}{\gamma_0 m v_0^2}\right)^{1/2}. \tag{5.289b}$$

The corresponding normalized emittance is

$$\tilde{\epsilon}_n = \tilde{x}\left(\frac{k_B T_\perp \gamma_0}{mc^2}\right)^{1/2}, \qquad \text{or} \qquad \epsilon_n = 2a\left(\frac{k_B T_\perp \gamma_0}{mc^2}\right)^{1/2} \tag{5.290a}$$

and in terms of the beam-frame temperature,

$$\tilde{\epsilon}_n = \tilde{x}\left(\frac{k_B T_{b\perp}}{mc^2}\right)^{1/2}, \qquad \text{or} \qquad \epsilon_n = 2a\left(\frac{k_B T_{b\perp}}{mc^2}\right)^{1/2}. \tag{5.290b}$$

Note the absence of the relativistic energy factor γ_0 when the normalized emittance is expressed in terms of the beam-frame temperature, $k_B T_{b\perp}$.

From Equations (5.288a) and (5.288b), we see that Ka^2/ϵ^2 used in Table 5.2 defines the ratio of space charge to emittance in the envelope equations and that it can be written in terms of the rms quantities as

$$\frac{Ka^2}{\epsilon^2} = \frac{K\tilde{x}^2}{4\tilde{\epsilon}^2}. \tag{5.291}$$

To exhibit the scaling with regard to the experimental parameters more clearly, we will use the definition (4.27a) for the generalized perveance K. Furthermore, we will introduce the normalized emittance, which is more useful than the unnormalized emittance, since it remains constant in the ideal case. We then obtain for (5.291) the alternative relation

$$\frac{Ka^2}{\epsilon^2} = \frac{I}{I_0}\frac{2a^2}{\beta_0\gamma_0\epsilon_n^2} = \frac{I}{I_0}\frac{\tilde{x}^2}{2\beta_0\gamma_0\tilde{\epsilon}_n^2},\tag{5.292}$$

where I is the beam current and I_0 the characteristic current defined in Equation (4.17).

Equations (5.289a) to (5.292) constitute the desired scaling relationships that allow us to analyze the behavior of the stationary beam when parameters such as transverse temperature, kinetic energy, current, emittance, or radius are changed adiabatically so that the beam remains matched. The beam radius $a = 2\tilde{x}$ depends, of course, on the focusing strength of the channel as defined by the wave constant k_0. If k_0, perveance K, and emittance ϵ are given, we can calculate it from Equation (5.288). The solution is found in Equations (4.91) to (4.93b), and the scaling of the radius with k_0, K, and ϵ is in general not very transparent from these equations. However, we can use the approximation

$$a \approx \left(\frac{K}{k_0^2} + \frac{\epsilon}{k_0}\right)^{1/2},\tag{5.293}$$

which shows the scaling more clearly. This relation is exact at both ends of the parameter range (i.e., when either $K = 0$ or $\epsilon = 0$), and in between it slightly overestimates the radius, with a maximum error of about $+12\%$ at $K/k_0\epsilon = 1.5$. For a space-charge-dominated beam, when emittance can be neglected ($\epsilon = 0$), we obtain the exact relation

$$a = 2\tilde{x} = \frac{\sqrt{K}}{k_0} = \left(\frac{I}{I_0}\frac{2}{\beta_0^3\gamma_0^3}\right)^{1/2}\frac{1}{k_0}.\tag{5.294a}$$

On the other hand, when emittance dominates and space charge is negligible ($K = 0$), we obtain [see Equation (4.91)]

$$a = \left(\frac{\epsilon}{k_0}\right)^{1/2} = \left(\frac{\epsilon_n}{\beta_0\gamma_0 k_0}\right)^{1/2},\tag{5.294b}$$

or, in terms of rms quantities,

$$\tilde{x} = \left(\frac{\tilde{\epsilon}_n}{\beta_0\gamma_0 k_0}\right)^{1/2}.\tag{5.294c}$$

We can of course adjust the focusing strength to obtain a desired beam size and then simply use the radius a as a measured or given quantity in the above scaling relations.

Let us now return to the Boltzmann density profiles of a thermal beam in Figure 5.12. From Table 5.2 we can determine which of the profiles is closest to the desired or experimentally known beam parameters. If we have a computer code we can of course calculate the profile by solving Equation (5.275) numerically. To illustrate the scaling and parametric dependence explicitly, let us assume that we have a space-charge-dominated beam near the zero-temperature limit so that the profile is practically uniform. From the relationships given above we have several possibilities of changing the profile toward the more Gaussian "high-temperature" shape:

1. *Increase in Transverse Beam Temperature $T_\perp$, and Hence Emittance $\tilde{\epsilon}_n$.* A number of effects, such as beam mismatch, nonstationary density profile, instabilities, and collisions, cause emittance growth and hence temperature rise; these effects are discussed in Chapter 6.

2. *Increase of Particle Kinetic Energy.* The acceleration of the beam decreases Ka^2/ϵ^2 in general; for instance, $Ka^2/\epsilon^2 \propto (\beta_0\gamma_0)^{-1}$ according to Equation (5.292) if current I, beam radius a, and normalized emittance remain constant. We discuss this change in an example below.

3. *Increase of Focusing Strength.* This reduces the beam radius and hence decreases Ka^2/ϵ^2; it leaves the normalized emittance unchanged, but in view of (5.290) it increases the temperature as $k_B T_\perp = \text{const}/a^2$.

4. *Longitudinal Debunching or Expansion.* This reduces the beam current in the bunch and also the beam radius if the transverse focusing strength remains constant; the result is a decrease in Ka^2/ϵ^2 and an increase in temperature, which moves the profile in the direction of a more Gaussian shape.

It is interesting to note that not all four effects cause an increase in the temperature. Particle acceleration in case 2 may, in fact, decrease the transverse laboratory temperature as $T_\perp \propto \gamma_0^{-1}$ according to (5.292) if the radius remains constant. However, the net effect is still a decrease in $Ka^2/\epsilon^2 \propto (\beta_0\gamma_0)^{-1}$ and hence a change of the density $n(r)$ toward a more Gaussian shape. The transverse beam-frame temperature $T_{b\perp}$ remains, of course, unaffected by acceleration.

To illustrate the effect of acceleration on the beam profile, let us consider the hypothetical case of a high-current linear accelerator of the type being considered for radioactive waste transmutation [15], spallation neutron sources [16], and other applications requiring high energy and beam power. We will assume a proton beam of 100 mA average current injected by an RFQ accelerator into a drift-tube linac (DTL) at an energy of 2 MeV ($\beta_0\gamma_0 \approx 0.065$) accelerated to some intermediate energy in the DTL and then further accelerated to a full energy of, say, 1000 MeV ($\beta_0\gamma_0 \approx 1.81$) in a coupled-cavity linac (CCL). Suppose that the bunching factor at 2 MeV is $B_f = \bar{I}/I = 0.1$, so that the current in the bunch is $I = 1$ A and that the beam has an rms width of $\tilde{x} = 2.0$ mm at this energy. The acceleration in the DTL shortens the phase width of the bunch with respect to the rf period $\beta_0\lambda$ and thereby increases the bunch current, and the CCL is usually designed for a higher frequency, say twice the DTL frequency. However, we are not interested in the details of the accelerator design and the coupling between longitudinal and transverse bunch sizes. For the purpose of our calculation we will simply assume that the transverse focusing strength is varied along the linac system in such a way that the product of $I\tilde{x}^2$ is the same at 1000 MeV as at 2 MeV. Furthermore, we assume that the normalized rms emittance remains constant in the acceleration process and is given by $\tilde{\epsilon}_n = 2 \times 10^{-7}$ m-rad. Using the above numbers we obtain from Equation (5.292) with $I_0 = 3.1 \times 10^7$ A,

$$\frac{Ka^2}{\epsilon^2} = \frac{1 \times (2 \times 10^{-3})^2}{3.1 \times 10^7 \times 2 \times 0.065 \times (2 \times 10^{-7})^2} = 24.8.$$

This value corresponds to curve 7a in Figure 5.12, which is a space-charge-dominated beam with a tune depression of $k/k_0 \sim 0.2$ and a relatively uniform density profile.

At the final energy of 1000 MeV, we obtain

$$\frac{Ka^2}{\epsilon^2} = \frac{24.8 \times 0.065}{1.81} = 0.89$$

and $k/k_0 = 0.73$, which corresponds to curve 4 in Figure 5.12 and is in the region where emittance begins to dominate over space charge (i.e., $Ka^2/\epsilon^2 < 1$ or $k/k_0 > \sqrt{0.5} = 0.707$). Thus, during the acceleration process in this hypothetical linac system the stationary beam profile $n(r)$ changes from a nearly uniform, sharp-edged shape at 2 MeV to a more Gaussian-like shape with a significant tail at 1000 MeV. In view of the high average power of about 100 MW of such a linac, particle losses must be kept extremely low to avoid activation of the machine. Prevention of halo formation is therefore of utmost importance (see Section 6.2.2). But as we see from our calculation, the stationary beam profile develops during the acceleration process a natural tail that is comparable to a halo. This feature of the thermal distribution must be taken into consideration in the design of such a linac. The parameter Ka^2/ϵ^2 should be kept as large as possible, and emittance growth, which would move the density profile even more toward a Gaussian shape, must be avoided.

What is particularly interesting in the example above is that the transverse beam temperature $T_\perp$ actually does not change very much. If the bunch current I increases by a factor of 2, for instance, the rms width $\tilde{x}$ decreases by $\sqrt{2}$ (since we assume that $I\tilde{x}^2 = $ const). With $\tilde{\epsilon}_n \propto \tilde{x}(k_B T_\perp \gamma_0) - $ const, one then finds that the laboratory temperature $T_\perp$ rises by only about 10% while the beam-frame temperature $T_{b\perp}$ would increase by a factor of 2 according to Equation (5.290). This shows that the *high-temperature* or *low-temperature* limits of the thermal distribution are better defined as the *emittance-dominated* or *space-charge-dominated* limits. The temperature, which is an appropriate parameter for a plasma, is obviously inadequate to describe the behavior of the density profile in a charged particle beam. The above example was chosen deliberately to be somewhat simplistic, to illustrate the energy scaling. A more realistic example of a high-current, high-brightness linac design, which includes the coupling between longitudinal and transverse bunch size, is presented in Appendix 4.

A similar argument can be made with regard to emittance. Even though it involves the product of two quantities, beam width and $\sqrt{T_\perp}$, emittance alone, like temperature, is not sufficient to characterize the beam profile. On the one hand, in the space-charge-dominated regime where the beam is nearly uniform, the effective emittance $\epsilon = 4\tilde{\epsilon}$ is a very useful quantity and includes nearly 100% of the beam particles. However, as the effect of space charge decreases and the density profile becomes more Gaussian in shape, neither the rms emittance nor the effective emittance provide a sufficient description of the particle distribution. Indeed, a significant fraction of the beam intensity may be outside the effective emittance area. To get a more quantitative estimate of the tail effect, let us consider the emittance-dominated limit of the Maxwell–Boltzmann distribution. Using cartesian coordinates and neglecting the space-charge potential ϕ_s, we can write Equation (5.274b) in the form

$$f(x, x', y, y') = f_0 \, \exp\left[-\frac{x^2 + x'^2(1/k_0^2) + y^2 + y'^2(1/k_0^2)}{2\delta^2} \right], \qquad (5.295a)$$

or in terms of the betatron function $\hat{\beta} = 1/k_0$ as

$$f(x, x', y, y') = f_0 \, \exp\left(-\frac{x^2 + \hat{\beta}_0^2 x'^2 + \hat{\beta}_0^2 y'^2}{2\delta^2} \right), \qquad (5.295b)$$

where

$$\delta = \tilde{x} = \tilde{y} = \frac{\tilde{r}}{\sqrt{2}} = \sqrt{\hat{\beta}_0 \tilde{\epsilon}} = \hat{\beta}_0 \left(\frac{k_B T_\perp}{\gamma_0 m v_0^2} \right)^{1/2} \qquad (5.296)$$

is the rms width of the Gaussian distribution.

By integrating (5.295b) over y, y', we obtain the density in the two-dimensional x–x' trace space

$$f(x, x') = f_{0x} \, \exp\left(-\frac{x^2 + \hat{\beta}_0^2 x'^2}{2\delta^2} \right). \qquad (5.297)$$

This relation allows us to calculate the emittance for any part of the beam inside of a given boundary defined by $x^2 + \hat{\beta}_0^2 x'^2 = \text{const}$. Thus, if f denotes a fraction of the beam ($0 \le f \le 1$), ϵ_f the emittance occupied by this fraction in x–x' trace space, and $\tilde{\epsilon}_m$ the rms emittance of the beam, one can show that

$$\epsilon_f = -2\tilde{\epsilon} \, \ln(1 - f). \qquad (5.298)$$

From this relation we see that the fraction of the beam within the rms emittance (i.e., $\epsilon_f = \epsilon$), is $f = 0.3935$. For the effective emittance, defined by $\epsilon = \epsilon_f = 4\tilde{\epsilon}$, one finds that $f = 0.8647$, while $\epsilon = 6\tilde{\epsilon}$ yields $f = 0.9502$. Of course, for the entire beam the total emittance would be infinite ($\epsilon_f \rightarrow \infty$ for $f = 1$) in view of the exponential tail of the distribution. In practice, however, the tail will always be cut off and the total emittance will always be finite, since the tail particles are lost to the wall of the beam pipe.

These results explain why different laboratories or researchers are using different definitions of emittance. At low energies where space-charge forces dominate and the density profile is nearly uniform in a linear focusing channel, the *effective emittance,* defined as *four-times rms emittance,* includes practically 100% of the beam. This definition was adopted by CERN, where it was first proposed by Lapostolle [5]. It was also used in our previous chapters on linear beam optics, which assumed a K–V distribution for the beam where $\epsilon = 4\tilde{\epsilon}$ exactly. In high-energy accelerators, synchrotrons, and storage rings, space-charge forces are usually small compared to the applied forces, and beams tend to have Gaussian profiles, as expected for a Maxwell–Boltzmann distribution. Some laboratories, including Fermilab, have introduced a *six-times emittance,* $\epsilon_f = 6\tilde{\epsilon}$, to define the phase-space area of the beam since it contains 95% of the particles. Needless to say, these varying definitions of effective emittance are a source of much confusion. The problem is compounded by the fact that many publications use the term *rms emittance* for the effective *four-times rms emittance,* a definition that dates back to Lapostolle's original proposal (see our comment in Reference 5). Different notations concerning the factor π were discussed in Section 3.1. At some places π is factored out in the transverse emittance but included in the longitudinal emittance. If a nominal emittance of $6\tilde{\epsilon}$ is adopted, it may be adequate for the experimentalists at the high-energy end of an accelerator chain. However, at low energies, near the ion source or linac of a proton or H$^-$ machine, most of the beam is space-charge-dominated and contained within the effective emittance of $\epsilon = 4\tilde{\epsilon}$, and $6\tilde{\epsilon}$ includes a large piece of empty phase space. No wonder that communication between workers within the same organization is often just as difficult as it is between personnel working in different laboratories.

The only emittance that plays a uniquely defined role in the physics, theory, and simulation of beams is the true rms emittance, which correlates with the mean kinetic energy per particle, measured by the temperature in the Maxwell–Boltzmann distribution, and the rms beam width. Other definitions, such as our effective emittance, $\epsilon = 4\tilde{\epsilon}$, or the six-times rms emittance, $6\tilde{\epsilon}$, are undoubtedly useful. But the range of applications should be clearly defined, and to avoid misunderstandings one should always be aware of the shape of the Maxwell–Boltzmann distribution in the various parameter regimes as shown in Figure 5.12 and Table 5.2. For proper comparison of beams at different energies or in different facilities the normalized emittance, ϵ_n, should be used, or else the beam energy should be listed so that ϵ_n can be readily calculated. Furthermore, the beam current should be mentioned since the figure of merit is usually not the emittance by itself but the two-dimensional phase-space density, $I/\epsilon_n \pi$ or the normalized brightness $\mathcal{B}_n = 2I/\epsilon_n^2 \pi^2$ [see

Equation (3.22)]. Finally, it should be noted in this context that the above relationships concerning a matched beam in a uniform focusing channel are valid not only for the transverse Maxwell–Boltzmann distribution, where the local temperature $T_\perp$ is the same everywhere in the beam. They also apply for other theoretical or experimental beam distributions, where the local temperature may vary within the beam (i.e., where the beam is not in transverse thermal equilibrium). The primary information is contained in the velocity distribution versus position, which may not necessarily be a Maxwellian. However, we can always define an "effective" temperature for any distribution by using the relation (5.270) and describe the distribution in terms of an analytical equivalent beam having the same rms parameters (or second moments).

5.4.5 Transverse Temperature and Beam-Size Variations in Nonuniform Focusing Channels

In the preceding section we treated a matched transverse Maxwell–Boltzmann distribution in a uniform focusing channel where the rms trace-space ellipse, the rms beam width, and the temperature $T_\perp$ remain constant. If the beam passes through a sequence of lenses and drift spaces, such as in a matching section or in a periodic-focusing channel, the stationary state is no longer defined by the equations given previously, such as (5.274b) or (5.295a). After passing through a focusing lens, for instance, the beam experiences first a compression in its transverse size and then expands again until it reaches the next lens, as illustrated in Figure 3.12. If the lens system is periodic, the stationary state is characterized by a periodic variation of the beam radius. The associated trace-space ellipse is tilted in general and oscillates in shape between the upright positions at the crests and waists of the transverse amplitude function $\hat{\beta}(z)$. By analogy with the compression and expansion of a gas, the transverse beam temperature, $T_\perp$, heats up during compression and cools during expansion. This temperature variation can be correlated with the corresponding orientation of the rms trace-space ellipse.

To analyze this correlation, let us consider the general Courant–Snyder equation (3.345) as it applies to the rms emittance, that is,

$$\hat{\gamma}x^2 + 2\hat{\alpha}xx' + \hat{\beta}x'^2 = \tilde{\epsilon}. \tag{5.299a}$$

Using the relation $\hat{\gamma} = (1 + \hat{\alpha}^2)/\hat{\beta}$ [Equation (3.343c)] between the three Courant–Snyder parameters, we can eliminate $\hat{\gamma}$ and write this equation in the alternative form

$$x^2 + (\hat{\alpha}x + \hat{\beta}x')^2 = \tilde{\epsilon}\hat{\beta}. \tag{5.299b}$$

The second term on the left-hand side of Equation (5.299a), which was zero in our previous discussion, shows that there is a correlation between x and x', or x and v_x, that depends on the parameter $\hat{\alpha}$.

The rms emittance of a general particle distribution $f(x, x')$ is defined in terms of the moments of the distribution by [see Equation (5.205)]

$$\tilde{\epsilon}^2 = \overline{x^2}\,\overline{x'^2} - \overline{xx'}^2. \tag{5.300a}$$

The corresponding equation in terms of the particle velocities is

$$\tilde{\epsilon}^2 = \overline{x^2}\frac{\overline{v_x^2}}{v_0^2} - \frac{\overline{xv_x}^2}{v_0^2}. \tag{5.300b}$$

For an equivalent K–V beam the particle density inside the emittance ellipse is constant and the moments are readily evaluated from Equation (5.299). One finds that

$$\tilde{x}^2 = \overline{x^2} = \hat{\beta}\,\tilde{\epsilon}, \tag{5.301}$$

$$\tilde{x}'^2 = \overline{x'^2} = \hat{\gamma}\,\tilde{\epsilon}, \tag{5.302}$$

$$\overline{xx'} = -\hat{\alpha}\epsilon. \tag{5.303}$$

If we divide the rms emittance equations (5.300) by $\overline{x^2} = \tilde{x}^2$, we obtain

$$\frac{\tilde{\epsilon}^2}{\tilde{x}^2} = \overline{x'^2} - \frac{\overline{xx'^2}}{\tilde{x}^2} = \frac{\overline{v_x^2}}{v_0^2} - \frac{\overline{xv_x}^2}{v_0^2\tilde{x}^2}. \tag{5.304}$$

This relation can be written in terms of rms transverse velocities and kinetic energies as

$$\tilde{v}_{x,\text{th}}^2 = \tilde{v}_x^2 - \tilde{v}_{x,\text{fl}}^2, \tag{5.305a}$$

or

$$\gamma_0 m\tilde{v}_x^2 = \gamma_0 m\tilde{v}_{x,\text{th}}^2 + \gamma_0 m\tilde{v}_{x,\text{fl}}^2. \tag{5.305b}$$

The physical interpretation of this relation is that the rms transverse velocity, or rms kinetic energy, of the particle distribution consists of a thermal (i.e., random) component, indicated by the subscript "th," and a flow component, indicated by the subscript "fl." The latter is due to the correlation (xv_x) between the velocity and the position of the particle in regions where the beam size contracts or expands.

Using the relations (5.301) to (5.305), we find that

$$\tilde{x}'_{\text{fl}} = \frac{\tilde{v}_{x,\text{fl}}}{v_0} = -\frac{\overline{xx'}}{\tilde{x}} = -\frac{\hat{\alpha}\tilde{\epsilon}}{\tilde{x}}, \tag{5.306a}$$

or, in view of $\tilde{\epsilon} = \tilde{x}^2 / \hat{\beta}$,

$$\tilde{x}'_{\text{fl}} = -\frac{\hat{\alpha}}{\hat{\beta}} \tilde{x}. \tag{5.306b}$$

Furthermore,

$$\tilde{x}'^2_{\text{th}} = \frac{\tilde{\epsilon}^2}{\tilde{x}^2} = \hat{\gamma}\tilde{\epsilon} - \left(\frac{\hat{\alpha}}{\hat{\beta}}\right)^2 \tilde{x} = \frac{\hat{\beta}\hat{\gamma} - \hat{\alpha}^2}{\hat{\beta}} \tilde{\epsilon},$$

or

$$\tilde{x}'_{\text{th}} = \frac{\tilde{\epsilon}}{\tilde{x}} = \left(\frac{\tilde{\epsilon}}{\hat{\beta}}\right)^{1/2}. \tag{5.307}$$

We note that the ratio of the flow divergence $\tilde{x}'_{\text{fl}}$ to the thermal divergence, $\tilde{x}'_{\text{th}}$, is given by $\tilde{x}'_{\text{fl}}/\tilde{x}'_{\text{th}} = -\hat{\alpha}$. When we compare these results with Figure 3.26 we recognize that $\tilde{x}'_{\text{th}}$ corresponds to the point of intersection of the tilted ellipse with the x'-axis (i.e., $\tilde{x}'_{\text{int}} = \tilde{x}'_{\text{th}} = \sqrt{\tilde{\epsilon}}/\hat{\beta}$), while $\tilde{x}'_{\text{fl}}$ relates to the dashed line with slope $-\hat{\alpha}/\hat{\beta}$ [i.e., $\tilde{x}'_{\text{fl}} = -(\hat{\alpha}/\hat{\beta})\tilde{x}$]. These relationships are thus readily identified from plots of the rms trace-space ellipse.

From (5.307) we see that the rms emittance can always be expressed by the product of the rms width, $\tilde{x}$, and the rms thermal velocity, $\tilde{v}_{x,\text{th}} = \tilde{x}'_{\text{th}} v_0$, as

$$\tilde{\epsilon} = \tilde{x}\tilde{x}'_{\text{th}} = \tilde{x}\left(\frac{\tilde{v}_{x,\text{th}}}{v_0}\right), \tag{5.308a}$$

which is identical with the relation given in Equation (3.2b). The corresponding normalized rms emittance is

$$\tilde{\epsilon}_n = \gamma_0 \beta_0 \tilde{x}\tilde{x}'_{\text{th}} = \gamma_0 \tilde{x}\left(\frac{\tilde{v}_{x,\text{th}}}{c}\right). \tag{5.308b}$$

In view of these results, the definition (5.270b) for the transverse laboratory temperature can now be generalized as

$$k_B T_\perp = k_B T_x = \gamma_0 m \tilde{v}^2_{x,\text{th}} = \gamma_0 m(\tilde{v}^2_x - \tilde{v}^2_{x,\text{fl}}). \tag{5.309}$$

For completeness, we include the general relation for the longitudinal laboratory temperature, where we obtain

$$k_B T_\parallel = k_B T_z = \gamma_0^3 m \, \widetilde{\Delta v}^2_{z,\text{th}} = \gamma_0^3 m\left(\widetilde{\Delta v}^2_z - \widetilde{\Delta v}^2_{z,\text{fl}}\right). \tag{5.310}$$

The flow terms in these equations are defined by the correlation terms (i.e., $\tilde{v}_{x,\text{fl}} = -\overline{xv_x}/\tilde{x}$ in the transverse direction), and with $z = s - s_0$, $\Delta v_z = v_z - v_0$, by $\widetilde{\Delta v}_{z,\text{fl}} = -\overline{z\Delta v_z}/\tilde{z}$ in the longitudinal direction.

The stationary state of a beam in a nonuniform focusing system is thus characterized by a variation in the rms beam width which correlates with the generation of an rms flow velocity and a variation of the thermal velocity and beam temperature.

If Coulomb collisions can be neglected or affect the beam temperature on a time scale that is long compared to the particle travel time between the focusing lenses, the normalized emittance is invariant. We then obtain from (5.308b) and (5.309) the relation

$$\tilde{x}^2 \gamma_0 k_B T_\perp = mc^2 \tilde{\epsilon}_n^2 = \text{const}, \tag{5.311}$$

which corresponds to Equation (4.4). It implies that the transverse beam temperature, which measures the average kinetic energy per particle due to the random part in the velocity distribution, is inversely proportional to the square of the rms beam width. For the stationary (matched) beam in a uniform focusing system that was treated in the preceding section, the rms radius, and hence the temperature, are constant (i.e., independent of distance z). In a nonuniform focusing system (e.g., a periodic channel or a matching section), the stationary state is characterized by a variation of the rms beam width with distance [i.e., $\tilde{x} = \tilde{x}(z)$]. The velocity distribution then consists of a flow part characterized by the rms flow velocity $\tilde{v}_{x,\text{fl}}$ and a thermal part defined by $\tilde{v}_{x,\text{th}}$. When the beam is diverging (i.e., when the radius expands), the flow velocity has an outward direction [$\hat{\alpha} < 0$, $\tilde{v}_{x,\text{fl}} > 0$, from Equation (5.306)] and the thermal velocity decreases. When the beam is converging, the flow is inward ($\hat{\alpha} > 0$, $\tilde{v}_{x,\text{fl}} < 0$) and the transverse temperature increases. During such expansion and compression of the beam radius the shape of the density profile changes. Although the curves in Figure 5.12 represent stationary states, they can still be used as a guide for the behavior in nonuniform systems. At the maxima and minima (waists) of the beam envelope, the rms emittance ellipse is upright ($\hat{\alpha} = 0$), so that the density profiles correspond to stationary Boltzmann profiles having the same upright ellipses. Thus, by calculating the parameter Ka^2/ϵ^2, one can use Table 5.2 and Figure 5.12 to identify the shapes of the profile at these positions. One can thereby visualize how the "operating point" wanders from profile to profile during an expansion and/or compression cycle. Further aspects of beam behavior in matching, focusing, and imaging systems are discussed in Section 5.4.11.

5.4.6 The Longitudinal Distribution and Beam Cooling due to Acceleration

In this section we want to turn our attention to the longitudinal part of the Maxwell–Boltzmann distribution and discuss the general parameters that are used to characterize its properties in a straight channel or linear accelerator. The stationary states and the longitudinal envelope equations in such straight channels are analyzed in Sections 5.4.7 and 5.4.8. The behavior in circular machines is discussed in Section 5.4.9. We will assume that coupling due to the transverse motion can be neglected and that the longitudinal part of the distribution has the form [see Equation (5.271) and subsequent comment]

$$f_\| = f_{\|0} \exp\left(-\frac{H_\|}{k_B T_\|}\right),$$

or

$$f_\| = f_{\|0} \exp\left[-\frac{\gamma_0^3 m (\Delta v_z)^2}{2 k_B T_\|} - \frac{q \phi_\|}{k_B T_\|}\right], \qquad (5.312)$$

where we dropped the subscript l in Δv_z and $\phi_\|$ for simplicity and where $\phi_\|$ is the longitudinal potential function that represents the focusing action of the applied longitudinal forces as well as the defocusing space-charge forces.

In rf linear accelerators each bunch of particles passes through the sequence of rf gaps in a phase interval during which the sinusoidally varying accelerating electric force is rising in time. The so-called "synchronous particle" at the center of the bunch always passes the gaps at the same rf phase (i.e., it is in synchronism with the rf field). Particles arriving earlier experience a smaller, those arriving later a greater accelerating force than that of the synchronous particle. In the beam frame (i.e., to an observer traveling at the velocity of the synchronous particle), the particles at the head of the bunch experience a force in the negative z-direction; those in the rear (behind the synchronous phase) experience a force in the positive z-direction. In the linear regime these forces are proportional to the difference in distance between the particle position and the bunch center defined by the synchronous particle. In a traveling-wave accelerator, often employed for electrons, the bunch is placed slightly ahead of the crest of the wave so that the forces act continuously. The net result is the same as in the case of periodically spaced rf cavities with acceleration gaps, namely a focusing effect that keeps the bunch from spreading longitudinally. The induction linear accelerators employed for high-current beams also can provide time-dependent longitudinal focusing. The gap voltage is increased with time along the pulse so that the early particles at the front of the pulse gain less energy than those in the rear. This force differential can either prevent the bunch from spreading or result in longitudinal compression farther down the beam line which is required for some applications. The particle oscillations due to these linear longitudinal forces are known as the *synchrotron oscillations* since, historically, the phase stability resulting from this focusing effect was crucial to the successful operation of the high-energy synchrotrons. We should note that the synchrotron oscillation frequency is usually much lower than the betatron oscillation frequency governing the transverse motion. The longitudinal dynamics, including the nonlinear motion in rf fields, is discussed in Section 5.4.8.

Let us now proceed with the general parameter characterization of the longitudinal distribution. Of particular interest is the longitudinal temperature and its changes due to acceleration and the relations between longitudinal temperature, emittance, and energy spread. The basic definition of normalized longitudinal emittance ϵ_{nz}

as the product of the longitudinal width and momentum spread of the particle distribution is the same as in the transverse case. However, the longitudinal phase space of a moving relativistic bunch can be characterized in several different ways, and accordingly, there are different definitions of the normalized and unnormalized longitudinal emittance, as we discuss below.

If we denote by $s(t)$ the distance of travel along the direction of beam propagation, then $z(t) = s(t) - s_0(t)$ is the difference in position, $\Delta P_z = P_z - P_0$ is the difference in longitudinal momentum, and $\Delta v_z = v_z - v_0$ is the difference in longitudinal velocity between a particle in the distribution and the center-of-momentum particle ("beam centroid"), indicated by the subscript "0." In terms of the associated rms quantities, the normalized rms emittance for a longitudinally matched beam is defined by

$$\tilde{\epsilon}_{nz} = \tilde{z}\frac{\widehat{\Delta P}_z}{mc} = \tilde{z}\gamma_0^3\frac{\widehat{\Delta v}_z}{c}. \tag{5.313}$$

The unnormalized longitudinal rms emittance is commonly defined in terms of the relative rms momentum spread as

$$\tilde{\epsilon}_z = \tilde{z}\frac{\Delta\tilde{P}_z}{P_0}, \tag{5.314}$$

so that $\tilde{\epsilon}_{nz} = \beta_0\gamma_0\tilde{\epsilon}_z$, as in the transverse case.

If we introduce $z' = dz/ds$ in place of the momentum or the velocity, we obtain

$$z' = \frac{dz}{ds} = \frac{\Delta v_z}{v_0} - \frac{1}{\gamma_0^2}\frac{\Delta P_z}{P_0}, \tag{5.315}$$

for a beam in a straight channel. With these relations we can define an unnormalized rms emittance $\tilde{\epsilon}_{zz'}$ in longitudinal trace space as

$$\tilde{\epsilon}_{zz'} = \tilde{z}\tilde{z}' = \frac{1}{\gamma_0^2}\tilde{\epsilon}_z \tag{5.316}$$

for straight beams. This relation differs from the conventional unnormalized emittance (5.314) by the factor γ_0^{-2}. Only in a nonrelativistic straight beam ($\gamma_0 = 1$) are the two emittances the same ($\tilde{\epsilon}_{zz'} = \tilde{\epsilon}_z$).

Using Equations (5.313), (5.314), (5.316), and (5.270d), we can relate $\tilde{\epsilon}_{nz}$ to $\tilde{\epsilon}_z$, $\tilde{\epsilon}_{zz'}$, and to the temperature $k_B T_\parallel$ in the laboratory frame as

$$\tilde{\epsilon}_{nz} = \beta_0\gamma_0\tilde{\epsilon}_z = \beta_0\gamma_0^3\tilde{\epsilon}_{zz'}, \tag{5.317a}$$

and

$$\tilde{\epsilon}_{nz} = \tilde{z}\left(\frac{\gamma_0^3 k_B T_\parallel}{mc^2}\right)^{1/2}. \tag{5.317b}$$

The above definitions have to be modified in circular machines, as discussed in Section 5.4.9. Note that in the beam frame, where $z_b = \gamma_0 z_l$, the normalized emittance has the same value as in the laboratory frame, $\tilde{\epsilon}_{bnz} = \tilde{\epsilon}_{nz}$. This is also true for the transverse rms emittances. Thus the normalized emittance is a Lorentz-invariant quantity [10]. In accelerator physics it has become customary to introduce energy E and time t as the conjugate canonical variables in place of P_z and z. Using $z = v_0 \Delta t = \beta_0 c \Delta t$, $\Delta P_z \simeq \Delta P = \Delta \gamma mc/\beta_0 = \Delta E/\beta_0 c$ and denoting this emittance by *, we have

$$\tilde{\epsilon}_{nz}^* = \widetilde{\Delta E}\, \widetilde{\Delta t}[\text{eV} \cdot \text{s}], \tag{5.318a}$$

which is measured in electronvolt-seconds. In rf accelerators this emittance is often expressed in terms of the rf phase φ and radian frequency ω_{rf} as

$$\tilde{\epsilon}_{nz}^* = \widetilde{\Delta E}\, \frac{\widetilde{\Delta \varphi}}{\omega_{rf}}. \tag{5.318b}$$

where $\widetilde{\Delta \varphi}$ defines the rms phase width of the particle distribution with respect to the accelerating rf field.

The relationship between $\tilde{\epsilon}_{nz}^*$ and $\tilde{\epsilon}_{nz}$ is given by

$$\tilde{\epsilon}_{nz}^* = \tilde{\epsilon}_{nz}\, \frac{mc^2}{c}, \tag{5.319}$$

where mc^2 is the particle rest energy in electronvolts and c is the speed of light in m/s. For protons one has $(mc^2)/c = 3.13$ eV-s/m, and the conversion factor for electrons is $(m_e c^2)/c = 1.70 \times 10^{-3}$ eV-s/m.

All of the above relations for the longitudinal emittance imply a finite beam length. The obvious application is a bunched beam with a longitudinal bunch size characterized by an rms width $\tilde{z}$. However, the definitions can also be applied to a continuous beam, where an emittance can be assigned to a slice of the beam containing a given number of particles. Take as an example the continuous beam which is extracted from a dc ion source and which, after being chopped and/or bunched, is injected into an rf linear accelerator. The bunch in each rf cycle, containing N_b particles, can be traced back to a slice of the continuous beam. This slice would contain N_b particles and would have the same emittance as the bunch in the linac if nonlinear forces during the chopping/bunching process have a negligible effect and if there are no particle losses. The bunch, of course, has a smaller rms size than the slice in the continuous beam from which it was formed.

But the rms momentum spread in the bunch is correspondingly larger than that in the slice, so that the normalized emittance is the same [see Equation (5.313)].

The relationships between the rms energy spread, rms momentum spread, and the longitudinal temperature in either continuous or bunched beams are given by

$$\Delta \tilde{E} = \beta_0 c \, \Delta \tilde{P}, \tag{5.320}$$

$$\frac{\Delta \tilde{P}}{P_0} = \frac{1}{\beta_0^2} \frac{\Delta \tilde{E}}{E_c} = \frac{\Delta \tilde{\gamma}}{\beta_0^2 \gamma_0}, \tag{5.321}$$

$$\Delta \tilde{P} = mc \left(\frac{\gamma_0^3 k_B T_\parallel}{mc^2} \right)^{1/2}, \tag{5.322}$$

$$\frac{\Delta \tilde{E}}{E_c} = \beta_0 \left(\frac{\gamma_0 k_B T_\parallel}{mc^2} \right)^{1/2}, \tag{5.323}$$

$$\frac{\Delta \tilde{P}}{P_0} = \frac{1}{\beta_0} \left(\frac{\gamma_0 k_B T_\parallel}{mc^2} \right)^{1/2}. \tag{5.324}$$

where $E_c = \gamma_0 mc^2$ is the center-of-momentum energy, $P_0 = \beta_0 \gamma_0 mc$, and $T_\parallel$ is the longitudinal temperature measured in the laboratory frame. All of the above relationships can of course be expressed in terms of the beam temperature $T_{b\parallel}$ by making the substitution $T_\parallel = T_{b\parallel}/\gamma_0$. In the extreme-relativistic case ($\beta_0 = 1$) we obtain $\Delta \tilde{E}/E_c = \Delta \tilde{P}/P_0 = (k_B T_{b\parallel}/mc^2)^{1/2}$.

In the nonrelativistic case, using the mean kinetic energy qV_0 instead of the total energy E_c, we have

$$\Delta \tilde{E} = (2qV_0 k_B T_\parallel)^{1/2}, \tag{5.325a}$$

or

$$\frac{\Delta E}{qV_0} = \left(\frac{2k_B T_\parallel}{qV_0} \right)^{1/2}. \tag{5.325b}$$

Let us now discuss the effect of acceleration on the longitudinal beam temperature in a continuous beam that corresponds to the distribution (5.312) with $\phi_\parallel = 0$. We will find that acceleration decreases the random velocity spread and hence cools the beam longitudinally. To see how this comes about, let us first consider the nonrelativistic situation as it may exist in or near an electron gun or ion source. Suppose that two particles, one (A) with initial velocity v_1, and the other (B) with velocity $v_1 + \Delta v_1$, are passing through an acceleration gap where they gain an energy of qV_0. After acceleration the two particles will have kinetic energies of (A)

$$\frac{m}{2} v_2^2 = \frac{m}{2} v_1^2 + qV_0 \tag{5.326}$$

and (B)

$$\frac{m}{2}(v_2 + \Delta v_2)^2 = \frac{m}{2}(v_1 + \Delta v_1)^2 + qV_0. \tag{5.327}$$

Expanding both sides of Equation (5.327) and assuming that $\Delta v \ll v$ so that the quadratic terms involving $(\Delta v_1)^2$ and $(\Delta v_2)^2$ can be neglected, we find that

$$\frac{m}{2}v_2^2 + mv_2\Delta v_2 = \frac{m}{2}v_1^2 + m\Delta v_1 + qV_0 = \frac{m}{2}v_2^2 + m\Delta v_1,$$

and hence

$$mv_2\Delta v_2 = mv_1\Delta v_1,$$

or

$$\Delta v_2 = \Delta v_1 \frac{v_1}{v_2}. \tag{5.328}$$

Thus the velocity difference between the two particles, which initially was Δv_1, is reduced by the ratio v_1/v_2 of the velocities before and after acceleration.

There is an inverse effect of the acceleration with regard to the separation Δz of the two particles. Let particle A with velocity v_1 be a distance Δz_1 behind when particle B with velocity $v_1 + \Delta v_1$ passes the gap. When A arrives at the gap after a time interval $\Delta t = \Delta z_1/v_1$, particle B will have traveled a distance $\Delta z_2 = (v_2 + \Delta v_2)\Delta t \simeq v_2\Delta t$. The relation between the two distances after and before acceleration is

$$\Delta z_2 = \Delta z_1 \frac{v_2}{v_1} \tag{5.329}$$

(i.e., the particles' separation increases by the velocity ratio v_2/v_1). Combining both equations, we see that

$$\Delta z_2 \, \Delta v_2 = \Delta z_1 \, \Delta v_1. \tag{5.330a}$$

We recognize that this result correlates with the invariance of the normalized longitudinal emittance in the nonrelativistic version. Indeed, if we average both sides of (5.330a) over the entire particle distribution prior to and after transversal of the acceleration gap, we obtain

$$\widetilde{\Delta z_2} \, \widetilde{\Delta v_2} = \widetilde{\Delta z_1} \, \widetilde{\Delta v_1}; \tag{5.330b}$$

since, nonrelativistically, $\Delta P/mc = \Delta v/c$, this is identical to

$$\widetilde{\Delta}_{z2}\frac{\widetilde{\Delta P}_{z2}}{mc} = \widetilde{\Delta}_{z1}\frac{\widetilde{\Delta P}_{z1}}{mc} = \tilde{\epsilon}_{nz},$$

as claimed. Of course, since we are dealing here with a continuous beam, $\widetilde{\Delta}_z$ and $\tilde{\epsilon}_{nz}$ are defined as the rms length and rms emittance of a given slice of the beam containing a fixed number of particles, as discussed earlier.

The decrease in the velocity spread according to (5.328) occurs when there are no longitudinal focusing forces acting on the particle distribution (i.e., in a continuous beam or in a drifting bunched beam). In the first case, both the applied and the space-charge force are zero in the longitudinal direction [i.e., $\phi_{\parallel} = 0$ in Equation (5.312)]. For the second case, only the applied longitudinal forces are zero, and the space-charge forces increase the bunch length.

After this general discussion, we now determine the temperature change, or *longitudinal cooling*, of the beam due to the decrease in the velocity spread by acceleration in a mathematically more rigorous form. For this purpose we assume a continuous beam (i.e., $\phi_{\parallel} = 0$) in the nonrelativistic regime where $\gamma_0 = 1$ and where the temperatures in the beam frame and in the laboratory frame are identical. Suppose that the initial state, which we will denote by i, corresponds to the distribution emerging from the particle source (e.g., thermionic cathode or ion source) and is given in the standard Maxwellian form as

$$f_i(v_i) = f_0 \exp\left(\frac{m v_i^2}{2k_B T_{\parallel i}} \right), \tag{5.331}$$

where v stands for $v_z = v_{\parallel}$.

The state of the distribution after acceleration, which we denote by the subscript f, is characterized by a new velocity v_f and temperature $k_B T_{\parallel f}$ as

$$f_f(v_f) = f_0 \exp\left[-\frac{m(v_f^2 - v_0^2)}{2k_B T_{\parallel f}} \right], \tag{5.332}$$

where

$$v_f^2 = v_i^2 + v_0^2 \tag{5.333}$$

and

$$\frac{m}{2} v_0^2 = q V_0. \tag{5.334}$$

We note that the longitudinal velocity distribution is significantly contracted while the transverse distribution and temperature remain unaffected by acceleration. The new longitudinal temperature $k_B T_{\|f}$ is defined in terms of the second moments of the velocity distribution as [from (5.310), with $\gamma_0 = 1$]

$$k_B T_{\|f} = m\left(\overline{v_{\|f}^2} - \overline{v_{\|f}}^2\right) = m\left(\overline{v_f^2} - \overline{v_f}^2\right). \tag{5.335}$$

Using (5.334) and (5.335), we find that

$$\overline{v_f^2} = \overline{(v_i^2 + v_0^2)} = \overline{v_i^2} + v_0^2,$$

$$\overline{v_f}^2 = \overline{\sqrt{v_i^2 + v_0^2}}^2 = v_0^2\left(1 + \frac{1}{2}\frac{\overline{v_i^2}}{v_0^2} - \frac{1}{8}\frac{\overline{v_i^4}}{v_0^4} + \cdots\right)^2, \tag{5.336}$$

$$\overline{v_f}^2 = v_0^2 + \overline{v_i^2} + \frac{1}{4}\frac{\overline{v_i^2}^2}{v_0^2} - \frac{1}{4}\frac{\overline{v_i^4}}{v_0^2} + \cdots \tag{5.337}$$

Thus, substituting (5.336) and (5.337) in (5.335), we find that

$$k_B T_{\|f} = \frac{m}{4v_0^2}\left(\overline{v_i^4} - \overline{v_i^2}^2\right). \tag{5.338}$$

To evaluate the fourth moments of the initial distribution we must first multiply Equation (5.332) by $v_i = v_{\|i}$ to get the particle current leaving the source. The integration then is from $v_i = v_{\|i} = 0$ to $v_i = v_{\|i} = \infty$ since only particles with velocities in the positive z-direction can escape. Carrying out the calculations, one finds that

$$k_B T_{\|f} = \frac{(k_B T_{\|i})^2}{2qV_0}. \tag{5.339}$$

The cooling effect predicted by relation (5.339) is very dramatic. Take, for instance, an electron beam with an initial temperature given by the cathode temperature of $k_B T_{\|i} = k_B T_c = 0.1$ eV. After acceleration to $qV_0 = 10$ keV, the longitudinal temperature drops to $k_B T_{\|f} \simeq 6 \times 10^{-7}$ eV; that is, for all practical purposes the beam is essentially "cold" in the longitudinal direction while the transverse temperature remains unaffected by the acceleration ($k_B T_{\perp f} = k_B T_{\perp i} = k_B T_c = 0.1$ eV). Also, in view of our discussion in connection with Equation (5.330), the longitudinal emittance as well as the transverse emittance remain the same. Likewise, the energy spread $\widetilde{\Delta E}$ should remain unaffected by the acceleration process. It is given by the initial temperature, and by solving (5.339) for $k_B T_{\|i}$ one obtains

$$k_B T_{\|i} = \widetilde{\Delta E}_i = (2qV_0 k_B T_{\|f})^{1/2} = \widetilde{\Delta E}_f, \tag{5.340}$$

in agreement with Equation (5.325a).

As we have now seen, the particle distribution comprising the beam is no longer in a three-dimensional thermal equilibrium state after it has been accelerated. Also, the energy spread is no longer identical to the temperature when acceleration has taken place. However, Coulomb scattering, instabilities, or other random processes coupling the longitudinal and transverse motion of the particles will have a tendency to restore equilibrium. The associated thermal energy transfer from the transverse to the longitudinal direction will increase the longitudinal emittance and energy spread. This is known as the *Boersch effect*, which is discussed in Section 6.4.1.

Although the longitudinal cooling effect due to acceleration is most pronounced at low energies near the particle source, it also occurs at relativistic particle energies. The evaluation of the temperature change in this case is essentially analogous to the nonrelativistic derivation. Instead of Equations (5.333) and (5.334), one uses

$$\beta_i^2 - 1 - \frac{1}{\gamma_i^2} = 1 - \frac{1}{(\gamma_{0i} + \Delta\gamma_i)^2} = 1 - \frac{1}{\gamma_{0i}^2}\frac{1}{(1 + \Delta\gamma_i/\gamma_{0i})^2}, \quad (5.341)$$

$$\beta_f^2 - 1 - \frac{1}{\gamma_f^2} = 1 - \frac{1}{(\gamma_{0f} + \Delta\gamma_i)^2} = 1 - \frac{1}{\gamma_{0f}^2(1 + \Delta\gamma_i/\gamma_{0f})^2}, \quad (5.342)$$

$$\gamma_f mc^2 = \gamma_i mc^2 + qV_0, \quad (5.343)$$

where $\gamma_0^2 - (1 - \beta_0^2)^{-1}$ defines the center-of-momentum energy and velocity and where we used $\Delta\gamma_f - \Delta\gamma_i$. Evaluating the moments using $\Delta\gamma_i = \frac{1}{2}\gamma_{0i}^3(\beta_i^2 - \beta_{0i}^2)$, one obtains

$$\overline{\beta_f^2} - \overline{\beta_f}^2 = \frac{1}{4}\frac{\gamma_{0i}^6}{\beta_{0f}^2\gamma_{0f}^6}\left(\overline{\beta_i^4} - \overline{\beta_i^2}^2\right). \quad (5.344)$$

Using the definition (5.310) and the result for the fourth moments analogous to the nonrelativistic calculation, one finds for the laboratory-frame temperature

$$k_B T_{\|f} = \frac{\gamma_{0i}^3 (k_B T_{\|i})^2}{\beta_{0f}^2\gamma_{0f}^2 mc^2}, \quad (5.345)$$

which in the nonrelativistic limit ($\gamma_0 = 1, \beta_0^2 mc^2 = mv_0^2 = 2qV_0$) agrees with the result (5.339). As an example, assume that an electron beam with initial energy of 10 keV ($\gamma_{0i} \simeq 1$) and initial laboratory temperature $k_B T_{\|i} = 0.5$ eV is accelerated to 1 MeV ($\beta_{0f}\gamma_{0f} = 2.783$). The temperature is then reduced to $k_B T_{\|f} = 6.32 \times 10^{-8}$ eV, which represents a drop by seven orders of magnitude.

The above cooling effect always occurs in electrostatic acceleration systems where no longitudinal focusing or bunching forces are present (i.e., in most electron guns, ion sources, dc acceleration columns, van de Graaff accelerators, etc.). In the bunched beams of rf accelerators and some induction linacs, the longitudinal forces generated by the time-varying fields tend to prevent the longitudinal expansion and hence to reduce the cooling effect. To illustrate the effect of bunching, consider a proton beam having an initial temperature of $k_B T_{\|i} = 0.5$ eV at the ion source. Now assume that this beam is accelerated to 40 keV and then injected into an RFQ accelerator, where it is bunched by a factor of 10. From Equation (5.339), the acceleration cools the longitudinal temperature to $k_B T_{\|f} = 3 \times 10^{-6}$ eV. However, the bunching process increases this value again, by a factor of 100, to 3×10^{-4} eV. This follows from Equation (5.317), which implies that $\tilde{z}^2 k_B T_\| = $ const if $\gamma_0 = 1$ and if the normalized emittance does not change. Of course, nonlinear beam dynamics effects in the RFQ accelerator may increase the normalized longitudinal emittance, which would then further increase the temperature.

5.4.7 Stationary Line-Charge Density Profiles in Bunched Beams

Let us now proceed with a more detailed analysis of the properties of bunched beams. The stationary longitudinal Maxwell–Boltzmann distribution in this case is of the general form (5.312). The function $\phi_\|$ consists of the applied potential $\phi_{\|a}$ that provides the longitudinal focusing force and the self-field potential $\phi_{\|s}/\gamma_0^2$ that produces a defocusing force. With $\phi_\| = \phi_{\|a} + \phi_{\|s}/\gamma_0^2$, Equation (5.312) can be written in the alternative form

$$ f_\| = f_{\|0} \exp\left[-\frac{\gamma_0^3 m (\Delta v_z)^2}{2 k_B T_\|} - \frac{q\phi_{\|a} + q\phi_{\|s}/\gamma_0^2}{k_B T_\|} \right]. \tag{5.346} $$

The factor γ_0^{-2} that occurs with the electrostatic space-charge potential $\phi_{\|s}$ is due to the Lorentz transformation, as explained below (p. 411). The two potentials are commonly taken to be functions of the distance z of a particle from the center of the bunch [i.e., $\phi_{\|a} = \phi_{\|a}(z)$, $\phi_{\|s} = \phi_{\|s}(z)$]. The applied potential is assumed to be a known function of z. Thus, if the focusing and acceleration is provided by rf cavities, $\phi_{\|a}$ is a periodically varying function of time, or of phase $\omega_{rf} \Delta t$ with respect to the passage of the bunch center through the cavities. Since Δt correlates with $z = s - s_0$ by $z = v_0 \Delta t$, where v_0 is the centroid velocity, one can define $\phi_{\|a}$ as a function of z. If the bunch length is small compared to the acceptance of the rf field (see the next section), one can approximate $\phi_{\|a}(z)$ by a harmonic oscillator potential ($\propto z^2$), which for convenience will be written in the form $\phi_{\|a}(z) = \phi_{\|0}[(z/z_0)^2 - 1]$, where $\phi_{\|a} = 0$ at $z = z_0$ and $\phi_{\|a}(0) = -\phi_{\|0}$.

The bunch usually travels through many rf cavities, as in a linear accelerator, or many times through the same cavities, as in a circular machine. In this general case, the applied potential amplitude would actually be a periodic function of the bunch travel distance s [i.e., $\phi_{\|0} = \phi_{\|0}(s)$]. On the other hand, if the longitudinal force acts continuously on the beam, as in a traveling-wave electron linac, where the bunch rides on the rf wave like a surfer on a wave in the ocean, $\phi_{\|0}$ is constant. For the general case of periodically spaced acceleration gaps it is customary to use the smooth-approximation theory whenever possible. This is analogous to the treatment of transverse focusing, and in this approximation the periodic gap system is identical to the longitudinal focusing in a traveling-wave accelerator.

With a constant or "smooth" applied harmonic-oscillator potential as discussed above, the stationary longitudinal distribution function, representing a perfectly matched bunch, takes the form

$$f_{\parallel} = f_{\parallel 0} \exp\left\{ -\frac{\gamma_0^3 m (\Delta v_z)^2}{2 k_B T_{\parallel}} - \frac{q\phi_{\parallel 0}[(z/z_0)^2 - 1] + q\phi_{\parallel s}(z)/\gamma_0^2}{k_B T_{\parallel}} \right\}. \qquad (5.347)$$

If the beam energy and the applied potential are constant, the temperature $T_{\parallel}$ will be constant. Moreover, if changes in energy, focusing potential, or temperature occur adiabatically, the distribution will remain in equilibrium although its density profile may change. If we multiply Equation (5.347) by the charge q and integrate over the velocities Δv_z, we obtain the longitudinal Boltzmann relation for the line charge density $\rho_L(z)$, that is,

$$\rho_L(z) = q \int f_{\parallel}(z, \Delta v_z) \, d(\Delta v_z), \qquad (5.348a)$$

yielding

$$\rho_L(z) = C \exp\left\{ -\frac{q\phi_{\parallel 0}[(z/z_0)^2 - 1] + q\phi_{\parallel s}(z)/\gamma_0^2}{k_B T_{\parallel}} \right\}, \qquad (5.348b)$$

where C is constant.

Let us now analyze the space-charge potential $\phi_{\parallel s}$ in the nonrelativistic limit (which is equivalent to treating the problem in the beam frame). As we will see, $\phi_{\parallel s}$ can be defined as a function of the line-charge density ρ_L. We will assume that the bunch has a cylindrically symmetric shape, and that it is inside of a cylindrical conducting tube of radius b. The total electrostatic potential due to the space charge of the bunch, ϕ_s, which is a function of radius r and axial position z, must obey the Poisson equation

$$\nabla^2 \phi_s(r, z) = -\frac{\rho(r, z)}{\epsilon_0}, \qquad (5.349)$$

where $\rho(r, z)$ is the volume charge density in the bunch. In general, this equation must be solved numerically. However, we can obtain useful analytical results for the special model where the bunch is represented by a well-defined ellipsoid with radial semiaxis a and longitudinal semiaxis z_m and with uniform charge density ρ_0, as illustrated in Figure 5.13. The potential inside the ellipsoidal boundary of the bunch in this case can be written as

$$\phi_s(r, z) = \phi_{fs}(r, z) + \phi_i(r, z) + \phi_0, \qquad (5.350a)$$

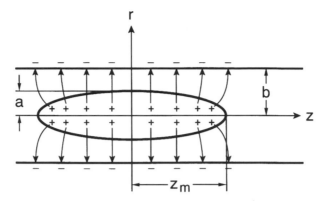

Figure 5.13. Geometry of ellipsoidal bunch in a cylindrical conducting drift tube.

where ϕ_{fs} is the free-space potential that can be found analytically [17,18] and is given by

$$\phi_{fs}(r,z) = -\frac{\rho_0}{2\epsilon_0}\left(\frac{1 - M_E}{2}r^2 + M_E z^2\right). \tag{5.350b}$$

$\phi_i(r,z)$ is the potential due to the image charges on the wall of the conducting tube, and the constant ϕ_0 is chosen to satisfy $\phi_s(b,z) = 0$ at the wall. The parameter M_E is defined as [17]

$$M_E = \frac{1 - \xi^2}{\xi^2}\left(\frac{1}{2\xi}\ln\frac{1 + \xi}{1 - \xi} - 1\right), \tag{5.351}$$

where $\xi = \sqrt{1 - (a/z_m)^2}$. When $z_m = a$, one finds that $M_E = \frac{1}{3}$, and over the range $0.8 < z_m/a < 4$, M_E can be approximated fairly well by [18] $M_E \approx a/(3z_m)$.

For the line-charge density $\rho_L(z)$ in the uniformly populated ellipsoid, one obtains

$$\rho_L(z) = 2\pi \int_0^{a\sqrt{1 - z^2/z_m^2}} \rho_0 r\, dr = \rho_0 a^2 \pi\left(1 - \frac{z^2}{z_m^2}\right),$$

or

$$\rho_L(z) = \rho_{L0}\left(1 - \frac{z^2}{z_m^2}\right). \tag{5.352}$$

Here

$$\rho_{L0} = \rho_L(0) = \rho_0 a^2 \pi = \frac{3}{4}\frac{Q}{z_m} \tag{5.353}$$

is the line-charge density at the center ($z = 0$) of the bunch, $Q = qN$ the total charge, and N the total number of particles in the bunch.

The axial gradient of the free-space part of the potential (5.350b) inside the bunch,

$$\frac{\partial \phi_{fs}}{\partial z} = -\frac{\rho_0}{\epsilon_0} M_E z, \tag{5.354a}$$

is a linear function of z and can be related to the derivative of the line-charge density by

$$\frac{\partial \phi_{fs}}{\partial z} = \frac{g_0}{4\pi\epsilon_0} \frac{\partial \rho_L(z)}{\partial z}, \tag{5.354b}$$

or

$$\frac{\partial \phi_{fs}}{\partial z} = -\frac{g_0}{4\pi\epsilon_0} \frac{2\rho_{L0}}{z_m^2} z - -\frac{\rho_0}{2\epsilon_0} \frac{g_0 a^2}{z_m^2} z. \tag{5.354c}$$

For the radial gradient of the free-space potential we obtain

$$\frac{\partial \phi_{fs}}{\partial r} = -\frac{\rho_0}{2\epsilon_0} (1 - M_E) r = -\frac{\rho_0}{2\epsilon_0} \left(1 - \frac{g_0}{2} \frac{a^2}{z_m^2} \right) r. \tag{5.354d}$$

Equation (5.354b) defines the *geometry factor* g, which plays an important role in the longitudinal beam dynamics. In this case we are dealing with the free-space potential gradient of a uniformly populated ellipsoid, where we define $g = g_0$. Below [see Equation (5.366)] we make use of an identical relationship in the more general case where image effects from the conducting boundary are included and where the line-charge density $\rho_L(z)$ may not be parabolic. The geometry factor g is then different from g_0 and defined by appropriate averaging over the charge distribution.

The free-space geometry factor g_0 depends on the eccentricity z_m/a of the ellipsoid and can be related to the parameters M_E or ξ by

$$g_0 = 2\frac{z_m^2}{a^2} M_E = \frac{2}{\xi^2} \left(\frac{1}{2\xi} \ln \frac{1 + \xi}{1 - \xi} - 1 \right). \tag{5.355}$$

For small eccentricities ($0.8 < z_m/a \le 4$), one can use $g_0 \approx 2z_m/3a$, which yields for the free-space gradients the approximate relations

$$\frac{\partial \phi_{fs}}{\partial z} \approx -\frac{\rho_0}{3\epsilon_0} \frac{a}{z_m} z = -\frac{Q}{4\pi\epsilon_0 a^2 z_m} z \tag{5.356a}$$

and

$$\frac{\partial \phi_{fs}}{\partial r} \approx -\frac{\rho_0}{2\epsilon_0}\left(1 - \frac{1}{3}\frac{a}{z_m}\right)r = -\frac{3}{2}\frac{Q}{4\pi\epsilon_0 a^2 z_m}\left(1 - \frac{1}{3}\frac{a}{z_m}\right)r, \qquad (5.356b)$$

where we introduced the total charge $Q = qN$ of the bunch from Equation (5.353).

In the free-space environment considered so far, the electric field lines originating from the charges in the bunch look at large distances like those from a point charge. This picture changes significantly when the conducting tube is present. The field lines will terminate at the image charges on the wall surface and will be pointing predominantly in the radial direction, as illustrated schematically in Figure 5.13. The electric field and hence also the total potential of Equation (5.350) will therefore essentially be concentrated in the region that is defined by the length $2z_m$ of the bunch. The image fields will reduce the axial defocusing space-charge force and increase the radial defocusing space-charge force on the particles. Mathematically, the image charge problem must be solved by numerical methods. Such numerical calculations were done recently [19] for ellipsoidal bunches with uniform charge density in a conducting cylindrical tube of radius b where the potential distribution is of the form (5.350). In these calculations the bunch eccentricity z_m/a and the ratio of tube radius to beam radius, b/a, were varied, covering a large number of cases in the range $1 \leq z_m/a \leq 20$ and $1.5 \leq b/a \leq 5$. Figure 5.14 shows the potential distribution $\phi_s(0, z)$ and the electric field $E_{sz}(0, z)$ along the axis for three bunches with (a) $z_m/a = 1$, (b) $z_m/a = 5$, and (c) $z_m/a = 10$, for $b/a = 2$. In addition, the electric field gradient $E'_{sz}(z)$ is shown for the case $z_m/a = 5$ in (d). These curves illustrate the general pattern in the electric field distribution. A careful analysis of the computer results and field plots obtained for all the cases leads to the following conclusions:

1. The axial electric field is practically linear with distance along the entire bunch as long as the bunch length does not exceed the tube diameter (i.e., if $z_m \leq b$). This is due to the fact that the nonlinear image field is relatively small compared to the linear free-space field.
2. With increasing bunch length the electric field becomes more and more nonlinear from the image effects. The data computed for $E_{sz}(z)$ can be fitted quite accurately with an analytical expression of the form

$$E_{sz}(z) = E'_{sz}(0)z\left[1 + A\left(\frac{z}{z_m}\right)^2 + B\left(\frac{z}{z_m}\right)^4\right], \qquad (5.357)$$

where $E'_{sz}(0)$ is the electric field gradient at the center ($z = 0$), and the coefficients A, B define the strength of the third- and fifth-order correction terms, respectively. For a self-consistent treatment of the equilibrium bunch distribution with images see remark and references at the end of Appendix 4.

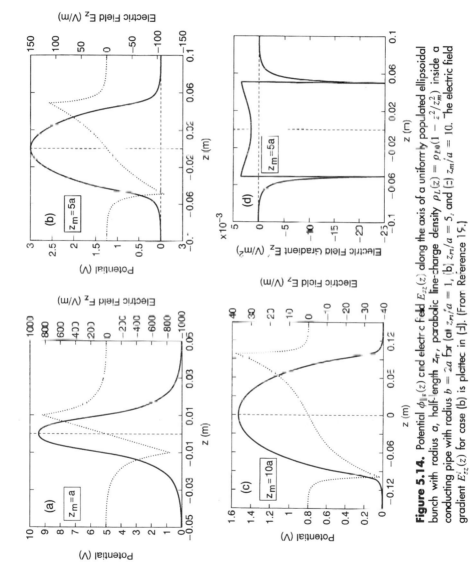

Figure 5.14. Potential $\phi_{\parallel s}(z)$ and electric field $E_{zz}(z)$ along the axis of a uniformly populated ellipsoidal bunch with radius a, half-length z_m, parabolic line-charge density $\rho_L(z) = \rho_{L0}(1 - z^2/z_m^2)$ inside a conducting pipe with radius $b = 2a$ for (a) $z_m/a = 1$, (b) $z_m/a = 5$, and (c) $z_m/a = 10$. The electric field gradient $E_{zz}'(z)$ for case (b) is plotted in (d). (From Reference 15.)

3. The linear electric field near the center or in short bunches where the nonlinear terms are negligible ($A \approx 0, B \approx 0$) can be related to the derivative of the line-charge density, in analogy to the free-space relation (5.354b), by

$$E_{sz}(z) = \frac{g(0)}{4\pi\epsilon_0} \frac{\partial \rho_L(z)}{\partial z} = \frac{g(0)}{4\pi\epsilon_0} \frac{2\rho_{L0}}{z_m^2} z , \tag{5.358}$$

where by comparison with (5.357) we have

$$E'_{sz}(0) = \frac{g(0)}{4\pi\epsilon_0} \frac{2\rho_{L0}}{z_m^2} . \tag{5.359}$$

Using the computed value for $E'_{sz}(0)$ and the relation (5.353), we can calculate the geometry factor $g(0)$ as

$$g(0) = \frac{2}{3} \frac{4\pi\epsilon_0}{Q} E'_{sz}(0) z_m^3 . \tag{5.360}$$

4. In the general nonlinear case, we can define a geometry factor g that describes the average behavior of the longitudinal distribution. The proper way of doing this is to take the average $\overline{zE_{sz}}$ of the computed electric field and compare it with the average $\overline{zE_{sz}}$ for the equivalent linear field. Now, by definition,

$$\overline{zE_{sz}} = \frac{2\pi \int_0^{z_m} \int_0^{a\sqrt{1-(z/z_m)^2}} zE_{sz}\rho(r,z)r\,dr\,dz}{Q/2} , \tag{5.361a}$$

where in our case the volume charge density in the ellipsoid is constant [i.e., $\rho(r,z) = \rho_0$]. The computer results show that the axial electric field E_{sz} is essentially independent of the radius r, so that (5.361a) can be integrated in r and the average $\overline{zE_{sz}}$ can be related to the integral over the longitudinal line charge density $\rho_L(z)$ by

$$\overline{zE_{sz}} = \frac{\int_0^{z_m} zE_{sz}(z)\rho_{L0}(1 - z^2/z_m^2)\,dz}{Q/2} . \tag{5.361b}$$

The equivalent linear electric field can be written in terms of a general geometry factor g as

$$E_{sz}(z) = \frac{g}{4\pi\epsilon_0} \frac{\partial \rho_L(z)}{\partial z} = \frac{g}{4\pi\epsilon_0} \frac{2\rho_L(z)}{z_m^2} z . \tag{5.362}$$

By evaluating the integral (5.361b) using (5.357) for $E_{sz}(z)$ and comparing the result with (5.362), one obtains for the geometry factor g the relation

$$g = g(0)\left(1 + \frac{3}{7}A + \frac{5}{21}B\right). \tag{5.363}$$

5. The longitudinal potential function $\phi_{\|s}(z)$ can be obtained by integrating $E_{sz}(z)$ in Equation (5.357). However, it will be more useful to relate $\phi_{\|s}$ to the line-charge density by integrating Equation (5.362), which yields

$$\phi_{\|s}(z) \simeq \frac{g}{4\pi\epsilon_0}\rho_L(z) + \text{const} = \frac{g}{4\pi\epsilon_0}\rho_{L0}\left(1 - \frac{z^2}{z_m^2}\right) + \text{const}. \tag{5.364}$$

For short bunches when $g = g(0)$ this expression is exact, whereas for long bunches it is an approximation based on the equivalent linear field.

6. The geometry factors g_0 for free space [Equation (5.355)], $g(0)$ [Equation (5.360)], and g [Equation (5.363)] for the various cases that were calculated are listed in Table 5.3 and plotted versus the eccentricity z_m/a of the bunch in Figure 5.15. As can be seen, $g(0)$ and g rise from values that are close to g_0 at $z_m/a = 1$ and then level off as z_m/a increases. The asymptotic values in the flat region of the curves are found to relate to the ratio b/a as [19]

$$g_{max}(0) = 2\ln\frac{b}{a}, \tag{5.365a}$$

$$g_{max} \approx 0.67 + 2\ln\frac{b}{a}. \tag{5.365b}$$

Table 5.3 Geometry parameters $g_0(0)$, and g for different values of z_m/a and b/a

Eccentricity	$b/a = 1.5$		$b/a = 2$		$b/a = 3$		$b/a = 5$		Free Space
	$g(0)$	g	$g(0)$	g	$g(0)$	g	$g(0)$	g	g_0
1	0.58	0.59	0.63	0.63	0.66	0.66	0.66	0.66	0.67
1.5	0.80	0.85	0.93	0.94	1.01	1.01	1.04	1.04	1.05
2	0.91	1.02	1.14	1.18	1.31	1.31	1.37	1.37	1.39
3	0.94	1.21	1.35	1.48	1.73	1.76	1.90	1.90	1.96
4	0.89	1.30	1.41	1.65	1.98	2.05	2.29	2.30	2.41
5	0.85	1.38	1.40	1.74	2.12	2.24	2.57	2.60	2.79
7.5	0.81	1.38	1.39	1.86	2.19	2.52	2.96	3.08	3.52
10	0.81	1.40	1.39	1.93	2.19	2.63	3.11	3.34	4.06
15	0.81	1.40	1.39	1.97	2.20	2.72	3.19	3.58	4.84
20	0.81	1.41	1.39	1.97	2.20	2.77	3.21	3.68	5.40

Source: Reference 19.

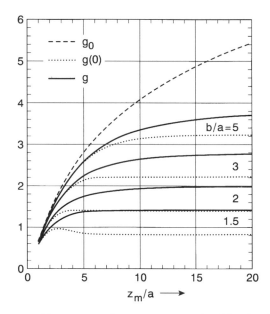

Figure 5.15. Variation of the geometry factors g_0, $g(0)$, and g defined in the text with the ratio of bunch length to radius, z_m/a, for different ratios of tube radius to bunch radius, b/a. (From Reference 19.)

The asymptotic result $g_{max}(0) = 2 \ln(b/a)$ can be derived analytically [19], and these analytical values were used in the flat regions of the curves for $g(0)$.

In summary, we can qualitatively distinguish between three regions with regard to the ratio of the bunch length $2z_m$ and the diameter $2b$ of the conducting tube:

Region 1 ($z_m \lesssim b$). The axial electric field is linear and hence satisfies the relation (5.358) for a parabolic line-charge density; the two geometry factors are essentially identical [i.e., $g = g(0)$], slightly lower than the free-space parameter g_0, and increase with z_m/a.

Region 2 ($b \lesssim z_m \lesssim 3b$). The axial electric field becomes increasingly nonlinear, the parabolic bunch relation (5.358) is no longer satisfied, the average geometry factor, g, becomes greater than $g(0)$, but the rate of increase with z_m/a shill begins to level off.

Region 3 ($z_m > 3b$). The nonlinearity of the axial electric field increases further as the fifth-order term ($\sim Bz^5$) becomes more and more significant, and the parabolic bunch relation (5.358) is even less satisfied than in region 2; the geometry factors $g(0)$ and g are essentially independent of the bunch length, or z_m/a, and can be approximated by the values given in (5.365a) and (5.365b).

We note that this behavior of the parabolic bunch with image fields differs significantly from the description found in the literature (see, e.g., [C.17, p. 181, or D.11, p. 24]), where the parabolic line-charge profile is assumed to be valid for any bunch length that is much larger than the tube diameter ($z_m \gg b$) and where the g-factor is assumed to have the constant value $g = 1 + 2 \ln(b/a)$. By contrast, we find that the parabolic profile, and hence the linearity of the electric field gradient, is valid only for short bunches ($z_m < b$); that the g-factor is a function of z_m/a and b/a, which increases at first with the bunch length and then levels off to an average value of $g \approx 0.67 + 2 \ln(b/a)$; and that the space-charge electric field becomes increasingly nonlinear toward the edges as the bunch becomes longer. The problem concerning the correct value of the g-factor will surface again in Chapter 6 when we deal with longitudinal space-charge waves and instabilities [see the discussion in connection with Equations (6.68), (6.69a), (6.69b), and Figure 6.18]. As we will see there, the g-factors for bunched beams and the g-factors for perturbations in continuous beams are different.

The preceding investigation of the behavior of an ideal ellipsoidal bunch in a cylindrical tube was intended to given us some physical insight into the effects of image charges. After this detour we return to the problem of finding the self-consistent longitudinal line-charge profiles for the stationary thermal distribution, as stated in Equation (5.348b). To make further analytic progress with our model, we assume that the relations $E_{sz} \sim \partial\rho_L/\partial z$ in Equation (5.362) and $\phi_{\|s}(z) \sim \rho_L(z)$ in Equation (5.364) are also satisfied in an approximate sense (and with the caveats presented in the preceding discussion) when $\rho_L(z)$ is not exactly parabolic. This assumption is necessary to reduce the three-dimensional problem to a one-dimensional problem whereby the axial space-charge field E_{sz} is related to the derivative of the line-charge density and thus is only a function of z. The approximation involved is usually quite satisfactory, and the relation $E_{sz} \sim \partial\rho_L/\partial z$ is widely used in the literature (see also Sections 6.2 and 6.3). We note that the one-dimensional approximation for the longitudinal motion is also an implicit assumption in our two-temperature model (5.274a) for the Maxwell–Boltzmann distribution.

The above analysis so far has been nonrelativistic. To extend it to relativistic beams, we must transform the electrostatic potential and field from the beam frame to the laboratory frame by multiplying with the factor γ_0^{-2}, as in the transverse case. This result can be obtained by applying a Lorentz transformation to $\partial\rho_L/\partial z$ in the beam frame yielding $(\partial\rho_L/\partial z)_{\text{beam}} = (\partial\rho_L/\partial z)_{\text{lab}}/\gamma_0^2$ due to the relativistic contraction of longitudinal dimensions. Thus one obtains in lieu of (5.362) and (5.364) the relativistic relations

$$E_{sz} = -\frac{g}{4\pi\epsilon_0\gamma_0^2}\frac{\partial\rho_L}{\partial z}, \tag{5.366}$$

$$\frac{\phi_{\|s}}{\gamma_0^2} = \frac{g}{4\pi\epsilon_0\gamma_0^2}\rho_L(z). \tag{5.367}$$

An alternative way is to express the laboratory electric field of the bunch in the form of Equation (2.35) that includes the inductive field due to the time-varying magnetic vector potential A_z (Faraday's law), that is,

$$E_{sz} = -\frac{\partial \phi_{\|s}}{\partial z} - \frac{\partial A_z}{\partial t}. \tag{5.368a}$$

As will be discussed in the next chapter [see Equation (6.68)], this relation may be written as

$$E_{sz} = -\frac{g}{4\pi\epsilon_0}\left(\frac{\partial \rho_L}{\partial z} + \frac{1}{c^2}\frac{\partial I}{\partial t}\right), \tag{5.368b}$$

and with $\partial I/\partial t \approx -v_0^2 \partial \rho_L/\partial z$, one obtains Equation (5.366) (i.e., the same result as with the Lorentz transformation).

By substituting the self-field potential (5.367) into the Boltzmann equation (5.348b) we obtain an integral equation from which the line-charge density profile $\rho_L(z)$ can be self-consistently calculated, namely

$$\rho_L(z) = \rho_L(0)\,\exp\left\{-\frac{q\phi_{\|0}}{k_B T_\|}\frac{z^2}{z_0^2} + \frac{qg\rho_L(0)}{4\pi\epsilon_0\gamma_0^2 k_B T_\|}\left[1 - \frac{\rho_L(z)}{\rho_L(0)}\right]\right\}. \tag{5.369}$$

This equation can be solved numerically in a straightforward manner [13]. The results for eight different temperatures are given in Figure 5.16. For zero temperature ($T_\| \to 0$) we find that

$$\rho_L(z) = \rho_{L0}\left(1 - \frac{z^2}{z_0^2}\right), \tag{5.370}$$

with

$$\rho_{L0} = \rho_L(0) = \frac{4\pi\epsilon_0\gamma_0^2 q\phi_{\|0}}{qg}, \tag{5.371}$$

which represents a parabolic line-charge density. Since $\phi_{\|s} \propto \rho_L(0)[1 - z^2/z_0^2]$, the space-charge force E_{sz} is linear in z, and it exactly balances the applied longitudinal force in this laminar-flow limit. The behavior of a parabolic line-charge profile in a low-temperature beam with linear external focusing has recently been studied experimentally [20].

In the high-temperature case ($k_B T_\| \gg q\phi_{\|0}$), where the space-charge forces can be neglected, the longitudinal density profile approaches the Gaussian form

$$\rho_L(z) = \rho_L(0)\,\exp\left(-\frac{z^2}{2\delta_z^2}\right), \tag{5.372}$$

where $\delta_z = \tilde{z} = (\overline{z^2})^{1/2}$ is the rms width of the distribution. The Gaussian profile is a good approximation when space-charge forces are negligible, while the parabolic

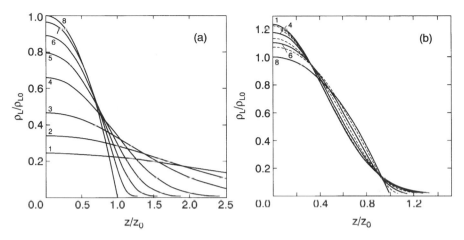

Figure 5.16. Line-charge density profiles for eight different longitudinal temperatures: (a) longitudinal focusing force is kept constant; (b) longitudinal focusing force is increased to keep rms radius constant. (From Reference 13.)

profile represents a bunched beam near the space-charge limit. We should note, however, from our preceding discussion that the self-consistent inclusion of image forces could yield a nonparabolic profile in the latter case.

Table 5.4 lists the parameter values associated with each of the eight density profiles. The column for $\tilde{z}/\tilde{z}_0$ shows the ratio of the rms width $\tilde{z}$ of each curve to the rms width $\tilde{z}_0$ of the parabolic zero-temperature profile in case (a) of Figure 5.16. Each density profile can be correlated with the equivalent parabolic beam of half-length z_m having the same perveance K_L, longitudinal rms width $\tilde{z}$, and rms emittance $\tilde{\epsilon}_{zz'} = \tilde{z}^2 k_z$, where $z_m = \sqrt{5}\,\tilde{z}$. As will be shown in the next section, K_L

Table 5.4 Parameter values for the eight longitudinal charge density profiles in Figure 5.16

Curve	$k_B T_\parallel/q\phi_{\parallel 0}$	$\rho_L(0)/\rho_{L0}$	$\tilde{z}/\tilde{z}_0$	k_z/k_{z0}	$\rho_L(0)/\rho_{L0}\ (\tilde{z}=\tilde{z}_0)$
1	10	0.237	5.03	0.994	1.19
2	5	0.332	3.59	0.985	1.19
3	2.5	0.455	2.59	0.965	1.18
4	1	0.645	1.76	0.898	1.14
5	0.5	0.777	1.41	0.793	1.10
6	0.25	0.873	1.21	0.653	1.06
7	0.1	0.947	1.09	0.459	1.03
8	0	1	1	0	1

is defined by

$$K_L = \frac{3}{2} \frac{gNr_c}{\beta_0^2 \gamma_0^5} = z_m^3 (k_{z0}^2 - k_z^2), \tag{5.373}$$

where k_z and k_{z0} are the focusing wave constants with and without space charge, N is the total number of particles in the bunch, and $r_c = q^2/4\pi\epsilon_0 mc^2$ is the classical particle radius. One finds that

$$\frac{k_B T_\parallel}{q\phi_{\parallel 0}} = \frac{2}{5} \frac{k_z^2}{k_{z0}^2} \frac{\bar{z}^2}{\bar{z}_0} \quad \text{or} \quad \frac{k_z}{k_{z0}} = \left(\frac{5}{2} \frac{k_B T_\parallel}{q\phi_{\parallel 0}}\right)^{1/2} \frac{z_0}{z_m}, \tag{5.374}$$

since $\bar{z}_0/\bar{z} = z_0/z_m$.

The results for the tune depression k_z/k_{z0} for the longitudinal particle oscillations shown in Table 5.4 have been calculated from Equation (5.374). They can be correlated with the longitudinal perveance parameter K_L by Equation (5.373).

5.4.8 Longitudinal Motion in rf Fields and the Parabolic Bunch Model

In the preceding section we analyzed the properties of the stationary state of the longitudinal distribution in a smooth, linear focusing system where the applied potential $\phi_{\parallel a}(z)$ is a quadratic function of the position z from the bunch center. We now proceed to derive the longitudinal equation of motion, which governs the general behavior of the bunch and which yields the stationary state, or matched beam, as a special solution. Since most machines employ rf fields for the acceleration and longitudinal bunching of the beam, we will analyze the particle motion in such fields. The forces acting on the particles in an electromagnetic field are, in general, nonlinear. We start with this general situation and then consider the special case where the bunch length is small compared to the rf wavelength, so that the linear approximation for the applied focusing force is valid. After that we will revisit the ellipsoidal bunch with parabolic line-charge density of the preceding section. We show that there exists a longitudinal phase-space distribution that satisfies the steady-state Vlasov equation, has a parabolic line-charge profile, and has linear forces over the entire range of possible emittance and space-charge parameters. This model can serve as an equivalent linear beam for the generally nonlinear longitudinal Maxwell–Boltzmann distribution. It thus plays the same role for the longitudinal beam physics as the K–V distribution does for the transverse phase space. Unfortunately, the K–V distribution cannot be extended self-consistently to six-dimensional phase space, as such an extension leads to a nonlinear space-charge force in the longitudinal direction (see Problem 5.12).

Let us now start with the derivation of the longitudinal equation of motion in rf fields. We will accomplish this task by considering the acceleration and longitudinal focusing process for a bunch of charged particles moving in a traveling electromagnetic wave. We ignore the details of mode structure, geometry, and radial variation as represented by the Bessel functions for cylindrical waveguides, and assume a TM wave having a longitudinal electric field component, which we write in the simple form

$$E_{az} = E_m \cos \varphi .$$
(5.375)

E_m is the peak electric field and φ represents the phase of the particle with respect to the peak field and is defined by the relation

$$\psi - \omega_{\rm rf} t - \omega_{\rm rf} \int_0^s \frac{ds}{v(s)} ,$$
(5.376a)

or

$$\varphi = \omega_{\rm rf} t - \frac{\omega_{\rm rf} s}{v_0}$$
(5.376b)

when $v(s) \approx v_0$ is approximately constant. Here $\omega_{\rm rf}$ is the angular frequency of the wave and $v(s)$ is the particle velocity, which is an increasing function of distance s when the particle is accelerated. We will assume that the phase velocity of the wave, v_p, increases with distance so that the particle at the center of the bunch moves in synchronism with the wave. Hence the velocity v_0 of this "synchronous" particle is the same as the phase velocity [i.e., $v_0(s) = v_p(s)$], and its phase φ_0 remains constant.

This simple picture of a traveling wave, which is illustrated in Figure 5.17, provides a good general description of the "smooth" longitudinal motion in rf linacs and synchrotrons with rf cavities. Factors such as the gap transit time in drift-tube linacs or the harmonic number $h = \omega_{\rm rf}/\omega_0$ between the rf and the orbital frequency ω_0 in synchrotrons can easily be taken into account by appropriate changes of the field amplitude E_m. For details we must refer to the various books on particle accelerators listed in the bibliography. In this section we consider the longitudinal motion in a linear accelerator. The extension to circular machines is treated in Section 5.4.9.

Figure 5.17(a) shows the accelerating field versus phase. The synchronous particle has phase φ_0 and momentum P_0, and its momentum change is defined by the equation of motion

$$\frac{dP_0}{dt} = mc \frac{d(\beta_0 \gamma_0)}{dt} = qE_m \cos \varphi_0 ,$$
(5.377)

where φ_0 remains constant and the space-charge force is zero at the center of the bunch. A nonsynchronous particle with phase φ and momentum $P = \beta \gamma mc$ will

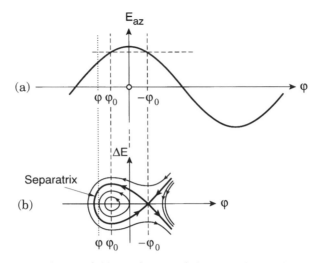

Figure 5.17. (a) Accelerating field as a function of phase; φ_0 denotes the synchronous phase (bunch centroid), φ a nonsynchronous particle; (b) particle trajectories in the $\Delta E - \varphi$ phase plane, including the "separatrix" which separates the stable from the unstable regions.

experience both the applied force $qE_m \cos \varphi$ and the space-charge force qE_{sz}, and its equation of motion is given by

$$\frac{dP}{dt} = mc \frac{d(\beta\gamma)}{dt} = qE_m \cos \varphi + qE_{sz}. \tag{5.378}$$

In Figure 5.17 the phases φ_0 and φ of the two particles considered are located to the left of the crest of the wave where the electric field is rising with time t. As we will see below, this is the region where the motion is stable. To a stationary observer, the nonsynchronous particle passes earlier in time than the synchronous particle. If we plot the wave as a function of distance, as is done in Figure 5.19, the two particles are both located to the right of the peak, and the nonsynchronous particle is ahead of the synchronous particle. These two different pictures regarding the phase φ of a particle as defined in Equation (5.376) must be kept in mind since we will now change from time t to distance s as the independent variable. It should be noted in this context that our definitions of phase (5.376) and electric field (5.375) are not unique. Quite often the phase is defined as $\varphi^* = \omega_{rf} s/v_0 - \omega_{rf} t$ or the electric field as $E_{az}^* = E_m \sin \varphi^*$, where we use the asterisk (*) to distinguish the two cases. The relevant equations can be readily converted to our notation by making the transformation $\varphi^* = -\varphi$ in the first case, or $\varphi^* = \varphi + (\pi/2)$ in the second case.

To proceed now with our analysis, it will be convenient to use energy E instead of momentum P. With $v_0 = ds/dt$, $dP/dt = (dE/dt)/v_0 = dE/ds$ from

Equation (2.25), we can write Equation (5.377) in the form

$$\frac{d}{ds}\frac{E_c}{mc^2} = \frac{d\gamma_0}{ds} = \frac{qE_m}{mc^2}\cos\varphi_0,$$

(5.379)

which determines the increase in energy E_c of the synchronous particle. The energy of a nonsynchronous particle differs from that of the synchronous particle by $\Delta E = E - E_c$, and the rate of change is obtained by subtracting (5.378) from (5.377), which yields

$$\frac{d}{ds}\frac{\Delta E}{mc^2} = \frac{d}{ds}(\gamma - \gamma_0) = \frac{qE_m}{mc^2}(\cos\varphi - \cos\varphi_0) + \frac{qE_{sz}}{mc^2}.$$

(5.380)

The phase difference $\Delta\varphi = \varphi - \varphi_0$ changes with the velocities $v = \beta c$ and $v_0 = \beta_0 c$ according to the relation

$$\frac{d}{ds}(\Delta\varphi) = \frac{d}{ds}(\varphi - \varphi_0) = \frac{\omega_{rf}}{c}\left(\frac{1}{\beta} - \frac{1}{\beta_0}\right).$$

(5.381)

Let us now assume that the energy difference ΔE is always very small compared to the energy of the synchronous particle (beam centroid) (i.e., $\Delta E = \Delta\gamma mc^2 \ll \gamma_0 mc^2$). Then, using the relation $\beta = (\gamma^2 - 1)^{1/2}/\gamma$, we obtain by Taylor expansion the result

$$\frac{1}{\beta} - \frac{1}{\beta_0} = -\frac{\gamma - \gamma_0}{(\gamma_0^2 - 1)^{3/2}} = -\frac{\Delta\gamma}{\beta_0^3\gamma_0^3}$$

(5.382)

to first order in the energy difference $\Delta\gamma mc^2$. Thus Equation (5.381) can be written in the alternative form

$$\frac{d}{ds}(\Delta\varphi) = -\frac{\omega_{rf}}{c}\frac{\Delta\gamma}{\beta_0^3\gamma_0^3} = -\frac{2\pi}{\lambda}\frac{\Delta E}{mc^2}\frac{1}{\beta_0^3\gamma_0^3} = -\frac{2\pi}{\lambda}\frac{\Delta P/P_0}{\beta_0\gamma_0^2},$$

(5.383)

where we introduced the wavelength $\lambda = 2\pi c/\omega_{rf}$ of the wave. By solving (5.383) for ΔE and substituting in (5.380), we obtain a single equation for the phase difference between a given particle and the bunch centroid:

$$\frac{d}{ds}\left[\beta_0^3\gamma_0^3\frac{d}{ds}(\Delta\varphi)\right] = -\frac{2\pi}{\lambda}\frac{qE_m}{mc^2}(\cos\varphi - \cos\varphi_0) + \frac{2\pi}{\lambda}\frac{qE_{sz}}{mc^2}.$$

(5.384)

If the motion is stable, the particles will oscillate in phase and energy about the synchronous particle (bunch centroid). To examine the stability conditions we assume that the energy $\gamma_0 mc^2$ changes adiabatically. The factor $\beta_0^3\gamma_0^3$ can then be treated as approximately constant during one phase oscillation period, so that Equation (5.384) becomes

$$\beta_0^3 \gamma_0^3 \frac{d^2}{ds^2} (\Delta\varphi) = - \frac{2\pi}{\lambda} \frac{qE_m}{mc^2} (\cos\varphi - \cos\varphi_0) + \frac{2\pi}{\lambda} \frac{qE_{sz}}{mc^2}. \tag{5.385}$$

Let us now temporarily neglect the space-charge force by setting $qE_{sz} = 0$. By multiplying both sides of Equation (5.385) with $d(\Delta\varphi)/ds$, we can integrate once and obtain

$$\frac{1}{2} \beta_0^3 \gamma_0^3 \left[\frac{d(\Delta\varphi)}{ds} \right]^2 = - \frac{2\pi}{\lambda} \frac{qE_m}{mc^2} (\sin\varphi - \varphi\cos\varphi_0 + C), \tag{5.386}$$

where C is an integration constant.

By substituting for $d(\Delta\varphi)/ds$ from Equation (5.383), we get

$$\frac{2\pi(\Delta E)^2}{2\beta_0^3 \gamma_0^3 \lambda mc^2} + qE_m(\sin\varphi - \varphi\cos\varphi_0 + C) = 0. \tag{5.387}$$

The constant C depends on the initial conditions $(\Delta E_i, \varphi_i)$ and is readily evaluated for any given set of the parameters $\beta_0\gamma_0$, λ, E_m, φ_0, and q/m. For each value of C, Equation (5.387) gives a possible trajectory in the $\Delta E - \varphi$ phase plane. Several such trajectories are shown in Figure 5.17. With the choice of the synchronous phase $\varphi_0 < 0$ in the figure we see that the particle motion is stable provided that the initial conditions are within the so-called *separatrix*. Inside the separatrix, particles move on closed curves in a counterclockwise direction, as illustrated in the figure. Particles whose initial phase and/or energy values are outside the separatrix will not be trapped and accelerated by the wave. They move on unstable trajectories similar to the one shown in Figure 5.17. Thus the separatrix, also known in the literature as the *rf bucket*, separates the stable from the unstable trajectories. As shown in Figure 5.17, the separatrix intersects the positive side of the φ-axis at the point $\varphi_{max} = -\varphi_0$, where $\varphi_0 < 0$ represents the synchronous phase. Setting $\varphi = -\varphi_0$, $\Delta E = 0$ in Equation (5.383) yields the value

$$C = \sin\varphi_0 - \varphi_0\cos\varphi_0 \tag{5.388}$$

that defines the trajectory for the separatrix. The value φ_{min} where the separatrix intersects the φ-axis on the negative side is found by numerical integration. For small values of the synchronous phase (i.e., $|\varphi_0| \ll \pi/2$), one finds that $\varphi_{min} \approx -2|\varphi_0|$, so that the phase difference $\Delta\varphi = \varphi - \varphi_0$ is to good approximation given by

$$-2|\varphi_0| \le \Delta\varphi \le |\varphi_0| \qquad \text{for } \varphi_0 \ll \frac{\pi}{2}. \tag{5.389}$$

As an example, for $\varphi_0 = -30°$, one finds $-60° \leq \Delta\varphi \leq 30°$, where $\varphi_{max} = 30°$ is exact and $|\varphi_{min}| - 60°$ is about 3% greater than the exact value.

When $\varphi_0 = -90°$, there is no net acceleration of the bunch, the constant C has the value $C = -1$, and the separatrix extends over the entire period of the rf wave, that is,

$$-\pi \leq \Delta\varphi \leq \pi. \tag{5.390}$$

The limiting values in the energy ΔE of the separatrix occur at the synchronous phase. By substituting $\varphi = \varphi_0$ and the relation (5.388) in (5.387), one obtains

$$\Delta E_{max} = -\Delta E_{min} = 2\left[\beta_0^3\gamma_0^3\frac{\lambda}{2\pi}mc^2qE_m(\varphi_0\cos\varphi_0 - \sin\varphi_0)\right]^{1/2}. \tag{5.391}$$

When $\varphi_0 = 0$, the area of the separatrix in the ΔE–φ plane shrinks to a point (i.e., there is no stable motion for a nonsynchronous particle). On the other hand, when $\varphi_0 = -90°$, the bucket size reaches a maximum with $\Delta\varphi$ defined by (5.390) and ΔE_{max} having the value

$$\Delta E_{max} = 2\left[\beta_0^3\gamma_0^3\frac{\lambda}{2\pi}mc^2qF_m\right]^{1/2} \quad \text{for } \varphi_0 = -\frac{\pi}{2}. \tag{5.392}$$

Figure 5.18 shows the shape of the separatrix and particle trajectories in the ΔE–$\Delta\varphi$ phase plane for two values of the synchronous phase. The case $\varphi_0 = -30°$ [Figure 5.18(a)] is typical for the acceleration regime. The separatrix in this case has a total width of $\Delta\varphi_{max} \approx 90°$, as discussed above, but the particle bunch usually occupies a much smaller phase-space area, like the little circle at the center

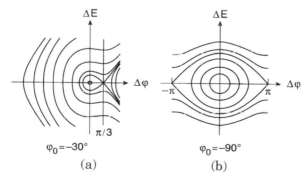

Figure 5.18. Separatrix and particle trajectories in ΔE–$\Delta\varphi$ phase plane for two values of the synchronous phase: (a) $-30°$; (b) $-90°$. Particle trajectories inside of the separatrix are closed stable orbits; those outside the separatrix are unstable.

of the separatrix. When $\varphi_0 = -90°$ [Figure 5.18(b)], the separatrix spans the entire phase range $-\pi \leq \Delta\varphi \leq \pi$, as shown in the picture on the right side. This regime is used to capture the beam from an injector, and the particles fill the entire 2π phase interval of the separatrix. Usually, the energy spread of the injected beam is considerably smaller than the height ΔE_{max} of the rf bucket. However, as is evident from the figure, there will always be some particles near the endpoints of the bucket $(-\pi, \pi)$ that will not be trapped if their energy ΔE is outside the separatrix boundary. A typical rf accelerator starts with $\varphi_0 = -90°$ to capture the injected beam, and then the synchronous phase is shifted adiabatically toward the acceleration point (e.g., $\varphi_0 = -30°$).

The shapes of the stable orbits inside the separatrix reflect the fact that the applied force is in general nonlinear with regard to the phase difference $\Delta\varphi$. However, if $\Delta\varphi \ll 1$, we can use the approximation $\cos(\varphi_0 + \Delta\varphi) - \cos\varphi_0 \approx -\Delta\varphi \sin\varphi_0$ and linearize the equation of motion. Thus, with $E_{sz} = 0$, Equation (5.385) becomes

$$\beta_0^3\gamma_0^3 \frac{d^2\Delta\varphi}{ds^2} = \frac{2\pi}{\lambda}\left(\frac{qE_m}{mc^2}\sin\varphi_0\right)\Delta\varphi. \tag{5.393}$$

This may be written in the harmonic-oscillator form

$$\frac{d^2(\Delta\varphi)}{ds^2} + k_l^2\,\Delta\varphi = 0, \tag{5.394}$$

where

$$k_l = \left[-\frac{2\pi qE_m\sin\varphi_0}{\lambda mc^2\beta_0^3\gamma_0^3}\right]^{1/2}, \tag{5.395a}$$

or

$$\omega_l = k_l\beta_0 c = \left[-\frac{\omega_{rf}qE_m\sin\varphi_0}{\beta_0\gamma_0^3 mc}\right]^{1/2}. \tag{5.395b}$$

Here k_l represents the wave number for the longitudinal phase oscillation, with wavelength $\lambda_l = 2\pi/k_l$; the corresponding oscillation frequency $\omega_l = k_l v_0$ is known as the *synchrotron frequency*. As can be seen from Equation (5.395), stable oscillations occur only for $\sin\varphi_0 < 0$, or $\varphi_0 < 0$, in agreement with our earlier discussion.

Equation (5.394) has the solution

$$\Delta\varphi = A\cos(k_l s + \alpha), \tag{5.396}$$

where A and α are defined by the initial conditions. The corresponding oscillations in the energy are obtained by differentiation of (5.396), which in view of (5.383) yields

$$\frac{d(\Delta\varphi)}{ds} = -Ak_l \sin(k_l s + \alpha) = -\frac{2\pi}{\lambda}\frac{\Delta E}{\beta_0^3 \gamma_0^3 mc^2},$$

or

$$\Delta E = B \sin(k_l s + \alpha), \qquad (5.397)$$

where

$$B = \frac{Ak_l \lambda \beta_0^3 \gamma_0^3 mc^2}{2\pi}. \qquad (5.398)$$

A particle with given phase amplitude A or given initial conditions $(\Delta E_i, \Delta\varphi_i)$ traces out an ellipse in ΔE–$\Delta\varphi$ phase space given by the equation

$$\frac{(\Delta\varphi)^2}{A^2} + \frac{(\Delta E)^2}{B^2} = 1, \qquad (5.399)$$

and the ratio of the semiaxes $(\Delta\varphi_m, \Delta E_m)$ of this ellipse is

$$\frac{\Delta E_m}{\Delta\varphi_m} = \frac{B}{A} = \frac{k_l \lambda \beta_0^3 \gamma_0^3 mc^2}{2\pi} = \left[\frac{qE_m \sin\varphi_0 \lambda \beta_0^3 \gamma_0^3 mc^2}{2\pi}\right]^{1/2}. \qquad (5.400)$$

We conclude from the above analysis of particle acceleration with rf fields that the linearity of the applied force is assured only if the bunch size is small compared to the phase width of the rf bucket. This linearity condition is usually satisfied in most rf machines during the acceleration regime. The notable exceptions are the injection from the source into an rf linac or from the rf linac into a synchrotron and the debunching (and rebunching) cycles in storage rings where the buckets can be completely filled. In these cases the longitudinal beam dynamics is highly nonlinear and becomes even more complicated when space-charge forces play a major role. Particles in the high-energy tail of the Maxwell–Boltzmann distribution with such a nonlinear potential function are then no longer confined and leak out of the rear of the separatrix. Furthermore, in the low-temperature limit the line-charge density profile is no longer parabolic with displacement from the beam centroid. This follows from the laminar-flow equilibrium condition, where the space-charge field completely cancels the applied field (i.e., $E_{sz} = E_{az}$) and the net potential is zero. We note that the applied potential in an rf bucket is reduced by the space-charge potential. As a result, the net bucket height shrinks and goes toward zero in the laminar-flow limit, and the net bucket area in ΔE–$\Delta\varphi$ phase space becomes a straight line in this limit. Further discussion of this topic is beyond the scope of this book. However, before we return to the initially stated goal of developing a linear model with space charge in this section, it will be useful to transform the above results from the $\Delta\varphi$, ΔE variables to the z, z' variables in which we described the longitudinal distribution earlier.

From the definition of phase in Equation (5.376b) and assuming adiabatic motion, we obtain the relation

$$\Delta\varphi = \varphi - \varphi_0 = -\frac{\omega_{rf}}{v_0}(s - s_0) = -\frac{\omega_{rf}}{v_0}z. \tag{5.401}$$

For the energy difference one finds that

$$\Delta E = mc^2\,\Delta\gamma = mc^2\frac{\Delta\beta}{\beta_0^2\gamma_0^3} = mc^2\frac{1}{\beta_0\gamma_0^3}\frac{dz}{ds}. \tag{5.402}$$

The difference in the applied force between a particle at position z in the bunch and the bunch centroid is given by

$$\Delta F_{az} = qE_m[\cos(\varphi_0 + \Delta\varphi) - \cos\varphi_0] = qE_m\left[\cos\left(\varphi_0 - \frac{\omega_{rf}z}{v_0}\right) - \cos\varphi_0\right]. \tag{5.403}$$

This equation may be integrated to get the potential function $\Delta U = -\int\Delta F_{az}\,dz$ for the applied force, which yields, with $\Delta U = 0$ at $z = 0$, the expression

$$\Delta U = qE_m\frac{v_0}{\omega_{rf}}\left[\sin\left(\varphi_0 - \frac{\omega_{rf}z}{v_0}\right) - \sin\varphi_0 + \frac{\omega_{rf}}{v_0}z\cos\varphi_0\right]. \tag{5.404}$$

The function is plotted schematically in Figure 5.19(b) versus the displacement $\omega_{rf}z/v_0$ from the synchronous phase. The accelerating field is shown on top (a) with $-\varphi$ pointing in the positive z-direction (i.e., the synchronous phase φ_0 is ahead of the crest of the wave). Figure 5.19(c) shows the separatrix and typical particle trajectories in a z–z' trace-space diagram. This diagram is basically a mirror image of the ΔE–φ plot in Figure 5.17, with distance z, rather than time t, being the independent variable on the abscissa. Figure 5.19(b) and (c) illustrate the longitudinal focusing potential and particle motion, respectively, as seen by an observer moving in a coordinate frame in which the synchronous particle is at the origin. Clearly, only particles whose total energy is less than a maximum value ΔU_{max} are trapped and accelerated by the wave. Note from Equation (5.404) that the slope $z' = \beta_0\gamma_0^3\,\Delta E/mc^2$ of a nonsynchronous particle in the moving frame

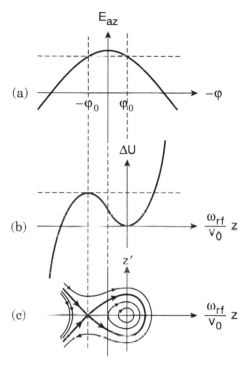

Figure 5.19. Accelerating electric field (a); potential (b) and trajectory diagram in $z-z'$ plane (c) as seen by an observer moving with a frame where the synchronous particle is at the origin.

Is proportional to the particle's difference in kinetic energy ΔE with respect to the synchronous particle as measured in the laboratory frame.

The general shape of the potential function $\Delta U(z)$ is highly nonlinear. However, near the origin (i.e., for particles whose oscillation amplitude is significantly smaller than the half width of the separatrix), the potential is harmonic ($\sim z^2$) and the focusing force is linear in z. By linear expansion of (5.403) we obtain for the applied force

$$\Delta F_{az} = \left(\frac{\omega_{rf}}{\beta_0 c} qE_m \sin \varphi_0 \right) z = -qE'_{az}z, \qquad (5.405)$$

where

$$E'_{az} = \frac{2\pi}{\lambda \beta_0} E_m | \sin \varphi_0 | \qquad (5.406)$$

is the field gradient defined as a positive quantity and $2\pi/\lambda = \omega_{rf}/c$.

With the aid of the relations (5.401) to (5.406) we can transform our equations of motion in an rf accelerator to the coordinates z, z' in the moving frame. Thus Equation (5.393) becomes

$$\beta_0^2 \gamma_0^3 \frac{d^2 z}{ds^2} = -q E'_{az} z ,$$
(5.407)

which is linear in the coordinate z.

Let us now assume that the space-charge force is also linear with a constant gradient E'_{sz}. By adding this linear space-charge force, Equation (5.407) can be written as

$$\frac{d^2 z}{ds^2} = z'' = - \frac{q E'_{az} z}{mc^2 \beta_0^2 \gamma_0^3} + \frac{q E'_{sz} z}{mc^2 \beta_0^2 \gamma_0^3} .$$
(5.408)

As we discussed in connection with the zero-temperature limit of the thermal distribution [Equation (5.369)], a parabolic line-charge density profile produces a linear space-charge force. We therefore assume for the desired linear-beam model that $\rho_L(z)$ is given by

$$\rho_L(z) = \rho_{L0} \left(1 - \frac{z^2}{z_m^2} \right) ,$$
(5.409)

so that in view of (5.366) we obtain

$$E'_{sz} = \frac{g}{4 \pi \epsilon_0 \gamma_0^2} \frac{2 \rho_{L0}}{z_m^2} .$$
(5.410)

The line-charge density $\rho_{L0} = \rho_L(0)$ at the bunch center can be related to the total number N of particles, or charge $Q = qN$, in the bunch and the half length z_m by Equation (5.353), and the rms width is given by

$$\tilde{z} = (\overline{z^2})^{1/2} = \frac{1}{\sqrt{5}} z_m = 0.447 z_m .$$
(5.411)

Using these relations one obtains for the equation of motion (5.408)

$$z'' = - \frac{q E'_{az}}{mc^2 \beta_0^2 \gamma_0^3} z + \frac{3}{2} \frac{g N r_c}{\beta_0^2 \gamma_0^5 z_m^3} z ,$$
(5.412)

where r_c is the classical particle radius. This equation has the desired linearity with the displacement z of the particle from the bunch center. We can write it in a form that is similar to the transverse force equations (4.111), namely

$$z'' + \kappa_{z0} z - \frac{K_L}{z_m^3} z = 0, \tag{5.413}$$

where

$$\kappa_{z0} = \frac{q E'_{az}}{mc^2 \beta_0^2 \gamma_0^3} \tag{5.414}$$

and

$$K_L = \frac{3}{2} \frac{g N r_c}{\beta_0^2 \gamma_0^5}. \tag{5.415}$$

Note that the longitudinal focusing function κ_{z0} has units of m^{-2}, as in the transverse case, while the *longitudinal perveance* parameter K_L has the unit of length [m], in contrast to the dimensionless perveance K for the transverse motion.

The equation for the longitudinal beam envelope z_m can be obtained by a procedure that is analogous to the transverse case [see Equation (4.178)]. Basically, this amounts to replacing z by z_m and adding an emittance term in the trajectory equation (5.413); one gets

$$z_m'' + \kappa_{z0} z_m - \frac{K_L}{z_m^2} - \frac{\epsilon_{zz'}^2}{z_m^3} = 0, \tag{5.416}$$

where $\epsilon_{zz'}$ represents the unnormalized total longitudinal emittance of the bunch (in the moving frame) enclosing the entire distribution of the particles given in (5.409). The applied longitudinal force κ_{z0} as well as the beam envelope z_m are in general functions of the distance s, usually of a periodic form that reflects the periodic traversal of acceleration and/or bunching gaps by the beam, as in the transverse case. For a continuously acting force, as when the bunch is propagating in a traveling wave or in the smooth approximation of a periodic system, we can replace κ_{z0} by the constant $k_{z0}^2 = (2\pi/\lambda_{z0})^2$, where $k_{z0} = k_l$. We then obtain from (5.416) for the matched-beam solution with $z_m'' = 0$ the fourth-order algebraic equation for the envelope

$$k_{z0}^2 z_m - \frac{K_L}{z_m^2} - \frac{\epsilon_{zz'}^2}{z_m^3} = 0, \tag{5.417a}$$

or

$$z_m^4 - \frac{K_L}{k_{z0}^2} z_m - \frac{\epsilon_{zz'}^2}{k_{z0}^2} = 0. \tag{5.417b}$$

When the space-charge force can be neglected ($K_L = 0$), the solution is

$$z_m = z_{m1} = \left(\frac{\epsilon_{zz'}}{k_{z0}}\right)^{1/2}.$$ (5.418)

On the other hand, when the emittance is negligible ($\epsilon_{zz'} = 0$) one gets

$$z_m = z_{m2} = \left(\frac{K_L}{k_{z0}^2}\right)^{1/3}.$$ (5.419)

The general solution of (5.417b) can be approximated by the relation [21]

$$z_m \approx \left(\frac{\epsilon_{zz'}^{3/2}}{k_{z0}^{3/2}} + \frac{K_L}{k_{z0}^2}\right)^{1/3},$$ (5.420)

which yields the correct results z_{m1} for $K_L = 0$ and z_{m2} for $\epsilon_{zz'} = 0$, and for the case where both terms are nonzero, it deviates by no more than $+3.4\%$ from the exact solution. Most important, this relation exhibits the scaling of the beam envelope with the beam parameters ($\epsilon_{zz'}$, K_L) and the longitudinal focusing strength (k_{z0}).

The effects of space charge on the longitudinal focusing of the bunch can be described by introducing the wave constant k_z, defined as

$$k_z = \left(k_{z0}^2 - \frac{K_L}{z_m^3}\right)^{1/2},$$ (5.421)

so that (5.417a) becomes

$$k_z^2 z_m = \frac{\epsilon_{zz'}^2}{z_m^3}.$$ (5.422)

In analogy to the transverse motion, we can define a longitudinal tune depression by

$$\frac{k_z}{k_{z0}} = \left(1 - \frac{K_L}{z_m^3 k_{z0}^2}\right)^{1/2},$$ (5.423a)

or

$$\frac{k_z}{k_{z0}} = \left(1 + \frac{K_L z_m}{\epsilon_{zz'}^2}\right)^{-1/2}.$$ (5.423b)

For a system with a constant or smooth force, where the last nine equations hold, the longitudinal Hamiltonian $H_\parallel$ is a constant of the motion, and hence, any distribution that is a function of $H_\parallel$ satisfies the stationary longitudinal Vlasov equation. Defining the Hamiltonian as

$$H_\parallel = \frac{1}{2}\left(k_{z0}^2 - \frac{K_L}{z_m^3}\right)z^2 + \frac{1}{2}z'^2 = \frac{1}{2}k_z^2 z^2 + \frac{1}{2}z'^2, \tag{5.424}$$

Neuffer [22] showed that

$$f(H_\parallel) = f_{\parallel 0}\sqrt{2(H_{\max} - H_\parallel)} = f_\parallel\sqrt{k_z^2 z_m^2 - k_z^2 z^2 - z'^2} \tag{5.425}$$

for $0 < H_\parallel < H_{\max}$ and zero elsewhere produces the line-charge density profile (5.409) and hence the desired linear equation of motion (5.413).

The corresponding distribution for the nonuniform (e.g., periodic) case, where the focusing function is $\kappa_{z0}(s)$ and varies with distance s so that $H_\parallel$ is no longer a constant of the motion, is given by [22]

$$f(z, z', s) = \frac{3N}{2\pi\epsilon_0}\sqrt{1 - \frac{z^2}{z_m^2} - \frac{z_m^2}{\epsilon_{zz'}^2}\left(z' - \frac{z_m'}{z_m}z\right)^2}. \tag{5.426}$$

The quadratic function under the square root is just a special form of the equation of the emittance ellipse in z–z' trace space. Thus, in analogy with the transverse K V beam, this longitudinal distribution is a function of the emittance $\epsilon_{zz'}$, and since $\epsilon_{zz'}$ is an invariant when there is no acceleration and the forces are linear, the distribution satisfies the time-independent Vlasov equation. From the customary Courant–Snyder relation

$$\hat{\gamma}z^2 + 2\hat{\alpha}zz' + \hat{\beta}z'^2 = \epsilon_{zz'} \tag{5.427}$$

one obtains by multiplication with $\hat{\beta}$ and with the substitutions $\hat{\beta}\hat{\gamma} = 1 + \hat{\alpha}^2$, $\hat{\beta}\epsilon_{zz'} = z_m^2$,

$$\frac{z^2}{z_m^2} + \frac{\hat{\beta}^2}{z_m^2}\left(z' + \frac{\hat{\alpha}}{\hat{\beta}}z\right)^2 = 1,$$

and finally, with $z_m' = -(\hat{\alpha}/\hat{\beta})z_m$ (see Figure 3.26),

$$\frac{z^2}{z_m^2} + \frac{z_m^2}{\epsilon_{zz'}^2}\left(z' + \frac{z_m'}{z_m}z\right)^2 = 1. \tag{5.428}$$

The longitudinal beam envelope z_m and its slope z_m' can be determined by solving the envelope equation (5.416) for given initial conditions and parameter values.

We see that the Neuffer distribution plays the same role for the longitudinal motion as the K–V distribution for the transverse motion. Like the K–V distribution, it yields linear forces over the entire parameter regime from a laminar beam ($\epsilon_{zz'} = 0$) to an emittance-dominated beam ($K_L = 0$). It therefore can be used as an equivalent analytical beam to model the longitudinal behavior of laboratory beams or of the nonanalytical Maxwell–Boltzmann distribution.

For comparison of different distributions it is desirable to use rms quantities such as rms beam envelope and rms emittance, as in the transverse case. With $z_m = \sqrt{5}\,\tilde{z}$, $\epsilon_{zz'} = 5\tilde{\epsilon}_{zz'}$ the envelope equation (5.416) takes the rms form

$$\tilde{z}'' + \kappa_{z0}\tilde{z} - \frac{K_L}{5\sqrt{5}\,\tilde{z}^2} - \frac{\tilde{\epsilon}_{zz'}^2}{\tilde{z}^3} = 0 \tag{5.429}$$

(i.e., one has a factor of $5\sqrt{5} \approx 11.18$ in the denominator of the space-charge term). By comparison, the space-charge term in the transverse envelope equation is $K/4\tilde{x}$, from Equation (5.218), for a round beam ($\tilde{x} = \tilde{y}$). Furthermore, in the longitudinal case, the space-charge term varies with the inverse square of the rms width ($\sim \tilde{z}^{-2}$), whereas it is inversely proportional to the width ($\sim \tilde{x}^{-1}$) in the transverse case.

5.4.9 Longitudinal Beam Dynamics in Circular Machines

The preceding analysis of the properties of the longitudinal distribution and the longitudinal beam dynamics (Sections 5.4.6 to 5.4.8) was based on the propagation of the beam in a straight channel such as a linear accelerator. In this section we extend our model to a circular accelerator such as a synchrotron or a storage ring. As we know from the discussion of the negative-mass effect in Section 3.6.4, there is a fundamental difference in the longitudinal dynamics between circular beams and straight beams. We will show, however, that this difference can be accounted for readily by introducing the effective mass $m^* = -\gamma_0 m/\eta$ and the associated slip factor η into the equations and relationships that we have derived so far for straight beams. As will be seen, the mathematical form remains the same, the relations are simply generalized to include circular motion, and the straight-beam results are recovered as a special case of the more general theory.

According to Equation (3.261) in Section 3.6.4, two particles with different momentum orbiting in a circular machine have different angular frequencies and revolution times. Consequently, their relative position will change with time in a way that depends on the slip factor η. This is illustrated schematically in Figure 5.20 for a machine that operates in the negative-mass regime. Let $\dot{\theta}_0 = \omega_0$ be the angular frequency, P_0 the momentum of the synchronous reference particle (A) and $\dot{\theta}_0 + \Delta\dot{\theta}$, and $P_0 + \Delta P$ the angular frequency and momentum of a nonsynchronous particle (B). Equation (3.261) may then be written in the form

$$\frac{\Delta\dot{\theta}}{\dot{\theta}_0} = -\eta \frac{\Delta P}{P_0}. \tag{5.430}$$

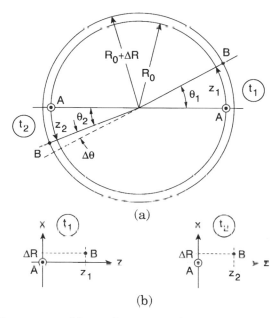

(a)

(b)

Figure 5.20. Relative motion of the synchronous particle (A) and a nonsynchronous particle (B) in a circular machine operating in the negative mass regime: (a) orbital motion; (b) relative position in the moving frame.

The slip factor is defined as $\eta = \alpha - 1/\gamma_0^2$ or $\eta - 1/\gamma_t^2 - 1/\gamma^2$ according to (3.262a) and (3.262b), where $\gamma_t mc^2$ is the transition energy and α the momentum compaction factor related to the horizontal betatron tune by $\alpha \approx 1/\nu_x^2$. This equation may be integrated with time to yield the change $\Delta\theta$ in the relative angular position of the two particles, namely

$$\Delta\theta = -\dot{\theta}_0 t \eta \frac{\Delta P}{P_0} = -\theta_0 \eta \frac{\Delta P}{P_0}, \qquad (5.431)$$

where we introduced the angular position of the synchronous particle $\theta_0(t) = \dot{\theta}_0 t$. When the slip factor η is positive, we are dealing with negative-mass behavior; that is, $\Delta\theta$ is negative, and the angular separation between the two particles decreases. Figure 5.20 shows the change in the relative angular position of the two particles after one half revolution for this case. Suppose that $s_0(t) = \dot{\theta}_0 t = R_0\theta_0$ is the distance traveled by the synchronous particle (A) and $s(t) = (\dot{\theta}_0 + \Delta\dot{\theta})t = R_0(\theta_0 + \Delta\theta)$ is the corresponding distance traveled by the nonsynchronous particle (B) in time t. In a coordinate system moving with the velocity $\dot{s}_0 = R_0\dot{\theta}_0$ of the synchronous particle and centered at A, particle B has the longitudinal position

$$z(t) = s(t) - s_0(t) = R_0\,\Delta\theta(t) \qquad (5.432)$$

and velocity

$$\dot{z}(t) = \dot{s}(t) - \dot{s}_0(t) = R_0 \Delta\dot{\theta} = \Delta v_z. \qquad (5.433)$$

Note that this moving coordinate system is not identical to the beam frame used in Section 5.4.3 for relativistic particles. The coordinate $z(t)$ and all other quantities ($\dot{z}$, ΔP_z, etc.) are measured in laboratory units, whereas the position z_b in the true beam frame, for instance, is related to z by the Lorentz transformation, $z_b = \gamma_0 z$, with $\gamma_0 = [1 - (v_0/c)^2]^{1/2}$ and $v_0 = \dot{s}_0$. Nonrelativistically, of course, there is no difference between z_b and z since $\gamma_0 = 1$ in this case.

In view of (5.432), the relationship between the relative velocity difference $\Delta v_z/v_0$ and the relative momentum difference $\Delta P_z/P_0$ between a nonsynchronous particle B and the reference particle A is given by

$$z' = \frac{\Delta v_z}{v_0} = -\eta \frac{\Delta P_z}{P_0}, \qquad (5.434)$$

where $z' = dz/ds$ is the slope of the trajectory. Since $P_0 = \gamma_0 m v_0$, this may be written as

$$\dot{z} = v_0 z' = \Delta v_z = -\frac{\eta}{\gamma_0 m}\Delta P_z. \qquad (5.435)$$

Introducing the effective mass $m^* = -\gamma_0 m/\eta$ from Equation (3.265), we obtain the relationship

$$\Delta P_z = m^* \Delta v_z, \qquad (5.436)$$

between ΔP_z and Δv_z, the momentum and velocity, respectively, of the nonsynchronous particle in the moving frame; this is identical to Equation (3.264). In the negative-mass regime ($\eta > 0$, $m^* < 0$) the distance between a particle with greater momentum and the reference particle decreases. This is shown in Figure 5.20, where after one half revolution the longitudinal position of particle B has changed by the amount $\Delta z = z_2 - z_1 = -R_0 \Delta\theta$, where $\Delta\theta = \Delta\dot{\theta}(t_2 - t_1)$. Clearly, particle B has a negative velocity $\dot{z} = \Delta v_z < 0$ in the moving frame [which can be calculated from Equation (5.435)], even though its momentum is greater than that of the synchronous particle A.

The above relations governing the longitudinal motion in a circular machine are readily applied to a straight beam. By setting $\alpha = 0$, or $\gamma_t \to \infty$, we recover the relations $m^* = \gamma_0^3 m$, $\Delta P_z = \gamma^3 m \Delta v_z$, $\Delta P_z/P_0 = (1/\gamma_0^2)\Delta v_z/v_0$, and so on, that we used in the preceding sections. The only question for which the answer is not so obvious is how to define the longitudinal temperature in a circular machine.

However, if we go back to the basic definition of temperature as a measure of the random (thermal) part of the velocity distribution, we are led to the relation

$$k_B T_\| = m^* \overline{\Delta v_{z,\text{th}}^2} = -\frac{\gamma_0 m}{\eta} \overline{\Delta v_{z,\text{th}}^2}.$$

(5.437)

In a stationary beam the thermal rms velocity spread, $\widetilde{\Delta v}_{z,\text{th}} = [\overline{\Delta v_{z,\text{th}}^2}]$ is identical to the total rms velocity spread, $\widetilde{\Delta v}_z$. Otherwise, it is defined by the difference between the total rms velocity spread and the rms flow velocity $\widetilde{\Delta v}_{z,\text{fl}}$ in the moving frame; that is,

$$\overline{\Delta v_{z,\text{th}}^2} = \widetilde{\Delta v}_{z,\text{th}}^2 = \widetilde{\Delta v}_z^2 - \widetilde{\Delta v}_{z,\text{fl}}^2,$$

(5.438)

in accordance with Equation (5.310).

Our generalized definition (5.437) implies that the effective longitudinal temperature is a negative quantity when the effective mass is negative ($m^* < 0$), which occurs for $\eta > 0$. This has the consequence that there is no three-dimensional thermal equilibrium in the negative-mass regime, as we discuss in connection with intrabeam scattering in Section 6.4.2. Under negative-mass conditions, the space-charge force is actually focusing, thereby increasing the longitudinal bunching and the negative longitudinal temperature of the beam. This is the source of the negative-mass instability discussed in Section 6.3.3.

Another peculiarity of particle motion in a circular machine occurs at the transition energy where $\gamma_0 = \gamma_t$ or $\gamma_0 = 1/\sqrt{\alpha}$. In this case all particle orbits are isochronous; that is, they have the same revolution time regardless of the relative momentum. The longitudinal particle motion thus "freezes," the flow is laminar, and the longitudinal temperature is zero, which follows from Equations (5.437) and (5.438) with $m^* \to \infty$. The beam is extremely sensitive to small perturbations in this regime, which is why special design features are implemented to pass very quickly through the transition point in a circular machine. In the positive mass regime below transition, the behavior of the particle distribution is, of course, similar to that in a straight channel.

Let us now turn our attention to the longitudinal equation of motion in a circular machine, with rf cavities providing the acceleration and longitudinal focusing. Since the motion is usually adiabatic (i.e., changes occur very slowly compared with the revolution time), we can treat the electric field as a smooth, continuously acting function. If V_m is the peak voltage gain per turn and $\overline{R}$ the average orbit radius, we can express the peak longitudinal electric field as

$$E_{zm} = \frac{V_m}{2\pi \overline{R}}.$$

(5.439)

In place of (5.414) or (5.395), we obtain for the synchrotron oscillation constant $k_{z0} = (\overline{\kappa_{z0}})^{1/2}$ in a circular machine the relation

$$k_{z0} = \frac{v_{z0}}{R} = \frac{\omega_{z0}}{v_0} = \left(\frac{2\pi}{\lambda} \frac{q E_m \eta \sin \varphi_0}{\beta_0^3 \gamma_0 mc^2} \right)^{1/2}, \tag{5.440}$$

while the space-charge parameter is given by

$$K_L = -\frac{3}{2} \frac{g N r_c}{\beta_0^2 \gamma_0^3} \eta. \tag{5.441}$$

With the last two relations the linearized trajectory equation (5.412) may be written as

$$z'' = \frac{2\pi}{\lambda} \frac{q E_m \eta \sin \varphi_0}{\beta_0^3 \gamma_0 mc^2} z + \frac{3}{2} \frac{g N r_c \eta}{\beta_0^2 \gamma_0^3 z_m^3} z, \tag{5.442a}$$

or

$$z'' + k_{z0}^2 z - \frac{K_L}{z_m^3} z = 0, \tag{5.442b}$$

while the longitudinal envelope equation for a parabolic bunch in a circular machine becomes

$$z_m'' + \frac{2\pi}{\lambda} \frac{q E_m \eta \sin \varphi_0}{\beta_0^3 \gamma_0 mc^2} z_m - \frac{3}{2} \frac{g N r_c \eta}{\beta_0^2 \gamma_0^3 z_m^2} - \frac{\epsilon_{zz'}^2}{z_m^3} = 0, \tag{5.443a}$$

or

$$z_m'' + k_{z0}^2 z_m - \frac{K_L}{z_m^2} z - \frac{\epsilon_{zz'}^2}{z_m^3} = 0. \tag{5.443b}$$

Here $\epsilon_{zz'}$ is the unnormalized total emittance of the bunch in the moving frame, as in the case of a straight channel. It is related to ϵ_z and ϵ_{nz} by the relation

$$\epsilon_{zz'} = \epsilon_z |\eta| = \frac{\epsilon_{nz} |\eta|}{\beta_0 \gamma_0}. \tag{5.444}$$

When $\eta = 0$ (transition energy), $\epsilon_{zz'}$ is zero, as expected for the laminar flow in this case.

We note that our previous straight-beam results can be recovered from the above relations by setting $\eta = -1/\gamma_0^2$; the synchronous phase must be negative ($\varphi_0 < 0$) in this case to get focusing ($k_{z0} > 0$), as expected. Below transition ($\gamma_0 < \gamma_t$) the slip factor is negative ($\eta < 0$); hence $k_{z0}^2 > 0$ provided that $\varphi_0 < 0$ and $K_L > 0$, so that mathematically the situation is perfectly analogous to that in a linear accelerator. However, in the negative mass regime above transition ($\gamma_0 > \gamma_t$) the slip factor is positive ($\eta > 0$), so that longitudinal focusing ($k_{z0}^2 > 0$) requires a shift of the synchronous phase from a negative to a positive value ($\varphi_0 > 0$). Furthermore, we get the interesting result mentioned earlier that the space-charge force is focusing since the perveance term is negative ($K_L < 0$) in this case. The accelerating force experienced by a particle at the front end of the bunch due to the space-charge electric field increases the particle kinetic energy. This in turn increases the particle's orbit radius, which slows down its angular motion and hence decreases its distance z from the bunch center. This longitudinal focusing effect of the space-charge force in the negative-mass regime is opposite to the usual defocusing action by the space charge below transition in linear accelerators and in the transverse direction.

The relation for the synchrotron oscillation wave constant with space charge is formally the same as Equation (5.421). However, we can call the ratio k_z/k_{z0} "tune depression" only below transition. In the negative-mass regime, $k_z/k_{z0} = \nu_z/\nu_{z0}$ is in fact greater than unity. The synchrotron tune in the presence of space charge is given by

$$\nu_z^2 = \nu_{z0}^2 - \frac{K_L \overline{R}^2}{z_m^2} = \nu_{z0}^2 + \frac{3}{2} \frac{gNr_c \overline{R}^2 \eta}{\beta_0^2 \gamma_0^3 z_m^2}. \tag{5.445}$$

For a small difference $|\nu_z^2 - \nu_{z0}^2|$ we obtain the tune-shift relation

$$\Delta \nu_z - \frac{3}{4} \frac{gNr_c R^2 \eta}{\beta_0^2 \gamma_0^3 z_m^2 \nu_{z0}}. \tag{5.446}$$

Note that the longitudinal tune shift due to space charge has the same sign as η; that is, it is negative below transition and in linear accelerators, and positive in the negative-mass regime above transition. All of the equations and relationships presented so far in this section are valid only for the parabolic beam model, where both the space-charge force and the focusing force are linear functions of the particle position z. When the space-charge force is nonlinear we must use the rms envelope equation (5.429) and rms values for all relevant quantities. The stationary longitudinal Maxwell–Boltzmann (thermal) distribution (5.312) for circular machines can be written in terms of the effective mass m^* as

$$f_\parallel(z, \Delta v_z) = f_{\parallel 0} \exp\left[-\frac{m^*(\Delta v_z)^2 + q\phi_\parallel(z)}{k_B T_\parallel} \right], \tag{5.447a}$$

or in view of (5.434) as

$$f_\parallel(z, z') = f_{\parallel 0} \exp\left[-\frac{m^* v_0^2 z'^2 + q\phi_\parallel(z)}{k_B T_\parallel} \right], \tag{5.447b}$$

where $\phi_\parallel(z)$ includes both the applied focusing as well as the space-charge potential.

The unnormalized longitudinal rms emittance for the stationary distribution in $z-z'$ space is given by

$$\tilde{\epsilon}_{zz'} = \tilde{z}\tilde{z}', \tag{5.448}$$

and $\tilde{z}'$ can be related to the longitudinal temperature $T_\parallel$ with the aid of (5.437) by

$$\tilde{z}' = \left(\frac{\overline{\Delta v_z^2}}{v_0^2}\right)^{1/2} = \left(\frac{|k_B T_\parallel|}{|m^*|v_0^2}\right)^{1/2} = \left(\frac{|k_B T_\parallel \eta|}{\gamma_0 \beta_0^2 mc^2}\right)^{1/2}, \tag{5.449}$$

so that

$$\tilde{\epsilon}_{zz'} = \tilde{z}\left(\frac{|k_B T_\parallel \eta|}{\gamma_0 \beta_0^2 mc^2}\right)^{1/2}. \tag{5.450}$$

The relation between $\tilde{\epsilon}_{zz'}$, $\tilde{\epsilon}_z$, and $\tilde{\epsilon}_{nz}$ is analogous to (5.444), that is,

$$\tilde{\epsilon}_{zz'} = |\eta|\tilde{\epsilon}_z = \frac{|\eta|}{\beta_0 \gamma_0}\tilde{\epsilon}_{nz}, \tag{5.451}$$

and the normalized longitudinal rms emittance $\tilde{\epsilon}_{nz}$ can be expressed in terms of the temperature $T_\parallel$ as

$$\tilde{\epsilon}_{nz} = \tilde{z}\left(\frac{\gamma_0 |k_B T_\parallel|}{|\eta|mc^2}\right)^{1/2}. \tag{5.452}$$

Note that when $\eta = -1/\gamma_0^2$ we recover the earlier relations (5.316) and (5.317) for a linear accelerator.

For a stationary (matched) beam we have $\tilde{z}' = k_z\tilde{z}$ and get with (5.422) for the emittance the alternative expressions

$$\tilde{\epsilon}_{zz'} = k_z\tilde{z}^2 \tag{5.453}$$

and

$$\tilde{\epsilon}_{nz} = \frac{\beta_0 \gamma_0}{|\eta|}k_z\tilde{z}^2. \tag{5.454}$$

When space charge is negligible so that $K_L = 0$ and $k_z = k_{z0}$, the normalized emittance becomes

$$\tilde{\epsilon}_{nz} = \frac{\beta_0 \gamma_0}{|\eta|}k_{z0}\tilde{z}^2, \tag{5.455}$$

or with (5.440),

$$\tilde{\epsilon}_{nz} = \tilde{z}^2 \left[\frac{2\pi}{\lambda} \frac{qE_m \gamma_0 |\sin \varphi_0|}{\beta_0 |\eta| mc^2} \right]^{1/2}. \tag{5.456}$$

Since $\tilde{\epsilon}_{nz}$ = const under ideal conditions we find for a circular accelerator where E_m, λ, and $\sin \varphi_0$ are constant that the rms width of the bunch in this case scales as

$$\tilde{z} = \text{const} \left[\frac{|\eta| \beta_0}{\gamma_0} \right]^{1/4}. \tag{5.457}$$

In an rf linac operating under the same conditions, we have $|\eta| = 1/\gamma_0^2$, so that $\tilde{z}$ scales as

$$\tilde{z} = \text{const} \frac{\beta_0^{1/4}}{\gamma_0^{3/4}}. \tag{5.458}$$

For an induction linac one must use the average field gradient E'_{az} and $k_{z0} = \sqrt{\kappa_{z0}} = (qE'_{az}/mc^2\beta_0^2\gamma_0^3)^{1/2}$ as defined in Equation (5.414). If E'_{az} = const, the rms bunch length in this case then scales as

$$\tilde{z} = \text{const } \gamma_0^{-5/4}. \tag{5.459}$$

The relative rms momentum spread $\widetilde{\Delta P}/P_0$ scales as

$$\frac{\widetilde{\Delta P}}{P_0} = \frac{\tilde{\epsilon}_{nz}}{\tilde{z} \beta_0 \gamma_0} = \frac{\text{const}}{|\eta|^{1/4} \beta_0 \gamma_0^{3/4}} \tag{5.460}$$

in the circular machine, and as

$$\frac{\widetilde{\Delta P}}{P_0} = \frac{\text{const}}{\beta_0 \gamma_0^{1/4}} \tag{5.461}$$

in the rf linac.

From Equations (5.452) and (5.457) we find that the longitudinal temperature for the assumed parameter regime scales as

$$k_B T_{\parallel} = -\text{const} \frac{\eta}{\gamma_0^{1/2}} \tag{5.462}$$

for a circular machine, and as

$$k_B T_\| = \text{const} \, \frac{1}{\gamma_0^{5/2}} \tag{5.463}$$

for a linear accelerator.

One must keep in mind that these scaling laws [Equations (5.455) to (5.463)] apply only for the case where the space charge is negligible ($K_L = 0$). Note that in the circular machine the longitudinal temperature decreases with energy from positive values below transition, passes through zero at the transition point, and then becomes negative, in agreement with our earlier discussion.

The above scaling relations for a circular machine are, strictly speaking, valid only when the energy is not close to the transition point (i.e., when $|\eta| \neq 0$). At the transition energy we would get $\tilde{z} \to 0$ and $\widetilde{\Delta P}/P_0 = \to \infty$, which is unphysical since the normalized emittance $\tilde{\epsilon}_{nz}$ must remain constant (unlike $\tilde{\epsilon}_{zz'}$, which does go to zero, as discussed earlier). Proper treatment of the problem for the case where the beam passes through the transition point or remains at transition, as in the isochronous cyclotron, reveals that $\tilde{z}$ and $\widetilde{\Delta P}/P_0$ remain finite.

In the space-charge-dominated regime of the longitudinal motion, where $\epsilon_{zz'}$ is negligible, we have from (5.443) for a matched parabolic beam ($z_m'' = 0$) the relation

$$K_L = k_{z0}^2 z_m^3, \tag{5.464}$$

which yields the scaling

$$z_m = \left(\frac{K_L}{k_{z0}^2} \right)^{1/3} = \left[\frac{3}{2} \, \frac{g N r_c}{\gamma_0^2} \, \frac{\lambda \beta_0 mc^2}{2\pi q E_m |\sin \varphi_0|} \right]^{1/3}. \tag{5.465}$$

For $q E_m |\sin \varphi_0| \gamma_0^2 / \lambda \beta_0 = \text{const}$, we find from this relation that the beam envelope z_m scales with the number of particles in the bunch, N, as

$$z_m = \text{const} \, N^{1/3} \tag{5.466a}$$

or, alternatively, since the average beam current $\bar{I}$ is proportional to N,

$$z_m = \text{const} \, \bar{I}^{1/3}, \tag{5.466b}$$

a result that was recently confirmed experimentally [23].

5.4.10 Effects of Momentum Spread on the Transverse Distribution

In our analysis of the behavior of the transverse and longitudinal distributions in the preceding section we tacitly assumed that these distributions are independent of each other. In fact, however, there is mutual coupling between the two distributions via momentum spread, the collective space-charge forces, and Coulomb collisions. In this section we study only the effects of longitudinal momentum spread on the transverse motion for both continuous or bunched beams. The coupling through space charge, which occurs in the case of bunched beams, is treated in Section 5.4.11, and Coulomb collisions are reviewed in Section 6.4.

The first effect caused by momentum spread is known as *chromatic aberration*. It is due to the fact that the strength of transverse focusing, that is, the focal length f of the discrete lenses or the focusing functions $\kappa_0(z)$ of arrays of lenses and periodic lattices, depend on the momentum of the particles. The chromatic aberration in a single lens was discussed in Section 3.4.6 and the change in the betatron oscillation frequency due to momentum spread in a circular machine with gradient n in Section 3.6.4. We can generalize the results obtained there to any focusing channel, whether straight or circular, by defining the relative *chromaticity parameter* ξ in terms of the general focusing function $\kappa_0(z)$ as

$$\xi = \frac{d\sqrt{\kappa_0}/\sqrt{\kappa_0}}{dP/P_0}.$$
(5.467a)

For a uniform channel with focusing strength defined by the wave number k_0 this relation becomes

$$\xi = \frac{dk_0/k_0}{dP/P_0},$$
(5.467b)

which for circular machines may be written in terms of the tune ν_0 in the form of Equation (3.268), that is,

$$\xi = \frac{d\nu_0/\nu_0}{dP/P_0}.$$
(5.467c)

It should be noted that this definition of chromaticity is not unique. Many authors prefer to define the chromaticity in terms of the absolute, rather than the relative, change of focusing parameter. Thus, in lieu of Equation (5.467c), one has

$$\xi^* = \frac{d\nu_0}{dP/P_0},$$
(5.468)

where we used the asterisk (*) to indicate the difference in the definition. Equation (3.271) gives the relations for the chromaticity parameters in a constant-gradient field. For $n = 0.5$, one obtains the values $\xi_r = -1.5$ for the radial and $\xi_z = 1.5$ for the vertical chromaticity parameters. In the case of a solenoid channel, where according to (4.86b)

$$
k_0 = \frac{\omega_L}{v} = \frac{qB}{2mc\beta\gamma} = \frac{qB}{2P},
$$

one gets

$$
\xi = \frac{dk_0/k_0}{dP/P_0} = -1. \tag{5.469}
$$

For magnetic quadrupoles, which make by far the largest contributions to chromaticity in a ring lattice, one obtains the relations

$$
\xi_x^* = -\frac{1}{4\pi} \int_0^C \hat{\beta}_{x0}(s)\kappa_{x0}(s)\, ds \tag{5.470a}
$$

for the horizontal motion and

$$
\xi_x^* = -\frac{1}{4\pi} \int_0^C \hat{\beta}_{y0}(s)\kappa_{y0}(s)\, ds \tag{5.470b}
$$

for the vertical motion. The betatron function $\hat{\beta}_0(s)$ is always positive, while the focusing function κ_0 varies periodically between positive (focusing) and negative (defocusing) values. In the focusing plane where $\hat{\beta}_0$ reaches a maximum, the function κ_0 is positive and also at its maximum, so that the chromaticity is negative. On the other hand, in the defocusing plane, $\hat{\beta}_0$ has its minimum while κ_0 is negative, so that ξ^* is positive at these positions. Since the large negative chromaticity values at the focusing planes outweigh the small positive values at the defocusing planes, the integrals in (5.470) representing the average chromaticity over one revolution are always negative.

Although the change of the betatron oscillation frequency due to momentum spread is relatively small, it can cause emittance growth. More important, however, in high-energy storage rings, the effect is responsible for the *head–tail instability*. The variation in betatron frequencies and the accumulated phase difference between head and tail particles drive this instability, whose growth rate is proportional to ξ and the number N of particles in the bunch (for a discussion of the effect, see [D.10, Sect. 6.4.3]). Thus there is a strong reason to reduce the chromaticity effect and ideally, to avoid it altogether. The method for compensating the chromaticity effect is to use sextupole magnets, which are usually placed at the locations of the quadrupoles in the FODO system of the typical accelerator lattice. In cylindrical coordinates the field components of a magnetic sextupole vary as

$$B_x = B''r^2 \sin 3\theta, \quad\quad B_y = B''r^2 \cos 3\theta, \quad\quad (5.471a)$$

and in cartesian coordinates as

$$B_x = B''xy, \quad\quad B_y = \frac{1}{2}B''(x^2 - y^2). \quad\quad (5.471b)$$

Here $B'' = (\partial^2 B_y/dx^2)$ is the second derivative, which in an ideal sextupole field can be equated with the ratio of the pole tip field B_0 and the pole tip "radius" squared, a_q^2, by analogy with the quadrupole case in Section 3.5.

Another longitudinal-transverse effect that occurs only in circular systems, where it is even more important than chromatic aberration, is *dispersion*. As discussed in Section 3.6.4, particles with a momentum differing from that of the synchronous particle, P_0, by an amount ΔP, have a different closed (equilibrium) orbit. The horizontal (radial) displacement of this equilibrium orbit from that of the synchronous particle can be written as

$$x_e(s) = D_e(s)\frac{\Delta P}{P_0}, \quad\quad (5.472)$$

where $D_e(s)$ is referred to as the *dispersion function*. The total displacement of a particle from the central orbit can be expressed as the sum of x_e and the betatron oscillation amplitude x_b [see (Equation (3.273)], that is,

$$x(s) = x_b(s) + x_e(s), \quad\quad (5.473a)$$

$$x'(s) = x_b'(s) + x_e'(s), \quad\quad (5.473b)$$

where $x_e' = D_e'\Delta P/P_0$ is proportional to the derivative $D_e'(s)$ of the dispersion function with respect to distance s.

Since the two effects are statistically uncorrelated, they add quadratically (i.e., $\tilde{x}^2 = \tilde{x}_b^2 + \tilde{x}_e^2$), so that the total rms width of the beam is

$$\tilde{x} = (\tilde{x}_b^2 + \tilde{x}_e^2)^{1/2}. \quad\quad (5.474)$$

As mentioned in Section 3.6.4, the dispersion in the vertical direction is zero to first order.

In modern, strong focusing synchrotrons and storage rings the rms average dispersion function $\tilde{D}_e$ around the closed orbit is typically in the range of 1 to several meters. Thus if $\widetilde{\Delta P}/P_0 \simeq 10^{-3}$, the rms width $\tilde{x}_e = \tilde{D}_e\widetilde{\Delta P}/P_0$ is in the range of a few millimeters. Unlike the chromaticity effect, dispersion is reversible (i.e., it does not by itself generate emittance growth). However, it does play an important role in intrabeam scattering, as we discuss in Section 6.4.2. The dispersion function varies along the equilibrium orbit. Its amplitude as well as its rms average value, $\tilde{D}_e$, around the orbit can be chosen by the lattice designer within certain limits to satisfy the requirements for a particular machine. Thus, the lattice design for a circular collider must provide a dispersion function $D_e(s)$ whose local value is zero at the interaction points of the two beams, so that the "spot" size of the beam is determined only by the emittance.

As shown in Equation (3.276a), the combined action of betatron oscillations and dispersion can be represented by a 3×3 matrix for the total displacement x, the total divergence x', and the momentum spread. We can generalize the description given in Equation (3.276) if we replace k_r^2 by the gradient function κ_0 so that $k_r = k_x = \sqrt{\kappa_0}$. The most common elements in a ring are dipole magnets for bending the beam and quadrupole magnets for focusing. In the first case, $\kappa_0 = (1 - n)/R^2$, where n is the magnetic field index and R the radius of the beam centroid trajectory in the bending magnet. In the second case, $\kappa_0 = \pm q B'/\gamma_0 m v_0 = \pm q B'/P_0$, or alternatively, with $B' = B_0/a_q$ and $P_0/qB_0 = R$, $\kappa_0 = \pm 1/a_q R$, where B' is the field gradient, B_0 the pole tip field, and a_q the distance of the pole tip from the axis; the plus sign indicates a focusing plane, the minus sign a defocusing plane. For elements of length l with piecewise constant-gradient function κ_0, the matrix elements a_{ij} in (3.276a) then depend on whether κ_0 is positive, zero, or negative, as shown in Table 5.5. The column with $\kappa_0 = 0$ represents the bending magnets between quadrupoles, which are assumed to act like a drift space on the betatron motion but bend the off-momentum particles with respect to the central orbit. In this case, R represents the local cyclotron radius in the bending magnets.

The first four rows in the table are the matrix elements for betatron motion, so that in the case $\kappa_0 > 0$ one has

$$x_b(l) = x_b(0) \cos \sqrt{\kappa_0}\, l + x_b'(0)(1/\sqrt{\kappa_0}) \sin \sqrt{\kappa_0}\, l, \qquad (5.475a)$$

$$x_b'(l) = -x_b(0)\sqrt{\kappa_0} \sin \sqrt{\kappa_0}\, l + x_b'(0) \cos \sqrt{\kappa_0}\, l. \qquad (5.475b)$$

Table 5.5 Matrix elements a_{ij} for different κ_0

κ_0	>0	0	<0
a_{11}	$\cos \sqrt{\kappa_0}\, l$	1	$\cosh \sqrt{|\kappa_0|}\, l$
a_{12}	$(1/\sqrt{\kappa_0}) \sin \sqrt{\kappa_0}\, l$	L	$(1/\sqrt{|\kappa_0|}\, l) \sinh \sqrt{|\kappa_0|}\, l$
a_{21}	$-\sqrt{\kappa_0} \sin \sqrt{\kappa_0}\, l$	0	$\sqrt{|\kappa_0|} \sinh \sqrt{|\kappa_0|}\, l$
a_{22}	$\cos \sqrt{\kappa_0}\, l$	1	$\cosh \sqrt{|\kappa_0|}\, l$
a_{13}	$(1/R\kappa_0)(1 - \cos \sqrt{\kappa_0}\, l)$	$L^2/2R$	$(1/R|\kappa_0|)(\cos \sqrt{\kappa_0}\, l - 1)$
a_{23}	$(1/R\sqrt{\kappa_0}) \sin \sqrt{\kappa_0}\, l$	L/R	$(1/R\sqrt{|\kappa_0|}) \sinh \sqrt{|\kappa_0|}\, l$

The last two rows are the matrix elements for dispersion, and in the case $\kappa_0 > 0$, one gets the general solution for an off-momentum particle,

$$
\begin{bmatrix} x(s) \\ x'(s) \\ \dfrac{\Delta P}{P_0} \end{bmatrix} =
$$

$$
\begin{bmatrix} \cos\sqrt{\kappa_0}\,l & \dfrac{1}{\sqrt{\kappa_0}}\sin\sqrt{\kappa_0}\,l & \dfrac{1}{R\kappa_0}(1 - \cos\sqrt{\kappa_0}\,l) \\ -\sqrt{\kappa_0}\sin\sqrt{\kappa_0}\,l & \cos\sqrt{\kappa_0}l & \dfrac{1}{R\sqrt{\kappa_0}}\sin\sqrt{\kappa_0}\,l \\ 0 & 0 & 1 \end{bmatrix}
\begin{bmatrix} x(0) \\ x'(0) \\ \dfrac{\Delta P}{P_0} \end{bmatrix} \qquad (5.476)
$$

By multiplying the matrices for the different sections of a lattice, one can find the matrix for one period or for one turn consisting of an integral number of periods. The condition that the displaced equilibrium orbits be closed implies that the vector $x(s)$, $x'(s)$ must be the same at $s = C = 2\pi\overline{R}$ as at $s = 0$, that is,

$$
\begin{bmatrix} x(C) \\ x'(C) \\ \dfrac{\Delta P}{P_0} \end{bmatrix} = \widetilde{M}_{\text{turn}}
\begin{bmatrix} x(0) \\ x'(0) \\ \dfrac{\Delta P}{P_0} \end{bmatrix} =
\begin{bmatrix} x(0) \\ x'(0) \\ \dfrac{\Delta P}{P_0} \end{bmatrix}, \qquad (5.477a)
$$

where $C = 2\pi R$ is the circumference of the equilibrium orbit. Note that by factoring out $\Delta P/P_0$ we can apply the last two equations to the dispersion function $D_e(s)$ itself, so that (5.477a) may be written as

$$
\begin{bmatrix} D_e(C) \\ D_e'(C) \\ 1 \end{bmatrix} = \widetilde{M}_{\text{turn}}
\begin{bmatrix} D_e(0) \\ D_e'(0) \\ 1 \end{bmatrix} =
\begin{bmatrix} D_e(0) \\ D_e'(0) \\ 1 \end{bmatrix}. \qquad (5.477b)
$$

The matrix $\widetilde{M}_{\text{turn}}$ for one revolution and the initial condition must satisfy this relation to obtain a closed-orbit solution. Both the betatron function $\hat{\beta}_0(s)$ and the dispersion function $D_e(s)$ represent the characteristics of the focusing ring lattice while emittance ϵ and momentum spread $\Delta P/P_0$ define the properties of the beam.

As discussed in Section 3.6.4, the dispersion effect is also represented by the momentum compaction factor α. For axisymmetric, constant-gradient fields, we found that $\alpha = 1/v_r^2 > 1$ since $v_r < 1$ in that case. This relation is still approximately true in modern strong-focusing (alternating-gradient) synchrotrons, where $v_x = v_r > 1$ and $\alpha \approx 1/v_x^2$ and where α is usually related to the transition energy γ_t by $\alpha = 1/\gamma_t^2 < 1$, which is less than unity since $\gamma_t > 1$.

The smaller betatron oscillation amplitudes and smaller momentum compaction factors of alternating-gradient lattices have made it possible to build modern synchrotrons and storage rings with much smaller magnet gaps and hence lower costs than would have been required with the old constant-gradient machines.

The small gap size and low dispersion of modern circular machines, combined with the increasingly smaller emittances of the beams produced by advanced particle sources and injector linacs, have led to a significant increase in the possible current density in the rings. Higher current density increases the space-charge forces and hence aggravates the tune-shift problem discussed in Section 4.5.1. This problem is further compounded by the *non-Liouvillean injection* schemes into the rings employed in high-energy accelerators [24] and proposed for heavy-ion fusion [25]. In the first case, a beam from an H^- ion source is accelerated by the linac and injected into the ring through a foil. The two electrons of the H^- ions are stripped in the foil, and the resulting H^+ ions (protons) are then deflected into the circular orbit of the ring machine. This process does not obey Liouville's theorem, which states that the phase-space density of a particle distribution remains constant. Thus new proton bunches can be injected and overlapped in phase space with the circulating bunches that have been injected earlier. With the non-Liouvillean charge-stripping process in the foil, the phase-space density, and hence charge density in the circulating beam, can be increased by multiturn injection to a much higher level than that of a single bunch, while preserving the small emittance of a single bunch. Without such a technique, the bunches injected during subsequent turns would have to be placed adjacent to each other in phase space, which, of course, results in a correspondingly larger emittance.

As mentioned, a major obstacle standing in the way of achieving the substantial increases in phase-space density that are possible is the incoherent space-charge tune-shift limit. From Equations (4.252) and (4.253), the tune-shift relation may be written in terms of the normalized rms emittance $\tilde{\epsilon}_n$, rather than the effective emittance $\epsilon_n = 4\tilde{\epsilon}_n$, as

$$\Delta\nu = -\frac{\bar{I}\,\bar{R}}{4I_0\tilde{\epsilon}_n\beta^2\gamma^2 B_f} = -\frac{N_t r_c}{8\pi\tilde{\epsilon}_n\beta\gamma^2 B_f}, \tag{5.478}$$

where $\bar{I}$ is the average current, $\bar{R}$ the average ring radius, N_t the total number of particles in the ring, B_f the bunching factor, and r_c the classical particle radius, as defined in Equation (5.244). The unperturbed tune ν_0 of a machine is usually designed to fall between a half-integral and an integral resonance (e.g., $\nu_0 = 6.7$ in the Fermilab booster synchrotron). If the space-charge tune shift gets large enough, it will push particles into the nearest resonance, say $\nu = 6.5$ in the Fermilab example. Traversal through the resonance will increase the amplitudes of the particle distribution. This amplitude growth is an incoherent process that will increase the emittance as well. The process will saturate when the emittance growth and the loss of particles that may occur are large enough that $|\Delta\nu|$ decreases and the resonance is no longer encountered. From Equation (5.478) we can see that for given energy, machine radius, and bunching factor, the requirement that $|\Delta\nu| \leq |\Delta\nu|_{\max}$ implies that the phase-space density has an upper limit that is defined by the relations

$$\frac{\bar{I}}{\bar{\epsilon}_n} = \frac{4I_0\beta^2\gamma^2 B_f}{\bar{R}}|\Delta\nu|_{\max},$$
(5.479a)

or

$$\frac{N_t}{\bar{\epsilon}_n} = \frac{8\pi\beta\gamma^2 B_f}{r_c}|\Delta\nu|_{\max},$$
(5.479b)

where $|\Delta\nu|_{\max}$ is typically in the range 0.3 to 0.5, depending on the machine design. The net result of the tune-shift limit is that one cannot take full advantage of the high-brightness beams being produced by modern ion sources or of the full potential offered by non Liouvillean injection. From the scaling given in Equation (5.479), it is obvious that one way out of this dilemma is to increase the injection energy and hence the length of the linear accelerator delivering the beam to the ring machine. Thus, the Fermilab upgrade project included an increase in the linac energy from 200 MeV ($\gamma = 1.21$, $\beta = 0.57$) to 400 MeV ($\gamma = 1.43$, $\beta = 0.71$), which results in a theoretical increase of $N_t/\bar{\epsilon}_n$ by a factor of 1.74.

The horizontal spread of the beam due to dispersion, which was neglected in the above relations, can also have a significant effect on the tune shift and increase the space-charge limit [26]. To include dispersion we will return to our original derivation for $\Delta\nu$ given in Equation (4.247). Due to dispersion the beam will have a larger width in the horizontal than in the vertical direction. With $x_{\max} = a$, $y_{\max} = b$, and $R = \bar{R}$ we obtain in place of (4.247) the relation

$$\Delta\nu_x = -\frac{K\bar{R}^2}{\nu_x a(a + b)}$$
(5.480a)

and

$$\Delta\nu_y = -\frac{K\bar{R}^2}{\nu_y b(a + b)},$$
(5.480b)

from which we recover the formula (4.247) when $a = b$, $\nu_x = \nu_y = \nu_0$. This result follows from the fact that in an elliptical beam with uniform density the space-charge electric fields are

$$E_x \propto \frac{K}{a(a + b)}x, \qquad E_y \propto \frac{K}{b(a + b)}y,$$

as discussed in Section 4.4.2 [see Equations (4.174) and (4.175)].

The above relations for the space-charge tune shift assume a beam with uniform density (K–V beam). To present them in a form that is independent of the distribution we introduce the rms widths $\delta_x = \tilde{x} = (\overline{x^2})^{1/2} = a/2$, $\delta_y = \tilde{y} = (\overline{y^2})^{1/2} = b/2$. Then we can write

$$\Delta\nu_x = -\frac{\overline{K}\,\overline{R}^2}{4\nu_x \delta_x (\delta_x + \delta_y) B_f}, \tag{5.481a}$$

$$\Delta\nu_y = -\frac{\overline{K}\,\overline{R}^2}{4\nu_y \delta_y (\delta_x + \delta_y) B_f}, \tag{5.481b}$$

where we added the bunching factor B_f and replaced K by the average perveance $\overline{K}$, which is proportional to the average beam current in the ring. When dispersion is present the horizontal rms width will consist of the contribution due the emittance of the beam, defined by $\delta_{xb} = \tilde{x}_b$, and the contribution due to the momentum spread, defined by $\delta_{xe} = \tilde{x}_e$. In view of (5.474), the total rms width of the beam is then given by

$$\delta_x = (\delta_{xb}^2 + \delta_{xe}^2)^{1/2} = \delta_{xb}(1 + \Delta_D^2)^{1/2}, \tag{5.482}$$

where

$$\Delta_D = \frac{\delta_{xe}}{\delta_{xb}}. \tag{5.483}$$

For the vertical direction we will assume that dispersion is zero, so that $\delta_y = \delta_{yb}$. Note that δ_{yb} will not be the same as δ_{xb} unless $\epsilon_y = \epsilon_x$ and $\nu_y = \nu_x$. We will introduce the parameter

$$\Delta_\delta = \frac{\delta_{xb}}{\delta_{yb}} \tag{5.484}$$

to define the ratio between the two quantities. The geometric terms in the denominator of Equation (5.481) then become

$$\frac{\delta_x(\delta_x + \delta_y)}{2} = \delta_{xb}^2 g_x, \tag{5.485a}$$

$$\frac{\delta_y(\delta_x + \delta_y)}{2} = \delta_{yb}^2 g_y, \tag{5.485b}$$

where

$$g_x = \frac{(1 + \Delta_D^2)^{1/2}}{2\Delta_\delta} [\Delta_\delta(1 + \Delta_D^2)^{1/2} + 1], \tag{5.486a}$$

$$g_y = \frac{1}{2} [\Delta_\delta(1 + \Delta_D^2)^{1/2} + 1]. \tag{5.486b}$$

For a matched beam, the normalized emittances in both directions will be given by the relations

$$\tilde{\epsilon}_{nx} = \frac{\beta\gamma\delta_{xb}^2 \nu_x}{R}, \tag{5.487a}$$

$$\tilde{\epsilon}_{ny} = \frac{\beta\gamma\delta_{yb}^2 \nu_y}{R}. \tag{5.487b}$$

Substituting Equations (5.485) and (5.487) in Equation (5.481) and introducing the average current $\bar{I}$ or the total number of particles in the ring N_t, we obtain

$$\Delta\nu_x = -\frac{\bar{I}\,R}{4I_0\tilde{\epsilon}_{nx}\beta^2\gamma^2 B_f g_x} = -\frac{N_t r_c}{8\pi\tilde{\epsilon}_{nx}\beta\gamma^2 B_f g_x}, \tag{5.488a}$$

$$\Delta\nu_y = -\frac{\bar{I}\,R}{4I_0\epsilon_{ny}\beta^2\gamma^2 B_f g_y} = -\frac{N_t r_c}{8\pi\tilde{\epsilon}_{ny}\beta\gamma^2 B_f g_y}, \tag{5.488b}$$

These equations have the same form as Equations (4.252) and (4.253) except that the normalized rms emittance $\tilde{\epsilon}_n$ is used in place of the effective emittance $\epsilon_n = 4\tilde{\epsilon}_n$. Furthermore, they have in the denominator the geometry factors g_x, g_y which define the decrease in tune shift due to dispersion and unequal tunes ($\nu_x \neq \nu_y$) or unequal emittances ($\tilde{\epsilon}_{nx} \neq \tilde{\epsilon}_{ny}$). Table 5.6 shows the values of the geometry parameters g_x and g_y for different ratios of $\Delta_D = \delta_{xe}/\delta_{xb}$ and $\Delta_\delta = \delta_{xb}/\delta_{yb}$. The examination of the results in the table and of the equations for g_x and g_y for the case where $\epsilon_{nx} = \tilde{\epsilon}_{ny}$ shows the following:

1. For symmetric focusing ($\nu_x = \nu_y$, $\Delta_\delta = 1$), dispersion decreases both tune shifts; however, $g_y < g_x$, hence $|\Delta\nu_y| > |\Delta\nu_x|$, so that the space-charge limit is determined by $\Delta\nu_y$ [Equation (5.488b)] and hence is increased by

Table 5.6 g_x and g_y for different values of $\Delta_D = \delta_{xe}/\delta_{xb}$ and $\Delta_\delta = \delta_{xb}/\delta_{yb}$

	$\Delta_\delta = 1$		$\Delta_\delta = 1.5$		$\Delta_\delta = 2$	
Δ_D	g_x	g_y	g_x	g_y	g_x	g_y
0	1.000	1.000	0.833	1.250	0.500	1.500
0.50	1.184	1.059	0.998	1.339	0.905	1.618
1.00	1.707	1.207	1.471	1.561	1.354	1.914
1.50	2.526	1.401	2.226	1.852	2.076	2.303
2.00	3.618	1.618	3.245	2.177	3.059	2.736
2.50	4.971	1.846	4.522	2.519	4.298	3.193
3.00	6.581	2.081	6.054	2.872	5.791	3.662

Source: Reference 26.

the factor g_y. As an example, for $\Delta_D = 2$, $\Delta_\delta = 1$, one finds from Table 5.5 that due to dispersion the phase-space density $N/\tilde{\epsilon}_n$ can be increased by a factor of $g_y = 1.618$ compared with the case where dispersion is negligible.

2. Asymmetric focusing ($\nu_x < \nu_y$, $\Delta_\delta > 1$) further enhances the g_y factor provided that $\Delta_D > (\Delta_\delta^2 - 1)^{1/2}$. For the above example ($\Delta_D = 2$), if $\Delta_\delta = 1.5$ (i.e., $\nu_y = 2.25\nu_x$) one obtains $g_y = 2.177$, which is significantly higher than in the symmetric case. In the region below the limit $\Delta_D = (\Delta_\delta^2 - 1)^{1/2}$, where $g_x < g_y$, the tune shift is controlled by g_x, which is less interesting from a practical point of view.

These examples show that the effect may be quite significant and much stronger than the Laslett tune-shift correction, due to image forces [see the example following Equation (4.277)] that we are neglecting in the present analysis. The image factors shown in brackets in Equations (4.276), (4.282), and (4.283) must, of course, be added to our results here to obtain the most general expressions for the tune shifts.

Since the beam profile in high-energy circular machines tends to have a Gaussian shape, there is a spread of the betatron oscillation frequencies (i.e., a particle's betatron tune depends on its radial amplitude, and hence its transverse kinetic energy). The above tune-shift relations [e.g., (5.488b)] represent rms averages over the particle distribution. They are appropriately called *rms tune shifts*. Particles with large betatron amplitudes scanning the thin tail of the Gaussian distribution have a smaller tune shift. Those with small amplitudes stay near the center in the beam core and experience a larger tune shift. The linear part of the space-charge force in the center of a Gaussian distribution is two times stronger than the rms force used in the above equation. Accordingly, the tune shift in the core of the Gaussian is a factor of 2 greater than the rms tune shift [e.g., $(\Delta\nu_y)_{\text{core}} = 2\Delta\nu_y$].

Let us now take a closer look at the dispersion effect represented by the parameter $\Delta_D = \delta_{xe}/\delta_{xb}$. For the lattice configuration of modern synchrotrons the dispersion function D_e varies periodically with path length s, as discussed above. In an ideal FODO lattice, which is uniformly occupied by bending magnets and quadrupole lenses and has no long straight sections, $D_e(s)$ is always positive and the rms value $\tilde{D}_e = (\overline{D^2})^{1/2}$, obtained from the integral over the closed orbit with average radius $\overline{R}$, is given by

$$\tilde{D}_e = \frac{\overline{R}}{\nu_x^2}, \tag{5.489}$$

as in the classical axisymmetric field. However, it should be noted that lattice designers can significantly enhance or decrease the average dispersion compared with this simple relation.

The rms width δ_{xe} of the beam due to dispersion is obtained from Equation (5.472) by averaging over the distributions in x_e and $\Delta P/P_0$ around the equilibrium orbit. Using (5.489) for the rms average dispersion, one can write

$$\delta_{xe} = \widetilde{D}_e \frac{\widetilde{\Delta P}}{P_0} = \frac{\overline{R}}{\nu_x^2} \frac{\widetilde{\Delta P}}{P_0}.$$ (5.490)

Although these relations for the dispersion effect are good approximations only for the ideal FODO lattice, they do show the general trend toward smaller dispersion when the tune is increased. From Equations (5.487a) and (5.490) one obtains for the parameter Δ_D the result

$$\Delta_D = \frac{\delta_{xe}}{\delta_{xb}} = \frac{1}{\nu_x^{3/2}} \left(\frac{\beta\gamma\overline{R}}{\tilde{\epsilon}_{nx}} \right)^{1/2} \frac{\widetilde{\Delta P}}{P_0}.$$ (5.491)

The scaling displayed by this relation implies that one should operate at a low tune to maximize the dispersion effect and hence the geometry factor g_y in the tune-shift formula. Low-tune operation increases the horizontal beam size and therefore requires a large beam pipe aperture. This conflicts with the historical trend toward stronger focusing (higher tune) and smaller apertures to minimize costs. However, in some specific cases a low-tune, large-dispersion design may provide a more attractive option to achieve the desired phase-space density or luminosity than other alternatives. Equation (5.491) shows, for instance, that the dispersion effect is the more pronounced the smaller the emittance and would therefore be particularly useful for the non-Liouvillean injection schemes discussed earlier. As already mentioned, a lattice with unequal tunes (i.e., $\nu_y \gg \nu_x$) also helps in increasing the space-charge limit, as is evident from the parameter Δ_s in Equation (5.486b). A practical upper limit to the achievable aspect ratio δ_x/δ_y of the beam is given by the size of the beam-pipe aperture, which cannot be too large, for various reasons. Also, one must allow enough space between the rms width δ_x of the beam and the wall of the beam pipe to accommodate the Gaussian particle distribution.

So far, the large-dispersion effect to increase the space-charge limit described here has not been used in existing circular machines. It remains to be seen whether it can provide a cost-effective option for future designs or upgrades of existing machines.

5.4.11 Coupled Envelope Equations for a Bunched Beam

In this section we attempt to integrate the models for the transverse and longitudinal distributions into a coherent theoretical description of a bunched beam that includes the transverse-longitudinal coupling through the space-charge forces. To simplify the analysis we make use of the smooth approximation, that is, we will neglect the usually very small envelope ripple due to the periodic-focusing structures. Furthermore, we assume that the average focusing forces and the emittances in the two orthogonal transverse directions are the same and that the bunch propagates in a cylindrical tube of radius b. In short, the system will have axial symmetry so that the particle density in the bunch will only be a function of radius r and axial displacement z from the bunch center [i.e., $n = n(r,z)$]. We will be concerned primarily with the properties of the quasi-stationary state of the bunch in a straight channel or linear accelerator, where both the transverse and the longitudinal distributions are perfectly matched and where the applied focusing forces are linear. Our analysis can also be applied to a circular machine with a symmetric lattice and negligible dispersion by incorporating the slip factor η into the longitudinal equations.

As we know from previous discussions, our two-temperature Maxwell–Boltzmann distribution (5.271) and the associated Boltzmann density profiles for the radial and axial directions, given by Equations (5.275) and (5.369), respectively, generate in general nonlinear space-charge forces, except for the zero-temperature case. However, we can model the bunch by an ellipsoid with uniform charge density, radius a, and axial half width z_m, in which the self forces are linear. This model is consistent with a zero-temperature Maxwell–Boltzmann distribution but not with a *hot* beam having finite emittances in the three phase-space projections. As we discussed at the beginning of Section 5.4.8, the extension of the K–V distribution to six-dimensional phase space leads to a nonlinear space-charge force in the longitudinal direction (see Problem 5.12). Unfortunately, no distribution exists that yields linear space-charge forces in *both* the transverse and longitudinal directions for a beam with nonzero average temperatures. Basically, the ellipsoidal model is consistent with a K–V distribution of the form (5.160) in transverse phase space and a Neuffer distribution of the form (5.425) in longitudinal phase space which cannot be derived from a single phase-space distribution. These two distributions are, however, adequate approximations for modeling of the bunch, and they can be correlated with the thermal distribution or any other particle distribution having the same rms width and emittance by using the concept of equivalent beams described in Section 5.3.4.

Since we are dealing with a bunched beam, it will be necessary to redefine the generalized perveance K for the transverse space-charge force. First, we must include the radial geometry factor due to the image force in the relativistic form $1 - (ga^2)/(2\gamma_0^2 z_m^2)$ [see Equation (5.354d), which represents the free-space situation where $g = g_0$], where $\gamma_0 z_m$ must be used in place of z_m to account for the longitudinal Lorentz contraction of the bunch in relativistic beams. Second, we will use the total number of particles in the bunch N in lieu of the peak current I. For our ellipsoidal bunch we find from Equation (5.353) that

$$I = \rho_{L0} v_0 = \frac{3}{4} \frac{qN\beta_0 c}{z_m}, \tag{5.492}$$

and hence we obtain for the perveance

$$K = \frac{qI}{2\pi\epsilon_0 mc^3 \beta_0^3 \gamma_0^3} \left(1 - \frac{g}{2} \gamma_0^2 \frac{a^2}{\gamma_0^2 z_m^2} \right) = \frac{3}{2} \frac{Nr_c}{\beta_0^2 \gamma_0^3} \frac{1}{z_m} \left(1 - \frac{g}{2} \frac{a^2}{\gamma_0^2 z_m^2} \right). \tag{5.493}$$

Here, r_c is the classical particle radius [see Equation (5.244)], and the term in brackets represents the geometry factor due to the radial image force. For the longitudinal perveance parameter K_L we use the definition (5.415).

With these modifications we can write the transverse and longitudinal envelope equations for the bunched beam in the form

$$a'' + k_{x0}^2 a - \frac{3}{2} \frac{Nr_c}{\beta_0^2 \gamma_0^3} \frac{1}{az_m} \left(1 - \frac{g}{2} \frac{a^2}{\gamma_0^2 z_m^2} \right) - \frac{\epsilon_x^2}{a^3} = 0 \qquad (5.494)$$

and

$$z_m'' + k_{z0}^2 z_m - \frac{3}{2} \frac{Nr_c}{\beta_0^2 \gamma_0^5} \frac{g}{z_m^2} - \frac{\epsilon_{zz'}^2}{z_m^3} = 0. \qquad (5.495)$$

Note that ϵ_x and $\epsilon_{zz'}$ can be related to the respective normalized emittances by $\epsilon_{nx} = \beta_0 \gamma_0 \epsilon_x$ and $\epsilon_{nz} = \beta_0 \gamma_0^3 \epsilon_{zz'}$. In the nonrelativistic limit ($\gamma_0 = 1$), and for free space ($g = g_0$) our last two equations agree with the equations used by Chasman for linear accelerator design studies in the late 1960s [27].

It is readily apparent that these two equations are coupled to each other via the space-charge term. As discussed in Section 5.4.7, the geometry factor g is in general a function of the semiaxes a and z_m of the ellipsoidal bunch and of the tube radius b (see Figure 5.15 and Table 5.3) [i.e., $g = g(z_m/a; b/a)$]. Thus, for a given number of particles N, emittances ϵ_x and $\epsilon_{zz'}$, external focusing forces, as represented by the wave numbers k_{x0} and k_{z0}, tube radius b, and energy $\gamma_0 mc^2 = mc^2(1 - \beta_0^2)^{-1/2}$, these coupled nonlinear equations must be solved numerically to find the radius a and half-length z_m of the bunch. The geometry factor g has a nonanalytic form except for free space ($g = g_0$) and for the long-bunch limit [$g \approx 0.67 + 2 \ln(b/a)$]. Hence, one must use approximate values for $g = g(z_m/a; b/a)$ from Figure 5.15 or interpolate numerically between the given curves. Note that z_m must be replaced by $\gamma_0 z_m$ for relativistic beams.

If the bunch is perfectly matched in both directions, then $a'' = 0$ and $z_m'' = 0$, and the semiaxes a and z_m can be calculated for a given set of parameters, including the beam energy. Moreover, in an accelerator, if the rate of energy change occurs adiabatically, as is usually the case, the change in the bunch radius a and half-length z_m can also be calculated from the matched envelope equations. However, it is then better to use the normalized emittances to exhibit the scaling with the velocity and energy parameters β_0, γ_0. Thus, the matched coupled envelope equations take the form

$$k_{x0}^2 a - \frac{3}{2} \frac{Nr_c}{\beta_0^2 \gamma_0^3} \frac{1}{az_m} \left(1 - \frac{g}{2} \frac{a^2}{\gamma_0^2 z_m^2} \right) - \frac{\epsilon_{nx}^2}{\beta_0^2 \gamma_0^2 a^3} = 0 \qquad (5.496)$$

and

$$k_{z0}^2 z_m - \frac{3}{2} \frac{Nr_c}{\beta_0^2 \gamma_0^5} \frac{g}{z_m^2} - \frac{\epsilon_{nz}^2}{\beta_0^2 \gamma_0^6 z_m^3} = 0. \qquad (5.497)$$

For the beam physics, the wave numbers k_x and k_z that include the space-charge defocusing effect on the betatron and synchrotron wavelengths $\lambda_x = 2\pi/k_x$ and $\lambda_z = 2\pi/k_z$ are very important. They are defined by

$$k_x^2 = k_{x0}^2 - \frac{3}{2} \frac{Nr_c}{\beta_0^2 \gamma_0^3} \frac{1}{a^2 z_m} \left(1 - \frac{g}{2} \frac{a^2}{\gamma_0^2 z_m^2} \right), \qquad (5.498)$$

$$k_z^2 = k_{z0}^2 - \frac{3}{2} \frac{Nr_c}{\beta_0^2 \gamma_0^5} \frac{g}{z_m^3} . \qquad (5.499)$$

In terms of these quantities the matched envelope equations may be written as

$$k_x^2 a - \frac{\epsilon_{nx}^2}{\beta_0^2 \gamma_0^2 a^3} = 0, \qquad (5.500)$$

$$k_z^2 z_m - \frac{\epsilon_{nz}^2}{\beta_0^2 \gamma_0^6 z_m^3} = 0. \qquad (5.501)$$

The above set of equations (5.496) to (5.501) allow us to calculate the properties of the bunch, in particular the semi-axes a and z_m and the physics parameters k_x and k_z for any given set of input parameters (N, k_{x0}, k_{z0}, ϵ_{nx}, ϵ_{nz}, β_0, γ_0, and b). Furthermore, by introducing the rms quantities $\tilde{x} = a/\sqrt{5}$, $\tilde{z} = z_m/\sqrt{5}$, $\tilde{\epsilon}_{nx} = \epsilon_{nx}/5$, $\tilde{\epsilon}_{nz} = \tilde{\epsilon}_{nz}/5$, we can determine the properties of any other equivalent bunched beam having the same rms widths and emittances as the ellipsoidal bunch considered here. Note that the relationships for the *transverse rms width* and *rms emittance of the ellipsoidal bunch* differ from those in a continuous beam where $\tilde{x} = a/2$, $\tilde{\epsilon} = \epsilon/4$. (See Problem 5.21.)

The matched envelope equations (5.496) and (5.497) represent a quasi-stationary state of the bunch in which the applied focusing force (first term), the space-charge force (second term), and the emittance (third term) are balanced, but not necessarily in three-dimensional thermal equilibrium. [See the comments at the beginning of this section and following Equation (5.271).] When $T_\perp \neq T_\parallel$ and space-charge forces are strong, there will be rapid change and emittance growth towards an equipartitioned state as discussed in Appendix 4. With regard to practical application and mathematical solution of these coupled equations, we can distinguish the following regimes:

1. The bunch is space-charge dominated in both directions, so that the emittance terms can be neglected ($\epsilon_{nx} = \epsilon_{nz} = 0$) for the calculation of a and z_m. This occurs in high-intensity linacs, and we discuss this case further below and in Appendix 4.

2. Space charge dominates in one direction but not in the other. In a circular machine, for instance, the transverse space-charge effect is usually small compared to the emittance, but the bunch could well be space-charge dominated longitudinally. Of course, we would have to use the slip factor η in

the longitudinal envelope equation, as discussed in Section 5.4.9. The radius a is then readily determined analytically from Equations (5.496), namely $a = (\epsilon_{nx}/\beta_0\gamma_0 k_{x0})^{1/2}$. This result can be substituted into the longitudinal equation to find z_m.

3. Space charge is negligible compared to emittance. This case is trivial, and the semiaxes are found analytically as $a = (\epsilon_{nx}/\beta_0\gamma_0 k_{x0})^{1/2}$ and $z_m = (\epsilon_{nz}/\beta_0\gamma_0^3 k_{z0})^{1/2}$.

4. The bunch length is large compared to the radius (i.e., $\gamma_0 z_m/a \gg 1$). In this case the geometry factor is $g \approx 0.67 + 2\,\ln(b/a)$ [from Equation (5.365b)], the image-force term $ga^2/\gamma_0^2 z_m^2$ in the radial envelope equation can be neglected, and the solution is then simplified.

5. The eccentricity of the bunch is small, say $\gamma_0 z_m/a \lesssim 4$, and the tube radius is significantly larger than the beam radius, say $b/a \gtrsim 5$. From the graph in Figure 5.15 we can see that the g-factor in this case does not differ significantly from the free-space value g_0. Thus we can use the approximation $g \approx g_0$. Furthermore, in this regime the free space geometry factor can be approximated by $g_0 \approx 2\gamma_0 z_m/3a$ [see Equations (5.356a) and (5.356b)]. Thus, in this case the envelope equations can be written in the simpler form

$$k_{x0}^2 a - \frac{3}{2}\frac{Nr_c}{\beta_0^2\gamma_0^3}\frac{1}{az_m}\left(1 - \frac{1}{3}\frac{a}{\gamma_0 z_m}\right) - \frac{\epsilon_{nx}^2}{\beta_0^2\gamma_0^2 a^3} = 0, \quad (5.502)$$

$$k_{z0}^2 z_m - \frac{Nr_c}{\beta_0^2\gamma_0^4}\frac{1}{az_m} - \frac{\epsilon_{nz}^2}{\beta_0^2\gamma_0^6 z_m^3} = 0, \quad (5.503)$$

which can be solved more easily than the general equations, where the g-factor must be found by interpolation from Table 5.3.

It is apparent from this discussion of the various regimes that analytic solutions for a and z_m can be obtained only in case 3, which is trivial, and in case 5, which is more involved since we are dealing with a set of fourth-order, coupled algebraic equations (see Problem 5.20). In case 5, if the bunch is space-charge dominated, we can neglect the emittance terms ($\epsilon_{nx} = \epsilon_{nz} = 0$) and hence obtain from (5.502), (5.503) the simpler equations

$$k_{x0}^2 a - \frac{3}{2}\frac{Nr_c}{\beta_0^2\gamma_0^3}\frac{1}{az_m} + \frac{1}{2}\frac{Nr_c}{\beta_0^2\gamma_0^4}\frac{1}{z_m^2} = 0, \quad (5.504)$$

$$k_{z0}^2 z_m - \frac{Nr_c}{\beta_0^2\gamma_0^4}\frac{1}{az_m} = 0, \quad (5.505)$$

which can be solved without difficulty; one finds that

$$a = \left(\frac{3}{2}\right)^{2/3}\frac{(Nr_c)^{1/3}}{\beta_0^{2/3}\gamma_0^{2/3}}\frac{1}{k_{z0}^{2/3}}\left(\frac{k_{x0}^2}{k_{z0}^2} + \frac{1}{2}\right)^{-2/3}, \quad (5.506)$$

$$z_m = \left(\frac{2}{3}\right)^{1/3} \frac{(Nr_c)^{1/3}}{\beta_0^{2/3}\gamma_0^{5/3}} \frac{1}{k_{z0}^{2/3}} \left(\frac{k_{x0}^2}{k_{z0}^2} + \frac{1}{2}\right)^{1/3}. \tag{5.507}$$

These relations reveal very clearly the scaling of the bunch size with the number of particles N, the wave numbers k_{x0} and k_{z0}, representing the applied focusing forces, and the kinetic energy through the factors β_0 and γ_0. Of particular interest, and somewhat unexpected, is the fact that the ratio of the semiaxes of the ellipsoidal bunch in this space-charge-dominated parameter regime is independent of the particle number N and hence the beam current. This ratio is given by the simple relation

$$\frac{z_m}{a} = \frac{2}{3\gamma_0}\left(\frac{k_{x0}^2}{k_{z0}^2} + \frac{1}{2}\right). \tag{5.508}$$

In a high-current rf linac, for instance, the ratio k_{x0}/k_{z0} usually increases with increasing energy, so that, at least in the nonrelativistic regime ($\gamma_0 \approx 1$), the eccentricity z_m/a of the bunch also increases with energy. Of course, we must keep in mind that our results (5.506) to (5.508) are valid only as long as $\gamma_0 z_m \lesssim 4$, $b/a \gtrsim 5$, and the space-charge terms in the envelope equations dominate over the emittance terms. Note that a bunch of half length z_m in the laboratory frame will be elongated by the factor γ_0 in the beam frame, i.e., an almost spherical bunch in the lab frame can still have large image forces and long-bunch behavior if γ_0 is high enough.

In Appendix 4 we apply the above results to a specific example of a high-intensity rf drift-tube linac. We also investigate the relationship between longitudinal and transverse temperature and the question of equipartitioning, which is of great importance for such devices.

From a mathematic point of view, the easiest way of solving the general coupled envelope equations (5.496) and (5.497) is to specify desired values for the bunch radius a and half-length z_m and the emittances as well. This approach is very useful in the design phase of a linear accelerator. One can then readily solve the envelope equations for the number of particles N in the bunch, or the equivalent average beam current $\bar{I}$. For an rf linac with frequency f and wavelength $\lambda = c/f$, one has the relation

$$\bar{I} = Qf = \frac{qNc}{\lambda}, \tag{5.509}$$

where $Q = qN$ is the total charge in the bunch.

By specifying a, z_m and k_{x0}, k_{z0}, one defines the acceptance of the linac in the transverse and longitudinal directions. If ϵ_{nx} or ϵ_{nz} is given or can be neglected since the beam is space-charge dominated, and if the tube radius b is given so that the g-factor can be determined from Figure 5.15, one can solve the envelope equations for the particle number N or the equivalent average current $\bar{I}$. The results obtained from the two equations will, in general, be different; that is, one will get a transverse current limit $\bar{I}_t$ and a longitudinal current limit $\bar{I}_l$ [28].

However, two different values for $\bar{I}_t$ and $\bar{I}_l$ imply that the beam is not matched in both directions. Even if it were initially matched transversely, for instance, with the choice $\bar{I} = \bar{I}_t < \bar{I}_l$, it would not be matched longitudinally, and space-charge coupling would immediately mismatch it transversely as well. Thus, in this approach, one would have to change the bunch-size parameters a and z_m and/or the emittances until a single solution for N, and hence $\bar{I}$, is found. (See also Appendix 4.)

5.4.12 Matching, Focusing, and Imaging

The self-consistent theory of beams developed in this chapter—laminar flow, Vlasov equation, and thermal distribution—has been applied mainly to determining the properties of the transverse or longitudinal meta-equilibrium states. In Chapter 6 we deal with the emittance growth that occurs when the beam is not in a stationary state (thermal equilibrium) or when instabilities and other effects, such as collisions, perturb the particle distribution. However, before proceeding to this next stage we review briefly in this section the topics of matching, focusing, and imaging within the context of a self-consistent description. These topics were, of course, discussed to some extent in Chapters 3 and 4. But with the exception of the short overview of aberrations in Section 3.4.6, we always assumed a uniform beam model in which both the applied focusing force and the space-charge force are linear.

The main objective of our discussion in this section is to obtain some physical insight and a qualitative picture of the role of particle distribution and nonlinear forces (aberrations) when a beam is focused by discrete lenses. To simplify the analysis, we consider only axisymmetric, thin lenses, such as electrostatic einzel lenses or short solenoids, as shown schematically in Figure 5.21. Figure 5.21(a) illustrates the matching by a single lens of a beam into a periodic-focusing channel. Ordinarily, one needs two lenses to match a beam into an axisymmetric channel since both the radius R and the slope $R' = dR/dz$ need to be changed. However, this task can, in principle, also be accomplished with a single lens that can be moved in position until the desired matching conditions are met. As another simplification for the purpose of our discussion we consider the matching transformation from a waist $R_1(R_1' = 0)$ to a waist $R_2(R_2' = 0)$, where we assumed in the figure that $R_2 < R_1$. Note that R_2 also corresponds to the waists between the lenses of the periodic array. Finally, for describing the effects and changes of the distribution through the lens system we assume a thermal beam having a transverse Boltzmann density profile, as illustrated in Figure 5.12.

We begin our analysis by considering an ideal, aberration-free lens that matches the beam into a periodic array of ideal lenses. There are two aspects to this problem. One is the behavior of the rms radius of the thermal beam, and the other one is the change in bunch profile and temperature in the focusing process. To evaluate the rms average behavior of the particle distribution, we will use the *equivalent uniform beam* having the same second moments as the thermal beam, following the description given in Section 5.3.4. For changes in the shape of the charge density and in the temperature of the distribution we refer to Figure 5.12 and Section 5.4.5, where some essential features of matching have already been discussed.

(a)   Matching into a periodic channel

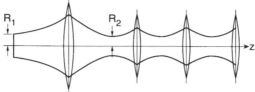

(b)   Focusing to a small spot size

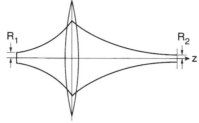

Figure 5.21. Schematic illustration of matching an axisymmetric beam into a periodic-focusing channel (a), and of focusing such a beam to a small spot size (b).

Consider now the beam at the waist upstream from the matching lens. Let $\tilde{x}_1 = \tilde{y}_1$ denote the rms width in the two transverse directions, $\tilde{r}_1 = \sqrt{2}\,\tilde{x}_1$ the rms radius, and $R_1 = \sqrt{2}\,\tilde{r}_1 = 2\tilde{x}_1$ the full radius of the equivalent uniform beam. If K is the generalized perveance, $\tilde{\epsilon}_1$ the rms emittance in x or y, and $\epsilon = 4\tilde{\epsilon}$ the total emittance of the equivalent uniform beam, the parameter $KR_1^2/\epsilon_1^2 = K\tilde{x}_1^2/4\tilde{\epsilon}_1^2$ will determine the ratio $\overline{\lambda_D}/R_1$ and hence the temperature and shape of the Boltzmann profile according to Figure 5.12 and Table 5.1. Since the rms ellipse of the distribution at the waist is upright, we have $\tilde{\epsilon} = \tilde{x}\tilde{x}' = \tilde{x}\tilde{v}_x/v_0$, where $\tilde{x}'$ is the rms divergence, $\tilde{v}_x$ the rms velocity, and v_0 the mean axial velocity. Thus if $R_1 = 2\tilde{x}_1$ and $\epsilon = 4\tilde{\epsilon}$ are given, we find for the rms velocity at the waist

$$\tilde{v}_{x1} = v_0 \frac{\tilde{\epsilon}}{\tilde{x}_1} = v_0 \frac{\epsilon}{2R_1}, \tag{5.510}$$

and for the transverse temperature from (5.289b),

$$k_{\mathrm{B}}T_{\perp 1} = \gamma_0 m\tilde{v}_{x1}^2 = \gamma_0 m v_0^2 \frac{\tilde{\epsilon}^2}{\tilde{x}_1^2} = \gamma_0 m v_0^2 \frac{\epsilon^2}{4R_1^2}. \tag{5.511}$$

The variation in rms radius as the beam propagates from the initial waist through the matching lens and the periodic-focusing channel is described by the rms envelope equation (5.218) with $k_{x0}^2 = 0$ and $\tilde{x} = \tilde{y}$, namely

$$\tilde{x}'' - \frac{K}{4\tilde{x}} - \frac{\tilde{\epsilon}^2}{\tilde{x}^3} = 0, \tag{5.512a}$$

or in terms of the effective radius $R = 2\tilde{x}$ and effective emittance $\epsilon = 4\tilde{\epsilon}$ of the equivalent uniform beam by

$$R'' - \frac{K}{R} - \frac{\epsilon^2}{R^3} = 0. \tag{5.512b}$$

At each lens the slope of the envelope is changed by $\Delta\tilde{x}' = -\tilde{x}/f$ or $\Delta R' = -R/f$, where f is the focal length of the lens. We assume that the emittance remains constant, so that for given initial conditions ($R = R_1$, $R_1' = 0$ in our case), the envelope can be calculated at any position along the system by integrating (5.512). If $R = R_2$, $R_2' = 0$ at the first waist downstream from the lens, we obtain for the transverse rms velocity and temperature at this position the relations

$$\tilde{v}_{r2} = v_0 \frac{\tilde{\epsilon}}{\tilde{v}_2} = v_0 \frac{\epsilon}{2R_2} \tag{5.513}$$

and

$$k_B T_{\perp 2} - \gamma_0 m \tilde{v}_{x2}^2 = \gamma_0 m v_0^2 \frac{\tilde{\epsilon}^2}{\tilde{x}_2^2} = \gamma_0 m v_0^2 \frac{\epsilon^2}{4R_2^2}. \tag{5.514}$$

By comparing the temperatures at the two waists we get

$$\frac{k_B T_{\perp 2}}{k_B T_{\perp 1}} = \frac{\tilde{x}_1^2}{\tilde{x}_2^2} = \frac{R_1^2}{R_2^2} \tag{5.515}$$

in agreement with Equation (5.311) for $\gamma_0 = $ const.

At a waist, the temperature $k_B T_\perp$ is identical to the average transverse kinetic energy per particle. Since $R_2 < R_1$ our relation (5.515) states that the transverse kinetic energy of the beam has increased by a factor of $(R_1/R_2)^2$ after passing through the lens. This additional transverse energy comes from the longitudinal energy of the particles. A lens transforms longitudinal momentum into transverse momentum, and vice versa. Consider a particle with velocity $v = v_{z1}$ entering the lens at radius r_1 with zero slope ($r_1' = 0$) (i.e., on a trajectory parallel to the axis). After passing through the lens it has a slope of $r_2' = -r_1/f$, hence a transverse velocity of $v_{r2} = r_2' v_{z2} = -r_1 v_{z2}/f$. Its axial velocity has been reduced to

$$v_{z2} = \left(v^2 - v_{r2}^2\right)^{1/2} = \frac{v}{[1 - (r_1/f)^2]^{1/2}}. \tag{5.516}$$

This decrease in the axial velocity is zero for a particle on the axis ($r_1 = 0$) and is a maximum for a particle passing through the lens at the outermost radius, $R_{1,max}$. The focusing action of a lens thus introduces a spread in the longitudinal energy distribution. This spread is reversible in the case of ideal lenses, but it may become irreversible if nonlinear forces from lens aberrations or space-charge nonuniformities are present.

The momentum transfer between longitudinal and transverse motion due to focusing changes the center-of-momentum velocity v_0 of the distribution. Thus, we should have used v_{01} and v_{02} in the above equations for the transverse velocities and temperatures in the two waists. Note that $v_{01} = v_{02}$, $\tilde{v}_{x1} = \tilde{v}_{x2}$, and $k_B T_{\perp 1} = k_B T_{\perp 2}$ when $\tilde{x}_2 = \tilde{x}_1$ (i.e., when the rms widths or corresponding radii at the two waists are the same).

Our analysis of the transverse energy variation due to focusing is incomplete so far, as we considered only the kinetic part. We also need to include the average potential energy per particle associated with the electric and magnetic forces due to the beam's space charge and current. This can be done by calculating the field energy per unit length of the beam and dividing by the number of particles per unit length, N_L. Since the transverse Boltzmann profiles in Figure 5.12 are nonanalytic, this calculation would have to be done numerically. However, for our purpose it will be adequate to use the equivalent uniform-beam model to obtain an analytic approximation that exhibits scaling with the pertinent parameters. This approach gives the correct result in the low-temperature limit where space charge dominates; and at higher temperatures, where the profiles become more Gaussian, the error is found to be relatively small.

The field energy per unit length for a uniform beam is given in Equation (4.68). However, since the self force is the difference between the repulsive Coulomb force and the attractive magnetic force [i.e., $F_r = qE_r - qv_z B_\theta = qE_r(1 - \beta_0^2)$], we must subtract the magnetic field energy from the electrostatic field energy. This yields [from Equation (4.68), with $f_e = 0$, $f_m = 0$] for the field energy per unit length

$$w = \frac{I^2(1 - \beta_0^2)}{16\pi\epsilon_0 c^2 \beta_0^2}\left(1 + 4\ln\frac{b}{R}\right), \tag{5.517}$$

where $R = 2\tilde{x}$ is the radius of the equivalent uniform beam. The field energy per particle w/N_L is identical to the potential energy qV_s due to the self forces. Since $N_L = I/qv_0$, we obtain

$$qV_s = \frac{w}{N_L} = \gamma_0 m v_0^2 \frac{K}{8}\left(1 + 4\ln\frac{b}{R}\right), \tag{5.518}$$

where we introduced the generalized perveance K defined in (4.127a).

The total average transverse energy per particle is then the sum of the kinetic energy $E_k = \gamma_0 m(\tilde{v}_x^2 + \tilde{v}_y^2)/2$ and the potential energy $E_s = qV_s$, or with $\tilde{v}_x^2 + \tilde{v}_y^2 = 2\tilde{v}_x^2 = 2v_0^2\epsilon^2/4R^2$ from (5.513),

$$E = E_k + E_s = \gamma_0 m v_0^2 \frac{\epsilon^2}{4R^2} + \gamma_0 m v_0^2 \frac{K}{8}\left(1 + 4\ln\frac{b}{R}\right). \tag{5.519}$$

Using this relation and assuming that $v_{01} \approx v_{02} \approx v_0$, we obtain for the total energy difference between the two waists,

$$E_2 - E_1 = \gamma_0 m v_0^2 \left[\frac{\epsilon^2}{4} \left(\frac{1}{R_2^2} - \frac{1}{R_1^2} \right) + \frac{K}{2} \ln \frac{R_1}{R_2} \right]. \tag{5.520}$$

If $R_2 < R_1$, as is the case in the example shown in Figure 5.21, we can conclude that the focusing action of the matching lens increases both the kinetic energy and the space-charge-related potential energy by an amount that can be calculated from Equation (5.519). The longitudinal energy of the beam, the velocity v_0, and the energy factor γ_0 are then reduced correspondingly. Since the corrections in v_0 and γ_0 are usually very small, we neglected them in Equation (5.519) [see also the discussion following Equation (5.516)].

In the periodic channel following the matching lens the total transverse beam energy remains constant and equal to E_2. This is also true for the emittance ϵ, which remains conserved for an ideally matched beam. By contrast, the transverse Hamiltonian for the motion of a single particle in the beam is not a constant, due to the periodic variation in the focusing potential.

If the beam is not perfectly matched, the energy will be greater than for the matched (stationary) case. The excess amount will constitute free energy that can thermalize and hence lead to emittance growth, as discussed in Section 6.2.

Let us now consider case (b) of Figure 5.21, which illustrates the focusing of a beam to a small spot size. With an ideal aberration-free lens there would be no fundamental difference to the matching case (a), except that the radius at the focused beam is usually much smaller. However, with a real lens the aberrations have a much stronger effect in the focusing system than in the matching system. For beams where space-charge forces are not very significant, these nonlinear effects are well understood and well documented in the literature (see our brief review in Section 3.4.6). When space charge is dominant, on the other hand, as in the focusing of very intense, high-brightness beams, the situation is much more complicated. We will therefore limit our discussion to the latter case and use as an illustrative example the experimental investigation of the effects of space charge and lens aberrations in the magnetic focusing of an electron beam by Loschialpo et al. [29]. In this experiment, a 5-keV 190-mA electron beam is focused by a short solenoid whose axial magnetic field can be approximated analytically by an expression of the form (4.127). The lens and beam geometry were deliberately designed to exhibit the effects of the inherent nonlinearity of the lens. Since the spherical aberration was the dominant effect, we will for the purpose of this discussion approximate the solenoid by a thin lens whose action can be described by the equation

$$r_2' - r_1' = -\frac{1}{f} r - \alpha_s r^3, \tag{5.521}$$

where $r = r_1$ in the thin-lens approximation. The third-order term is defined by the positive parameter α_s and has a focusing effect. Figure 5.22 shows the results of the trajectory calculation, which illustrate the focusing of an initially parallel uniform beam by the lens for the two extreme cases where space charge is zero (top) and where the temperature is zero (bottom). On the left side are the trajectories without aberration ($\alpha_s = 0$) and on the right side are the trajectories when the aberrations are present ($\alpha_s \neq 0$). The case without space charge shows the well-known axis crossing of the trajectories at the focal point ($z \approx 7$ cm) when $\alpha_s = 0$ (a) and the spreading of the crossing points when aberration is present (b), as discussed in Section 3.4.6. The behavior of the zero-temperature laminar beam is fundamentally different. Without aberration (c) the trajectories do not cross the axis but form a waist that occurs at a distance that is significantly greater than the focal length ($z_w \approx 10$ cm). When aberration is present (d), the beam breaks up into an inner core whose trajectories behave as in (c) and an outer part whose trajectories cross the axis. This effect can be explained by comparing the applied focusing force from the lens with the defocusing force due to the space charge. In the ideal linear case ($\alpha_s = 0$) these forces are acting in such a way that all trajectories are similar. Furthermore, the transverse kinetic energy acquired by the particles in the passage through the lens is fully converted into potential energy at the waist where the slope of each trajectory is zero ($r' = v_r/v_0 = 0$). When a nonlinear force is present ($\alpha_s \neq 0$), on the other hand, the particles gain additional transverse energy, which increases rapidly with the radius r. Thus, there will be a critical radius r_c beyond which this additional transverse kinetic energy is greater than the potential

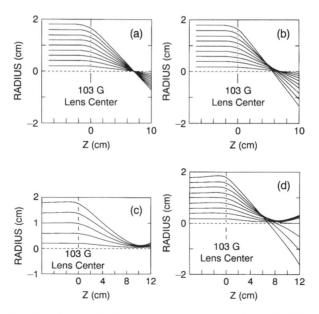

Figure 5.22. Focusing of a parallel beam without space charge by an ideal lens (a) and a lens with spherical aberrations (b); focusing of a laminar beam by an ideal lens (c) and a lens with spherical aberrations (d). (From Reference 29.)

energy at the waist. Therefore, the particles with $r > r_c$ at the lens will cross the axis. As a result, the beam profile, which was uniform initially, will become hollow downstream from the lens. The degree of nonuniformity will depend on the focusing strength f, the spherical aberration parameter α_s and the width of the beam in the lens. The measurements by Loschialpo shown in Figure 5.23 illustrate this effect very graphically. Note that the dip in the profile is most pronounced at the waist, and farther downstream it shows a tendency to flatten out. At higher fields ($B_0 = 147$ G) a single peak develops at $z \geq 20$ cm. As the focusing strength is increased further, the waist becomes smaller and a triple-peak profile develops when the beam expands again beyond the waist. Computer simulations yielded excellent agreement with these experimental observations [29]. Although the beam is not in thermal equilibrium during the focusing process, it is still useful to compare it with the stationary Boltzmann profiles. When the aberrations are absent ($\alpha_s = 0$), the curves in Figure 5.12 will give us a good idea of how the profiles change with temperature increases as the beam is focused down to the waist. The sharp edge of the low-temperature space-charge-dominated initial beam will become more fuzzy when the temperature effects, and hence the emittance term in the envelope equation (5.512), become important or even exceed the space-charge force.

The development of a hollow profile when aberrations are present is also consistent with the stationary Boltzmann density distribution. Consider, for instance, a periodic channel consisting of short solenoids with spherical aberrations, as described by Equation (5.521). The applied focusing potential in the Hamiltonian will then have the form $\phi_1(r, z) = A(z)r^2 + B(z)r^4$, which includes the fourth order aberration term. In the smooth approximation, where the potential function is averaged over z, the zero-temperature density profile will have the parabolic form $n(r) = n(0)[1 + C(r/R)^2]$, where the constant C depends on the aberration coefficient and R is the beam radius. At higher temperatures, the dip in the profile will be washed out (see Problem 5.17).

Spherical aberrations in the electrostatic potential distribution also explain why the electron beams from high perveance guns, such as the gun pictured in Figure 1.1, tend to have a hollow profile when the anode hole is not covered by a mesh. This is true even with the standard Pierce-type electrode geometry [C.3, Chap. 10.1]. The assumption of a uniform density profile made in Pierce's theory is not correct. But this does not affect the electrode design, in which the radial focusing force component balances the space-charge force at the beam edge; from Gauss's law, the latter depends only on the total current and not on the density profile.

The deviation of the density profile in space-charge-dominated beams from a uniform distribution may cause emittance growth, as discussed in Section 6.2. With regard to focusing a high-intensity, high-brightness beam to a small spot size, it is important that lens aberrations be minimized.

As a final topic in this section, let us now briefly discuss the problem if imaging in electron microscopy, ion-beam projection lithography, and other applications. To form an undistorted image of an object it is essential that all types of aberrations be minimized. This includes the effect of space charge, which tends to act like a spherical aberration, as discussed in Section 3.4.6 (Figure 3.14). For imaging

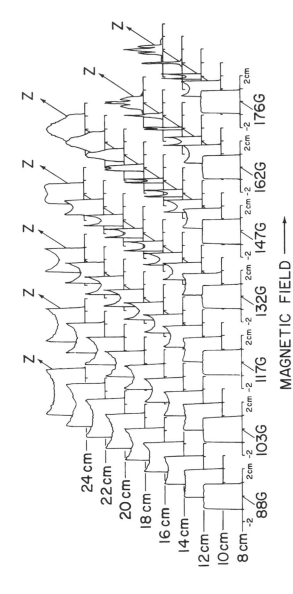

Figure 5.23. Focusing of an initially uniform space-charge-dominated electron beam by a solenoidal lens with spherical aberrations. The curves show the measured current density profiles versus distance for different magnetic field strengths. The lens center is at $z = 8.8$ cm. (From Reference 29.)

purposes the particle source (cathode, plasma) must by necessity have a very small diameter, and the beam current must be relatively low. The beam is therefore temperature dominated, so that the Boltzmann density profile has the Gaussian shape of curve 1 in Figure 5.12. The finite temperature, or emittance, causes chromatic aberrations but also spherical aberrations by radial spreading of the particle distribution into the nonlinear regions of the lenses. These detrimental effects can be minimized by the use of apertures, just as in a camera. The beam current involved in the image formation process is therefore always smaller than the total current emitted from the source. Indeed, the higher the required image resolution, the lower the usable current and current density. If J_s, J_i and r_s, r_i denote the current densities and radii at the object (source) and image, respectively, and $M_i = r_i/r_s$ is the magnification, ideally one would expect that $J_i = M_i^2 J_s$. However, this ideal value can never be reached in practice since current must be sacrificed with the aid of apertures to reduce the aberrations and achieve the desired resolution. This problem is discussed in Pierce's book [C.3, Chap. VIII] and reviewed by Lawson [C.17, Sec. 4.8].

REFERENCES

1. M. Reiser, *Phys. Fluids* **20**, 477 (1977).

2. A. A. Vlasov, *J. Phys. USSR* **9**, 25 (1945).

3. I. M. Kapchinsky and V. V. Vladimirsky, *Proc. International Conference on High Energy Accelerators*, CERN, Geneva, 1959, p. 274 ff. For the generalized treatment presented in Section 5.3.2, see G. P. Saraph and M. Reiser, "Transverse Energy Distribution of Asymmetric or Mismatched K–V Beams", CPB Technical Report 94-002, February 14, 1994, Institute for Plasma Research, University of Maryland, College Park, MD, to be published.

4. D. A. Hammer and N. Rostoker, *Phys. Fluids* **13**, 1831 (1970).

5. P. M. Lapostolle, *IEEE Trans. Nucl. Sci.* **NS-18**, 1101 (1971). In this paper, Lapostolle actually defined $\epsilon_x = 4\bar{\epsilon}_x$ as the "rms emittance." This has caused confusion in the literature, which we try to avoid by calling ϵ_x the "effective emittance."

6. F. J. Sacherer, *IEEE Trans. Nucl. Sci.* **NS-18**, 1105 (1971).

7. J. Struckmeier, J. Klabunde, and M. Reiser, *Part. Accel.* **15**, 47 (1984).

8. P. Debye and E. Hückel, *Z. Phys.* **24**, 185 (1923).

9. G. H. Jansen, *Coulomb Interactions in Particle Beams*, Academic Press, New York, 1990.

10. This problem was discussed in a report by M. Reiser, G. Schmidt, and T. Wangler, "Relativistic Beam Relationships," *CPB Technical Report 91-015*, December 20, 1991, Institute for Plasma Research, University of Maryland, College Park, MD, unpublished.

11. See, for example, R. C. Tolman, *Relativity, Thermodynamics and Cosmology*, Clarendon Press, Oxford, 1934.

12. S. R. DeGroot, W. A. van Leeuwen, and Ch. G. van Weert, *Relativistic Kinetic Theory*, North-Holland/Elsevier, Amsterdam, 1980.

13. M. Reiser and N. Brown, *Phys. Rev. Lett.* **71**, 2911 (1993).

14. I. Hofmann and J. Struckmeier, *Part. Accel.* **21**, 69 (1987).

15. R.A. Jameson, "Advanced High-Brightness Ion rf Accelerator Applications in the Nuclear Arena," in *AIP Conf. Proc.* **253**, 139 (April 1992) ed. W. W. Destler and S. K. Guharay.

16. *Proc. Workshop on Accelerator for Future Spallation Neutron Sources,* Santa Fe, NM, February 16–20, 1993; *Los Alamos National Laboratory Report LA-UR-93-1356.*

17. I. M. Kapchinsky, *Theory of Resonance Linear Accelerators,* Harwood Academic Publishers, New York, 1985, Chap. 3.1.

18. R. L. Gluckstern, in *Linear Accelerators* (ed. P. Lapostolle and A. Septier), North-Holland, Amsterdam, 1967, p. 827.

19. C. K. Allen, N. Brown, and M. Reiser, *Part. Accel.* **45**, 149 (1994).

20. D. X. Wang, J. G. Wang, and M. Reiser, *Appl. Phys. Lett.* **62**, 3232 (1993).

21. N. Brown and M. Reiser, *Part. Accel.* **43**, 231 (1994).

22. D. Neuffer, *IEEE Trans. Nucl. Sci.* **NS-26**, 3031 (1979).

23. T. J. P. Ellison, S. S. Nagaitsev, M. S. Ball, D. D. Caussyn, M. J. Ellsion, and B. J. Hamilton, *Phys. Rev. Lett.* **70**, 790 (1993).

24. For a historical account of non-Liouvillean injection, see R. L. Martin, *AIP Conf. Proc.* **253**, 232 (April 1992), ed. W. W. Destler and S. K. Guharay.

25. C. Rubbia, *Nucl. Instrum. Methods* A **278**, 253 (1989).

26. M. Reiser, "Use of Dispersion to Increase the Space-Charge Limit in Circular Accelerators," *CPB Technical Report 92-003,* Institute for Plasma Research, University of Maryland, College Park, MD, April 1992.

27. R. Chasman, *Proc. 1968 Proton Linear Accelerator Conference,* BNL 50120 (C-54), p. 378.

28. M. Reiser, *J. Appl. Phys.* **52**, 555 (1981).

29. P. Loschialpo, W. Namkung, M. Reiser, and J. D. Lawson, *J. Appl. Phys.* **57**, 10 (1985).

PROBLEMS

5.1 Consider a planar diode formed by two infinite parallel planes separated a distance d with potentials $V = 0$ at $x = 0$ and $V = V_0$ at $x = d$. The plane at $x = 0$ forms a cathode from which a steady stream of electrons is emitted, and as a result, a negative space charge of density $\rho(x)$ is building up in the gap between anode and cathode. If the thermal velocities of the electrons are neglected, a steady-state situation develops in which $dV/dx = 0$ at $x = 0$ and the electron current density reaches an upper limit, J_{max}. The general approach to finding the steady-state solution for $V(x)$ and J_{max} for relativistic electron velocities leads to an equation for V which is not integrable in terms of elementary functions.

   (a) Carry out the analysis relativistically correct and find the (nonintegrable) equation $\int f(V)\,dV = Cx$. Determine the constant C. Explain why the self-magnetic field of the electron stream can be neglected.
   (b) Solve $\int f(V)\,dV = Cx$ for the nonrelativistic limit; obtain $V(x)$ and J_{max} in terms of V_0 and d.

(c) Current flow across the diode can be impeded by applying a uniform magnetic field $\mathbf{B} = B\mathbf{a}_z$ perpendicular to the electric field. Above a critical value B_c, no electron leaving the cathode with zero initial velocity will reach the anode. Derive an expression for B_c (in terms of V_0, d, and other parameters) that is relativistically correct.

5.2 Determine $n(r)$, $v(r)$, and $E_r(r)$ and plot as functions of radius ($0 \le r \le a$) for the laminar beam treated in Section 5.2.1.

5.3 The rigid-rotor equilibrium beam is characterized by the solution $\omega = $ const of the equations of state for all particles in the nonrelativistic energy regime.

 (a) Show that $\omega = \omega_L \pm \omega_L \left[1 - \omega_p^2/2\omega_L^2\right]^{1/2}$ by solving the equations of state.

 (b) Show how ω relates to the magnetic field configuration (B_s — field at the source, $B = $ field in the equilibrium region).

 (c) Find the axial velocity $v_z(r)$ for the entire range of ω values.

 (d) Evaluate and discuss the results (a) to (c) for the cases $\omega = 0$, $\omega = 0.5\omega_L$, $\omega = \omega_L$, and $\omega = 2\omega_L$.

5.4 Consider a cold relativistic electron beam with a total current of 10 kA that is emitted from a magnetically shielded diode with a cathode–anode voltage of 1 MV. The initial beam profile is defined by a radius of $a = 1$ cm and zero slope.

 (a) Determine the distance at which the beam radius doubles when the beam propagates in a field-free drift tube.

 (b) Suppose that the beam is injected into a tube of radius $b = a$. Calculate the magnetic field B_0 necessary to achieve uniform focusing using paraxial theory, with $\gamma = \gamma_a$ determined by the diode voltage.

 (c) Determine the variation with radius of the energy parameter $\gamma = \gamma(r)$ assuming that $\beta \approx 1$ and density $n = n_0 = $ const. Using the value $\gamma_0 = \gamma(0)$ on the axis rather than $\gamma_a = \gamma(a)$, recalculate the magnetic field B_0 necessary to achieve uniform focusing.

 (d) Compare the paraxial result with the exact self-consistent theory of relativistic Brillouin flow equilibrium of Section 5.2.3 by calculating the equilibrium current that corresponds to the magnetic field B_0 obtained in the two cases (b) and (c). Explain why the results differ.

5.5 Show that the relations (5.45b), (5.46a), and (5.51a) for a nonrelativistic solid Brillouin beam can be obtained from the corresponding relativistic equations (5.72), (5.70), and (5.69).

5.6 Derive equations (5.105) and (5.106).

5.7 With q_i denoting the three space variables and p_i the three conjugate canonical momenta, Liouville's theorem may be stated in the alternative forms

$$\frac{df(q_i, p_i)}{dt} = 0, \qquad \iint d^3q\, d^3p = \text{const.}$$

Prove that the theorem also holds in q, P space, where P is the mechanical momentum, that is,

$$\iint d^3q\, d^3p = \iint d^3q\, d^3P.$$

5.8 Prove that the generalized K–V distribution $f = f_0\delta(G - 1)$ represents in the x–y plane a beam with elliptic cross section (semiaxes X and Y) and uniform charge density $\rho = I/\pi v XY$, where I is the total beam current, v the particle velocity in the z-direction, $X = \sqrt{\hat{\beta}_x \epsilon_x}$, and $Y = \sqrt{\hat{\beta}_y \epsilon_y}$. [*Hint:* It will be helpful to introduce new variables α, ψ by the transformations

$$w_x x' - w'_x x = \alpha \cos\psi$$
$$w_y y' - w'_y y = \alpha \sin\psi$$

and to make use of the properties of the Dirac delta function, which in this case takes the form $\delta(\alpha^2 - \alpha_0^2)$, where α_0^2 represents a function that is constant with regard to the integration.]

5.9 Consider a K–V beam whose projection in x–x' trace space corresponds to a tilted ellipse in the Courant–Snyder form of Equation (3.22) (i.e., $\hat{\gamma}x^2 + 2\hat{\alpha}xx' + \hat{\beta}x'^2 = \epsilon_x$ with $\hat{\alpha} \neq 0$).

(a) Evaluate the rms emittance $\tilde{\epsilon}_x$ as defined in Equation (5.205) and show that $\epsilon_x = 4\tilde{\epsilon}_x$.
(b) Calculate the first moments $\bar{x}$ and $\bar{x'}$ of this distribution.

5.10 Prove that the brightness of a K–V beam is given by

$$B = \frac{2I}{\pi^2\epsilon^2},$$

where $\epsilon = \epsilon_x = \epsilon_y$ is the 100% emittance of the beam in each transverse direction.

5.11 (a) Carry out the relativistic transformation of the Maxwell–Boltzmann distribution (5.266) from the beam frame to the laboratory frame using the covariant relations (5.261) to (5.265). *Hint:* Define the laboratory velocity components by $\beta_{l\perp}^2 = \beta_{lx}^2 + \beta_{ly}^2$, $\beta_{lz} = \beta_0 + \Delta\beta_{lz}$; assume that $\beta_{l\perp} \ll \beta_0$, $\Delta\beta_{lz} \ll \beta_0$, and expand $\gamma_l(\beta_{l\perp}, \Delta\beta_{lz})$ about the center-of-momentum value $\gamma_0 = (1 - \beta_0^2)^{-1/2}$ up to second order in $\beta_{l\perp}$, $\Delta\beta_{lz}$.

(b) Show that the laboratory distribution (5.268a) in a relativistic beam can be obtained directly from the beam-frame distribution (5.266) by applying the Lorentz transformations for the velocities and the scalar potential and then using the temperature relation (5.260).

5.12 Consider a K–V type distribution where the particles occupy uniformly the surface of a hyperellipsoid in six-dimensional phase space. Calculate the longitudinal charge density profile $\rho_L(z)$ and show that it does not yield a linear force in the variable z.

5.13 Show that the rms kinetic energy per particle in the nonrelativistic Boltzmann distribution

$$f(v) = f_0 \, \exp\left[-\frac{m(v_x^2 + v_y^2 + v_z^2)}{2k_B T} \right]$$

is $\frac{3}{2}k_B T$. Determine $\tilde{v}_x$, $\tilde{v}_\perp = (v_x^2 + v_y^2)^{1/2}$, and $\tilde{v}$ as functions of the temperature $k_B T$.

5.14 Prove that the longitudinal distribution function $f(H_\parallel) = f_{\parallel 0}\sqrt{2(H_{max} - H_\parallel)}$ of Equation (5.425) yields the longitudinal line-charge density variation $\rho_L(z)$ given in Equation (5.409).

5.15 Solve the longitudinal envelope equation (5.416) for a cold ($\epsilon_{zz'} = 0$) drifting beam with initial conditions at $s = 0$ of $z_m(0) = z_0$, $z_m'(0) = z_0'$. Find the distance s_w where the beam envelope goes through a minimum (waist) defined by z_w and determine the compression ratio z_0/z_w as a function of the beam parameters.

5.16 Perform the integration (5.348a) of the longitudinal Maxwell Boltzmann distribution that yields the density profile (5.348b), and determine the constant C. What is the value of C in the case where the space-charge potential $\phi_{\parallel s}$ is negligible?

5.17 Consider the stationary transverse Boltzmann distribution in a smooth-focusing channel with an applied potential function of the form $\phi_{\perp a}(r) = \gamma_0 m_0 v_0^2 k_0^2 (r^2 + Ar^4)$, where k_0 is the focusing constant for the linear part of the force and A represents the spherical aberration. Determine and sketch the density profiles for the zero-temperature case ($k_B T_\perp = 0$) and for the high-temperature case, where the space-charge potential can be neglected ($\phi_{\perp s} = 0$). Choose the constant A so that $n(0) = 0.6n_1$, where $n_1 = n(R)$ and R is the radius of the beam in the case $k_B T_\perp = 0$.

5.18 Consider an axisymmetric beam with an arbitrary density profile $n(r)$, a radial electric field $E_r(r)$, and a number of particles per unit length N_L.

(a) Prove that the average $\overline{rE_r}$ has a value that is independent of the shape of the radial density profile.

(b) Prove that $\overline{xE_x} = \frac{1}{2}\overline{rE_r}$, where E_x is the x-component of the electric field.

(c) Show that $\overline{xE_x}/\tilde{x} = K/4\tilde{x}$, as stated in (5.216).

5.19 Find the solutions (5.506), (5.507) for the coupled envelope equations (5.504), (5.505).

5.20 By analogy with Equations (5.293) and (5.420), the general solutions of the coupled envelope equations (5.502) and (5.503) can be approximated by

$$
a \approx \left[\left(\frac{3}{2} \right)^2 \frac{Nr_c}{\beta_0^2 \gamma_0^2} \frac{1}{k_{z0}^2} \left(\frac{k_{x0}^2}{k_{z0}^2} + \frac{1}{2} \right)^{-2} + \left(\frac{\epsilon_{nx}}{\beta_0 \gamma_0 k_{x0}} \right)^{3/2} \right]^{1/3}
$$

$$
z_m \approx \left[\frac{2}{3} \frac{Nr_c}{\beta_0^2 \gamma_0^5} \frac{1}{k_{z0}^2} \left(\frac{k_{x0}^2}{k_{z0}^2} + \frac{1}{2} \right) + \left(\frac{\epsilon_{nz}}{\beta_0 \gamma_0 k_{z0}} \right)^{3/2} \right]^{1/3}
$$

Evaluate the accuracy of these expressions.

5.21 Consider the ellipsoidal bunch with uniform volume charge density ρ_0 and semi-axes a and z_m discussed in Sections 5.4.7 and 5.4.11.

(a) Show that the rms widths and emittances are given by $\tilde{x} = a\sqrt{5}$, $\tilde{z} = z_m/\sqrt{5}$, $\tilde{\epsilon}_x = \epsilon_x/5$, $\overline{\epsilon_z} = \tilde{\epsilon}_z/5$.

(b) Prove that in the long-bunch limit where $z_m \to \infty$, Equation (5.494) becomes identical to the transverse K–V envelope equation (4.85a), as one would expect.

CHAPTER 6

Emittance Growth

6.1 CAUSES OF EMITTANCE CHANGE

In the self-consistent theory of Chapter 5 we limited our analysis for the most part to stationary or quasistationary beams where the applied focusing forces are linear and the emittances associated with each direction are constant. These beams are best described by a Maxwell–Boltzmann distribution with different transverse and longitudinal temperatures. The forces arising from the space charge of such stationary beams are in general nonlinear except at very low temperatures, where the perveance dominates over the emittance and where the transverse density profile tends to be uniform. However, in the equilibrium state the nonlinear space-charge forces do not, by definition, cause any changes in temperature and emittance.

Real laboratory beams are usually not in perfect equilibrium, and there are a large number of effects that can cause the temperature and emittance to increase. The most important causes of emittance growth are the following:

- Nonlinearities in the applied forces
- Chromatic aberrations
- Nonlinear forces arising from nonstationary beam density profiles
- Beam mismatch causing oscillations of the rms radius
- Beam off-centering causing coherent oscillations around the optical axis or central orbit
- Misalignments of the focusing and accelerating elements
- Collisions between the beam particles (Coulomb scattering) and between the beam and a background gas or a foil
- Instabilities, including unstable interactions with applied or beam-generated electromagnetic fields
- Nonlinear single-particle resonances and nonlinear coupling between longitudinal and transverse motion (especially important in circular accelerators)
- Beam–beam effects in the interaction regions of high-energy colliders

• Random kicks due to rf noise, mechanical vibrations of the magnets, and other sources of statistical fluctuations (limiting the lifetime of beams in storage rings)

There are also effects that cause the emittance to decrease. An example of this type is *equipartitioning,* where Coulomb collisions or collective forces tend to drive a beam with an anisotropic temperature distribution toward three-dimensional thermal equilibrium. Thus, if the temperature is high in the transverse direction, it will fall, while the low temperature in the longitudinal direction will rise until both temperatures are equal. As a result, the transverse emittance in this case will become smaller while the longitudinal emittance will increase. Such equipartitioning will be discussed in connection with the *Boersch effect,* intrabeam scattering, and instabilities being treated in this chapter. A brief section is devoted to *beam cooling* schemes in storage rings where the six-dimensional phase-space volume of a beam (i.e., transverse and longitudinal emittance) is reduced. The three schemes that have been employed most successfully are electron beam cooling of ions, stochastic cooling, and radiation cooling of electrons.

However, the major topic of this chapter is emittance growth, which is one of the most fundamental issues in beam physics. Many advanced accelerator applications, such as high-energy colliders, heavy-ion inertial fusion, and free electron lasers, require beams with very small emittance and high beam intensity. As discussed in Section 1.3, some modern particle sources produce beams with high intrinsic phase-space density $I/\tilde{\varepsilon}_n$ (or brightness $2I/\pi^2\tilde{\varepsilon}_n^2$), which is often more than adequate for a particular application. Near the source and in the low-energy part of the accelerator system, such beams are dominated by the space-charge forces, which depend strongly on the shape of the particle distribution. As we will see in the next section, any deviation from the nearly uniform density profile of a space-charge-dominated Maxwell–Boltzmann distribution will cause emittance growth. This is true even if the rms radius is matched to the acceptance of the focusing channel (where the emittance remains constant when the space charge is negligible).

As an example, consider the gas focusing of high-brightness proton or H^- beams discussed in Section 4.6.2. The collisions with the gas molecules and the resulting charge neutralization of the beam will produce a Boltzmann distribution with a Gaussian density profile of the form (5.316b), where $n(0) = n_0$ if the charge is fully neutralized. When the beam enters the radio-frequency-quadrupole (RFQ) accelerator, where the focusing is entirely by electromagnetic forces, it experiences a rapid change toward the uniform density profile of the ideal Maxwell–Boltzmann distribution. The Gaussian beam has more electrostatic field energy than the ideal uniform beam. During the charge homogenization process this energy difference will be converted into thermal energy, which will cause emittance growth. The energy conversion is driven by the nonlinear space-charge forces associated with the nonuniform initial density profile.

On the other end of the spectrum is the emittance growth caused by nonlinear external focusing forces in beams where space charge is negligible. To illustrate this effect let us consider the propagation of a beam through a periodic channel consisting of axisymmetric lenses with spherical aberration. For simplicity we assume that the lenses are thin, so that the change of the slope of a particle's trajectory at each lens crossing is given by

$$\Delta r' = -a_1 r - a_3 r^3, \tag{6.1}$$

where $a_1 = 1/f$ is defined by the focal length f and a_3 by the spherical aberration coefficient C_3 (see Section 3.4.6). Without aberration ($a_3 = 0$) the beam can be perfectly matched, as discussed in Section 3.8.1, and the trace-space ellipse will rotate, keeping the area constant and maintaining an elliptic shape. On the other hand, when the spherical aberration is present ($a_3 \neq 0$), the trace-space ellipse will be distorted, as shown in Figure 6.1. The area enclosed by the trace-space boundary remains constant, in agreement with Liouville's theorem. However, the illumentation due to the aberration becomes progressively worse. After a sufficient number of periods the particle distribution fills a diluted trace-space area that is bounded by an ellipse of larger size than the initial ellipse. The increase in the effective trace-space area can be measured by evaluating the rms emittance of the distorted distribution and comparing it with the rms emittance at the beginning of the channel.

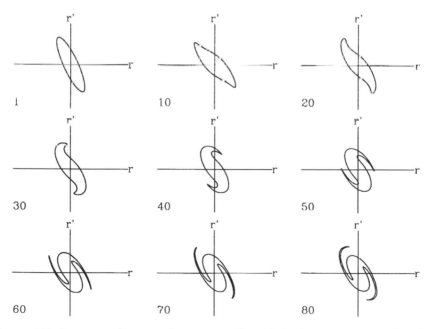

Figure 6.1. Progressive distortion of trace-space ellipse during beam propagation through a periodic channel of thin lenses with spherical aberrations. The numbers associated with each figure indicate the lens periods that have been traversed.

A problem with the rms emittance is that it tends to give more weight to particles with large amplitudes (x, x') since it is based on an evaluation of the second moments of the distribution [see Equations (5.240) through (5.246)]. Thus the protrusion from the ellipse developing in the first few lens crossings of Figure 6.1 causes an rms emittance growth that is considerably larger than the percentage of particles involved in this effect. By placing collimators with appropriately small apertures into the focusing channel, one can intercept these particles and prevent the rms emittance increase that would otherwise occur.

In the more general case, where nonlinear forces from both the applied fields as well as from the space-charge fields are present, the theoretical analysis is exceedingly difficult, and computer simulations become indispensable. Analytical modeling of beams with space charge is by and large restricted to relatively simple systems such as the uniform or linear periodic channels discussed in Chapter 5. To evaluate emittance growth in nonlinear periodic channels, beam matching systems, transfer lines, and so on, one must rely almost exclusively on computer simulation and experiment.

We begin our formal discussion in the next section with an investigation of the transverse emittance growth in linear focusing channels when the beam is not in the equilibrium state corresponding to a Boltzmann density profile.

6.2 FREE ENERGY AND EMITTANCE GROWTH IN NONSTATIONARY BEAMS

6.2.1 Analytical Theory

In Sections 5.3.3 and 5.4.4 we discussed the stationary state of a continuous beam in a linear, uniform focusing channel. According to the smooth-approximation theory, the uniform channel is also a good model for a linear periodic focusing system. From a thermodynamic point of view, the equilibrium state of such a beam is best described by a transverse Maxwell–Boltzmann distribution. The temperature-dependent Boltzmann profiles are shown in Figure 5.12, and in the space-charge-dominated regime these profiles tend to be uniform.

Let us now examine what happens when the beam does not satisfy the stationary-state requirements at injection into the focusing channel. The three most important examples of such a "nonstationary" initial beam are mismatch in the density profile (e.g., the beam is not uniform in the low-temperature, space-charge-dominated case), mismatch in the rms radius, and off-centering, or a combination of these three effects. As we know from thermodynamics, a nonstationary initial beam has a higher total energy per particle than that of the corresponding stationary beam. The energy difference ΔE between the nonstationary and the stationary beam represents *free energy* that can be thermalized by nonlinear space-charge forces, instabilities, or collisions. This produces emittance growth as the beam relaxes toward a final stationary state at the higher energy per particle [1].

Since the Boltzmann profile is nonanalytic in general, it will not be possible to model the system in a mathematically exact form. Instead, we use the concept of equivalent beams introduced in Section 5.3.4 to obtain an approximate description following the theory developed in Reference 1. This concept implies that the behavior of a general nonuniform particle distribution can be modeled with good approximation by using the equivalent uniform K–V beam having the same second moments (rms width, rms divergence, rms emittance), current, and kinetic energy. The theory compares the initial nonstationary beam with the equivalent stationary distribution to determine the free energy ΔE. It then assumes that the beam relaxes into a stationary distribution at the higher energy $E + \Delta E$. Using force-balance and energy-conservation relations, the change in beam radius and emittance is then calculated as a function of the free energy and the tune depression k_i/k_0 defining the ratio of the betatron wave constant k_i for the initial stationary beam with space charge and the betatron wave constant k_0 without space charge.

Consider a continuous round beam with current I, particle kinetic energy $(\gamma - 1)mc^2$, rms width $\tilde{x} = \tilde{y}$, transverse rms velocity $\tilde{v}_x = \tilde{v}_y$, and (unnormalized) rms emittance $\tilde{\epsilon}_x = \tilde{\epsilon}_y = \tilde{\epsilon}$ in a linear focusing channel and surrounded by a conducting tube of radius b. Assume that $v \simeq v_z \gg \tilde{v}_x$. For a periodic focusing channel the smooth-approximation theory (see Sections 4.4.1 and 4.4.2) relates the wave number without space charge, k_0, to the phase advance per period, σ_0, and the period length, S, by $k_0 = \sigma_0/S$. The presence of the beam's self field will reduce the net focusing force acting on the particle, and the wave number, oscillation wavelength, and phase advance *with* space charge will be defined by k, λ, and σ, respectively, so that $k = 2\pi/\lambda = \sigma/S$.

It will be convenient to use the effective quantities $a = 2\tilde{x}$, $v_x = 2\tilde{v}_x$, $\epsilon = 4\tilde{\epsilon}$, and the generalized perveance $K = (I/I_0)(2/\beta^3\gamma^3)$, where $I_0 = 4\pi\epsilon_0 mc^3/q$ is the characteristic current. According to the theory developed in Section 4.3.2, the stationary state of a beam in a linear focusing channel is characterized by a constant effective radius and perfect balance between external focusing force, $k_0^2 a$, self force, K/a, and the emittance term, ϵ^2/a^3. The relevant equations (4.88) and (4.89) will be repeated here for convenient reference:

$$k_0^2 a - \frac{K}{a} - \frac{\epsilon^2}{a^3} = 0, \tag{6.2}$$

which may be written in terms of the wave number *with* self fields, k, as

$$k^2 a - \frac{\epsilon^2}{a^3} = 0, \qquad \text{or} \qquad \epsilon = ka^2, \tag{6.3}$$

with

$$k^2 = k_0^2 - \frac{K}{a^2} = k_0^2 - \frac{K}{4\tilde{x}^2}. \tag{6.4}$$

For such a stationary particle distribution, the total energy is a minimum, and the density profile is practically uniform when the beam is space-charge domin-

ated (i.e., when $Ka^2 \gg \epsilon^2$). The average transverse kinetic energy per parti-
cle is $E_k = \frac{1}{2}\gamma m(\tilde{v}_x^2 + \tilde{v}_y^2) = \gamma m \tilde{v}^2 \tilde{x}'^2$, where $x' = dx/dz$ and nonrelativistic
transverse velocities are assumed. Since $\tilde{x}' = k\tilde{x}$, we have $E_k = \gamma m v^2 k^2 \tilde{x}^2$.
The average potential energy per particle due to the external focusing force is
$E_p = \gamma m v^2 k_0^2 \tilde{x}^2$. The average energy per particle associated with the self forces
of the beam E_s was calculated in Equation (5.518). Using $K = (k_0^2 - k^2)4\tilde{x}^2$ from
(6.4), one gets $E_s = \gamma m v^2 (k_0^2 - k^2)\tilde{x}^2[1 + 4\ln(b/2\tilde{x})]/2$. With $a = 2\tilde{x}$, one thus
obtains for the total energy per particle in a stationary beam,

$$E = E_k + E_p + E_s = \frac{\gamma m v^2}{4}\left[k^2 a^2 + k_0^2 a^2 + \frac{1}{2}[k_0^2 - k^2]a^2\left(1 + 4\ln\frac{b}{a}\right)\right].$$

$$(6.5)$$

Let us now suppose that the beam injected into the focusing channel is not
perfectly matched and that the total energy per particle is E_n, while E_i represents
the energy in the equivalent matched (stationary) beam. The free energy per particle
is then $\Delta E = E_n - E_i$. The possible emittance growth can be calculated if we
assume that the nonstationary initial beam with energy E_n will relax into a stationary
state with final energy $E_f = E_n = E_i + \Delta E$ due to the action of nonlinear space
charge forces or other effects. Since both the initial stationary state (denoted by the
subscript i) and the final stationary state (subscript f) must obey Equation (6.5),
we obtain the following total energy relation:

$$\frac{\gamma m v^2}{4}\left[k_f^2 a_f^2 + k_0^2 a_f^2 + \frac{1}{2}\left(k_0^2 - k_f^2\right)a_f^2\left(1 + 4\ln\frac{b}{a_f}\right)\right]$$
$$= \frac{\gamma m v^2}{4}\left[k_i^2 a_i^2 + k_0^2 a_i^2 + \frac{1}{2}(k_0^2 - k_i^2)a_i^2\left(1 + 4\ln\frac{b}{a_i}\right)\right] + \Delta E. \quad (6.6)$$

It will be convenient to write ΔE in the form

$$\Delta E = \frac{1}{2}\gamma m v^2 k_0^2 a_i^2 h,\qquad(6.7)$$

where h is a dimensionless parameter that can be calculated for each effect,
producing free energy and emittance growth. Furthermore, from (6.4) we have $k_f^2 = k_0^2 - K/a_f^2$ and $k_i^2 = k_0^2 - K/a_i^2$, hence $k_f^2 = k_0^2 - (a_i/a_f)^2(k_0^2 - k_i^2)$. Using
this result for k_f^2 and substituting (6.7) into (6.6), we find the following relation:

$$\left(\frac{a_f}{a_i}\right)^2 - 1 - \chi\ln\frac{a_f}{a_i} = h,\qquad(6.8)$$

where

$$\chi = 1 - \frac{k_i^2}{k_0^2}.\qquad(6.9)$$

Since a_i, χ, and h are known from the initial beam, we can calculate the final effective beam radius u_f from Equation (6.8). Figure 6.2 shows a_f/a_i versus h for various ratios k_i/k_0. For $a_f - a_i \ll a_i$, one obtains from (6.8) the first-order relation

$$\frac{a_f}{a_i} \simeq 1 + \frac{h}{2 - \chi} = 1 + \frac{h}{1 + (k_i/k_0)^2}, \qquad (6.10)$$

which is sufficient for most cases of practical interest.

Next, using Equation (6.3), we obtain for the emittance difference between the final and initial stationary beam

$$\Delta \epsilon^2 = \epsilon_f^2 - \epsilon_i^2 - k_f^2 a_f^4 - k_i^2 a_i^4, \qquad (6.11)$$

With $k_f^2 = k_0^2 - (a_i/a_f)^2(k_0^2 - k_i^2)$ and $\epsilon_i^2 = k_i^2 a_i^4$, we find for the final emittance

$$\epsilon_f = \frac{a_f}{a_i}\left\{\epsilon_i^2 + k_0^2 a_i^4\left[\left(\frac{a_f}{a_i}\right)^2 - 1\right]\right\}^{1/2}, \qquad (6.12a)$$

or

$$\frac{\epsilon_f}{\epsilon_i} = \frac{a_f}{a_i}\left\{1 + \frac{k_0^2}{k_i^2}\left[\left(\frac{a_f}{a_i}\right)^2 - 1\right]\right\}^{1/2}. \qquad (6.12b)$$

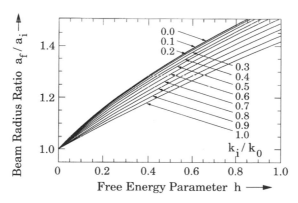

Figure 6.2. Ratio a_f/a_i of final and initial stationary beam radius versus the free-energy parameter h for different values of k_i/k_0, where $k_0 = 2\pi/\lambda_0 = \sigma_0/S =$ external focusing wavenumber, $k_i = 2\pi/\lambda = \sigma/S =$ initial focusing wavenumber with self fields. (From Reference 1.)

The laminar beam case is obtained from (6.12a) by setting $\epsilon_i = 0$. The emittance increase, calculated by substituting a_f/a_i from (6.8) into (6.12), is plotted in Figure 6.3 versus h for different initial tune depressions k_i/k_0.

If $a_f - a_i \ll a_i$, we can use (6.10) and get the first-order approximation

$$\frac{\epsilon_f}{\epsilon_i} = \left(1 + 2\frac{k_0^2}{k_i^2}h\right)^{1/2}. \tag{6.13}$$

Note that the above formulas can be applied to a periodic-focusing channel by substituting σ_0/σ_i for k_0/k_i.

It is very important to recognize that the above relations define the theoretically *possible* increase in beam radius and emittance due to free energy. The predicted change will occur only if nonlinear external or space-charge forces or stochastic effects (e.g., rf noise, Coulomb collisions, etc.) act on the beam to thermalize the free energy. Take as an example the case where space-charge forces are negligibly small, so that $k_i/k_0 \approx 1$, and where the applied focusing force is perfectly linear. If the beam is mismatched (case 2 below) or off-centered (case 3 below) it will have a higher total energy compared to the ideally matched and centered beam, due to the additional kinetic energy associated wtih the coherent envelope and centroid oscillations. This excess amount of energy constitutes free energy that could in principle be thermalized. The above formulas correctly calculate this free energy and the possible emittance growth. However, since no nonlinear forces are present, this energy would not be thermalized, and the beam would continue to perform mismatch or off-centering oscillations. In storage rings, of course, where the beam lifetime is very long, stochastic effects such as rf noise or Coulomb collisions would eventually thermalize the free energy.

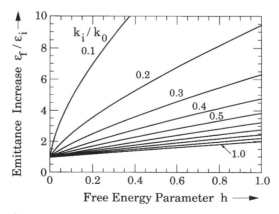

Figure 6.3. Emittance growth ϵ_f/ϵ_i versus free-energy parameter h for different values of the initial space-charge tune depression k_i/k_0. (From Reference 1.)

Let us now evaluate the free-energy parameter h for nonuniform, mismatched, and off-centered beams.

Case 1 (Nonuniform Charge Distribution)   If $U = w_n - w_u$ denotes the field energy difference per unit length between the nonuniform and the uniform (stationary) initial beam, one can show that

$$h = h_s = \frac{1}{4}\left(1 - \frac{k_i^2}{k_0^2}\right)\frac{U}{w_0}, \tag{6.14}$$

where $w_0 = I^2/(16\pi\epsilon_0\beta^2 c^2)$ and U/w_0 is a dimensionless parameter. If the linear approximations (6.10) and (6.13) are valid, we find that

$$\frac{\epsilon_f}{\epsilon_i} = \left[1 + \frac{1}{2}\left(\frac{k_0^2}{k_i^2} - 1\right)\frac{U}{w_0}\right]^{1/2}. \tag{6.15}$$

As an example, for a Gaussian distribution one has $U/w_0 = 0.154$. If $k_i/k_0 = 0.2$, one finds $h_s = 0.037$ and $\epsilon_f/\epsilon_i = 1.688$.

Historically, the emittance growth in space-charge-dominated beams having nonuniform (nonstationary) density profiles was first identified in connection with computer simulation studies by Struckmeier, Klabunde, and Reiser, and an equation of the form (6.15) was derived which showed good agreement with the simulation results [2]. Wangler then derived the differential equation for the emittance change [3]

$$\frac{d(\epsilon)^2}{dz} = -a^2 K \frac{d}{dz}\left(\frac{U}{w_0}\right), \tag{6.16}$$

and showed that it yields the solution (6.15) if the radius remains constant or does not change significantly. (This differential equation had been derived earlier by Lapostolle [4], who, however, at that time thought that the effect was not very significant.) Detailed computer simulations of nonuniform beams injected into a uniform focusing channel confirmed the theoretical predictions and revealed that the emittance grows very rapidly in a distance z_m that corresponds to a quarter of a beam plasma period given by [3]

$$z_m = \frac{\lambda_p}{4} = \frac{\pi v}{2\omega_p} = \frac{\pi a}{2\sqrt{2K}}. \tag{6.17}$$

Anderson obtained the same result by analytically modeling the dynamic evolution of a nonuniform laminar sheet beam [5]. Hofmann and Struckmeier extended the theory to three-dimensional bunched beams [6].

Case 2 (Mismatched Beam) In x–x' phase space a mismatched beam is represented by a tilted ellipse that rotates clockwise as the beam propagates along the focusing channel, and the effective radius oscillates between the minimum and maximum values, which are denoted by a_0 and a_1 in Figure 6.4. We choose the upright (waist) position of the ellipse with semiaxes a_0 and a_0', as indicated in Figure 6.4, to evaluate the free energy associated with the mismatch. For an initially tilted ellipse in the Courant–Snyder form [Equation (5.160b)], $\hat{\gamma}_0 x^2 + 2\hat{\alpha}_0 xx' + \hat{\beta}_0 x'^2 = \epsilon_i$, the two radii a_0 and a_1 are given by

$$
a_\pm^2 = \frac{\epsilon_i}{2}\left[\left(\frac{\hat{\gamma}_0}{k_i^2} + \hat{\beta}_0\right) \pm \sqrt{\left(\frac{\hat{\gamma}_0}{k_i^2} + \hat{\beta}_0\right)^2 - \frac{4}{k_i^2}}\right], \qquad (6.18)
$$

where $a_1 = a_+$, $a_0 = a_-$, and $\hat{\gamma}_0, \hat{\alpha}_0, \hat{\beta}_0$ are the usual Courant–Snyder parameters.

The customary assumption is that nonlinear external forces will eventually cause the beam to fill the enclosing ellipse with width a_1 and slope a_0' so that the effective emittance increase due to mismatch is then simply calculated as

$$
\frac{\epsilon_{\text{eff}}}{\epsilon_i} = \frac{a_1}{a_0}. \qquad (6.19)
$$

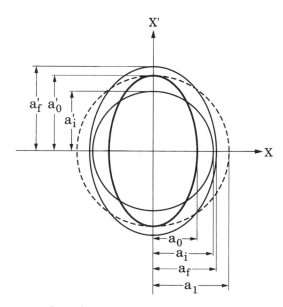

Figure 6.4. Trace-space ellipses for the initial mismatched beam (radius a_0, maximum slope a_0') and the corresponding "effective" emittance $(a_1 a_0')$, the initial stationary beam (a_i, a_i'), and the final stationary beam $(a_f a_f')$. (From Reference 1.)

We note in this context that P. Lapostolle in a 1970 CERN report [7] proposed another empirical relation, which applies to both the symmetrical (in-phase) and the antisymmetrical (180° out-of-phase) mismatch and which was found very useful and accurate in computer simulation studies.

In our model we compare the mismatched beam with the stationary (matched) beam having the same emittance but semiaxes a_i, a_i', as indicated in Figure 6.4. For convenience we assume that the mismatch oscillations in the x and y directions are in phase, so that according to Equation (4.204) the envelope oscillation wave number $k_e = 2\pi/\lambda_e$ is given by $k_e = (2k_0^2 + 2k_i^2)^{1/2}$.

The energy difference per particle between the initial mismatched beam (subscript 0) and the initial matched beam (subscript i) is calculated to be

$$\Delta E = \frac{\gamma m v^2}{4} \left[a_0'^2 - a_i'^2 + k_0^2(a_0^2 - a_i^2) + 2(k_0^2 - k_i^2)a_i^2 \ln \frac{a_i}{u_0} \right]. \tag{6.20}$$

Since the initial emittance of the two beams is assumed to be the same, we have $a_0 a_0' = a_i a_i'$; furthermore, $a_i' = k_i a_i$ and $a_0' = a_i a_i'/a_0 = k_i a_i^2/a_0$. Using these relations, we obtain for the free-energy parameter $h = h_m$ due to mismatch

$$h_m = \frac{1}{2} \frac{k_i^2}{k_0^2} \left(\frac{a_i^2}{a_0^2} - 1 \right) - \frac{1}{2} \left(1 - \frac{a_0^2}{a_i^2} \right) + \left(1 - \frac{k_i^2}{k_0^2} \right) \ln \frac{a_i}{a_0}. \tag{6.21}$$

The mismatch leads to a possible final emittance of $\epsilon_f = a_f a_f'$ (indicated in Figure 6.4) that can be calculated for any given value of h_m from Equations (6.8) and (6.12) or, if $h_m \ll 1$, from (6.13). As an example, we find for a mismatched beam with $a_0/a_i = 0.8$ and $k_i/k_0 = 0.2$ from (6.21) $h_m = 0.0455$, hence a radius increase of $u_f/u_i = 1.0413$ and an emittance growth of $\epsilon_f/\epsilon_i = 1.836$. By comparison, the effective emittance growth factor from (6.19), with $a_1 = a_0 + 2(a_i - a_0)$, is found to be $\epsilon_{eff}/\epsilon_i = 2(a_i/a_0) - 1 = 1.5$.

Case 3 (Off-Centered Beam) Let us assume that the beam is properly matched but off-centered in the x-direction, as shown in Figure 6.5. The centroid of the beam will perform coherent oscillations about the axis of the ideal focusing channel and move on the small ellipse having semiaxes x_c and x_c'. The wave number $k_c = 2\pi/\lambda_c$ of the coherent oscillations is given by $k_c = k_0$, or if image forces are present, by [see Equations (4.238) to (4.240)]

$$k_c = (k_0^2 - k_{im}^2)^{1/2}, \tag{6.22}$$

where

$$k_{im}^2 = \begin{cases} \dfrac{K}{b^2} & \text{for} \quad \tau < \tau_m \\[2mm] \dfrac{K\gamma^2}{b^2} & \text{for} \quad \tau > \tau_m. \end{cases} \tag{6.23}$$

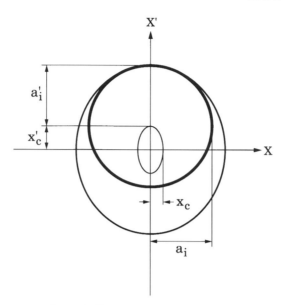

Figure 6.5. Trace-space ellipse of off-centered beam. The centroid moves clockwise along the small ellipse while the outermost point in the beam traces the large outer ellipse. (From Reference 1.)

τ_m is the magnetic diffusion time defined by $\tau_m = 4d^2\sigma\mu/\pi^2$, where d is the wall thickness of the conducting tube, σ the conductivity, and μ the magnetic permeability of the wall material. Evaluating the total energy of the off-centered beam where the centroid coincides with the beam axis (Figure 6.5), we obtain

$$\Delta E = \frac{\gamma m}{2}v_c^2 = \frac{\gamma m}{2}v^2 x_c'^2. \tag{6.24}$$

Since $x_c' = k_c x_c$, we get for the free-energy parameter

$$h = h_c = \left(\frac{x_c}{a_i}\right)^2 \frac{k_c^2}{k_0^2}. \tag{6.25}$$

As an example, if $k_c = k_0$, $x_c = 0.2a_i$, we have $h_c = 0.04$, and for $k_i = 0.2k_0$, we find from (6.13) that $\epsilon_f/\epsilon_i = 1.732$.

The examples given indicate that each of these three effects can cause considerable emittance increase. In practice, all three effects can be present, and the associated free-energy terms add linearly:

$$h = h_s + h_m + h_c. \tag{6.26}$$

While the uniform *equivalent K–V beam* was used to model the behavior of an initially nonstationary distribution in a linear focusing channel it is important to recognize that an *ideal* nonstationary K–V beam would not exhibit any emittance growth. Since all forces are linear in this case, the beam envelope would oscillate indefinitely and the excess free energy would not thermalize, leaving the emittance constant. Real beams, however, differ from the K–V distribution, and any perturbation in the equilibrium density profile gives rise to nonlinear space-charge forces, which may lead to thermalization of the free energy and hence emittance growth. The nonlinear collective forces have the same effect as collisions in thermalizing a particle distribution [8]. The available free energy is, however, not entirely thermalized. Some of the energy will be converted to potential energy, due to the change in beam radius and density profile. Relation (6.12) of the theory will therefore tend to slightly overestimate the emittance growth since it does not take into account the nonuniform part of the field energy in the final stationary state. Furthermore, a small fraction of the particles with large transverse energy may form a "halo" surrounding the thermal core of the beam and contributing a disproportionate amount to the rms emittance growth, as discussed in the next section.

Another important observation mentioned earlier, but worth repeating, is that the theory calculates the *possible* emittance growth. Whether for a given situation all of the theoretically possible emittance growth will actually occur depends on the time scale and the dynamical details of the nonlinear effects. In the case of a nonuniform beam (case 1), most of the possible emittance growth occurs in a quarter of a plasma wavelength, $\lambda_p/4 = 2\pi v/\omega_p$, as mentioned. On the other hand, analytical studies for a mismatched laminar beam in a uniform focusing channel [9] and computer simulation of off-centered beams in a periodic focusing channel [10,11] show that the associated emittance growth is a slow process that can take place over a large number of focusing periods. Specifically, the time scale for the mismatched beam is defined by the betatron oscillation period, since it takes one or more betatron oscillations to get the phase mixing leading to the randomization of the velocity distribution. By contrast, the coherent oscillations due to beam off-centering may persist for a very large number of betatron oscillations [11], since the beam centroid is affected mainly by the linear external force, which preserves the coherence in the beam. In high-energy synchrotrons and storage rings the acceleration and storage times are always long enough that even relatively small nonlinearities in the transverse focusing forces lead to phase mixing and thereby convert all of the coherent energy due to off-centering at injection into emittance growth. This effect is discussed in the book by Edwards and Syphers (Sect. 7.1, pp. 222 to 238). These authors define our off-centering as "injection steering error"; and with regard to mismatch, they distinguish between "betatron function mismatch" and "dispersion function mismatch," as is appropriate for circular machines.

6.2.2 Comparison of Theory, Simulation, and Experiment

The theoretical model described in the preceding section assumes a round, continuous beam in an axisymmetric channel with uniform focusing. However, as already pointed out, the basic results should also apply to periodic-focusing channels in the regime where the smooth approximation is valid (i.e., for a phase advance of

$\sigma_0 < 90°$). According to the results of Section 4.4, the smooth-approximation relations for transportable beam current, beam radius, space-charge-depressed phase advance σ, and so on, are accurate to within a few percent for the range $\sigma_0 < 90°$. Furthermore, the envelope instabilities treated in Section 4.4.3 prohibit the transport of space-charge-dominated beams in the region above $\sigma_0 = 90°$ where the smooth approximation fails. Thus we would expect that the theory of emittance growth in nonstationary, uniformly focused beams can also be used to predict the behavior of periodically focused beams in axisymmetric (e.g., solenoid) or quadrupole (FODO) channels.

In this section we compare the theory with numerical simulation results for beams in a uniform focusing channel, a magnetic quadrupole channel of the FODO type, and a periodic solenoid channel. Furthermore, for the solenoid case, both theory and numerical simulation are compared with experimental results. It will be shown that the interplay of theory, simulation, and experiment reveals important details of beam behavior that would be missed if either simulation studies or experiment were done alone.

Let us begin with the computer simulation studies by Struckmeier, Klabunde, and Reiser [2] that were mentioned in the preceding section following Equation (6.15). These studies were aimed at obtaining an understanding of the behavior of different types of distributions in a magnetic quadrupole channel. Specifically, the goal was to investigate theoretically predicted instabilities and to find out if the growth rate for these instabilities depended on the form of the charge distribution. The part of this work that relates to the instabilities is described in Section 6.3.1. Here we limit the discussion to the discovery in these simulation studies of the very rapid and unexpected initial emittance growth due to charge nonuniformity, which appeared to be unrelated to the instability problem. This emittance growth was found to depend on the form of the distribution. It was strongest in the case of a Gaussian distribution, weakest in the Waterbag case, and practically absent in the K–V beam. In all cases it was found that the emittance rises rapidly to a peak within approximately one FODO period, and then oscillates with relatively small amplitudes about a constant mean value over the 50 FODO periods for which the simulation runs were made. The FODO lattice was the same as that in Figure 4.11(b) in Section 4.4.3. Since the emittance growth was most pronounced in the Gaussian distribution and was not observed in the K–V distribution, it was concluded that the effect was caused by the nonuniformity of the particle distribution.

As we now know from our discussion of the Boltzmann distribution in Section 6.5, the stationary state for a space-charge-dominated beam is characterized by a uniform charge-density profile. A nonuniform distribution has a greater amount of field energy than that of the equivalent uniform beam. The energy difference represents free energy that thermalizes as the beam evolves toward a new steady state having a higher temperature and a more uniform density profile. This effect causes the emittance growth observed in the computer simulation. Another way of describing the process is that the nonlinear space-charge forces associated with the nonuniform initial density profile produce a rapid redistribution of the particles toward a more uniform density. The charge homogenization effect converts the potential energy of the particles into thermal kinetic energy, which results in the observed emittance growth. In the computer simulation, both charge homogenization and emittance growth occur simultaneously on a very fast time scale: within one FODO period. As was found in later studies, this corresponds approximately to a quarter of a plasma wavelength.

To determine the effects of beam intensity on emittance growth, Gaussian and K–V distributions were studied systematically for the case $\sigma_0 = 60°$, with decreasing phase advance σ (i.e., increasing values of the space-charge parameter Ka^2/ϵ^2). The results are shown in Figure 6.6, where the ratio of final to initial emittance after 50 FODO periods is plotted.

As can be seen, the K–V distribution has practically no emittance growth except at very low values of σ below about 10°. The small increase in this high-intensity region is probably due to an instability caused by a fourth-order resonance of the type described in Section 6.3.1. The behavior of the Gaussian distribution, on the other hand, is markedly different. It, too, shows essentially no emittance growth in the region of $\sigma > 40°$. But then, as the intensity increases (i.e., as σ becomes smaller), the emittance rises rapidly to large values. This computer result is consistent with the predictions of the theory. Using Equations (6.8), (6.12b), and (6.14), with $U/w_0 = 0.154$ for the Gaussian beam from Table 5.2, one obtains the dashed curve that is shown in Figure 6.6. The agreement between theory and simulation is remarkably good with regard to general behavior. However, the theory overestimates the emittance growth by about 15 to 20%. But this is not surprising since the analytical model assumes that all of the free energy is thermalized, when in fact some of it remains as potential energy in the final equilibrium state of the beam.

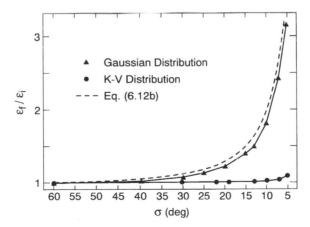

Figure 6.6. Computer simulation results of emittance growth in a FODO channel with $\sigma_0 = 60°$ versus σ for a Gaussian distribution and comparison with Equation (6.12b) of theory (dashed curve). (From Reference 2.)

Indeed, if one keeps track of the actual variation of the field energy factor U/w_0 in the simulation and uses it in the theoretical formula, the agreement between theory and simulation becomes almost perfect. This was done in the study by Wangler et al. that was mentioned earlier [3], where the behavior of a Gaussian distribution in a uniform focusing channel was investigated. The tail of the Gaussian was truncated at two standard deviations and it was assumed that the initial tune depression of this space-charge-dominated beam was $k_i/k_0 = \omega_i/\omega_0 = 0.02$. The variation with distance z (in units of plasma periods, λ_p) of the parameter $\rho = 1 - \bar{a}/\bar{a}_u$, the field energy factor U/w_0, and the emittance ratio ϵ/ϵ_i are shown in Figure 6.7. The parameter ρ is positive for a peaked charge distribution, zero for a uniform beam, and negative for a hollow beam. It compares the behavior of the average radius $\bar{a}$ of the Gaussian distribution with the average radius $\bar{a}_u$ of the equivalent uniform beam. As one can see, $\bar{a}$ oscillates about $\bar{a}_u$ with a period length that is approximately equal to the plasma period, λ_p. On the other hand, the field energy factor U/w_0 and the emittance oscillate with a frequency that is twice as fast as the plasma frequency. The square symbols in the emittance curve correspond to emittance calculations from the second moments of the distribution according to Equation (5.246) at each step. The triangles correspond to emittance calculations using the nonlinear field energy factor U/w_0 and Equation (6.15) at each step. Note the excellent agreement between analytical theory and simulation if the evolution of the remaining field energy is taken into account as the beam propagates along the channel. On the other hand, if one just uses Equation (6.15) with $k_i/k_0 = 0.02$ and the initial value of $U/w_0 \simeq 0.043$ for a Gaussian truncated at two standard deviations, one obtains $\epsilon_f/\epsilon_i = 7.4$. This value is about 9% higher than the numerical result of $\epsilon_f/\epsilon_i \simeq 6.8$ from Figure 6.7. Thus, as expected, the theory overestimates the emittance growth effect by a small amount even in the case of a uniform focusing channel.

Following these early simulation studies, the concept of free-energy conversion into thermal motion and emittance growth was investigated experimentally at the University of Maryland [12,13]. In the experiments, a 5-keV electron beam from a gun with thermionic cathode was passed through an aperture plate outside the anode and matched with the aid of two short solenoids into a 5-m-long periodic channel consisting of an array of 36 solenoid lenses. The injector part of the system, consisting of the electron gun, aperture plate, and the two matching lenses, is shown in Figure 6.8. With the aid of small holes in the aperture plate a nonuniform beam consisting of a configuration of five beamlets was created, as illustrated in the figure. The evolution of this multiple-beam distribution through the injector and down the periodic-focusing channel was observed on a movable fluorescent screen. The images on the screen were photographed with a CCD camera and stored in an Apple Macintosh II computer for further analysis. Emittance measurements with a slit-pinhole type of meter [14] suitable for axisymmetric beams were made at the end of the long channel. The initial emittance of the five-beamlet configuration was inferred from measurements of the full round beam produced by the gun. This beam is converging initially and has a waist at the position of the aperture plate. Thus the initial effective emittance is defined by Equation (5.318b), which can be

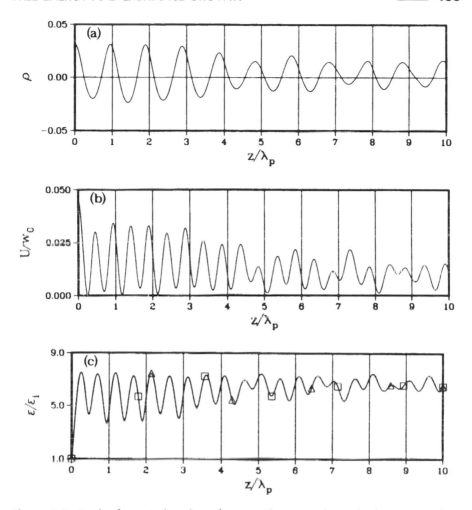

Figure 6.7. Results of numerical simulation for an initial Gaussian charge distribution in a uniform focusing channel. Details are described in the text. (From Reference 3; © 1985 IEEE.)

written in the form

$$\epsilon_i = 4\tilde{\epsilon}_i = R_i \left(\frac{2k_B T}{eV_0} \right)^{1/2}. \tag{6.27}$$

V_0 is the beam voltage, which is 5 kV in this experiment, and $k_B T$ the temperature at the waist. R_i is the effective initial radius, which is defined by the geometry of the beamlet distribution. Assuming a uniform density in each beamlet, one finds from Figure 6.8 that

$$R_i = (a^2 + 1.6\delta^2)^{1/2} = 3.924a = 4.67 \text{ mm}, \tag{6.28}$$

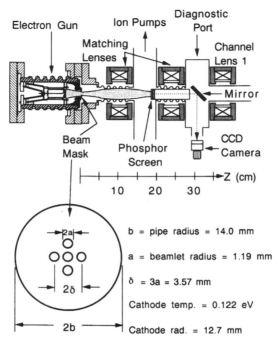

Figure 6.8. Schematic of multiple-beam experiment showing electron gun, beam mask (aperture plate), two matching lenses, first of the 36 lenses of periodic solenoid channel, and diagnostics.

where $a = 1.19$ mm is the beamlet radius and $\delta = 3a = 3.57$ mm defines the beamlet separation as indicated in the figure. The beam radius at the waist was approximately one-half of the cathode radius, and the cathode temperature was typically 0.12 eV. In view of Equation (5.343), one then obtains a waist temperature of

$$k_B T_{\text{waist}} = \left(\frac{R_{\text{cathode}}}{R_{\text{waist}}}\right)^2 k_B T_{\text{cathode}} = 0.48 \text{ eV}. \tag{6.29}$$

Substituting (6.28) and (6.29) in (6.27), one finds for the initial effective emittance of the five-beamlet configuration

$$\epsilon_i = 6.48 \times 10^{-5} \text{m-rad} = 64.8 \text{ mm-mrad}. \tag{6.30}$$

The total current in the five beamlets was $I = 44$ mA or 8.8 mA per beamlet, yielding a generalized perveance of [see Equation (4.27a)]

$$K = \frac{I}{I_0}\frac{2}{\beta^3\gamma^3} = 1.88 \times 10^{-3}. \tag{6.31}$$

The calculation of the nonuniform field energy factor U/w_0 that determines the free-energy parameter h according to Equation (6.14) is straightforward but tedious and yields [12]

$$\frac{U}{w_0} = 0.16\left\{5 - \ln\left[\left(\frac{\delta}{a}\right)^5\left(\frac{1 - s^8}{4}\right)^4\left(\frac{t^2 + 1.6s^2}{s^2}\right)^{12.5}\right]\right\} = 0.2656, \quad (6.32)$$

where $s = \delta/b$, $t = a/b$, and $b = 14$ mm is the radius of the conducting beam tube. The axial magnetic field produced by the solenoidal lenses used for matching and periodic focusing of the beam in the experiments can be approximated by the analytical relation given in Equation (4.127). The period length is $S = 13.6$ cm. In the latest series of experiments performed by Kehne [15], the peak field B_0 of the 36 lenses in the periodic channel was set to give a zero-current phase advance of $\sigma_0 = 77°$. Using this number one finds from the smooth-approximation theory discussed in Section 4.4.1 [Equations (4.144) to (4.147)] for the effective average radius of the initial matched beam

$$a_i = 4.61 \text{ mm} \qquad (6.33)$$

and

$$\frac{\sigma_i}{\sigma_0} = \frac{k_i}{k_0} = 0.31. \qquad (6.34)$$

The free-energy parameter for the initial five-beamlet distribution is given by Equation (6.14), and using the results (6.32) and (6.34), one obtains

$$h = h_s = \frac{1}{4}\left(1 - \frac{\sigma_i^2}{\sigma_0^2}\right)\frac{U}{w_0} = 0.06. \qquad (6.35)$$

To check the predictions of the theory, numerical simulation studies and experiments were performed for two cases. In the first case, the five-beamlet configuration was rms-matched to the periodic channel; that is, with the aid of the two matching lenses the beam was injected so as to produce a matched periodic envelope with initial average radius $a_i = 4.61$ mm. In the second case, the beam was deliberately mismatched into the channel so that the initial radius was only half of the matched radius (i.e., $a_0/a_i = 0.5$). The results of the two experiments and the numerical simulation studies are summarized below.

Case 1 (rms Matched Beam) The free-energy parameter for the initial five-beamlet configuration in this case is due entirely to the nonuniformity of the distribution and given by Equation (6.35). For the small increase in beam radius, one can use Equation (6.10) and finds that

$$\frac{a_f}{a_i} \approx 1 + \frac{h_s}{1 + (\sigma_i/\sigma_0)^2} = 1.055. \qquad (6.36)$$

The emittance increase due to thermalization of the nonuniform field energy is calculated from (6.12), which yields

$$\frac{\epsilon_f}{\epsilon_i} = 1.56 , \tag{6.37a}$$

or

$$\epsilon_f = 101 \text{ mm-mrad} . \tag{6.37b}$$

From Equation (6.17) one would expect that the emittance growth and the correlated charge homogenization takes place in a distance of

$$z_m = \frac{\pi R}{2\sqrt{2K}} \approx 12 \text{ cm} . \tag{6.38}$$

The simulation results for the variation of the effective beam radius and of the effective emittance with distance z are shown in Figure 6.9. They are in remarkably good agreement with the theoretical expectations. Thus, the radius oscillates about an average value corresponding to the prediction of Equation (6.36), and the emittance grows in a distance of about one lens period (13 to 14 cm) to a value close to the predicted result of Equation (6.37).

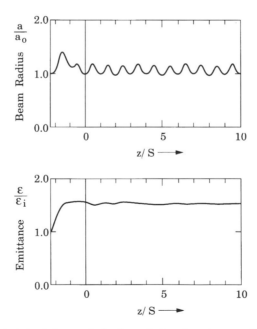

Figure 6.9. Numerical simulation results for the variation of rms beam radius and emittance with distance z along the solenoidal transport channel in the rms matched case of the initial five-beamlet configuration. S is the solenoid periodicity (13.6 cm in the experiment), $z = 0$ is the periodic channel entrance, and the curves start at the location of the mask upstream from the two matching lenses (see Figure 6.8).

The fluorescent screen pictures taken in the experiment at many locations along the channel revealed that the beam retains a rather intricate dynamical structure over a distance of more than 1 m. However, at the end of the channel the profile is perfectly round, with a slight peak at the center as would be expected for a Maxwell–Boltzmann distribution. It took some effort to obtain graphic displays of the beam profile that conform with the fluorescent-screen images. The experimental and numerical beam images at six different locations are shown in Figure 6.10. They agree in many details to a remarkable degree, except for a slight variation in the rotation angle which was due to a small difference between the measured magnetic field and that used in the simulation. Of particular interest is the formation of an image of the initial beam configuration at a distance of about 1 m. It can be attributed to the fact that a large group of particles in each beamlet core is not yet

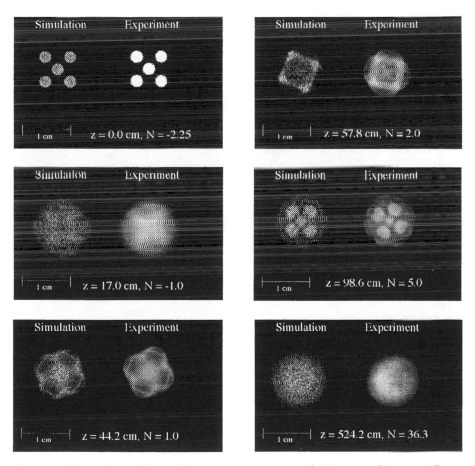

Figure 6.10. Simulation plots and fluorescent-screen pictures of the beam profiles at six different locations along the transport channel for the rms matched beam. (Courtesy of D. Kehne.)

affected by the developing turbulence caused by the nonlinear space charge forces in the initial part of the transport channel. These particles perform quasiperiodic betatron oscillations with a wavelength

$$\lambda_b = \frac{2\pi S}{\sigma_i},$$ (6.39)

corresponding to the initial depressed phase advance of $\sigma_i = 0.31\sigma_0 = 24°$. In turn, these oscillations lead to image formation at a distance of $z_{image} = \lambda_b/2 \simeq$ 1 m, in good agreement with the observations in Figure 6.9. Further downstream, however, this coherence in the beam structure disappears, and no images are observed at subsequent half periods of the betatron oscillations. The beam entropy increases and the final beam profile at 524 cm has the expected axisymmetric structure of a Boltzmann distribution at a higher temperature than the initial state. The emittance measurements [15] with a slit-pinhole system at the end of the channel yielded a value of about 110 mm-mrad, in relatively good agreement with theory and simulation. The fact that this value is slightly higher (by about 9%) than the theoretical prediction can be attributed to scattering in the residual gas of the vacuum system. [See the example following Equation (6.187) in Section 6.4.3.]

Case 2 (rms Mismatched Beam) In this case the beam was overfocused by the two matching lenses to produce a mismatch ratio of $a_0/a_i = 0.5$ at the entrance of the periodic channel. The free-energy parameter due to this mismatch can be calculated from Equation (6.21) using (6.34), and one obtains

$$h_m = 0.396.$$ (6.40)

The total free energy is defined by the sum of the nonuniform field energy (6.35) and the mismatch energy (6.40), that is,

$$h = h_s + h_m = 0.456.$$ (6.41)

Using (6.34) and (6.41), one obtains for the ratio of the final beam radius a_f to the initial radius a_i from (6.8) the result

$$\frac{a_f}{a_i} \simeq 1.3.$$ (6.42)

The emittance growth predicted by the theory is then found from Equation (6.12b) as

$$\frac{\epsilon_f}{\epsilon_i} = 3.72.$$ (6.43)

The simulation results for the variation of beam radius and emittance growth with axial distance are shown in Figure 6.11. As can be seen, after an initial increase

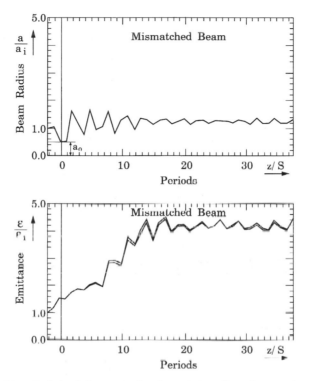

Figure 6.11. Numerical simulation results for the variation of rms beam radius and emittance with distance z along the solenoidal transport channel in the rms-mismatched case of the initial five-beamlet configuration.

the radius oscillates about a mean value that is close to the theoretical prediction of Equation (6.42). The emittance, on the other hand, shows the small increase due to the charge nonuniformity in the first two lens periods. It then grows further, over a distance of about 12 lens periods, where it settles down and oscillates about a mean value of $\epsilon_f/\epsilon_i \simeq 4.2$. This emittance increase, most of which is attributable to the beam mismatch, is approximately 16% higher than the theoretical estimate of Equation (6.43). To obtain some insight into possible causes of this unexpected discrepancy, we need to examine the simulation and fluorescent-screen images shown in Figure 6.12. This series of pictures begins at a distance of 17 cm from the beam aperture and shows an image at about 44 cm, less than half the distance of the image location in case 1. An important new feature is the formation of a ring at 126 cm corresponding to 7 lens periods. At 194 cm (i.e., after a total of 12 lens periods), the ring develops into a large halo surrounding the beam core. This location corresponds to the position where the emittance reaches its maximum value (see Figure 6.11). The halo persists through the remaining length of the focusing channel, although it is not visible in the reproduction of the images at $z = 524$ cm.

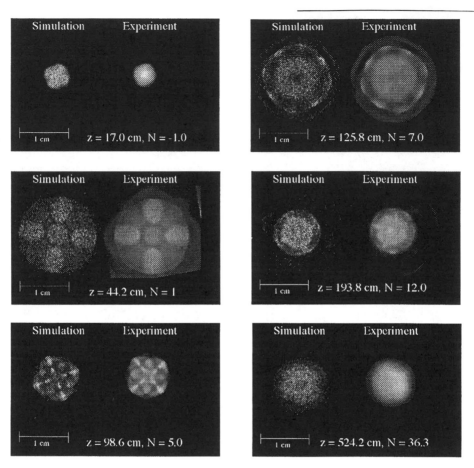

Figure 6.12. Simulation plots and fluorescent-screen pictures of the beam profiles at six different locations along the transport channel for the rms mismatched beam. (Courtesy of D. Kehne.)

The final state of the beam at the end of the channel is characterized by a well-behaved axisymmetric beam core resembling a Boltzmann distribution and a halo. A detailed analysis shows that the halo comprises about 20% of the beam current and is responsible for almost all of the emittance growth due to the mismatch, while the well-behaved beam core has an emittance growth of about 1.5, which corresponds to the nonuniform field energy. Only the core approaches a thermal equilibrium state, while the free energy due to the very large mismatch studied here produces a cloud of particles having considerably higher energy and oscillation amplitudes than those in the core. Further studies show that the large discrepancy between the theoretical prediction of emittance growth and the numerical result becomes smaller when the mismatch ratio a_0/a_i is reduced [15]. Numerical simulation studies of round mismatched beams launched into a uniform focusing channel reveal that in these cases the theory predicts an emittance growth that is actually higher than the numerical value [16]. This is true even for a large mismatch where halo formation occurs. One is therefore led to the conclusion that the low estimate of the theory in the five-beamlet case is attributable to the fact that there is more free energy available in this case due to the asymmetry than in a comparable axisymmetric beam configuration. In any case, these findings point out that there is need for further studies of emittance growth and halo formation in nonstationary beams.

Although some details, such as halo formation, are not yet fully understood, the above studies have provided valuable information on the emittance growth and on the time scales for the various effects [17]. Considering the assumptions made in the model, the theoretical predictions with regard to emittance growth are remarkably good and very useful for practical applications. Also, the information obtained from simulation studies and experiments on the time scales of the various effects are extremely important. Thus the fastest process is the emittance growth associated with the thermalization of free energy in an rms-matched beam with a nonstationary density profile. It occurs in a quarter of a plasma period. The conversion of free energy into emittance growth due to rms mismatch, on the other hand, occurs in a distance corresponding to a betatron oscillation, λ_b, and is almost exclusively associated with the formation of a halo at this distance. These observations are in general agreement with studies by Anderson on the dynamic evolution of laminar sheet beams, which are initially nonuniform in the density profile or mismatched [9]. Thermodynamically, the conversion of free energy into thermal energy and emittance growth corresponds to an increase in the entropy of the particle distribution. The mathematical correlation between emittance and entropy was discussed in a paper by Lawson, Lapostolle, and Gluckstern [18] in 1973.

6.3 INSTABILITIES

6.3.1 Transverse Beam Modes and Instabilities in Periodic Focusing Channels

The theoretically possible stationary states of a beam in a linear focusing channel were described in Section 5.3 by distribution functions that satisfy the time-independent Vlasov equation. Specifically, for a uniform (continuous) focusing channel, all distributions that are functions of the transverse Hamiltonian $H_\perp$ (which is a constant of the motion in this case) are stationary. However, in the case of periodic-focusing channels, $H_\perp$ is no longer constant, and the only stationary state for which an analytic representation could be found is the K–V distribution.

A key question in the theory of particle beams is whether a particular distribution is stable or unstable against various types of perturbations. From the thermodynamic arguments presented in Section 5.4, we would expect that in the presence of collisions or, in general, due to the actions of nonlinear forces of a stochastic nature, all distributions will eventually relax into thermodynamic equilibrium (i.e., into a Maxwell–Boltzmann distribution). One thermalization mechanism that we discussed in the preceding section is the conversion of the free energy associated with nonstationary distributions into random motion and hence emittance growth. Another mechanism that can lead to emittance growth is instability, which is the subject of this section. Instabilities can affect both stationary and nonstationary initial distributions. The most powerful analytic technique for investigating the instability problem involves the use of the Vlasov equation and hence the neglect of collisional effects. In this technique the Vlasov equation is linearized by expanding the perturbation about the known stationary solution and determining the perturbed electromagnetic fields with the aid of Maxwell's equations. If the perturbations consist of simple electrostatic charge-density oscillations, which is the case with the problem that we are discussing in this section, the fields can be found by solving Poisson's equation for the associated space-charge potentials. For a beam transport channel consisting of a periodic array of discrete solenoid or quadrupole lenses, the linearized Vlasov theory is limited to the stability analysis of the K–V distribution. For a more general investigation one must rely on computer simulations.

Historically, this stability problem was first examined in 1970 for stationary distributions in a continuous focusing channel (such as a long solenoid) by Gluckstern [19], who analyzed the K–V distribution, and by Davidson and Krall [20], who showed that a large class of stationary solutions of the Vlasov equation is stable against arbitrary charge-density fluctuations. As in a plasma, local density perturbations in beams can produce collective modes of oscillations. Gluckstern described these modes for the K–V distribution in terms of the oscillations of the space-charge potential associated with the density fluctuations. The solutions for the perturbed potentials that satisfy the Vlasov equations and the beam's boundary condition can be expressed in the form $V_n \propto e^{i\omega t} G(r, \phi)$, where the time t can be related to the propagation distance s and the particle velocity v by $t = s/v$ and where $G(r, \phi)$ describes the geometric dependence of the potentials on the cylindrical coordinates r and ϕ. Basically, $G(r, \phi)$ consists of terms such as $r^n \cos m\phi$ and $r^n \sin m\phi$, where the integer n denotes the "order" of the mode and the integer $m \leq n$ the azimuthal variation. Gluckstern found that the stability of the K–V distribution in a continuous focusing channel depends on the numbers m and n and on the tune depression $k/k_0 = \nu/\nu_0$ of the particle oscillations due to space charge. For $\nu/\nu_0 \gtrsim 0.4$ all modes are stable, while in the region below this value unstable behavior occurs if ν/ν_0 falls below a certain value, which depends on the structure of the mode. In simulation studies performed later [21] it was found that these "instabilities" of the K–V distribution in uniform transport channels manifest themselves only as redistributions of the charge density with no actual emittance growth—in agreement with much earlier simulation studies by Lapostolle [7]. Thus, in essence, the K–V distribution in a uniform channel is stable.

In 1983, Hofmann et al. extended Gluckstern's stability analysis of the K–V distribution to periodic solenoid and quadrupole channels [21]. They found that many of the Gluckstern-type modes become unstable when the associated frequencies interact resonantly with the periodicity of the focusing system ("structure resonances") or when the frequencies of two modes converge ("confluence of eigenvalues"). The regions of instability depend predominantly on the zero-current phase advance σ_0 and the tune depression σ/σ_0 due to space charge and to a lesser extent on the "filling factor," or the ratio l/L of lens width l to drift space L in the focusing lattice. The stability analysis and computer simulation studies showed that the modes of lowest order, the quadrupole ($n = 2$) and sextupole ($n = 3$) modes, are the most dangerous and generate large emittance growth. Those of high order ($n > 3$) are generally less pronounced and their effects on the beam emittance appear to decrease rapidly with increasing mode order.

By far the most destructive modes are those of the quadrupole type, which are identical to the envelope instabilities studied by Struckmeier and Reiser [22] that were discussed in Section 4.4.3. What happens in this case is illustrated in Figures 4.12 (for a solenoid channel) and 4.13 (for a quadrupole channel) in that section. Shown on the left side of these figures are the phase advance Φ per lattice period of the two fundamental quadrupole oscillations, the "in-phase" mode, where $V_2 \propto r^2$, and the "out-of-phase" mode, where $V_2 \propto r^2 \sin 2\phi$. For a zero-current phase advance of σ_0, both modes start at $\Phi = 2\sigma_0$ and then decrease on separate curves as the intensity is increased and, correspondingly, the phase advance with space charge, σ, is decreased. When the phase advance Φ of either mode passes through 180°, a resonant interaction occurs with the periodic structure that is analogous to the $\sigma_0 = 180°$ threshold for stable single-particle motion [see Equation (3.302)]. As can be seen in Figure 4.12, for the solenoid case the modes become phase-locked to the structure period, so that rather than single resonance points, one has extended regions in $\sigma - \sigma_0$ space where the beam is unstable. The growth rates of the instabilities $|\lambda|$ are plotted on the right side of Figures 4.12 and 4.13. In the quadrupole channel, the dominant effect is confluence of the two modes near and slightly below the 180° threshold line (Figure 4.13). Computer simulation with an initial K–V distribution at $\sigma_0 = 120°$, $\sigma = 35°$ in the quadrupole channel of Figure 4.11 are plotted in Figure 6.13 and show the destructive effect of the envelope instability. The conclusion, stated in Section 4.4.3, is that periodic-focusing channels for transport of high-intensity beams should be designed to operate at a zero-current phase advance of $\sigma_0 \leq 90°$ to avoid these quadrupole-type instabilities.

The analysis of third-order (sextupole) modes shows that instabilities due to structure resonances at 180° and confluence of modes occur when $\sigma_0 \geq 60°$. In this case the phase advance of the mode starts at $\Phi = 3\sigma_0$ when the current is zero, and if $\sigma_0 > 60°$, the curves of the eigenmodes pass through the 180° line, at which point structure instability occurs analogous to the envelope modes. Figure 6.14 shows the growth rate of the confluent and 180° instabilities as a function of σ for a quadrupole channel with $\sigma_0 = 90°$ and filling factor $\eta = l/(L + l) = 0.1$ (see Figure 3.27). For a solenoid channel, the studies in Reference 21 show that the third-order instabilities are much less pronounced than in the quadrupole channel and depend more strongly on the filling factor, as illustrated in Figure 6.15. Indeed, one can see the trend whereby the instabilities become negligibly small when the filling factor goes toward unity and one obtains a continuous long solenoid. Computer simulations presented in Reference 2 for an initial K–V distribution in

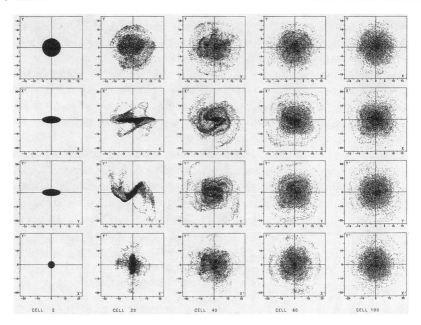

Figure 6.13. Computer simulation of an initial K–V distribution in the quadrupole channel of Figure 4.11 at $\sigma_0 = 120°$, $\sigma = 35°$, showing the evolution of the envelope instabilities and a state of saturation reached after about 100 periods. The computer results are transformed to correspond to an upright phase-space ellipse. (Courtesy of J. Struckmeier.)

the quadrupole channel of Figure 4.11 (Section 4.4.3) at $\sigma_0 = 90°$ and $\sigma = 41°$ are shown in Figure 6.16. The transverse density oscillations and the third-order structure of the modes are quite apparent although higher-order modes may also be present.

The question arises whether these instabilities are a peculiarity of the K–V distribution, or whether they also occur in other, more realistic distributions. For

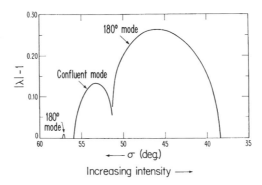

Figure 6.14. Growth rate of the confluent and 180° instabilities as a function of σ for a quadrupole channel with $\sigma_0 = 90°$ and filling factor $\eta = l/(L + l) = 0.1$. (From Reference 21.)

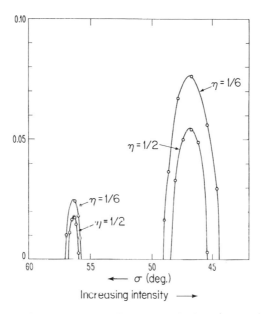

Figure 6.15. Behavior of third-order mode for interrupted-solenoid system for which $\eta = 1/2$ or $\eta = 1/6$ and $\sigma_0 = 90°$. (From Reference 21.)

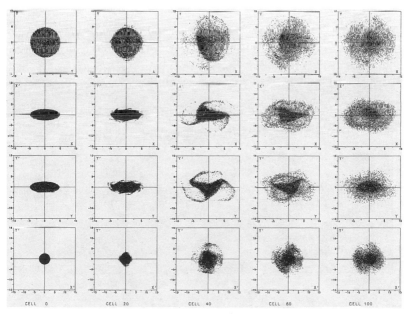

Figure 6.16. Computer simulation of an initial K–V distribution in the quadrupole channel of Figure 4.11 at $\sigma_0 = 90°$, $\sigma = 41°$, showing the evolution of the third-order instability in the upright phase-space ellipse position. (Courtesy of J. Struckmeier.)

envelope-type instabilities, computer simulations by Struckmeier and Reiser [22] and experiments [23] show that they are quite universal. The third-order modes are also observed in computer simulation studies of various distributions. However, the effects of these instabilities decrease substantially when the distribution approximates a realistic beam. This is illustrated in Figure 6.17, which is identical to Figure 2 in Reference 2 except that the semi-Gaussian distribution (uniform in space, thermal in the velocities) is included. The conditions are the same as in Figure 6.16 (i.e., $\sigma_0 = 90°$ and $\sigma = 41°$), and the emittance growth is plotted versus the number of quadrupole periods (cells). Substantial emittance growth that is attributable to the third-order instability occurs for the K–V, waterbag, and parabolic distributions and to a lesser extent for the conical distribution. (Note that the various distributions are defined in Table 5.2.) The Gaussian distribution shows a rapid initial emittance growth within the first cell and then a rather slow, almost linear increase over the entire length of the channel (100 cells). The conical, parabolic, and waterbag distributions also show a rapid initial emittance increase before third-order instability sets in. All of these initial emittance changes are due to the fact that these distributions have nonstationary density profiles, and they are explained by the conversion of the free energy into thermal energy discussed in Section 6.2. The semi-Gaussian is a notable exception in that the initial emittance actually decreases at first before it rises and eventually converges with the Gaussian curve after about 75 cells. The reason for the initial decrease is that

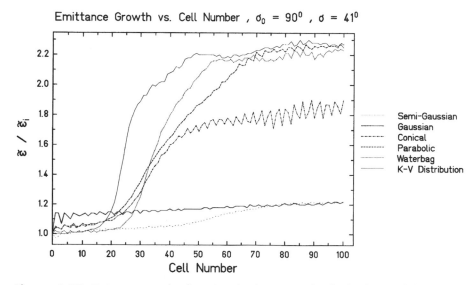

Figure 6.17. Emittance growth of various distributions in the third-order instability regime ($\sigma_0 = 90°$, $\sigma = 41°$) of the quadrupole channel of Figure 4.11. (From Reference 2.)

the semi-Gaussian must convert some of its thermal energy into field energy as it assumes a more stationary density profile. Of particular interest is the fact that all distributions do not reach a plateau of constant emittance. Instead, there is a small but fairly linear increase of the emittance with distance that was observed already by Lapostolle et al. [24] but is not fully understood. This small increase may, of course, be due to computational error. However, it is only observed in the simulation of quadrupole channels and not in periodic solenoid channels [25]. Hence, there exists the possibility that a genuine physical mechanism occurring only in the quadrupole case may be responsible for it.

As we have seen, the emittance growth due to third-order modes is very small in the more realistic Gaussian and semi-Gaussian distributions, and the effects have so far not been detected in quadrupole transport experiments [26, 27]. One therefore concludes that it should be possible to operate in the parameter range above $\sigma_0 = 60°$ (where the third-order modes occur) but below $\sigma_0 = 90°$ to avoid the envelope instabilities. Within this regime of operation the experiments show that there appears to be no lower limit in σ for stable beam transport in long periodic channels [12, 26, 27].

From the computer simulation results, especially the plots in Figure 6.13 showing a K–V distribution in the envelope instability regime of a quadrupole channel, one infers that the beam eventually becomes stable. The final stationary state is characterized by a larger radius, an increased emittance, and a density profile with a more Gaussian-like shape than that for the uniform-density initial K–V beam. Apparently, the beam reaches a Boltzmann-type equilibrium, having greater random velocity distribution, and hence higher temperature, than the original distribution. Where does the energy that causes this increase of emittance and temperature come from? As we discussed in Section 5.4.12, the total transverse energy of a perfectly matched beam in an ideal periodic channel remains constant (i.e., there is no coupling between the longitudinal and transverse energies in the channel). We conclude, therefore, that the increase in emittance and temperature caused by the instability is due to a conversion of potential energy into random kinetic energy. The envelope instability, for instance, is caused by the resonant interaction between the collective force due to charge perturbations defined by the plasma frequency ω_p and the external periodic-focusing force characterized by the particle oscillation frequency without space charge, $\omega_0 = (\sigma_0/S)v_0$, and the frequency $\omega_f = (2\pi/S)v_0$ due to the periodicity S of the lattice. As the instability effect increases the rms widths $\tilde{x}$, $\tilde{y}$, and thus the mean radius of the beam, the particle density n decreases. Hence, the plasma frequency $\omega_p \propto \sqrt{n}$ decreases until the resonance condition is no longer satisfied. Another way of looking at the problem is that the velocity spread in the beam is increased until *Landau damping* occurs (see the discussion at the beginning of Section 6.3.3). In principle, one could calculate the radius and emittance increase due to the envelope instability in a fashion similar to that done in Section 6.2. However, in the region $\sigma_0 > 90°$, the smooth approximation is no longer valid, and therefore such analytical estimates of the emittance growth become very difficult.

It should also be noted in this context that the unstable modes discussed in this section (i.e., essentially the envelope instabilities) occur only in straight, linear transport systems for intense beams, which includes rf and induction linacs at high beam current. In circular machines the space-charge tune shift severely limits the beam intensity, so that the tune depression ν/ν_0 is always close to unity and the unstable space-charge modes cannot develop.

The instabilities due to the interaction between the space charge and the periodic-focusing force discussed in this section assumed a beam with transverse symmetry (i.e., identical emittances and energies, or temperatures, in the two orthogonal phase-space areas). I. Hofmann showed that additional instabilities occur in beams with anisotropic distributions having different energies, or temperatures, in the two directions of motion [28]. While these collective instabilities are of general importance and may, for instance, occur in the transport of sheet beams, they are particularly relevant to high-current linear accelerators. The bunched beams in rf linacs usually have anisotropic energy distributions in the longitudinal and transverse directions. This leads to equipartitioning and emittance growth of the beams via the collective space-charge forces, as was shown by R. Jameson in computer simulation studies and analytical considerations [29]. We treat the topic of equipartitioning in a high-current rf linac in Appendix 4.

6.3.2 Longitudinal Space-Charge Waves and Resistive-Wall Instability

Perturbations in the longitudinal charge distribution can create an instability that poses a serious problem for both high-current linear accelerators and transport systems, as well as for circular machines. This longitudinal instability adversely affects the longitudinal particle distribution, limits the beam intensity, and may also increase the transverse emittance by generating changes in the transverse distribution. To understand the physical mechanisms driving this instability, it will be useful first to discuss how perturbations of the longitudinal charge density propagate along the beam as space-charge waves. We start with a simple, one-dimensional, nonrelativistic beam model where boundary effects are ignored. Next, we present an analysis of a cylindrical beam inside a perfectly conducting boundary, which includes relativistic effects. This analysis is then extended to the case where the beam tube has a finite resistivity, which causes the *resistive wall instability*. Temperature effects are neglected, i.e., we treat the beam as a cold (laminar) fluid.

As we know from basic theory (see, for instance, B.1., Chapter 1), local charge perturbations in a plasma generate plasma oscillations with frequency ω_p. This is also the case in a beam that can be described as a nonneutal plasma (see discussion in Section 4.1). Thus, if a charged particle is displaced longitudinally from its equilibrium position due to a perturbation, it will perform longitudinal oscillations with frequency ω_p. If $s(t)$ denotes the displacement from the equilibrium position in the moving beam frame as a function of time t, the equation of motion of the particle is of the simple harmonic-oscillator form

$$\ddot{s} + \omega_p^2 s = 0. \tag{6.44}$$

The solution of this equation can be expressed as

$$s(t) = C_1 e^{i\omega_p t} + C_2 e^{-i\omega_p t}, \tag{6.45}$$

and

$$\dot{s}(t) = i\omega_p C_1 e^{i\omega_p t} - i\omega_p C_2 e^{-i\omega_p t}, \tag{6.46}$$

where C_1 and C_2 are complex constants determined by the initial conditions.

Let us a now consider the case where the perturbation is caused by un external force acting on the beam. To be specific, we assume that the beam passes through a small gap in an rf cavity where a periodic electric field with frequency ω produces a velocity modulation (as in a klystron). Suppose that the gap width is infinitesimally small so that a particle passing through it will receive an instantaneous "kick" that changes its velocity but not its position in the beam at that point. If $t = t_0$ denotes the time of gap crossing, we can express the total initial velocity after the kick as

$$v(t_0) = v_0 + v_1 \cos \omega t_0, \tag{6.47}$$

where v_0 is the unperturbed velocity and v_1 the amplitude of the velocity modulation. The velocity change leads to a displacement of the particle from its equilibrium position downstream from the gap, in turn produces the plasma oscillation described by Equations (6.45) and (6.46). The constants C_1 and C_2 can be evaluated by using the initial conditions

$$s(t_0) = 0, \tag{6.48a}$$

$$\dot{s}(t_0) = v_1 \cos \omega t_0, \tag{6.48b}$$

which yields

$$C_1 = \frac{v_1}{2i\omega_p} e^{i(\omega - \omega_p)t_0}, \tag{6.49a}$$

$$C_2 = -\frac{v_1}{2i\omega_p} e^{i(\omega + \omega_p)t_0}, \tag{6.49b}$$

By substituting these results into Equation (6.45) we obtain

$$s(t, t_0) = \frac{v_1}{2i\omega_p} e^{i(\omega - \omega_p)t_0} e^{i\omega_p t} - \frac{v_1}{2i\omega_p} e^{i(\omega + \omega_p)t_0} e^{-i\omega_p t}. \tag{6.50}$$

This equation describes the displacements of the particles from their equilibrium positions in the moving beam in terms of the times t_0 when they cross the gap. Since the distance z of travel from the gap is given by $z = v_0(t - t_0)$, and hence

$$t_0 = t - \frac{z}{v_0}, \tag{6.51}$$

we can eliminate t_0 and express the displacement as a function of t and z, rather than t and t_0. By using relation (6.51) we can write Equation (6.50) in the form

$$s(t, z) = \frac{v_1}{2i\omega_p} e^{i(\omega t - k_f z)} - \frac{v_1}{2i\omega_p} e^{-i(\omega t - k_s z)}. \tag{6.52}$$

The two wave numbers in this equation are given by

$$k_f = \frac{\omega - \omega_p}{v_0}, \tag{6.53}$$

$$k_s = \frac{\omega + \omega_p}{v_0}. \tag{6.54}$$

They satisfy the *dispersion relation* between ω and k, which applies for such perturbations in a cold beam and which is given by

$$(\omega - kv_0)^2 = \omega_p^2 \tag{6.55a}$$

$$\omega - kv_0 = \pm\omega_p. \tag{6.55b}$$

Equation (6.52) represents the sum of two traveling waves called *space-charge waves,* one with wave number k_f and wavelength $\lambda_f = 2\pi/k_f$, the other with wave number k_s and wavelength $\lambda_s = 2\pi/k_s$. The corresponding phase velocities are

$$v_f = \frac{\omega}{k_f} = \frac{v_0}{1 - (\omega_p/\omega)} \approx v_0\left(1 + \frac{\omega_p}{\omega}\right), \tag{6.56}$$

$$v_s = \frac{\omega}{k_s} = \frac{v_0}{1 + (\omega_p/\omega)} \approx v_0\left(1 - \frac{\omega_p}{\omega}\right), \tag{6.57}$$

where we assume that $\omega_p/\omega \ll 1$. In the first case, the phase velocity is seen to be greater than the beam velocity ($v_f > v_0$), and we call this wave the *fast wave.* In the second case, the phase velocity is less than the beam velocity ($v_s < v_0$), and we call this wave the *slow wave.*

The displacements of the particles from their equilibrium positions can be correlated with perturbations of the velocity $v(t, z)$, the space-charge density $\rho(t, z)$ and the current density $J(t, z)$ which are connected by the continuity equation

$J = \rho v$. The superposition of the fast and slow space-charge waves leads to distinct patterns of bunching and debunching of J or ρ along the beam. Thus J reaches its first peak at a distance of $\lambda_p/4 = \pi v_0/\omega_p$ from the "input" cavity (where the initial velocity modulation occurs). In a *klystron*, microwaves are generated in a second cavity (the "output" cavity), which is located at this first current maximum. If we examine the two space-charge w.ves in the beam frame, we obtain for the phase velocities the results

$$v_+ = v_f - v_0 = v_0 \frac{\omega_p}{\omega - \omega_p}, \tag{6.58a}$$

$$v_- - v_s - v_0 = -v_0 \frac{\omega_p}{\omega - \omega_p}. \tag{6.58b}$$

Thus an observer moving with the beam velocity v_0 would see the two waves moving in opposite directions with frequency ω_p and unequal phase velocities $|v_+| = |v_-|$. Although the phase velocities of the two waves differ, the group velocity v_g in the laboratory frame is the same, namely

$$v_g = \frac{\partial \omega}{\partial k} = v_0, \tag{6.59}$$

as can be verified from Equation (6.55b).

Accordingly, energy and information will travel with the velocity of the beam, as one expects from the well-known arguments of classical physics. We also note that Equation (6.55) represents the classical Doppler shift for the frequency ω_p measured by an observer in the beam frame to the frequency ω measured in the laboratory frame.

Our analysis in this section as well as in Section 6.3.3 treats the disturbances that are producing the space-charge waves as periodically acting harmonic forces with frequency ω. But it should be pointed out in this context that such perturbations and the associated space-charge waves can occur also in the form of localized single pulses. Usually the space-charge waves are created as a pair of slow and fast waves. However, recent experiments with localized, single perturbations on an electron beam and theoretical analysis have demonstrated that one can create only one or the other as a single wave by controlling the conditions at the gridded cathode of an electron gun [30]. We note in this context that the interaction of these two waves with external electromagnetic fields plays the key role in either acceleration or microwave generation or in causing longitudinal beam instabilities. Take as an example a beam propagating in a waveguide together with an externally launched electromagnetic (EM) wave traveling in the same direction as the beam. If the phase velocity of the fast space-charge wave is in synchronism with the phase velocity of the traveling EM wave, the beam will gain energy from the EM wave (i.e., it will be accelerated). The fast space-charge wave is called a *positive-energy wave* since it gains energy from the EM wave as its amplitude is increasing. On the other hand, when the slow wave is in synchronism with the phase velocity of the EM wave, as is the case in a *slow-wave structure*, the beam will give up energy to the EM wave. We are dealing in this case with a traveling-wave rf generator in which the beam kinetic energy is transformed into microwave energy. The amplitude of the slow space-charge wave will grow as it gives up energy to the EM wave, which is why it is called a *negative-energy wave*. As we will see later in this section and in the next section, it is the interaction of the slow wave with an external circuit impedance that produces longitudinal instability when the slow-wave amplitude is growing.

The above analysis of the space-charge waves was somewhat academic in that we assumed a strictly one-dimensional geometry in which the beam was assumed to be infinitely large in the transverse direction. We can obtain a more realistic description of the problem by taking the beam to be an infinitely long cylinder of line-charge density ρ_L and radius a inside a conducting drift tube of radius b. To simplify the use of subscripts, we use the symbol Λ for the line-charge density ρ_L (i.e., we set $\rho_L = \Lambda$). Let us assume that the unperturbed beam has a constant line-charge density Λ_0, which implies that the longitudinal dc electric field is zero. If a local perturbation develops, it will emit two space-charge waves according to our discussion above. In the steady state, all quantities associated with this perturbation will consist of a dc value and a wavelike ac perturbation. Specifically, the line-charge density Λ, velocity v, and beam current I will be of the form

$$\Lambda(z,t) = \Lambda_0 + \Lambda_1 e^{i(\omega t - kz)}, \tag{6.60a}$$

$$v(z,t) = v_0 + v_1 e^{i(\omega t - kz)}, \tag{6.60b}$$

$$I(z,t) = \bar{I} + I_1 e^{i(\omega t - kz)}, \tag{6.60c}$$

where

$$I = \Lambda v. \tag{6.61}$$

Note that we use the symbol $\bar{I}$ for the dc (average) current to avoid confusion with the characteristic current I_0.

To analyze the propagation properties of the perturbed waves, we will use a linearized, cold, one-dimensional fluid model that consists of the continuity equation and the momentum transfer equation (which is identical with the longitudinal equation of motion). Linearization requires that the perturbations be small compared to the dc quantities (i.e., $\Lambda_1 \ll \Lambda_0$, etc.). The cold-beam approximation implies that we neglect the longitudinal momentum spread of the beam. (We generalize the analysis in Section 6.3.3 by including momentum spread and using the Vlasov equation instead of the cold-fluid model.)

The continuity equation yields

$$\frac{\partial(\Lambda v)}{\partial z} + \frac{\partial \Lambda}{\partial t} = 0, \tag{6.62a}$$

or

$$\frac{\partial I}{\partial z} + \frac{\partial \Lambda}{\partial t} = 0, \tag{6.62b}$$

Keeping only the first-order terms, one obtains from (6.60) and (6.61)

$$\bar{I} = \Lambda_0 v_0, \tag{6.63a}$$

$$I_1 = \Lambda_0 v_1 + v_0 \Lambda_1 \tag{6.63b}$$

and from (6.62),

$$-ikI_1 = -i\omega\Lambda_1, \tag{6.64a}$$

or

$$\Lambda_1 = \frac{kI_1}{\omega}. \tag{6.64b}$$

The space-charge waves will produce a longitudinal electric field $E_z(z,t)$ which will exert a force on the particles in the beam. This force changes the velocity in accordance with the longitudinal equation of motion, which has the relativistic form

$$\gamma_0^3 m \frac{dv}{dt} = qE_z, \tag{6.65a}$$

or

$$\frac{\partial v}{\partial z} v_0 + \frac{\partial v}{\partial t} = \frac{q}{\gamma_0^3 m} E_z, \tag{6.65b}$$

if the perturbed velocity amplitude v_1 is small compared to the dc velocity v_0 so that the relativistic energy factor $\gamma_0 = (1 - v_0^2/c^2)^{-1/2}$ remains essentially constant. With $E_z = E_s \exp[i(\omega t - kz)]$ and using (6.60b), one obtains from (6.65b)

$$(-ikv_0 + i\omega)v_1 = \frac{q}{\gamma_0^3 m} E_s, \tag{6.66a}$$

or

$$v_1 = -i\frac{qE_s}{\gamma_0^3 m(\omega - kv_0)}. \tag{6.66b}$$

Substitution of (6.64b) and (6.66b) in Equation (6.63b) yields the following relationship between the perturbed electric field and current amplitudes:

$$E_s = i\frac{\gamma_0^3 m}{q\Lambda_0} \frac{(\omega - v_0 k)^2}{\omega} I_1. \tag{6.67}$$

The longitudinal electric field E_z must also obey Maxwell's equations. We will use for our analysis the low-frequency approximation where the displacement current term $\partial\mathbf{D}/\partial t = \epsilon_0\partial\mathbf{E}/\partial t$ in Maxwell's equations can be neglected. The longitudinal electric field can then be calculated by applying Faraday's law, $\oint\mathbf{E}\cdot d\mathbf{l} = -\partial/\partial t\int\mathbf{B}\cdot d\mathbf{S}$ and Ampère's circuital law, $\oint\mathbf{B}\cdot d\mathbf{l} = \mu_0\int\mathbf{J}\cdot d\mathbf{S}$, to the closed rectangular loop shown in Figure 6.18. For later use we have assumed in the figure that the wall of the beam tube has a finite resistivity, so that the induced current I_w generates an axial electric field E_{zw} in the wall. E_{zs} denotes the axial field associated with the space-charge perturbation in the beam itself. The evaluation of the integrals involved can be simplified by using in place of the generally nonuniform radial density profile the equivalent uniform-beam model defined in Section 5.3.4 (see also Figure 5.12 and Table 5.2 and related discussion in the text). Carrying out the calculation, which is left as a problem (6.4), yields the following result:

$$E_{sz} = -\frac{g}{4\pi\epsilon_0}\left(\frac{\partial\Lambda}{\partial z} + \frac{1}{c^2}\frac{\partial I}{\partial t}\right) + E_{wz}. \qquad (6.68a)$$

or, since $\partial I/\partial t \approx -v_0^2\partial\Lambda/\partial z$ and $1 - v_0^2/c^2 = 1/\gamma_0^2$,

$$E_{sz} = -\frac{g}{4\pi\epsilon_0\gamma_0^2}\frac{\partial\Lambda}{\partial z} + E_{wz}. \qquad (6.68b)$$

The parameter g in Equation (6.68) is a geometry factor that defines the proportionality between the longitudinal electric field E_{sz} and the derivatives $\partial\Lambda/\partial z$, $\partial I/\partial/t$ of the charge and current perturbations, respectively. The situation is analogous to the relation (5.366) between the electric field and the line-charge density gradient in a bunched beam. However, in the present situation, we are dealing with line-charge density perturbations in a continuous round beam, and hence the geometry factor g should not be the same as for the bunched beam. Mathematically, the solution for perturbations in a cylindrical beam with radius a inside a conducting pipe with radius b involves Bessel functions, and the geometry factor depends on the wave constant k or wavelength $\lambda = 2\pi/k$ of the perturbation. Only in the long-wavelength limit being considered here, where $\lambda \gg a$ does one get an asymptotic value for g that is independent of λ. This asymptotic expression depends on the relationship between the perturbed line-charge density Λ, the volume charge density ρ, and the beam radius a which for a uniform beam is given by $\Lambda(z) = \rho(z)a^2(z)\pi$. The radius, in turn, depends on the wave constant k_0 (or betatron function $\hat{\beta}_0 = 1/k_0$) of the focusing channel, the beam perveance K, and the emittance ϵ, as defined by the approximate relation (5.293), which is more useful for our purpose than the exact result in Equation (4.93). Since K is proportional to the beam current I and $I(z) \approx \Lambda(z)v$, we find from (5.293) that $a^2(z) = \Lambda(z)/C + \epsilon/k_0$, where $C = 2\pi\epsilon_0(mc^2/q)\beta^2\gamma^3k_0^2$. If the emittance term dominates, i.e., if $\epsilon/k_0 \gg \Lambda(z)/C$, the radius is essentially constant ($a = $ const), i.e., $\Lambda(z) = \rho(z)a^2\pi$, and the line-charge perturbation manifests itself as a longitudinal variation (bunching and debunching) of the volume charge density $\rho(z)$. This is the situation in circular accelerators and storage rings. On the other hand, if space charge dominates, i.e., if $\Lambda(z)/C \gg \epsilon/k_0$, one has $a^2(z) \approx \Lambda(z)/C$; hence, the volume charge density remains constant ($\rho = $ const), and the line-charge perturbation produces a variation of the beam radius. This is the case in high-current linear accelerators and beam transport systems. The calculation yields for the g-factor in these two limiting cases the following result (see Problem 6.4):

$$g = 1 - \frac{\overline{r^2}}{a^2} + 2 \ln \frac{b}{a} = \frac{1}{2} + 2 \ln \frac{b}{a} \qquad (6.69a)$$

for the emittance-dominated (high temperature) beam where $\epsilon/k_0 \gg \Lambda/C$;

$$g = 2 \ln \frac{b}{\overline{a}} \qquad (6.69b)$$

for the space-charge dominated (low temperature) beam where $\Lambda/C \gg \epsilon/k_0$.

In the first case, the axial electric field E_{sz} varies with radius r while a is constant, and we calculated the g-factor by averaging over r^2, with $\overline{r^2} = a^2/2$. In the second case, E_{sz} is constant across the beam, while a varies with distance z, and we took the average radius $\overline{a}$ to define the g-factor. Note that both results differ from the asymptotic average value of the g-factor for a bunched beam given in Equation (5.365b), as expected [see also our discussion following Equation (5.365b)]. For the general case where both space charge and emittance affect the beam radius, the values of the g-factor are of course in the range between the above two limits, i.e., $2 \ln (b/a) \le g \le 0.5 + 2 \ln (b/\overline{a})$.

We should point out in this context that there is some confusion in the literature regarding the proper expression for the g-factor. Many authors (see, for instance, D.10, Section 6.2.1), use the relation $g = 1 + 2 \ln (b/a)$, which corresponds to the axial electric field on the axis ($r = 0$) of an emittance-dominated beam, rather than the average value (6.69a), which is more appropriate. Similarly, there has been disagreement in the past on the formula for g in a space-charge dominated beam. A recent experiment has confirmed for the first time the validity of Equation (6.69b) in this case [31]. Finally, we should point out that all our calculations here assume a cylindrical beam in a cylindrical conducting tube. In other geometries, the g-factor will be different. For a round beam of radius a between two parallel conducting plates of separation $2b$, the term $2 \ln (b/a)$ in the formula for g must be replaced by $2 \ln (4b/\pi a)$ [see Reference 37, Equation (9)]. This is also a good approximation for a rectangular pipe with height $2b$ and width $2w$ when $w \gg b$. On the other hand, when the width is comparable to the height ($w \approx b$), one can use $2 \ln (b/a)$ as a good approximation.

After this brief detour on the g-factor, let us return to Equation (6.68a) and consider first the case of a perfectly conducting tube where $E_{wz} = 0$. The amplitude of the total axial electric field perturbation is then

$$E_s = -\frac{g}{4\pi\epsilon_0}\left(-ik\Lambda_1 + \frac{i\omega}{c^2}I_1\right), \qquad (6.70a)$$

which, in view of (6.64b) may be written in the form

$$E_s = i \frac{g}{4\pi\epsilon_0} \left(\frac{k^2}{\omega} - \frac{\omega}{c^2} \right) I_1 . \tag{6.70b}$$

We now have derived two relationships between the axial electric field and current perturbations, E_s and I_1. The first one, Equation (6.67), was obtained from the continuity and force equations; the second one, Equation (6.70b), was derived from Maxwell's equations. Clearly, E_s and I_1 must satisfy both equations, and hence the two terms associated with I_1 on the right-hand side of the equations must be equal. This yields the desired dispersion relation between the frequency ω and the propagation constant k, namely

$$\frac{\gamma_0^3 m}{q \Lambda_0} \frac{(\omega - v_0 k)^2}{\omega} = -\frac{g}{4\pi\epsilon_0} \frac{\omega^2 - k^2 c^2}{\omega c^2} ,$$

or

$$(\omega - v_0 k)^2 - \gamma_0^2 c_s^2 k^2 \left(1 - \frac{\omega^2}{k^2 c^2} \right) = 0 . \tag{6.71}$$

Here we introduced the parameter c_s, defined as

$$c_s = \left(\frac{qg\Lambda_0}{4\pi\epsilon_0 \gamma_0^5 m} \right)^{1/2} , \tag{6.72}$$

which corresponds to the speed of sound in the mathematically equivalent problem of the propagation of a perturbation in a nonrelativistic cold fluid.

In many cases the difference between the phase velocities of the two space-charge waves and the beam velocity v_0 is very small. Hence we can make the approximation $\omega \simeq k v_0$ in the second term of Equation (6.71) and obtain the simpler dispersion relation usually found in the literature:

$$(\omega - k v_0)^2 - c_s^2 \gamma_0^2 k^2 \left(1 - \frac{v_0^2}{c^2} \right) = 0 ,$$

or

$$(\omega - k v_0)^2 - c_s^2 k^2 = 0 . \tag{6.73}$$

The solution of Equation (6.71) is

$$\omega = \frac{k \left[v_0 \pm c_s \sqrt{1 + \gamma_0^4 c_s^2 / c^2} \right]}{1 + \gamma_0^2 c_s^2 / c^2} , \tag{6.74a}$$

while (6.73) yields

$$\omega = k(v_0 \pm c_s). \tag{6.74b}$$

As we see, when the condition $\gamma_0^4 c_s^2/c^2 \ll 1$ is satisfied, Equation (6.74a) becomes identical with Equation (6.74b). This condition can be stated in terms of the average current as $\bar{I} \ll I_0 \beta_0 \gamma_0 / g$.

Thus we obtain, as in the previous case, two space-charge waves whose phase velocities for the simpler version of the dispersion relation are obtained from (6.74b) as

$$v_f = \frac{\omega}{k_+} = v_0 + c_v = v_0 \left(1 + \frac{c_s}{v_0} \right), \tag{6.75a}$$

$$v_s = \frac{\omega}{k_-} = v_0 - c_s = v_0 \left(1 - \frac{c_s}{v_0} \right). \tag{6.75b}$$

By introducing the plasma frequency

$$\omega_p = \left(\frac{q^2 n_0}{\epsilon_0 m \gamma_0^3} \right)^{1/2} = \left(\frac{q \Lambda_0}{\epsilon_0 m \gamma_0^3 a^2 \pi} \right)^{1/2} = \left(\frac{4 \gamma_0^2 c_s^2}{g a^2} \right)^{1/2}, \tag{6.76}$$

the velocity ratio c_s/v_0 can be expressed in terms of ω_p and Λ_0 as

$$\frac{c_s}{v_0} = \frac{\omega_p a}{v_0} \frac{\sqrt{g}}{2\gamma_0} = \left(\frac{q \Lambda_0 g}{4 \pi \epsilon_0 \gamma_0^5 m v_0^2} \right)^{1/2}, \tag{6.77}$$

where the g-factor is as defined in Equation (6.69).

It is interesting to compare the last four equations for the phase velocities of the waves in the pencil beam with Equations (6.55) and (6.56) for the infinite beam. We see that in place of ω_p/ω in the infinite-beam case we have c_s/v_0, and in view of (6.77) we have the correlation

$$\frac{\omega_p}{\omega} \leftrightarrow \frac{\omega_q}{\omega},$$

where

$$\omega_q = \frac{\omega_p \sqrt{g} \, \omega a}{2 \gamma_0 v_0} = \frac{\omega_p \sqrt{g} \, ka}{2 \gamma_0} \tag{6.78}$$

is the "reduced" plasma frequency due to screening by the wall of the vacuum tube in the pencil-beam case. Note that c_s corresponds to the phase velocity in the beam frame since $v_\pm = \pm c_s$ for $c_s \ll v_0$.

As can be seen from the above results density perturbations in a pencil beam surrounded by a perfectly conducting drift-tube wall travel along the beam as fast and slow space-charge waves. The frequencies for the two waves are real, hence there is no change (growth or decay) of the wave amplitudes. The beam neither loses nor gains energy and there is no instability.

This situation changes if we consider the case where the drift-tube wall has a finite resistance per unit length defined by $R_w^*[\Omega/m]$. The electric field E_{zw} along the wall surface is then no longer zero, as with the perfect conductor. The ohmic losses due to the image currents in the wall lead to growth of the slow-wave amplitude, and the associated *resistive-wall instability* was first investigated by Birdsall and Whinnery in 1953 for the possibility of microwave generation [32]. More recently, Smith and others [33–36] studied the effects of this instability on high-current beams in induction linacs, where it poses a threat to the beam quality. To analyze the problem we note that the electric field along the wall surface has a dc value that is defined by the product of the dc image current $\bar{I} = \Lambda_0 v_0$ and R_w^* (i.e., $E_{w0} = -\bar{I}R_w^* = -\Lambda_0 v_0 R_w^*$) and an ac component E_w determined by the product of the perturbed current $I_1 = \Lambda_0 v_1 + \Lambda_1 v_0$ and R_w^*:

$$E_w = -R_w^* I_1 = -R_w^*(\Lambda_0 v_1 + \Lambda_1 v_0). \tag{6.79}$$

For the behavior of the space-charge waves only the ac component E_w is relevant. From Figure 6.18 and Equation (6.68), E_w must be added to the space-charge field E_s of Equation (6.70b) to give a total longitudinal electric field of

$$E_s + E_w = \left(-i \frac{g}{4\pi\epsilon_0} \frac{\omega^2 - k^2 c^2}{\omega c^2} - R_w^*\right) I_1. \tag{6.80}$$

This total field must also satisfy Equation (6.67), and by equating $(E_s + E_w)/I_1$ from the two equations we obtain the dispersion relation

$$i \frac{\gamma_0^3 m}{q\Lambda_0} \frac{(\omega - v_0 k)^2}{\omega} = -i \frac{g}{4\pi\epsilon_0} \frac{\omega^2 - k^2 c^2}{\omega c^2} - R_w^*,$$

or

$$(\omega - v_0 k)^2 - \gamma_0^2 c_s^2 k^2 \left(1 - \frac{\omega^2}{k^2 c^2}\right) - i\omega \frac{q R_w^* \Lambda_0}{\gamma_0^3 m} = 0. \tag{6.81}$$

For $R_w^* = 0$, we recover our previous result of Equation (6.71). To simplify the analysis we will use the approximation $\omega \approx k v_0$ in the second term, as before, and introduce the frequency parameter ω_1 defined as

$$\omega_1 = \frac{q R_w^* \Lambda_0}{\gamma_0^3 m} \tag{6.82}$$

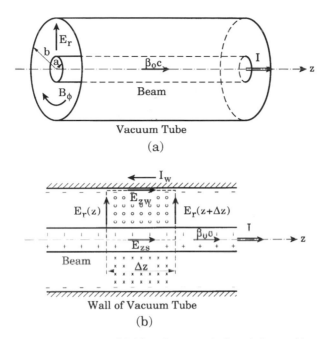

Figure 6.18. Beam geometry (a) and field configuration (b) for calculation of longitudinal electric field and impedance.

and representing the resistive wall effect. This yields the dispersion relation

$$(\omega - kv_0)^2 - c_s^2 k^2 - i\omega\omega_1 = 0. \tag{6.83}$$

This equation can be solved for the wave number k, and one obtains

$$k = \frac{\omega v_0}{v_0^2 - c_s^2}\left\{1 \pm \left[\frac{c_s^2}{v_0^2} + i\frac{\omega_1}{\omega}\left(1 - \frac{c_s^2}{v_0^2}\right)\right]^{1/2}\right\}. \tag{6.84}$$

Using the fact that $c_s^2 \ll v_0^2$ (i.e., $v_0^2 - c_s^2 \approx v_0^2$) and introducing the parameter $k_0 = \omega/v_0$, we can rewrite (6.84) in the approximate form

$$k = k_r + ik_i \approx k_0\left[1 \pm \frac{c_s}{v_0}\left(1 + i\frac{\omega_1 v_0^2}{\omega c_s^2}\right)^{1/2}\right]. \tag{6.85}$$

When $\omega_1 = 0$, the imaginary part is zero ($k_i = 0$) and we recover our previous result in Equation (6.74b). On the other hand, when $\omega_1 \neq 0$, we can solve (6.85) for the real and imaginary parts of k and obtain

$$k_r = k_0 \left\{ 1 \pm \left[\frac{1}{2} \frac{c_s^2}{v_0^2} \left(\sqrt{1 + \left(\frac{\omega_1 v_0}{c_s^2 k_0} \right)^2} + 1 \right) \right]^{1/2} \right\}, \qquad (6.86a)$$

$$k_i = \pm k_0 \left[\frac{1}{2} \frac{c_s^2}{v_0^2} \left(\sqrt{1 + \left(\frac{\omega_1 v_0}{c_s^2 k_0} \right)^2} - 1 \right) \right]^{1/2}. \qquad (6.86b)$$

For typical experimental parameters one finds that $\omega_1 v_0 \ll c_s^2 k_0$, so that one gets the approximate expressions

$$k_r \simeq k_0 \left(1 \pm \frac{c_s}{v_0} \right) \approx \frac{\omega}{v_0 \mp c_s}, \qquad (6.87a)$$

$$k_i \simeq \pm \frac{1}{2} \frac{\omega_1}{c_s}. \qquad (6.87b)$$

The result (6.87a) for the real part k_r of the propagation constant k is identical with Equation (6.74b) for the case where $R_w^* = 0$. The imaginary part k_i in (6.87b) is due to the wall resistivity R_w^*. It indicates that the amplitude of the slow wave will grow exponentially with distance as $\exp[(\omega_1/2c_s)z]$; that is, the beam will lose energy via dissipation in the resistive wall. This effect is known in the literature as the *resistive-wall instability*. It limits the beam current and causes the beam quality to deteriorate.

It will be useful for the general analysis and interpretation of the various dispersion relations for the space-charge waves to introduce the *space-charge impedance* $Z_s^*[\Omega/\text{m}]$, which is defined as the ratio of the voltage per meter $V^* = -E_s$ and the current amplitude I_1. From Equation (6.70b) we get

$$Z_s^* = \frac{V^*}{I_1} = -\frac{E_s}{I_1} = -i \frac{g}{4\pi\epsilon_0} \left(\frac{k^2}{\omega} - \frac{\omega}{c^2} \right). \qquad (6.88)$$

We see that Z_s^* is complex if ω and k are complex. However, for our analysis we will treat ω and k in the space-charge impedance as real quantities; in this case Z_s^* has only an imaginary, or reactive, part which we will define by $-X_s^*$. Furthermore, there is a capacitive component, $(i\omega C^\dagger)^{-1}$, and an inductive component, $i\omega L^*$. The capacitance $C^\dagger$ is associated with the perturbed longitudinal charge density and electric field, and it has units of F-m, which is why we use the superscript† rather than*. This is in contrast to the capacitance C^* per unit length associated with the transverse electric field due to the space charge of the beam, which is in units of F/m (see Problem 6.1). L^* is the inductance per unit length [H/m] associated with the perturbed current. In terms of the two parameters $C^\dagger$ and I^* we can write the space-charge impedance as

$$Z_s^* = -iX_s^* = \frac{1}{i\omega C^\dagger} + i\omega L^*, \qquad (6.89a)$$

or

$$Z_s^* = -\frac{i}{\omega C^\dagger}(1 - \omega^2 L^* C^\dagger). \qquad (6.89b)$$

By comparing the last three equations, we obtain

$$X_s^* = \frac{gk^2}{4\pi\epsilon_0\omega}\left(1 - \frac{\omega^2}{k^2 c^2}\right) \qquad [\Omega/\text{m}], \qquad (6.90)$$

$$C^\dagger = \frac{4\pi\epsilon_0}{gk^2} \qquad [\text{F-m}], \qquad (6.91)$$

$$L^* = \frac{g}{4\pi\epsilon_0 c^2} = \frac{g\mu_0}{4\pi} \qquad [\text{H/m}], \qquad (6.92)$$

$$L^* C^\dagger = \frac{1}{k^2 c^2} \qquad [\text{s}^{-2}]. \qquad (6.93)$$

In the approximation $\gamma_0^4 c_s^2/c^2 \ll 1$, where $\omega = k(v_0 \pm c_s)$ [see Equation (6.74b)], and $c_s \ll v_0$, we can use $\omega \simeq kv_0$ and obtain the relation

$$Z_s^* = -i\frac{g}{4\pi\epsilon_0}\frac{k^2}{\omega}\left(1 - \frac{v_0^2}{c^2}\right) = -i\frac{g}{4\pi\epsilon_0}\frac{k^2}{\omega\gamma_0^2}, \qquad (6.94a)$$

or

$$Z_s^* = -iX_s^* = -i\frac{gk}{4\pi\beta_0\gamma_0^2}Z_0 = -i\frac{g\omega}{4\pi c\beta_0^2\gamma_0^2}Z_0, \qquad (6.94b)$$

where $Z_0 = (\epsilon_0 c)^{-1} = (\mu_0/\epsilon_0)^{1/2} \simeq 377~\Omega$ is the free-space impedance.

We conclude from these relations that the space-charge impedance is always negative imaginary (i.e., the capacitive part is always greater than the inductive part). The ratio of the two impedances varies as $\omega L^*/(\omega C^\dagger)^{-1} = \omega^2 L^* C^\dagger \approx v_0^2/c^2$; that is, the inductive part is negligible at nonrelativistic velocities and becomes more and more comparable to the capacitive part at highly relativistic energies. The net effect is that the space-charge impedance Z_s^* is essentially capacitive and decreases with increasing kinetic energy as $(\beta_0\gamma_0^2)^{-1}$. Furthermore, Z_s^* is proportional to the geometry factor $g = \alpha + 2\ln(b/a)$ and inversely proportional to the wavelength $\lambda = 2\pi/k$ of the perturbation where $0 \le \alpha \le 1$ [see (6.69) and related discussion].

If we introduce the space-charge impedance X_s^* from Equation (6.90) and the generalized perveance $K = (\bar{I}/I_0)(2/\beta_0^3 \gamma_0^3)$, the dispersion relation (6.81) may be written in the alternative form

$$\omega - kv_0 = \pm k_0 v_0 \left[i \frac{2\pi \beta_0 K}{k_0 Z_0} (R_w^* - iX_s^*) \right]^{1/2}. \tag{6.95}$$

By defining

$$\Delta \omega = \omega - kv_0 = \Delta \omega_r + i \Delta \omega_i, \tag{6.96}$$

we can write the general wave solution in terms of the amplitude and phase factors as

$$u = u_0 + u_1 e^{-\Delta \omega_i t} e^{i(\omega_r t - k_r z)}, \tag{6.97}$$

where $\Delta \omega_r t = \omega_r t - k_r v_0 t$ and $v_0 t = z$ was used. The amplitude factor $e^{-\Delta \omega_i t}$ measures the exponential growth of the slow wave or the decay of the fast wave, depending on whether the sign of $\Delta \omega_i$ is negative or positive. This notation for the wave amplitude as a whole is preferable over the solutions for either ω_i or k_i alone, as we did in our analysis so far and which we still can get separately from (6.83) if we wish [see Equations (6.84) to (6.87)]. In terms of $\Delta \omega$, the dispersion relation (6.95) may be written as

$$\Delta \omega = \Delta \omega_r + i \Delta \omega_i = \pm k_0 v_0 \left[\frac{2\pi \beta_0 K}{k_0 Z_0} (X_s^* + iR_w^*) \right]^{1/2}, \tag{6.98}$$

and the solutions for the real and imaginary parts of $\Delta \omega$ are

$$\Delta \omega_r = \pm k_0 v_0 \left[\frac{\pi \beta_0 K}{k_0 Z_0} \left(\sqrt{R_w^{*2} + X_s^{*2}} + X_s^* \right) \right]^{1/2}, \tag{6.99a}$$

$$\Delta \omega_i = \mp k_0 v_0 \left[\frac{\pi \beta_0 K}{k_0 Z_0} \left(\sqrt{R_w^{*2} + X_s^{*2}} - X_s^* \right) \right]^{1/2}, \tag{6.99b}$$

where the upper signs indicate the fast wave ($\Delta \omega_r > 0, \Delta \omega_i < 0$) and the lower signs the slow wave ($\Delta \omega_r < 0, \Delta \omega_i > 0$), and where $k_0 = \omega/v_0$ was used.

If $R_w^* \ll X_s^*$, one obtains for the growth rate of the slow wave the approximate result

$$\Delta \omega_i = k_0 v_0 \left(\frac{\pi \beta_0 K R_w^{*2}}{2 k_0 Z_0 X_s^*} \right)^{1/2}, \tag{6.100a}$$

or by substituting for K and X_s^* [Equation (6.94b)],

$$\Delta\omega_i \simeq 2\pi \frac{R_w^*}{Z_0}\beta_0 c\left(\frac{\bar{I}}{I_0}\frac{1}{g\beta_0\gamma_0}\right)^{1/2}. \qquad (6.100b)$$

This may be written in terms of the imaginary wave number $k_i = \Delta\omega_i/v_0$ as a spatial growth rate,

$$k_i = 2\pi \frac{R_w^*}{Z_0}\left(\frac{\bar{I}}{I_0}\frac{1}{g\beta_0\gamma_0}\right)^{1/2}. \qquad (6.100c)$$

For electrons ($I_0 \simeq 1.70 \times 10^4$ A) the last relation becomes

$$k_i = 1.28 \times 10^{-4}R_w^*\left(\frac{\bar{I}}{g\beta_0\gamma_0}\right)^{1/2}. \qquad (6.101a)$$

For ions ($I_0 \simeq 3.13 \times 10^7 A/Z$ amperes) with mass number A, charge state Z, and average particle current $\bar{I}_p = \bar{I}/Z$, one gets

$$k_i = 2.98 \times 10^{-6}R_w^* Z\left(\frac{\bar{I}_p}{g A \beta_0\gamma_0}\right)^{1/2}, \qquad (6.101b)$$

which in the nonrelativistic regime ($\gamma_0 \simeq 1$, $\beta_0 = \sqrt{2T/mc^2}$) may be written in the form

$$k_i = 1.39 \times 10^{-5}R_w^* Z\left(\frac{\bar{I}_p}{g A}\right)^{1/2}\left(\frac{1}{T/A}\right)^{1/4}, \qquad (6.101c)$$

where R_w^* is in Ω/m, I_p in amperes, and T/A in MeV/nucleon.

The growth of the slow space-charge wave predicted by the above theory implies that energy is lost by the beam to the external resistance and that the beam quality deteriorates. As we will see in the next section, the growth of the resistive-wall instability is damped by momentum spread, $\widetilde{\Delta P}/P$, in the beam. If we start with a cold beam ($\widetilde{\Delta P}/P = 0$), as in the above analysis, the instability will cause a momentum spread to develop which eventually will become large enough to saturate the growth. In turn, this momentum spread may cause excessive chromatic aberrations which make it impossible to focus the beam to a desired spot size (i.e., the instability produces in effect an increase of the emittance).

The resistive wall instability is of concern for high-current electron and heavy-ion linear accelerators and transport systems and for circular machines. The latter have generally lower currents but many revolutions and hence a longer interaction time than that of the linear machines. We treat the instability in circular machines in the next section.

To illustrate the application of the above theory to laboratory beams, let us consider as a first example a relativistic 10-kA electron beam in an induction linac with $R_w^* = 10\ \Omega/\text{m}$, $g \simeq 2$, and $\beta_0\gamma_0 \simeq 5$. From (6.101a) we find $k_i = 0.04\ \text{m}^{-1}$ or $z = k_i^{-1} = 25\ \text{m}$. Clearly, in this case one would expect problems with the instability since the length of such an induction linac would be greater than the e-folding growth distance k_i^{-1}.

As a second example, let us take the case of a 10-GeV $_{137}\text{Ba}^{2+}$ beam for possible use in heavy-ion inertial fusion, with $A = 137$, $Z = 2$, $I_p = 10^4\ \text{A}$, $g = 2$, $R_w^* = 100\ \Omega/\text{m}$, and $\beta_0\gamma_0 \simeq 0.4$. From (6.101b) we get $k_i = 5.75 \times 10^{-3}$ m^{-1} or $z = k_i^{-1} = 174\ \text{m}$. Since the final transport line of the beam would be considerably longer than this distance, the resistive-wall instability may pose a severe problem for heavy-ion inertial fusion drivers.

The above analysis can be readily extended from a purely resistive wall to the general case of a complex impedance $Z_w^* = R_w^* + iX_w^*$. By including the space-charge impedance $Z_s^* = -iX_s^*$ we then can define a total longitudinal impedance $Z_\parallel^*$ as

$$Z_\parallel^* = Z_w^* + Z_s^* = Z_r^* + iZ_i^*, \qquad (6.102a)$$

where the real part is given by

$$Z_r^* = R_w^* \qquad (6.102b)$$

and the imaginary part by

$$Z_i^* = X_w^* - X_s^*. \qquad (6.102c)$$

Note that all impedances are in general functions of the frequency ω and wave number k. This is also true for the space-charge impedance X_s^*, as can be seen from Equations (6.88) or (6.94b). In terms of the total longitudinal impedance $Z_\parallel^*$, the dispersion relation (6.95) may be written as

$$\Delta\omega = \omega - kv_0 = \pm k_0 v_0 \left[i\frac{2\pi\beta_0 K}{k_0 Z_0} Z_\parallel^* \right]^{1/2}, \qquad (6.103a)$$

or

$$\Delta\omega = \pm k_0 v_0 \left\{ \frac{2\pi\beta_0 K}{k_0 Z_0} [(X_s^* - X_w^*) + iR_w^*] \right\}^{1/2}. \qquad (6.103b)$$

If Z_w^* represents a lossy transmission-line model, where a resistance R^* is in series with a distributive inductance L^* and both are connected to ground by a distributive capacitance C^*, one has

$$Z_w^*(k, \omega) = \frac{k^2(R^* + i\omega L^*)}{k^2 - \omega^2 L^* C^* + i\omega R^* C^*} = R_w^* + iX_w^*. \qquad (6.104)$$

The linear growth rate for the slow-wave amplitude is then found to be

$$\Delta\omega_i = -k_0 v_0 \left\{ \frac{\pi\beta_0 K}{k_0 Z_0} \left[\sqrt{Z_r^{*2} + Z_i^{*2}} \mid Z_i^* \right] \right\}^{1/2}, \qquad (6.105a)$$

or alternatively, for the imaginary wave constant

$$k_i = k_0 \left\{ \frac{\pi\beta_0 K}{k_0 Z_0} \left[\sqrt{R_w^{*2} + (X_w^* - X_s^*)^2} + (X_w^* - X_s^*) \right] \right\}^{1/2}. \qquad (6.105b)$$

The analysis of these relations shows that an inductive wall impedance enhances the growth rate, while a capacitive impedance decreases the growth rate. In the first case, instability can arise even if $R_w - 0$ (see Problem 6.9).

6.3.3 Longitudinal Instability in Circular Machines and Landau Damping

In Sections 3.6.4 and 5.4.9 we pointed out that the negative-mass behavior of charged particle beams in circular accelerators above the transition energy $\gamma_t mc^2$ can cause longitudinal bunching and instability. This *negative mass instability* was first identified and analyzed theoretically by Nielsen, Sessler, and Symon [37] and independently by Kolomenskij and Lebedev [38] in 1959. Later studies [39] showed that instability also occurs below transition energy and hence is not restricted to the negative-mass regime when the finite wall resistivity is taken into account. In fact, the underlying physical mechanism in circular machines is basically the same as in the resistive-wall instability discussed in the preceding section. Perturbations of the beam's line-charge density produce electromagnetic fields via the image charges flowing through the surrounding walls and these fields act back on the beam. If the wall impedance has a resistive component as in our previous case, there will be unstable growth of the slow space-charge wave, which in turn may result in beam deterioration and particle loss. Since the effect is frequency dependent and shows a resonant-like behavior at high frequencies, it is also known in the literature as the *longitudinal microwave instability*. As mentioned in the preceding section, momentum spread in the beam can decrease the growth rate or prevent the instability from developing in the first place. What happens in this case is that the phase spread in the particle oscillations due to the different momenta offsets the bunching that is otherwise produced by the instability. This effect is known as

Landau damping since it is mathematically analogous to the damping of unstable electromagnetic perturbations in an infinite plasma that was first investigated by Landau [40].

To analyze the longitudinal instability with damping due to momentum spread we have to use the Vlasov equation and a proper longitudinal distribution function for the beam. In our description of the problem we follow the review given by Hofmann [41], except that we use somewhat different notation consistent with the preceding section. Let us assume that the distribution of the particles in longitudinal phase space depends on the energy E, the distance s along the circumference $C = 2\pi\overline{R}$ of the circular accelerator, and time t as

$$f(E, s, t) = f_0(E) + f_1(E)e^{i(\omega t - ks)}. \tag{6.106}$$

Here $f_0(E)$ is the unperturbed beam, assumed to be continuous (unbunched) along the circumference, and $f_1(E)$ is the amplitude of the perturbation, $\overline{R}$ the average orbit radius, k the wave number, and ω the frequency of the perturbation, which can be complex in general. It is customary to introduce the angle $\theta = s/\overline{R}$ and the number of wavelengths n of the perturbation within the circumference of the ring. With $2\pi\overline{R} = n\lambda$, we get $ks = n\theta$, so that the distribution function can be written in the alternative form

$$f(E, \theta, t) = f_0(E) + f_1(E)e^{i(\omega t - n\theta)}. \tag{6.107}$$

If Λ_0 is the unperturbed line charge density, the total number of particles is $N = 2\pi\overline{R}\Lambda_0/q$, which leads to the normalization relation

$$\int_0^{2\pi} \int_0^\infty f_0(E)\, dE\, d\theta = \frac{2\pi\overline{R}\Lambda_0}{q} \tag{6.108}$$

for the unperturbed distribution function.

The distribution (6.107) must satisfy the Vlasov equation

$$\frac{\partial f}{\partial t} + \frac{\partial f}{\partial \theta}\dot{\theta} + \frac{\partial f}{\partial E}\dot{E} = 0, \tag{6.109}$$

or

$$(i\omega - in\dot{\theta})f_1 + \frac{\partial f_0}{\partial E}\dot{E} = 0, \tag{6.110}$$

where we assumed that $\partial f/\partial E \approx \partial f_0/\partial E$. $\dot{E}$ is the rate of change of the particle energy in the distribution due to the electric field produced by the perturbation, $E_\| = E_s + E_w$. Here E_s is the space-charge field, and E_w is the field generated by the perturbed current I_1 due to the impedance of the wall. It is customary in the theory of circular machines to introduce the voltage drop through one revolution, $V_1 = -2\pi \overline{R}E_\parallel$, and express it as the product of the total impedance $Z_\parallel$ and the perturbed current I_1, that is,

$$V_1 = -2\pi\overline{R}E_\parallel = -2\pi\overline{R}(E_s + E_w) = Z_\parallel I_1 . \tag{6.111}$$

If $\omega_0 = \overline{R}/v_0$ denotes the angular revolution frequency, the rate of change (decrease) of the particle energy due to the perturbation can then be written as

$$\dot{E} = \frac{dE}{dt} = -qV_1\frac{\omega_0}{2\pi} = -qZ_\parallel I_1\frac{\omega_0}{2\pi} . \tag{6.112}$$

The longitudinal impedance $Z_\parallel$ consists of the space-charge impedance, Z_s, and the wall impedance, Z_w, with all contributions of individual elements (such as drift-tube sections, rf gaps, diagnostic ports, etc.) summed up along the entire circumference of the machine. With (6.94b), the total space-charge impedance is given by

$$Z_s = -2\pi\overline{R}i\frac{gkZ_0}{4\pi\beta_0\gamma_0^2} , \tag{6.113}$$

and since $k = n\theta/s = n/R$, we can write this relation in the form

$$Z_s = -iX_s = -\frac{ingZ_0}{2\beta_0\gamma_0^2} . \tag{6.114}$$

The total wall impedance Z_w will, in general, have a resistive part, R_w, and a reactive part, X_w (i.e., $Z_w = R_w + iX_w$), so that we have

$$Z_\parallel = Z_s + Z_w = R_w + i(X_w - X_s) . \tag{6.115}$$

By substituting (6.112) into (6.110), we obtain

$$(i\omega - in\dot{\theta})f_1(E) = \frac{\partial f_0(E)}{\partial E}\frac{q\omega_0}{2\pi}Z_\parallel I_1 . \tag{6.116}$$

Since the perturbed current I_1 is related to the perturbed distribution function by

$$I_1 = q\omega_0 \int f_1(E)\,dE , \tag{6.117}$$

we can write the dispersion relation (6.117) as

$$1 = \frac{-iq^2\omega_0^2}{2\pi}Z_\parallel \int \frac{\partial f_0(E)}{\partial E}\frac{dE}{\omega - n\dot{\theta}} . \tag{6.118}$$

To proceed further it will be helpful to change variables from E to $\dot\theta$ so that

$$\frac{\partial f_0(E)}{\partial E} dE = \frac{\partial f_0(\dot\theta)}{\partial \dot\theta} \frac{d\dot\theta}{dE} d\dot\theta. \tag{6.119}$$

From (3.261) and (5.351b) we have

$$\frac{d\dot\theta}{\omega_0} = -\eta \frac{dP}{P_0} = -\frac{\eta}{\beta_0^2} \frac{dE}{E_0} = -\frac{\eta}{\beta_0^2} \frac{dE}{\gamma_0 mc^2},$$

or

$$\frac{d\dot\theta}{dE} = -\frac{\eta}{\beta_0^2} \frac{\omega_0}{E_0} = -\frac{\eta\omega_0}{\beta_0^2 \gamma_0 mc^2}, \tag{6.120}$$

where $E_0 = \gamma_0 mc^2$ and $\eta = (1/\gamma_t^2 - 1/\gamma_0^2)$, as defined in (3.262b).

Using (6.119) and (6.120), we can write the dispersion relation (6.118) in the alternative form

$$+i\frac{q^2\omega_0^3\eta Z_\|}{2\pi\beta_0^2\gamma_0 mc^2} \int \frac{\partial f_0(\dot\theta)}{\partial\dot\theta} \frac{d\dot\theta}{\omega - n\dot\theta} = 1. \tag{6.121}$$

Let us first consider the cold-beam limit (zero energy spread) by assuming a delta function for the distribution, that is,

$$f_0(\dot\theta) = \frac{\overline{R}\Lambda_0}{q} \delta(\dot\theta - \omega_0) \tag{6.122}$$

using the normalization (6.108). Then

$$\int \frac{\partial f_0(\dot\theta)}{\partial\dot\theta} \frac{d\dot\theta}{\omega - n\dot\theta} = -\frac{n\overline{R}\Lambda_0}{q(\omega - n\omega_0)^2}, \tag{6.123}$$

and with $\overline{R}\Lambda_0 = \overline{R}\overline{I}/v_0 = \overline{I}/\omega_0$, Equation (6.123) becomes

$$(\omega - n\omega_0)^2 = -i\frac{q\eta\omega_0^2 n Z_\|\overline{I}}{2\pi\beta_0^2\gamma_0 mc^2}. \tag{6.124}$$

It is left as a problem (6.11) to show that (6.124) converts to (6.85) if one makes the correct transition from circular to straight beam. By introducing $\Delta\omega = \Delta\omega_r + i\Delta\omega_i = \omega - n\omega_0$, we can write the last relation in the form

$$\omega - n\omega_0 = \Delta\omega = \pm\left(-i\frac{q\eta\omega_0^2 n\overline{I}Z_\|}{2\pi\beta_0^2\gamma_0 mc^2}\right)^{1/2}, \tag{6.125}$$

which is convenient for a stability analysis. Obviously, when $\Delta\omega$ is imaginary (i.e., $\Delta\omega = i\Delta\omega_i$), the exponential wave factor of the perturbation will have the form $e^{-\Delta\omega_i t}$, which indicates unlimited exponential growth of the perturbation amplitude for $\Delta\omega_i < 0$. We see immediately that such an instability will occur when $\eta > 0$, or $\gamma_0 > \gamma_t$ (above transition), and when $Z_{\parallel}$ is negative imaginary. The simplest case for which this can happen is for zero wall impedance (i.e., $Z_w = R_w + iX_w = 0$), so that $Z_{\parallel} = -iX_s$ is entirely determined by the reactive space-charge impedance, X_s, and with (6.114), relation (6.125) becomes

$$\Delta\omega = \pm i\Delta\omega_i = \pm i\omega_0 \left(\frac{q\eta n^2 \bar{I} g Z_0}{4\pi\beta_0^3 \gamma_0^3 mc^2} \right)^{1/2}. \tag{6.126}$$

For $\Delta\omega_i < 0$, this may be written in terms of the generalized perveance $K = (\bar{I}/I_0)(2/\beta_0^3\gamma_0^3)$ as

$$-\Delta\omega_i = \frac{1}{\tau} = \omega_0 \left(\frac{\eta n^2 g K}{2} \right)^{1/2}, \tag{6.127}$$

or, with $\omega_0 = \beta_0 c/\overline{R}$, in the alternative form

$$-\Delta\omega_i = \frac{1}{\tau} - \left[\frac{\eta n^2 \bar{I} g Z_0 c^2}{4\pi\overline{R}^2 (mc^2/q)\beta_0\gamma_0^3} \right]^{1/2}. \tag{6.128}$$

The unstable situation defined by these relations is known as the *negative-mass instability* [29]. It occurs only in circular machines above transition energy ($\gamma_0 > \gamma_t$) and is attributable entirely to the negative mass behavior ($m^* = -\gamma_0 m/\eta$) discussed in Section 3.6.4. In this negative-mass regime, a local density increase in the particle distribution will grow with time, leading to bunching of the beam. If the space-charge forces associated with these bunches become large enough, emittance growth and particle loss will occur when the tune shift $\Delta\nu$ exceeds the threshold for a resonance. One should note that the negative-mass instability occurs with perfectly conducting walls (i.e., under conditions where a straight beam is stable). The straight-beam case can be obtained from (6.125) by letting $\gamma_t \to \infty$ and hence $\eta \to -1/\gamma_0^2$, so that $\Delta\omega$ becomes real and the perturbation remains stable, in agreement with our discussions in the preceding section. Of course, in the circular machine the negative-mass instability does not occur below transition when $\eta < 0$, as is evident from the last four equations.

Let us now discuss the case when the wall has a finite resistivity so that the impedance $Z_{\parallel}$ is complex. To be as general as possible, we will define the impedance as

$$Z_{\parallel} = Z_r + iZ_i, \tag{6.129}$$

where Z_r is the real part (e.g., $Z_r = R_w$), and Z_i is the imaginary part (e.g., $Z_i = X_w - X_s$), as in Equation (6.115). The dispersion relation (6.125) then becomes

$$\Delta\omega = \pm\omega_0\left\{-i\frac{q\eta n\bar{I}}{2\pi\beta_0^2\gamma_0 mc^2}[Z_r + iZ_i]\right\}^{1/2}. \tag{6.130}$$

With the impedance given in Equation (6.115) we can write

$$\Delta\omega = \pm\omega_0\left\{-i\frac{q\eta n\bar{I}}{2\pi\beta_0^2\gamma_0 mc^2}[R_w + i(X_w - X_s)]\right\}^{1/2}. \tag{6.131}$$

It can be seen that $\Delta\omega$ always has a nonzero imaginary part no matter what the sign of η is or whether or not $X_w = 0$. This case is known as the *resistive-wall instability* in circular machines. The only difference with respect to the straight-beam case discussed in the preceding section is that the growth rate depends on η and the sign of η. To further analyze the dispersion relation (6.131), let us introduce the parameter Δ defined as

$$\Delta = \frac{q|\eta|n\bar{I}}{2\pi\beta_0^2\gamma_0 mc^2}. \tag{6.132}$$

The growth rate $\Delta\omega_i$ then depends on the sign of η. When η is positive [i.e., in the negative-mass regime (above transition)], one obtains

$$\Delta\omega_i = \pm\omega_0\left[\frac{1}{2}\Delta\left(\sqrt{Z_r^2 + Z_i^2} - Z_i\right)\right]^{1/2} \qquad \text{for } \eta > 0, \tag{6.133a}$$

and when η is negative (below transition) one finds that

$$\Delta\omega_i = \pm\omega_0\left[\frac{1}{2}\Delta\left(\sqrt{Z_r^2 + Z_i^2} + Z_i\right)\right]^{1/2} \qquad \text{for } \eta < 0. \tag{6.133b}$$

Note that the result for the last case ($\eta < 0$) has the same form as in Equation (6.105a) for a straight beam. These cold-beam results can be summarized by stating that there is always instability when $R_w \neq 0$ (resistive-wall instability) and, furthermore, for $R_w = 0$, the beam is unstable above transition ($\eta > 0$, negative-mass instability). Also, it is interesting to evaluate how the capacitive space-charge impedance and inductive ($X_w > 0$) or capacitive ($X_w < 0$) wall impedances affect the instability growth rate in the general case. Above transition ($\eta > 0$) in the negative-mass regime, both the space charge and a capacitive wall increase the growth rate, while an inductive wall decreases $\Delta\omega_i$. The fact that an inductive wall impedance tends to stabilize the negative-mass behavior was first pointed out by Briggs and Neil [42]. Below transition ($\eta < 0$) the opposite is true: space charge and capacitive wall lower the growth rate, while an inductive wall increases it. [See the discussion following Equation (6.105b).]

To determine the effect of Landau damping on the longitudinal instability, one must use an appropriate distribution function in energy E or rotation frequency $\dot{\theta}$ and evaluate the integral in the dispersion relation (6.121). If the frequency ω of the perturbation lies within the frequency distribution $n\dot{\theta}$ of the particles, the denominator of the integral will become zero at $\omega = n\omega_0$. In this case the integral can be split into two parts, the principal value (P.V.) and the residue term, and one obtains

$$\int \frac{\partial f_0(\dot{\theta})/\partial \dot{\theta}}{\omega - n\dot{\theta}}\, d\dot{\theta} = \int_{\text{P.V.}} \frac{\partial f_0(\dot{\theta})/\partial \dot{\theta}}{\omega - n\dot{\theta}}\, d\dot{\theta} \pm i\pi \frac{\partial f_0(\dot{\theta})}{\partial \dot{\theta}}\bigg|_{\dot{\theta}=\omega/n}, \qquad (6.134)$$

so that (6.121) can be written as

$$-\frac{q^2 \omega_0^3 \eta Z_\parallel}{2\pi \beta_0^2 \gamma_0 mc^2}\left[\pm \pi \frac{\partial f_0(\dot{\theta})}{\partial \dot{\theta}}\bigg|_{\dot{\theta}=\omega/n} - i \int_{\text{P.V.}} \frac{\partial f_0(\dot{\theta})/\partial \dot{\theta}}{\omega - n\dot{\theta}}\, d\dot{\theta} \right] = 1. \qquad (6.135)$$

To solve this dispersion relation for different distributions it will be convenient to introduce [41] the half width $S = \Delta\dot{\theta}/2$ of the angular frequency distribution $f_0(\dot{\theta})$ measured at half-height and relate the frequencies in the dispersion integral to S by means of two dimensionless variables x and x_1. The first is defined by $\dot{\theta} = \omega_0 - xS$, or

$$n\dot{\theta} - n\omega_0 = xnS, \qquad (6.136)$$

and describes the angular frequencies of the particles in the beam. The second is defined by

$$\Delta\omega = \omega - n\omega_0 = x_1 nS, \qquad (6.137)$$

and gives the frequency ω with which the instability is driven. As mentioned earlier, ω_0 is the revolution frequency of the central-orbit particle. Furthermore, the distribution function $f_0(\dot{\theta})$ is expressed in terms of x as

$$f(x) = \frac{2\pi S f_0(\dot{\theta})}{N}, \qquad (6.138)$$

where N is the total number of particles in the ring and

$$\int f(x)\, dx = 1. \qquad (6.139)$$

Finally, the half-width S is related to the full momentum spread ΔP at half-height of the distribution via

$$2S = -\eta\omega_0 \frac{\Delta P}{P}. \tag{6.140}$$

With these definitions one can express the dispersion relation (6.121) in the form

$$-\text{sign}\left(\frac{d\dot{\theta}}{dE}\right) \frac{2\bar{I}qZ_\|}{\pi mc^2\beta_0^2\gamma_0|\eta|(\Delta P/P_0)^2 n} I_D' = 1, \tag{6.141}$$

where I_D' is the normalized dispersion integral given by

$$I_D' = \pm\pi \frac{dE}{dx}(x_1) - i\int_{\text{P.V.}} \frac{df/dx}{x - x_1}\,dx. \tag{6.142}$$

The $\text{sign}(d\dot{\theta}/dE)$ is $+1$ below transition energy and -1 above transition energy.

The factor in front of I_D' in (6.141) is originally defined as a complex quantity [37] $V' + iU'$ that can be related to the complex impedance $Z_\| = Z_r + iZ_i$ by

$$V' + iU' = \frac{2\bar{I}q}{\pi mc^2\beta_0^2\gamma_0|\eta|(\Delta P/P)^2 n}(Z_r + iZ_i), \tag{6.143}$$

so that (6.141) may be written as

$$-\text{sign}\left(\frac{d\dot{\theta}}{dE}\right)(V' + iU')I_D' = 1. \tag{6.144}$$

This equation defines a relation between x_1 and V', U'. The quantity x_1 is related to the real and imaginary frequency shifts as

$$\Delta\omega_r = nS\,\text{Re}(x_1), \tag{6.145}$$

$$|\Delta\omega_i| = \frac{1}{\tau} = nS\,\text{Im}(x_1), \tag{6.146}$$

where $\tau = |\Delta\omega_i^{-1}|$ is the growth rate of the instability. The stability limit is defined by $\Delta\omega_i = 0$, or by the curve

$$\text{Im}(x_1) = 0 \tag{6.147}$$

in a U' versus V' stability diagram. The region inside the curve $\text{Im}(x_1) = 0$ is stable and the region outside it is unstable. Figure 6.19 shows these curves for several distributions $f(x)$ investigated by Ruggiero and Vaccaro [43]. For high-energy accelerators, where $\gamma_0 \gg 1$ and where the space-charge impedance is small compared to the reactive part of the wall impedance (i.e., $X_s \ll |X_w|$, or $|Z_s| \ll |Z_\parallel|$), one can establish a very conservative stability criterion by approximating the stability limit $\text{Im}(x_1) = 0$ with a circle that fits inside all these curves. Using this circle, shown in Figure 6.19 with dashed interior, one obtains from (6.141) the *Keil–Schnell stability criterion* [44]

$$\left|\frac{Z_\parallel}{n}\right| \leq F \frac{mc^2\beta_0^2\gamma_0|\eta|(\Delta P/P_0)^2}{q\overline{I}} = F \frac{mc^2|\eta|[\Delta(\beta_0\gamma_0)]^2}{q\overline{I}\gamma_0} . \tag{6.148}$$

The form factor F is determined by the radius of the circle. In Figure 6.19 this radius is 0.6 and gives a form factor of $F \simeq 1$. Relation (6.148) can be used in many ways. Thus it shows the absolute value of the longitudinal impedance $|Z_\parallel|$ divided by the harmonic number n of the perturbation that is necessary to obtain stability for a given beam distribution with average current $\overline{I}$, energy $\gamma_0 mc^2$, and momentum spread $(\Delta P/P_0)$. Conversely, one can calculate the current threshold $\overline{I}$ for given $|Z_\parallel|$, $(\Delta P/P_0)$ and γ_0, and so on.

Note that the effect of Landau damping is given by the momentum spread $\Delta P/P_0$. The smaller $\Delta P/P_0$, the smaller is the beam current that can be circulated in the ring. If $\Delta P/P_0 = 0$, there is no stability, and we recover the previous cold-beam results where $f_0(E)$ was a delta function.

The Keil–Schnell criterion is very conservative and deliberately underestimates the stability threshold to provide a margin of flexibility. As we discussed in Section 5.4, laboratory beams tend to have a Maxwell–Boltzmann distribution as represented by the curve $e^{-x^2/2a^2}$ in Figure 6.19. Thus, in practice, the region of stability is much larger than the Keil–Schnell limit implies, especially with regard to the imaginary part of the impedance ($U' \propto Z_i$), which can be many times greater than the Keil–Schnell value. In the stability diagram of Figure 6.19, which applies for a high-energy machine above transition, one could therefore tolerate a high net inductive impedance ($Z_i = X_w - X_s > 0$) that exceeds the Keil–Schnell limit (i.e., $U' > U'_{K.S.}$) if this were practical. Below transition, the stability curves in Figure 6.19 should be flipped over since the stable region in this case extends toward the negative U' direction where the net impedance is capacitive ($Z_i < 0$). If the particle energies are not highly relativistic, as in some heavy-ion synchrotrons or in low-energy proton machines, the space-charge impedance may be considerably greater than the wall impedance (i.e., $X_s > |X_w|$) and the operating point could be well outside the Keil–Schnell circle in the long neck of the stable region. An example of this type is the beam behavior in a heavy-ion storage ring discussed by Hofmann [45]. In high-current induction linacs, proposed as drivers for heavy-ion inertial fusion, one would always operate in such a space-charge-dominated regime. The boundaries of the stable region then depend not only on the momentum spread but also on the beam current or generalized perveance K. This is illustrated in Figures 6.20 and 6.21, which show the stability diagrams for a Gaussian momentum

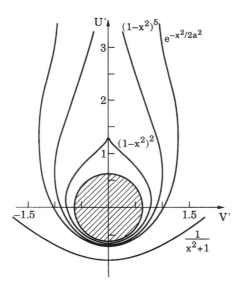

Figure 6.19. Ruggiero–Vaccaro stability diagram for several distributions $f(x)$. (From Reference 43.)

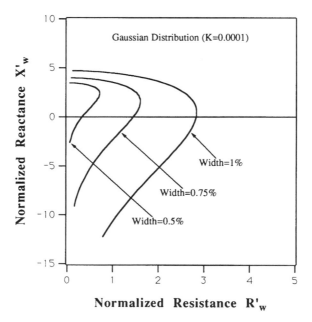

Figure 6.20. Boundary of stable regions for a Gaussian distribution in the half $R'_w - X'_w$ plane for three different momentum spreads at a fixed beam perveance $K = 10^{-4}$, where $\beta_0 = 0.3$, and $g = 2$ were used in the calculation. (From Reference 36.)

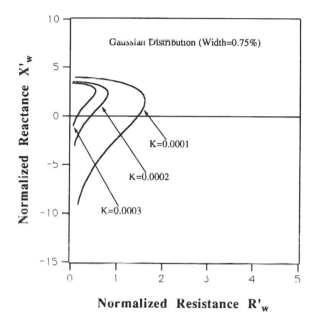

Figure 6.21. Stable-region boundary of the Gaussian distribution of Figure 6.20 for three values of the perveance K and a momentum spread of 0.75%. (From Reference 36.)

distribution $f_0(p) = (\sqrt{\pi}\,\alpha\, p_0)^{-1} \exp\{-[(p - p_0)/\alpha\, p_0]^2\}$ with different values of α and K, and with $\beta_0 = v_0/c = 0.3$ and $g = 2$ (see Reference 36). The two axes correspond to the normalized resistive and reactive parts of the longitudinal wall impedance defined as $R'_w = R^*_w(\lambda_0/Z_0)$ and $X'_w = X^*_w(\lambda_0/Z_0)$, where $\lambda_0 = 2\pi/k_0$ is the wavelength of the perturbation and $Z_0 = 377\ \Omega$ the free-space impedance.

Finally, we note that there are many other instabilities, such as the transverse resistive-wall instability, and instabilities of bunched beams. However, in many circular accelerators the most important limits for the beam current are the space-charge tune shift treated in Sections 4.5.3 and 5.4.7 and the longitudinal instability.

6.4 COLLISIONS

6.4.1 The Boersch Effect

In Section 5.4.6 we showed that acceleration produces a rather dramatic cooling of the longitudinal beam temperature, while it leaves the transverse temperature in the beam frame unchanged. The beam is therefore not in three-dimensional thermal equilibrium. However, Coulomb collisions or other effects of a random nature, such as instabilities, will tend to drive the beam toward thermodynamic equilibrium so that the longitudinal temperature increases while the transverse temperature decreases. In the final stationary state—if it could be reached—the temperatures in all three degrees of freedom would be the same (i.e., the beam would be *equipartitioned*). Unfortunately, the time constant for Coulomb collision is much too long to achieve this equilibrium state in typical linear transport channels or electrostatic accelerators. However, instabilities, beam mismatch, longitudinal-transverse coupling of the space-charge forces in bunched beams (see Appendix 4), and other nonlinear effects may shorten the relaxation time considerably and play a major role in equipartitioning.

In this section we consider only the effects of Coulomb collisions between the particles in a continuous beam that propagates through a smooth focusing channel. We adopt the theory of Ichimaru and Rosenbluth [46] for a nonrelativistic plasma with initially unequal longitudinal and transverse temperatures, $T_\parallel$ and $T_\perp$, confined by an axial magnetic field. Specifically, we consider the case where the magnetic field has no effect on the relaxation toward equilibrium. This relaxation is defined by the equation [Equation (71) in Reference 46]

$$\frac{dT_\perp}{dt} = -\frac{1}{2}\frac{dT_\parallel}{dt} = -\frac{T_\perp - T_\parallel}{\tau}, \tag{6.149}$$

where the factor $\frac{1}{2}$ is due to the fact that $T_\parallel$ changes twice as fast as $T_\perp$. The relaxation time τ is given by the relation [Equation (76) in Reference 46]

$$\frac{1}{\tau} = \frac{8\pi^{1/2}nq^4}{15(4\pi\epsilon_0)^2 m^{1/2}(k_B T_{\text{eff}})^{3/2}}\ln\Lambda, \tag{6.150}$$

where n is the particle density, $\ln\Lambda$ is the Coulomb logarithm defined in Equations (5.247) and (5.248), and where the effective temperature T_{eff} is obtained from the integral [Equation (77) in Reference 46]

$$\frac{1}{(T_{\text{eff}})^{3/2}} = \frac{15}{4}\int_{-1}^{1}\frac{\mu^2(1-\mu^2)\,d\mu}{[(1-\mu^2)T_\perp + \mu^2 T_\parallel]^{3/2}}. \tag{6.151}$$

When equilibrium is reached ($T - T_{\text{eq}}$), the three temperatures are the same (i.e., $T_{\text{eff}} = T_\parallel = T_\perp = T_{\text{eq}}$). By introducing the classical particle radius $r_c = q^2/(4\pi\epsilon_0 mc^2)$ we can write (6.150) in the alternative form

$$\frac{1}{\tau} = \frac{8\pi^{1/2}nr_c^2 c}{15(k_B T_{\text{eff}}/mc^2)^{3/2}}\ln\Lambda. \tag{6.152}$$

If we apply these relations to our case of the accelerated beam and assume initial temperatures of $T_{\parallel 0} = 0$ and $T_{\perp 0} \neq 0$, we obtain for the effective initial temperature from (6.151) the relation

$$\frac{1}{T_{\text{eff},0}^{3/2}} = \frac{15\pi}{8T_{\perp 0}^{3/2}}$$

or

$$T_{\text{eff},0}\left(\frac{8}{15\pi}\right)^{2/3} T_{\perp 0} = 0.307 T_{\perp 0}. \tag{6.153}$$

The initial relaxation time is then defined by

$$\frac{1}{\tau_0} = \frac{\pi^{3/2} n r_c^2 c}{(k_B T_{\perp 0}/mc^2)^{3/2}} \ln \Lambda. \tag{6.154}$$

If we assume that the total thermal energy in the beam is constant and given by $T_{\perp 0}$, then with $T_{\text{eq}}/2$ for each degree of freedom, the final equilibrium temperature T_{eq} is

$$\frac{3}{2} T_{\text{eq}} = T_{\perp 0}, \qquad \text{or} \qquad T_{\text{eq}} = \frac{2}{3} T_{\perp 0}. \tag{6.155}$$

With these initial and final conditions, the integration of Equation (6.149) and (6.151) using Equation (6.150) yields the temperature changes as a function of time shown in Figure 6.22. The parallel and perpendicular temperatures are plotted in units of the equilibrium temperature and the time is in units of the relaxation time τ_{eq} at equilibrium. It can be shown that τ is increasing with time, reaching $\tau_{\text{eq}} = 3.20\tau_0$ at the equilibrium temperature. The two curves in Figure 6.22 can be approximated by exponential functions as

$$T_\perp = \frac{2}{3} T_{\perp 0}\left(1 + \frac{1}{2}e^{-3t/\tau_{\text{eff}}}\right), \tag{6.156a}$$

$$T_\| = \frac{2}{3} T_{\perp 0}\left(1 - e^{-3t/\tau_{\text{eff}}}\right), \tag{6.156b}$$

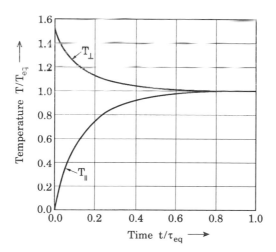

Figure 6.22. Relaxation of transverse and longitudinal beam temperatures in a uniform focusing channel when initial longitudinal temperature $T_{\| 0} = 0$. Temperatures are in units of the equilibrium temperature T_{eq}, and the time is in units of the equilibrium value of the relaxation constant τ_{eq}. (Courtesy of N. Brown.)

where the best fit is obtained with $\tau_{\text{eff}} = 1.34\tau_0 = 0.42\tau_{\text{eq}}$, which correlates with an effective temperature of $T_{\text{eff}} = 0.373T_{\perp 0} = 0.56T_{\text{eq}}$. The theoretical model of Ichimaru and Rosenbluth has been confirmed in recent experiments with a nonneutral electron plasma [47].

Relations (6.149) to (6.156b) for a nonrelativistic plasma apply directly to a nonrelativistic beam propagating in a focusing channel since in this case there is no difference between particle densities and temperature in the beam frame and the laboratory frame. However, they also apply to a relativistic beam with nonrelativistic transverse and longitudinal velocities in the beam frame. To express the above relations in terms of laboratory parameters, one must use the Lorentz transformations $n_l = \gamma_0 n$, $T_l = T/\gamma_0$, and $\tau_l = \gamma_0 \tau$ for the density, temperature, and relaxation time, respectively.

Although the same physics applies to charged particle beams as seen by an observer in the beam frame, the propagation time in a focusing channel of typical length is much shorter than the relaxation time, so that the beam will not reach thermal equilibrium. To illustrate this point, let us consider a 5-keV electron beam with a current of 200 mA launched from a thermionic cathode with radius $r_c = 6$ mm and temperature $k_B T_c = 0.1$ eV and then focused by a long solenoid in which the beam radius is $a = 0.6$ mm. Since the beam is compressed by a factor of 10, the transverse temperature in the solenoid will be [from (5.343)] $k_B T_{\perp 0} = k_B T_c (r_c/a)^2 = 0.1 \times 10^2 = 10$ eV. From (6.152) one obtains for electrons in the beam frame (or nonrelativistically in both beam and lab frame)

$$\tau_{\text{eff}} = 4.44 \times 10^{20} \frac{(k_B T_{\text{eff}}/mc^2)^{3/2}}{n \ln \Lambda}. \tag{6.157}$$

For the Coulomb logarithm one obtains from Equation (5.247)

$$\ln \Lambda = \ln\left[5.66 \times 10^{21} \frac{(k_B T/mc^2)^{3/2}}{n^{1/2}} \right]. \tag{6.158}$$

Since $I = ena^2 \pi v$, one has

$$n = \frac{I}{ea^2 \pi v}. \tag{6.159}$$

With the given parameters and $\beta = v/c \simeq 0.139$ one gets $n = 2.65 \times 10^{16}$ m^{-3}. Since the thermal energy remains constant in our case, we use $k_B T = \frac{2}{3} k_B T_{\perp 0}$ in the Coulomb logarithm, giving $\ln \Lambda = 14.3$. For the effective temperature, on the other hand, we choose the exponential-fit value $k_B T_{\text{eff}} = 0.373 k_B T_{\perp 0} = 3.73$ eV.

With these numbers we find $\tau_{eff} = 2.31 \times 10^{-5}$ s. The corresponding distance is $L = v\tau_{eff} = 925$ m. Thus our solenoid would require a length of 925 m to achieve thermal equilibrium via Coulomb collisions for the 5-keV electron beam. One would therefore tend to conclude that Coulomb collisions play no role at all in conventional laboratory experiments with straight beams, short transport lines, or even linear accelerators. However, this conclusion is not correct. It turns out that even in short distances on the order of 1 m, the collisions produce a significant increase in the beam's energy spread, ΔE. This phenomenon was first observed experimentally in 1954 by Boersch [48] and is since known as the *Boersch effect*. To understand this effect, let us first calculate the initial longitudinal beam temperature $T_{\parallel 0}$ after acceleration of the above 5-keV electron beam. With an initial temperature of 0.1 eV at the cathode, one obtains from Equation (5.339) a longitudinal temperature of $k_B T_{\parallel 0} = (0.1)^2/2 \times 5 \times 10^3 = 1 \times 10^{-6}$ eV in the accelerated beam. Next, let us determine what the longitudinal temperature $T_{\parallel}$ will be after the beam propagates a distance of $L = 1$ m in the solenoid. Using $\tau_{eff} = 2.31 \times 10^{-5}$ s, $t - L/v = 2.4 \times 10^{-8}$ s, and $k_B T_{\perp 0} = 10$ eV, one finds from Equation (6.156b) a value of $k_B T_{\parallel} = 2.1 \times 10^{-2}$ eV. This implies that in the short distance of 1 m the longitudinal temperature has increased from 1×10^{-6} eV by four orders of magnitude. While it is still far from equilibrium, this temperature is large enough to cause a significant increase in the longitudinal energy spread, ΔE. Initially, this energy spread is defined by the cathode temperature and hence has the value of $\Delta E_c = k_B T_c = 0.1$ eV. Acceleration does not change this energy spread—it changes only the temperature, the part of the kinetic energy that is related to the thermal motion of the particles. However, after the temperature increases due to the Coulomb collisions, we have from Equation (5.340) an rms energy spread of

$$\widetilde{\Delta E} = (2qV_0 k_B T_{\parallel})^{1/2} = (2 \times 5 \times 10^3 \times 2.1 \times 10^{-2})^{1/2} = 14.5 \text{ eV}.$$

This represents a significant increase in the initial energy spread by a factor of 145. Since $\tau \propto n^{-1}$, $\widetilde{\Delta E}$ increases with beam density or current. As an example, doubling the beam current to 400 mA and leaving all other parameters the same yields a value of $\tau_{eff} = 1.11 \times 10^{-5}$ s and a longitudinal temperature of $k_B T_{\parallel} = 4.3 \times 10^{-2}$ eV at a distance of 1 m. The energy spread then increases to $\widetilde{\Delta E} = 20.8$ eV. This sensitivity of the energy spread with beam current was observed by Boersch in his original experiments, which, however, were quite different from our example here. Boersch measured the energy distribution as a function of beam current for a 27-keV focused electron beam from a thermionic cathode. The energy spread measured downstream from the crossover point (waist) of the beam showed anomalous broadening that increased with the current density at the waist. Boersch did not attribute this energy broadening to Coulomb collisions, which are now generally considered to be the cause of this effect.

An interesting consequence of the Boersch effect is that the longitudinal emittance increases while the transverse emittance decreases (albeit at a much smaller rate). As an example, take a beam of finite longitudinal rms width, $\widetilde{\Delta z}$, or a given number of particles occupying a slice of a continuous beam with width $\widetilde{\Delta z}$. The normalized longitudinal rms emittance is proportional to $(k_B T_\parallel)^{1/2}$ according to Equation (5.317). Hence, for our 5-keV electron beam it will increase by a factor of $(2.1 \times 10^{-2}/1 \times 10^{-6})^{1/2} = 145$ in the 100-mA case and $(4.3 \times 10^{-2}/1 \times 10^{-6})^{1/2} = 207$ in the 400-mA case. This is a rather significant effect, while the associated decrease in the transverse temperature $T_\perp$ and emittance is relatively small. If one could reach thermodynamic equilibrium, these effects would be even more pronounced, and the transverse emittance would decrease to $(\frac{2}{3})^{1/2} = 0.816$ of its initial value (i.e., by about 18.4%). However, the long relaxation times make it impractical to achieve equilibrium in a straight beam. Only in storage rings where particles are confined for long times can Coulomb collisions produce full equipartitioning of a beam, as discussed in the next section.

Before proceeding to this topic, one should note that the treatment of the Boersch effect given in this section is somewhat simplistic. A very thorough review that deals with the rather complicated physical and theoretical details can be found in Jansen's book (Reference 10 in Chapter 5). Thus, for example, the observed energy distributions may differ significantly from the Maxwellian shape assumed here. Furthermore, one must differentiate between the smooth uniform beam in the long solenoid treated here and the beam that is focused to a small waist or, more generally, a beam whose radius varies strongly, as in a matching section. In the first case (smooth beam) the total thermal energy remains constant (i.e., $2k_B T_\perp + k_B T_\parallel = \text{const}$). In the second case, however, the temperature increases as coherent longitudinal kinetic energy becomes thermalized in large-angle collision so that $2k_B T_\perp + k_B T_\parallel \neq \text{const}$. We pursue this point further in the next section.

6.4.2 Intrabeam Scattering in Circular Machines

The effects of Coulomb collisions between the particles in circular machines are commonly referred to in the literature as *intrabeam scattering*. The lifetimes of the beams in circular machines are much longer than in linear devices; this is especially true for storage rings and circular colliders, where the beams can be trapped for many hours. Consequently, intrabeam scattering plays an important role in these machines and may, in fact, impose an upper limit for the luminosity, brightness and beam lifetime that can be achieved.

As we know from Section 5.4.9, the particle dynamics in a circular focusing lattice differs significantly from that in a linear focusing channel, and hence, the effects of intrabeam scattering also differ substantially. The two most important differences with regard to Coulomb collisions are negative-mass behavior of the particles in a circular machine above transition energy and dispersion.

Let us first consider the ideal machine with a smooth-focusing lattice below transition and negligible dispersion. Such a machine behaves essentially like a linear focusing channel except that the beam goes around in a circle and that the current is limited by the space-charge tune shift. But even in this ideal case there is a subtle difference, as will be shown now. As we discussed in the preceding section, the total thermal energy per particle in a smooth linear beam channel is conserved; that is, one has (in the beam frame as well as in the lab frame) for a beam with constant energy (γ_0 = const)

$$2k_B T_\perp + k_B T_\| = \text{const}, \tag{6.160a}$$

or if x and y denote the two transverse directions,

$$k_B T_x + k_B T_y + k_B T_\| = \text{const}. \tag{6.160b}$$

Coulomb collisions drive the beam toward an isotropic thermal equilibrium, in which case the three temperatures would be the same, that is,

$$k_B T_x = k_B T_y = k_B T_\| = k_B T_{eq}. \tag{6.161}$$

In view of the relations (5.270) between temperature and rms velocity spread, we can put the conservation law (6.160b) into the laboratory form

$$\gamma_0 m \overline{v_x^2} + \gamma_0 m \overline{v_y^2} + \gamma_0^3 m \overline{(\Delta v_z)^2} = \text{const}, \tag{6.162}$$

or in terms of the slopes $x' = v_x/v_0$, $y' = v_y/v_0$, and relative momentum spread $\Delta P_z/P_0 = \gamma_0^2 \Delta v_z/v_0$ from (5.315):

$$\overline{x'^2} + \overline{y'^2} + \frac{1}{\gamma_0^2} \overline{\left(\frac{\Delta P_z}{P_0}\right)^2} = \text{const}. \tag{6.163}$$

This relation holds for a straight beam. However, in a circular machine we must replace $1/\gamma_0^2$ in the third term on the left side of Equation (6.163) by $-\eta = 1/\gamma_0^2 - 1/\gamma_t^2$ [see Equation (5.434)], which yields

$$\overline{x'^2} + \overline{y'^2} - \eta \overline{\left(\frac{\Delta P}{P_0}\right)^2} = \text{const}. \tag{6.164}$$

This relationship is essentially identical to the invariant for intrabeam scattering derived in 1974 by Piwinski [49]. We recognize that it is just another form of the conservation law (6.160) for the beam temperature. However, we see immediately that there is a significant difference between a linear and a circular beam that is represented by the factor η. For a linear beam ($\gamma_t \to \infty$, $\eta = -1/\gamma_0^2$) Equation (6.164) is identical with (6.163), as expected. For a circular beam the behavior of the system depends on the sign of η [i.e., whether we are below transition ($\gamma_0 < \gamma_t$) or above ($\gamma_0 > \gamma_t$)]. Below transition, η is negative, and from Equation (5.437) and the discussion following Equation (5.438), we find that the longitudinal temperature $k_B T_\parallel$ is a positive quantity. This means that for the smooth, dispersion-free lattice below transition, thermal equilibrium can be reached. However, in the negative-mass regime above transition, η is positive and $k_B T_\parallel$ becomes negative. This implies that thermal equilibrium is not possible. An increase in momentum spread $\Delta P/P_0$ or negative temperature must be offset by a corresponding increase in the transverse temperatures to maintain the "conservation law" (6.160b). Physically, negative temperature means that there is a source of energy that continuously drives up the transverse temperature. This source is basically the coherent longitudinal kinetic energy of the beam, which is thermalized via the negative-mass effect.

What are the consequences of the above analysis for the transverse and longitudinal emittance of the beam? First, in the linear beam case and below transition in the ideal circular machine, there will only be emittance change if the beam initially is not in three-dimensional thermal equilibrium. If, for instance, $k_B T_\parallel < k_B T_\perp$, as is usually the case, there will be longitudinal emittance growth and the transverse emittance may actually decrease slightly until equilibrium is reached, as discussed in the preceding section. Second, above transition, there will be continuous emittance growth in transverse and longitudinal directions, and equilibrium will never be established.

The ideal circular machine with smooth focusing that we just described almost never exists in the real world, where the effects of dispersion must be taken into account and where the lattice is not smooth but often has a rather strong variation around the circumference. This variation is described by the betatron function, $\hat{\beta}_x(s)$, and the dispersion function, $D_e(s)$, and their derivatives, $\hat{\beta}_x'(s)$ and $D_e'(s)$. Without scattering, the emittance of the beam remains preserved in a dispersive lattice, as discussed in Section 5.4.7. However, Coulomb collisions will change a particle's momentum or slope x', whether it is dispersed or not; that is, a particle having position $x_1 = x_{b1} + D_e(\Delta P/P_0)$ and slope x_1' prior to a collision will have a changed slope x_2' after the collision. These changes of the slopes of the particles in the beam distribution cause an irreversible increase in the corresponding emittance. The theory of intrabeam scattering by Bjorken and Mtingwa [50] shows that the emittance will always grow in a lattice when the combination of lattice functions defined by the parameter $\phi_l = D_e' - D_e \hat{\beta}_x'/2\hat{\beta}_x$ does not vanish. This condition ($\phi_l \neq 0$) is always satisfied along large fractions of the lattices of modern strong-focusing rings.

In summary, the behavior of a circular machine with regard to intrabeam scattering (as compared with an equivalent linear transport channel of sufficient length) is defined by the two parameters

$$\eta = \frac{1}{\gamma_t^2} - \frac{1}{\gamma_0^2} \approx \overline{\left(\frac{D_e^2}{\hat{\beta}_x^2}\right)} - \frac{1}{\gamma_0^2} \tag{6.165a}$$

and

$$\phi_l = D'_e - \frac{D_e \hat{\beta}'_x}{2\hat{\beta}_x},$$ (6.165b)

where the relation $\gamma_t \approx \hat{\beta}_x/\tilde{D}_e$ used in Equation (6.165a) follows from (5.489), with $\gamma_t \approx \nu_x$ and $\hat{\beta}_x = \bar{R}/\nu_x$. Three-dimensional thermal equilibrium can be achieved only if $\eta < 0$ ($\gamma_0 < \gamma_t$ or $D_e\gamma_0 > \hat{\beta}_x$), as first shown by Piwinski [49], and if, in addition, $\phi_l = 0$, as pointed out by Bjorken and Mtingwa [50]. In principle, these two conditions can be satisfied simultaneously only in an ideal smooth-focusing machine below transition energy. In modern strong-focusing machines the condition $\phi_l = 0$ is never fully satisfied, so that in practice three-dimensional equilibrium is never achieved, and the six-dimensional phase-space volume defined by the product of the three emittances, $\epsilon_x \epsilon_y \epsilon_l$, always increases. As mentioned in the discussion following Equation (6.164), this increase in beam temperature occurs at the expense of the total kinetic energy of the beam, which is orders of magnitude larger than the energy of the betatron and synchrotron oscillations. While this observation concerning the parameter ϕ_l is correct, the computations for existing machines show that the contributions from $\phi_l \neq 0$ to the growth rates are almost negligibly small in many cases [51, 52], so that the results from the simpler smooth-lattice calculations are adequate. The error made by neglecting ϕ_l should be tolerable if one keeps in mind that the Coulomb logarithm ln Λ is often taken to have a constant value when in fact it may vary appreciably and is only an approximate statistical parameter anyway.

The theory of intrabeam scattering in circular machines is rather complicated mathematically. The calculation of the growth rates in each degree of freedom involves integration and averaging procedures that must be done by computer and are rather lengthy if the lattice parameter ϕ_l is included. Besides, it appears that there are still significant differences between the various models that have not been explained in a satisfactory manner. Thus the parameter H in the theory of Bjorken and Mtingwa [50] is not exactly identical to Piwinski's invariant [Equation (6.164)]; the slip factor η does not appear explicitly, and hence the fact that equilibrium cannot be achieved for $\gamma_0 > \gamma_t$ even if $\phi_l = 0$ does not follow from their theory. Conte and Martini [53] found that the Bjorken and Mtingwa model applies mainly to high-energy rings ($\gamma_0 \gtrsim 10$), and they revised this model to give more satisfactory results for low energies ($\gamma_0 < 10$) as well. An excellent general review of intrabeam scattering was given by Sørensen [54]. By using appropriate reduced variables, Sørensen showed that the computer results for different values of the transverse emittances and momentum spread in a given machine lattice can all be represented by a single universal curve. Moreover, the regime where the horizontal growth rate dominates is distinctly separated from the regime where the longitudinal growth rate dominates.

As an example of this interesting result, Sørensen's universal curve for a coasting proton beam in the former ICE storage ring at CERN is shown in Figure 6.23. Note

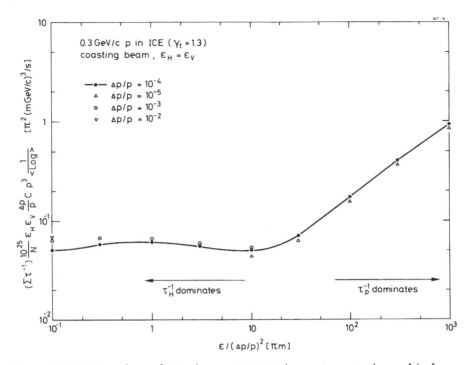

Figure 6.23. Universal curve for intrabeam scattering in the coasting proton beam of the former ICE storage at CERN. (Courtesy of A. Sørensen; see Reference 54.)

that the data points shown on this plot cover a large range of rms momentum spreads ($\Delta \tilde{P}/P$ in our notation) from 10^{-5} to 10^{-2}. ($\sum \tau^{-1}$) on the abscissa represents the sum of the growth rates in the three degrees of freedom, C is the circumference of the ring, $\langle \text{Log} \rangle = \ln \Lambda$ is the average Coulomb logarithm, N is the total number of particles in the ring, and the ϵ-parameters represent normalized rms values for the emittances (i.e., $\tilde{\epsilon}_n$ in our notation). On the right side in Figure 6.23 the growth rate τ_p^{-1} for the momentum spread dominates. Here the universal curve represents essentially the longitudinal temperature increase of a collapsed thermal distribution, as in the Boersch effect. On the left side, where the horizontal growth rate τ_H^{-1} dominates, the curve is more sensitive to the specific lattice design (see Reference 54 for details).

The growth rate for intrabeam scattering in high-energy circular machines can be written in the simple relativistic form [50]

$$\frac{1}{\tau_j} = \frac{1}{\tau_0} \langle H_j \rangle = \frac{\pi^2 c r_0^2 m^3 N \ln \Lambda}{\gamma_0 \Gamma} \langle H_j \rangle, \qquad (6.166)$$

where N is the total number of particles, Γ the six-dimensional phase-space volume occupied by N, and where the function H_j depends on γ_0, the emittances

$\tilde{\epsilon}_x$, $\tilde{\epsilon}_y$, $\tilde{\epsilon}_z$, and the lattice parameters $\hat{\beta}_x$, D_e, $\hat{\beta}'_x$, D'_e, and $\hat{\beta}_y$. The function H_j is averaged over a lattice period and the subscript j denotes the three orthogonal directions [i.e., j = horizontal (x), vertical (y), and longitudinal (s)].

For bunched beams, N represents the number of particles in a single bunch and the six-dimensional volume Γ of the bunch is given by

$$\Gamma_b = (2\pi)^3 \frac{P_0^3}{c^3} \widetilde{\Delta x} \frac{\widetilde{\Delta P_x}}{P_0} \widetilde{\Delta y} \frac{\widetilde{\Delta P_y}}{P_0} \widetilde{\Delta z} \frac{\widetilde{\Delta P_z}}{P_0}, \tag{6.167a}$$

or

$$\Gamma_b = (2\pi)^3 \frac{P_0^3}{c^3} \tilde{\epsilon}_x \tilde{\epsilon}_y \tilde{\epsilon}_z = (2\pi)^3 (\beta_0 \gamma_0)^3 m^3 \tilde{\epsilon}_x \tilde{\epsilon}_y \tilde{\epsilon}_z. \tag{6.167b}$$

For unbunched beams, N denotes the number of particles in the circumference $2\pi \bar{R}$ of the ring and Γ is given by

$$\Gamma_u = 4(\pi)^{5/2} (\beta_0 \gamma_0)^3 m^3 \tilde{\epsilon}_x \tilde{\epsilon}_y \frac{\widetilde{\Delta P}}{P_0} 2\pi \bar{R}. \tag{6.168}$$

The factor γ_0 in the denominator of Equation (6.166) is due to the Lorentz transformation from the beam frame to the laboratory frame, which yields $\tau = \gamma_0 \tau_b$ (time dilation) for the relaxation time in the laboratory frame.

In the periodic-focusing systems of modern circular machines the temperature is not a constant, and it is customary to use the rms emittance $\tilde{\epsilon} \sim \delta \sqrt{k_B T}$ and the six-dimensional phase-space volume Γ, which is invariant when scattering is neglected. The relaxation times are then defined by the increase of the emittances rather than the temperatures as in the preceding section. Thus τ_j in Equation (6.166) is defined as

$$\frac{1}{\tau_j(\epsilon)} = \frac{1}{\tilde{\epsilon}_j} \frac{d\tilde{\epsilon}_j}{dt}. \tag{6.169}$$

For a smooth lattice where the rms beam width $\delta \simeq$ const, one can use the temperature relaxation time given by

$$\frac{1}{\tau_j(T)} = \frac{1}{T_j} \frac{dT_j}{dt}. \tag{6.170}$$

Since $\epsilon \sim \delta \sqrt{k_B T}$, one has the relation

$$\frac{1}{\tau_j(\epsilon)} = \frac{1}{2} \frac{1}{\tau_j(T)}, \tag{6.171}$$

that is, the emittance relaxation time $\tau_j(\epsilon)$ is a factor of 2 longer than the temperature relaxation time $\tau_j(T)$.

It is interesting to compare the result (6.166) with Equation (6.152) for the case of a nonrelativistic ($\gamma_0 = 1$) unbunched beam in a linear smooth-focusing channel. For the density n one has

$$n = \frac{N}{4\pi\delta_x\delta_y 2\pi\overline{R}}, \tag{6.172}$$

where $\delta_x = \widetilde{\Delta x}$, $\delta_y = \widetilde{\Delta y}$ denote the rms widths of the beam. Using (6.168) and (6.172), one obtains for the factor $1/\tau_0$ in Equation (6.166),

$$\frac{1}{\tau_0(\epsilon)} = \frac{\sqrt{\pi}\, cr_c^2 n \ln \Lambda}{(k_B T_x/mc^2)^{1/2}(k_B T_y/mc^2)^{1/2}(k_B T_z/mc^2)^{1/2}}. \tag{6.173}$$

In comparison with (6.152), this equation exhibits the same scaling except that the constants are different and that in place of $T_{\text{eff}}^{3/2}$ one has the product of the square roots of the three temperatures. The definition (6.173) requires that the three temperatures not differ drastically. If one of them, say T_z, goes toward zero, as was the case in the Boersch effect with $T_{\parallel}$, the growth rate $1/\tau_0$ becomes infinitely large, which is unphysical. The problem is with the definition of the growth rate (6.170), which does not allow for an equilibrium to exist. If an equilibrium temperature T_{eq} can be reached, as in the ideal smooth system below transition, the relaxation time should be defined by $\tau_j^{-1} = (T_j - T_{\text{eq}})^{-1} dT_j/dt$, as in Equation (6.149).

In the theory of intrabeam scattering in circular machines the Coulomb logarithm $\ln \Lambda$ is defined as

$$\ln \Lambda = \ln \frac{r_{\max}}{r_{\min}}, \tag{6.174}$$

where $r_{\max}$ is taken to be the smaller of the Debye length λ_D or the rms beam width δ_x and $r_{\min}$ is the classical impact parameter b [Equation (5.287)]. We note that this definition of $\ln \Lambda$ differs somewhat from that given in Equations (5.286) to (5.289) in that the effective beam radius a in Equation (5.289) is replaced by the rms width δ_x. However, in practice, $\ln \Lambda$ is a large number between 10 and 30 and the various definitions differ at most by a factor of 2. Usually, a constant value of $\ln \Lambda \simeq 20$ is used in the computer codes on intrabeam scattering. Despite the discrepancies that exist between the various models, as discussed above, the computational results appear to be generally within a factor of 2 or so of the experimental observations [50, 55].

Finally, we want to mention a special phenomenon caused by intrabeam scattering in bunched beams which was first analyzed correctly by Touschek [56] and is since known as the *Touschek effect*. In a relativistic storage ring, Coulomb collisions lead to a momentum transfer from the transverse into the longitudinal direction that is amplified by the Lorentz factor γ_0. This is illustrated in the diagram of Figure 6.24, which is shown in Sørensen's review article [54] and can be

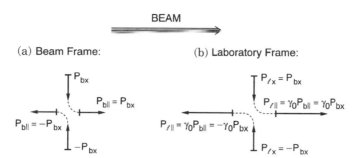

Figure 6.24. Elastic Coulomb collision between two particles as seen in the beam frame (a) and the laboratory frame (b). (See References 54 and 57.)

attributed to Derbenev [57]. The figures portray an elastic collision between two particles, as seen in the beam frame (a) and in the laboratory frame (b). While the total momentum in the collision is preserved, the two particles emerge from this collision with opposite longitudinal momentum components that are larger by the factor γ_0 than the original transverse momentum component before the collision. If the longitudinal momentum acquired in such a collision is greater than the momentum acceptance of the rf bucket that keeps the beam longitudinally bunched, the two particles involved in such a collisions will be lost. For, after the collision process, the forward-scattered particle will have too much, and the backward-scattered particle too little energy to be contained within the stable region (*bucket*) of the rf voltage acting on the beam. The net result is that the lifetime of the stored beam is reduced [56].

6.4.3 Multiple Scattering in a Background Gas

The collisions of the beam particles with the atoms or molecules of the residual gas in a vacuum tube can cause a large variety of effects, all of which depend on the particles' kinetic energy. We have already treated the ionization of the gas due to such collisions in Section 4.6.1. Other effects are the excitation of the gas atoms or molecules, charge exchange between gas and beam particles, and at higher energies the many types of nuclear reactions whose cross sections are considerably smaller than the atomic effects. All of the above interactions involve energy loss of the beam particles and are therefore characterized as *inelastic collisions*.

The most frequent events in the encounters between beam particles and gas molecules are, however, the *elastic collisions*, which change a particle's momentum without energy loss. The deflections or angular scattering of the beam particles by such elastic multiple collisions in the gas cause an irreversible increase of the emittance, which is the subject of this section.

The theoretical treatment of elastic scattering of fast particles by atoms is analogous to that of Coulomb scattering within a beam discussed in the two previous sections. It differs only in the fact that we are dealing with two particle species having a large relative velocity with respect to each other. For a detailed discussion of the subject we will refer to Jackson [A.4, Secs. 13.7 and 13.8] or Lawson [C.17, Secs. 5.2 and 5.3]. According to the theory, a fast particle with momentum $P = \gamma m v$ and charge Ze passing an atom with nuclear charge $Z_g e$ at a distance defined by the impact parameter b, will experience an angular deflection given by the polar angle θ in spherical coordinates. The probability that a particle will be deflected into a solid angle $d\Omega = \sin\theta \, d\theta \, d\phi$ is determined by the cross section for nuclear scattering, which for small angles θ obeys the famous *Rutherford formula*

$$\frac{d\sigma_s}{d\Omega} = \frac{b}{\theta}\left|\frac{db}{d\theta}\right| = \left(\frac{2ZZ_g e^2}{4\pi\epsilon_0 P v}\right)^2 \frac{1}{\theta^4}. \tag{6.175}$$

Thus, in view of the θ^{-4} dependence, the probability for a small-angle deflection is much greater than for a large-angle deflection. As discussed by Jackson, the cross section will actually flatten off at small angles, and one has the more general form

$$\frac{d\sigma_s}{d\Omega} = \left(\frac{2ZZ_g e^2}{4\pi\epsilon_0 P v}\right)^2 \frac{1}{\left(\theta^2 + \theta_{min}^2\right)^2}, \tag{6.176}$$

where θ_{min} is a cutoff angle. The mean-square angle for single scattering is defined by

$$\overline{\theta^2} = \frac{\int \theta^2 (d\sigma_s/d\Omega) \, d\Omega}{\int (d\sigma_s/d\Omega) \, d\Omega} = 2\theta_{min}^2 \, \ln\left(\frac{\theta_{max}}{\theta_{min}}\right), \tag{6.177}$$

where θ_{max} represents an upper bound for the scattering angle and where, according to Jackson,

$$\ln\left(\frac{\theta_{max}}{\theta_{min}}\right) \approx \ln\left(204 Z_g^{-1/3}\right). \tag{6.178}$$

If the beam traverses a gas region with density n_g and length s, the particles will undergo multiple collisions. Since these collisions are statistically independent events, the *central limit theorem* states that the distribution in angles will be approximately Gaussian with a mean-square angle $\overline{\Theta^2} = N_s \overline{\theta^2}$. N_s is the number of collisions given by $N_s = n_g \sigma_s' s$, where $\sigma_s' = \int (d\sigma_s/d\Omega) \, d\Omega$ is the total cross section. Following Jackson, one obtains for the mean-square angle due to multiple scattering the result

$$\overline{\Theta^2} = 16\pi n_g \left(\frac{ZZ_g e^2}{4\pi\epsilon_0 mc^2 \gamma \beta^2}\right)^2 \ln\left(204 Z_g^{-1/3}\right) s, \tag{6.179a}$$

which may be written in the alternative form

$$\overline{\Theta^2} = 16\pi n_g \frac{Z_g^2 r_c^2}{\beta^4 \gamma^2} \ln\left(204 Z_g^{-1/3}\right) s.$$ (6.179b)

Here Z, γmc^2, $v = \beta c$, and r_c denote the charge state, relativistic energy, velocity, and classical radius of the beam particles, respectively, and Z_g is the nuclear charge number of the gas atoms.

The above derivation implicitly assumes that the beam enters the gas region with the particles having initially straight trajectories, corresponding to zero initial emittance. In practice, the beam has, of course, a finite initial rms emittance, and $\overline{\Theta^2}$ defines the change of the mean-square slope according to the relation

$$\Delta \overline{x'^2} = \frac{1}{2} \overline{\Theta^2},$$ (6.180)

where the factor $\frac{1}{2}$ results from the projection of the deflection angles into the $x-s$ plane. The associated increase of the rms emittance $\tilde{\epsilon}$ is readily calculated if we assume that the beam propagates through a smooth channel characterized by a wave number $k = 1/\hat{\beta} = 2\pi/\lambda$ and that the change is adiabatic. In this case an initially matched beam remains matched and since $\tilde{\epsilon} = \tilde{x}\tilde{x}' = \tilde{x}'^2/k$ (using $\tilde{x}' = k\tilde{x}$), the emittance change is given by

$$\Delta \tilde{\epsilon} = \frac{\Delta \tilde{x}'^2}{k} = \frac{\overline{\Theta^2}}{2k}.$$ (6.181)

This result can be expressed as a differential change per unit length $d\tilde{\epsilon}/ds$ along the distance s of propagation through the background gas as

$$\frac{d\tilde{\epsilon}}{ds} = \frac{1}{2k} \frac{d(\overline{\Theta^2})}{ds} = \frac{8\pi}{k} n_g \frac{Z_g^2 r_c^2}{\beta^4 \gamma^2} \ln\left(204 Z_g^{-1/3}\right),$$ (6.182a)

or, in terms of the normalized rms emittance $\tilde{\epsilon}_n = \beta\gamma\tilde{\epsilon}$, as

$$\frac{d\tilde{\epsilon}_n}{ds} = \frac{8\pi}{k} n_g \frac{Z_g^2 r_c^2}{\beta^3 \gamma} \ln\left(204 Z_g^{-1/3}\right).$$ (6.182b)

For electron beams, the classical radius is $r_c \approx 2.8 \times 10^{-15}$ m; for ion beams we can show the explicit dependence on the charge state Z and mass number A by expressing r_c as

$$r_c = \frac{Z}{A} r_p,$$ (6.183)

where $r_p \approx 1.5 \times 10^{-18}$ m is the classical proton radius. In storage rings it is more convenient to express the emittance increase due to gas scattering as a change per unit time rather than unit length; that is, one has with $ds/dt = v = \beta c$,

$$\frac{d\tilde{\epsilon}_n}{dt} = \frac{d\tilde{\epsilon}_n}{ds}\beta c = \frac{8\pi c}{k}n_g\frac{Z_g^2 r_c^2}{\beta^2 \gamma}\ln\left(204Z_g^{-1/3}\right). \qquad (6.184)$$

Furthermore, we can relate the gas density n_g to the pressure using Equation (4.286) so that (6.184) may be written in the form

$$\frac{d\tilde{\epsilon}_n}{dt} = 2.67 \times 10^{32}\hat{\beta}_{[m]}P_{[torr]}\frac{Z_g^2 r_{c[m]}^2}{\beta^2 \gamma}\ln\left(204Z_g^{-1/3}\right), \qquad (6.185)$$

where $\hat{\beta} = 1/k$ is the average betatron function.

As an example, consider a proton beam ($Z = 1$, $A = 1$) with air as the residual gas in the vacuum chamber. Taking $Z_g \simeq 8$ for oxygen atoms, we obtain

$$\frac{d\tilde{\epsilon}_n}{dt} = 0.18 \times \hat{\beta}_{[m]}\frac{P_{[torr]}}{\beta^2 \gamma} \qquad \left[\frac{\text{m-rad}}{\text{s}}\right]. \qquad (6.186)$$

In a hypothetical proton storage ring with a radius of $\overline{R} = 50$ m, a tune of $\nu_x = \nu_y = 4$ (i.e., $\hat{\beta} = \overline{R}/\nu_x = 12.5$), a kinetic energy of 300 MeV ($\gamma = 1.32$, $\beta = 0.65$), and a pressure of 10^{-9} torr, the normalized rms emittance would increase at a rate of about 4×10^{-9} m-rad/s, or 14.5×10^{-6} m-rad per hour. Since emittances are typically in the range of a few mm-mrad, this increase would be a significant factor in limiting the storage time. The emittance growth would, of course, be a factor of 10 lower if one could operate at a background pressure of 10^{-10} torr.

As can be seen from the above formulas, the emittance increase due to scattering is most severe at low energies. Thus, for a 50-keV proton beam ($\gamma \approx 1$, $\beta \approx 0.01$) propagating through a transport channel having a betatron function of, say, $\hat{\beta} = \lambda/2\pi = 0.5$ m and a pressure of 10^{-5} torr, the rms emittance growth is 9×10^{-3} m-rad/s or $9 \times 10^{-3}/(0.01 \times 3 \times 10^8) = 3 \times 10^{-9}$ m-rad per meter of travel. For a short channel this is not very significant, although the rate of change is more than six orders of magnitude higher than in the above storage-ring example.

The above formulas also show that electrons are much more strongly scattered than ions of the same velocity since $\overline{\Theta^2} \propto r_c^2 \propto (Z/m)^2$. For the rate of emittance change per meter of an electron beam, one obtains with $r_c = 2.8 \times 10^{-15}$ m:

$$\frac{d\tilde{\epsilon}_n}{ds} = 7 \times 10^{-6}\hat{\beta}P_{[torr]}\frac{Z_g^2}{\beta^3 \gamma}\ln\left(204Z_g^{-1/3}\right) \qquad [\text{m-rad/m}]. \qquad (6.187)$$

As an example, let us consider the 5-keV electron beam ($\beta = 0.14$, $\gamma \approx 1$) discussed in Section 6.2.2, case 1 (rms matched beam). It propagated through a periodic solenoid channel over a distance of about $s = 5.2$ m. In the smooth approximation, the $\hat{\beta}$-function relates to the particles' betatron wavelength λ by

$\hat{\beta} = \lambda/2\pi = S/\sigma$, with $\lambda = 2\pi S/\sigma$ [Equation (6.39)], where $S = 0.136$ m is the period length and σ the phase advance with space charge. At the beginning, one has $\sigma_i = 0.31\sigma_0 \approx 24°$ since $\sigma_0 = 77°$, and after the rapid emittance growth one finds from (4.147) that $\sigma_f = 0.43\sigma_0 \approx 33°$. Using the latter value, the average betatron function can be approximated by $\hat{\beta} \approx 0.24$ m. The average residual gas pressure in the beam tube was around $p \approx 3.5 \times 10^{-7}$ torr (D. Kehne, private communication). With the above numbers and taking air ($Z_g \approx 8$) as the background gas, one obtains from Equation (6.187) an rms emittance increase of $\Delta\tilde{\epsilon}_n \approx 3.3 \times 10^{-7}$ m-rad. This corresponds to an effective emittance change of $\Delta\epsilon = 4\Delta\tilde{\epsilon}_n/\beta\gamma = 9.4 \times 10^{-6}$ m-rad. Adding this value to the theoretically predicted emittance of 101 mm-mrad [Equation (6.37b)], one obtains $\epsilon = 110.4$ mm-mrad, in remarkably good agreement with the measurement. Thus, gas scattering appears to explain why the measured value of the emittance was consistently about 9% higher than expected in the experiment of case 1, Section 6.2.2. Gas scattering would, of course, also have affected the case 2 (rms mismatched beam) experiment described in Section 6.2.2. However, the halo formation prevented an accurate emittance measurement in that experiment, so that a quantitative evaluation is not possible.

Returning now to the general discussion of gas scattering, it will be useful for us to make a comparison with the intrabeam Coulomb collisions treated in the preceding two sections. First, it should be pointed out that both mechanisms are *elastic* collision processes (i.e., the particles involved suffer no energy loss). In the Coulomb collisions between the beam particles we are dealing, on the one hand, with relaxation of initially different longitudinal and transverse temperatures toward thermal equilibrium. On the other hand, we have a continuous transformation of coherent longitudinal kinetic energy into thermal energy when an equilibrium does not exist. Scattering in a background gas is related to the latter case; that is, it is a nonequilibrium process in which the coherent, center-of-momentum energy is gradually converted into random, incoherent transverse motion and hence thermal energy. This process continues in principle until all coherent kinetic beam energy is thermalized. This extreme case occurs when the beam is stopped completely, as happens at high gas pressure or in a solid material. Of course, in these extreme cases there are also many collisions involving *inelastic* processes where true energy loss or *dissipation* occurs so that the kinetic energy of the beams is completely transformed into heat and radiation or other forms of energy.

6.5 BEAM COOLING METHODS IN STORAGE RINGS

6.5.1 The Need for Emittance Reduction

For many applications of charged particle accelerators, such as high-energy colliders, special nuclear physics studies, short-wavelength free electron lasers, and so on, the inherent emittances and momentum spreads of the beams are too large to satisfy the experimental requirements. The best examples are the antiproton beams used in high-energy proton–antiproton ($p\overline{p}$) colliders and the positrons used in electron–positron (e^-e^+) colliders. These beams of antiparticles are produced by bombarding special targets with primary beams of sufficiently high energy, and they have therefore inherently large emittances and momentum spreads. It was the need to reduce the phase-space volumes of the antiproton beams for successful high-energy collision experiments that led to the invention of *electron cooling* by Budker [58] at Novosibirsk and *stochastic cooling* by van der Meer and his co-workers [59] at CERN. Stochastic cooling of antiprotons was instrumental in the discovery of the W and Z particles (vector bosons) at CERN by Rubbia and his team. Electron beam cooling, on the other hand, plays an important role in a number of lower-energy facilities, such as the storage ring at Bloomington mentioned in Section 6.3 (see Reference 13 at the end of Chapter 3).

For high-energy lepton machines such as the e^-e^+ linear collider at SLAC, *radiation cooling* is the method of choice. In view of the very stringent emittance requirements for achieving high luminosity in the interaction point, both positrons and electrons require cooling in special damping rings before they are accelerated to full energy. This technique utilizes synchrotron radiation to dampen the amplitudes of the particles' betatron oscillations and also to reduce the longitudinal momentum spread.

All of the three cooling methods (electron, stochastic, and radiation) require long interaction times that can only be achieved in storage rings over thousands of revolutions. A technique for rapid cooling of a beam in a straight transport line (rather than an expensive storage ring) has yet to be found. In the subsections below we discuss briefly each of the three successful methods employed in ring machines.

6.5.2 Electron Cooling

If a low-temperature electron beam is combined with a high-temperature ion beam traveling in the same direction and at the same speed, Coulomb collisions between the two particle species will lead to temperature relaxation. The electron beam will heat up while the ion beam cools down as the two-beam system is driven toward thermal equilibrium. As a result of this thermal energy exchange the emittance of the electron beam increases while that of the ion beam is reduced.

In practice, the electron beam interacts with the ion beam only along a short straight section of length L_e built into the ion storage ring, whose circumference C is usually much larger than L_e . During each traversal of the cooling section, the ion beam imparts a small amount of its thermal energy to the electron beam. The latter is produced by an electron gun with thermionic cathode. It thus has a transverse temperature on the order of 0.1 eV and a longitudinal temperature that is several orders of magnitude lower due to acceleration, as discussed in Sections 5.4.6 and 6.4.1. The electron beam is extracted from the cooling section after the interaction with the circulating ion beam and hence carries the transferred thermal energy from the ions out of the system. In each pass through the cooling section the ions encounter a fresh group of cold electrons from the gun so that, in principle, they could be cooled to the intrinsic electron temperature. However, intrabeam scattering in the ion beam may prevent such high cooling rates to be achieved. For instance, if the storage ring operates in the negative-mass regime there will be a continuous thermalization of coherent ion-beam energy, as discussed in Section 6.4.2, which will tend to reduce the cooling effect of the electron beam. Likewise, if the lattice function ϕ defined in Equation (6.165b) is not zero, there will also be a lower limit to the ion-beam temperature that can be reached. In either case, this limit will be defined by the equilibrium state in which the rate of ion temperature increase due to intrabeam scattering is just balanced by the cooling rate due to the interaction with the cold electron beam. It is found that the cooling effect can be greatly enhanced by providing a longitudinal magnetic field B_z that confines the electrons to helical orbits of small radius while leaving the heavier ions essentially unaffected.

The theory of electron cooling is, like that of intrabeam scattering, rather involved as there are many different regimes of operation and parameters to be taken into account. However, if one assumes a simple electron-ion plasma model where three-dimensional equilibrium can be achieved, one can derive an approximate relation for the relaxation time, which is analogous to Equations (6.152) and (6.173) and given by [60]

$$\tau_e = \frac{C}{L_e} \frac{F_1 \gamma_0^2}{r_e r_i n_e c \ln \Lambda} \left[\left(\frac{k_B T_{be}}{m_e c^2} \right)^{3/2} + \left(\frac{k_B T_{bi}}{m_i c^2} \right)^{3/2} \right]. \tag{6.188}$$

Here n_e is the electron density, assumed to be the same as the ion density n_i, r_e and r_i are the classical electron and ion radii, F_1 is a constant that for a smooth focusing system has the value $F_1 = 3/4\sqrt{2\pi} \approx 0.3$, and γ_0 is the relativistic energy factor (identical for both beams); the electron and ion temperatures are measured in the beam frame. L_e/C is the fraction of the storage ring occupied by the cooling section, and $\ln \Lambda$ is the Coulomb logarithm as defined in Equations (5.247) and (5.248), with $k_B T/mc^2 \approx k_B T_{he}/m_e c^2$ since $k_B T_{hi}/m_i c^2 \ll k_B T_{he}/m_e c^2$. When equilibrium is reached, the two beam temperatures are the same (i.e., $T_{bi} = T_{be}$). Assuming that both beams have identical transverse cross sections, one obtains an emittance ratio of $\epsilon_i/\epsilon_e \approx (m_e/m_i)^{1/2}$; that is, the ion-beam emittance would be considerably smaller than that of the electron beam in view of the inverse square-root mass ratio. As an example, consider the electron cooling of a 200-MeV proton beam so that $\gamma_0 \approx 1.21$ and the electron energy is about 109 keV. Assuming a density of $n_e \approx 10^{15}$ m^{-3}, $C/L_e \approx 50$, $F_1 \approx 0.3$, and $k_B T_e = 0.1$ eV, one finds from (6.158) $\ln \Lambda = 16.5$ and from (6.188) an approximate relaxation time of $\tau_e \approx 0.1$ s. The equilibrium temperature of the protons would be $k_B T_i \approx k_B T_e = 0.1$ eV, and the transverse emittance of the proton beam would be $\epsilon_i \approx 0.02\epsilon_e$ if identical beam size is assumed. In practice, the lattice design and other parameters come into play, as mentioned above, which change the factor F_1. Furthermore, Coulomb scattering between the protons in the ring works against the electron cooling. The final equilibrium is reached when the cooling rate due to the electron beam and the heating rate due to intrabeam scattering are equal, and the proton temperature is always higher than the 0.1-eV electron temperature. At higher energies (i.e., above $\gamma_0 \simeq 1.25$) the relaxation times become too long and the electron cooling systems too bulky to be practical, and the stochastic cooling method described in the next section is superior.

6.5.3 Stochastic Cooling

To obtain a conceptual understanding of stochastic cooling, let us first consider the radial betatron oscillation of a single particle about the ideal equilibrium orbit in the midplane of a storage ring. Assume that a pickup probe consisting of two electrodes is located at some position along the ring. One electrode is inside and the other outside the central orbit. If the particle trajectory coincides with the equilibrium orbit, there will be no betatron oscillation and no signal will be induced in the pickup plates. On the other hand, if the particle deviates from the equilibrium orbit it will perform betatron oscillations and induce an electric signal in the pickup probe. This signal is proportional to the displacement from the central orbit. It can be amplified and fed to a "kicker" consisting of two electrodes and located an odd number of quarter-wavelengths of the betatron oscillation downstream of the pickup. The kicker then provides a deflection to the particle that is proportional to the displacement sensed at the pickup probe and has a polarity that tends to reduce the betatron amplitude. The signal path between pickup and kicker must, of course, be sufficiently shorter than the orbital path length between the two locations so that the signal reaches the kicker at the same time as the particle. As this process is repeated during several successive revolutions, the particle gradually loses all its transverse energy and will move along the ideal equilibrium orbit without deflections. In a sense, the particle has been "cooled" and its initial transverse energy has been dissipated in the kicker system.

In a real beam, there will be many particles performing betatron oscillations with a random distribution in phase. The signals induced in the pickup probe by the group of particles being sampled will, however, not cancel each other completely. Due to the finite number of particles and the stochastic nature of the oscillations there will in general be fluctuations of the sampled group's centroid position with respect to the equilibrium orbit. Suppose that the number of particles in the sample is N_s and that the mean displacement of this group of particles from the equilibrium orbit at the pickup probe is $\bar{x}$. The corresponding signal in the pickup probe will be amplified and fed to the kicker. There, each particle in the sample will receive a deflection Δx of its trajectory that is proportional to the mean displacement $\bar{x}$ at the pickup, say $\Delta x = \alpha \bar{x}$. Hence, after the kick, each individual particle's displacement from the equilibrium orbit will be $x = x_k - \alpha \bar{x}$, where x_k is the position before the kick. The net result is that the mean square $\overline{x^2}$, and hence the rms width $\tilde{x} = (\overline{x^2})^{1/2}$ of the particle distribution after the kick, is reduced compared with the value $\overline{x_k^2}$ before the kick. Thus, the emittance is also reduced. It can be shown that the change of $\overline{x^2}$ per revolution (i.e., per passage through the pickup-kicker system) is to good approximation given by (see, e.g., [D.10, Sec. 7.3.1])

$$\frac{d\overline{x^2}}{dn} = -\frac{2\alpha - \alpha^2}{N_s} \overline{x_k^2}. \tag{6.189}$$

The corresponding rate of emittance change is then

$$\frac{1}{\epsilon}\frac{d\epsilon}{dn} = -\frac{2\alpha - \alpha^2}{N_s}. \tag{6.190}$$

If $\tau_{\text{rev}} = C/v$ is the revolution time in the ring, the characteristic time constant τ for the cooling process can be defined by

$$\frac{1}{\tau} = -\frac{1}{\epsilon}\frac{d\epsilon}{dt} = -\frac{1}{\epsilon}\frac{d\epsilon}{dn}\frac{1}{\tau_{\text{rev}}} = \frac{2\alpha - \alpha^2}{N_s\tau_{\text{rev}}}. \tag{6.191}$$

Let us assume that the entire beam comprises a total number of particles N that are uniformly distributed around the circumference of the storage ring. If Δt_s is the pulse length of the slice consisting of the N_s particles being sampled by the pickup probe, then

$$N_s = N\frac{\Delta t_s}{\tau_{\text{rev}}} \tag{6.192}$$

and the cooling rate (6.191) may be written as

$$\frac{1}{\tau} = \frac{1}{N\,\Delta t_s}(2\alpha - \alpha^2). \tag{6.193}$$

Thus, the time constant τ for the cooling process is seen to be proportional to the total number of particles N and the sampling time Δt_s and inversely proportional to the function $2\alpha - \alpha^2$ of the signal amplification factor α. The emittance decreases with time t as

$$\epsilon(t) = \epsilon_i e^{-t/\tau}, \tag{6.194}$$

where ϵ_i is the initial value prior to the onset of stochastic cooling.

By using pickup and kicker probes with vertical as well as horizontal electrode configurations, one can cool the emittances of the beam in both transverse directions simultaneously. The above analysis applies, of course, for either direction.

The stochastic cooling technique can also be employed to reduce the longitudinal momentum spread of the beam. Momentum differences are detected by the related difference in the revolution times or orbital frequencies. A synchronous particle having the ideal momentum and orbital frequency will remain unaffected. Non-synchronous particles will receive a longitudinal kick from an appropriately designed sensing and feedback system so that the momentum difference is reduced. The mathematical analysis for this longitudinal cooling technique is beyond the scope of our brief review of the subject. An excellent general review can be found in the book by Edwards and Syphers [D.10] mentioned earlier. A more comprehensive treatment of the theory was given by Möhl [61], and a good introduction to both electron and stochastic cooling was given by Cole and Mills [62].

6.5.4 Radiation Cooling

As is well known from classical electrodynamics, a charged particle will emit electromagnetic radiation when it is accelerated or decelerated. If the particle moves on a straight path and the acceleration is in the direction of the particle's velocity, as is the case in linear accelerators, the radiation effect is generally insignificant. By contrast, if the acceleration is perpendicular to the velocity, as in the bending magnets of synchrotrons and other circular accelerators, the effect is very pronounced. The radiated power rises strongly with the particle's energy $E = \gamma mc^2$ as E^4, and it is inversely proportional to the square of the mass. Thus, the radiation plays a significant role only in the case of highly relativistic electrons and other light particles (*leptons*). Indeed, in circular electron machines, *synchrotron radiation*, as the effect is known in the literature, poses an upper limit to the achievable energy that is in the range of about 100 GeV. On the other hand, it is negligible in existing high-energy *hadron* colliders (protons and antiprotons) where the energy is below 1 TeV, although it would be significant at energies above 10 TeV.

A beneficial effect of synchrotron radiation in high-energy rings is the damping of the amplitude of the incoherent particle oscillations about the beam centroid. This can be understood intuitively by considering the transverse betatron oscillations. A particle performing an oscillation about the equilibrium orbit has a higher energy and hence emits a larger amount of radiation power than the equilibrium particle. Synchrotron radiation is a dissipative non-Liouvillean process and thus it can be employed to reduce the transverse emittance and the longitudinal momentum spread of electron or positron beams. Radiation cooling in special damping rings, for instance, is a necessity in a linear e^+e^- collider. To achieve the desired luminosity at the interaction point, the transverse dimensions, and hence the emittances, of the two colliding beams have to be extremely small.

Let us now take a brief look at the existing theory of synchrotron radiation and radiation cooling. We will not give any detailed derivations, but merely present and discuss the major relations that describe these effects. Following Jackson [A.4, Chap. 14], the power radiated by an accelerated particle of charge q can be expressed in the form

$$\mathcal{P} = \frac{q^2\gamma^2}{6\pi\epsilon_0 m^2 c^3}\left[\left(\frac{d\mathbf{P}}{dt}\right)^2 - \frac{1}{c^2}\left(\frac{dE}{dt}\right)^2\right], \tag{6.195a}$$

or, with $E = \gamma mc^2 = (d\mathbf{P}/dt)\cdot\mathbf{v}$ [Equation (2.25)],

$$\mathcal{P} = \frac{q^2\gamma^2}{6\pi\epsilon_0 m^2 c^3}\left[\left(\frac{d\mathbf{P}}{dt}\right)^2 - \frac{1}{c^2}\left(\frac{d\mathbf{P}}{dt}\cdot\mathbf{v}\right)^2\right]. \tag{6.195b}$$

For a given applied force $d\mathbf{P}/dt = \mathbf{F}$, this formula shows the inverse dependence on the square of the mass mentioned above. Furthermore, the radiated power depends very strongly on the direction of the applied force relative to the particle velocity **v**. Consider first the case of a linear accelerator where the accelerating force $\mathbf{F}_\parallel = (d\mathbf{P}/dt)_\parallel$ is parallel to the direction of the particle velocity **v**. Since in this case $(d\mathbf{P}/dt) \cdot \mathbf{v} = v\, dP/dt$, one obtains

$$\mathcal{P} = \frac{q^2}{6\pi\epsilon_0 m^2 c^3}\left(\frac{d\mathbf{P}}{dt}\right)^2, \tag{6.196a}$$

or

$$\mathcal{P} = \frac{2}{3}\frac{r_c c}{mc^2}\left(\frac{dE}{dz}\right)^2, \tag{6.196b}$$

where we used $\gamma^2(1 - \beta^2) = 1$, $dP/dt = dE/dz$, and introduced the classical particle radius r_c.

We can compare the radiated power with the energy gain per second in the linac, $dE/dt = (dE/dz)v$, by writing the last equation in the alternative form

$$\frac{\mathcal{P}}{dE/dt} = \frac{2}{3}\frac{r_c}{mc^2}\frac{1}{\beta}\frac{dE}{dz}. \tag{6.197}$$

From this relation we can see that the radiation loss of an electron will be unimportant unless the rate of energy gain is on the order of $mc^2 = 0.511$ MeV in a distance corresponding to the classical radius of $r_c = 2.82 \times 10^{-15}$ m (i.e., about 1.8×10^{14} MeV/m). In linear accelerators, the electric field gradients, and hence the rates of energy gain, are severely limited by electrical breakdown and other effects. Typical field gradients are in the range 10 to 100 MV/m. Thus, radiation losses are completely negligible in linear machines.

The situation is quite different in circular accelerators, where the Lorentz force $F_\perp = qvB$ is perpendicular to the direction of motion. With $d\mathbf{P}/dt \perp \mathbf{v}$ the second term in brackets in Equation (6.195b) is zero, and the radiated power becomes

$$\mathcal{P} = \frac{q^2\gamma^2}{6\pi\epsilon_0 m^2 c^3}\left(\frac{d\mathbf{P}}{dt}\right)^2 = \frac{q^2\gamma^2}{6\pi\epsilon_0 m^2 c^3}(\mathbf{F}_\perp)^2. \tag{6.198}$$

Now $|d\mathbf{P}/dt| = |\mathbf{F}_\perp| = \gamma m v^2/R = \omega|\mathbf{P}| = \omega\beta\gamma mc$; hence,

$$\mathcal{P} = \frac{q^2 c\gamma^4\beta^4}{6\pi\epsilon_0 R^2}, \tag{6.199}$$

where R is the radius of curvature of the particle orbit. Clearly, the radiated power in this case can be very high, as it increases with the fourth power of the energy, E. In practice, the radiation losses must be compensated by increasing the energy provided by the rf cavities that are located along the circumference of a circular machine. Since high rf power is difficult to achieve and expensive, an energy limit is reached where electron synchrotrons are no longer feasible or cost-effective. This is the motivation for the development of linear colliders [63], where these radiation losses are insignificant.

The energy loss due to synchrotron radiation per revolution of a circulating particle can be expressed as

$$\Delta E_{\text{turn}} = \frac{1}{\beta c} \int_0^C \mathcal{P} \, ds = \frac{q^2 \gamma^4 \beta^3 \overline{R}}{3\epsilon_0} \overline{\left(\frac{1}{R^2}\right)}, \tag{6.200}$$

where $C = 2\pi \overline{R}$ is the circumference, $\overline{R} = C/2\pi$ the average radius of the equilibrium orbit, and $\overline{(1/R^2)}$ represents the square of the local curvature radius averaged over the circumference. Most high-energy rings consist of straight sections and bending magnets. If the bending magnets all have the same magnetic field, so that the local orbit radii are the same and η_b is the fraction of the circumference occupied by bending magnets, then $\overline{(1/R^2)} = \eta_b(1/R^2)$.

In a betatron, of course, where the orbit is perfectly circular, we have $\eta_b = 1$ and $R = \overline{R}$. For highly relativistic electrons ($\beta = 1$) the energy loss per turn will be

$$\Delta E_{\text{turn}} = \frac{4\pi r_e E^4}{3(mc^2)^3} \frac{1}{R} \tag{6.201a}$$

or, numerically,

$$\Delta E_{\text{turn}_{[\text{MeV}]}} = 8.85 \times 10^{-2} E_{[\text{GeV}]}^4 \frac{1}{R_{[\text{m}]}}. \tag{6.201b}$$

It has been shown by Richter [64] that the cost of a circular machine rises as the square of the energy, E^2, whereas that of a linear collider is proportional to E. The crossover point for the two curves is near 100 GeV, and a linear collider becomes less expensive than a ring above this point.

As an example, consider a 10-GeV electron synchrotron with $C = 700$ m and $R = 0.8 \, C/2\pi \approx 90$ m. According to the preceding equations, at 10 GeV, each electron will lose an energy of 9.83 MeV/turn. Thus the rf system must provide an acceleration rate of 9.83 MeV/turn to make up for the radiated power. At an average electron beam current $\overline{I}$, the total rf power required to maintain the electron energy would be

$$\mathcal{P}_{\text{rf}[\text{MW}]} = \Delta E_{\text{turn}[\text{MeV}]} \cdot \overline{I}_{[\text{A}]}; \tag{6.202}$$

that is, for $\overline{I} = 10$ mA one would need an rf power of 98.3 kW to maintain the electron energy. At higher energies and higher beam currents the rf power requirements quickly become excessive, and 100 GeV is considered an upper limit for electron synchrotrons, as mentioned above.

Proceeding now to the topic of radiation cooling, we first note that from Equation (6.200) the average power radiated by an electron can be expressed as

$$\overline{\mathcal{P}} = \frac{1}{C} \int_0^C \mathcal{P}\, ds = \frac{v}{C} \Delta E_{\text{turn}} = f_0\, \Delta E_{\text{turn}}, \tag{6.203}$$

where $f_0 = v/C$ is the revolution frequency. It will be convenient to define the damping rates of the particle oscillations in terms of the characteristic time τ_0 in which a particle radiates all its energy, that is,

$$\tau_0 = \frac{E}{\overline{\mathcal{P}}}. \tag{6.204}$$

Due to the effect of momentum dispersion, D_e, the horizontal particle oscillations in a synchrotron differ from the vertical oscillations, as discussed in Section 5.4.10. Likewise, the longitudinal dynamics is not the same as that in the two transverse directions. As a result, the radiation damping rates for the particle oscillations in each direction are different. A formal derivation of radiation damping was given by Robinson [65], who also referenced earlier work in this field. Later, Sands [66] presented a detailed physical discussion of the effects. More recent reviews of the topic can be found in the books by Lawson [C.17, Sec. 5.10] and by Edwards and Syphers [D.10, Secs. 8.1 to 8.3]. The latter includes derivations of the key relations, and in the following we quote the results in the form presented by these authors.

According to the theory, the effects of dispersion on the radiation damping rates of the particle oscillations amplitudes can be described by the function

$$\mathcal{D} = \int_0^C \frac{D_e}{R^2}\left(\frac{1}{R} + 2\frac{B'}{B} \right) ds \bigg/ \int_0^C \frac{ds}{R^2}, \tag{6.205}$$

where B' is the magnetic field gradient and $B = \gamma m v/qR$.

Since the vertical motion is usually dispersion-free for all practical purposes, the time constant for damping of the vertical oscillation amplitudes is independent of the function $\mathcal{D}$ and given by

$$\tau_y = 2\tau_0. \tag{6.206}$$

The vertical betatron amplitude thus decreases exponentially as

$$y(t) = y_0 e^{-t/\tau_y} = y_0 e^{-t/2\tau_0}. \tag{6.207}$$

For the horizontal oscillations, which consist of the betatron and dispersion part according to Equation (5.416a), one finds a radiation damping time of

$$\tau_x = \frac{2}{1 - \mathcal{D}} \tau_0, \tag{6.208}$$

while the damping time for the synchrotron oscillations is given by

$$\tau_s = \frac{2}{2 + \mathcal{D}} \tau_0. \tag{6.209}$$

The three time constants obey *Robinson's theorem* [65], which states that

$$\frac{1}{\tau_x} + \frac{1}{\tau_y} + \frac{1}{\tau_s} = \frac{2}{\tau_0}. \tag{6.210}$$

Thus when two of the three damping constant are known, the third can be calculated directly from Robinson's relation.

The above relations for the time constants indicate that the horizontal or longitudinal oscillation amplitudes may actually grow, rather than damp, depending on the value of $\mathcal{D}$. Thus, one can see from Equation (6.208) that τ_x becomes negative, that is, the horizontal oscillation amplitudes will grow exponentially, when $\mathcal{D} > 1$. On the other hand, it follows from (6.209) that there is amplitude growth of the synchrotron oscillations when $\mathcal{D} < -2$. The vertical motion is always damped since the time constant does not depend on $\mathcal{D}$. For damping to occur in all three degrees of freedom, $\mathcal{D}$ must satisfy the relation

$$-2 < \mathcal{D} < 1. \tag{6.211}$$

For a weak-focusing machine with axial symmetry (no straight sections) like a betatron, one finds that

$$\mathcal{D} = \frac{1 - 2n}{1 - n}. \tag{6.212}$$

The field index $n = -RB'/B$ must satisfy the condition $0 < n < 1$ [Equation (3.204)] to assure focusing in both transverse directions. Thus the vertical and horizontal oscillation amplitudes are always damped. However, the synchrotron oscillations are damped only if $n < 0.75$, which, in practice, is readily achieved.

Modern strong-focusing synchrotrons and storage rings are built with "separated-function" lattices, where the bending occurs in uniform-field magnets and the focusing in straight sections with magnetic quadrupoles. In these machines it is found that $\mathcal{D}$ is positive and small compared with unity, so that there is always damping in all three degrees of freedom. If $\mathcal{D}$ can be neglected, one simply gets $\tau_x = \tau_y = 2\tau_0$ and $\tau_s = \tau_0$.

The above classical theory of radiation cooling predicts that the transverse and longitudinal beam temperatures and emittances would exponentially go toward zero with time. However, this is not the case, as the classical model must be corrected by taking into account the quantum mechanical description of the radiation effect. According to modern theory, radiation occurs in the form of discrete photon emission, which is essentially a stochastic process. The emission of a photon changes the momentum of the particle and hence the phase and amplitude of its oscillation about the beam centroid. According to the quantum-statistical description of the process there is a large spread in the photon energies emitted by the electrons in a beam. The photon energy $w = \hbar\omega$ is usually expressed in terms of the *critical energy* w_c, defined as

$$w_c = \frac{2}{3}\gamma^3\hbar\omega_0, \tag{6.213}$$

where $\hbar = h/2\pi = 1.0545 \times 10^{-34}$ J-s $= 6.5906 \times 10^{-16}$ eV-s, and ω_0 is the instantaneous angular frequency of the particle motion in the circular machine. Figure 6.25 shows the energy distribution $S(w/w_c)$ of the synchrotron radiation. The mean and mean-square values of the energy spectrum are found to be

$$\overline{w} = \frac{8}{15\sqrt{3}}w_c, \tag{6.214}$$

$$\overline{w^2} = \frac{11}{27}w_c^2. \tag{6.215}$$

The random fluctuations of the emitted photon energies produce a spread of the particle oscillation amplitudes that leads to emittance growth for the horizontal

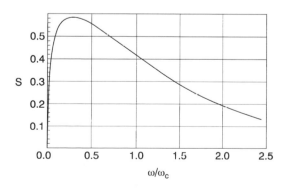

Figure 6.25. Energy spectrum of synchrotron radiation, S versus $\omega/\omega_c = w/w_c$.

and longitudinal motion which opposes the radiation damping effect. Growth in the vertical direction is negligible since there is no vertical dispersion. If N_ν is the number of photons emitted per revolution defined by

$$N_\nu = \frac{\overline{\mathcal{P}}}{f_0 \overline{w}} = \frac{\Delta E_{\text{turn}}}{\overline{w}}, \tag{6.216}$$

one obtains the following differential equations for the horizontal and vertical rms beam widths, δ_x, δ_y and the rms energy spread δ_E:

$$\frac{d\delta_x^2}{dt} = -\frac{2}{\tau_x} \delta_x^2 + \frac{1}{2} N_\nu f_0 \hat{\beta}_x \overline{\mathcal{H} \frac{w^2}{E^2}}, \tag{6.217}$$

$$\frac{d\delta_y^2}{dt} = -\frac{2}{\tau_y} \delta_y^2, \tag{6.218}$$

$$\frac{d\delta_E^2}{dt} = -\frac{2}{\tau_s} \delta_E^2 + \frac{1}{2} N_\nu f_0 \overline{w^2}. \tag{6.219}$$

The function $\mathcal{H}$ represents the effect of dispersion and is defined as

$$\mathcal{H} = \hat{\gamma} D_e^2 + 2\hat{\alpha} D_e D_e' + \hat{\beta} D_e'^2. \tag{6.220}$$

$\overline{\mathcal{H}}$ is the average value of $\mathcal{H}$ over the closed orbit, $\hat{\alpha}$, $\hat{\beta}$, $\hat{\gamma}$ are the Courant–Snyder parameters for the horizontal motion, and $\overline{\mathcal{H}} \approx 0$ for the vertical motion.

The three equations (6.217) to (6.219) can readily be integrated, yielding

$$\delta_x^2(t) = \delta_x^2(0)e^{-2t/\tau_x} + \frac{1}{4} N_\nu f_0 \hat{\beta}_x \tau_x \overline{\mathcal{H} \frac{w^2}{E^2}} (1 - e^{-2t/\tau_x}), \tag{6.221}$$

$$\delta_y^2(t) = \delta_y^2(0)e^{-2t/\tau_y}, \tag{6.222}$$

$$\delta_E^2(t) = \delta_E^2(0)e^{-2t/\tau_s} + \frac{1}{4} N_\nu f_0 \tau_s \overline{w^2} (1 - e^{-2t/\tau_s}). \tag{6.223}$$

Assuming that all three damping time constants are positive, we see that equilibrium can be reached within a few damping times (strictly speaking, as $t \to \infty$). Since the transverse rms emittances are defined by $\tilde{\epsilon}_x = \delta_x^2/\hat{\beta}_x$ [see Equation (5.334)] and $\tilde{\epsilon}_y = \delta_y^2/\hat{\beta}_y$, one obtains for the normalized rms emittance $\tilde{\epsilon}_n = \gamma\tilde{\epsilon}$ (in the relativistic limit where $\beta = 1$) and the relative rms energy spread δ_F/E the following equilibrium values:

$$\tilde{\epsilon}_{nx} = C_1 \left| \frac{\mathcal{H}}{1 - \mathcal{D}} \right| \frac{w_c}{mc^2}, \tag{6.224}$$

$$\tilde{\epsilon}_{ny} = 0, \tag{6.225}$$

$$\frac{\delta_E}{E} = \left[C_1 \frac{1}{2 + \mathcal{D}} \frac{w_c}{\gamma mc^2} \right]^{1/2}. \tag{6.226}$$

The constant C_1 is given by

$$C_1 = \frac{55\sqrt{3}}{2^4 3^2} = 0.66. \tag{6.227}$$

Note that the vertical equilibrium emittance is approximately zero in this model. In reality one has to take into account other effects, such as intrabeam scattering, which transfers thermal energy from the transverse and longitudinal directions to the vertical phase space. However, even though the vertical emittance will not be zero, it will still remain significantly smaller than the horizontal emittance. Thus, the beams in electron synchrotrons always have a rectangular ribbon-like cross section with $\delta_x \gg \delta_y$. However, the energy distribution is Maxwellian. To illustrate the radiation cooling effect, let us consider the 10-GeV electron machine discussed earlier [following Equation (6.201b)]. For convenience we assume a smooth lattice where $D'_e = 0$ and $\mathcal{D} \approx 0$. The critical energy can be written in the form

$$w_c = \frac{9}{8\pi} \frac{\hbar c}{r_e} \left(\frac{R}{\overline{R}} \right) \frac{\Delta E_{\text{turn}}}{E} = 2.51 \times 10^7 \left(\frac{R}{\overline{R}} \right) \frac{\Delta E_{\text{turn}}}{E}. \tag{6.228}$$

Using $\Delta E_{\text{turn}} = 9.83$ MeV and $R/\overline{R} = 0.8$, one obtains $w_c = 19.7$ keV. The function $\mathcal{H}$ is given by

$$\overline{\mathcal{H}} = \hat{\gamma} D_e^2 = \frac{D_e^2}{\hat{\beta}} = \frac{D_e^2 \nu_x}{R} \tag{6.229}$$

since $\hat{\beta}\hat{\gamma} = 1$, and $\hat{\beta} = \overline{R}/\nu_x$. Taking $D_e \approx 2$ m, $\nu_x = 4.8$, one gets $\overline{\mathcal{H}} = 0.21$. With these values one obtains $\tilde{\epsilon}_{nx} = 5.36 \times 10^{-3}$ m-rad for the normalized rms emittance, $\tilde{\epsilon}_x = \tilde{\epsilon}_{nx}/\gamma = 2.7 \times 10^{-7}$ m-rad for the unnormalized rms emittance, and $\delta_E/E = 8.1 \times 10^{-4}$ for the relative rms energy spread. In an actual machine, these values would, of course, differ somewhat since D'_e and $\mathcal{D}$ are not exactly zero. But, more important, intrabeam scattering provides coupling between the three degrees of freedom and may introduce additional thermal energy, as discussed in Section 6.4.2.

6.6 CONCLUDING REMARKS

The topics of emittance growth, emittance preservation, and emittance reduction by cooling techniques discussed in this chapter are of fundamental importance for the design and application of advanced particle accelerators and other devices. We reviewed three major causes of emittance growth—beam mismatch, instabilities, and collisions—but our list of effects in each category is by no means complete.

In Section 6.2 we discussed the thermodynamic concept of free energy in nonstationary, or mismatched, beams and its possible conversion into thermal energy and associated emittance increase. This topic is relatively new, as most of the research results obtained during the last few years have not yet been reviewed in other books. The emphasis in this section was on the role of space charge and the shape of the particle distribution. Even if a beam is rms-matched into a focusing channel or accelerator, emittance growth can occur if the initial density profile differs from that of the stationary Maxwell–Boltzmann distribution. Still, our analysis was limited to a symmetrical beam in transverse phase space. In some applications the beam cross section may be asymmetric with different rms widths and rms emittances in the two orthogonal directions (e.g., in a "sheet beam") and the theory—both the Maxwell–Boltzmann distribution as well as the emittance growth formalism—needs to be extended accordingly. The work by Wangler, Lapostolle, and Lombardi is a first step in this direction [67].

More research is needed to correlate the time scales for emittance growth with the nonlinearities due to the applied focusing forces and due to the space-charge density perturbations from the stationary profile. Most likely, one needs to go back to the Fokker–Planck equation and try to obtain a better model for the diffusion coefficient and relaxation rate, as attempted by Bohn [68].

The formation of a *halo* in the mismatched beam is largely unexplained. We know that the halo is caused by the existence of free energy and the nonlinear interaction of the particles with the density oscillations and fluctuations in the beam. Recent studies by Jameson provide some insight into the mechanisms that cause individual particles to gain transverse energy and to become part of the halo [69]: The interaction of single particles with the time-varying collective fields due to the plasma oscillations in the beam core may lead to a net increase of the transverse energy and amplitude of particle excursion. In related work, the origin of the halo particles in computer simulation was traced [70]. It was found that a large fraction of the halo consisted of particles from the outer regions of the beam's phase space (i.e., their initial energy is considerably greater than the average energy). However, there are also many particles from the interior of the phase space that are kicked out, gaining sufficient energy to become part of the halo. Because of the stochastical nature of the interaction, such tracking of individual particles may not provide the definitive explanation of the halo effect. Why does some fraction of the available free energy increase the temperature of the beam core while the rest goes into the nonthermal high-energy tail? Can we develop a model that can predict this behavior quantitatively and allow us to determine which fraction of the initial mismatched particle distribution and/or free energy ends up in the halo? Does one reach a final steady state with a thermal beam core surrounded by a halo? Can the halo be removed by appropriately placed aperture plates without disturbing the stationary core? There is also the problem discussed in Section 5.4.4 that the Boltzmann density profile develops a natural Gaussian-like tail as the beam is accelerated to high energy even if the distribution is rms-matched adiabatically during the acceleration process so that no free energy is created.

All of these questions need more research. Furthermore, our analysis in Section 6.2 needs to be extended to the longitudinal direction of bunched beams where nonstationary line-charge density profiles, mismatch, and off-centering lead to longitudinal emittance growth and halo formation, just as in the transverse phase planes. The situation may even be worse than in the transverse case because of the highly nonlinear nature of the longitudinal forces in the buckets of rf accelerators. Ultimately, the goal must be to understand fully the behavior of the three-dimensional particle distribution with space charge when free energy is created that increases the beam's temperature and its three-dimensional phase-space volume. This relates also to the problem of equipartitioning discussed in Appendix 4.

In Section 6.3 we limited our discussion essentially to three topics. The first was the transverse instabilities caused by the resonant interaction between density perturbations oscillating with the plasma frequency and focusing forces in a periodic channel. These instabilities in space charge dominated beams are closely related to the emittance growth effects discussed in Section 6.2.

Our second topic was concerned with the nature and behavior of longitudinal space-charge waves, which are created by perturbations of a beam's line charge-density profile. Space-charge waves play the key role in the physics of longitudinal instabilities, which are the cause of beam degradation and emittance growth. As our third topic we chose the longitudinal instability that is created by the interaction of the slow space-charge wave with an external circuit (e.g., resistive wall) in a linear channel. In view of the general importance of circular accelerators and for historical reasons, we extended this discussion in Section 6.3.3 to circular machines. There the *negative-mass instability* is of particular interest since it occurs as a result of the particle dynamics in the negative-mass regime above transition energy discussed in Section 5.4.9.

These two examples of longitudinal beam instability were intended to serve as an illustration and introduction into the topic. An excellent, comprehensive treatment of collective instabilities in high-energy accelerators is given in the book by Chao [D.11]. At high, relativistic particle energies the long-wavelength electrostatic model that we used in our analysis of space-charge waves and instabilities is no longer sufficient. A fully electromagnetic treatment is required in which the wakefields created by bunched beams or by perturbations in unbunched beams are taken into account. These wakefields act back on the beam and cause transverse and longitudinal instabilities which depend on the transverse and longitudinal impedances $Z_\perp(\omega)$ and $Z_\parallel(\omega)$ of the beam's enviroment which are functions of the frequency ω. The mathematical treatment of these wakefield effects is quite complicated and requires a mixture of analysis and simulation. We will not attempt to go any further into this highly specialized topic, which is treated in great depth and detail by Chao. The book by Edwards and Syphers provides a good elementary introduction into the topic of wakefields, impedances, and instabilities in high-energy accelerators [D.10, Chap. 6]. A more general and excellent review of waves and instabilities in charged-particle beams can be found in Lawson's book [C.17, Chap. 6].

The third category of effects causing emittance growth, Coulomb collisions between particles in the beam or between beam particles and a background gas, was reviewed in Section 6.4. Our treatment of collisions in Section 6.4.1 (Boersch effect) is new. We applied the theory of Ichimara and Rosenbluth for a stationary, nonrelativistic, magnetically confined plasma to a charged particle beam in a uniform focusing channel. The beam's longitudinal temperature $T_\parallel$ was assumed to be much lower than the transverse temperature $T_\perp$, due to cooling by acceleration. Historically, the Boersch effect was observed in a focused electron beam where the transverse temperature $T_\perp$ increases with distance, reaching a maximum at the crossover point (waist). The theoretical treatments of the collisions in such a focused beam, reviewed in Jansen's book (see Reference 9 in Chapter 5), are rather involved, and scaling with physical parameters is not readily apparent. Our treatment, on the other hand, yields a relatively simple analytical relation for growth rate and parameter scaling. The numerical examples show that the temperature relaxation can increase the longitudinal energy spread by two orders of magnitude in short distances of 1 m even though the beam is far from thermal equilibrium. This result is in good agreement with Boersch's observations for a focused beam. A comparison of the results from our uniform-beam model with those from the models for a focused beam would be very interesting. Our results indicate that significant broadening of the energy spread should occur not only in the crossover point of a strongly focused beam, but also in the more smoothly focused beams of electrostatic accelerators, low-energy beam transport systems, induction linacs, and even in rf linacs, where, however, it may be masked by the energy spread due to bunching.

Our review of intrabeam scattering in circular machines (Section 6.4.2) provided some new insight by emphasizing the thermodynamic aspects (i.e., the relationships between momentum spreads and temperatures in the three degrees of freedom). Theoretically, thermal equilibrium is possible in a smooth channel with zero dispersion below transition energy, but no equilibrium exsits in the negative-mass regime above transition. In practice, nonzero dispersion and the variations of the betatron function along the circumference contribute additional energy, which prevents the attainment of equilibrium below transition as well. As in the negative-mass regime, this additional energy is due to thermalization of longitudinal kinetic energy of the beam (see our discussion in Section 5.4.12). However, the effect appears to be rather small in most cases studied, so that the simple smooth-focusing models should be adequate to calculate the growth rates. We found it somewhat puzzling that the more sophisticated computer models do not distinguish explicitly between the two fundamentally different regimes above and below transition. Clarification of this puzzle would be highly desirable.

Our review of gas scattering in Section 6.4.3 and beam cooling methods in Section 6.5 follows the standard treatment found in the literature, except for some changes in notation to maintain consistency with that in other sections of the book.

With regard to our discussion of emittance growth in this chapter, we have selected those effects that are fundamental to most beams, but we emphasized the role of space charge and the thermodynamic concept of free energy. We did not discuss many effects that are unique to a particular device, such as the instabilities in high-energy accelerators (except for the longitudinal instability. Other examples of this type are the special emittance preservation requirements in future linear colliders [71], synchrotron light sources, free-electron lasers [72], and inverse free electron laser applications [73].

For linear colliders, free electron lasers and other applications, the development of new electron guns with higher current density and brightness that can be achieved with thermionic cathodes is of great importance. The *rf photocathode gun*, also known as the *laser-driven rf electron gun*, mentioned in Section 3.1 is the leading candidate, with research and developmental work in progress at several laboratories (see References 3–5 in Chapter 1). Emittance growth in the high-density electron bunches produced by these rf guns is a major concern and we will present a brief general discussion of this problem in Appendix 5.

Finally, we want to mention that the emittance growth of beams in drift space, a special topic that was not treated in this book, was studied by Lee *et al.* [74], Wangler [75], and Noble [76]. An expanding or converging beam is of course not in thermal equilibrium as the temperature decreases or increases, respectively. If the particle density deviates from the ideal Boltzmann profile, emittance growth occurs, and this can be significantly stronger than in the uniformly focused beams discussed in Section 6.2. This applies not only to the continuous or long beams that have been studied so far, but also to bunched beams which, to our knowledge, have not yet been the subject of systematic theoretical and experimental investigations. This topic, which is of great importance for the behavior of beams in the various matching sections of accelerator systems and in final focusing of the beams, deserves further research in the future.

REFERENCES

1. M. Reiser, *J. Appl. Phys.* **70**, 1919 (1991).

2. J. Struckmeier, J. Klabunde, and M. Reiser, *Part. Accel.* **15**, 47 (1984).

3. T. P. Wangler, K. R. Crandall, R. S. Mills, and M. Reiser, *IEEE Trans. Nucl. Sci.* **NS-32**, 2196 (1985).

4. P. M. Lapostolle, *IEEE Trans. Nucl. Sci.* **NS-18**, 1101 (1971).

5. O. A. Anderson, *Part. Accel.* **21**, 197 (1987).

6. I. Hofmann and J. Struckmeier, *Part. Accel.* **21**, 69 (1987).

7. P. Lapostolle, *CERN Report ISR/D1-70-36,* CERN, Geneva, 1970 (unpublished); for a review of the early work on the effects of space charge in linear accelerators, see P. Lapostolle, *AIP Conf. Proc.* **253,** 11 (April 1992), ed. W. W. Destler and S. K. Guharay.

8. G. Schmidt, *Physics of High Temperature Plasmas,* Second Edition, Academic Press, New York, 1979, p 317.

9. O. A. Anderson, *AIP Conf. Proc.* **152,** 253 (1986); ed. M. Reiser, T. Godlove and R. Bangerter.

10. I. Haber, *AIP Conf. Proc.* **139,** 107 (1986); ed. G. Gillespie, Y. Y. Kuo, D. Keefe, and T. P. Wangler.

11. C. M. Celata, I. Haber, L. J. Laslett, L. Smith, and M. G. Tiefenback, *IEEE Trans. Nucl. Sci.* **NS-32,** 2480 (1985).

12. M. Reiser, C. R. Chang, D. Kehne, K. Low, T. Shea, H. Rudd, and I. Haber, *Phys. Rev. Lett.* **61,** 2933 (1988).

13. I. Haber, D. Kehne, M. Reiser, and H. Rudd, *Phys. Rev. A* **44,** 5194 (1991).

14. M. J. Rhee and R. F. Schneider, *Part. Accel.* **20,** 133 (1986).

15. D. Kehne, "Experimental Studies of Multiple Electron Beam Merging, Mismatch and Emittance Growth in a Periodic Solenoid Channel," Ph.D. Dissertation, Electrical Engineering Department, University of Maryland, College Park, MD, April 1992.

16. A. Cucchetti, M. Reiser, and T. Wangler, *Conference Record of the IEEE 1991 Particle Accelerator Conference,* 91CH3038-7, May 6–9, 1991, p. 251.

17. For an excellent review, see T. P. Wangler, in *AIP Conf. Proc.* **253,** 21 (April 1992), ed. W. W. Destler, S. K. Guharay.

18. J. D. Lawson, P. M. Lapostolle, and R. L. Gluckstern, *Part. Accel.* **5,** 61 (1973).

19. R. L. Gluckstern, *Proc. 1970 National Accelerator Laboraories Linear Accelerator Conference* (ed. M. R. Tracey), FNAL, Batavia, IL, 1970. Vol. 2, p. 811.

20. R. C. Davidson and N. A. Krall, *Phys. Fluids* **13,** 1543 (1970).

21. I. Hofmann, L. J. Laslett, L. Smith, and I. Haber, *Part. Accel.* **13,** 145 (1983).

22. J. Struckmeier and M. Reiser, *Part. Accel.* **14,** 227 (1984).

23. D. Kehne, K. Low, M. Reiser, T. Shea, C. R. Chang, and Y. Chen, *Nucl. Instrum. Methods Phys. Res. A* **278,** 194 (1989).

24. P. Lapostolle, *CERN Report ISR/78-13,* CERN, Geneva, 1978.

25. J. Struckmeier, private communication.

26. M. G. Tiefenback and D. Keefe, *IEEE Trans. Nucl. Sci.* **NS-32,** 2483 (1985).

27. J. Klabunde, P. Spädtke, and A. Schönlein, *IEEE Trans. Nucl. Sci.* **NS-32,** 2462 (1985).

28. I. Hofmann, *IEEE Trans. Nucl. Sci.* **NS-28,** 2399 (1981).

29. R. A. Jameson, *IEEE Trans.* **NS-28,** 2408 (1981).

30. J. G. Wang, D. X. Wang, and M. Reiser, *Phys. Rev. Lett.* **71,** 1836 (1993).

31. J. G. Wang, D. X. Wang, H. Suk, and M. Reiser, *Phys. Rev. Lett.,* **72,** 2029 (1994).

32. C. K. Birdsall and J. R. Whinnery, *J. Appl. Phys.* **24,** 314 (1953).

33. L. Smith, "ERDA Summer Study of Heavy Ions for Inertial Fusion," *Report LBL-5543,* Berkeley, California, 1976, p. 77.

34. S. Humphries, *J. Appl. Phys.* **53,** 1334 (1982).

35. J. Bisognano, I. Haber, and L. Smith, *IEEE Trans. Nucl. Sci.* **NS-30,** 2501 (1983).

36. J. G. Wang and M. Reiser, *Phys. Fluids B* **5**, 2286 (1993).

37. C. E. Nielsen, A. M. Sessler, and K. R. Symon, *Proc. International Conference on High Energy Accelerators,* CERN, Geneva, 1959, p. 239.

38. A. A. Kolomenskij and A. N. Lebedev, *Proc. International Conference on High Energy Accelerators,* CERN, Geneva, 1959, p. 115.

39. V. L. Neil and A. M. Sessler, *Rev. Sci. Instrum.* **36**, 429 (1963).

40. L. D. Landau, *J. Phys. (USSR)* **10**, 25 (1946).

41. A. Hofmann, *CERN Report 17-13,* CERN, Geneva, 1977, p. 139 ff.

42. R. J. Briggs and V. K. Neil, *J. Nucl. Energy* **9**, 209 (1967).

43. A. G. Ruggiero and V. G. Vaccaro, *CERN Report ISR-TH/68-33,* CERN, Geneva, 1968.

44. E. Keil, W. Schnell, *CERN Report ISR-TH-RF/69-48,* CERN, Geneva, 1969.

45. I. Hofmann, *Proc. 1991 IEEE Particle Accelerator Conference,* Conf. Rec. 91 CH 3038-7, p. 2492.

46. S. Ichimaru and M. N. Rosenbluth, *Phys. Fluids* **13**, 2778 (1970).

47. A. W. Hyatt, C. F. Driscoll, and J. H. Malmberg, *Phys. Rev. Lett.* **59**, 2975 (1987); B. R. Beck, J. Fajans, and J. H. Malmberg, *Phys. Rev. Lett.* **68**, 317 (1992).

48. H. Boersch, *Z. Phys.* **139**, 115 (1954).

49. A. Piwinski, *Proc. 9th International Conference on High Energy Accelerators,* SLAC, 1979, p. 405.

50. J. D. Bjorken and S. K. Mtingwa, *Part. Accel.* **13**, 115 (1983).

51. A. Piwinski, private communication.

52. M. Conte, private communication.

53. M. Conte and M. Martini, *Part. Accel.* **17**, 1 (1985).

54. A. H. Sørensen, *Proc. CERN Accelerator School,* CERN Report 87-10, CERN, Geneva, 1987, p. 135, A. H. Sørensen, in *Frontiers of Particle Beams: Intensity Limitations* (ed. M. Dienes, M. Month, and S. Turner), *Lecture Notes in Physics* **400**, Springer-Verlag, New York, 1992, pp. 467–487.

55. A. Piwinski, private communication.

56. C. Bernardini, G. Corazza, G. DiGiugno, G. Ghigo, J. Haissinski, P. Marin, R. Querzoli, and B. Touschek, *Phys. Rev. Lett.* **10**, 407 (1963). The theoretical interpretation of the effect is contained in this letter; there appears to be no separate paper or report by Touschek himself.

57. Ya. S. Derbenev, *Fermilab $\bar{p}$-Note 176* (1981); see also Reference 54.

58. G. I. Budker, *Proc. International Symposium on Electron and Positron Storage Rings,* Saclay, France, 1966, p. II-1-1.

59. D. Möhl, G. Petrucci, L. Thorndahl, and S. van der Meer, "Physics and Technique of Stochastic Cooling," *Phys. Rep.* **58** (2) (1980).

60. N. S. Dikansky, I. N. Meshkov, and A. N. Skrinskij, *Nature* **276**, 763 (1978).

61. D. Möhl, "Stochastic Cooling for Beginners," in *CERN Accelerator School: Antiprotons for Colliding Beam Facilities,* CERN Report 84-15, CERN, Geneva, 1984.

62. F. T. Cole and F. E. Mills, *Annu. Rev. Nucl. Sci.* **31**, 295 (1981).

63. B. Richter, *IEEE Trans. Nucl. Sci.* **NS-32**, 3828 (1985).

64. B. Richter, *Nucl. Instrum. Methods* **136**, 47 (1976).

65. K. W. Robinson, Phys. Rev. **111**, 373 (1958).

66. M. Sands, "The Physics of Electron Storage Rings: An Introduction," *Proc. International School 'Enrico Fermi', Course XLVI* (ed. B. Touschek), Academic Press, New York, 1971.

67. T. Wangler, P. Lapostolle, and A. Lombardi, *Conference Record of the IEEE 1993 Particle Accelerator Conference,* 93CH3279-7, p. 3606.

68. C. L. Bohn, *Phys. Rev. Lett.* **70**, 932 (1993).

69. R. A. Jameson, "Beam Halo from Collective Core/Single-Particle Interactions," *Los Alamos National Laboratory Report LA-UR-93-1209,* March 1993.

70. D. Kehne, M. Reiser, and H. Rudd, *Conference Record of the IEEE 1993 Particle Accelerator Conference,* 93CH3279-7, p. 65.

71. See, for instance, the review by R. D. Ruth, *Conference Record of IEEE 1991 Particle Accelerator Conference,* 91CH2669-0, May 6–9, 1991, p. 2037, or the article by J. T. Seeman, in *AIP Conf. Proc.* **253**, 129 (April 1992), ed. W. W. Destler and S. K. Guharay.

72. See, for instance, the review article by C. W. Roberson and P. Sprangle, *Phys. Fluids B* **1**, 3 (1989), and the paper on emittance and energy spread in an FEL by B. Hafizi and C. W. Roberson, *Phys. Rev. Lett.* **68**, 3539 (1992).

73. See, for instance, C. Pellegrini, J. Sandweiss, and N. Barov, *AIP Conf. Proc.,* **279**, 338 (1993), ed. J. S. Wurtele.

74. E. P. Lee, S. S. Yu, and W. Barletta, *Nucl. Fusion* **21**, 961 (1981).

75. T. P. Wangler, private communication.

76. R. J. Noble, *Conference Record of the 1989 IEEE Particle Accelerator Conference,* 89CH2669-0, p. 1067.

PROBLEMS

6.1 Consider a uniformly focused beam with constant particle density n_0, current I, velocity v, and radius a inside a cylindrical drift tube of radius b. It can be compared with a coaxial cylindrical transmission line, where the beam takes the place of the solid inner conductor.

(a) Calculate the electric and magnetic fields and the electric and magnetic field energies per unit length for both the beam and the equivalent transmission line (with the same current and "voltage").

(b) Calculate the capacitance C_b^* and inductance L_b^* per unit length of the beam configuration and compare the results with $C_t^\dagger$ and inductance L_t^* for the equivalent transmission line. (*Hint:* Use relations between capacitance and inductance and the appropriate field energy.)

6.2 The beams emerging from electron guns with thermionic cathodes (as in Appendix 1) often have a hollow density profile $n(r)$ due to nonlinear field configurations in the diode or nonuniform cathode emission. Suppose that the density profile has the form

$$n(r) = \begin{cases} n_0\left[1 + \delta\left(\frac{r}{a_0}\right)^2\right] & \text{for } 0 \leq r \leq a_0, \\ 0 & \text{for } a_0 \leq r \leq b, \end{cases}$$

where a_0 is the beam radius, b the conducting tube radius, and δ is a number in the range $0 < \delta < 1$. Assume that the beam is injected into a linear, uniform focusing channel and that the *equivalent* uniform beam (with the same current I and kinetic energy) has initial radius a_i and density $n = n_i = $ const for $r \leq a_i$, $n = 0$ for $a_i < r < b$. Calculate the following:

(a) a_i as a function of a_0 and δ
(b) n_i as a function of n_0 and δ
(c) Electric self field $E_r(r)$ for the nonuniform beam
(d) Electrostatic field energy per unit length, w_E, for the nonuniform beam
(e) Nonuniform field energy factor U/w_0
(f) Radius increase a_f/a_i, emittance growth ϵ_f/ϵ_i, and distance $z_p = \lambda_p/4$ if the electron beam current is 3 kA, the kinetic energy 1.5 MeV, the cathode radius $r_c = 6$ cm, the cathode temperature $k_B T = 0.1$ eV, the geometry factor $\delta = 0.5$, and the effective initial beam radius in the focusing channel $a_i = 5$ cm.

6.3 The nonstationary waterbag (WB) distribution listed in Table 5.1 has a nonuniform density profile of the form

$$n(r) = \begin{cases} n_1\left[1 - \left(\frac{r}{a_1}\right)^2\right] & \text{for } 0 \leq r \leq a_1, \\ 0 & \text{for } a_1 < r < b. \end{cases}$$

The beam radius a_1 and the drift-tube radius b are constant (i.e., independent of the axial coordinate z).

(a) Find the electric field, $E(r)$ and the electrostatic energy per unit length, w_E.
(b) For comparison, calculate $E(r)$ and w_E for a beam with uniform density

$$n(r) = \begin{cases} n_0 & \text{for } 0 \leq r \leq a_0, \\ 0 & \text{for } a_0 \leq r \leq b. \end{cases}$$

(c) Find the relation between n_1 and n_0, and a_1 and a_0 if both beams have the same number of particles per unit length (i.e., the same current) and the same rms radius $\tilde{r} = (\overline{r^2})^{1/2}$. Show that the difference in field energy between the two beams is

$$\Delta w_E = U = 0.0224 w_0,$$

where w_0 is defined in connection with Equation (6.14).

6.4 Derive the relations (6.68) for the longitudinal space charge field E_{sz} and (6.69a) for the g-factor in an emittance-dominated beam by applying Faraday's law and Ampere's circuital law to the beam configuration shown in Figure 6.18. Read the discussion preceding and following these equations.

6.5 Derive the solutions (6.86a) and (6.86b) from the relation (6.85).

6.6 Derive the expressions for the real part R_w^* and the imaginary part X_w^* of the transmission-line impedance Z_w^* given in Equation (6.104). Discuss the dependence on ω and k, draw a diagram ω versus k, and show qualitatively the lines $\omega(k)$ for the slow and fast space-charge waves.

6.7 Find the solutions (6.105a) and (6.105b) from the dispersion relation (6.103a).

6.8 Consider the relation (6.105a) or (6.105b) and show that an inductive transmission-line impedance enhances the growth rate while a capacitive impedance decreases the growth rate of the longitudinal instability. Under what conditions can the beam still be unstable (i.e., $k_i > 0$) even if $R_w^* = 0$?

6.9 Consider the dispersion relations (6.124) and (6.126) for the situation where $R_w = 0$. Discuss all possible combinations of $\eta > 0$, $\eta < 0$, $X_w > 0$ (inductive wall impedance), $X_w < 0$ (capacitive wall impedance), $|X_w| < X_s$, $|X_w| > X_s$, and compare the growth rates with each other and with the case $X_w = 0$.

6.10 In a beam vacuum tube with a smooth wall, the impedance Z_w^* is determined by the skin effect. As long as the wall thickness is larger than the skin depth δ_s we have a resistive (real) component and an inductive (imaginary) component of equal magnitude. The complex wall impedance is given by

$$Z_w^* = \frac{(1+i)}{2\pi b} \frac{1}{\sigma \delta_s} = R_w^* + i\omega L_w^*.$$

Here b is the radius of the vacuum tube, σ the conductivity of the wall material in Ω^{-1}/m, and δ_s the skin depth defined by

$$\delta_s = \left(\frac{2}{\omega \mu \sigma}\right)^{1/2},$$

where μ is the permeability of the material in H/m. (In nonmagnetic materials, $\mu \approx \mu_0 = 4\pi \times 10^{-7}$ H/m.)
Evaluate the longitudinal stability of a cold 50-MeV 1-kA electron beam with radius $a = 0.5$ cm in an aluminum tube ($\sigma = 3.26 \times 10^{-7}\sqrt{\omega/2\pi}\ \Omega^{-1}/\text{m}$) with radius $b = 2.5$ cm.

6.11 Show that the dispersion relation (6.124) is identical to (6.85) when the transition from circular to straight beam is made.

6.12 Using the proper definitions for the temperature in a circular machine, show that Equation (6.164) is identical to (6.160b).

6.13 Carry out the details of the derivations leading to Equation (6.52).

6.14 Derive the equations corresponding to Equation (6.52) for the perturbations (ac components) of the velocity, space-charge density, and current density using the first-order relations

$$v_{ac} = v_0 \frac{\partial s}{\partial z} + \frac{\partial s}{\partial t},$$

$$\rho_{ac} = -\rho_0 \frac{\partial s}{\partial z},$$

$$J_{ac} = \rho_0 v_{ac} + v_0 \rho_{ac}.$$

Plot the real part of J_{ac} as a function of distance z for one instant of time ($e^{i\omega t} = 1$) in the interval $0 < z < \lambda_p - 2\pi/\omega_p$, and show that the first peak occurs at $z = \lambda_p/4$.

Example of a Pierce-Type Electron Gun with Shielded Cathode

Electron guns with magnetically shielded cathodes producing high-perveance solid beams are used for high-power microwave devices such as klystrons and traveling-wave tubes, electron linacs, and many other applications. Since the effect of temperature or emittance on the beam radius is negligibly small compared to the space-charge force, such beams can be modeled with good accuracy by the Brillouin laminar-flow theory discussed in Section 5.2.2 [case (a), Brillouin solid beam]. In most of those guns it is desirable to compress the beam radially so that it fits into the aperture of the rf cavity structure being employed for microwave generation or for acceleration of the beam. The compression is achieved by shaping the electrodes in the gun to introduce a focusing transverse electric field near the cathode and by utilizing the focusing action of the magnetic fringe field. By choosing appropriate electrode angles, one can obtain a transverse focusing force at the beam edge that either exactly balances the space-charge force to keep the radius constant or exceeds the space charge to obtain a converging beam (for radial compression) in the gun region. This concept was first proposed and examined theoretically by Pierce (see [C.3, Chap. X]). It is used in all modern high-perveance, solid-beam electron guns as well as in the design of ion sources. In the literature these devices are known as electron guns or ion sources with *Pierce-type geometry*, or in the electron case, simply as *Pierce guns*. A schematic of an electron gun with Pierce-type geometry producing a parallel beam is given in Figure 1.1. In this particular case, a mesh covers the anode hole to suppress the transverse electric field components, which would defocus the beam in this aperture region if the mesh were absent.

564

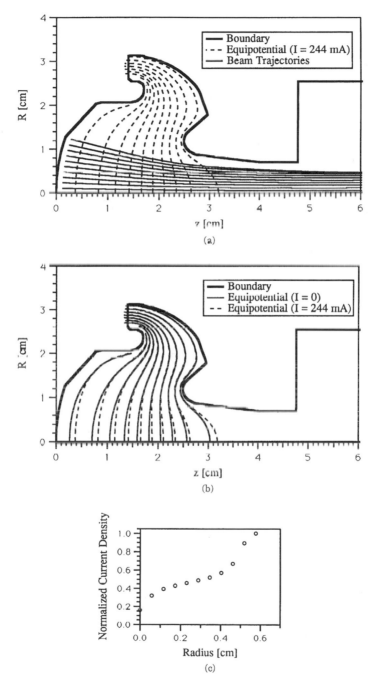

Figure A.1.1. (a) Electrode boundaries, equipotential contours, and particle trajectories in a compression-type Pierce gun; (b) equipotentials with and without space charge; (c) current density versus radius at the waist ($z = 6$ cm) of the beam. (Courtesy of D. Kehne.)

The electron gun employed in the beam transport experiments described in Section 6.2.2 and shown in Figure 6.8 represents an example of a compression-type Pierce gun. It is a reduced version (scale factor 1:5) of a SLAC-type klystron gun, designed with the help of W. Herrmannsfeldt and built by the Hughes Electron Dynamics Division in 1980. The dispenser-type cathode disk is concave and has a diameter of 1 inch. Beam current and voltage can be changed over a wide range, with 200 to 250 mA and 5 kV being typical for most experiments. The contours of the electrode geometry and equipotential surfaces as well as typical electron trajectories are shown in Figure A1.1. The potential and trajectory computations were performed for a 5-kV, 244-mA beam with Herrmannsfeldt's code, which is widely used in electron gun and ion source design [1]. Figure A1.1(a) shows the equipotentials and trajectories for the full beam (244 mA). In Figure A1.1(b) are the equipotential contours without the beam ($I = 0$) and with the beam ($I = 244$ mA). From the curvature of these contours one can envision the electric field lines, which have a focusing radial component E_r from the cathode, where E_r is a maximum, to a distance of $z \approx 2$ cm, where E_r becomes zero. In the region $z > 2$ cm of the anode aperture the radial electric field has a defocusing polarity, but the overall effect of the electric field in this gun is focusing, producing a converging beam that reaches a waist radius of about 0.6 cm at a distance of 6 cm from the cathode. On closer inspection one finds that the focusing electric field is nonlinear and has a strong third-order (spherical aberration) component, which is typical for these electrostatic lenses. A laminar beam adjusts its density profile such that the transverse space-charge force exactly balances the external focusing force. Consequently, the beam assumes the hollow shape shown in Figure A1.1(c), where the computed current density is plotted versus radius at the waist position.

Measurements confirm these theoretical expectations and computer results [2]. Indeed, most electron guns with such a Pierce-type geometry produce hollow beam profiles. When such beams are injected into a linear focusing channel where the equilibrium state has a uniform density profile (in the laminar limit), emittance growth occurs due to conversion of free energy (see Section 6.2.2). Note that Problems 5.17 and 6.2 relate to this hollow-beam phenomenon.

REFERENCES

1. W. B. Herrmannsfeldt, "Electron Trajectory Program," *SLAC Report 226*, November 1979.
2. D. Kehne, "Experimental Studies of Multiple Electron Beam Merging, Mismatch and Emittance Growth in a Periodic Solenoid Channel," Ph.D. Dissertation, Electrical Engineering Department, University of Maryland, College Park, MD, April 1992.

APPENDIX 2

Example of a Magnetron Injection Gun

Magnetron injection guns (MIGs) have been used successfully in a number of different applications, including switch tubes [1] and microwave sources (gyrotrons) [2]. For the former application, the equilibrium is essentially as described in this book (see Figure 5.5). For the latter application, MIGs are used to generate beams that give up energy in a microwave circuit via the cyclotron resonance instability [3]. To achieve efficient operation, the required equilibrium differs from the one described in this book in several ways. First, the beam is usually tenuous and is dominated by magnetic field effects. Second, although individual electrons essentially perform helical orbits, each center of gyration is sufficiently large that the electrons never encircle the axis. Furthermore, although some designs produce beams that are laminar near the cathode, all gyrotron MIG beams eventually evolve to a phase-mixed [4] state where the orbits cross. Finally, the emitter strips do not necessarily follow the magnetic flux lines.

The electrode configuration for a high-power coaxial gyrotron MIG is shown in Figure A2.1. The applied axial magnetic field profile and sample ray trajectories for the beam are also indicated in the figure. The design beam voltage and current are 500 kV and 480 A, respectively. The average cathode radius is 7.5 cm and its slant angle with respect to the axis is approximately 37°. The evolution of the average ratio of the beam electron's perpendicular to parallel velocity is depicted in Figure A2.2(a). When the flow has nearly reached the anode plane ($z = 16$ cm), the beam is laminar, most of the energy is in the axial motion, and the spread in axial velocity is nearly zero [see Figure A2.2(b)]. However, as the beam progresses through the increasing magnetic field, energy is adiabatically pumped into the perpendicular motion until the velocity ratio reaches about $v_\perp / v_z = 1.5$. The average beam radius also decreases toward its final value of 2.6 cm. Toward the end of the MIG ($z \gtrsim 40$ cm), the electron orbits begin to cross and space-charge effects fuel a spread in axial velocity. The final rms spread of nearly 6.5% is well in the suitable range for efficient microwave production.

567

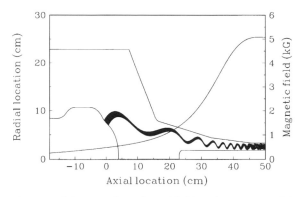

Figure A.2.1. Schematic of the MIG electrode and field configurations and the simulated beam trajectory. (Courtesy of W. Lawson.)

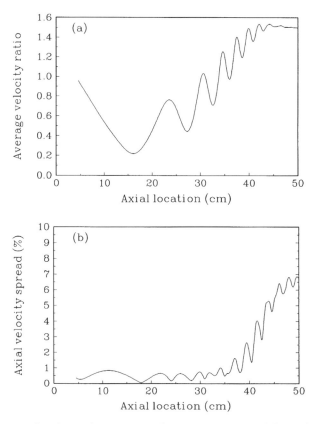

Figure A.2.2. Axial evolution of (a) average velocity ratio ($v_\perp/v_z$) and (b) axial velocity spread. (Courtesy of W. Lawson.)

REFERENCES

1. W. Lawson, private communication.

2. J. M. Baird and W. Lawson, "Magnetron Injection Gun Design for Gyrotron Applications," *Int. J. Electron.* **61,** 953–967 (1986).

3. J. Schneider, "Stimulated Emission of Radiation by Relativistic Electrons in a Magnetic Field," *Phys. Rev. Lett.* **2,** 504–505 (1959).

4. J. Neilson, M. Caplan, N. Lopez, and K. Felch, "Simulation of Space-Charge Effects on Velocity Spread in Gyro-Devices," *1985 IEDM Tech. Dig.,* 184–186 (1985).

Four-Vectors and Covariant Lorentz Transformations

The four-vector covariant form of Lorentz transformations is discussed in standard textbooks such as Panofsky and Phillips [A.1, Chaps. 17 and 18] or Jackson [A.4, Chap. 11]. For convenience, we will present here a few definitions and relations that are relevant to our work.

The four-momentum vector is defined as

$$p^i = \left(p_x, p_y, p_z, \frac{E}{c} \right), \tag{A3.1}$$

where $E = \gamma mc^2$. The Lorentz transformation from the laboratory frame (subscript l) to the beam (rest) frame (subscript b) for a beam moving in the positive z-direction is given by

$$p_b^i = Q_j^i p_l^j, \tag{A3.2}$$

where

$$Q_j^i = \begin{pmatrix} 1 & 0 & 0 & 0 \\ 0 & 1 & 0 & 0 \\ 0 & 0 & \gamma_0 & -\beta_0\gamma_0 \\ 0 & 0 & -\beta_0\gamma_0 & \gamma_0 \end{pmatrix}. \tag{A3.3}$$

Thus

$$
\begin{pmatrix} p_{bx} \\ p_{by} \\ p_{bz} \\ \dfrac{E_b}{c} \end{pmatrix} = \begin{pmatrix} 1 & 0 & 0 & 0 \\ 0 & 1 & 0 & 0 \\ 0 & 0 & \gamma_0 & -\beta_0\gamma_0 \\ 0 & 0 & -\beta_0\gamma_0 & \gamma_0 \end{pmatrix} \begin{pmatrix} p_{lx} \\ p_{ly} \\ p_{lz} \\ \dfrac{E_l}{c} \end{pmatrix}. \tag{A3.4}
$$

The center-of-momentum velocity of the beam measured in the laboratory frame is v_0, and $\beta_0 = v_0/c$, $\gamma_0 = (1 - \beta_0^2)^{-1/2}$.

The inverse transformation from beam frame to laboratory frame is

$$
p_{lj} = (Q_j^i)^{-1} p_{bj}, \tag{A3.5}
$$

with

$$
(Q_j^i)^{-1} = \begin{pmatrix} 1 & 0 & 0 & 0 \\ 0 & 1 & 0 & 0 \\ 0 & 0 & \gamma_0 & \beta_0\gamma_0 \\ 0 & 0 & \beta_0\gamma_0 & \gamma_0 \end{pmatrix}. \tag{A3.6}
$$

Hence

$$
p_{lx} = p_{hx}, \qquad p_{ly} = p_{hy}, \tag{A3.7a}
$$

$$
p_{lz} = \gamma_0 \left(p_{hz} + \beta_0 \frac{E_b}{c} \right), \tag{A3.7b}
$$

$$
E_l = \gamma_0 (E_h + \beta_0 c\, p_{hz}). \tag{A3.7c}
$$

Transformations of the type (A3.2) and (A3.5) are called *Lorentz covariant*. A four-vector quantity A_j that transforms as Equation (A3.5) is called a *covariant four-vector*. On the other hand, a quantity B^i that transforms as Equation (A3.2) is called a *contravariant four-vector*. By means of the relation

$$
B_i = g_{ij} B^j, \tag{A3.8}
$$

where

$$
g_{ij} = \begin{pmatrix} -1 & 0 & 0 & 0 \\ 0 & -1 & 0 & 0 \\ 0 & 0 & -1 & 0 \\ 0 & 0 & 0 & 1 \end{pmatrix}, \tag{A3.9}
$$

one can define for any contravariant four-vector B^j its covariant partner B_i, and vice versa. As we can see from the form of the matrix g_{ij}, the transformation (A3.8) merely changes the sign of the first three components of the four-vector.

The product of a contravariant four-vector and a covariant four-vector is *Lorentz invariant*. As an example, for the four-momentum one has

$$p^i p_i = \text{const},$$

or

$$-p_{bx}^2 - p_{by}^2 - p_{bz}^2 + \frac{E_b^2}{c^2} = -p_{lx}^2 - p_{ly}^2 - p_{lz}^2 + \frac{E_l^2}{c^2} = m^2 c^2, \quad \text{(A3.10)}$$

since $E^2/c^2 = p^2 + m^2 c^2$.

The four-vector potential A^j is composed of the three spatial components of the ordinary vector potential ($\mathbf{A}$) and the scalar potential divided by the speed of light (ϕ/c):

$$A^j = \left(A_x, A_y, A_z, \frac{\phi}{c} \right). \quad \text{(A3.11)}$$

The four-velocity is defined as

$$u^i = (\gamma v_x, \gamma v_y, \gamma v_z, \gamma c). \quad \text{(A3.12)}$$

Its covariant partner is

$$u_i = (-\gamma v_x, -\gamma v_y, -\gamma v_z, \gamma c), \quad \text{(A3.13)}$$

where γ is defined in terms of the total velocity $v = (v_x^2 + v_y^2 + v_z^2)^{1/2}$ as $\gamma = (1 - v^2/c^2)^{-1/2}$.

The Lorentz-invariant product $u^i u_i$ is

$$u^i u_i = -\gamma^2 v^2 + \gamma^2 c^2 = (-\gamma^2 \beta^2 + \gamma^2) c^2 = c^2 \quad \text{(A3.14)}$$

since $\gamma^2 - \gamma^2 \beta^2 = 1$.

The center-of-momentum four-velocity in the beam frame is

$$U_b^i = (0, 0, 0, c). \quad \text{(A3.15)}$$

The corresponding velocity in the laboratory frame is

$$U_l^j = (Q_j^i)^{-1} U_b^i = (0, 0, \beta_0 \gamma_0 c, \gamma_0 c), \quad \text{(A3.16)}$$

where $\gamma_0 = (1 - \beta_0^2)^{-1/2}$, $\beta_0 = v_0/c$, and v_0 is the center-of-momentum velocity along the z-direction in the laboratory frame. For completeness we also present the four-coordinate vector, which is defined as

$$x^i = (x, y, z, ct), \quad \text{(A3.17)}$$

and which also transforms like any other four-vector, such as p^i in (A3.2) or (A3.5).

APPENDIX 4

Equipartitioning in High-Current rf Linacs

As discussed in Section 6.3.1, collective instabilities due to coupling between the longitudinal and transverse direction via space-charge forces can cause emittance growth if the bunches in an rf linac have different longitudinal and transverse temperatures. This effect was first demonstrated in a theoretical study by Jameson [1]. Using a realistic model of a high-current deuteron rf linac and PARMILA code simulation of the particle dynamics, Jameson found that significant emittance growth occurred when the temperatures differed and that this growth became negligibly small when the beam was equipartitioned.

In related work, Hofmann analyzed the eigenmodes of an anisotropic K-V distribution due to coupling between two orthogonal directions [2]. He found that unstable collective modes occur if the tune depressions in both directions fall below a threshold curve that differs for each mode and depends on the ratio of the two-particle oscillation frequencies. This analysis is consistent with the results obtained by Jameson, who showed that the emittance growth observed in his computer simulation for anisotropic beams can be correlated with Hofmann's coupled instabilities.

These findings are also consistent with the thermodynamic description presented in this book. Most beams have different longitudinal and transverse temperatures. Various effects, such as mismatch, instabilities, and collisions, tend to drive the particle distribution toward three-dimensional thermal equilibrium. In space-charge-dominated beams, the relaxation times can be very short, as discussed in Section 6.2. The equipartitioning effect is therefore particularly strong in high-current rf linacs with anisotropic temperatures.

Theoretically, for a matched (stationary) bunch in a smooth-focusing system, the temperatures can be related to the rms beam widths and normalized rms emittances. From Equations (5.290a) and (5.317) one obtains for the ratio of the transverse

573

and longitudinal temperatures the relation

$$\frac{k_B T_\perp}{k_B T_\parallel} = \frac{\tilde{\epsilon}_{nx}^2}{\gamma_0 \tilde{x}^2} \frac{\gamma_0^3 \tilde{z}^2}{\tilde{\epsilon}_{nz}^2} = \gamma_0^2 \frac{\tilde{\epsilon}_{nx}^2}{\tilde{\epsilon}_{nz}^2} \frac{\tilde{z}^2}{\tilde{x}^2} . \tag{A4.1}$$

We can express this relation also in terms of the effective normalized emittances $\epsilon_{nx} = 5\tilde{\epsilon}_{nx}$, $\epsilon_{nz} = 5\tilde{\epsilon}_{nz}$ and the effective widths $a = \sqrt{5}\tilde{x}$, $z_m = \sqrt{5}\tilde{z}$ of the equivalent uniform-density ellipsoidal bunch (see Problem 5.21) as

$$\frac{T_\perp}{T_\parallel} = \gamma_0^2 \frac{\epsilon_{nx}^2}{\epsilon_{nz}^2} \frac{z_m^2}{a^2} . \tag{A4.2}$$

Alternatively, we can introduce the focusing wave numbers with space charge, k_x and k_z. Using $a' = k_x a$ and $z'_m = k_z z_m$, we get

$$\epsilon_{nx} = \beta_0 \gamma_0 \epsilon_x = \beta_0 \gamma_0 k_x a^2, \tag{A4.3}$$

$$\epsilon_{nz} = \beta_0 \gamma_0^3 \epsilon_{zz'} = \beta_0 \gamma_0^3 k_z z_m^2, \tag{A4.4}$$

so that Equation (A4.2) may be written in the form

$$\frac{T_\perp}{T_\parallel} = \frac{k_x \epsilon_{nx}}{k_z \epsilon_{nz}} . \tag{A4.5}$$

The beam is equipartitioned ($T_\perp = T_\parallel$) when

$$\gamma_0 \frac{\epsilon_{nx}}{\epsilon_{nz}} \frac{z_m}{a} = 1 , \tag{A4.6}$$

or, alternatively, when

$$\frac{\epsilon_{nx}}{\epsilon_{nz}} \frac{k_x}{k_z} = 1 . \tag{A4.7}$$

The wave numbers k_x and k_z depend on the beam widths a and z_m, as given in Equations (5.498) and (5.499), so that relation (A4.6) is more explicit than (A4.7). In any case, one must calculate a and z_m from the coupled envelope equations (5.496) and (5.497) for a given set of beam parameters.

If the beam is space-charge dominated and the conditions mentioned in Section 5.4.11 are satisfied, one obtains the analytical approximations (5.506) to (5.508) for a, z_m, and z_m/a, which are repeated here for easy reference:

$$a = \left[\left(\frac{3}{2} \right)^2 \frac{N r_c}{\beta_0^2 \gamma_0^2} \frac{1}{k_{z0}^2} \left(\frac{k_{x0}^2}{k_{z0}^2} + \frac{1}{2} \right)^{-2} \right]^{1/3} , \tag{A4.8}$$

$$z_m = \left[\frac{2}{3} \frac{N r_c}{\beta_0^2 \gamma_0^5} \frac{1}{k_{z0}^2} \left(\frac{k_{x0}^2}{k_{z0}^2} + \frac{1}{2} \right) \right]^{1/3} , \tag{A4.9}$$

$$\frac{z_m}{a} = \frac{2}{3\gamma_0} \left(\frac{k_{x0}^2}{k_{z0}^2} + \frac{1}{2} \right) , \tag{A4.10}$$

Using (A4.10), one can express the equipartitioning condition (A4.6) in the form

$$\frac{\epsilon_{nx}}{\epsilon_{nz}}\frac{2}{3}\left[\frac{k_{x0}^2}{k_{z0}^2}+\frac{1}{2}\right]=1.\qquad\qquad(A4.11)$$

The wave numbers k_{x0} and k_{z0} represent the external focusing forces in the transverse and longitudinal directions. They can be controlled to a certain extent while the emittances depend on the history of the beam. Solving (A4.11) for k_{x0}/k_{z0}, we obtain the relation

$$\frac{k_{x0}}{k_{z0}}=\left(\frac{3}{2}\frac{\epsilon_{nz}}{\epsilon_{nr}}-\frac{1}{2}\right)^{1/2}\qquad\qquad(A4.12)$$

for a space-charge-dominated beam.

As an example, consider a high-current rf linac accelerating protons from a nonrelativistic initial energy of 2 MeV ($\gamma_0\approx1$, $\beta_0=0.065$) to a relativistic final energy of 938 MeV ($\gamma_0\approx2$, $\beta_0=0.866$). Assume that the normalized longitudinal emittance is twice as large as the normalized transverse emittance, so that $\epsilon_{nz}/\epsilon_{nx}=2$. To satisfy the equipartitioning conditions, the transverse and longitudinal focusing strengths must be designed so that $k_{x0}/k_{z0}=\sqrt{2.5}\approx1.58$ at injection (2 MeV). If this beam is to remain equipartitioned and the emittance ratio does not change, the focusing-strength ratio must remain constant through the linac system to satisfy Equation (A4.12) as the energy γ_0mc^2 increases. The bunch size ratio has the values $z_m/a-2$ at injection, $z_m/a\approx1.33$ at $\gamma_0=1.5$, and $z_m/a=1$ at full energy, in agreement with the condition (A4.6). Note that the bunch eccentricity z_m/a becomes smaller with increasing energy and that in our particular example the bunch shape becomes spherical ($z_m=a$) at full energy in the lab frame.

So far, these calculations have been rather general, and we need to examine whether the equipartitioning conditions for the focusing-strength ratio can in fact be satisfied in practice [4]. Since rf linacs employ magnetic quadrupoles for transverse focusing, there is, in principle, no difficulty in varying the value of k_{x0}. On the other hand, one does not have much flexibility with the longitudinal focusing strength, which, according to Equation (5.395a), is defined by

$$k_{z0}=\left(-\frac{2\pi qE_m\sin\varphi_0}{\lambda mc^2\beta_0^3\gamma_0^3}\right)^{1/2}.\qquad\qquad(A4.13)$$

The synchronous phase angle φ_0 and the maximum electric field strength E_m are usually fixed so that $E_m\sin\varphi_0$ is constant. Consequently, k_{z0} varies with increasing energy as

$$k_{z0}\propto\frac{1}{(\beta_0\gamma_0)^{3/2}}.\qquad\qquad(A4.14)$$

The preferred design method in such rf linacs is to keep the transverse phase advance without space charge, σ_{x0}, constant. Since $k_{x0} \propto \sigma_{x0}/\beta_0\lambda$, we then have the scaling

$$k_{x0} \propto \frac{1}{\beta_0}. \tag{A4.15}$$

Consequently, we find for this linac design scenario that the focusing-strength ratio varies as

$$\frac{k_{x0}}{k_{z0}} \propto \beta_0^{1/2}\gamma_0^{3/2}. \tag{A4.16}$$

Clearly, this variation with energy does not satisfy Equation (A4.12). The beam does not remain equipartitioned, the transverse temperature $T_\perp$ becomes higher than the longitudinal temperature $T_\parallel$, and longitudinal emittance growth occurs, as observed in the computer simulation studies of Jameson [1] and more recently by Wangler et al. [3]. Since we have no control over the longitudinal focusing strength, it is clear from the above analysis that we must change the transverse focusing conditions so that the ratio k_{x0}/k_{z0} meets the requirements for equipartitioning. For the space-charge-dominated equipartitioned beam being discussed here, one finds from (A4.12) and (A4.14) that the transverse wave number k_{x0} should obey the scaling

$$k_{x0} \propto \frac{1}{(\beta_0\gamma_0)^{3/2}}. \tag{A4.17}$$

This implies that the transverse phase advance σ_{x0} must decrease with energy as

$$\sigma_{x0} \propto \frac{1}{(\beta_0\gamma_0^3)^{1/2}}. \tag{A4.18}$$

and not remain constant, as is usually the case in many rf linac designs.

The required decrease with energy of the transverse focusing strength to keep the beam equipartitioned has the consequence that like z_m, the bunch radius a increases along the linac. By substituting (A4.11) into (A4.8), (A4.9) , one obtains the relations

$$a = \left[\frac{\epsilon_{nx}^2}{\epsilon_{nz}^2} \frac{1}{k_{z0}^2} \frac{Nr_c}{\beta_0^2\gamma_0^2} \right]^{1/3}, \tag{A4.19}$$

$$z_m = \left[\frac{\epsilon_{nz}}{\epsilon_{nx}} \frac{1}{k_{z0}^2} \frac{Nr_c}{\beta_0^2\gamma_0^5} \right]^{1/3}, \tag{A4.20}$$

for a space-charge-dominated equipartitioned linac. For a given emittance ratio and particle number, one then gets the scaling

$$a \propto \frac{1}{k_{z0}^{2/3}} \frac{1}{\beta_0^{2/3} \gamma_0^{2/3}} \tag{A4.21}$$

$$z_m \propto \frac{1}{k_{z0}^{2/3}} \frac{1}{\beta_0^{2/3} \gamma_0^{5/3}} \tag{A4.22}$$

or, in view of (A4.14),

$$u \propto \beta_0^{1/3} \gamma_0^{1/3} \tag{A4.23}$$

$$z_m \propto \frac{\beta_0^{1/3}}{\gamma_0^{2/3}} . \tag{A4.24}$$

Thus, in the linac example above, the bunch size increases from 2 MeV to 938 MeV by the factors

$$\frac{u(938 \text{ MeV})}{a(2 \text{ MeV})} = 2.98$$

in radius and by

$$\frac{z_m(938 \text{ MeV})}{z_m(2 \text{ MeV})} = 1.49$$

in the axial length.

The increase in beam radius is particularly troublesome. The bore radius b of the drift tubes is usually fixed, and one wants to maintain a safe ratio of say $b/a \geq 5$ to avoid particles from striking and (if the energy is high enough) activating the tube walls. This problem can be alleviated somewhat by making a transition at an appropriate energy to a linac operating at twice the frequency (i.e., half the wavelength) as the injector linac.

Since the electrical breakdown threshold in rf systems increases with frequency, one can increase the accelerating field accordingly, and the scaling is approximately given by $E_m \propto (1/\lambda)^{1/2}$. Combined with the factor λ in the denominator of (A4.13), one thus finds that k_{z0} scales with the rf wavelength as

$$k_{z0} \propto \frac{1}{\lambda^{3/4}}, \tag{A4.25}$$

so that the radius varies as

$$a \propto \lambda^{1/2}. \tag{A4.26}$$

Thus by decreasing the wavelength by a factor of 2, the increase in the bunch radius can be reduced to

$$\frac{a(938 \text{ MeV}, \lambda/2)}{a(2 \text{ MeV}, \lambda)} = \frac{2.98}{\sqrt{2}} = 2.11.$$

The bunch length is, of course, also reduced by the factor $1/\sqrt{2}$. In such a scenario a bunch radius of, say, $a = \sqrt{5}\tilde{x} = 2$ mm at 2 MeV would then grow to $a = 4.22$ mm at full energy. In view of (A4.23), most of the bunch-size increase occurs in the low-energy part of the linac system. Thus it would be important to make the transition to the high-frequency linac at low enough energy that the radius does not exceed the above limit. In our case this transition point occurs at an energy of about 162 MeV. Nevertheless, it is not clear whether such a significant increase' in bunch radius is tolerable in practice. As already mentioned, the customary design philosophy is to increase the transverse focusing and hence to reduce the radius with rising energy. Consequently, the temperature anisotropy $T_\perp/T_\parallel$ and the bunch eccentricity z_m/a become larger. The price one pays is an increase in longitudinal emittance. For the future high-power linacs being considered, this conventional approach may not be acceptable, and a design of a system that is either equipartitioned, if this is feasible, or at least closer to thermal equilibrium may be required.

In closing, we note again that the above analysis is based on three assumptions: (1) the bunch is perfectly matched in the transverse and longitudinal directions and the acceleration process is adiabatic; (2) the beam is space-charge dominated through the entire linac; (3) the values for the bunch eccentricity z_m/a and for the ratio b/a of drift-tube radius b to bunch radius a are in the range where the approximation $\overline{g} \approx g_0 \approx 2\gamma_0 z_m/3a$ for the geometry factor is valid. If assumption (1) is not satisfied, the free energy associated with mismatch will lead to emittance growth in both directions due to the space-charge coupling forces. If assumptions (2) and (3) are not satisfied, the envelope equations for a and z_m must be solved numerically, as discussed in Section 5.4.11. On the other hand, if the effect of emittance on beam size is not negligible, so that assumption (2) is not met, but (1) and (3) hold, we find the approximate solutions

$$a \approx \left[\left(\frac{3}{2}\right)^2 \frac{N r_c}{\beta_0^2 \gamma_0^2} \frac{1}{k_{z0}^2} \left(\frac{k_{x0}^2}{k_{z0}^2} + \frac{1}{2}\right)^{-2} + \left(\frac{\epsilon_{nx}}{\beta_0 \gamma_0 k_{x0}}\right)^{3/2} \right]^{1/3}, \qquad (A.4.27)$$

$$z_m \approx \left[\frac{2}{3} \frac{N r_c}{\beta_0^2 \gamma_0^5} \frac{1}{k_{z0}^2} \left(\frac{k_{x0}^2}{k_{z0}^2} + \frac{1}{2}\right) + \left(\frac{\epsilon_{nz}}{\beta_0 \gamma_0^3 k_{z0}}\right)^{3/2} \right]^{1/3}. \qquad (A.4.28)$$

By dividing these two equations and using the equipartitioning condition (A4.6), one obtains a single equation that defines the parameter space for which the beam is in thermal equilibrium (i.e., $T_\perp = T_\parallel$). Alternatively, we can solve either of the two equations to get the relation for the particle number N or average current $\bar{I} = qNf = qNc/\lambda$. Thus, we obtain from (A4.27) for the average beam current

$$\bar{I} = \left(\frac{2}{3}\right)^2 I_0 \frac{a^3}{\lambda} \beta_0^2 \gamma_0^2 k_{z0}^2 \left(\frac{k_{x0}^2}{k_{z0}^2} + \frac{1}{2}\right)^2 \left[1 - \left(\frac{\epsilon_{nx}}{\beta_0\gamma_0 k_{x0}}\right)^{3/2} \frac{1}{a^3}\right], \qquad (A4.29)$$

where $I_0 = 4\pi\epsilon_0 mc^3/q$, as defined in Equation (4.17). The bunch radius a and half length z_m are of course coupled by the two envelope equations and cannot be chosen independently. In the space-charge dominated case, where the emittance term in the bracket is negligibly small compared to unity and the current has its maximum, we can replace a by z_m from Equation (4.10), and use $z_m = \beta_0\lambda(\Delta\varphi_m/2\pi)$ [Equation (5.401)] to introduce $\Delta\varphi_m$ and (A4.13) for k_{z0}^2. This yields the relation

$$\bar{I}_{\max} = \frac{\pi}{10}\lambda\beta_0^2\gamma_0^2\left(\frac{\Delta\varphi_m}{2\pi}\right)^3 \frac{E_m|\sin\varphi_0|}{(k_{x0}^2/k_{z0}^2) + 1/2}. \qquad (A4.30)$$

If, furthermore, the beam is equipartitioned we can use (A4.12) and write this relation for the maximum current in the form

$$\bar{I}_{\max} = \frac{2\pi}{30}\lambda\beta_0^2\gamma_0^2\left(\frac{\Delta\varphi_m}{2\pi}\right)^3 E_m|\sin\varphi_0|\frac{\epsilon_{nx}}{\epsilon_{nz}}. \qquad (A4.31)$$

Since $E_m \propto 1/\sqrt{\lambda}$, we get for fixed values of $\Delta\varphi_m$ and φ_0, the scaling $I_{\max} \propto \lambda^{1/2}\beta_0^2\gamma_0^2(\epsilon_{nx}/\epsilon_{nz})$. If φ_0 and radius a are fixed, we have $\bar{I} \propto \lambda^{-5/2} \times (\beta_0\gamma_0)^{-1}(\epsilon_{nz}/\epsilon_{nx})^2$, $\Delta\varphi_m \propto (\lambda\beta_0\gamma_0)^{-1}(\epsilon_{nz}/\epsilon_{nx})$. Thus, the allowable maximum values for $\Delta\varphi_m$ and a determine the maximum current for an equipartitioned beam.

In the parameter regime where image forces become important, the equilibrium bunch shape must be determined self-consistently by numerical methods [5,6]. One finds, for instance, that in the zero-temperature limit the bunch assumes a non-ellipsoidal equilibrium boundary. This restores the linearity of the total space-charge forces (including images) and the balance with the applied linear forces that is required for the $T = 0$ equilibrium state while keeping the charge density ρ_0 constant [5]. The equivalent linear ellipsoid bunch in Section 5.4.11, and the parabolic line-charge model and g-factor calculations (Fig. 5.15) in Section 5.4.7 are good approximations that can be used to determine the average (rms) bunch behavior and dimensions for the general case with images and different temperatures. Figure A4.1 shows the self-consistent, numerically calculated Boltzmann density profile for a spherically symmetric, space-charge dominated (low-temperature) bunch (for $\gamma_0 = 1$) in thermal equilibrium (see figure for details). It illustrates that the uniform-density ellipsoid is a good model for such bunches.

If thermal equilibrium is not possible within the constraints of a particular design, one should try to minimize the deviation and the associated emittance growth, as was done in the computer studies mentioned above [1, 3]. In any case, our analysis and the relations derived in this section should serve as a useful guide for designers of rf linacs.

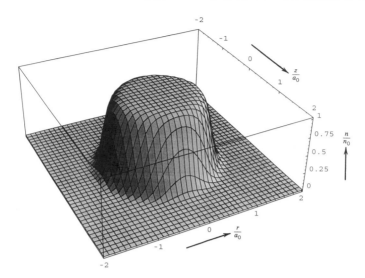

Figure A4.1. Self-consistent three-dimensional Boltzmann density profile of a spherically symmetric bunch in thermal equilibrium in the space-charge dominated (low-temperature) regime with tune depression of $k_x/k_{x0} = k_z/k_{z0} = 0.3$. The density $n(r,z)$ and coordinates r, z are in units of the density n_0 and the radius a_0 of the zero-temperature case (ideal uniform ellipsoid), respectively. Boundary is at $b = 2a_0$, but has no significant effect in this case. (Courtesy N. Brown.)

REFERENCES

1. R. A. Jameson, *IEEE Trans. Nucl. Sci.* **NS-28,** 2408 (1981).

2. I. Hofmann, *IEEE Trans. Nucl. Sci.* **NS-28,** 2399 (1981).

3. T. P. Wangler, T. S. Bhatia, G. H. Neuschaefer, and M. Pabst, *Conference Record of the 1989 IEEE Particle Accelerator Conference,* 89CH2669-0, March 20–23, 1989, p. 1748.

4. R. A. Jameson, *AIP Conference Proceedings* **279,** 969 (1993), ed. J. S. Wurtele.

5. C. K. Allen and M. Reiser, "Zero-Temperature Equilibrium for Bunched Beams in Axisymmetric Systems," CPB Technical Report #94-005, March 25, 1994, Institute for Plasma Research, University of Maryland, College Park, MD 20742; to be published.

6. N. Brown, "Three-Dimensional Thermal Distribution for Bunched Beams," CPB Technical Report #94-006, April 4, 1994, Institute for Plasma Research, University of Maryland, College Park, MD 20742; to be published.

Radial Defocusing and Emittance Growth in High-Gradient rf Structures (Example: The rf Photocathode Electron Gun)

In Sections 1.3 and 6.6 we mentioned the laser-driven photocathode electron gun as a new high-brightness electron beam source for linear colliders, free electron lasers (FELs) and other advanced accelerator applications [1–3].

The photocathode is located inside of a high-gradient (20–100 MV/m) rf resonant cavity structure operating in a TM_{01} mode and consisting of $(n + 1/2)\lambda/2$ cells, where n is an integer. Most rf guns are designed with $n = 1$ and $n = 2$; at Los Alamos the rf gun is an integral part of a high-gradient rf linac for FEL experiments with $n = 10$. Figure A5.1 shows the schematic of a two-and-a-half cell design, with the electron bunch and lines of force during the accelerating half-cycle in the first cell. The laser beam is focused on the photocathode at an angle with

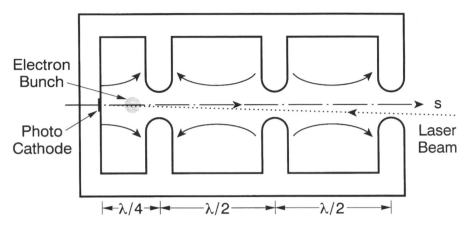

Figure A5.1. Schematic of a laser-driven rf photocathode electron gun in a 5λ/4 rf structure.

respect to the cavity axis through the apertures in the rf structure from downstream or through a special port in the cavity wall. The timing and pulse length of the laser is designed to produce short electron bunches during the rising part of the accelerating field. Resonator frequencies are typically in the range of 0.4 to 3 GHz.

A major problem in these rf guns is emittance growth due to rf defocusing and space charge. This problem was first analyzed theoretically by Kim [4] and Kim and Chen [5], by McDonald [6], Sarafini [7], and others; there has also been extensive computer simulation work. (See Reference [3] for a review of recent studies.) There is, in general, reasonably good agreement between simulation and emittance measurements while the analytical theory tends to overestimate the emittance growth significantly [3]. A major reduction of emittance growth can be achieved by using solenoidal focusing, first suggested by Carlston [8] and experimentally verified at Los Alamos [9,10].

A more detailed description of rf guns and related ongoing research is beyond the scope of this book. We will instead present a brief and more general discussion that extends the theoretical concepts developed in Sections 5.4 (especially 5.4.7, 5.4.8, and 5.4.11), 6.2 and Appendix 4 to the behavior of intense bunched electron beams in high-gradient, high-frequency rf linac structures like the one depicted in Figure A5.1. The situation here differs from the self-consistent treatments given in these preceding sections in several ways:

1 The strong electric fields accelerate the electrons very rapidly to relativistic energies; the motion can in general no longer be treated as adiabatic, and the equations governing the beam physics must be solved numerically or by simulation.

2 The radially defocusing rf forces due to the high electric fields, which we tacitly ignored in our preceding discussions, may cause significant emittance growth unless strong external magnetic focusing is applied.

3 Thermodynamically, the electron bunches are far from three-dimensional equilibrium; the drive towards an equipartitioned state via the coupled space-charge forces is offset by the rapid acceleration.

4 At increasingly relativistic energies wakefield effects come into play. These are treated in Chao's book (D.11) and are included in the simulation codes. We are limiting ourselves here to electron acceleration in the rf injector systems (i.e., energies in the range of a few MeV). This is where space-charge effects are most pronounced and wakefields can be neglected.

Let us now begin our theoretical analysis with a review of radial defocusing in rf fields. We will consider an axisymmetric TM_{01}-type standing wave in a structure such as the one depicted in Figure A5.1. Space-charge effects will be ignored first, but included subsequently. Let the axial electric field be of the form

$$E_z(s, t) = E_m(s) \cos ks \cos \omega_{rf} t. \tag{A5.1}$$

Here s denotes the distance of travel along the system and $z = s - s_0$ will be used to define the relative position of a particle with respect to the bunch center, s_0, as in Section 5.4.8. Furthermore, $k = 2\pi/\lambda = \omega_{rf}/c$ is the wavenumber, λ the wavelength, and ω_{rf} the radian frequency. The peak field $E_m(s)$ is essentially constant except near the apertures and at the exit of the cavity structure (where it falls to zero). From Maxwell's equations one obtains for the radial electric and azimuthal magnetic field the first-order relations

$$E_r(s, t) = -\frac{r}{2} \frac{\partial E_z}{\partial s}, \quad B_\theta(s, t) = \frac{r}{2c^2} \frac{\partial E_z}{\partial t}. \tag{A5.2}$$

The radial force is given by

$$F_r = \frac{dP_r}{dt} = qE_r - qvB_\theta, \tag{A5.3}$$

and, using the first two equations, we obtain (with $\beta = v/c$)

$$\frac{dP_r}{dt} = -\frac{qr}{2}$$

$$\left[\frac{\partial E_m}{\partial s} \cos ks \cos \omega_{rf} t - E_m k \sin ks \cos \omega_{rf} t - \frac{\omega_{rf}}{c} \beta E_m ks \cos \omega_{rf} t \right]. \tag{A5.4}$$

By introducing the phase of the particle with regard to the rf wave,

$$\varphi = \omega_{rf} t - ks, \tag{A5.5}$$

we an write Equation (A5.4) in the alternative form

$$\frac{dP_r}{dt} = \frac{r}{2}kqE_m$$

$$\cdot\left[-\frac{1}{E_m}\frac{\partial E_m}{\partial s}\cos ks\,\cos(ks + \varphi) + \frac{1}{2}(1 - \beta)\,\sin(2ks + \varphi) - \frac{1}{2}(1 + \beta)\,\sin \varphi\right].$$

$$(A5.6)$$

This equation shows that the radial momentum change depends on the phase φ with respect to the wave, i.e., on the relative position of a particle within the bunch. It can be integrated if the variation of the velocity βc and the phase φ of the particle are obtained by solving independently the longitudinal momentum and phase equations, (5.378) and (5.376a). The form of Equation (A5.6) is sufficiently general that it can be applied to different types of rf systems. A detailed analysis shows that the radial rf forces are generally defocusing in the phase interval $-\pi/2 < \varphi < 0$ required for longitudinal focusing (see the discussion in Section 5.4.8). Furthermore, this effect is significant only in low-energy ion linacs and in the high-gradient structures of electron injector linacs, rf guns and bunching systems. We will limit our discussion to electron-beam defocusing in high-gradient structures where the electrons are rapidly accelerated to relativistic velocities so that $\beta \approx 1$ can be assumed. The term involving $\partial E_m/\partial s$ is usually not very important in this case, and we will neglect it. With these assumptions we obtain from (A5.6) to good approximation the result

$$\frac{dP_r}{dt} = -\frac{r}{2}keE_m\,\sin \varphi,\tag{A5.7}$$

where we introduced the electron charge $(q = e)$.

If φ_0 denotes the phase of the bunch centroid, $\varphi = \varphi_0 + \Delta\varphi$ the phase of any other particle, and if the bunch width is short compared to the rf period, we can express Equation (A5.7) as

$$\frac{dP_r}{dt} = -\frac{r}{2}\,keE_m\sin(\varphi_0 + \Delta\varphi) \approx -\frac{r}{2}keE_m[\sin\,(\varphi_0 + \Delta\varphi)\,\cos\,\varphi_0],$$

$$(A5.8)$$

This relation shows very clearly that the rf force is defocusing radially ($\sin\,\varphi_0 < 0$) in the phase interval $-\pi/2 < \varphi < 0$ where the accelerating force is rising with time, and furthermore, that the defocusing force depends on the particles relative position $\Delta\varphi$ within the bunch. Using $dt = ds/v \approx ds/c$, and assuming $r \approx \text{const}$, $P_r = 0$ at $s = 0$, we can integrate Equation (A5.7) over the length of a cavity structure, say from $s = 0$ to $s = 5\lambda/4$ in the case of Figure A5.1, and obtain

$$P_r = \frac{erE_m}{2c}\,\cos\,\varphi_1,\tag{A5.9}$$

where φ_1 denotes the phase of the particle at the cavity entrance. In a more accurate calculation taking into account that $\beta < 1$ near the entrance (cathode of rf gun), one finds that it is better to take φ_1 as the phase at the cavity exit [5]. We can express φ_1 as $\varphi_1 = \varphi_0 + \Delta\varphi$ and $\Delta\varphi \approx -k(s - s_0) = -kz$ and write (A5.9) in terms of P_x, x in place of P_r, r. One then obtains from (A5.9) for the momentum difference $\Delta P_x = P_x(x, s) - P_x(x, s_0)$ between a particle at position (x, s) in the bunch and the bunch centroid (x, s_0) the result

$$\Delta P_x - -\frac{eE_m k}{2c}(\sin \varphi_0)xz, \tag{A5.10a}$$

or

$$\Delta P_x = \frac{eE_m k}{2c}|\sin \varphi_0|xz \quad \text{for} \quad 0 < \varphi_0 < \frac{\pi}{2}. \tag{A5.10b}$$

This relation for the rf defocusing effect shows the transverse-longitudinal coupling (xz) that causes an undesirable increase in the effective emittance. By averaging over the entire particle phase-space distribution in the bunch, we obtain for the increase of the normalized rms emittance the expression [5]

$$\Delta \tilde{\epsilon}_{nx}^{rf} - \frac{\left[\overline{x^2 \Delta P_x^2} - \overline{x\Delta P_x}^2\right]^{1/2}}{mc} - \frac{eE_m k}{2mc^2}\tilde{x}^2\tilde{z}|\sin \varphi_0|, \tag{A5.11}$$

where $\tilde{z} - (\overline{z^2})^{1/2}$ denotes the longitudinal rms width of the beam and $\tilde{x}.\tilde{P}_x$ are the rms width and rms momentum spread in the x-direction.

If $\tilde{\epsilon}_{nx}^{th}$ denotes the intrinsic (thermal) normalized rms emittance, we obtain for the total emittance in an rf gun (not including space-charge effects)

$$\tilde{\epsilon}_{nx} = \left[(\tilde{\epsilon}_{nx}^{th})^2 + (\Delta\epsilon_{nx}^{rf})^2\right]^{1/2}. \tag{A5.12}$$

In view of the high peak fields in rf guns the effective emittance increase due to rf defocusing can be very significant. As an example, suppose that $\tilde{\epsilon}_{nx}^{th} = \tilde{x}(kT/mc^2)^{1/2} = 3 \times 10^{-6}$ m-rad, $E_m = 50$ MV/m, $\tilde{x} = 2$ mm, $\tilde{z} = \lambda/50 = 2\pi/50k$, $\varphi_0 = -30°$. With these numbers one gets from (A5.12) an effective emittance increase due to rf defocusing of $\tilde{\epsilon}_{nx}/\tilde{\epsilon}_{nx}^{th} = 4.1$, which is quite dramatic. In practice, this effect could be much worse in view of radial beam expansion due to rf defocusing and space charge forces. The obvious answer is that the beam must be confined radially by strong magnetic focusing forces (from solenoid or quadrupole lenses) which significantly exceed the rf defocusing forces. In a solenoid, for instance, the radial focusing force is given by

$$\frac{dP_r}{dt} = -r\frac{eB^2}{2\gamma m}, \tag{A5.13}$$

and the ratio of the rf defocusing force and solenoidal focusing force is from (A5.7) with $\varphi = \varphi_0$ and (A5.13) found to be

$$\frac{F_r(\text{rf})}{F_r(\text{sol})} = \frac{E_m \pi \gamma mc^2 |\sin \varphi_0|}{\lambda e B^2 c^2} \tag{A5.14}$$

For a system with $E_m = 50$ MV/m, $\lambda = 0.015$ m, $|\sin \varphi_0| = 0.5$ and an electron energy of 1 MeV ($\gamma = 3$) a magnetic field strength of $B = 0.4$ T makes this ratio less than 0.1.

How can we now incorporate the rf defocusing and rapid acceleration in high-gradient structures into our self-consistent theory of bunched beams with space charge? In view of what has been said in points 1–4 above, the adiabatic coupled envelope equations of Section 5.4.11 and Appendix 4 cannot be used without qualification even if rf defocusing is added. We must instead turn to the more general envelope equations that include the effect of acceleration and rf defocusing. Thus, one obtains for the radial motion with $ds/dt = \beta_0 c$

$$\frac{dP_r}{dt} = \frac{ds}{dt}\frac{dP_r}{ds} = mc^2(\gamma_0' r' + \beta_0^2 \gamma_0 r''), \tag{A5.15}$$

and hence [see Equation (4.78)]

$$r'' + \frac{\gamma_0' r'}{\beta_0^2 \gamma_0} + \kappa_0 r - \frac{keE_m}{mc^2}\frac{|\sin \varphi_0|}{\beta_0^2 \gamma_0}\frac{r}{2} + \frac{K}{a^2}r = 0, \tag{A.16}$$

where the third term represents the rf defocusing effect. Using relation (5.493) for the perveance K, we then obtain in place of (5.494) the radial envelope equation

$$a'' + \frac{\gamma_0' a'}{\beta_0^2 \gamma_0} + k_{x0}^2 a - \frac{\pi e E_m |\sin \varphi_0|}{\lambda mc^2 \beta_0^2 \gamma_0}a$$

$$-\frac{3}{2}\frac{Nr_c}{\beta_0^2 \gamma_0^3}\frac{1}{az_m}\left(1 - \frac{g}{2}\frac{a^2}{\gamma_0^2 z_m^2}\right) - \frac{\epsilon_{nx}^2}{\beta_0^2 \gamma_0^2 a^3} = 0. \tag{A5.17}$$

For the longitudinal motion we must retain the general form (5.384) without the adiabatic assumption. The corresponding envelope equation replacing (5.495) is then

$$\frac{d}{ds}\left[\beta_0^2 \gamma_0^3 \frac{dz_m}{ds}\right] + \beta_0^2 \gamma_0^3 k_{z0}^2 z_m - \frac{3}{2}\frac{gNr_c}{\gamma_0^2 z_m^2} - \frac{\epsilon_{nz}^2}{\gamma_0^3 z_m^3} = 0. \tag{A5.18}$$

The corresponding adiabatic equations are obtained by setting $\gamma_0' = 0$ in (A5.17) and $(\beta_0^2 \gamma_0^3 z_m')' = \beta_0^2 \gamma_0^3 z_m''$ in (A5.18). Note that the semi-axes of the elliptical bunch arc related to the rms width by $a = \sqrt{5}\tilde{x}$, $z_m = \sqrt{5}\tilde{z}$, and the emittances by $\epsilon_n = 5\tilde{\epsilon}_n$.

To integrate the general coupled envelope equations we must solve simultaneously the two equations for the change of the energy $\gamma_0 mc^2$ (5.379) and phase φ_0(5.376) for the beam centroid. The set of the four equations [(A5.17), (A5.18), (5.376), and (5.379)] determines self-consistently the bunch widths a and z_m for a given number of particles, N, radial and longitudinal focusing forces, emittances, rf structure parameters E_m, λ, g-factor curves (Figure 5.15 and Table 5.3), and initial conditions. These equations must be solved numerically. They can serve as a guide for computer simulation studies. More importantly, they have an advantage in that scaling with the various parameters is much more transparent than with simulation, where many runs are required to obtain such information in an empirical way.

For the space-charge dominated beams desired in rf guns and injector linacs the emittance terms in (A5.17) and (A5.18) can be neglected. Furthermore, since the electron bunches are usually very short, so that the eccentricity z_m/a is close to unity or even less than unity, the free space expression for the geometry factor ($g = g_0$) and the relation $g_0 = 2\gamma_0 z_m/3a$ can be used to good approximation. The two coupled envelope equations then take the simpler form

$$a'' + \frac{\gamma_0' a'}{\beta_0^2 \gamma_0} + k_{x0}^2 a - \frac{\pi e E_m |\sin \varphi_0|}{\lambda mc^2 \beta_0^2 \gamma_0} a - \frac{3}{2} \frac{N r_c}{\beta_0^2 \gamma_0^3} \frac{1}{a z_m} \left(1 - \frac{1}{3} \frac{1}{\gamma_0 z_m} \right) = 0$$

(A5.19)

and

$$(\beta_0^2 \gamma_0^3 z_m')' + \beta_0^2 \gamma_0^3 k_{z0}^2 z_m - \frac{N r_c}{a \gamma_0 z_m} = 0$$

(A5.20)

Finally, we need to discuss the problem of emittance growth if free energy is created due to deviations of the density profiles from the ideal Boltzmann distribution, beam mismatch and off-centering, or if the beam is not in three-dimensional thermal equilibrium. From our analysis of the Maxwell-Boltzmann distribution in Section 5.4 it is clear that for a space-charge dominated (i.e., low-temperature) bunch in a linear focusing system the uniform-density ellipsoidal bunch with parabolic line-charge profile is a good approximation for the stationary state. Thus, if the beam is launched with this shape there should be little or no emittance growth provided that the transverse and longitudinal temperatures do not differ too much. (See the discussion in Appendix 4 on equipartitioning in rf linacs when there is a significant difference between $T_{\parallel}$ and $T_{\perp}$.)

Emittance growth occurs if the three-dimensional density profile of the bunch deviates from the uniform ellipsoid. To obtain an upper limit for this growth one must determine the free energy, which is defined as the difference of the total energy per particle between the nonstationary (e.g., Gaussian) and the stationary ellipsoidal distribution. The procedure is similar to the continuous-beam case discussed in

Section 6.2; but the actual calculations are more involved because of the three-dimensional bunch geometry. We note that the space-charge model used widely to explain emittance growth in rf guns relates the growth to the total rms space-charge forces in both directions [4,5] and not to the energy difference between the nonstationary and the stationary case. This explains why the theoretical predictions from this model always overestimate by a significant factor the emittance growth observed in simulation studies and experiments [3].

A more accurate theoretical description based on the thermalization of the field-energy difference $U = W_n - W_u$ between a nonlinear distribution and the linear uniform ellipsoidal distribution yields for a spherically symmetric bunch ($z_m = a$) the approximate emittance-growth relation [11,12]

$$\frac{\tilde{\epsilon}_{nf}}{\tilde{\epsilon}_{ni}} = \left[1 + \frac{N r_c \tilde{x}}{15\sqrt{5}\gamma_0 \tilde{\epsilon}_{ni}^2} \frac{U}{W_1} \right]^{1/2}. \tag{A5.21}$$

Here $\tilde{\epsilon}_{ni}$ and $\tilde{\epsilon}_{nf}$ are the initial and final normalized rms emittances, $W_1 = Q^2(40\pi\epsilon_0 a_0)$, $a_0 = \sqrt{5}\tilde{x}$ is the radius of the uniform-density ellipsoid, $Q = eN$ is the charge, N is the number of particles in the bunch, r_c is the classical particle radius. For the dimensionless quantity U/W_1 one finds [11] $U/W_1 = 0.308$ for a Gaussian distribution and $U/W_1 = 0.0368$ for a parabolic distribution $n(r) = n_0[1 - (r/a_0)^2]$. When the bunches are not spherically symmetric one must use a more general formula [11] that also includes the emittance growth due to equipartitioning when the bunch is not in thermal equilibrium. As an example, take an electron bunch ($r_c = 2.82 \times 10^{-15}$ m) with $N = 3 \times 10^{10}$, $\tilde{x} = 3 \times 10^{-3}$ m, $\gamma_0 = 3$ (1 MeV), $\tilde{\epsilon}_{ni} = 3 \times 10^{-6}$ m-rad. From (A5.21) one then obtains an emittance growth of $\tilde{\epsilon}_{nf}/\tilde{\epsilon}_{ni} = 9.3$ for a Gaussian distribution and $\tilde{\epsilon}_{nf}/\tilde{\epsilon}_{ni} = 3.4$ for a parabolic distribution.

It is obvious from the general discussion in this appendix that much more work is needed to obtain a better understanding of the beam physics in rf guns or high-current electron injector linacs and to explain the empirical scaling of emittance growth with beam current deduced from experimental observations [10]. Only when this is accomplished and the parametric dependences are apparent can one determine the ultimate fundamental limits to the particle number per bunch and to the achievable brightness in such devices.

REFERENCES

1. J. S. Fraser, R. L. Sheffield, E. R. Gray, and G. W. Rodenz, *IEEE Trans. Nucl. Sci.* **NS-32**, 1791 (1985); J. S. Fraser and R. L. Sheffield, *IEEE J. Quantum Electron.* **23**, 1489 (1987).

2. P. O'Shea and M. Reiser, *AIP Conf. Proc.* **279**, 579 (1993), ed. J. S. Wurtele.

3. I. Ben-Zvi, *Conference Record of the 1993 IEEE Particle Accelerator Conference*, 93CH3279-7, p. 2964.

4. K.-J. Kim, *Nucl. Instr. and Meth.* **A275**, 201 (1989).
5. K.-J. Kim and Y.-J. Chen, 1988 Linear Accelerator Conference Proceedings, Newport News, VA, CEBAF-Report-89-001, June 1989, p. 322.
6. K. T. McDonald, *IEEE Trans. Elect. Dev.* **ED-35**, 2052 (1988).
7. L. Serafini, *AIP Conf. Proc.* **279**, 645 (1993), ed. J. S. Wurtele.
8. B. E. Carlsten, *Nucl. Instr. and Meth.* **A285**, 313 (1989).
9. P. G. O'Shea et al., *Nucl. Instr. and Meth.* **A331**, 62 (1993).
10. P. G. O'Shea et al., *Phys. Rev. Lett.* **71**, 3661 (1993).
11. T. P. Wangler, F. W. Guy, and I. Hofmann, *1986 Linear Accelerator Conference Proceedings,* SLAC-Report-303, Stanford, CA, September 1986, p. 340.
12. M. Reiser, *AIP Conf. Proc.* **193**, 311 (1989), ed. C. Yoshi.

List of Frequently Used Symbols

The list presented here contains those symbols that are frequently used in this book. Symbols that are used only within a particular context of one section are defined locally and not listed here. Where appropriate, the units of measurements are given and reference is made to the section(s) and/or equation(s) where the symbol is discussed.

A	Ampere
A	Atomic mass number
$\mathbf{A}$	Vector potential
A	Trace-space area [m-rad]
A	Amplitude in phase-amplitude variables [Equation (3.337); Section 4.4.1]
a	Radius of cylindrical beam
a	Radius of ellipsoidal bunch (semiaxis in transverse direction)
a	Semiaxis in x-direction of continuous beam with elliptical cross section (Section 4.5.3), also denoted by X (e.g., Section 4.5.3; Section 5.3.2)
$\mathbf{B}, B$	Magnetic flux density [T]
B	Brightness [A/(m-rad)^2]
B_n	Normalized brightness
b	Radius of conducting tube surrounding beam
b	Semiaxis in y-direction of continuous beam with elliptical cross section (Section 4.5.3), also denoted by Y (e.g., Section 4.5.3; Section 5.3.2)
C	Circumference of orbit in a circular accelerator
C_c	Chromatic aberration coefficient (Section 3.4.6)
C_s	Spherical aberration coefficient (Section 3.4.6)
C	Coulomb
C^*	Capacitance per unit length [C/m] (Problem 6.1)

C^+	Capacitance associated with longitudinal space-charge impedance [C · m] (Section 6.3.2)
c	Speed of light
c_s	Phase velocity of space-charge wave in beam frame, also called "speed of sound" (Section 6.3.2)
D, D	Electric flux density [C/m²]
D	Diffusion coefficient (Section 5.4.2)
D_e	Dispersion function [m] in circular accelerator lattice (Section 5.4.10)
d_1 (d_2)	Distance of object (image) side principal planes from center of lens (Section 3.4.2, Figure 3.4)
E, E	Electric field intensity [V/m]
E	Energy
e	Electron charge
F, F	Force
f	Focal length of a lens
$f(\ )$	Particle distribution function, with variables defined locally in text
f_e	Electric (charge) neutralization fraction of a partially charge-neutralized beam (Section 4.2.1; Section 4.6.2; Section 4.6.6)
f_m	Magnetic (current) neutralization fraction of beam, e.g., electron beam with comoving ions (Section 4.2.3)
G	Ripple factor of matched beam in a periodic focusing channel [Eq. (4.157)]
g	Geometry factor associated with the longitudinal field of bunched beams [Section 5.4.7, Equations (5.354) to (5.365), Figure 5.15, Table 5.3] or line-charge perturbations on continuous beams [Section 6.3.2; Section 6.3.3; Equation (6.68) and subsequent discussion].
H, H	Magnetic field intensity [A/m]
H	Hamiltonian (Section 2.3.4)
h	Planck's constant
h	Free-energy parameter (Section 6.2.1)
I	Current
I_Λ	Alfvén current (Section 4.2.3)
I_L	Space-charge current limit (Section 4.2.3)
I_0	Characteristic current, $4\pi\epsilon_0 mc^3/q$ [Equation (4.17)]
i	$\sqrt{-1}$
J, J	Current density [A/m²]
J_i	Action integral (Section 2.3.4)
K	Generalized dimensionless perveance [Equation (4.24)]

K_L Longitudinal perveance parameter [m], (Section 5.4.8)

k_B Boltzmann's constant

k_0 Wave number, $2\pi/\lambda_0$, associated with particle oscillations in a smooth focusing channel *without* space charge: transverse ("betatron") oscillations (Section 4.3.2), and longitudinal ("synchrotron") oscillations (Section 5.4.8; Section 5.4.9)

k Wave number, $2\pi/\lambda$, of particle oscillations *with* space charge (Section 4.3.2; Section 5.4.8; Section 5.4.9)

k Wave number of space charge waves (Section 6.3.2)

k Wave number of rf wave, $2\pi/\lambda = \omega_{rf}/c$.

k_1 Intensity parameter associated with waterbag distribution [Equation (5.171)]

k_e, k_m Electric and magnetic field indices in $\mathbf{E} \times \mathbf{B}$ fields (Section 3.6.3), $k_m = -n$

$\overline{k}$ Average field index in sector-focusing cyclotrons (Section 3.8.4), $\overline{k} = \overline{k}_m = -\overline{n}$

L Lagrangian (Section 2.3.1)

L Length of drift space between lenses of periodic focusing channel (Section 3.8.3; Section 4.4)

L^* Inductance per unit length [H/m] (Section 6.3.2, Problem 6.1)

$L_1 (L_2)$ Distance between object (image) side focal point and the respective principal planes (Section 3.4.2, Figure 3.4)

l Length of lens in "hard-edge" approximation of periodic focusing channel (Section 3.8.3; Section 4.4)

$\tilde{M}$ Transfer matrix in a linear focusing system (Equation (3.85); Section 3.4.1)

m Mass of particle (Section 2.1)

m^* Effective mass of particle [Equations (3.625) to (3.267) in Section 3.6.4]

N Number of particles in a bunch

N_L Number of particle per unit length of beam $[m^{-1}]$

N Number of focusing periods in a circular accelerator lattice

n Particle density $[m^{-3}]$

n Field index [Equation (3.193)]

P Momentum

P_0 Momentum of centroid particle (Section 5.4.3; Section 5.4.6; Sections 5.4.8–5.4.10)

$\mathcal{P}$ Radiated power [Equation (6.195)]

p Canonical momentum

p_θ Canonical angular momentum

p	Pressure [Torr] or [Pa], defined in Equation (4.286)
Q	Total charge of a bunch
q	Charge of a particle
q	Generalized coordinates (Section 2.3.1)
R	Radius or envelope of uniform cyclindrical beam, also "effective radius" ($R = \sqrt{2}\,\tilde{r} = 2\tilde{x}$) of beam with nonuniform density profile
R	Dimensionless beam radius, r_m/r_0 [Equation (4.30)]
R	Reduced variable for radial coordinate, $(\beta\gamma)^{1/2}r$ (Section 3.3.3)
R	Orbit radius in circular accelerator
$\overline{R}$	Average orbit radius ($C/2\pi$) in circular accelerator, $C/2\pi$
R_0	Orbit radius of "centroid" particle in circular accelerator
R_w^*	Beam tube wall resistance per unit length [Ω/m] [Equation (6.79)]
r	Radial variable
r_c	Classical particle radius, $q^2/4\pi\epsilon_0 mc^2$, [Equation (5.244)]
r_m	Envelope radius of uniform beam [Equations (4.48), (4.49)]
S	Length of one period in a periodic focusing channel (Section 3.81; Section 4.4.1; Section 4.4.2)
s	Coordinate along direction of beam propagation in curved systems (e.g., circular machines) for bunched beams and in some cases for straight focusing systems in place of z
s	Longitudinal displacement of particle from equilibrium position due to perturbation (plasma oscillations, space charge waves, Section 6.3.2)
T	Kinetic energy
T	Temperature, usually in the combination k_BT where k_B is Boltzmann's constant
$T_\perp$	Transverse temperature of a beam
$T_\parallel$	Longitudinal temperature of a beam (in direction of propagation)
t	Time
U	Potential energy (Section 2.2)
U	Nonuniform field energy per unit length of beam [Equation (6.14)]
U'	Dimensionless parameter related to imaginary part of longitudinal impedance in circular machines [Equation (6.143)]
u, v	Independent principle solutions of paraxial ray equation (Section 3.3.3)
u	Space-charge parameter [Equation (4.92)]
V	Volt
V	Potential on the axis of electrostatic focusing system (Section 3.3.2; Section 3.4.3)
V'	Dimensionless parameter related to real part of longitudinal impedance in circular machines [Equation (6.143)]

v, v	Particle velocity
W	Wronskian determinant of paraxial ray equation (Section 3.3.3)
W	Total field energy per unit length of beam [Equations (4.67), (4.68)]
w	Amplitude function in a linear focusing channel; in Chapter 3 where beams without space charge are treated (Section 3.8.3); when space charge is included in beam dynamics, as in Chapter 4 and elsewhere, w denotes amplitude function *with* space charge (Section 4.4).
w_0	Amplitude function in linear focusing channel *without* space charge in Chapter 4 and elsewhere where beam theory includes space charge (Section 4.4).
w_0	Field energy parameter, $w_0 = I^2/16\pi\epsilon_0\beta^2 c^2$ [Equation (6.14)]
X, Y	Transverse beam envelopes in a quadrupole focusing channel (Section 4.4.2)
X_s	Space-charge impedance [Ω] in circular accelerator [Equation (6.114)]
X_w	Imaginary part of wall impedance [Ω] in circular accelerators [Equation (6.115)]
$\tilde{x}$	rms width of beam density profile in transverse x-direction, i.e. $\tilde{x} = (\overline{x^2})^{1/2}$; for uniform-density beam $\tilde{x} = X/2$
$\tilde{x}'$	rms divergence defined as $\tilde{x}' = (\overline{x'^2})^{1/2}$, where x' is the slope of a particle trajectory
$\tilde{y}'$	rms width of beam density profile in transverse y-direction, i.e. $\tilde{y} = (\overline{y^2})^{1/2}$; for uniform-density beam $\tilde{y} = Y/2$
Z	Charge state of ion
$Z_\parallel$	Longitudinal impedance [Ω] in circular machines [Equation (6.115)]
$Z_\parallel^*$	Longitudinal impedance per unit length [Ω/m] in linear accelerators or beam transport systems [Equation (6.102)]
Z_s^*	Space-charge impedance per unit length of beam [Ω/m] [Equations (6.88) to (6.94)]
z	Coordinate in the direction of beam propagation for axisymmetric and quadrupole focusing systems (occasionally s is also used in place of z)
z	Relative coordinate of particle in a bunch with respect to centroid position s_0, i.e. $z = s - s_0$ (Section 5.4.6 to Section 5.4.9; Section 5.4.11)
z_m	Longitudinal half length (semiaxis) of ellipsoidal bunch [Equation (5.416)]
$\tilde{z}$	rms width of bunch in longitudinal direction, i.e. $\tilde{z} = (\overline{z^2})^{1/2}$; for a uniform-density ellipsoidal bunch, $\tilde{z} = z_m/\sqrt{5}$ [Equation (5.411)]

$\tilde{z}'$	Longitudinal rms divergence defined as $\tilde{z}' = (\overline{z'^2})^{1/2}$; where $z' = dz/ds$ is the slope of the longitudinal trajectory in the frame moving with the bunch centroid
α	Momentum compaction factor [Equation (3.256)]
α	Acceptance (or admittance) of a focusing system [Equations (3.311), (3.353), (4.94)]
$\hat{\alpha}, \hat{\beta}, \hat{\gamma}$	*Courant–Snyder* (or *Twiss*) *parameters* describing trace-space ellipse in linear focusing systems in Chapter 3 where beams without space charge are treated (Section 3.8.3); when space charge is included in the beam dynamics, as in Chapter 4 and elsewhere, these symbols denote the ellipse parameters *with* space charge.
$\hat{\alpha}_0, \hat{\beta}_0, \hat{\gamma}_0$	Courant–Snyder parameters *without* space charge in linear focusing systems in Chapter 4 and elsewhere where beam theory includes space charge.
$\hat{\beta}$	Amplitude (betatron) function with space charge, $\hat{\beta} = 1/w^2$ (see w)
$\hat{\beta}_0$	Amplitude (betratron) function without space charge, $\hat{\beta}_0 = 1/w_0^2$ (see w_0)
β	Particle velocity divided by speed of light, $\beta = v/c$
γ	Total energy of particle, $E = E_0 + T$, divided by rest energy, $E_0 = mc^2$, i.e. $\gamma = 1 + T/mc^2 = (1 - \beta^2)^{-1/2}$; also known as *Lorentz factor*
γ_0	Value of γ on beam axis (Section 5.2.3), or for "center of momentum" particle [Equation (5.259)] in a beam with momentum distribution
γ_a	Value of γ at beam edge (Section 5.2.3)
γ_b	Value of γ at injection into beam tube (Section 5.2.3)
ϵ_0	Permittivity (dielectric constant) of free space: 8.854×10^{-12} F/m $\simeq (1/36\pi) \times 10^{-9}$ F/m
ϵ_x, ϵ_y	Emittance in x or y direction; defined as total emittance of a K-V beam (uniform charge density) [Equations (3.4), (3.5c), (5.138)], or as the "effective emittance," $\epsilon = 4\tilde{\epsilon}$, of a nonuniform beam [Equation (5.206)]
$\tilde{\epsilon}_x, \tilde{\epsilon}_y$	rms emittance in x or y direction [Equations (3.2a), (3.2b); Section 5.3.4]
ϵ_n	Normalized emittance, defined as $\epsilon_n = \beta\gamma\epsilon$ [Equation (3.21b)]
$\tilde{\epsilon}_n$	Normalized rms emittance, defined as $\tilde{\epsilon}_n = \beta\gamma\tilde{\epsilon}$ [Equation (3.21a)]
$\epsilon_{zz'}$	Longitudinal emittance of an ellipsoidal bunch with parabolic line charge profile (uniform volume charge density) or "effective longitudinal emittance," $\epsilon_{zz'} = 5\tilde{\epsilon}_{zz'}$, of a bunch with nonuniform volume charge density [Equations (5.416), (5.427), (5.429)]
$\tilde{\epsilon}_{zz'}$	Longitudinal rms emittance, $\tilde{z}\tilde{z}'$ [Equation (5.316)]

$\tilde{\epsilon}_z$	Longitudinal rms emittance defined in terms of the momentum spread: $\tilde{\epsilon}_z = \tilde{z}\Delta\tilde{P}_z/P_0 = \gamma_0^2\tilde{\epsilon}_{zz'}$ [Equations (5.314), (5.316)]
$\tilde{\epsilon}_{nz}$	Normalized longitudinal rms emittance; $\tilde{\epsilon}_{nz} = \beta_0\gamma_0\tilde{\epsilon}_z = \beta_0\gamma_0^3\tilde{\epsilon}_{zz'}$ in linear accelerators [Equation (5.317a)], and $\tilde{\epsilon}_{nz} = \beta_0\gamma_0\tilde{\epsilon}_z = \beta_0\gamma_0\tilde{\epsilon}_{zz'}/|\eta|$ in circular accelerators [Equation (5.444)]
ϵ_{nz}	Normalized longitudinal emittance defined as $\epsilon_{nz} = 5\tilde{\epsilon}_{nz}$
η	Slip factor in a circular accelerator [Equation (3.262)]
θ	Angular coordinate
θ	Focusing-strength parameter, $\theta = \sqrt{\kappa}\, l$ [Equation (3.357)]
θ_r	Angle rotation in a solenoidal lens [Equations (3.140) to (3.142)]
κ	Focusing function [Equation (3.312)]; when space charge is included in beam dynamics, κ and κ_0 denote the focusing functions *with* and *without* space charge, respectively [Equations (4.111) to (4.116)]
Λ	ln Λ = Coulomb logarithm [Equations (5.242) to (5.248)]
Λ	Line-charge density (Section 6.3.2), defined as ρ_L in Chapters 4 and 5
λ	Wavelength of electromagnetic wave (Section 5.4.8)
λ	Wavelength of betatron oscillation with space charge (see k)
λ_0	Wavelength of betatron oscillation without space charge (see k_0)
λ_D	Debye length [Equation (4.1)]
μ_0	Permeability of free space: $4\pi \times 10^{-7}$ H/m
ν	Betatron tune, betatron oscillation frequency normalized to orbital frequency in a circular accelerator in Chapter 3; in the context of beam theory that includes space charge (Chapter 4 and elsewhere), ν also denotes the betratron tune *with* space charge.
ν_0	Betatron tune *without* space charge in Chapter 4 and elsewhere where beam theory includes space charge (Section 4.5).
ν_B	Budker parameter [Equation (4.18)]
ρ	Volume-charge density [C/m^3]
ρ_L	Line-charge density [C/m], defined as Λ in Chapter 6
σ	Phase advance of betatron oscillations in one period of a periodic focusing channel in Chapter 3 where beams without space charge are discussed (Section 3.8.2); when space charge is included in beam dynamics, as in Chapter 4 and elsewhere, σ denotes phase advance *with* space charge (Section 4.4)
σ_0	Phase advance of betatron oscillations in one period of a periodic focusing channel *without* space charge in Chapter 4 and elsewhere where beam theory includes space charge (Section 4.4)
σ	Conductivity [Equation (4.240)]
σ_i	Ionization cross section (Section 4.6.1)
τ	Relaxation time in Coulomb collision (Section 6.4)

τ_N	Neutralization time (Section 4.6)
ϕ	Potential [Equation (2.13)]
ϕ	Phase, in phase-amplitude variables [Equation (3.337); Section 4.4.1]
φ	Phase of particle in rf field [Section 5.4.8, Equations (5.375, (5.376)]
ψ	Magnetic flux [T/m^2], [Equation (2.77)]
ψ	Phase function in a linear focusing channel [Equation (3.337); Section 4.4.1]
ω	Angular frequency (often simply referred to as "frequency")
ω_c	Cyclotron frequency [Equation (2.81)]
ω_L	Larmor frequency [Equation (2.83)]
ω_l	Synchrotron frequency [Equation (5.395b)]; also ω_{z0} [Equation (5.440)]
ω_p	Plasma frequency [Equation (4.2)]
ω_{rf}	Angular frequency of rf field (Section 5.4.8)

Bibliography (Selected List of Books)

A. Classical Mechanics and Electrodynamics

A.1. W. K. H. Panofsky and M. Phillips, *Classical Electricity and Magnetism,* Addison-Wesley, Reading, Mass., 1962.

A.2. W. R. Smythe, *Static and Dynamic Electricity,* McGraw-Hill, New York, 1968.

A.3. H. Goldstein, *Classical Mechanics,* Addison-Wesley, Reading, Mass., 1965.

A.4. J. D. Jackson, *Classical Electrodynamics,* Second Edition, Wiley, New York, 1975.

B. Plasma Physics

B.1. N. A. Krall and A. W. Trivelpiece, *Principles of Plasma Physics,* McGraw-Hill, New York, 1973; also available in paperback from San Francisco Press, 1986.

B.2. G. Schmidt, *Physics of High Temperature Plasmas,* Second Edition, Academic Press, New York, 1979.

B.3. R. C. Davidson, *Physics of Nonneutral Plasmas,* Addison-Wesley, Reading, Mass., 1990.

C. Charged Particle Dynamics and Beam Physics

C.1. V. K. Zworykin et al., *Electron Optics and the Electron Microscope,* Wiley, New York, 1945.

C.2. O. Klemperer, *Electron Optics,* Cambridge University Press, Cambridge, 1953.

C.3. J. R. Pierce, *Theory and Design of Electron Beams,* Second Edition, Van Nostrand, New York, 1954.

C.4. P. A. Sturrock, *Static and Dynamic Electron Optics,* Cambridge University Press, Cambridge, 1955.

C.5. *Korpuskularoptik,* in *Handbuch der Physik,* Bd. 33, Springer, Berlin, 1956 (Elektronen und Ionenoptik, W. Glaser).

C.6. B. Lehnhart, *Dynamics of Charged Particles,* North-Holland, Amsterdam, and Wiley, New York, 1964.

C.7. K. G. Steffen, *High Energy Beam Optics*, Academic Press, New York, 1970.

C.8. A. P. Banford, *The Transport of Charged Particle Beams*, E. & F.N. Spon, London, 1966.

C.9. P. T. Kirstein, G. S. Kino, and W. E. Waters, *Space Charge Flow*, McGraw-Hill, New York, 1967.

C.10. P. W. Hawkes, *Quadrupoles in Electron Lens Design*, Academic Press, New York, 1970.

C.11. P. Grivet, *Electron Optics*, Pergamon Press, Elmsford, NY, 1972.

C.12. C. L. Hemenway, R. W. Henry, and M. Caulton, *Physical Electronics*, Second Edition, Wiley, New York, 1967.

C.13. A. Septier, Editor, *Focusing of Charged Particles*, two volumes, Academic Press, New York, 1968.

C.14. A. B. El-Kareh and J. C. J. El-Kareh, *Electron Beams, Lenses, and Optics*, two volumes, Academic Press, New York, 1970.

C.15. R. G. Wilson and G. R. Brewer, *Ion Beams with Application to Ion Implantation*, Wiley, New York, 1973.

C.16. L. Vályi, *Atom and Ion Sources*, Wiley, New York, 1977.

C.17. J. D. Lawson, *The Physics of Charged-Particle Beams*, Oxford University Press, Oxford, 1977; Second Edition, 1988.

C.18. R. B. Miller, *An Introduction to the Physics of Intense Charged Particle Beams*, Plenum Press, New York, 1982.

C.19. A. Septier, Editor, *Applied Charged Particle Optics, Advances in Electron Optics and Electron Physics*, Supplements 13A and 13B, 1980, and Supplement 13C (*Very-High-Density Beams*), 1983, Academic Press, New York.

C.20. H. Wollnik, *Optics of Charged Particles*, Academic Press, New York, 1987.

C.21. M. Szilagyi, *Electron and Ion Optics*, Plenum Press, New York, 1988.

C.22. P. W. Hawkes and E. Kaspar, *Principles of Electron Optics*, Vols. 1 and 2, Academic Press, New York, 1989.

C.23. Ian G. Brown, Editor, *The Physics and Technology of Ion Sources*, Wiley, New York, 1989.

C.24. S. Humphries, Jr., *Charged Particle Beams*, Wiley, New York, 1990.

D. Charged Particle Accelerators

D.1. J. J. Livingood, *Principles of Cyclic Particle Accelerators*, D. Van Nostrand, Princeton, NJ, 1961.

D.2. M. S. Livingston and J. P. Blewett, *Particle Accelerators*, McGraw-Hill, New York, 1962.

D.3. M. Reiser, "Particle Accelerators," Section 8j in *American Institute of Physics Handbook*, Third Edition, McGraw-Hill, New York, 1972.

D.4. C. L. Olson and U. Schumacher, *Collective Ion Acceleration*, Springer-Verlag, New York, 1979.

D.5. N. Rostoker and M. Reiser, Editors, *Collective Methods of Acceleration*, Harwood Academic Publishers, Chur, Switzerland, 1979.

D.6. S. Humphries, Jr., *Principles of Charged Particle Acceleration,* Wiley, New York, 1986.

D.7. M. Month, Editor, *Physics of High Energy Particle Accelerators, AIP Conf. Proc.* **105** (1983), **153** (1987), **184** (1989), and **249** (1992).

D.8. *Advanced Accelerator Concepts, AIP Conf. Proc.* **156** (1986), ed. F. E. Mills; **193** (1989), ed. C. Joshi; **279** (1992), ed. J. S. Wuertele.

D.9. M. Month and S. Turner, Editors, *Frontiers of Particle Beams, Lecture Notes in Physics* **296,** Springer-Verlag, Berlin, 1988.

D.10. D. A. Edwards and M. J. Syphers, *An Introduction to the Physics of High Energy Accelerators,* Wiley, New York, 1992.

D.11. A. W. Chao, *Physics of Collective Beam Instabilities in High Energy Accelerators*, Wiley, New York, 1993.

Index

Aberrations:
    chromatic, 106, 108, 437
    geometrical, 106
    spherical, 106–107
Acceleration:
    adiabatic, 172, 417–418
    collective, 287
    wakefield, 289
Acceptance, 143–144, 153, 215
    of FODO channel, 156
Action integrals, 35
Admittance, see Acceptance
Alfvén current (or Alfvén-Lawson current),
    6, 204–205, 310
Alternating-gradient principle, 144, 162–
    163
Amplitude function, 148–149, 153, 179,
    223–228, see also Betatron function;
    Courant–Snyder parameters
    maximum in FODO channel, 156
    maximum in periodic solenoid channel,
    232

Barber's rule, 179
Beam(s):
    bunched, see Bunched beams
    charge-neutralized relativistic electron,
    203
    emittance-dominated, 217
    equivalent, see Equivalent beams
    intense relativistic electron (IREB), 5,
    285, 323–331
    laminar, mismatched, 331–334
    laminar, in uniform magnetic field, 306
    matched, in a FODO channel, 235–239

matched, in a periodic solenoid channel,
    143, 153, 223–233
    matched, in a uniform channel, 213–217
    mismatched, emittance growth, 476–477,
    488–491
    mismatched, in periodic channel, 240–
    251
    mismatched, in uniform channel, 217–
    218
    off centered, 252–260, 477
    space-charge dominated, 217
Beam breakup instability, 260
Beam centroid, 252, 375
Beam cooling:
    electron beam, 468, 542–543
    longitudinal, due to acceleration, 399–402
    radiation, 468, 542, 546–553
    stochastic, 468, 542, 544–545
Beam current:
    average, 436, 452
        in rf linac, 452, 579
    maximum, in rf linac, 579
    maximum, in transport channel, 215
    peak, in ellipsoidal bunch, 448
Beam frame, 365, 375
Beam matching, 217, 302, 453–454
Beam radius (continuous beams):
    approximate relation in uniform (or
    smooth) channel, 385
    in drift space, 197–203
    effective, 360
    matched, average, in periodic channel,
    228
    matched, in uniform channel, 215
    rms, 359

Bennett pinch, 209
Betatron, 54 (Problem 2.13), 116
  modified, 265
  plasma, 265
Betatron function, 148, 153, 388, 438, 532, 540–541, *see also* Amplitude function,
Betatron oscillations, 116-121
  coherent, 252
  incoherent, 270
Betatron tune, 118
  coherent, 267, 271
  forbidden values of, 170
  incoherent, 267
Betatron wavelength, 118
Betatron wave number, 118
Boersch effect, 401, 468, 525–530
Boltzmann density profile:
  longitudinal (or line-charge), 412–413
  transverse (or radial), 189, 380, 383, 386
Boltzmann equation, 366
Boltzmann relation, 188, 340, 379
  longitudinal, for line-charge density, 403
Brightness, 13, 61
  average, 61
  normalized, 14, 62, 65, 389
Brillouin beams:
  hollow, 318, 321
  solid, 317, 319
Brillouin flow, 214, 216, 315
Budker condition of self-focusing, 196
Budker parameter, 195
Bunched beams:
  approximate solutions for semi-axes, 449–452, 466 (Problem 5.20), 578
  coupled envelope equations, 449–451
  ellipsoidal model, 403–404, 448, *see also* Ellipsoidal bunch; Parabolic bunch model
  half-length of ellipsoid, 403, 449–452, 466 (Problem 5.20), 578
  image effects, *see* Geometry factor
  radius of ellipsoid, 403, 449–452, 466 (Problem 5.20), 578
Busch's theorem, 34, 313

Canonical angular momentum, 29
  conservation of, 33
  relation to normalized emittance, 211
Canonical momentum, 28
Center-of-momentum energy, 397
Center-of-momentum frame, 375

Center-of-momentum particle, 395, *see also* Beam centroid
Central limit theorem, 372, 538
Charge-neutralization effects:
  in intensive relativistic electron beams, 285–289, 295–296
  linear beam model, 278–280
  in low-energy p and H$^-$ beams, 281–284
  in storage rings, 289
Charge-neutralization factor, 192–197, 206
Charge-neutralization time, 274
Charged-particle beam lithography, 5, 459, 461
Characteristic current, 195
Child's law (or Child–Langmuir Law), 6, 10, 45–46
Chromaticity parameters, 133, 437
Classical particle radius, 195, 370
Collisions:
  Coulomb, 364–368, 526
  elastic, 537
  inelastic, 537
Cooling, *see* Beam cooling
Conjugate momentum, *see* Canonical momentum
Conservation:
  of canonical angular momentum, 33
  of energy, 20
  of magnetic moment, 37
Conservative systems, 19, 24, 30
Coulomb collisions, *see* Collisions
Coulomb parameter (or Coulomb logarithm), 370–371
Courant–Snyder parameters, 148, 219
Current neutralization, 206–207
Cusped magnetic field, 53
Cyclotron:
  classical, 116–117
  isochronous, 132, 158
  sector-focusing, 157
  superconducting, 162

Debye length, 64, 185, 366–367
  definition of, 185
  in a relativistic beam, 186
Debye shielding, 184, 364–368
Debye sphere, 368
Diamagnetic field, 324
Differential algebra, 171
Diode, *see* Planar diode
Dispenser cathodes, 9

Dispersion, 439
  effect on tune shift, *see* Tune shift, incoherent
Dispersion function, 439
Dispersion relation:
  for longitudinal space-charge waves, 500
  for negative mass instability, 519
  for relativistic beam with trapped particles of opposite charge, 291–292
  for resistive wall instability, 509, 520
Distribution:
  conical (nonstationary), 360–361
  Gaussian, 357, 388
    nonstationary, 360–361
  Kapchinsky–Vladimirsky (K–V), 6, 61, 190, 341–347, 350–351, 448
    stability in periodic solenoid and quadrupole channels, 492–497
    stability in a uniform focusing channel, 492
  Maxwell–Boltzmann, 188, 340, 373, 379
    longitudinal, 402–403
    longitudinal for circular machines, 433
    for a relativistic beam, 372–377
    two-temperature, 378
  Maxwellian velocity, 11, 370
  microcanonical, 341
  Neuffer, 427, 448
  parabolic (nonstationary), 360
  thermal, 185–186, 188, 340, *see also* Distribution, Maxwell-Boltzmann
  two-temperature, 378
  waterbag, 61, 352
    nonstationary, 360–361
Divergence:
  effective, 59
  flow, 392
  rms, 58
  thermal, 392

Effective divergence, 59
Effective emittance, 59, 360, 389
Effective width, 59
Electrical breakdown in rf systems, 577
Electron gun:
  Pierce-type, 7, 564–566
  rf photocathode, 12, 581–582
  thermionic cathode, 7
Electron microscope, 4
Electron ring accelerator, 281
Electrostatic analyzer, 48

Ellipsoidal bunch: *see also* Bunched beams; Parabolic bunch model
  longitudinal half-width, 449–451, 478
  longitudinal rms width, 412–413
  rms emittance, 450, 466
    geometry factor, 405–411
    potential due to image charges, 404
  transverse half-width (radius), 449–451, 478
  transverse rms width, 450, 466
Emittance, 57–59
  effective, 59, 360, 389
  and entropy, 491
  four-times rms, *see* Emittance, effective
  longitudinal, 395–396
  unnormalized, 423
  normalized, 11, 62, 65
    relation to canonical angular momentum, 211
  rms, 57, 65, 358–359, 389, *see also* rms emittance
  six-times rms, 389
  unnormalized, 13
Emittance growth:
  in drift space, 557
  and equipartitioning, 498
  and free energy, 470
  in a mismatched beam, 176 177, 188, 491
  due to nonuniform charge distribution, 475
  in off-centered beams, 477
  due to rf defocusing 581, 585
  in spherical bunch, 588
  time scales of, 491
Energy:
  conservation of, 20
  difference between mismatched and matched beam, 477
  difference between nonuniform and uniform beam, 475
  free, 189, 470, 477
  of off-centered beam, 478
  rest (definition), 20
  total (kinetic and potential) average per particle, 456
  total field (electric and magnetic) of beam, 207
  transition, 131, 165, 431
  voltage-equivalent of kinetic, 70

Envelope equations:
in a circular machine, 432
coupled for a bunched beam, 449–451, 587
in drift space, 105, 197, 203
longitudinal, 425
longitudinal rms, 428
nonadiabatic for a bunched beam, 586
in a periodic focusing system, 149, 222, 236
rms, 363
in a uniform focusing system, 125, 212
Envelope oscillations due to mismatch:
fundamental modes, 242
in-phase mode, 218, 242
instabilities in a periodic channel, 243–251
out-of-phase mode, 219, 242
Equipartitioning 378, 468, 526
and emittance growth in rf linacs, 498, 573
Equivalent beams, 184, 359, 362
and rms emittance, 358
Equivalent linear beam for longitudinal distribution, 414
Equivalent ellipsoidal model for bunched beams, 448
Euler equations, 39

Field index, 118
Floquet functions, 148
Floquet's theorem, 148
Focal points of a lens, 80
Focusing:
edge, 135–136
gas, 190, 281, 294, 468
periodic, 139
of intense beams, 221
with thin lenses, 139
strong, 144, 346
weak, 144, 346
Fokker–Planck equation, 368–369
Four-coordinate vector, 572
Four-momentum vector, 374, 570
Four-times rms emittance, 59, 389
Four-vector potential, 374, 572
Free electron lasers, 5–6, 557, 581
Free energy, 189, 470
and emittance growth, 470
Frequency slip factor, 131, 165

Gabor lens, *see* Lenses
Gas focusing, *see* Focusing
Gaussian density profile, 380

Geometry factor or *g*-factor:
definition for bunched beams, 405, 408
definition for line-charge perturbations, 505
for ellipsoidal bunch in free-space, 405
for emittance dominated beams, 505
for space-charge dominated beams, 505
Generalized coordinates and velocities, 23
Generalized momenta, 28
Generalized potential, 25
Grids, 96–98
Gyrotron 6, 567

H⁻ beams, 281
Halo, 479, 489–491, 554
Hamiltonian for relativistic particle, 33
Hamilton's equations, 30
Hamilton's variational principle, 23
Hard-edge approximation, 231
Harmonic oscillator, 49, 370
Heavy-ion inertial fusion, 5, 442, 514
Herrmannsfeldt's code, 566
High-voltage breakdown in ion sources, 12

Image forces:
in bunched beams, *see* Geometry factor
effects on betatron tune, 267, *see also* Tune shift
in off-centered beams, 250–259
Image formation, 83–86, 459, 461
Immersed flow, 318
Impact parameter, 370
Instability:
in beams with trapped particles of opposite charge, 290–293
head–tail, 438
longitudinal microwave, 515
negative-mass, 7, 132, 519
in periodic focusing systems, 243–251, 491–498
resistive-wall, 508, 510, 513–514
in circular machines, 520
Instability stop band, 161
Intense relativistic electron beam (IREB), 5, 323
charge neutralization effects in, 285–289
Interparticle distance, 186
Intrabeam scattering, 530–537
emittance change by, 532
invariant for, 531
theory of, 532

Ion implantation, 5
Ionization cross sections, 273–278
  for electron and proton beam in hydrogen
    gas, 276
Ion propulsion, 206
Ion shaking, 294
Ion sources, 9–10

Kapchinsky–Vladimirsky distribution, *see*
    Distribution, Kapchinsky–Vladimirsky
Keil–Schnell stability criterion, 523
Kerst–Serber equations, 120
Kinetic equation, *see* Vlasov equation
Klystron, 501

Lagrange equations of motion, 24
Lagrange function (or Lagrangian), 23
  generalized, 25
  for relativistic particle, 27
Laminar flow, 45
  beams in uniform magnetic field, 306
  model, 189
Landau damping, 294, 497, 516, 521
Larmor frame, 73, 75
Laser beat-wave acceleration, 289
Laslett tune shift, *see* Tune shift, Laslett
Lattice in synchrotrons, 163
Lenses:
  bipotential (or immersion), 88, 90–91,
    176 (Problems 3.4–3.5)
  cathode, 91, 95
  cinzel (or unipotential), 88, 90, 176
  electrostatic, 86
  Gabor (or space-charge), 294
  plasma, 294
  quadrupole, 111–116
  sextuple, 439
  single-aperture, 90
  solenoidal magnetic, 98–103
Lie operators, 171
Linear aperture, 215
Linear colliders, 12, 294, 302 (Problem
    4.15), 546, 557, 581
Linear induction accelerator, 5
Line-charge density, 403, 412–413, 502
  parabolic profile, 424
  of uniformly populated ellipsoid, 404
Liouville's theorem, 62–64
Longitudinal envelope equation in a circular
    machine, 432

Longitudinal equation of motion in a circular
    machine, 431–432
Lorentz factor, 16
Lorentz force, 15
Lorentz transformations, 570–572

Magnetic diffusion time, 258–259
Magnetic neutralization, 207
Magnetron, 54 (Problem 2.14)
Magnetron injection gun, 318, 567–568
Magnification (image), 83–86
Mass:
  effective, 16, 130–132
  longitudinal, 17, 132
  negative, 132
  relativistic, 17
  rest, 17
  transverse, 17
Matching, *see* Beam matching
Mathieu–Hill equation, 145, 148
Mathieu stability diagram, 169
Maxwell's equations, 15
Mean free path, 274
Mechanical momentum, 16, 28–29
Meridional plane, 74
Meta-equilibrium state, 453
Method of images, 267–268
Micromachining, 5
Mismatch, *see* Beams, mismatched; Enve-
    lope oscillations
Momentum compaction factor, 130
Momentum-energy four-vector, *see* Four-
    momentum vector
Multiple scattering, 537
  emittance growth, 539–541

Necktie diagram, 181 (Problem 3.24)
Negative-mass instability, *see* Instability
Negative-mass regime, 433
Newton's equation, 15
Nonlinear beam optics, 2
  experimental investigation, 171
Non-Liouvillean injection, 442
Nonneutral plasma, 2, 185
Nonuniform charge distribution, *see* Emit-
    tance growth

Oscillations:
  beam mismatch, 217, 219
  betatron, *see* Betatron oscillations
  coherent due to injection errors and mis-

Oscillations (*Continued*)
 alignments, 252–260
 envelope, *see* Envelope oscillations
 incoherent, 252
 synchrotron, *see* Synchrotron oscillations

Parabolic bunch model, 412 414, 424, *see also* Ellipsoidal bunch
Parabolic line-charge density, *see* Parabolic bunch model
Paraxial ray equation, 69–75
Periodic solenoid channel, 221–234
Perveance, 197, 309
 generalized, 196–197
 longitudinal, 425
 maximum, in uniform focusing channel, 215
  hollow electron beam, 322
  solid electron beam, 320
Phase advance, 142, 152
 with/without space charge, 223
 zero current, 230
Phase-amplitude form of solution to Mathieu–Hill equation, 149, 222
Phase shift, *see* Phase advance
Phase space, 11, 49, 62
Phase-space density, 389
Phase stability, 165
Photocathode rf electron gun, *see* Electron gun
Pierce-type electron gun, *see* Electron gun
Pierce-type geometry, 564
Piwinski invariant, *see* Intrabeam scattering, invariant
Planar diode, 7, 43–46, 54 (Problem 2.10), 55, 462 (Problem 5.1)
Plasma channel, 295
Plasma frequency, 184–185, 195
Plasma lenses, *see* Lenses
Plasma wave number, 219
Power-balance equation, 208
Principal planes of a lens, 80–81
Principle of least action, 37

Quadrupole doublets, 114
Quadrupole lenses, *see* Lenses
Quadrupole triplets, 116

Radioactive waste transmutation, 386
Radio-frequency (rf) bucket, 418–420, *see also* Separatrix

Radio-frequency (rf) linear accelerators (or rf linacs), 5, 415, 573
Radio-frequency quadrupole (RFQ) accelerators, 6, 281
Random walk, 255, 372
Relaxation time:
 for Coulomb collisions, 526–527
 for electron beam cooling, 543
 for intrabeam scattering, 534–535
Resonances:
 in circular accelerators, 166–171
 confluent and parametric in K–V beam instabilities, 246
Rest energies of some isotopes and ions, 22
Richardson–Dushman equation, 8
Rigid-rotor equilibrium, 312, 315
Ripple factor:
 in an axisymmetric periodic focusing channel, 230, 233–234
 in a FODO channel, 239
rms average dispersion, 447
rms beam width, 358
rms emittance, 57, 65, 358–359, 389
 normalized, 359
 normalized longitudinal, 395, 434
 total or 100%, 359
 unnormalized longitudinal, 395, 434
rms energy spread, 397
rms momentum spread, 397
Robinson's theorem, 550
Ruggiero–Vaccaro stability diagram, 524
Rutherford formula, 538

Sector magnets, 135–136
Separatrix, 418–420, *see also* Radio-frequency (rf) bucket
Sextuple lenses, *see* Lenses
Shielded source, 315, 325
 magnetically, 315
Six-times rms emittance, 389
Slip factor, *see* Frequency slip factor
Slow-wave structure, 502
Smooth approximation, 155, 221, 224, 339
Space-charge current limit, 310, 328
Space-charge impedance (longitudinal), 510–511
Space-charge waves:
 fast and slow, 500
 growth rate of, 513, 515
 negative-energy, 502
 positive-energy, 501

Spallation neutron sources, 386
Stochastic cooling, *see* Beam cooling
Stochastic effects, 372, 414, 554
Stop bands, 170
Storage rings, 369, 428
Strong-focusing principle, *see* Alternating-gradient principle
Synchrocyclotrons, 157
Synchronous particle, 394, 415
Synchrotron(s), 157, 428
    Fermilab booster, 264, 442, 443
    strong-focusing, 162
Synchrotron frequency, 420
Synchrotron oscillations, 394
Synchrotron principle, 157
Synchrotron radiation, 546
    energy loss, 548
    energy spectrum, 551
Synchrotron tune with space charge, 433

Temperature:
    beam-frame, 186, 375, 378
    laboratory, 186, 375, 378
        longitudinal, 378, 392
        transverse, 378, 392
    longitudinal, 397
        in a circular machine, 430–431
        cooling, due to acceleration, 399–401
        effective, 431
        negative, 431, 532
        and rms energy spread, 397
    relativistic definition of, 186, 375, 378
Tevatron, 4, 19
Theorem of adiabatic invariance, 35
Theorem of magnetic flux conservation, 36
Thermionic cathode, 8
Thermonuclear fusion, 5
Thin-lens approximation, 83, 155
Total field energy, 207
Touschek effect, 536–537
Trace space, 57
    area, 58, 65, 79
    ellipse 103–105, 151
        in a betatron-type field, 121–123
    longitudinal emittance, 395

Tracking codes, 171
Transfer matrix, 78–83
Transition energy, 131, 165, 431
Transverse wakefield effect, 260
Triplet, *see* Quadrupole triplets
Tune, 153, 163
Tune depression, due to space charge:
    longitudinal, 426, 433
    transverse, 214, 216, 383
Tune shift, coherent, due to images, 271
Tune shift, incoherent:
    above and below transition, 433
    due to charge neutralization, 289–290
    Laslett, 7, 273
    longitudinal, due to space charge, 433, *see also* Synchrotron tune
    resonance traversal due to space charge, 264
    rms 446
    due to space charge, 260–264, 442–443
    due to space charge and dispersion, 444–445
    due to space charge and images, 270, 272
Twiss parameters, *see* Courant–Snyder parameters

Velocity analyzer, 48
Velocity of space-charge waves:
    group, 501
    phase (fast and slow wave), 500–501, 507
Virial theorem, 51
Virtual cathode, 285–288, 310
Vlasov equation, 335–336, *see also* Vlasov–Maxwell equations
    relativistic, 336
Vlasov–Maxwell equations, 337, 344
Voltage-equivalent of kinetic energy, 70, 322

Wakefields, 3, 555
Weak-lens approximation, *see* Thin-lens approximation
Welding with particle beams, 5

Z-pinch, 294